Project Editor: *Larry Olsen*
Copy Editor: *Sean Cotter*
Designer: *Sharon H. Smith*
Production Coordinator: *Bill Murdock*
Illustration Coordinator: *Cheryl Nufer*
Compositor: *Bi-Comp, Inc.*
Printer and Binder: *Arcata Book Group*
Cover: The cover shows the outer layers of the Earth, where interacting thermal and gravitational energy drive mantle convection, plate subduction and rifting, magma generation, volcanism, and many other geologic processes that are responsible for the creation of magmatic and metamorphic rocks. Design by Michael C. Graves.

Library of Congress Cataloging in Publication Data
Best, Myron G.
  Igneous and metamorphic petrology.

  (A Series of books in geology)
  Includes bibliographies and index.
  1. Rocks, Igneous. 2. Rocks, Metamorphic. I. Title.
II. Series.
QE461.B53   552′.1   81-17530
ISBN 0-7167-1335-7   AACR2

Printed in the United States of America

4 5 6 7 8 9 KP 0 8 9 8 7 6 5 4

Igneous and Metamorphic Petrology

Myron G. Best

Brigham Young University

Line Drawings by the Author

W. H. Freeman and Company
New York

*Igneous and
Metamorphic
Petrology*

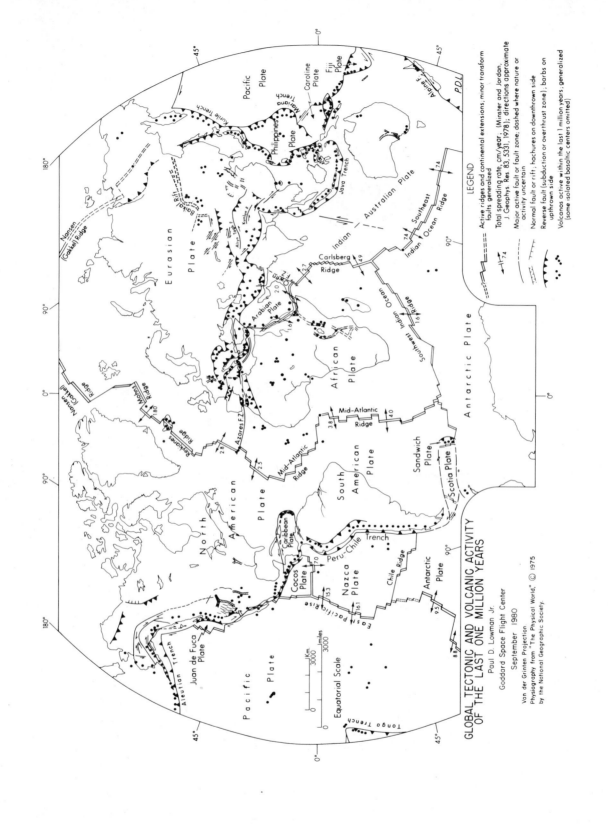

GLOBAL TECTONIC AND VOLCANIC ACTIVITY OF THE LAST ONE MILLION YEARS

Paul D. Lowman Jr.
Goddard Space Flight Center
September 1980

Van der Grinten Projection
Physiography from "The Physical World," © 1975
by the National Geographic Society.

LEGEND

Active ridges and continental extensions, minor transform faults generalized

Total spreading rate, cm/year, (Minster and Jordan, J. Geophys Res. 83, 533), 1978); directions approximate

Major active fault or fault zone, dashed where nature or activity uncertain

Normal fault or rift, hachures on downthrown side

Reverse fault (subduction or overthrust zone), barbs on upthrown side

Volcanos active within the last 1 million years, generalized (some isolated basaltic centers omitted)

To Ken,
  who led the way, and
To Viv,
  who made it possible

Contents in Brief

Extended Contents

Part III

*Metamorphic Bodies
and Systems   341*

Preface

Many petrology texts have appeared during the revolutionary growth of geology in the past two decades. Most of these texts are specialized state-of-the-art presentations designed for the advanced student or research worker. I have attempted in this text to present reasonably comprehensive coverage of the fundamental concepts, principles, and factual data that form the foundation of igneous and metamorphic petrology for the undergraduate student. At the same time, I have emphasized the uncertainties and limitations in our current knowledge, in the hope that the reader can discern lines of fruitful research in the future. The text is designed to be a balanced instructional tool, showing which properties of rock bodies are relevant to an understanding of their origin and how these properties can be specifically interpreted to provide genetic information on ancient geologic and planetary systems. Descriptive material on composition, fabric, and field relations on all scales of observation—from global to submicroscopic—is presented along with basic thermodynamic principles and experimental data. Abundant illustrations encourage the student to make direct observations of real rock bodies. The intimate association between the origin and evolution of rock bodies and tectonic processes and environments, especially on a global scale, is a pervasive theme of the text.

The level of treatment varies somewhat among chapters but is generally intended for the undergraduate geology major who has had introductory courses in physical and historical geology and mineralogy. Some knowledge of basic physics and calculus would be helpful but is not necessary, as I realize that many students in the United States enter geology with uneven backgrounds in science and mathematics. The organization of topics and the level of treatment have been designed for flexibility in classroom use. Some beginning courses of a more classical, descriptive nature can omit Chapters 7, 8, 9, 13, and 14. But the more advanced topics in these chapters could be considered in a subsequent intermediate-level course as a springboard toward graduate studies. Another option for a two-semester sequence is to cover Chapters 1 through 9, essentially on igneous petrology, followed by Chapters 10 through 16, on metamorphic, global, and planetary topics. Many of the essentially descriptive chapters (3, 4, 5, 6, 12) are more or less self-contained and could be omitted without unduly jeopardizing an understanding of the remainder of the text. Numerous cross-references link factual or descriptive material in one chapter with theoretical concepts in another.

In writing and illustrating this text, I became acutely aware of the necessity for balance between clarity and precision and depth and breadth. Because of my commitment to the undergraduate student, I tried to opt for clarity and breadth; I hope there are not too many errors of fact or principle that will have to be unlearned in subsequent in-depth studies.

Numerous persons helped to make this text a reality. W. Kenneth Hamblin contributed countless hours of discussion and guidance in the formative stages and set me on the right track with his persistent question, "Why is it important?" John Merrill, Bevan Ott, Paul Baclowski, Ronald Kistler, Alan White, Keith Cox, Herbert Shaw, J. G. Liou, Harry Green, Robert Baragar, Gordon Goles, Ronald Emslie, Michael Drake, Allan Cox, and Lehi Hintze reviewed individual chapters, and Dana Griffen, Paul Robinson, Howard Day, and especially Douglas Coombs courageously critiqued major parts of the entire text. Their encouragement, skill in ferreting out many errors, and constructive comments are greatly appreciated. Copy editing by Sean Cotter was remarkably astute and insightful. Without the skillful help of Keith Rigby, Revell Phillips, Stephen Leedom, Daniel Harris, Gary Harris, and Steven Knowles, the photography never would have been done. Richard Best assisted in assembling the figures. The forbearance of my colleagues in the Geology Department at Brigham Young University during the lengthy and trying gestation of the book was greatly appreciated. The positive attitude of John Staples, Inez Burke, and everyone else at W. H. Freeman and Company encouraged me when I needed it most; in the final stretch, the burden of reading a flood of proofs was lightened by the cheerful patience of Larry Olsen. Cheryl Nufer, Kathy Monahan, and Nancy Warner assisted in the final preparation of the line drawings. Finally, my special thanks to my wife Vivian, who typed countless drafts of the manuscript while keeping things running at home, and to Karen, Jenny, Karl, Teresa, Katrina, Richard, Laura, and Tyler, who missed many special outings but didn't complain.

November 1981   Myron G. Best

Symbols

A	Area; also a constant in various equations	J	Time rate of diffusion of an atom or molecule
a	Activity	K	Rate of a chemical reaction; also designates thermal diffusivity
C	The number of components in a thermodynamic system		
c	Specific heat	K_D	Distribution coefficient
D	Diffusion coefficient	m	Mass
E	Energy	P	Confining, load, or lithostatic pressure
E_a	Activation energy	P_f	Fluid pressure
E_i	Internal energy	P_i	Partial equilibrium vapor pressure of a component i, which may be H_2O, CO_2, O_2, S, and so on.
E_s	Surface energy		
E_t	Thermal energy		
F	The variance, or the degrees of freedom of a thermodynamic system	$\Delta Q, q$	Quantity of heat transferred
f	Fugacity	R	Gas constant; also the dimensionless Rayleigh number
G	Gibbs free energy	r	Radius
g	Acceleration of gravity	S	Entropy
h	Height of a column of rock	S_m	Entropy of melting
		T	Temperature

T_e	Temperature of melting or freezing	$\dot{\epsilon}$	Strain rate
t	Time	γ	Radioactive decay constant
V	Volume	μ_i^β	Chemical potential of a component i in a phase β
v	Velocity		
dw	Increment of mechanical energy, or work	η	Coefficient of Newtonian viscosity
X	Chemical composition expressed in terms of the mole fraction of a component	ϕ	The number of phases in a thermodynamic system
α	Coefficient of thermal expansion	ρ	Density
δ	A function that expresses the ratio of oxygen isotopes per mil in a sample; see Equation 2.1	σ	Normal stress
		σ_1	Maximum, or compressive, stress
ϵ	Strain	σ_3	Minimum, or tensile, stress
		τ	Shear stress

*Igneous and
Metamorphic
Petrology*

1
Overview of Fundamentals

Bodies of rock possess a wide variety of physical, chemical, spatial, and chronological properties that reflect the geologic processes responsible for their formation. **Petrology** is the study of these properties and processes.

In this chapter, we will consider how particular properties of bodies of rocks can provide specific clues to their origins. The petrologically most meaningful properties are the mineralogical and chemical compositions of the body and its fabric and field relations. The petrologist can interpret these four properties in terms of geologic events—that is, in terms of the way volcanoes, mountains, continents, the Earth, and even other planets originated.

All natural processes, including those by which bodies of rock are created, involve transfer and transformation of different forms of energy and movement of matter, establishing new states of more stable equilibrium. A particular state of mineralogical equilibrium is governed by the pressure, temperature, and chemical composition of the geologic system and is reflected in the mineralogical composition of the rock body. Thus, if a petrologist knows the mineralogical and chemical composition of a rock body, he can draw certain inferences about the pressure and temperature that prevailed when the minerals formed. Changing states of equilibrium print new mineralogical compositions over old. Petrologists can determine the history of such changes by analyzing the fabric of the rock body and comparing the relations of one body to another in space and time.

The most significant transfers and transformations of energy and movements of matter for geologic processes occur along boundaries of lithospheric plates. Thus, global tectonics and rock genesis are closely linked; specific types of rocks originate in specific plate environments.

1.1 THERMODYNAMICS

The most meaningful properties of a body of rock to the petrologist trying to understand its origin are its chemical and mineralogical compositions, its fabric, and its relationships to other bodies of rock in the field. To comprehend fully how and why these four properties are petrologically significant, we must first consider some basic relations between energy, matter, and states of equilibrium as governed, for example, by temperature and pressure. Fundamentally, all natural processes, including geologic processes of rock formation, involve movement of matter and transfer and transformation of energy; in brief, there is flow of matter and energy. The field of study embodying these concepts is **thermodynamics,** which takes its name from the flow of the most important form of energy, thermal energy.

A part of the Earth or Universe that is set aside in our mind for special attention is called a **system.** A geologic system is a group of mutually interactive geologic processes, such as occurs in a magmatic intrusion or along the margin of a continent where a lithospheric plate is being subducted. Systems can be large or small, momentary or lasting over millions of years. Two or more systems might overlap in space and time. Compared to the well-defined systems of the experimental chemist or physicist working in the laboratory, natural geological systems tend to be large, poorly definable, and ever changing, so that states of equilibrium are more commonly approached rather than firmly established. Everything outside of a particular system can be called the **surroundings.**

Forms of Energy

Energy is commonly defined in physics as the capacity for doing work. Energy exists in various forms and is manifest in terms of motion, or potential for motion, by the temperature and physical state of matter. A moving automobile, a tankful of gas, and a hot lava flow all have energy, but in different forms. In dealing with energy, it is far easier to determine the change in some form of energy associated with a process than it is to quantify the absolute amount in a static situation.

Kinetic energy is associated with the motion of matter. If a body has a mass m and a velocity v, its kinetic energy is given by the formula $\frac{1}{2}mv^2$. The movement of a lava flow, the movement of molecules in a gas, the fall of a meteor through the Earth's atmosphere, and the motion of volcanic ejecta thrown from an explosive vent are all examples of kinetic energy.

Potential energy is energy of position; it is potential in the sense that it can be converted, or transformed, into kinetic energy. A body held above the ground has potential energy because it gains kinetic energy when dropped. Potential energy can be equated with the amount of work required to move a body from one position to another in a potential field, in this instance, the gravitational field of the Earth. In lifting a body of mass m through a vertical distance h in the Earth's gravitational field, whose acceleration is g, the amount of **work** expended is

$$\text{work} = \text{force} \times \text{distance} = mgh$$

In other words, the **gravitational potential energy** of the mass above some reference level is given by mgh. It should be noted that h is calculated to be positive outward from the Earth. Thus, the farther the body is from the Earth, the greater the potential it possesses to do work, or the greater the kinetic energy that can be gained by it.

As Rumford realized two centuries ago, work performed on a system can cause changes other than increases in potential or kinetic energy. He noted that the work of boring cannons for the Bavarian army caused the cannons to become very hot, and he reasoned that the temperature of the cannon must reflect some form of internal energy created by frictional resistance to the drilling tool during the boring operation. Mechanical work seemed to be transformed into another, internal form of energy. In 1849, Joule established the equivalence between internal and other forms of energy, and today the fundamental unit of energy is called the joule (see Appendix A). The **internal energy** of a body includes both the potential energy and the kinetic energy of the particles within it. Potential energy derives from interatomic force fields, and kinetic energy derives from motion of the particles. An increase in the internal energy of a solid body, which is manifest by an increase in temperature, is associated with greater kinetic energy (faster motion) of particles. This motion can become sufficiently vigorous to break bonding forces between particles so that the body becomes a liquid, or even a gas. Because of the manifestation of internal energy by the temperature of the body, it is common practice to refer to internal energy in the context of temperature changes as thermal energy. For example, we say that as **thermal energy** is imparted to a body, its temperature will rise. The term **heat** is sometimes used synonymously with thermal energy in this context, but heat refers to transferred thermal energy. For example, heat moves from a person's body into the surroundings on a cold day and the body temperature falls.

In the more abstract mathematical formulations of thermodynamics, internal energy represents the total energy of the system. In this sense, changes in chemical (or if you like, electrical) energy associated with transfer of outer valence electrons of atoms in chemical reactions represent changes in internal energy. Likewise, nuclear energy, involving changes in the nuclei of radioactive atoms, is internal energy.

Still other forms of energy, such as electrical and magnetic energy, are important, but chiefly in man-made systems, not in geologic systems, where natural electrical and magnetic potential fields are too weak to cause movement of matter.

Transformation and Conservation of Energy

Transformation, or conversion, from one form of energy to another occurs in all natural processes. Trade-offs from one form of energy to another are rigorously and quantitatively conserved, and this conservation has been formulated as a universal law, verified by innumerable tests.

The Law of Conservation of Energy, also called the **First Law of Thermodynamics,** states that the total amount of energy in a system is conserved in any process. In some natural processes, such as occur in the nuclei of radioactive atoms, matter is converted into energy; consequently, mass has energy. A more general statement of the First Law therefore is as follows: The total amount of energy and mass in a system is conserved in any process.

A meteor falling to the Earth illustrates the First

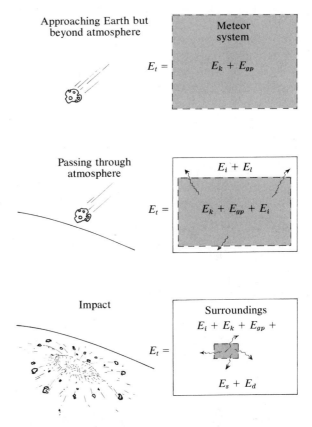

Figure 1-1 Conservation of energy between a meteor system, represented by the shaded boxes, and its surroundings. E_t, total energy; E_k, kinetic energy; E_{gp}, gravitational potential energy; E_i, internal energy; E_l, "light" energy; E_s, sound energy; E_d, mechanical work of deformation.

Law. From some point in space, a meteor is accelerated toward Earth by its gravitational field, gaining kinetic energy and losing gravitational potential energy, but conserving the total, as shown in Figure 1-1. As the meteor enters the Earth's atmosphere, its velocity decreases due to frictional drag imparted by air molecules; some kinetic energy of the meteor is thus transformed into a little thermal energy in the meteor and surrounding air molecules. Some electromagnetic energy, in the form of light, may also be produced by transformation of kinetic energy as the meteor reaches incandescent temperatures. If the meteor survives its flight through the atmosphere and impacts the ground, there is further transformation and transference of energy within the meteorite and with its surroundings. Overall, energy is conserved, but its flow is complex. Immediately before impact, the total energy is kinetic and a little internal. At impact, for a large meteorite, some of the total energy is converted into thermal energy, causing the meteorite and impacted rocks to become hotter; some becomes kinetic and potential energy as pieces of meteorite and Earth material are thrown out from the point of impact; and some is consumed in the mechanical work of compressing mineral grains, causing them to break and/or to transform to denser phases (for example, quartz → coesite → stishovite).

This qualitative example shows how the First Law helps one to analyze the flow of energy in natural systems. It emphasizes that processes in natural systems always involve flow of energy—both transfers and transformations—and that the total energy is always conserved.

Global Energy Budget

A quantitative inventory of the global energy budget of the Earth system serves to emphasize the flow of energy in geologic processes (Oort, 1970; Hubbert, 1971). The largest source of energy driving terrestrial processes, nearly 5,000 times all other sources, is radiant thermal energy from the Sun (see Figure 1-2). About 30% of this energy is directly reflected away from the Earth, and most of the remainder is consumed in the global hydrologic system of evaporation, movement of air masses, stream transport, and so on. Surface temperatures of the Earth are governed by the radiant influx from the Sun. Although thermal energy flowing out from the interior of the Earth—its **heat flow**—is minute compared to the solar influx, it is 100 times greater than all of the energy expended in volcanism, which also dissipates the Earth's internal energy, and earthquake activity. Even if

the energy flow in other major internal geologic processes, such as metamorphism and permanent deformation of rocks by folding and faulting, were considered, it is unlikely that the global heat flow would ever be surpassed. Apparently, then, there is still ample thermal energy within the Earth, 4.6 Gy (billion years; see Appendix A) after its creation, to drive major geologic processes, thus assuring geologists of reasonable employment opportunities for the near future!

With all the solar energy feeding into the Earth's atmosphere (plus a trickle from the interior), it may be wondered why the surface of the Earth is not hotter. In the operation of natural processes, a vast amount of low-temperature, nonusable thermal energy is dissipated into the atmosphere and the ground. This energy is radiated back out into space. The record in the rocks of the Earth indicates that the surface of our planet generally has remained within a relatively narrow temperature interval (0°C to 100°C) for all of decipherable geologic time, almost 4 Gy.

The origin of the thermal energy within the Earth is not completely understood, but some definite possibilities may be reviewed. Current views of the formation of the Earth involve accretion of solid particles condensed from a hot gaseous solar nebula some 4.6 Gy ago (see Chapter 16). As these particles accreted into a proto-earth, their gravitational potential energy and kinetic energy were transformed to thermal energy. Compression of these particles by additional accretion of more solids on top added a little more thermal energy (see box on page 6). If the matter accreted was homogeneous, a large amount of thermal energy was generated as the metallic core segregated from the silicate mantle. The decay of some short-lived radioactive elements, such as $^{26}Al \rightarrow$ ^{26}Mg with a half-life of 0.7 My (million years), during or shortly after accretion contributed additional thermal energy. It is debatable how much of the energy from these and possibly other primeval sources is still manifest in the elevated temperatures of the interior of the Earth.

In contrast to these primeval sources, there are others that have functioned continually throughout

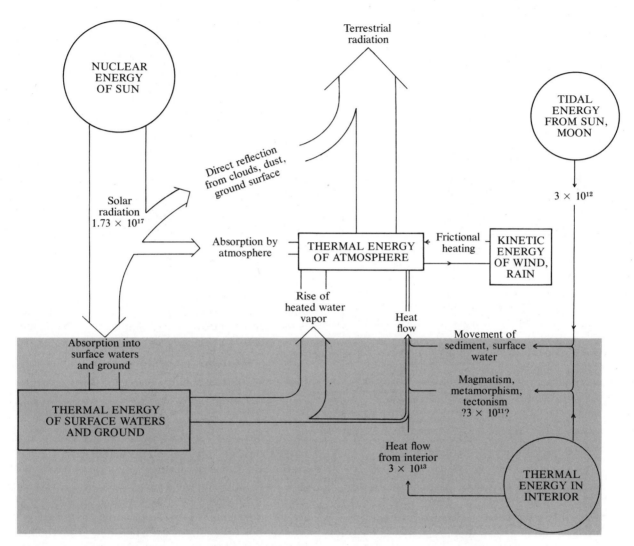

Figure 1-2 Flow of energy into and out of the Earth. Sources of energy are labeled in capital letters, energy flows in lowercase letters. Widths of energy flow paths are proportional to the amount of transferred energy. Numerical values are cited in joules per second. The shaded area represents the solid Earth.

the Earth's history. Tidal deformation of the solid Earth due to gravitational pull of the Sun and Moon is dissipated as thermal energy in the interior. A major source of thermal energy has been the decay of the long-lived radioactive isotopes, especially ^{40}K, ^{238}U, and ^{232}Th, whose half-lives are on the order of billions of years. As a parent atom decays to its daughter (for example, ^{40}K $\rightarrow$ ^{40}Ar), high-speed nuclear particles fly out of the atom, and their kinetic energy is transformed into thermal energy in the surrounding material. Although the exact distribution of these isotopes within the

Adiabatic Compression

Adiabatic compression of a body causes its temperature to rise; work done on the body is transformed into internal or thermal energy. The term **adiabatic** refers to an isolated system in which thermal energy neither leaves nor enters. For example, a hand bicycle pump gets hot mostly because of the compression of air during the pumping operation. But rocks are much less compressible than air, and hence the temperature rise within the Earth due to the compressional effect of the overlying mass is very small. Calculations indicate the temperature rise in the Earth is only a few tenths of a degree per kilometer depth. Assuming this **adiabatic gradient** is 0.3°C/km, at a depth of 100 km the temperature would be 30°C, plus 20°C for the mean surface temperature, or 50°C. This value is only a small fraction of the actual temperature at this depth (see Figure 1-15), judging from all geophysical and petrological data, such as melting temperatures of ~1500°C for magma production. Thus, within at least a few hundred kilometers of the surface, self-compression can only account for a fraction of the internal temperature of the Earth.

core, mantle, and crust of the Earth is controversial (because of limited facts), there is a remarkable coincidence between the measured heat flow at the surface and heat flow that is calculated by assuming that the bulk chemical composition of the Earth approximates a chondritic meteorite. (Evidence will be reviewed in Chapter 16 indicating that the so-called carbonaceous chondrite meteorites likely approximate the overall chemical composition of the Earth. The concentrations of radioactive K, U, and Th have been determined in the laboratory for this class of meteorites, and this gives their approximate concentrations in the Earth.) We have, then, an apparent paradox in that none of today's heat flow from the interior need be derived from the primeval thermal energy sources discussed in the previous paragraph. The rate of radioactive decay is exponential, so the thermal energy produced by Th, U, and K 3 Gy ago was roughly three times today's production, and 4.6 Gy ago, when the Earth was born, the production was roughly six times greater.

Although the Earth is a giant heat engine that both generates and consumes thermal energy (by melting of rock to form magma, deformation of rocks, and so on), the operation of geologic processes cannot function solely on this form of energy. Without gravity, matter would be dispersed indefinitely by the thermal process of expansion, melting, and vaporization. Gravity pulls matter together, and in so doing causes matter to give up gravitational potential energy.

The constant interaction between thermal and gravitational energy drives internal processes of magmatism, metamorphism, and tectonism (deformation of rocks). This significant concept will be amplified in the last section of this chapter.

Geologic Processes and Energy Flow

Heat is thermal energy moving between bodies by virtue of a difference in temperature between them. The transfer or flow of heat is one of the most significant phenomena in geologic as well as man-made systems. Everyday experience shows that heat flows spontaneously from a hotter to a colder body. Thus, ice cubes melt as heat is transferred into them from surrounding warm water; coolant in a car radiator transfers heat from the hot engine to the cooler air; and a hot magmatic intrusion transfers heat to its wall rocks. Measurements of the transfer of heat show that different substances in different states—solid, liquid, or gas—possess different **specific heats** (weight basis), or **heat capacities** (molar basis). The heat transferred, ΔQ, the mass of the substance, m, the specific heat, c, and the change in the temperature, ΔT, due to the transfer are related by

$$\Delta Q = mc \, \Delta T \qquad (1.1)$$

This equation shows that substances with large specific heats have the capacity to store more thermal energy and that large transfers of heat into them cause somewhat smaller rises of temperature. Water has a greater specific heat than rocks and magmas. Equation (1.1) is important in showing the relation between incremental changes in the heat brought into or out of a body and corresponding changes in the temperature of the body.

Heat is transferred in three different ways—by conduction, convection, and radiation. **Conduction** involves transfer between neighboring bodies of material in physical contact with one another by virtue of the temperature difference between them. The initial temperature difference or gradient between the two bodies is smoothed and can eventually be eliminated by the conduction process. A metal pan on a hot stove gets hot due to conduction. Wall rocks surrounding a magmatic intrusion become hot by conduction. Metals are far better thermal conductors than silicate minerals and hence than most rocks, which are better considered as thermal insulators. The **thermal conductivity** of metals is approximately 100 times that of rocks. **Convection** involves physical movement of fluid material that, having absorbed heat in one place, mixes with cooler fluid and gives up heat in another place. If the motion of the fluid is caused by density differences associated with differential expansion of the fluid at contrasting temperatures, it is **natural convection.** Water in a pan on a hot stove convects naturally as it warms and expands at the bottom, flows upward, and mixes with and displaces cooler water at the top of the pan. In geologic systems natural convection occurs in subterranean magma chambers and possibly in the Earth's mantle. **Forced convection** occurs if the fluid is made to move by some external influence. A fan blowing in a room forces air to convect; in the Earth, subducting lithosphere may force the adjacent mantle to convect. In some geologic systems a third type of convection, **penetrative convection,** occurs when fluid, such as hot magma, moves through openings in solid rock, transferring heat by conduction as the fluid moves. **Radiation** involves emission of electromagnetic energy from the surface of a hot body into its cooler surroundings. As the amount of thermal radiation emitted is proportional to the fourth power of the temperature (T^4), this mode of heat transfer is most effective for very hot bodies. Within the Earth's interior, the chief mode of heat transfer is probably by convection, even though the rate of movement of matter may be as low as 1 mm/y.

The three modes of heat flow—by conduction, convection, and radiation—are **thermal processes.** Work is a **mechanical process** in which force acts on a body as it is displaced; examples include folding, faulting, and breaking a body into smaller pieces, compressing it into a smaller volume, or causing phase transformations such as andalusite $\rightarrow$ kyanite. If an energy transfer involves a change in mass or concentration of chemical elements, it is a **chemical process.** Figure 1-3 illustrates some examples of these three types of processes, which collectively may be called **thermodynamic processes.** Those in geologic settings may also be called **geologic processes.**

King (1962, pp. 5–7) describes some of these processes as follows:

A change in internal energy often is accompanied by a change in temperature. That the temperature of a system may be increased by mechanical and chemical processes, as well as by thermal processes, is demonstrated in a simple experiment. A closed cylinder with a piston in one end is filled with a mixture of dry hydrogen and oxygen gases in the ratio of two parts by volume of hydrogen to one of oxygen. Initially the system is at room temperature, but...heat can readily flow through...[the walls of the cylinder] into the system. If the cylinder is immersed in a warm bath and the volume of gas in it is held constant, the temperature of the system is observed to rise by purely thermal processes. Suppose now that the cylinder is thoroughly insulated to prevent heat exchanges with the near-surroundings, that is, is made adiabatic.... When the gas in the insulated cylinder is compressed by means of the piston, again the temperature of the system is observed to

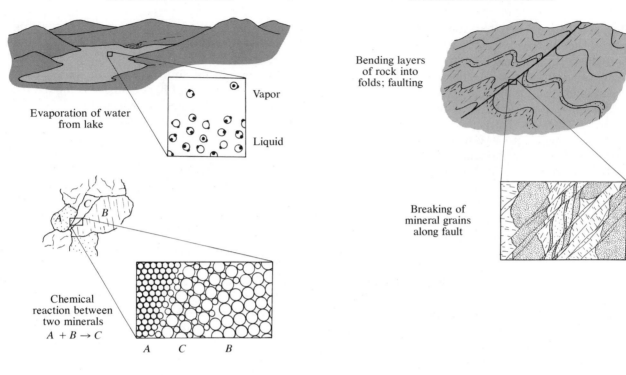

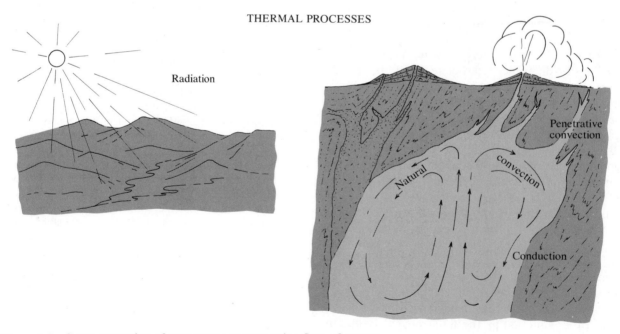

Figure 1-3 Some examples of processes accompanying flow of energy.

increase, this time by a purely mechanical process. Finally, suppose the cylinder is not only thermally insulated, but also has its piston held immovable to prevent any work exchange with the near-surroundings. By means of an electrical spark or a suitable catalyst such as platinized asbestos the gases are induced to react and form water, and the temperature of the system increases by a purely chemical process. Clearly temperature is a physical property of the system which can be changed by any one or more of the energy processes, not just by heating alone.

State Properties

The preceding paragraph quoted from King introduces us to the more general topic of states of systems and changes in states. We saw how changes in thermal energy caused certain changes in the system, not only in temperature but also in volume (after compression by the piston) and in density (after the oxygen and hydrogen had reacted to form water). In his illustration, we can recognize four different **states** of the system—the initial state of oxygen and hydrogen gas, the warmed state of the gases, the warmed compressed state of diminished volume, and the state of water. Each of these states can be characterized, or defined, by a particular set of **state properties,** sometimes called **variables of state,** including temperature (T), pressure (P), volume (V), mass (m), density (ρ), internal energy (E_i) and chemical composition (X). The chemical variable of state is the molar fraction of chemical constituents comprising the system, $X_{SiO_2}, X_{FeO}, X_{H_2O}$, and so on. For brevity, we simply denote it as X. Heat and work are not state properties, as are P, T, and X, but energy in transit whose manifestation depends on the path followed.

Properties of state may be classified as either intensive or extensive. **Intensive variables** have definite values at each point within a system; temperature, pressure, and concentrations of particular chemical species are intensive. In contrast, **extensive properties** depend on the total quantity of matter in the system; hence internal energy, mass, and volume are extensive.

How many independent properties must be specified in order to define a particular state completely and unambiguously? Could only three be specified, all others being fixed or derived from these three? Or could there be a fourth that could vary independently, changing the state of the system? To answer this question, we must determine what properties are of greatest concern, which ones can be ignored, and the nature of the system (Turner and Verhoogen, 1960, p. 25). Is the system **open** or **closed** to the exchange of matter? Is the system **isolated** with respect to transfer of energy across its boundary with the surroundings? It will be shown later that the following statement holds true: The intensive variables P, T, and X are one important set of state properties that can completely define a homogeneous system in chemical equilibrium.

Many geologic systems are so complex (varying significantly in both space and time) as to make definition of these variables quite difficult. Nonetheless, a reasonable first approximation, or **model,** can often be adopted that adequately represents the system.

An important feature of systems that have attained a state of no change, or equilibrium, is that the intensive variables of the system such as P, T and X are independent of any prior states and of the paths between prior and final states. In the cylinder of oxygen and hydrogen gas previously described, the two states of warmed gas could have identical P, T, and X and could therefore be the identical state; yet they arrived at that state by either of two process paths—insertion of thermal energy or compression by a piston—from prior states that could have been quite different.

As will be shown later, the P, T and X of the final state of a system is recorded in the mineralogical composition of a rock. Suppose a particular rock is comprised of quartz, feldspar, and biotite (see Figure 1-4). Whether it was previously a volcanic tuff, a rhyolite flow, a granitic intrusion, or an arkosic sandstone cannot be answered only from the minerals present, which only record the intensive variables P, T, and X.

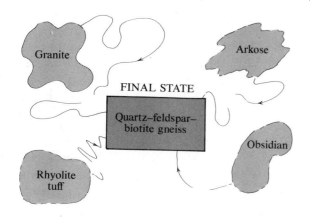

Figure 1-4   It is impossible to decide only on the basis of the mineralogical composition, which manifests the *P*, *T* and *X* of the final equilibrium state of a system, what the prior state was. Some possible prior states are shown as irregularly shaped areas.

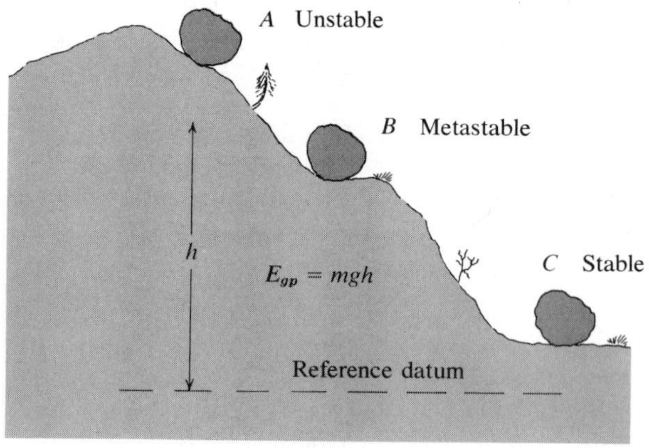

Figure 1-5   Concept of stability illustrated by three boulders possessing different amounts of gravitational potential energy.

Direction of Changes in State

The Earth is a dynamic, ever-changing set of interacting systems. Internal geologic processes build up surface features, such as mountains and volcanoes, and hydrologic processes wear them down. In what direction do changes in states of systems proceed? How may we predict the direction of change? This problem is so important that we will consider it at some length, first in general qualitative terms using some familiar examples, and then more quantitatively, using thermodynamic functions.

Stability and equilibrium.   Changing states are related to concepts of stability and equilibrium, and these can be introduced by a familiar illustration from nature.

Let each of the three boulders in Figure 1-5 represent different possible states of a system. Which state is the most stable? The least stable? Common sense tells us the state represented by boulder *C* in the valley is the most stable, and boulder *A* highest on the hill slope is the least stable. These boulders have different amounts of gravitational potential

energy, the most stable one in the valley having the least. The most **stable state,** relative to nearby possible states, is the one of lowest energy.

The most stable boulder in the valley may be rocked from side to side (say, by an earthquake) but will return spontaneously to its original position. In contrast, any slight disturbance of boulder *A* will grow, and recovery to its original **unstable state** is not spontaneous. Boulder *B* represents a **metastable state** whose energy is less than immediately adjacent possible states but more than some distant state. The metastable state will return spontaneously to its original configuration if subject to only a small disturbance—for example, slight rocking of boulder *B* in the declivity on the hillslope. But a larger disturbance will cause it to move toward a more stable state—once the boulder is over the lip of the declivity, it moves downhill without stopping to stable state *C*.

Each of the boulders in Figure 1-5 is in equilibrium because none is accelerating due to unbalanced forces. State *A* is an unstable equilibrium; any small disturbance or applied force will cause it to move out of that state. State *B* is a metastable equilibrium, and *C* is a stable equilibrium. A system is said to be in **equilibrium** when the net result of forces acting on the system at rest is zero. Thus, equilibrium is a state of no change.

Although this illustration of stability and equilibrium refers to a special mechanical state governed by gravitational potential energy, the concepts are equally valid for other types of changes in states. Thus mechanical equilibrium as a whole prevails when there is no unbalanced force; chemical equilibrium prevails when no chemical reactions (diffusion, solution, and so on) are occurring; and thermal equilibrium prevails when all parts of the system are at the same temperature and no heat is flowing. The general condition of thermodynamic equilibrium is satisfied when all conditions for each of the three types of equilibrium are satisfied.

In evaluating the stability and equilibrium of any system, the nature of possible flows of energy that could cause a change in state must be carefully specified. Thus, a particular system, such as boulder *A* in Figure 1-5, might be in a state of stable chemical and thermal equilibrium but unstable with respect to gravitational potential energy. Boulder *C* would be chemically unstable if it were limestone sitting in slightly acid water.

Moderation theorem. We can now specify, qualitatively, a criterion for determining the direction in which a particular state of thermodynamic equilibrium will change because of some sort of process (flow of energy) that changes a state property. The criterion is sometimes called the moderation theorem, but a more familiar term for it is **LeChatelier's Principle:** If a system in a state of equilibrium is perturbed, it will respond in such manner as to moderate the effect.

An example of this principle is afforded by the endothermic (consuming thermal energy) dissociation at 1200 K (degrees Kelvin; see Appendix A), $P = 1$ atm, of calcium carbonate yielding lime and carbon dioxide:

$$1.24 \times 10^5 \text{ J/mole} + CaCO_{3_\text{solid}} =$$
$$CaO_\text{solid} + CO_{2_\text{gas}}$$

If the three compounds are initially in a state of equilibrium (forward reaction is just balanced by the reverse), insertion of thermal energy into the system promotes dissociation of more $CaCO_3$ into $CaO + CO_2$, consuming thermal energy and thereby moderating the temperature increase. For another example, consider a system in which the two silica polymorphs, quartz and coesite, coexist in stable equilibrium. If the system is mechanically squeezed, pressure is increased and the denser coesite will grow at the expense of quartz, diminishing the volume of the system and moderating the pressure increase.

One of the aims of thermodynamics is to find mathematical functions that can be applied to systems in which flows of energy are complex and cause more than one state property to change. Different functions have been formulated for different purposes, but the most useful one for us is the Gibbs free energy. Its formulation depends on several concepts, the first of which concerns entropy and the Second Law.

Entropy and the Second Law of Thermodynamics. Suppose that, along one of the paths leading to the assemblage of quartz, feldspar, and biotite in Figure 1-4, there is an infinitesimal change in internal energy, dE_i, brought about by adding an infinitesimal amount of heat, dq, to the system. P, T, and X all remain constant during this process. This type of change is common in geologic systems; for example, if the dissociation of $CaCO_3$ into CaO plus CO_2 previously cited proceeds slowly, the P and T remain fixed as long as both products and reactants coexist. As a result of absorbing heat at constant P and T, the system expands, doing work on the surroundings. From the First Law, the increase in internal energy due to heat absorbed is diminished by the amount of work done on the surroundings:

$$dE_i = dq - dw \qquad (1.2)$$

The work done in expansion against the surroundings is

$$dw = \text{force} \times \text{distance} =$$
$$P \times \text{area} \times \text{distance} = PdV \quad (1.3)$$

giving

$$dE_i = dq - PdV \quad (1.4)$$

where dV is the infinitesimal increase in volume.

Changes in state can be either reversible or irreversible and spontaneous. A **reversible** change involves only small departures from equilibrium; at any point in the path between states, the change can be made to go in the opposite direction; thus a reversible path is defined by many intermediate equilibrium states. If left to themselves, most natural changes are **irreversible** or spontaneous, proceeding quickly in finite steps. (One way to distinguish between these two paths would be to make a movie of the change and then run the movie backwards. If the backwards version does not ever happen spontaneously, or appears silly, the path is irreversible.) An ice cube in a glass of warm (20°C) water melts irreversibly, but at 0°C, any theoretically slight melting is reversible.

According to the First Law, energy is conserved in any change—whether it is reversible or irreversible—so why are virtually all processes in nature irreversible? Consider a bottle filled with neatly alternating layers of white and red sand. When shaken vigorously, the white and red sand grains become thoroughly mixed. Was the change in the arrangement of the contents of the bottle reversible or irreversible? The change was, of course, irreversible. Can further shaking organize the grains back into their original neat layers? No. What was irretrievably lost in the shaking process? Order.

The increase in disorder in many spontaneous processes is not always easy to perceive; it can be disarmingly subtle. Consider the organization of energy in the system of warm water (20°C) and ice. Higher-energy molecules of the water are initially separated from the lower-energy molecules of the ice; this separation or organization constitutes a degree of order. At the end of the irreversible path, when the ice has melted and the glass of water is

uniformly at a temperature somewhere between 0° and 20°C, the faster, higher-energy molecules have mixed with the slower ones, the concentration of energy has evened out, and order has been lost.

We may now ask: Can order ever be increased? The answer is yes. The sand grains can carefully be sorted and placed in their original neat layers and the glass of water can be frozen by placing it in a freezer. However, additional energy has been expended in the surroundings outside the system to do the ordering within it. The *total* disorder of the universe has increased in creating better order in a part of it. One statement of the **Second Law of Thermodynamics** is as follows: If a system is isolated, so no energy can flow into it, then any spontaneous process in it causes an increase in disorder in the system.

Another expression of the Second Law derives from the concept of energy distribution. In thermal processes, thermal energy always flows from hotter to cooler regions—hot magma intruded into cooler rocks causes a flow of heat into the wall rocks, where work is done in volumetric expansion and chemical reactions of recrystallization occur. Without this uneven concentration of thermal energy, or the opportunity for energy to flow, no work can be done. Water in an enclosed lake on a high plateau has lots of potential energy relative to sea level, but so long as it is isolated from sea level and cannot flow, its concentration is uniform and no work can be done. However, in the natural course of events, a river drains the lake into the sea, forming a process path between the high and low potential energy levels, so that work can be done—driving turbines to generate electricity, eroding the river channel, transporting sediment, and so on. These are but two examples of innumerable natural processes that, operating spontaneously by themselves, even out the concentrations of various forms of energy. Thus the Second Law can also be stated as follows: Natural processes, if left to themselves, eventually tend to even out the concentration of any form of energy.

In thermodynamics, there is another state property called the **entropy,** denoted by the symbol S, which can be equated with the disorder of particles in a system (such as the red and white sand grains in the bottle), or with the evenness in concentration of energy. The greater the particle disorder, the greater the entropy. The more even the concentration of some form of energy, the greater the entropy. The greater the entropy, the less opportunity there is to perform work.

The Second Law of Thermodynamics states that entropy increases in every spontaneous irreversible process in an isolated system. The stable state of a system is one of maximum entropy.

In thermodynamics, entropy is defined as follows: "(1) in any reversible process, the infinitesimal change in entropy, dS, is measured by the heat received by the system divided by its temperature and (2) in any irreversible spontaneous process dS is greater than this amount" (Turner and Verhoogen, 1960, p. 8). Symbolically,

reversible process $\quad\quad dS = dq/T \quad$ (1.5)

irreversible process $\quad\quad dS > dq/T \quad$ (1.6)

Gibbs free energy. The law of increasing entropy during a spontaneous change in an isolated system, reaching a maximum in the stable state, has great importance. However, as the systems of concern in geology are generally not isolated, it is perhaps a little impractical. For nonisolated systems in which energy is flowing in or out, it was previously suggested (see p. 10) that some form of energy is minimized as the system changes to a more stable state. For the general thermodynamic system, the drift toward a state of stable equilibrium must be governed by compound variations in both entropy and internal energy. As it turns out, mechanical work done on the system is a third factor.

These three factors can be combined to define a new mathematical energy function for thermodynamic systems of constant chemical composition:

$$G = E_i - TS + PV \quad\quad (1.7)$$

G is a new state variable, the **Gibbs free energy**, named after one of the pioneers of thermodynamics, J. W. Gibbs. G can be considered as thermodynamic potential energy that, like gravitational potential energy in systems involving only gravity, must be less than the energy of nearby states for stable equilibrium to prevail. Equation (1.7) is formulated in such a way that G is a minimum value if E_i or PV are as small as possible and S is a maximum. In the general case, however, one or two of the three factors may not be an optimum value, and there must then be a compromise. For example, to make entropy large, energy may have to be expended, but so long as S is larger than E_i, perhaps with PV small, then G can still be minimized in a certain change toward a state of stable equilibrium.

But as we are more interested in changes in states and the accompanying changes in state variables, let us differentiate this expression, obtaining

$$dG = dE_i - TdS - SdT + PdV + VdP \quad (1.8)$$

but from equation (1.4)

$$dE_i = dq - PdV$$

so

$$dG = dq - TdS - SdT + VdP \quad\quad (1.9)$$

Most important changes in geologic systems occur at constant P and T—for example, a crystal of pyroxene grows from a silicate melt at 1200°C and 1 atm pressure. If P and T are constant, then from calculus $dP = 0$ and $dT = 0$. Combining with equations (1.5) and (1.6) we find that

$$dG = 0 \quad\quad\quad (1.10)$$

for thermodynamic equilibrium at constant P and T, and

$$dG < 0 \qquad (1.11)$$

for any spontaneous process at constant P and T.

It happens that the state properties S, V, P, and T can be evaluated for geologic systems at least qualitatively and in many cases quantitatively, so that rigorous statements can be made predicting the direction that a geologic system will change, or whether it is in an equilibrium state of no change.

It may be noted in Equation (1.9) that P and T are the relevant intensive state variables governing the energy function. In systems where chemical composition can change, X is also a relevant intensive variable. Thus P, T, and X are all important in determining thermodynamic equilibrium.

1.2 KINETICS

The basic questions with regard to changing states are: Where is the system going, and how fast will it get there? So far we have answered only the first question and have found that systems move predictably toward a more stable state of lower free energy. With regard to the second, it has been found that there is usually a certain amount of inertia or resistance to the changing of a state. There is seldom an unrestricted mobility or access of a system to a more stable nearby state. Thousands of high-energy sports fans, for example, cannot immediately get out of a football stadium. Water may rest in a high mountain lake for a very long time, even though it is unstable relative to the ocean, to which it will ultimately move. Assemblages of coexisting minerals in igneous and metamorphic rocks formed at high temperatures are mostly unstable at the wet surface of the Earth; yet these rocks will not convert into stable minerals, such as clay, after sitting in museum drawers in our lifetime, and they had not done so for perhaps centuries before they were collected.

A limitation of the application of classical thermodynamics to any system is that it cannot predict *when* a particular change in state will occur— whether in 10 minutes, 10 years, or 10 million years. In other words, thermodynamics has noth-

ing to say about **kinetics,** or time rates of motion of matter or flow of energy.

Systems must usually surmount an energy barrier, called the activation energy, before a change in state can occur. In Figure 1-6, for the system in unstable state A to change toward stable state B, the system must first surmount, or overcome, the **activation energy** barrier, E_a. In the case of the boulder midway up the hill in Figure 1-5, the activation energy is associated with an unbalanced force—such as that due to an animal, another rolling boulder, or running water—that can push it over the lip of the small declivity in which it rests. Metastable volcanic glasses may be made to crystallize by adding thermal energy to the system, raising their temperature. Stable equilibrium in chemical systems is generally more readily attained at higher temperatures than at lower ones. This accords with our common experience that chemical processes happen faster at higher temperatures. Because of the greater motion of atoms in bodies at higher temperatures, the activation energy barrier is more easily exceeded. Magmatic and metamorphic systems reach states of stable thermodynamic equilibrium more readily than do low-temperature sedimentary systems.

The rate at which a state of stable thermodynamic equilibrium is reached from some prior state depends exponentially upon temperature, according to the relation

$$K = Ae^{-E_a/RT} \qquad (1.12)$$

where K is the rate, R is the gas constant, A is a constant, T is the absolute temperature, and E_a is the activation energy (Turner and Verhoogen, 1960, p. 48).

Any change of state, a chemical reaction, for example, has an associated rate, K. For geologic systems, a particular change in state generally will involve several rates K_1, K_2, K_3, and so on. One of these—the slowest one—determines the rate of the overall change. Similarly, the determining rate for exit of frenzied football fans from the stadium is

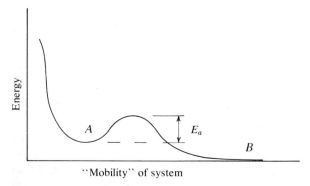

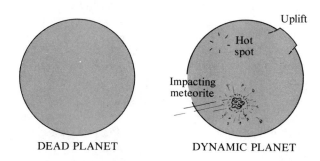

Figure 1-6 Concept of activation energy, E_a. For a system to move from state A to the more stable state B, it must first surmount an energy barrier, E_a.

Figure 1-7 Contrast between a dead planet (a) and a dynamic planet (b), in which there is unevenness in concentration of energy so that rock-forming processes can occur.

not how fast they can run, but how many can squeeze through the exit gates in a unit of time.

1.3 CHANGING STATES AND CREATION OF ROCK BODIES

Thus far in this chapter, we have dealt mostly with rather abstract thermodynamic concepts. Our objective now is to apply these concepts to specific rock-forming geologic systems.

All geologic processes—magmatism, metamorphism, sedimentation, tectonism—involve flow of energy and movement of matter that change the state of the geologic system. Suppose the Earth had no surface relief (no mountains, valleys, ocean trenches) and was covered to a uniform depth by a lifeless, world-encircling ocean over the solid interior (see Figure 1-7a). Suppose further that no heat was received from the Sun, the internal temperature of the Earth was uniform, and density did not vary laterally (only increased uniformly with depth). Could any new body of rock ever form on such an Earth in a state of maximum entropy? The answer is an unequivocal and emphatic No!

It will be left as an exercise for the reader to consider how the three processes shown in Figure 1-7b that have caused an uneven distribution of energy can lead to formation of rocks.

In many geologic systems, the large dimensions and the long duration of activity involved necessitate an arbitrary distinction between the initial state and the final state. Geologic systems tend to behave irreversibly; the ongoing continuum of processes and changing states in a dynamic Earth implies that one cannot always rigorously say when or where a particular state started changing, or whether a prior state had reached stable equilibrium. In nonequilibrium, irreversible processes, only local, temporary states of equilibrium can be recognized. For example, along margins of continents where subduction is occurring, multiple magmatic intrusion, metamorphism, and deformation constitute a continuum of processes extending over tens of millions of years. Local systems on a smaller scale might be recognized within this vast, complex system.

Significance of the Rock Body

Up to this point, our approach has been deductive, starting with general principles and illustrating them by specific examples. In practice, however, the geologist is faced with an entirely opposite, inductive problem: Given a particular rock body and its observable properties, what was the nature of the ancient geologic system in which it was created? What geologic processes of energy flow and movement of matter were involved? From the only tangible record of the system—the body of rock—we must work backwards to try to comprehend the system.

According to the principles of classical thermodynamics, the final equilibrium state of a system as defined by the intensive variables *P, T,* and *X* gives no indication of prior states. So, even if we could interpret *P, T,* and *X* from the mineralogical composition of the rock, that would be the end of our rock history. No earlier history, as for example, the nature of a metamorphic rock before its recrystallization, could be elucidated.

However, from a geological standpoint, we realize that there are other important aspects of any rock body in addition to its mineralogical composition that provide very significant information regarding its history. These other aspects include the fabric and the field relations of the body and its chemical composition.

The mineralogical and chemical composition, fabric, and field relations of the rock body provide all the essential information regarding its origin. This fundamental concept of petrology is one of the dominant themes of this book; it will be only briefly amplified in the following sections.

Significance of fabric and field relations of rock bodies. **Fabric** and **field relations** both encompass all of the geometric, noncompositional properties of the rock body. Although the distinction between them is not always clearly definable, fabric has to do mainly with internal features within the body, whereas things external to the body pertain to field relations. But, of course, any body larger than the size of a hand specimen must be investigated as it sits in place in the field, and so in this regard study of fabric becomes as much an exercise in field work as in laboratory work.

Fabric is an inclusive term for both texture and structure. **Texture** has to do with relationships between grains of a rock, specifically:

1. Degree of crystallinity—proportion of crystalline material to glass

2. Grain size
3. Shapes of grains

Structure embodies the mutual relations of groups or aggregates of grains such as bedding or folds in bedding.

Fabric furnishes information regarding the flows of energy that caused the change in state of the system—in other words, information regarding the geologic processes that produced the rock body. For example, many aspects of the texture of an igneous rock are determined by the manner of extraction of thermal energy, or of dissolved gas, from the magma body. Large grains testify to slow loss, whereas crystals embedded in natural glass suggest an early episode of slow cooling followed by rapid exit of energy or gas from the magma. Columnar amphiboles in a granite that are oriented like logs in a fast-moving river connote differential flow in the magma containing the amphiboles. Petrologic implications of many other types of fabric will be discussed in succeeding chapters.

Field relations pertain to the size, shape, and dimensions of the rock body, its chronologic setting relative to neighboring bodies, and the nature of its external contacts with neighboring bodies—whether sharp or gradational, parallel or cross-cutting; whether fabric varies in proximity to the contact; and so on. Study of these field relations in stratified sedimentary rock bodies is called **stratigraphy.** Conventionally, this term is not used in igneous and metamorphic contexts, although the principles of stratigraphy—superposition, cross-cutting relations, and correlation—are universal.

Field relations provide insight into the causes for changes in state, or why there was flow of energy. For example, it is reasonable to suppose that a body of volcanic glass overlying beds of volcanic tuff was emplaced on top of the ground on the tuffs and in contact with the atmosphere where it cooled very quickly, rather than being emplaced deep within the crust.

Significance of composition of rock bodies. The composition of any rock can be considered from different points of view, as shown in Figure 1-8.

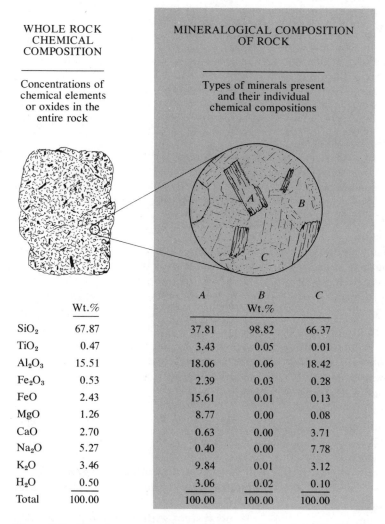

WHOLE ROCK CHEMICAL COMPOSITION		MINERALOGICAL COMPOSITION OF ROCK			MODAL COMPOSITION OF ROCK	
Concentrations of chemical elements or oxides in the entire rock		Types of minerals present and their individual chemical compositions			Proportions of the different types of minerals constituting the rock	

MODAL COMPOSITION OF ROCK

	Vol.%
A	13.7
B	15.9
C	70.4
Total	100.0

		A	B	C
	Wt.%		Wt.%	
SiO$_2$	67.87	37.81	98.82	66.37
TiO$_2$	0.47	3.43	0.05	0.01
Al$_2$O$_3$	15.51	18.06	0.06	18.42
Fe$_2$O$_3$	0.53	2.39	0.03	0.28
FeO	2.43	15.61	0.01	0.13
MgO	1.26	8.77	0.00	0.08
CaO	2.70	0.63	0.00	3.71
Na$_2$O	5.27	0.40	0.00	7.78
K$_2$O	3.46	9.84	0.01	3.12
H$_2$O	0.50	3.06	0.02	0.10
Total	100.00	100.00	100.00	100.00

Figure 1-8   The three compositional aspects of a rock body. Chemical composition of the whole rock and of its constituent minerals is conventionally expressed on the basis of weight percent of the oxides; other possible bases are molecular percent and percent of the elements. The modal composition is the proportion of constituent minerals, usually expressed on a percent volume basis.

The **bulk chemical** (or **whole rock**) **composition** is expressed in terms of weight concentrations on a percentage basis of chemical species such as SiO$_2$, Al$_2$O$_3$, H$_2$O, CaO, TiO$_2$, and so on. By **mineralogical composition** is meant not only the different types of minerals present—*A, B,* and *C* in Figure 1-8 could be biotite, quartz, and feldspar, for example—but also their individual chemical compositions. Still another aspect of the composition of the rock body is its **modal composition,** or **mode,** which is the volumetric proportions of the different minerals in the whole rock. As might be anticipated, there is a close correlation between these three different types of composition. Obviously, the more SiO_2 there is in a rock (bulk chemical composition), and if quartz is stable (mineralogical composition), there is likely to be more of it (modal composition). Rocks with abundant (modal composition) biotite (mineralogical composition) will have considerable potassium in the rock (bulk chemical composition).

The bulk chemical composition provides insight into the source of matter comprising the rock. Igneous rocks with high concentrations of Mg and Fe are likely to have formed from magmas originally derived from the mantle. A metamorphic rock comprised chiefly of CaO and CO_3 was probably a limestone before metamorphism.

The mineralogical composition of a rock body is a function of the *P, T,* and *X* of the last state of stable equilibrium in the system. The way in which the mineralogical composition of a rock body manifests the *P, T,* and *X* of the system when the minerals formed can be illustrated with reference to a system composed of the chemical constituents Na_2O, Al_2O_3, and SiO_2. What minerals would comprise the system? Albite, nepheline, jadeite, and quartz are possibilities, but which one(s) and in what proportions if more than one? Two reactions may be written relating these minerals:

$$NaAlSi_2O_6 + SiO_2 = NaAlSi_3O_8$$
$$\text{jadeite} \quad\;\; \text{quartz} \quad\;\; \text{albite}$$

$$NaAlSiO_4 + 2SiO_2 = NaAlSi_3O_8$$
$$\text{nepheline} \quad \text{quartz} \quad\;\; \text{albite}$$

Quartz and nepheline cannot be a stable pair of minerals under any conditions; hence, the direction of the second reaction is invariably to the right, forming stable albite. Now, if the relative concentrations of Na_2O, Al_2O_3, and SiO_2 are exactly $1:1:2$, then under certain *P–T* conditions, we may expect nepheline to form ($Na_2O + Al_2O_3 + 2SiO_2 \rightarrow 2NaAlSiO_4$). But if at the same *P* and *T* the concentration of SiO_2 in a system is twice as great as before, then nepheline and albite may

form:

$$Na_2O + Al_2O_3 + 4SiO_2 \rightarrow$$
$$NaAlSiO_4 + NaAlSi_3O_8$$
$$\text{nepheline} \quad\;\; \text{albite}$$

If an even greater concentration of SiO_2 were present at the same *P* and *T*, then the system might consist of albite plus quartz; for example,

$$Na_2O + Al_2O_3 + 10SiO_2 \rightarrow 2NaAlSi_3O_8 + 4SiO_2$$
$$\text{albite} \quad\;\;\; \text{quartz}$$

Hence the relative, not the absolute, concentrations of the different chemical constituents in a system dictate its mineralogical composition. Relative concentrations in thermodynamic studies are cast in the form of the molar fraction X_i of a particular chemical constituent *i*.

It might be wondered why, in the above reaction, jadeite did not form instead of albite plus nepheline. If *P* is sufficiently high, jadeite is stable because the molar volume of jadeite is less than albite plus nepheline. According to the moderation theorem, increasing the pressure on the equilibrium

$$2NaAlSi_2O_6 = NaAlSi_3O_8 + NaAlSiO_4$$
$$\text{jadeite} \qquad \text{albite} \qquad \text{nepheline}$$

will favor the lower volume state. Hence *P* can influence the mineralogical composition of a system.

If the system with the ten moles of SiO_2 considered above were at sufficiently high temperature, it might be comprised of a silicate melt plus quartz; no albite would be stable. Hence *T* also governs the mineralogical composition of a system.

Although these examples have dealt with the *types* of minerals occurring at different *P, T,* and *X*, many examples could be cited of how these three intensive variables control *compositions* of the individual minerals. Such examples, however, are a little more difficult to appreciate and will be deferred to later chapters.

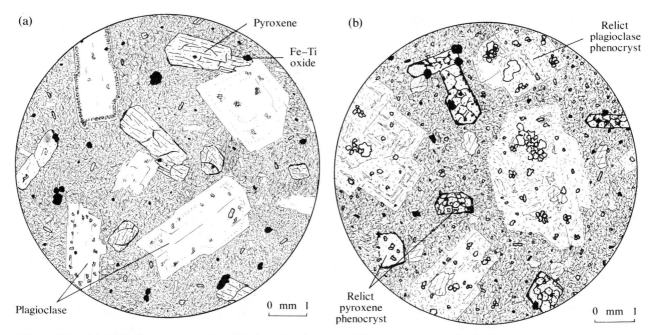

(a)

Pyroxene

Fe–Ti oxide

Plagioclase

0   mm   1

(b)

Relict plagioclase phenocryst

Relict pyroxene phenocryst

0   mm   1

Figure 1-9 (a) Andesite, as viewed under the microscope in thin section, comprised of phenocrysts of plagioclase and pyroxene in a fine-grained matrix of feldspars, oxides, and glass. (b) Metamorphosed andesite contains relict pyroxene and plagioclase phenocrysts, readily recognizable on the basis of their shape, even though the present rock is comprised of a low-grade metamorphic mineral assemblage.

Overprinting

It has been emphasized that specific processes in the changing state of a system leave specific imprints in fabric and that states of thermodynamic equilibrium are manifest in the mineralogical composition of the rock body. Prior states of a system and changes from them may be preserved in the same rock body in the form of superposed or overprinted fabrics and mineralogical compositions. Overprints are apparent provided the successive intermediate states did not come to complete equilibrium and geologic processes were not so intense as to totally eradicate prior fabrics.

Imprints of changing states preserved in the fabric and mineralogical composition of a rock body constitute a record of successive flows of energy. In the dynamic Earth, few geologic sys-

tems remain in one state forever—most change to a significant degree sometime during Earth history. The effects of each change will be overprinted on the previous effects, causing partial or complete obliteration of them. Metamorphic rocks best display superposition, though they are not unique in showing this phenomenon. Vestiges of the earlier fabric and composition can usually be discerned in any rock subjected to more than one state. By "unraveling" superposed effects, the petrologist can gain significant historical insight into the sequential changes in the states of the system.

As an example of **overprinting,** a fresh andesite and a metamorphosed andesite are shown in Figure 1-9. There is a striking similarity between the fabrics of the two rocks, even though their mineralogical compositions are quite different. As a consequence of a change in state, labradorite pheno-crysts in the second andesite have recrystallized to aggregates of albite, calcite, epidote, and white mica but retain the original form of the phenocryst. Pyroxene phenocrysts have recrystallized to "pyroxene-shaped" aggregates of Fe–Ti oxides, epidote, and chlorite. The fine matrix of feldspars and glass surrounding the phenocrysts in the original andesite has recrystallized to a confused aggregate of calcite, white mica, epidote, and chlorite. Hence, the minerals produced in the original high-temperature magmatic state have been completely obliterated in the final chemical state, but the magmatic fabric has not, and relicts of it remain.

1.4 CLASSIFICATION

Classification is a human endeavor that attempts to recognize and categorize common or contrasting features of related things. In petrology, classification is not an end in itself, but a means of seeing more clearly, simply, and unambiguously the interrelations of the different properties of different rocks. The focus should be on genetically significant, easily recognized aspects of the rocks, not trivial ones.

Field relations of the rock body are one basis for classification and provide information on the surroundings of the geologic system, or part of the system, in which the body originated and on the cause of the energy flow. Hence, useful information is provided in a classification that is based on categories such as volcanic, dike, marine, subaqueous, and contact metamorphic. Obviously, these categories are relatively broad and general, but are still very useful.

Internally, rock bodies exhibit a seemingly infinite variety of fabrics and chemical and mineralogical compositions indicative of an equally broad range and combination of geologic processes, sources of matter, and final states of chemical equilibrium. These variations comprise an essentially continuous spectrum, or continuum, with few well-defined discontinuities or breaks. In this respect, rocks differ from most minerals or organisms, which usually have natural breaks between species. Classifications based on fabric or composition attempt to subdivide this continuum in some meaningful way and attach labels or names to the subdivisions.

Fabric as a Basis of Classification

Ignoring minor exceptions, the more or less continuous spectrum of fabrics found in all different types of rocks consists of only five basic end-member types—glassy, sequential, clastic, strained, and crystalloblastic; all other fabrics are essentially combinations or variants of these. The five basic end-member types of fabric are most clearly distinguished in thin sections of rocks viewed under the microscope (see Figures 1-10 through 1-14).

Investigations of modern geologic processes, coupled with laboratory studies, provide insights into the origin of these five basic fabrics. Thus, genetic significance can be attached to these types, as indicated in the following discussion.

Solidification of a liquid solution.   Solidification of a liquid solution occurs by reducing T or by changing P and X. These changes in state variables are caused by flow of thermal energy out of the system, by work done on the system, or by flow of matter into or out of the system. Two of the five basic fabrics can be produced by this solidification process.

Glassy fabric consists of a silicate glass created by very rapid cooling of a high-T silicate melt; generally there are some enclosed crystals formed earlier that have well-developed faces due to freedom of growth in the liquid medium (see Figure 1-10). Glassy fabric is unique to magmatic (igneous) rocks and is most common in volcanic rocks.

Sequential fabric consists of interlocking mineral grains that precipitated one after another during crystallization of the solution; early formed and, therefore, more freely growing crystals tend to have better-developed crystal faces than later ones that fill in available remaining space. There may be exceptions to this control of relative development of crystal faces. Sequential fabric characterizes most magmatic rocks (see Figure 1-11a), but it is also found in precipitates from intermediate-*T* aqueous, or hydrothermal (literally, hot water), solutions and from low-*T* brines in evaporite basins and lake and ocean waters where biological agents may be active. Generally, sequential fabrics are random, or **isotropic,** having similar aspects in any direction, because the liquid media from which they crystallized were **hydrostatic,** pressures being equal in any direction. **Anisotropic** (or nonrandom or directional) fabrics, in which elongate mineral grains show a preferred orientation (see Figure 1-11b), can form under at least three special circumstances: (1) Flow of the solution during crystallization can orient the long dimension of grains in the direction of motion, like logs in a fast-moving river. (2) Settling of dense grains under the influence of gravity can orient the long dimensions parallel to the depositional interface, thereby minimizing the gravitational potential energy of each grain. (3) Elongate mineral grains can grow perpendicular to the boundary of the liquid body, like teeth in a comb; this is common in hydrothermal veins.

Disaggregation, dispersal, and deposition. Broken fragments, or **clasts,** of preexisting material are dispersed in the air or water and then are collected together (deposited) to yield a body with **clastic fabric** (see Figure 1-12). The clasts are usually denser than the dispersing and depositing media—air or water—and when they sink to a depositional base, their gravitational potential energy is minimized. Consolidation of the deposited particles is accomplished chiefly by subsequent precipitation of minerals from solutions percolating through or residing in pore spaces in the deposit, or by compaction. Clasts may be angular or rounded, depending upon the distance traveled and media of transport. Clastic fabric is most widespread in

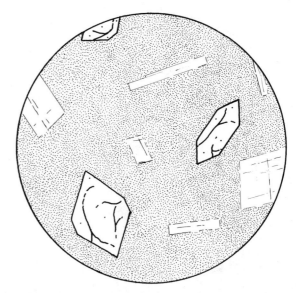

Figure 1-10 Glassy fabric as viewed under the microscope. Volcanic glass encloses crystals with well-developed faces.

sedimentary rock bodies but is found in some magmatic bodies as well.

Mechanical destruction. Mechanical destruction or deformation of mineral grains occurs as work is done on the rock body by tectonic forces such as in fault zones or by shock processes due to meteorite impact. The latter clearly involves transformation of kinetic energy, whereas the former involves more complex flows of energy during mountain building. The actual processes that shape the resulting **strained fabric** are complicated, but bent, distorted, faulted, and generally flattened grains are typical (see Figure 1-13). The fabric can be isotropic or anisotropic, depending on the nature of the applied forces.

Growth in the solid state. Growth of mineral grains by diffusion of ions in the solid state—essentially, recrystallization—can be caused by significant changes in any of the intensive variables *P, T,* and *X.* Such changes in state usually accompany flow

(a)

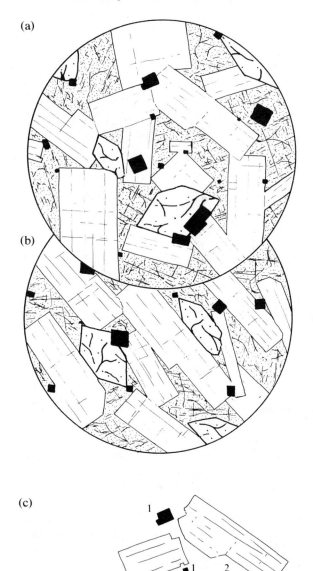

(b)

(c)

Sequence
of growth

1   First

2, 3

4   Last

◄*Figure 1-11*   Sequential fabric. (a) Isotropic sequential fabric. (b) Anisotropic sequential fabric. (c) Exploded view of central part of (a) showing contrasting shapes of earliest-formed (1) and latest-formed (4) mineral grains. Note crystal faces completely bounding mineral grains (1), partly bounding mineral grains (2) and (3), and reentrant angles (4).

(a)

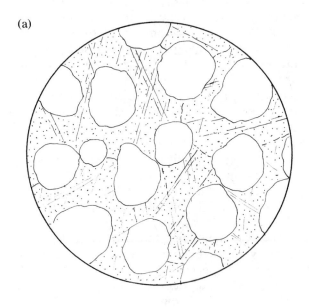

(b)

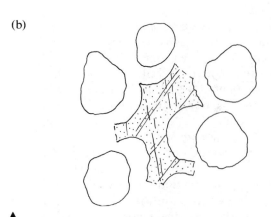

▲
Figure 1-12   (a) Clastic fabric. (b) Exploded view of central part of (a) showing the distinction between rounded clastic particles and secondary binding material precipitated after their deposition.

(a)

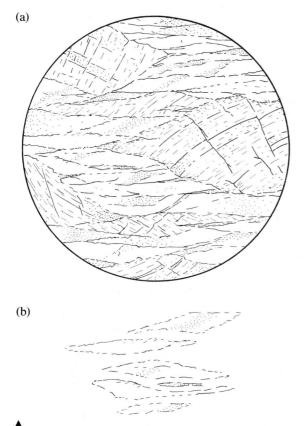

(b)

▲
Figure 1-13   Strained fabric. (b) is an exploded view of central part of (a). Note lenticular, bent grains.

(a)

(b)

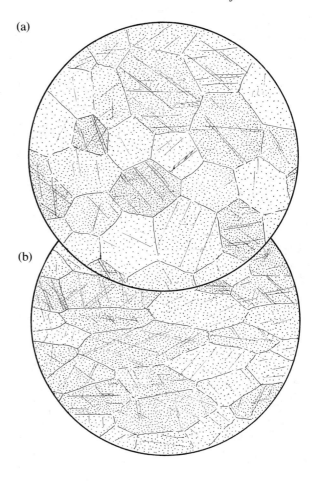

(c)

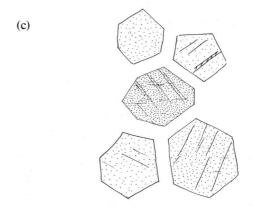

▲
Figure 1-14   Crystalloblastic fabric. (a) Isotropic. (b) Anisotropic. (c) Exploded view of central part of (a) showing polygonal shape of mineral grains.

of heat or matter into or out of a system. The resulting **crystalloblastic** fabric is characterized by a mutually interfering growth pattern in which polygonal grains meet at approximately 120° triple junctions (see Figure 1-14). Such fabric may be either isotropic or anisotropic, depending upon the forces acting on the rock body and upon the actual processes of fabric development. Recrystallization in rather passive environments, such as around a hot igneous intrusion after its emplacement or in diagenetically altered sedimentary rocks, yields an isotropic fabric. Recrystallization concurrent with deformation and solid flow of the body under tectonic forces results in the typical anisotropic fabric of schists and gneisses.

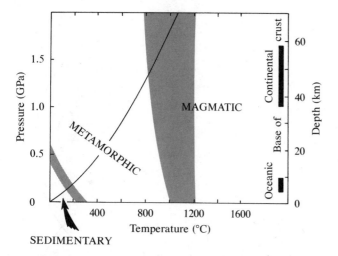

Figure 1-15 Approximate ranges of pressure and temperature over which the traditionally recognized sedimentary, metamorphic, and igneous (or magmatic) rock-forming processes operate. The indefinite boundaries between the three realms are real. Both *P* and *T* increase with depth in the Earth, but at different rates in different geologic environments. A geothermal gradient (see box on p. 25), which expresses the relation between *T* and depth in the Earth, such as might occur in an active mountain belt, is shown by the curved line. Heavy bars on the right of the diagram indicate the ranges of depths to the base of the oceanic and continental crusts.

P–T Realms of Rock-Forming Systems

There is an imperfect correlation between the five basic fabrics (and their four formative geologic processes) and the traditional threefold subdivision of rocks into igneous, sedimentary, and metamorphic, which was first proposed (in a slightly different form) in the nineteenth century. Solidification of liquid solutions is a widespread process in igneous systems, but it also occurs in sedimentary systems. Most clastic deposits are consolidated into solid rock due to the secondary precipitation of mineral material from percolating pore solutions. And precipitation of calcite, gypsum, and quartz from natural waters forms massive beds of sedimentary rocks such as limestones,

gypsum, and chert. Disaggregation, dispersal, and deposition are involved in the formation of sandstones, shales, and many other sedimentary rocks. But the so-called pyroclastic deposits form by the same processes, except that the disaggregated material is hot magma broken up by explosive expansion of gas; such deposits share igneous and sedimentary aspects in their origin. A similar duality of formative processes is found in what we may call magmatic sediments, which originate by gravitative settling of dense olivines and pyroxenes precipitated in slowly cooled bodies of basaltic magma. Mineral grain destruction and growth in the solid state typically occurs in metamorphic systems, but such growth also takes place in sedimentary systems, as in the diagenetic reconstitution of limestone into dolomite.

To some extent, the traditionally recognized igneous, sedimentary, and metamorphic rock-forming processes are more closely linked to the intensive state properties *T* and *P* than to geologic processes. Figure 1-15 shows the approximate range of *P* and *T* under which sedimentary, metamorphic, and igneous (or magmatic) processes operate.

1.5 HOW PETROLOGISTS STUDY ROCKS

The basic aim of petrology is to understand the geologic system in which a body of rock originated. Any petrologic study must rest on a firm foundation of facts. Only then can viable **petrogenetic interpretations** be made.

Factual data on a rock body are obtained from observations in the field and laboratory. Scales of observation range from continent to mountain range to outcrop to hand sample and finally to thin section under the microscope. Mapping, scrutiny of outcrops for chronologic, fabric, and mineralogical relations, and careful sampling of fresh rock in the field yield important rewards. Samples returned from the field are examined in thin section by the polarizing petrographic microscope for fabric and chronologic relations and for approximate mineralogical composition, using techniques

Geothermal and Geobaric Gradients

The change in temperature, ΔT, with respect to change in depth, ΔZ, is called the **geothermal gradient,** $\Delta T/\Delta Z$, or simply the **geotherm.** It is related to the heat flow from the surface of the Earth, Q, by

$$Q = k(\Delta T/\Delta Z)$$

where k is the thermal conductivity. The geothermal gradient is measured by means of holes bored into the Earth and ranges upward from 5°C/km with an average of approximately 30°C/km. Thermal conductivity can be determined in the laboratory from core samples recovered from the drill hole so that the heat flow, Q, can be calculated. Rock temperature, composition, and fabric (especially porosity and water content) dictate the measured value of k.

Pressure is the force acting over a particular area. Within the Earth, pressure caused by the weight of overlying rock is called load, lithostatic, or **confining pressure,** P. Its magnitude increases with depth in a predictable manner. To evaluate this, we first note that bodies of rock below several kilometers of the surface are hot and, over long periods of geologic time, behave approximately like fluids. Like water, which seeks its own level because it has no intrinsic strength, hot rocks have low strengths and flow readily. This concept is implied in mantle convection. The confining (or hydrostatic) pressure at the base of a vertical column of fluid with cross-sectional area A is

$$P = \text{force}/A = mg/A$$

from Newton's First Law, where m is mass and g is the acceleration of gravity. If the density, ρ, of the fluid is m/V where V is the volume of the mass, then

$$P = mg/A = \rho Vg/A = \rho gZ$$

where Z is the height of the column. In the Earth, the **geobaric gradient** is $\Delta P/\Delta Z = \bar{\rho}g$, where $\bar{\rho}$ is the mean density of the column of rock. In the crust, with $\bar{\rho} \sim 2.7$ g/cm^3, $\Delta P/\Delta Z = 27$ MPa/km; and in the upper mantle, with $\bar{\rho} \sim 3.3$ g/cm^3, the gradient is 33 MPa/km (refer to Appendix A for units).

of optical mineralogy. These microscopic observations are usually referred to as **petrography,** although the term is also used in a general sense for the descriptive aspects of petrology, as distinct from petrogenetic interpretations. Chemical analysis of the whole rock or its constituent minerals can be made by various spectrometric techniques. Mass spectrometric analyses of select stable and radioactive isotopes provide insight into the source of the rock and its age relations. X-ray diffractometry of minerals identifies them and may reveal additional information on such factors as cooling rates of the rock.

A careful, objective scrutiny of field relations and fabric can be the most rewarding aspect of a good many petrologic studies. In the field, on the outcrop, a petrologist can use imagination, intuition, experience, and familiar analogies to formulate interpretive petrogenetic models of what happened in the ancient geologic system; he or she should seek to understand each particular outcrop—the flows of energy and matter and the cause of these flows, the source of matter, chronologic relations, and perhaps even intensive variables and possible gradients in them. Multiple working hypotheses (Chamberlain, 1897) can be entertained during the course of a study, but these can be continually refined and a favored one eventually selected as facts accumulate. More expensive laboratory studies are an adjunct to field work and are used to refine interpretive models.

The above comments are not intended in any way to minimize the importance of experimental and theoretical investigations, which are assuming an increasingly significant role in petrology. These ''in-door'' endeavors provide, for example, valuable insights into the *P–T–X* conditions of mineral growth and the influence of tectonic forces on fabric development. Many discoveries of fundamental importance have been made by chemists and physicists with special training in experimental and theoretical techniques. Others with backgrounds in geology who have mastered "black boxes" and thermodynamic equations have also made important discoveries.

In his presidential address to the Geochemical Society in 1969, J. B. Thompson, Jr. (1970) had this to say regarding petrology:

> We who are concerned with such matters fall mainly in one or more of three categories: (1) those who collect rocks and study them; (2) those who try to duplicate rocks in the laboratory; and (3) those who worry about rocks. All three facets, the observational, the experimental and the theoretical, should clearly be integrated in a healthy science. Description as an end in itself is sterile, experiments that answer unasked questions are irrelevant, and theory unchecked by fact is useless. Theory, at its best, is in many ways a link between the other two. Good theory can thus be used to interpret or reinterpret the results of observation in order to ask more significant questions of experiment and vice versa.

1.6 PLATE TECTONICS AND PETROLOGY

Prior to the 1960s petrologists as well as structural geologists, stratigraphers, and solid-earth geophysicists had no unified Earth model for geological systems on a global scale. The underlying causes of magmatism, metamorphism, earthquake activity, mountain building, and epeirogenic movements were vigorously debated. But in the 1960s, largely through studies of the seafloor, a new scheme of global **plate tectonics** emerged— undoubtedly the greatest revolution in geologic thinking since the early beginnings of the science over two centuries ago (see Wilson, 1970; Cox, 1973). So petrologists, and their associates in other geologic disciplines, now view large-scale geologic systems in terms of plate motion, and with remarkable success. But much still remains to be learned in petrology, particularly on smaller scales of observation—the individual volcano, intrusive center, or metamorphic belt—where the connection to plate tectonics may be remote or totally obscure.

Plate tectonics has convinced us, more than ever before, of the intimate interrelation between tectonism, magmatism, and metamorphism and of the link all three have with thermal energy in the Earth. The connections between thermal energy and magmatism (via melting and magma production) and metamorphism (via the energy to drive destruction of existing rocks and recrystallization of new grains) are clear. The connection between thermal energy and tectonism is made apparent by the association of mountain building and seafloor spreading with magmatism and metamorphism.

Because of the inseparability of tectonism from magmatism and metamorphism, it is appropriate to consider plate tectonic regimes where rock-forming geologic systems occur. Remaining chapters of the book amplify this brief discussion.

Plate Boundaries and Motion

The relatively thin, rigid **lithosphere** of the Earth, encompassing the uppermost mantle and all of the crust, is divided into about a dozen separate plates (see frontispiece). Plate thickness ranges from only a few kilometers along oceanic ridges to over 200 km beneath old tectonically stable continental regions; horizontal dimensions are thousands of kilometers. Movement of these plates relative to one another is made possible because of an underlying substratum of soft, weak mantle material— the **asthenosphere**—that allows the lithospheric plates to be mechanically decoupled from the more rigid lower mantle.

It is commonly assumed that the seismic low velocity zone and the asthenosphere are the same. However, the slight degree of partial melting that may be responsible for the former does not appear to explain the kinetic behavior of the latter. Crystalline plasticity at high temperatures, as discussed in Chapter 13, probably has a more significant bearing on the weakness of the asthenosphere.

With respect to the relative motion of two neighboring plates, there are basically three types of plate margins (see Figure 1-16).

1. Divergent, spreading, or accretionary margins. Two plates, usually oceanic, move horizontally away from one another in a direction perpendicular to their common margin; new rock material is added to each.
2. Convergent, subduction, or consumptive margins. Two plates move toward each other and one plate of oceanic character slides beneath the other, descending into the mantle where it is consumed. The position of the descending plate is marked by the inclined **Benioff zone** of earthquake foci.
3. Passive, transform, or shear margins. Two plates slide horizontally past one another parallel to their boundary so that plate area is conserved, none being created or destroyed.

On an Earth of constant surface area, relentless growth of oceanic plates at accretionary boundaries—**sea-floor spreading**—requires their consumption elsewhere in subduction zones. This trade-off between spreading and consumption is the essence of plate tectonics—the **Wilson cycle** of opening and closing of the oceans.

Energy and Plate Tectonic Rock Associations

Plate tectonics is the surface manifestation of the need to maintain thermal and gravitational equilibrium within the earth. Plate motion and associated convection in the underlying mantle are the means of achieving this balance.

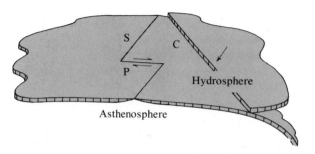

Figure 1-16 Lithospheric plate boundaries. S, spreading; P, passive; and C, consumptive plate edges. The hydrosphere lies above and the asthenosphere below the plates. Arrows indicate nature of motion between plates.

Rock-forming processes are concentrated along plate boundaries (and the lithosphere–hydrosphere interface) because that is where flows of matter and energy are focused due to plate motion. Particular boundaries are characterized by particular types and complexities of geologic processes affecting different rock materials. This favors development of specific types of rock bodies, or **plate tectonic rock associations.**

The following sections briefly review these associations (Dickinson, 1972), and further details are set out in the remainder of this book (see Figure 1-17). Only very minor rock-forming processes function along shear plate boundaries, chiefly mechanical destruction of preexisting rocks.

Divergent plate boundaries and associations. Along oceanic rifts, bouyant, partially melted mantle rises and continues to melt during decompression. Lithosphere and overlying sea water are heated convectively and conductively by contact with hot magma. Tensile forces produce subvertical fissures and steeply dipping fault blocks.

Melting of upwelling peridotitic mantle produces copious volumes of basaltic magma, which is extruded onto the surface as lava flows and is intruded into the crust where it freezes into bodies of gabbroic rocks or related differentiates (dunite,

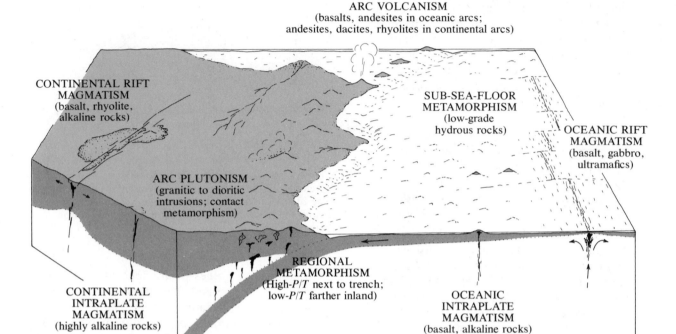

Figure 1-17   Idealized view of plate tectonic rock associations. Surface relief is somewhat exaggerated. Lithosphere is shaded dark gray.

pyroxenite, minor granite). In oceanic areas, newly solidified, still hot basaltic rocks near the rift interact with sea water that penetrates convectively through them, causing metamorphic reconstitution.

Convergent plate boundaries and associations. Convergent plate boundaries involve very complex transfers and transformations of energy and movement of matter associated with a wide range of geologic processes. Plate convergence causes not only mechanical work of deformation but also uplifting of the overriding plate, which increases its gravitational potential energy. The cool subducting plate serves as a heat sink as it descends into the hotter mantle, absorbing thermal energy and dehydrating the wet metamorphosed oceanic crust. Hot hydrous magmas move upward from the subducting plate and overlying mantle into the crust, heating and loading it. Hydrous

magmas erupt in subaerial volcanoes, adding water and other gaseous constituents to the hydrosphere and atmosphere.

Plate convergence is polar (directed to one side) and produces asymmetric patterns of metamorphism, magmatism, and tectonism in the overriding plate. Near trenches along continental margins, a tectonic mélange or mixture of rock types derived from both oceanic and continental plates is emplaced, including slices of dismembered ancient oceanic lithosphere, called **ophiolite,** and high-*P* low-*T* metamorphic rocks, called blueschists. Farther inland, mantle-derived magmas plus some originating in the deep crust erupt to the surface quiescently as well as explosively, forming a thick volcanic pile capped by lofty composite volcanoes; within the pile, arrested bodies of magma form plutons. Magma compositions range from basalt to granite, although andesite predominates in the volcanic pile, and chemically similar diorite and granodiorite in the plutons. Effects of magma mixing, contamination, and differentiation are widespread. Wall rocks heated by plutons suffer metamorphism, and in the deeper roots of this magmatic belt, pervasive, synchronous deformation and recrystallization produce foliated slates, schists, and gneisses; this inland metamorphism occurs on higher geothermal gradients so that temperatures are higher than in the near-trench metamorphism.

Where subduction occurs beneath the margin of an oceanic plate, an arc of volcanic islands is virtually the only expression of newly formed rock bodies at the surface. The volcanoes are built chiefly of basaltic to andesitic magmas.

Intraplate settings and associations. Although most magmatic activity occurs along convergent and divergent plate boundaries, some also is found apparently randomly distributed far from any boundary and well within a given plate. The reason for such activity is not understood; it may be prompted by aborted attempts at rifting a continental plate apart, or perhaps is related to plumes of deep, hotter mantle rising up to the base of the lithosphere.

Compositionally, intraplate magmatism is highly variable and includes many of the rock types found in convergent boundary associations plus a variety of much more alkaline rocks that are greatly enriched in Na and K relative to Si.

SUMMARY

1.1 Thermodynamics is concerned with the flows of energy and matter within a system and between a system and its surroundings. Kinetic, gravitational potential, and thermal energy are important in geologic systems, and their total can be called internal energy.

The Law of Conservation of Energy, also called the First Law of Thermodynamics, states that the total amount of energy in a system is conserved in any process. An extended version of this law involving mass states that the total of energy and mass is conserved in any process. Most of the

energy driving surface processes on Earth is derived from radiant solar heat, whereas internal processes of magmatism, metamorphism, and tectonism are driven by interacting gravitational and thermal energy from within the Earth.

Geologic processes are responses of geologic systems to flow (transfer and transformation) of energy and movement of matter. Thermal processes are some of the most significant aspects of geologic processes and are associated with conductive, convective, and radiant flow of heat.

Geologic processes cause a change in the state of a system. A system may be characterized by various state properties, among the most important of which are pressure (P), temperature (T), and chemical composition (X). These intensive variables, P, T, and X, in a system at equilibrium are independent of any previous state of the system.

The most stable state of a system, in terms of some possible flow of energy, and relative to nearby possible states, is the one having the lowest energy. A system is said to be in equilibrium when the net result of forces acting on the system at rest is zero. LeChatelier's Principle states that if a system in a state of equilibrium is disturbed, it will respond in such a manner as to moderate the effect of the disturbance. The Second Law of Thermodynamics states that entropy, or disorder, increases in every spontaneous irreversible process in an isolated system. In any spontaneous process moving to a more stable state, the free energy change is negative; that is, the free energy of the final state is less than the initial state. There is no difference in free energy between two states in thermodynamic equilibrium.

1.2 States of unstable and metastable equilibrium can persist indefinitely because of an energy barrier—the activation energy—between them and more stable states.

1.3. There can be no magmatism, metamorphism, tectonism, sedimentation, or any other geologic process without an unevenness of some form of energy that changes a system to a different state.

The mineralogical and chemical composition,

fabric, and field relations of a rock body provide all the essential information regarding its origin. The mineralogical composition indicates the P, T, and X of the last state of stable equilibrium in the system. The bulk chemical composition provides insight into the source of matter comprising the body. Fabric denotes the nature of geologic processes involved in the change of state accompanying formation of the body. Field relations provide clues regarding the chronology of the body and reasons for the changing state. Earlier states of a system may be perceptible in overprinted fabrics and mineralogical compositions.

1.4. The purpose of classification of rocks is to subdivide the continuum of fabric and composition in a meaningful way. Only five basic end-member types of fabric are found in rocks—glassy, sequential, clastic, strained, and crystalloblastic. Four basic processes—solidification of liquid solution; disaggregation, dispersal, and deposition; mechanical destruction; and growth in the solid state—produce these fabrics. These basic fabrics and processes do not correspond on a one-to-one basis with the traditional threefold division of rocks and formative processes into igneous, sedimentary, and metamorphic. These three processes, however, are correlatable with fairly distinct ranges of P and T.

1.5. Petrology is the science of the Earth that seeks to understand the geologic system in which a body of rock evolved. Any petrogenetic interpretation must be based upon an intelligent evaluation of facts.

1.6. Plate tectonics demonstrates an intimate interrelation between tectonism, magmatism, and metamorphism and is the surface manifestation of the need to maintain thermal and gravitational equilibrium within the Earth. Plate motion and associated convection in the underlying mantle are the means of achieving this balance.

The essence of plate tectonics is the spreading apart and consumption of lithospheric plates—the opening and closing of oceans. Rock-forming processes are concentrated along plate boundaries (and the lithosphere–hydrosphere interface) because that is where flows of matter and energy are focused due to plate motion. Particular boundaries are characterized by particular types and complexities of geologic processes affecting different rock materials; this favors development of specific types of rock bodies, or plate tectonic rock associations.

STUDY QUESTIONS

1. The major rock-forming processes on the Moon have been magmatism and shock metamorphism due to impacting meteorites. The youngest magmatic event has been dated radiometrically at about 3 Gy. Discuss possible fundamental reasons for the absence of lunar magmatic activity since then.

2. Draw several rectangular parallelepipeds representing tabular feldspar crystals, showing their orientation in a flowing magma and in a quiescent magma where they have settled under the influence of gravity on the floor of the chamber. What dictates their orientation?

3. Discuss and/or diagram how energy is conserved in a plutonic magma system.

4. Contrast geologic systems with the systems of the physical chemist in the laboratory.

5. How is kinetic energy important to the petrologist? Potential energy? Internal energy?

6. Discuss the creation of new bodies of rock in Figure 1-7b.

7. Discuss reasons for the lack of significant rock-forming processes along transform plate boundaries.

8. In what respects is the Earth a heat engine?

9. What is meant by a geologic model? Give an example.

10. Should a body of granitic magma in the crust be considered as an open, a closed, or an isolated system? Discuss.

11. Volcanic glasses will devitrify and crystallize over periods of hundreds to millions of years. Is volcanic glass stable, metastable, or unstable? Discuss.

12. What is activation energy? How is it related to the concept of minimum energy for stability?

13. What is involved in changing the state of a system?

14. In what respects is there an uneven concentration of energy near the surface of the Earth? Consider the different forms of energy. What is the fate of these uneven concentrations?

15. Compare the entropy of the crystalline and gaseous states for a system of constant chemical composition.

16. Discuss the real nature of the world's energy crisis in terms of the First and Second Laws.

17. From the standpoint of the Second Law, discuss the geologic processes of concentration of Cu atoms into an economically viable ore deposit, and its further concentration by man into usable materials such as wire.

18. What are the fundamental prerequisites for creation of a rock body?

19. If a system is stable thermally, is it necessarily stable chemically? Discuss.

20. Discuss the limitations of applying classical thermodynamics to the interpretation of geologic systems.

21. List the significant aspects of a rock body and indicate what each tells about its evolution.

22. Contrast modal, whole rock chemical, and mineralogical composition. Which can be determined in the field?

23. In what way is a jadeitic pyroxene an illustration of the theorem of moderation?

24. Give an example of overprinting in a sedimentary rock body.

25. Why is the color of a rock not a significant petrologic basis for classification?

26. In sequential growth fabric, why do early formed grains usually possess well-formed crystal faces?

27. Tuff is a consolidated deposit of volcanic ash blown explosively from a vent. What flows of energy are involved in the formation of bodies of tuff? Is the body igneous, sedimentary, or metamorphic? Discuss.

28. Contrast petrology, petrography, and petrogenesis.

29. The lithospheric plate of which Africa is a part has been a coherent entity since about Triassic time and is virtually encompassed by divergent plate boundaries. Comment on possible post-Triassic rock-forming processes in Africa.

30. Why might intraplate magmatism be subordinate in volume to subduction magmatism?

31. Discuss conservation of energy as dense metallic particles in a homogeneous protoearth segregate and sink to form a core, leaving an overlying silicate mantle.

REFERENCES

Angrist, S. W., and Hepler, L. G. 1967. *Order and Chaos: Laws of Energy and Entropy*. New York: Basic Books, 237 p. (An entertaining introduction to the basic fundamentals of thermodynamics without the rigorous mathematical language in which the subject is usually presented.)

Chamberlain, T. C. 1897. The method of multiple working hypotheses. *Journal of Geology* 5:837–848.

Cox, A. 1973. *Plate Tectonics and Geomagnetic Reversals*. San Francisco: W. H. Freeman and Company, 702 p.

Dickinson, W. R. 1972. Evidence for plate tectonic regimes in the rock record. *American Journal of Science* 272:551–576.

Dixon, J. R. 1975. *Thermodynamics I: An Introduction to Energy*. Englewood Cliffs, N.J.: Prentice-Hall, 458 p. (A problem-solving approach to thermodynamics using mathematics but also clearly presenting the basic concepts in simple language.)

Elder, J. 1976. *The Bowels of the Earth.* London: Oxford University Press, 222 p.

Hubbert, M. K. 1971. The energy resources of the earth. *Scientific American* 225:61–70.

King, A. L. 1962. *Thermophysics.* San Francisco: W. H. Freeman and Company, 379 p.

Oort, A. H. 1970. The energy cycle of the earth. *Scientific American* 223:54–63.

Thompson, J. B., Jr., 1970. Geochemical reaction and open systems. *Geochimica et Cosmochimica Acta,* 34:529–551.

Turner, F. J., and Verhoogen, J. 1960. *Igneous and Metamorphic Petrology,* 2d ed. New York: McGraw-Hill, 694 p. (Chapter 2 is a somewhat mathematical summary of the application of thermodynamics and kinetics to rock systems.)

Wilson, J. T. 1970. *Continents Adrift: Readings from Scientific American.* San Francisco: W. H. Freeman and Company, 172 p.

Part I
MAGMATIC ROCK BODIES

Petrology texts always consider magmatic rocks before metamorphic. This tradition follows from the primary nature of magmatic rocks; without them, there would be no sedimentary or metamorphic rocks.

Petrogenetic interpretations—the ultimate aim of petrology—must be founded upon intelligent evaluation of facts. The intent of this part of the book is to provide factual information regarding bodies of magmatic rocks—their composition, fabric, and field relations. Then, with this information for a foundation, Part II will explore the origin and behavior of magmatic systems from theoretical and experimental viewpoints.

A purely descriptive approach to rocks can be sterile and downright boring unless it is kept in mind that gathering facts and attaching names to things (classification) is not an end in itself, but rather is one step toward deciding how the rock body originated. To help maintain this viewpoint in Part I, questions will be asked regarding the data and brief comments made regarding interpretation to stimulate interest and indicate the petrogenetic relevance of particular information.

Organization of the subject material into chapters is according to composition. Chapter 2, however, presents an overall view of the fabric, composition, and classification of magmatic rocks.

2

Petrography and Composition of Magmatic Rocks

Magmatic rocks seemingly possess a bewildering array of fabrics and compositions that reflect a diverse combination of geologic processes acting on a variety of rock materials. Our task as petrologists is to extract from this array meaningful information that can enlighten our understanding of how bodies of rock form. Similarities and contrasts must be sought, correlations made, and a language devised for labeling and communication.

Rocks are usually classified according to some combination of fabric and composition—either mineralogical, modal, or chemical. Fundamental subdivisions of fabric are phaneritic, aphanitic, glassy, and clastic. Almost all magmatic rocks contain feldspar, in addition to some combination of quartz, micas, amphiboles, pyroxenes, olivines, Fe–Ti oxides, and other rarer mineral species; mineralogical-modal classifications are based chiefly on these seven major rock-forming minerals, the nature of the feldspar being of paramount importance. With the exception of some very rare magmatic carbonate rocks, most igneous rocks are composed chiefly of Si and O with lesser amounts of Al, Fe, Mg, Ca, Na, K, Ti, and so on. Hence, many chemical classifications depend upon silica (SiO_2), either in absolute concentration or relative to other constituents, especially Na_2O and K_2O; the latter basis distinguishes two major categories—alkaline and subalkaline, which is further split into tholeiitic and calc-alkaline.

The essentially continuous spectrum of magmatic rock chemical composition reflects the operation of a number of secondary processes of diversification such as magma mixing, contamination, and differentiation acting upon already variable magmas ascending from upper mantle and lower crustal source regions.

Trace elements and certain types of isotopes have become increasingly important as petrogenetic tracers that constrain the possible sources and evolutionary paths of magma systems.

The purpose of this chapter is to acquaint the reader with the array of fabrics and the compositional spectrum of magmatic rocks and the ways in which rock compositions are determined and compared.

2.1 CLASSICAL PETROGRAPHY OF MAGMATIC ROCKS IN HAND SAMPLE

No introductory petrology text would be complete without at least a brief discussion of how magmatic rocks can be described, in general terms, on the hand-sample scale using a hand lens. Extended descriptions are provided by Jackson (1970, chap. 5) and Dietrich and Skinner (1979). Such description is necessary in field mapping and for preliminary laboratory studies.

Fabric

Four principal types of fabric occur in magmatic rocks: phaneritic, aphanitic, glassy, and clastic. The first two refer to the dominant grain size in sequential fabrics (see Section 1.4). Each of these four connotes specific processes in the corresponding magmatic system.

Phaneritic Mineral grains are sufficiently large to be identifiable in hand sample. Intrusions and the cores of large extrusive bodies slowly lose heat to their surroundings and allow minerals to grow to large size.

Aphanitic Mineral grains are too small to be identified without a microscope. Aphanitic fabric usually originates in small magma bodies emplaced near or onto the Earth's surface, where loss of heat and gas is rapid.

Glassy Conspicuous amounts of volcanic glass visible in hand sample define this fabric. Crystals may be evident. Glassy fabric originates in magma bodies, such as lava flows and very shallow small intrusions, that lose heat and gas so rapidly that atoms in the silicate melt have insufficient opportunity to organize into the regular geometric arrays of crystals. Instead, the melt solidifies into a very viscous amorphous glass.

Clastic Aggregated clasts, or broken fragments of glass, rocks, or minerals, held together by some sort of binding material.

The most widespread clastic fabric in magmatic rocks is found in explosive volcanic deposits.

Phaneritic, aphanitic, and clastic fabrics are mutually exclusive; they do not occur in combination with one another. However, a rock whose fabric is clastic can be comprised of fragments of aphanitic or phaneritic rock, and if most of the clasts are glassy, the term **vitroclastic** is appropriate.

Many phaneritic and aphanitic rocks are **equigranular,** consisting of grains all about equal size. Other rocks, however, are **inequigranular,** consisting of conspicuously larger grains, called **phenocrysts,** in the finer-grained **matrix,** or **groundmass.** This **porphyritic** fabric generally originates as magma that has been slowly cooling and forming large crystals and is then more quickly cooled (by extrusion, for example), forming the finer-grained matrix. A rock can be porphyritic-aphanitic or porphyritic-phaneritic, depending upon the grain size of the matrix. A glassy rock with scattered crystals is **vitrophyric.**

Rock-Forming Minerals

Of the several hundred families and species of minerals now recognized, virtually all magmatic rocks are comprised chiefly of only seven:

Feldspars	Fe–Ti oxides
Quartz	Olivines
Pyroxenes	Amphiboles
Micas	

Most magmatic rocks have over 50% feldspar, and those totally lacking it are rare. Feldspars are thus **major** mineral constituents of most magmatic rocks, in contrast to minerals such as sphene, sulfides, and apatite, which nearly always comprise less than 5% of magmatic rocks and may therefore be designated as **accessory** or minor minerals. The 5% limit is arbitrary. A particular mineral—quartz, for example—can be a major constituent in one rock type (granite) yet be an accessory constituent in another type (diorite).

A brief review of the more important rock-forming minerals is presented in Appendix C. An extensive treatise is provided by Deer, Howie, and Zussman (1962; see also their abridged, single volume paperback).

Classification

A magmatic rock may be classified in a general way, depending upon easily recognized mineralogical aspects, as follows:

Leucocratic rock   Pale-colored minerals (for example, most rhyolites, granites)

Felsic rock   Rich in feldspar, feldspathoids, or quartz (for example, granite).

Silicic rock   Rich in silica, as manifest by an abundance of alkali feldspars, quartz, or glass rich in SiO_2 (for example, rhyolite, granite). The approximately synonymous terms *salic* or **sialic** are used for rocks rich in Si and Al (that is, with abundant feldspar) and especially with reference to the continental crust.

Mafic rock (or mineral)   High concentrations of Mg and Fe (for example, basalts, pyroxenes, biotite). Synonomous with the more cumbersome term *ferromagnesian*.

Ultramafic rock   Especially rich in Mg and Fe, with no feldspar or quartz (for example, dunite, peridotites).

Many classification diagrams have been published in geology texts subdividing the essentially continuous compositional spectrum of magmatic rocks (see Figure 2-6) into discrete compartments with specific labels such as granite, gabbro, and so on. However, it is impossible to represent in a very accurate and reliable manner this entire spectrum by means of any two-dimensional diagram; there are just too many variables. Nonetheless, by recognizing a **family** of many individual rock types, all related by some common compositional feature, a reasonably accurate classification diagram can be contrived. We recognize here four major families, each of which will be classified and discussed in

more detail in succeeding chapters, as follows:

1. **Calc-alkaline volcanic rocks** (Chapter 3)   Aphanitic, glassy, clastic, and porphyritic rocks with feldspar as the dominant mineral and lesser amounts of quartz, mica, amphibole, pyroxene, and other minerals.

2. **Calc-alkaline plutonic rocks** (Chapter 4)   Phaneritic and locally porphyritic rocks with compositions similar to calc-alkaline volcanic rocks.

3. **Mafic and ultramafic rocks** (Chapter 5)   Rocks of any fabric composed dominantly of some combination of calcic plagioclase, pyroxene, and olivine.

4. **Highly alkaline rocks** (Chapter 6)   Rocks of any fabric containing abundant Na and K relative to Si; diagnostic minerals include feldspathoids (nepheline, leucite), biotite, alkali feldspar; quartz is absent.

2.2 DESCRIPTIVE PETROCHEMISTRY

The field of petrology dealing with the chemical composition of rocks may be called **petrochemistry**. Obviously, it borders on the field of analytical chemistry because one cannot know the composition of a rock until appropriate analyses have been made, and this is usually an effort performed by an analytical chemist or by a petrologist using borrowed analytical techniques.

The validity of any chemical data depends upon the quality of the analysis itself as well as upon the quality of the sample collected in the field.

Sampling

The importance of this step is crucial, but commonly ignored. Always ask: What is the nature of the sample, or how good is it? For what purpose is it to be used?

If a chemical analysis is to be truly representative of the magmatic rock, the sample must be free of overprints of alteration and weathering; the chemical composition of the rock changes during these secondary processes, and it is very difficult to know in what way or to what extent. The sample must be fresh. Constituent minerals should have bright vitreous luster with clean grain boundaries and well-defined cleavage or fracture surfaces. The rock should be free of patchy discoloration due to films of secondary manganese or iron oxide, or a whitish clouding of feldspar caused by alteration to clay minerals. Except in high mountain terraines recently subjected to Pleistocene glaciation, outcrops are generally weathered; at lower elevations and in humid climates especially, rock tends to be deeply weathered. A hammer may be unable to break far enough into an outcrop to obtain fresh rock and a portable diamond core drill may be required. Unsatisfactory weathered samples crumble from the outcrop, whereas unweathered rock breaks with difficulty under the hammer and yields sharp-edged pieces.

Providing a fresh sample can be obtained, it must next be determined how large it should be, precisely where it should be taken from the outcrop or the rock body, and so on. Different sampling plans must be adopted to solve different problems. A common problem is to collect a sample from a rock exposure, say, of granite that is to be representative of the exposure. Just what is representative may not be easy to ascertain (see Baird and others, 1967; reference given at end of Chapter 4). Each sample must be many times larger than the dimensions of the coarsest grains. When all or part of the sample is pulverized for the analysis, it is necessary to demonstrate that the sample volume (or weight) is sufficiently large not to be influenced by grain heterogeneity; a sample consisting of only one feldspar grain is obviously too small to represent the composition of the entire granite. A small, unbiased aliquot of the pulverized rock may then be taken for the chemical analysis.

Analyses

Prior to about 1950 chemical analyses of rocks and minerals were generally performed by classical wet methods involving tedious precipitation, titration, and weighing techniques carried out by skilled chemical analysts. Today, however, many rapid instrumental methods are in wide use, such as atomic absorption spectrophotometry, X-ray fluorescence spectrometry, and neutron activation analysis. Chemical analysis of minerals has been revolutionized in the past couple of decades by the electron probe microanalyzer, which is essentially an X-ray fluorescence spectrometer attached to an electron microscope. A thin beam of electrons only a few microns in diameter impinging upon the surface of the mineral grain causes it to emit fluorescent X-rays. These instrumental methods detect the spectra of the elements in the sample and relate their intensities to intensities of a standard, giving finally the element concentration in the sample. Even though element concentrations are determined, major constituents are nonetheless reported as oxides of the elements, following the traditional presentation of the results of wet methods of analysis.

Chemical constituents as oxides of the elements are listed in weight percent (wt.%). In a 100-g sample, for example, if 55 g are SiO_2 and 8 g are MgO, then there is 55 wt.% SiO_2 and 8 wt.% MgO. Since each constituent is generally determined independently of the others, and since analytical errors are omnipresent, the sum of the constituents will usually differ from 100.00%. Furthermore, available time and limitations of cost and analytical technique do not allow the analyst to determine all elements. Hence, a sum of 99.53% recorded for a particular chemical analysis of a rock could mean that not all constituents were determined, or that all constituents present were analyzed for, but errors created a departure from 100.00%. A total over 100.00% doesn't necessarily mean all constituents were determined because the total analytical error could carry it over 100.00%. Generally speaking, if the analysis lists all major and minor constituents and the total lies between 99.75% and 100.50%, it is considered to be acceptable.

The compositional spectrum of most magmatic rocks (excluding the very rare carbonatites described in Chapter 6) can be defined by the specifications given in Table 2-1. It should be realized

Table 2-1 Chemical constituents of magmatic rocks (very rare carbonatites excluded).

Constituent	Range of concentration of constituent in magmatic rocks (in wt.%)	Rock-forming minerals (or volcanic glass) in which the constituent is present
1. Generally major constituents (>1.0 wt.%)		
SiO_2	30–78 wt.%	In glasses and silicate minerals
Al_2O_3	3–34	In glasses and silicate minerals, except olivine
Fe_2O_3	0–5	In glasses, pyroxenes, amphiboles, micas, Fe–Ti oxides (sometimes higher, especially in mafic volcanic rocks, where FeO has been oxidized after extrusion of the magma)
FeO	0–15	In glasses, olivines, pyroxenes, amphiboles, biotite, Fe–Ti oxides, garnets
MgO	0–40	In glasses, olivines, pyroxenes, amphiboles, biotite–phlogopite
CaO	0–20	In glasses, pyroxenes, amphiboles, feldspars, garnets, zeolites
Na_2O	0–10	In glasses, feldspars, feldspathoids, amphiboles, pyroxenes, zeolites
K_2O	0–15	In glasses, feldspars, feldspathoids, micas
2. Generally minor constituents (0.1 to 1.0 wt.% of rock)		
H_2O^+		Structural or combined water in glasses, amphiboles, all sheet silicates, zeolites
H_2O^-		"Dampness," or water adsorbed on grain surfaces and held in pore spaces that is driven off by heating to 110°C; also weakly bound water in zeolites and glass
TiO_2		In glasses, pyroxenes, amphiboles, biotite, Fe–Ti oxides, sphene, rutile; sometimes a major constituent
P_2O_5		In glasses, apatite
MnO		Same as for FeO
CO_2		In glasses and carbonates
Cl		In glasses, amphiboles, micas, sodalite
F		In glasses, amphiboles, micas, fluorite
S		In glasses, sulfides
3. Trace elements (<0.1 wt.%, equivalent to <1,000 ppm; concentrations are usually listed in ppm)		

that the concentration limits between major and minor constituents and trace elements are somewhat arbitrary and that a particular constituent can be a minor constituent in one rock and a major in another. K_2O, for example, is a major constituent in granites because of abundant alkali feldspar, but is a minor constituent of most basalts and a trace constituent of a dunite (olivine rock).

Table 2-2 gives an example of a carefully determined complete rock analysis, that of a basalt from the Columbia River Plateau. The grand total of this analysis of sample BCR-1, 100.14 wt.%, obviously involves some analytical errors.

Correlation Between Chemical, Mineralogical, and Modal Composition

As explained on page 17, any rock has three compositional aspects:

1. Bulk chemical composition: concentration of elements or oxides.

2. Mineralogical composition: types of constituent minerals and their chemical compositions.

3. Modal composition: volumetric proportions of minerals in the rock.

Table 2-2   Complete rock analysis of a basalt from the Columbia River Plateau.

Sample BCR-1

Chemical constituents (wt.%)				Trace elements (ppm)							
SiO_2	54.50	BaO	0.078	B	5	Ge	2	S	392		
TiO_2	2.22	Li_2O	0.01	Be	2	Hf	5	Sc	33		
Al_2O_3	13.41	Na_2O	3.23	Ce	54	Hg	11	Sm	7		
Fe_2O_3	3.28	K_2O	1.68	Cl	50	La	26	Sn	3		
FeO	9.17	Rb_2O	0.005	Co	38	Mo	1	Th	6		
MnO	0.19	P_2O_5	0.36	Cr	18	N	30	U	2		
MgO	3.41	H_2O^+	0.63	Cs	1	Nb	14	V	399		
CaO	6.98	H_2O^-	0.71	Cu	18	Nd	29	Y	37		
SrO	0.04	CO_2	0.03	Dy	6	Ni	16	Yb	3		
F	0.05	Total	99.98	Er	4	Pb	18	Zn	120		
				Eu	2	Pr	7	Zr	190		
				Ga	20			Total	1576 ppm		
				Gd	7						

Source: S. S. Goldich and others, 1967, Analysis of silicate rock and mineral standards, *Canadian Journal of Earth Science,* 4.

These three compositions are obviously interrelated. A high concentration of K implies the presence of a K-bearing mineral such as microcline or biotite; and conversely the presence of abundant microcline or biotite in the rock implies that K is a major chemical constituent. Let us compare these different compositional bases in more detail.

First, consider a hypothetical peridotite whose modal composition (volumetric basis) is olivine 70% and diopside 30%. The chemical composition of the olivine and diopside is listed in the second and third columns of Table 2-3. Oxide weights contributed to the whole rock by the 70% olivine (0.70 times the values listed in the second column) are listed in the fourth column, and contributions by the 30% diopside are shown in the fifth column. The whole rock chemical composition of this hypothetical peridotite, given in the sixth column, is calculated by summing columns four and five. As should have been expected from the chemical compositions of the constituent minerals, the peridotite is comprised essentially of four major oxides: SiO_2, MgO, FeO, and CaO.

Second, consider a hypothetical basalt whose mode is shown in Table 2-4. In this particular example, the constituent minerals have significantly different densities (density = molar weight ÷ molar volume), and the volumetric proportions of the mode cannot be used directly in calculating the chemical composition. (The olivine and pyroxene in the peridotite of the previous example had similar densities.) Instead, each of these volumetric proportions is multiplied by the respective density of the mineral, giving, after recalculation to 100%, the model composition in weight %, which then forms the basis of the calculation in Table 2-5.

There are more major chemical constituents in the basalt than in the peridotite. MgO is much lower and is contributed by the pyroxene and the small quantity of olivine. TiO_2 is contributed almost entirely by the small quantity of ilmenite. Plagioclase contributes essentially all of the Al_2O_3 and Na_2O. Plagioclase and pyroxene together contribute the CaO. FeO is contributed by pyroxene, olivine, ilmenite, and magnetite. Thus it can be seen that in rocks comprised of several different minerals, each of which contains several elements, the overall bulk chemical composition of the rock is built up in a rather complex way.

It is left for the reader to calculate the chemical

Table 2-3 Calculation of the bulk chemical composition of a hypothetical peridotite from its modal and mineralogical composition.

	Olivine[a]	Diopside[a]	70 vol.% Olivine[b]	+	30 vol.% Diopside[b]	=	Peridotite[b]
SiO_2	40.96 wt.%	52.34 wt.%	29 wt.%		16 wt.%		45 wt.%
TiO_2	0.01	0.18	0		0		0
Al_2O_3	0.21	2.72	0		1		1
Fe_2O_3	0.00	1.31	0		0		0
FeO	7.86	1.77	6		1		7
MnO	0.13	0.24	0		0		0
MgO	50.45	16.15	35		5		40
CaO	0.15	24.23	0		7		7
Na_2O	0.01	0.57	0		0		0
						Total	100

[a] Appendix C, Table C-4, No. 1, and Table C-5, No. 3.
[b] Weight percent values are rounded to two significant figures to accord with the modal values.

composition of a granite from its modal composition (in volume percent): perthite 65%, quartz 30%, and amphibole 5%.

It must be emphasized that these calculations of whole rock chemical compositions from modal and mineralogical compositions are for illustrative purposes only. The rock compositions obtained are not intended to be particularly representative or significant for these rock types. Moreover, this method ignores important compositional contributions of minute grains, such as apatite and zircon for P_2O_5 and ZrO. Few mineral grains in rocks are entirely homogeneous, and one can never be sure that analysis of one mineral grain is representative of those throughout the rock. For these reasons, *the only reliable way of obtaining the bulk chemical composition of a rock is the direct chemical analysis of the whole rock.*

Table 2-4 Recalculation of the modal composition of a hypothetical basalt.

	Mode (vol.%)	Density	Product of volume × density	Product recalculated to 100% Mode (wt.%)
Plagioclase (An_{59})	55	2.7	149	46
Augite	30	3.9	117	36
Olivine (Fo_{77})	10	3.4	34	11
Ilmenite	3	4.7	14	4
Magnetite	2	5.2	10	3
Total	100		324	100

Table 2-5 Calculation of the chemical composition of a hypothetical basalt (in Table 2-4) from its modal composition.

	46% Plagioclase[a] +	36% Augite[b] +	11% Olivine[c] +	4% Ilmenite[d] +	3% Magnetite[e] =	Basalt
SiO_2	25 wt.%	18 wt.%	4 wt.%	0 wt.%	0 wt.%	47 wt.%
TiO_2	0	0	0	2	0	2
Al_2O_3	14	1	0	0	0	15
Fe_2O_3	0	0	0	0	2	2
FeO	0	4	2	2	1	9
MnO	0	0	0	0	0	0
MgO	0	6	4	0	0	10
CaO	5	6	0	0	0	11
Na_2O	2	0	0	0	0	2
K_2O	0	0	0	0	0	0
					Total	98

[a] Table C-1, No. 5.
[b] Table C-5, No. 2.
[c] Table C-4, No. 2.
[d] Table C-3, No. 1.
[e] Table C-3, No. 2.

Chemical Classifications

The greatly increased availability of chemical analyses of rocks made possible by rapid and relatively inexpensive instrumental techniques has caused greater emphasis to be placed on purely chemical classifications of magmatic rocks. The advantages of chemical classifications are obvious for very fine-grained rocks, whose mineralogical compositions are obscure (except by X-ray diffractometry), or for glassy rocks. Magmatic rocks whose characterizing minerals have been obliterated by metamorphism can be analyzed to reveal their original nature, provided the recrystallization process has not severely opened the rock to exchange of chemical constituents with its surroundings. For these reasons, and because of the inherent precision of good chemical analyses, it is tempting to believe that very intricate and sophisticated schemes of classification are called for. However, the large number of variables, such as the chemical constituents themselves, and the lack

of agreement among petrologists as to the defining limits of specific rock types, such as andesite, hamper the use of chemical data.

Silica concentration. With few exceptions, silica (SiO_2) is the principal oxide constituent of igneous rocks. It is only natural then that it should serve as a basis for classification, either directly or indirectly. One widely used and relatively simple classification utilizes the concentration of SiO_2 as follows:

Silica concentration	Name
66 or more wt.%	**acid**
52 to 66	**intermediate**
45 to 52	**basic**
45 or less	**ultrabasic**

In this scheme, the peridotite in Table 2-3 is ultrabasic and the basalt in Table 2-5 is basic. These terms originated several decades ago when it was believed that silica combined with other oxides in minerals in the form of acids and bases. This notion has long since been abandoned, but the terms remain in use.

As defined here, acid and basic have no reference whatsoever to the hydrogen ion content, or pH, as used in chemistry. Also, these categories have no direct correlation with modal quantity of quartz in the rock, although as a general rule, acid rocks do contain quartz and ultrabasic ones do not. Two rocks having identical concentrations of silica may have widely different quantities of quartz, and two rocks of similar quartz content may have different silica concentrations, depending upon the composition and quantity of other minerals in the rock.

The degree of **silica saturation** depends upon the concentration of silica relative to the concentrations of other chemical constituents in the rock that combine with it to form silicate minerals. Let us illustrate using the relative concentrations of SiO_2 and Na_2O (see also page 18). The feldspathoid nepheline and quartz are unstable together so that the reaction

$$2NaAlSiO_4 \;+\; 4SiO_2 \;\rightarrow\; 2NaAlSi_3O_8$$

nepheline   quartz   albite

always goes to the right, forming stable albite. In a crystallizing magma, the two constituents on the left that are dissolved in the melt will combine to form albite, and the resulting rock is compositionally saturated with respect to silica. As the formula of nepheline may be rewritten as $Na_2O \cdot Al_2O_3 \cdot 2SiO_2$ and that of albite as $Na_2O \cdot Al_2O_3 \cdot 6SiO_2$, we note that the ratio $SiO_2/Na_2O = 6$ in the magma from which albite crystallized. If, on the other hand, the ratio $SiO_2/Na_2O < 6$ in the magma, there is insufficient silica to combine with all the Na_2O and the resulting crystalline material will consist of albite plus nepheline; or if the deficiency is such that $SiO_2/Na_2O < 2$, then no albite forms at all. Where nepheline occurs, the rock (and the magma) is said to be undersaturated in silica. Going in the opposite direction, if $SiO_2/Na_2O > 6$, then there is an excess of silica over

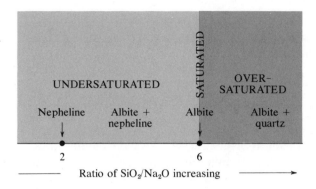

Figure 2-1 The degree of silica saturation depends upon the ratio of SiO_2/Na_2O and is manifest in the mineralogical composition.

that needed to form albite and the resulting oversaturated rock contains albite plus quartz. These relations are summarized in Figure 2-1.

Another pertinent mineral reaction illustrating the concept of silica saturation is

$$Mg_2SiO_4 \;+\; SiO_2 \;\rightarrow\; 2MgSiO_3$$

forsterite   enstatite
olivine   pyroxene

Still other reactions could be written involving Ca and K. Obviously, in natural systems with K, Na, Ca, Fe, Mg, and so on all competing for the available Si, the degree of silica saturation becomes complicated. These complications will be handled in the normative calculation (see below). Formally, we recognize two classes of minerals:

Silica-saturated minerals Compatible, or can exist at equilibrium, with quartz or its polymorphs; saturated minerals include feldspar, enstatite, hypersthene, amphiboles, micas, tourmaline, Fe-olivine, Fe–Ti oxides, sphene, and many others.

Silica-unsaturated minerals Incompatible and never found, except by accident, with quartz; unsaturated minerals include all feldspathoids, Mg-olivine, corundum, melilite.

And we recognize three classes of rocks:

Silica-oversaturated rocks Contain quartz or its polymorphs; granite is an oversaturated rock.

Silica-saturated rocks Contain neither unsaturated minerals nor quartz; diorite is an example.

Silica-undersaturated rocks Contain unsaturated minerals; nepheline syenite is an undersaturated rock.

Definition of standard rock types. Many schemes for more detailed chemical classification into discrete **rock types**—andesite, rhyolite, and so on— have been published in the petrologic literature. Most of these reflect the individual prejudices of their authors rather than widespread agreement among many petrologists. In an effort to sidestep this obstacle, LeMaitre (1976a) has compiled 26,373 published rock analyses that have been given a rock-type name by the original author. He then averaged all analyses with a particular label, for example, "andesite," irrespective of the classification scheme used. Hence, such an average represents a consensus of many petrologists and serves as a useful guide to the chemical composition of an andesite. Average compositions of some rock types in LeMaitre's compilation are given in Appendix D.

Chemical classification can also be achieved using data in graphic form, as for example in Irvine and Baragar (1971) and Cox and others (1979).

The CIPW normative composition. About the turn of the century, three petrologists (Cross, Iddings, and Pirsson) teamed up with a geochemist (Washington) and devised an elegant procedure for recalculating the chemical composition of a rock into a hypothetical assemblage of water-free standard minerals called the **CIPW normative composition,** or simply the **norm.** The merit of the norm, as contrasted with the actual modal and mineralogical composition of the rock, from which the norm may differ substantially, is that the limited number of standard normative minerals facilitates comparisons between rocks. The norm ignores the effects of geologic processes influencing fabric and of different P–T conditions influencing mineralogical composition, and so focuses upon the source of matter comprising the magmatic rock. Besides, of the three intensive variables (P, T, and X) governing the mineralogical composition of an igneous rock, the chemical composition (X) is the dominant variable. Glassy and crystalline rocks can be compared. Rocks with micas and amphiboles that crystallized from "wet" magmas can be compared with anhydrous rocks. Distinctions become apparent between superficially similar rocks in which there is extensive solid solution in the constituent minerals. For example, two basalts, both comprised of feldspar, olivine, pyroxene, and Fe–Ti oxides, could have significantly different chemical and, therefore, normative compositions. The norm emphasizes certain subtle distinctions in composition. (Additional details regarding the norm and its calculation are given in Appendix E.)

In its allocation of silica first to feldspars and then to mafic normative minerals, the norm calculation emphasizes the degree of silica saturation in the rock. Thus, on the basis of the norm, we can make the following redefinitions of the categories introduced above:

Silica-oversaturated rocks Contain $Q + Hy$ in the norm.

Silica-saturated rocks Contain Hy.

Silica-undersaturated rocks Contain $Ol \pm$ possible Ne.

2.3 THE ORIGIN AND EVOLUTION OF MAGMAS

Much more needs to be said regarding the chemical composition of magmatic rocks. However, before proceeding further, we must digress briefly to discuss some genetic aspects of magma systems that will make additional comments more meaningful. This discussion is a very short summary of Chapters 8 and 9.

From the information provided in this chapter to this point, it is obvious that magmatic rocks have a wide, yet definable, range of chemical compositions, matched, of course, by mineralogical variations. Why is there such a range? How does it originate? And why should it have certain bounds (no more than about 78 wt.% SiO_2, no more than 20 wt.% CaO, and so on)?

The title of this section suggests that magmas evolve. In the biological world, evolution depends upon internal genetic changes in offspring as compared with their parents and upon external environmental influences. These factors cause great numbers of new species to develop. Similarly, magmas evolve from some parent magma by various processes of diversification that are both internal to the magma system and external in the sense that the system is open to exchange of matter with its surroundings. Although it was once believed that there are only one or two parent magmas from which all others were produced by processes of diversification, this notion has been found to be faulty because a significant range of magmas can be generated right in their source regions.

The wide range in composition of igneous rocks reflects diversification processes acting on compositionally variable parent magmas. Diversification of open magma systems is an easily appreciated concept. Production of **hybrid magmas** can occur by liquid–solid interaction between the original magma and its contrastive wall rocks or by mixing of two (or possibly more) magmas of different composition. Evidence of **magma mixing** is manifest in igneous complexes with a history of sustained activity. Streaked rocks with complicated nonequilibrium phenocryst assemblages (such as Mg-olivine plus quartz) are one result. More complete homogenization and equilibration erases clues to the mixed origin. Because hot magmas spend so much of their history encased within older rocks, some degree of **contamination** by **assimilation** of wall rock is to be expected. In addition to strong thermal gradients across the contact there are generally significant compositional differences, both of which prompt the inter-change of matter across the contact in an effort to reach a more stable state. As interchange is facilitated by greater surface area of contact between the solid rock and magma, the assimilation of torn-off pieces of foreign wall rock, or **xenoliths,** can hasten the contamination of the magma. In some cases the assimilation is merely physical, whereas in other instances the ingested pieces of older rock are dissolved or melted and cause a distinct chemical contamination.

Internal diversification processes that split an initially homogeneous closed magma body into two or more contrasting parts is called **magmatic differentiation.** Such processes might occur in wholly liquid magma bodies consisting only of a high-T silicate melt, or in partly crystallized magmas consisting of crystals in a melt, or in magmas containing bubbles of gas.

In wholly liquid bodies, temperatures are too high for any crystals to have precipitated and pressures are too great to allow formation of bubbles. A rather surprising discovery from experimental studies, which has been confirmed in a few natural magmatic bodies, is that certain types of high-T silicate melts can split into two compositionally different immiscible portions. Familiar immiscible liquids that remain separate are oil and water. **Liquid immiscibility** is obviously a viable mechanism of magmatic differentiation.

Some magma bodies contain bubbles of gas that have separated out of the melt and have floated to the top of the magma body. The gas is chiefly H_2O and CO_2 but can contain small amounts of dissolved K, Na, and other chemical constituents that readily boil out of a melt at high T. The efficacy of **volatile transfer** in magmatic differentiation is uncertain but, if significant, would enrich the top of the magma body in K and Na relative to lower regions.

Bodies of magma that consist of both crystals and a high-T melt—that is, partly crystallized magmas—are probably the most common. Because the crystals always have a different chemical composition from the coexisting liquid, an effective mechanism of magmatic differentiation involves the separation of the crystals and liquid. Let us illustrate how this **crystal–liquid fractionation** can produce diverse rocks from an initially homogeneous parent magma. A basalt melt at a particular P and T can have crystals of olivine and pyroxene coexisting in stable equilibrium with it. Comparison of the hypothetical examples in Tables 2-3 and 2-5 shows that these crystals have a different composition from a basalt melt. A mixture of olivine and pyroxene crystals with basalt melt will obviously have an overall bulk chemical composition different from the individual solids and liquid; moreover, this bulk composition will depend upon the relative proportions of the solids and liquid. If the proportion of crystals to liquid is great, the bulk is actually peridotitic in composition; if the opposite, then it is basaltic. Moving crystal-laden magmas can experience **flowage differentiation** of crystals into the faster-moving portion of the body, leaving fewer crystals near the walls. In quiescent bodies of basaltic magma, perhaps sitting in a staging area beneath a volcano, crystals of olivine and pyroxene may settle and accumulate on the floor of the body because their densities are greater than the melt in which they are suspended. The crystal-rich base of the body formed by **crystal settling** assumes a peridotitic composition.

Fractionation of olivine crystals alone would leave the **residual melt** more enriched in Al, Ti, Ca, Na, and K, as these elements do not enter at all (or only in minute proportions) into olivine solid solutions. If the olivines were of the composition shown in Table 2-3 (second column), then the residual melt would also be enriched in SiO_2 and FeO, as these two constituents also have lower concentrations in the crystals than in an initially basaltic magma (Table 2-5, last column). The residual melt would be greatly depleted in MgO because the olivines are so rich in this oxide.

It is obvious that there are several processes of diversification possible in magma systems that could function alone or in combination to create a considerable range in chemical composition of derivative rocks. But this can only be part of the explanation for the spectrum of magmatic rocks, as there are crystal–liquid fractionation processes occurring in the **source region** of magma generation that are more or less the reverse of those just described. Such processes involving **anatexis,** or **partial melting,** and **melt segregation** yield variable magmas that can subsequently be diversified further by mixing, contamination, and magmatic differentiation.

Geophysical and petrologic investigations indicate that solid upper mantle peridotite, composed of olivine plus pyroxene, can partially melt locally, forming a small proportion of basalt liquid in equilibrium with it. During this incomplete melting, basaltic chemical constituents Si, Al, Fe, Na, and so on are preferentially dissolved out of the olivine and pyroxene solid solutions, leaving more refractory olivine and pyroxene solid-solution crystals that are enriched in Mg and other chemical constituents stable to higher temperatures. The exact chemical composition of the liquid depends upon the amount produced by melting—whether 2%, 12%, or whatever—and the pressure and composition of the peridotite source. For example, some peridotitic source regions appear to contain small amounts of phlogopite that, upon partial melting, yield liquids enriched in potassium. Hence, melts of variable chemical composition can be produced in the mantle source region. Segregation of the melts from their peridotite residuum, collection of small volumes of melt into larger, and their rise upward because of lower density are the initial stages in the evolution of bodies of basaltic magma.

In light of all these potential mechanisms causing variations in igneous rocks, it may be wondered why the range of observed chemical compositions has certain limits, as described earlier. Certainly, the composition of the source rocks that melt to form parent magmas has an important bearing; if source regions happened to be comprised of something other than peridotite, magmatic rocks might not be so dominated by mafic compositions. Open-system diversification processes are limited by the compositions of the two end-members that combine. Crystal–liquid fractionation processes are limited by the nature of early formed crystals. If quartz, for example, precipitated early in a basalt magma (actually it doesn't at all) and fractionated, residual melts could have a composition unlike anything observed.

Throughout our discussion of magmatic evolution (see Figure 2-2), there is the underlying concept of an initial, original, primary, or **parent magma** acted upon by secondary diversification processes that produce less primitive, more **evolved,** or **derivative magmas.** The fundamental importance of this concept should not be overlooked, even though the terminology might vary (for a discussion see Carmichael and others, 1974, pp. 44–46).

2.4 INTERPRETIVE PETROCHEMISTRY

An important record of the evolution and source of a magma is preserved in the chemical composition of the body, especially the specific nature of the interelement variations and trace-element and isotopic tracers.

Variation Diagrams

Significant insight into the origin and evolution of a magma is afforded by the compositional variations within the body itself and between it and other bodies that may be related in space and time. The value of such variations is in showing chemical trends that reflect evolutionary processes (Wilcox, 1979). For example, a group of basalt lavas forming a volcano may exhibit compositional variations explicable in terms of accumulation of different proportions of olivines by crystal–liquid fractionation. Or, a group of lavas may display linear elemental variations between two end-member parent magmas that mixed together.

An important first step in the interpretation of petrochemical data is to plot the concentration values on some sort of **variation diagram** so that any trends or patterns in variation are made more obvious.

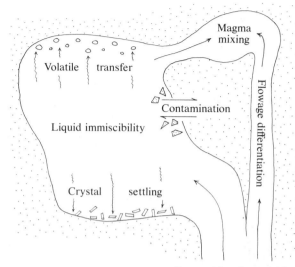

Figure 2-2 Schematic of possible processes of magmatic diversification that act on already variable magmas ascending from source regions. Wall rocks enclosing magma body are stippled.

Two basic types of diagrams are used by petrologists—a Cartesian graph of two variables (x and y) and a triangular diagram (see Figure 2-3). There are obvious weaknesses in the use of such plots to represent petrochemical data. The Cartesian diagram portrays best the absolute concentrations, but in order to represent all n constituents in a rock, $n - 1$ plots are required. The triangular plot presents one more constituent than can be represented in one Cartesian graph but ignores the absolute concentrations of the three variables; two rocks, one with FeO 2%, SiO_2 72%, and Na_2O 4%, the other with equivalent concentrations of 1%, 36%, and 2% would both plot as the same point. Triangular plots show only ratios of constituents.

In an effort to circumvent these weaknesses, some petrologists have resorted to combining two or more constituents into one variable, justifying

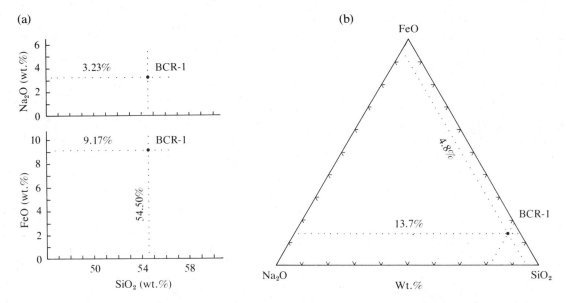

Figure 2-3   Plotting of compositional data on Cartesian (a) and triangular (b) diagrams. In the Cartesian diagram, weight percentages of FeO and Na₂O are plotted against weight percentage of SiO₂ from sample BCR-1 in Table 2-2. These three constituents can be summed ($54.50 + 9.17 + 3.23 = 66.90$), recalculated so the total is 100.00 (that is, $SiO_2 = 81.46$, $FeO = 13.71$, $Na_2O = 4.83$), and then plotted on the triangular diagram. This can be done as follows. First, draw on a piece of triangular graph paper (obtainable at most college bookstores) a line parallel to the leg of the triangle opposite the apex representing FeO and 13.7% of the way toward the apex; the leg opposite is the locus of points representing 0% FeO and the apex point itself represents 100% FeO. Next draw a line through the triangle representing 4.8% Na₂O. The intersection of these two lines is a point that represents the relative SiO₂, FeO, and Na₂O weight percentages in sample BCR-1. Note that it is only necessary to draw lines for any two of the three variables represented in the diagram because the third variable is dependent on the other two.

the combination on some geochemical grounds. Thus, MnO and FeO might be combined together, as Mn is a minor element and generally substitutes for Fe in most silicates and oxides. Another way to represent the constituents more completely is to plot in three dimensions and project the data points onto a two-dimensional graph. Still more sophisticated plotting procedures using a computer have also been employed (for example, LeMaitre, 1976a,b).

Chemical analyses of five representative samples from Mt. Lassen, a composite Quaternary volcano in the northern California Cascade Mountains, are given in Table 3-2, and most of the constituents are plotted in a **Harker diagram** in Figure

2-4. This variation diagram, named after the British petrologist Alfred Harker, plots the weight percentage of SiO₂ along the *x*-axis (abscissa) against concentrations of other oxides along the ordinate axis. The rocks display a relatively smooth sympathetic, or coherent, variation of many oxides with respect to SiO₂. The significant increase in SiO₂, from 54 to 74 wt.%, correlates with increasing K₂O. Because the oxides for each sample total 100%, other oxides must diminish as SiO₂ increases; these negatively correlated oxides include Fe₂O₃ and FeO (combined to minimize effects of secondary oxidation of the lavas after extrusion), MgO, and CaO. Other constituents—Al₂O₃ and Na₂O—vary little. These coherent chemical variations in rocks ranging from basalt (54% SiO₂) through andesite and dacite to rhyolite (74%) are clearly reflected in mineralogical variations; the more mafic silica-poor rocks are comprised principally of pyroxene and plagioclase, and as these Fe–Mg–Ca-rich minerals diminish, the more Si–K-rich assemblage of quartz and alkali feldspar becomes dominant in rhyolite.

If it is assumed that rocks related in space and time are cogenetic, then coherent chemical variations reflect an evolutionary path in the corresponding magma system. The exact nature of the path—involving processes of magmatic diversification and magma generation in the source region as outlined in the previous section—is the central problem facing the petrologist. Does, for example, the Mt. Lassen trend indicate a sequence of diversified magmas drawn from one source region? Or, one parent magma subjected to crystal–liquid fractionation? Or, mixing of a basaltic magma derived from the upper mantle with a rhyolitic magma generated from the sialic crust beneath the Cascades? What about the possibility of contamination? Although petrologists have yet to solve this particular question of the petrogenesis of Mt. Lassen lavas, our intent in succeeding chapters will be to consider fundamental concepts and information that can provide insight into this and allied problems.

It must be emphasized that a regular pattern of data points on a variation diagram does not prove the corresponding rocks are genetically related, though a coherent trend may be compatible with such a relationship. For example, data from such widely scattered rocks as a Precambrian andesite from Canada, a Paleozoic dacite from Australia, a Holocene obsidian from California, and so on could define the same pattern shown in Figures 2-4 and 2-5. The purpose of the variation diagram is to represent numerical data in an easily visualized manner. Only subsequent investigation will answer specific questions on the origin of the rock body (Wright, 1974).

Regardless of the exact petrogenetic processes involved, if that can ever be decided, it is sometimes convenient to devise some sort of evolu-

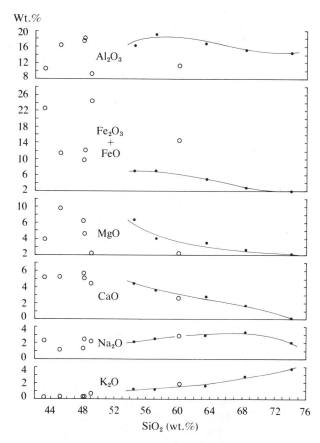

Figure 2-4 Harker variation diagram for representative samples from Mt. Lassen in the Cascade Mountains of northern California (filled small circles) and from the Skaergaard layered gabbroic intrusion in east Greenland (open circles). A line that best fits the trend of variation for Mt. Lassen is drawn for each set of points. Complete chemical analyses are given in Tables 3-2 and 5-4.

tionary index for groups of genetically related rocks. To be useful, such an index would have to be independent of any specific mechanism, such as magma mixing, and simply distinguish in a general manner individual rocks, or a whole group of rocks, that are more "evolved" from those that are closer to parent magmas, however these may be defined.

One index, popular several decades ago, is simply the SiO₂ concentration as displayed on a

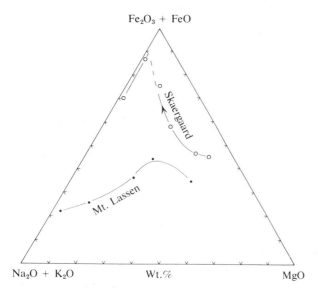

Figure 2-5 AFM diagram for representative samples from Mt. Lassen and the Skaergaard intrusion (see Tables 3-2 and 5-4).

Harker diagram such as that in Figure 2-4; higher SiO_2 values were alleged to indicate solidification from a more evolved magma. Although this may be true in some instances, possible confusion arises in the case of Mt. Lassen lavas, as some dacites with only 68 wt.% SiO_2 are younger than rhyolite with 74 wt.%. Also, in mafic rocks such as basalts and gabbros, SiO_2 may vary little or even decrease in residual liquids produced by crystal–liquid fractionation. This variation has been documented in solidifying Hawaiian lava lakes and in the Skaergaard intrusion, where layers of gabbroic differentiates stacked on top of each other are interpreted to have formed by differential gravitative settling of dense mafic minerals. Some representative samples from the Skaergaard plotted in Figure 2-4 show the scattered, noncoherent nature of the data at low SiO_2 concentrations. However, a late residual-liquid differentiate with 60 wt.% SiO_2 is present, but its oxide values do not always lie on the Mt. Lassen trend.

In view of the inadequacies of SiO_2 as a reliable index, petrologists have turned to other parameters. Because crystal–liquid fractionation processes are commonly believed to influence magma evolution strongly, parameters have been selected to emphasize the contrasting compositions of early formed crystals versus late residual liquids (Cox and others, 1979, pp. 26–40; Carmichael and others, 1974, p. 48). Thus, a **felsic index**, (Na_2O + K_2O)/(Na_2O + K_2O + CaO), has low values in accumulations of early crystallized calcic plagioclases and augite, such as in gabbroic rocks, whereas high values obtain in granitic rocks with abundant alkali feldspar and biotite. A **mafic index**, (Fe_2O_3 + FeO)/(Fe_2O_3 + FeO + MgO), can distinguish between accumulations of high-T Mg-rich olivines and pyroxenes and solidified residual melts enriched in Fe. These two indices can be illustrated in a Cartesian variation diagram. Or, what is virtually the same display can be shown on a triangular AFM diagram, in which the apices of the triangle represent (Na_2 + K_2O) = A, (Fe_2O_3 + FeO) = F, and MgO = M (see Figure 2-5). This plot clearly distinguishes the Mt. Lassen trend from the Skaergaard trend. The postulated crystal–liquid fractionation in the latter produced progressively more Fe-rich rocks, and only in the late stages of differentiation did more felsic rocks rich in alkalies appear. In contrast, the mafic index of the Mt. Lassen lavas varied much less and alkalies reach higher concentrations in most rocks.

Two other evolutionary indices that follow more or less from the above are the **solidification index** (SI) and the **differentiation index** (DI) (see Cox and others, 1979, p. 30). The SI, originally proposed by the Japanese petrologist H. Kuno and equal to the ratio

$$\frac{100 \text{ MgO}}{\text{MgO} + \text{FeO} + Fe_2O_3 + Na_2O + K_2O}$$

is simply a numerical expression of the AFM diagram. The DI of Thornton and Tuttle is the sum of the normative *Q, Or, Ab, Ne, Ks,* and *Lc* and reflects the tendency of magmatic differentiation to enrich residual liquids in these felsic minerals—the so-called **petrogeny's residua system.**

Rock Suites

Scrutiny of variation diagrams soon convinces us that many different chemical trends exist within magmatic rock bodies; compare, for example, the Mt. Lassen trend versus the more Fe-rich Skaergaard trend in Figure 2-5. A group or assemblage of rock types that share some common chemical property have been given many labels in the petrologic literature, including association, series, suite, lineage, and with the additional constraint of geography, petrographic province. Recognition of a **rock suite** implies that the members within it are somehow genetically related, but in what way is not always easy to answer. There are different hierarchies of suites; some are broad and general and embody others that are narrowly defined. Although sometimes clearly distinguishable, as in Figure 2-5, rock suites as a whole grade into one another and must be defined by specific chemical parameters, because there is an essentially continuous spectrum of rock compositions.

Figure 2-6 shows this spectrum in terms of three oxides; this diagram should dispel, once and for all, any notion that natural discontinuities exist in the compositions of magmatic rocks that can be "boxed-in" and given convenient labels. Any boundaries between individual rock types or between suites of rocks arbitrarily cut the compositional spectrum, not always in the best place.

The silica–alkalies variation diagram of Figure 2-6 can be used to subdivide all igneous rocks graphically into two categories—alkaline and subalkaline. **Alkaline rocks** have relatively high concentrations of alkalies (Na, K) relative to silica, so that usually they are undersaturated and have normative nepheline. These chemical properties are manifest in the presence of actual minerals such as alkali feldspar, feldspathoids (sometimes to the exclusion of feldspar in extremely undersaturated rock types), alkali-rich pyroxenes, and alkali-rich amphiboles. Ca-poor pyroxene (such as hypersthene) is never found in alkaline rocks. The most common alkaline rocks belong to the alkali olivine basalt suite, comprising basalts as well as more felsic rock types.

Subalkaline rocks have higher silica relative to alkalies and consequently generally have orthopyroxene or quartz in the norm and never feldspathoids. Subalkaline rocks may be further divided into the tholeiitic suite and the calc-alkaline suite (see Figure 2-7). The **tholeiitic suite,** comprised primarily of basalts but including more felsic rocks as well, shows a higher degree of iron enrichment on an AFM variation diagram and lower concentrations of aluminum. The **calc-alkaline** suite comprises the common volcanic basalt–andesite–dacite–rhyolite and plutonic gabbro–diorite–granodiorite–granite associations, which contain combinations of olivine, Ca-rich clinopyroxene, Ca-poor orthopyroxene, hornblende, biotite, feldspars, and quartz. Comparison of Figures 2-5 and 2-7 indicates that the Skaergaard rocks are tholeiitic and the Mt. Lassen lavas calc-alkaline.

Rarely, some magmatic rocks have molecular $(Na_2O + K_2O) > Al_2O_3$; the excess of alkalies relative to alumina needed to make feldspars or feldspathoids in these **peralkaline rocks** resides in alkali-rich mafic minerals. In the norm, there will be normative acmite $(NaFe^{3+}Si_2O_6)$.

Some geologists are tempted to make uncompromising correlations between rock suites and plate tectonic setting; however, there always seem to be exceptions. The *general* pattern is as follows:

Plate tectonic setting	*Magmatic rock suite*
Oceanic rift	Tholeiitic, some marginally alkaline
Continental rift	Tholeiitic, alkaline, peralkaline
Subduction zone	Calc-alkaline, tholeiitic, rare alkaline
Intraplate	All suites except(?) calc-alkaline

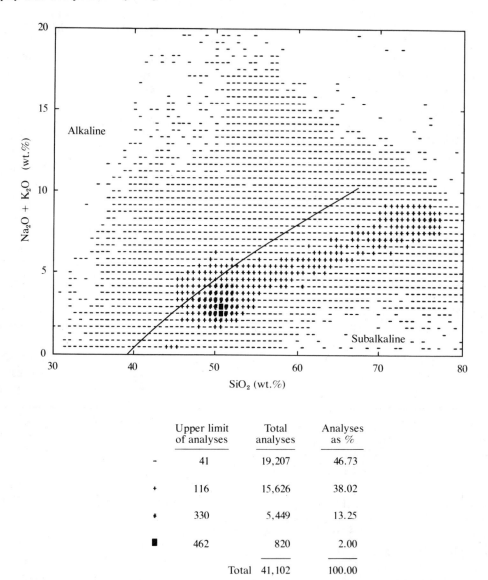

	Upper limit of analyses	Total analyses	Analyses as %
-	41	19,207	46.73
✦	116	15,626	38.02
✦	330	5,449	13.25
■	462	820	2.00
		Total 41,102	100.00

Figure 2-6 Chemical analyses of over 41,000 igneous rocks from around the world have been compiled and stored on magnetic tape by Roger W. LeMaitre of the University of Melbourne. In this variation diagram that he created with a computer plotter, each symbol represents a particular number of analyses falling within the range specified under "upper limit of analyses." About half of all igneous rocks are widely scattered over the diagram (dash symbol), whereas more than half are tightly clustered in a band mostly below and to the right of the boundary line chosen by Irvine and Baragar (1971) to discriminate between alkaline and subalkaline rocks. Most of the samples shown are, therefore, subalkaline. Note the somewhat higher density of rocks near 2.5 wt.% $(Na_2O + K_2O)$ and 50 wt.% SiO_2 and near 8% and 75%, corresponding roughly to basalt and granite (rhyolite), the two dominant magmatic rock types on Earth.

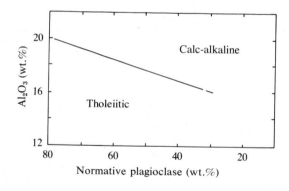

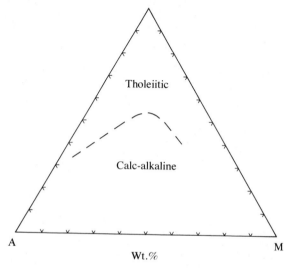

Figure 2-7 Dividing lines between the tholeiitic and calc-alkaline suites (Irvine and Baragar, 1971). Normative plagioclase composition is 100 $An/(An + Ab + 5/3\ Ne)$. A = $Na_2O + K_2O$; M = MgO; F = $FeO + 0.8998\ Fe_2O_3$.

Distribution of Elements

Because of differing bonding characteristics, atoms of different elements prefer different types of chemical environments in which to reside. This is demonstrated by the distribution of elements in the coexisting metallic, sulfide, and silicate (or oxide) minerals in meteorites and in the pig iron, copper matte, and slag produced by smelting of ores (see Table 2-6). Clearly, Fe, Ni, and Co prefer

to form a metallic phase, Cu and Zn prefer to combine with sulfur, and Si, Al, Mg, Ca, and K seek out oxygen. Other elements, perhaps Ti and Na, are a little more indifferent. In view of these affinities, several decades ago the geochemist V. M. Goldschmidt, classified elements as follows:

Siderophile (prefer metallic phase)	Fe, Co, Ni, Pd, Pt, Mo
Chalcophile (prefer sulfide phase)	S, Cu, Zn, As, Ag, Au, Hg, Bi
Lithophile (prefer silicate phase)	Li, Be, O, Na, Mg, Al, Si, K, Ca, Ti, Rb, Sr, Y, Zr, Nb, Cs, Ba, Th, U, and the rare earth elements (REE), which include La, Ce, Pr, Nd, Pm, Sm, Eu, Gd, Dy, Er, Yb, Lu

In the Earth as a whole, siderophile elements are most concentrated in the core, lithophile in the mantle and crust, and siderophile in sulfide ore deposits.

Although our emphasis thus far has been on major elements, it has become apparent in the past decade or so that trace elements can provide very significant insight into the evolution of many magmatic systems. Because of their very low concentrations (in the ppm range), they do not govern the appearance of major minerals such as feldspar, pyroxene, and so on but instead occur as **dispersed elements** in whatever phases happen to be stable under the prevailing *P, T,* and *X*. In this sense, some minor elements such as Mn and Ti can be considered as dispersed provided no mineral forms that has the element as a major stoichiometric constituent, such as ilmenite, $FeTiO_3$.

The importance of dispersed elements to petrogenetic models stems from the fact that their relative concentrations in different minerals in a

Table 2-6   Concentrations of some elements in coexisting minerals of meteorites and in the products of smelted Fe–Cu ores.

	Minerals in meteorites			Metallurgical products		
	Metallic	Sulfide	Silicate	Metal (pig iron)	Sulfide (copper matte)	Silicate (slag)
Si	0.015 wt.%	0 wt.%	21.6 wt.%	0.02 wt.%	0.05 wt.%	22.1 wt.%
Al				0.05	0	9.1
Fe	88.6	13.3		73.6	22.9	3.0
Mg				0	0.05	7.5
Ca				0.003	0.001	13.5
Na				0.1	0.1	0.6
K					0.5	3.3
P	1,800 ppm	3,000 ppm	700 ppm	18,400 ppm	0 ppm	300 ppm
Cr	300	1,200	3,900	0	0	40
Ni	84,900	1,000	3,300	17,200	2,800	500
Co	5,700	100	400	24,400	2,500	40
V				800	100	200
Ti	100	0	1,800	20	20	300
Mn	300	460	2,050	0	6,400	2,000
Cu	200	500	2	64,400	462,000	2,340
Pb	56	20	2	20	2,200	200
Zn	115	1,530	76	8	16,800	3,700

Source: After K. B. Krauskopf, 1967, *Introduction to Geochemistry* (New York: McGraw-Hill).

rock can differ significantly; the dispersed element composition of a magma can, therefore, place constraints on the nature of its source rock, especially what minerals were present and were dissolved to form a partial melt, and what minerals might have been involved in subsequent fractionation processes. Whereas variations in major element concentrations due to magmatic evolutionary processes are generally less than one order of magnitude (see pages 38–39), variations in trace elements can be over three orders of magnitude.

Petrogenetic models based on dispersed elements can be made quite quantitative (for example, Carmichael et al., 1974, pp. 631–639; Cox and others, 1979, chap. 14) because of the precision of

the analyses and because there is an increasing body of data regarding their distribution, or partition, coefficients. If a particular mineral is in chemical equilibrium with a silicate melt, the **distribution coefficient**, K_D, is the ratio of the concentrations of a specific dispersed element in these two phases:

$$K_D = \frac{\text{concentration in mineral}}{\text{concentration in melt}}$$

Coefficients shown in Table 2-7 depend upon the *P*, *T*, and *X* of the system. Dispersed elements can be classified as **incompatible** if $K_D \ll 1$ and **compatible** if $K_D > 1$. Incompatible elements are preferentially concentrated in the "looser" structure of a silicate

Table 2-7 Some approximate trace element partition coefficients (K_D) for minerals in equilibrium with a basalt melt.

				Rare earth elements				
	Rb	Sr	Ba	Ce	Sm	Eu	Yb	Ni
Olivine	0.001	0.001	0.001	0.001	0.002	0.002	0.002	10
Orthopyroxene	0.001	0.01	0.001	0.003	0.01	0.01	0.05	4
Clinopyroxene	0.001	0.07	0.001	0.1	0.3	0.2	0.3	2
Garnet	0.001	0.001	0.002	0.02	0.2	0.3	4.0	0.04
Plagioclase	0.07	2.2	0.2	0.1	0.07	0.3[a]	0.03	0.01
Amphibole	0.3	0.5	0.4	0.2	0.5	0.6	0.5	3

[a] Europium exists in both the di- and trivalent state in magmas, and the divalent ion is preferred by plagioclase. So under strongly reducing conditions, $Eu^{3+} \rightarrow Eu^{2+}$ and the partition coefficient can be as great as that for Sr.
Source: After K. G. Cox, J. D. Bell, and R. J. Pankhurst, 1979, *The Interpretation of Igneous Rocks* (London: Allen and Unwin).

melt coexisting with a crystalline solid. During partial melting of mantle peridotite (olivine + orthopyroxene + clinopyroxene ± garnet), for example, Rb, Sr, Ba, and rare earth elements (REE) are highly concentrated in the melt, by factors as great as 1,000. During crystal–liquid fractionation, the residual melt separated from crystals is similarly enriched in these incompatible elements. Compatible elements are enriched in crystalline phases and depleted in a coexisting melt. More Ni stays in peridotite than in a partial melt, and fractionation of olivines from a basalt magma will leave the residual melt impoverished in this element. It should be kept in mind that whether an element is compatible or incompatible depends entirely on what minerals exist in the magma system. For example, Sr is enriched as an incompatible element in the melt of a crystallizing basalt magma as long as olivines and pyroxenes are forming, but when plagioclase crystallizes, it takes Sr out of the melt as a compatible element.

Principal factors that account for the differing partition coefficients and whether an element is compatible or incompatible are the radii and charges of the ions (see any modern introductory mineralogy text for a discussion). In Table 2-8 it can be seen why Rb, Sr, Ba, and so on and REE are incompatible with respect to minerals in peridotite; their radii are too large and their charges too high to be tolerated in Mg, Fe, and Ca sites in the peridotite minerals, and their radii are too large for them to fit in Si and Al sites. The size and charge of Ni, however, is close to Mg and so Ni readily substitutes for Mg in mafic minerals and is a compatible element with respect to them.

It can be seen from Table 2-7 that partial melts derived from a source rock containing plagioclase that is not completely dissolved during melting will contain relatively less Eu than other REE because K_D for Eu is larger. A similar "europium anomaly" could develop in residual melts from which plagioclase has been fractionated. K_D values involving garnet increase from light through heavy REE. Therefore, a partial melt of a garnetiferous source rock, or a residual melt from garnet fractionation, will have a high Ce/Yb ratio. These are the sorts of constraints that can be applied to evolutionary processes in magmatic systems using dispersed elements.

Isotopes

Radioactive isotopes of K, Rb, Sm, U, and Th are critically important in establishing the chronology of magmatic events. These isotopes and certain stable ones are also valuable as petrogenetic trac-

Table 2-8 Ionic radius and charge of major elements in minerals occurring in mantle peridotite compared to dispersed elements.[a]

Major elements			Dispersed elements			
	Coordination	Radius		Coordination	Radius	Nomenclature
Si^{4+}	IV	0.34	P^{5+}	IV	0.25	
Al^{3+}	IV	0.47	Ti^{4+}	VI	0.69	
	VI	0.61	K^+	VIII	1.59	
Fe^{2+}	VI	0.69	Rb^+	VIII	1.81	
Mg^{2+}	VI	0.80	Sr^{2+}	VIII	1.33	
Ca^{2+}	VI	1.08	Cs^+	VIII	1.82	
Na^+	VI	1.10	Ba^{2+}	VIII	1.50	
			Th^{4+}	VIII	1.12	
			U^{4+}	VIII	1.08	
			La^{3+}	VIII	1.26	
			Ce^{3+}	VIII	1.22	
			Nd^{3+}	VIII	1.20	
			Sm^{3+}	VIII	1.17	
			Eu^{2+}	VIII	1.33	
			Eu^{3+}	VIII	1.15	
			Yb^{3+}	VIII	1.06	
			Co^{2+}	VI	0.73	
			Ni^{2+}	VI	0.64	

Nomenclature brackets: Large ion lithophile elements (LIL) and Rare earth elements (REE) — Light (LREE) and Heavy (HREE) — comprise the Incompatible elements; Co^{2+} and Ni^{2+} are Compatible elements.

[a] Radii (in Å) are a function of coordination. Nomenclature of dispersed elements is also shown; compatibility and incompatibility are relative to minerals in peridotite.
Source: After E. J. W. Whittaker and R. Muntus, 1970, Ionic radii for use in geochemistry, *Geochimica et Cosmochimica Acta,* 34.

ers in evaluating the evolution of magmas in much the same way as dispersed elements (Cox and others, 1979). Investigations of dispersed-element and isotope tracers go hand-in-hand because the important isotopes are of dispersed elements. **Isotopes** of an element are atoms whose nuclei contain the same number of protons but a different number of neutrons. Different isotopes provide insight into (1) the age of the rock or mineral, (2) the temperature at which minerals crystallized in a state of equilibrium, (3) the source of the magmatic or metamorphic rock, and (4) processes acting on the rock body during its history.

In petrology, isotopes of oxygen, rubidium, strontium, lead, uranium, thorium, samarium, and neodymium have proven to be especially significant (Faure, 1977). Oxygen isotopes are particularly valuable as geothermometers, and oxygen, strontium, lead, and neodymium isotopes are important as radiogenic tracers, or indicators of the source of material comprising the rock body.

Oxygen isotopes. Oxygen has three stable isotopes, not subject to radioactive decay, with the following determined abundances in ocean water:

^{16}O	99.756%
^{17}O	0.039
^{18}O	0.205

The superscripts denote the mass number, that is, number of neutrons plus protons in the nucleus.

Substantial variations in the abundance of these

three isotopes occur in natural waters, rocks, and minerals. The conventional manner of expressing the isotopic composition is by referring the $^{18}O/^{16}O$ ratio to the ratio of a standard, usually standard mean ocean water, $^{18}O/^{16}O_{SMOW}$

$$\delta^{18}O/^{16}O = \left[\frac{^{18}O/^{16}O}{^{18}O/^{16}O_{SMOW}} - 1 \right] 1,000 \quad (2.1)$$

The δ value is in parts per mil (thousand). Meteoric waters are enriched in ^{16}O relative to ^{18}O and thus have negative values ($^{18}O/^{16}O$ water $>$ $^{18}O/^{16}O_{SMOW}$). Rocks usually have positive δ values.

Variations in oxygen isotope abundances are caused by fractionation mechanisms whereby one isotope is preferentially incorporated into one material relative to another. As the vapor pressure, or escaping tendency, of an isotope is inversely proportional to its mass, during evaporation of sea water the atmospheric vapor and hence meteoric (rain) waters are enriched in the lighter ^{16}O isotope. Fractionation of light isotopes such as oxygen is temperature dependent, but is insensitive to pressure. Hence, in a system at equilibrium, the isotopic composition of two coexistent phases, such as calcite–water, quartz–water, or quartz–feldspar, is a function of temperature. Equilibrium isotopic compositions as a function of temperature have been determined experimentally. A pair of minerals formed in nature at equilibrium can thus be used as a **geothermometer**. Moreover, certain fractionation processes and interactions between materials of different isotopic composition can be evaluated.

Rubidium and strontium isotopes. Rubidium occurs in nature as the isotopes ^{86}Rb and ^{87}Rb; the latter is radioactive and decays by beta-emission to ^{87}Sr with a half-life of 50 Gy. The present relative abundance of the rubidium isotopes—72.17% ^{86}Rb and 27.83% ^{87}Rb—is the same in all rocks and minerals, regardless of age, indicating that these heavy isotopes were thoroughly mixed in the primeval Earth and have not experienced fractionation since then regardless of the geologic processes acting on them.

The ionic charge and electronegativity of rubidium and potassium ions are the same. Their ionic radii are 1.81 Å and 1.59 Å, respectively. Rb substitutes for K in micas and potassium-bearing feldspars. Rocks and minerals that have high concentrations of K also tend to have relatively high Rb, although the K/Rb ratio is not uniform in all materials.

The crystal chemistry of Sr is a little more complicated than that of Rb, but follows Ca (Faure, 1977), so that Sr is most highly concentrated in calcic plagioclase and apatite. Ca^{2+} sites in calcic pyroxenes are too small for the somewhat larger Sr^{2+} ions.

Strontium has four stable isotopes, ^{88}Sr, ^{87}Sr, ^{86}Sr, and ^{84}Sr, whose relative abundances are, respectively, about 10, 0.7, 1, and 0.07. But because ^{87}Sr is a decay product of ^{87}Rb, its actual abundance in a rock or mineral depends not only upon the amount of ^{87}Sr present when the material formed but also upon the concentration of Rb and the age. Materials rich in Rb, such as micas and alkali feldspars, will obviously contain considerable ^{87}Sr, especially if they are old. As isotopic ratios are more accurately measured by a mass spectrometer than absolute amounts, the abundance of ^{87}Sr is expressed as the ratio $^{87}Sr/^{86}Sr$, where the abundance of ^{86}Sr is constant because it is a stable isotope not formed as a decay product of any other naturally occurring radioactive isotope. The relationship between the present day $^{87}Sr/^{86}Sr$ ratio, the initial ratio $(^{87}Sr/^{86}Sr)_0$ when the rock or mineral formed at time zero, the Rb content, the age in t years, and the decay constant λ is expressed by the equation

$$^{87}Sr/^{86}Sr = (^{87}Sr/^{86}Sr)_0 + (^{87}Rb/^{86}Sr)\lambda t \quad (2.2)$$

This is a linear equation of the form $y = a + xb$, where a is $(^{87}Sr/^{86}Sr)_0$ and $b = \lambda t$. A plot (see Figure 2-8) of $x = {^{87}Rb/^{86}Sr}$ and $y = {^{87}Sr/^{86}Sr}$ measured on separated minerals from one rock, or on a group of genetically related whole rocks, in a body that has behaved as a closed system since $t = 0$, yields a straight line called an **isochron**. The intercept of the isochron on the y axis is the initial ratio

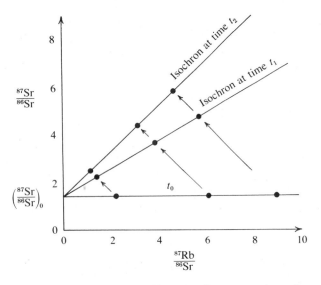

Figure 2-8 An isochron diagram of comagmatic rock or mineral samples (solid circles) represents a closed Rb–Sr system decaying through time according to equation (2.2). When the magma crystallizes at $t = 0$ (t_0), all minerals and rocks within the system are at equilibrium with identical $^{87}Sr/^{86}Sr$ ratios equal to $(^{87}Sr/^{86}Sr)_0$. Their $^{87}Rb/^{86}Sr$ ratios differ, however, depending upon Rb/Sr ratio; muscovite, for example, has a greater ratio than plagioclase. Subsequent to time $t = 0$, individual rocks or minerals move along straight lines having a slope of -1 due to decay of ^{87}Rb to ^{87}Sr, provided the system remains closed to rubidium and strontium. At any one time (t_1, t_2, and so on), points representing the minerals or rocks define a straight line isochron whose slope is dictated by the age of the system. The initial ratio $(^{87}Sr/^{86}Sr)_0$ is given by the intercept of the isochron with the vertical axis.

$(^{87}Sr/^{86}Sr)_0$ and the slope of the line is the age times the decay constant, λt. Because of the long half-life of ^{87}Rb, the present day $^{87}Sr/^{86}Sr$ measured on a mass spectrometer for samples only a few million years old is essentially the same as the initial ratio.

The initial ratio $(^{87}Sr/^{86}Sr)_0$ serves as a valuable petrogenetic tracer. Magmas derived by partial melting of source rocks with high Rb/Sr ratios, or contaminated by such material, such as old continental crustal sial, will inherit this geochemical property in a high initial ratio. Sources in the peridotitic mantle, where Rb/Sr is low, will yield magmas with low initial ratios.

Although discussions in later chapters will explore some specific implications of Rb–Sr isotopic compositions to magma genesis, it is pertinent here to present a generalized model of Sr-isotope evolution in the Earth. Figure 2-9a shows that after the origin of the Earth 4.6 Gy ago and assuming an initial chondritic meteorite composition, the peridotitic upper mantle and continental granitic crust have experienced an increasing $^{87}Sr/^{86}Sr$ ratio due to the slow decay of ^{87}Rb. Crustal ratios, however, have been increasing at a greater rate because of greater Rb/Sr. The model further assumes that granitic crust has been derived from the mantle over most of the history of the Earth and that older crustal segments now have the highest $^{87}Sr/^{86}Sr$ ratios. Magmas derived from the mantle have present day ratios generally <0.706, whereas those derived by melting or assimilation of continental crust have ratios >0.706.

Samarium and neodymium isotopes. There are many isotopes of these two light REEs, but the ones of most relevance to geochronology and petrology are ^{147}Sm, which decays by alpha emission to ^{143}Nd with a half-life of 106 Gy. The abundance of the daughter product is referred to by a ratio with the stable ^{144}Nd isotope—$^{143}Nd/^{144}Nd$. The Sm–Nd isotopic system is similar to the Rb–Sr system just described and is handled in much the same manner. Measurement of Sm–Nd isotopes (only since the mid-1970s) had to await the development of a new breed of very high-precision mass spectrometers and chemical separation techniques capable of distinguishing the very small differences in their ratios (see review of published work in Bickford and Van Schmus, 1979). Such small differences ($^{143}Nd/^{144}Nd$ ratios of 0.510 to 0.512) stem from the small variations in Sm/Nd ratios of 0.1 to 2.0 in rocks (compare Rb/Sr ratios of 0.005 to 3.0), which in turn are related to the similar chemical bonding of these two REEs (see Table 2-8). Because of this similarity, there are no reservoirs on Earth especially rich in Sm nor Nd like the Sr-rich

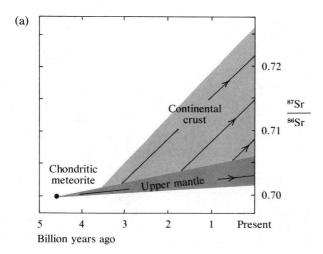

(a)

Chondritic
meteorite

Continental
crust

Upper mantle

$\frac{^{87}Sr}{^{86}Sr}$

0.72

0.71

0.70

5   4   3   2   1   Present

Billion years ago

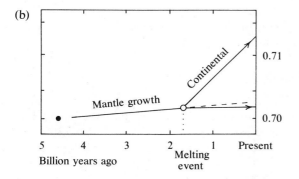

(b)

Mantle growth

Continental

0.71

0.70

5   4   3   2   1   Present

Billion years ago

Melting
event

Figure 2-9 Strontium isotope evolution diagram assuming the Earth has evolved from a chondritic meteorite composition 4.6 Gy ago. (a) The upper mantle shows only slight but somewhat variable growth in $^{87}Sr/^{86}Sr$ with time as a result of its initial low concentration of ^{87}Rb. Youthful mantle-derived magmas have $^{87}Sr/^{86}Sr$ ratios less than 0.706. Withdrawal of Rb-enriched partial melts from the upper mantle during the Earth's history has produced a broad spectrum of growth trends for the continental crust, yielding $^{87}Sr/^{86}Sr$ ratios greater than 0.706. (b) Detailed growth lines surrounding a melting event in the mantle. Removal of Rb-rich partial melt from the mantle causes its $^{87}Sr/^{86}Sr$ ratio to grow at a slower rate than undepleted mantle (dashed line).

oceans that produce Sr-rich marine carbonates and the Rb-rich old continental crust.

Despite the analytical challenges, Sm–Nd isotopes offer significant advantages over other systems such as Rb–Sr and U–Pb because Sm and Nd occur in major minerals such as pyroxenes and plagioclases and are not nearly so subject to chemical disturbances; thus, geologic processes such as intracrystalline deformation and recrystallization accompanying metamorphism that can greatly perturb and open Rb–Sr and U–Pb systems, erasing primary crystallization events, may not be felt in Sm–Nd systems. Comparison of Sr and Nd isotopic ratios can, therefore, constrain models of magmatic evolution—mantle partial melting, crustal contamination, interaction between oceanic basalt magma and sea water—to a better degree than Sr isotopes alone.

SUMMARY

2.1. Basic hand-sample petrography is essential to many petrologic investigations. Sequential magmatic fabrics are expressed in aphanitic, phaneritic, and porphyritic grain sizes that reflect differing relative rates of crystal formation in the magma. Glassy fabric originates by loss of heat or volatiles from the magma that is too rapid to allow crystal formation. Clastic fabric reflects a fragmented source material.

Magmatic rocks are comprised of only seven major rock-forming mineral groups: feldspars, quartz, pyroxenes, olivines, amphiboles, micas, Fe-Ti oxides. Magmatic rock bodies can be divided into the following four major families of rocks: (1) calc-alkaline volcanic rocks, (2) calc-alkaline plutonic rocks, (3) mafic and ultramafic rocks, and (4) highly alkaline rocks.

2.2. The validity of any chemical data depends upon the quality of the laboratory analysis as well as upon the quality of the sample collected in the field. Samples collected in the field to evaluate magmatic evolution must be fresh (free of secondary alteration or weathering) and sufficiently large to be representative of the composition of the body. Chemical analyses of igneous rocks, mainly by rapid spectrometric methods, reveal usually major amounts of O, Si, Al, Fe, Mg, Ca, Na, and K; other elements are present in minor or even trace amounts measurable in parts per million (ppm). Modal, mineralogical, and chemical compositions of a rock are correlated.

As there are few significant discontinuities in the compositional spectrum, most classification schemes for magmatic rocks must use arbitrarily defined limits or averages. Several different general chemical classifications exist that are based mainly on silica content. Specific classification into rock types is hampered by lack of agreement among petrologists. The normative composition of a rock calculated from its chemical composition is expressed in terms of readily compared hypothetical standard minerals.

2.3. Compositionally variable parent magmas ascending from source regions in the upper mantle and lower crust are subjected to subsequent processes of magmatic diversification including, but perhaps not limited to, mixing of two or more magmas, contamination with wall rock, liquid immiscibility, volatile transfer, crystal settling (or flotation), and flowage differentiation. The challenge facing the petrologist is to find which evolutionary process or combination of processes among the many possible were responsible for the origin of a rock body.

2.4. An important record of the evolutionary processes experienced by a magma, as well as clues to its source, is preserved in the chemical composition of the rock body, especially the specific nature of interelement variations and dispersed-element and isotopic tracers. Many different Cartesian and triangular variation diagrams and chemical indices can be employed to provide insight on the evolution of a magma system.

Rock suites consisting of many individual rock types and encompassing a significant part of the whole igneous compositional spectrum seemingly are genetically related by common evolutionary processes. The major suites are the silica-saturated to silica-oversaturated subalkaline suite and the mostly silica-undersaturated alkaline suite; subalkaline rocks include the tholeiitic, following more Fe-enriched evolutionary trends, and the calc-alkaline. There is an imperfect correlation between these magmatic rock suites and plate tectonic setting.

Dispersed, mostly trace, elements are partitioned in significantly different concentrations in different minerals and any coexisting silicate melt. As these concentrations can vary hundreds of times more than major elements, dispersed elements can indicate specific minerals with which partial melts equilibrated at their source and what minerals might have been subsequently fractionated.

Isotopes can be used to establish the chronology of magmatic processes and place constraints on models of magma sources and of interactions between the magma system and its surroundings; oxygen isotopes can also serve as geothermometers of mineral growth.

STUDY EXERCISE

1. Classify the rocks listed in Appendix D as either alkaline or subalkaline; if the latter, further classify them as either calc-alkaline or tholeiitic.

REFERENCES

Allègre, C. J., and Hart, S. R. (eds.), 1978. *Trace Elements in Igneous Petrology.* New York: Elsevier, 272 p.

Bickford, M. E., and Van Schmus, W. R. 1979. Geochronology and radiogenic isotope research. *Reviews of Geophysics and Space Physics* 17:824–839.

Carmichael, I. S. E., Turner, F. J., and Verhoogen, J. 1974. *Igneous Petrology.* New York: McGraw-Hill, 739 p.

Cox, K. G., Bell, J. D., and Pankhurst, R. J. 1979. *The Interpretation of Igneous Rocks.* London: George Allen and Unwin, 450 p.

Deer, W. A., Howie, R. A., and Zussman, J. 1962. *Rock-Forming Minerals*. London: Longman (5 volumes).

Dietrich, R. V., and Skinner, B. J. 1979. *Rocks and Rock Minerals:* New York: John Wiley and Sons, 319 p.

Faure, G. 1977. *Principles of Isotope Geology*. New York: Wiley, 464 p.

Irvine, T. N., and Baragar, W. R. A. 1971. A guide to the chemical classification of the common volcanic rocks. *Canadian Journal of Earth Sciences* 8:523–548.

Jackson, K. C. 1970. *Textbook of Lithology*. New York: McGraw-Hill, 552 p.

LeMaitre, R. W. 1976a. The chemical variability of some common igneous rocks. *Journal of Petrology* 17:589–637.

LeMaitre, R. W. 1976b. Some problems of the projection of chemical data in mineralogical classifications. *Contributions to Mineralogy and Petrology* 56:181–189.

Wilcox, R. E. 1979. The liquid line of descent and variation diagrams. In H. S. Yoder, Jr. (ed.), *The Evolution of the Igneous Rocks: Fiftieth Anniversary Perspectives*. Princeton, N.J.: Princeton University Press, pp. 205–232.

Wright, T. L. 1974. Presentation and interpretation of chemical data for igneous rocks. *Contributions to Mineralogy and Petrology* 48:233–248.

3

Calc-Alkaline Volcanic Rock Bodies

Volcanic rocks comprised dominantly of feldspar with lesser amounts of quartz, biotite, amphiboles, pyroxenes, and other accessory minerals occur in large volumes in volcanic arcs alongside oceanic trenches above subducting lithospheric plates. Fabrics are basically aphanitic, glassy, or clastic but are highly variable due to complex flows of energy in magmas that range widely in viscosity and gas content. Magma extrusions forming the volcanic bodies are from coherent localized lava flows and explosive pyroclastic eruptions of gas-charged magma that can create voluminous deposits covering thousands of square kilometers. Rapid evacuation of magma from shallow chambers allows the roof to collapse into the void, forming a bowl-shaped caldera. The familiar large, conical composite volcanoes are built of alternating layers of pasty lava and volcaniclastic deposits.

Although the main emphasis of this chapter is on the descriptive aspects of calc-alkaline volcanic rocks in arc settings, some generalizations are in order concerning their evolution and origin. Arc magmas appear to be polygenetic, originating by partial melting of hydrous peridotitic upper mantle, subducting oceanic crust, and continental crustal material and by magma mixing, differentia-

tion, and assimilation. Compositional differences in some volcanic sequences tend to reflect the basement on which they rest, whether oceanic or continental. The challenge facing the petrologist working on a particular body of rock is to decide, on the basis of its specific properties, how the magmas originated and evolved and how they were extruded and emplaced.

3.1 PETROGRAPHY

The rocks described in this chapter are chiefly aphanitic, glassy, or clastic and occur as surficial deposits or small shallow intrusions. Compositionally, they range from highly silicic rhyolites to basalts (see Appendix D), but all are characterized by abundant feldspar with lesser amounts, in various combinations, of quartz, biotite, hornblende, pyroxenes, or rarely olivine, sphene, and other accessory minerals. Although the emphasis of this chapter is on rocks of the calc-alkaline suite (see Figures 3-1 through 3-11), some rock types of the peralkaline and tholeiitic suites are intimately associated with the calc-alkaline suite and have similar fabrics and modes of occurrence.

(a)

(b)

Pyroxene
phenocryst

Plagioclase phenocryst
with corroded core

Hornblende
phenocryst

Microlites
of feldspar
in felty matrix

Hornblende phenocrysts
with decomposed rims
rich in opaque Fe–Ti oxides

0 cm 1

0 mm 0.5

Figure 3-1   The most common fabric in volcanic rocks is a combination of porphyritic and aphanitic, such as in this andesite shown at (a) hand-sample and (b) microscopic scales of view.

Aphanitic refers to crystalline aggregates whose grain size is so small that minerals are not identifiable with the naked eye. They form by rapid crystallization with considerable undercooling, so that constituent minerals have abundant nuclei from which to grow. This fabric also forms secondarily by devitrification of glasses. A rock is designated as aphanitic regardless of the concentration of any phenocrysts that might occur.

In the microscopic view of the andesite, the aphanitic matrix is comprised of small birefringent **microlites** of euhedral feldspar that are interwoven in an irregular fashion as **felty texture**. (Birefringence refers to the colors of light seen in a mineral in thin section under a petrographic microscope, due to the transmission of cross-polarized light.) If the microlites are disposed in a subparallel manner as a result of flow of the magma, the fabric is **trachytic**. **Phyric**, or **porphyritic**, refers to the presence of larger **phenocrysts** residing in an appreciably finer **matrix**, or **groundmass**. Combined porphyritic and aphanitic fabric implies rapid crystallization at numerous nucleation centers of a melt in which the phenocrysts had already grown. Phenocrysts that grew slowly beneath the surface prior to a second stage of rapid crystallization of the magma are sometimes called **intratelluric crystals**. Not all porphyritic fabrics necessarily reflect a two-stage cooling history (see pages 241–242). In some porphyritic rocks, phenocrysts are clustered together as **glomerophyric** fabric, which originates as suspended crystals in a melt that somehow become attached together, or as remnants of a partially disaggregated crystal segregation that collected on the floor or margin of the magma body. Wholly aphanitic rocks without phenocrysts are said to be **nonporphyritic** or **aphyric**; they form by rapid crystallization from hot melts lacking large suspended crystals.

Fabrics

The three major types of fabrics found in volcanic and shallow intrusive bodies form in response to different geologic processes. Crystalline aphanitic rocks reflect rapid loss of thermal energy, or gas, or both. For a given chemical composition, glassy rocks represent more rapid heat or volatile loss. Indirectly, chemical composition of the melt is also involved because the more siliceous

(a)

(b)

0 cm 2

0 mm 0.04

Figure 3-2 In **glassy** fabrics the amount of glass can range from traces to most of the rock, as seen in this obsidian with conchoidal fracture (a). However, even in apparently crystal-free glasses, high magnification will invariably reveal minute, nonbirefringent **crystallites** such as the spiderlike ones shown in the photomicrograph (b).

and therefore viscous the melt, the greater the tendency for glass formation. More viscous melts are less mobile on all scales down to the atomic scale. Clastic volcanic rocks result from fragmentation by gas explosions or by shattering due to thermal stresses in a rapidly cooling extruded lava exposed to the cool atmosphere or to bodies of water.

Some of the fabrics encountered in calc-alkaline volcanic rocks are more common in their plutonic counterparts and in basalts; these are treated in the next two chapters. The major types of fabrics found in calc-alkaline rocks are illustrated and briefly discussed in Figures 3-1 through 3-11.

Classification

Within each of the three major fabric groups, further subdivision into specific rock types is made according to composition or some other petrologically significant property.

Classification of aphanitic calc-alkaline volcanic rocks. Although coarsely crystalline phaneritic rocks such as granite (described in the next chapter) can be classified on the basis of modal proportions of quartz and feldspars, this approach is generally unrealistic for volcanic rocks because of the presence of glassy and microcrystalline

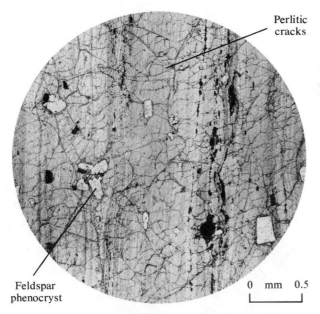

Perlitic
cracks

Feldspar
phenocryst

0 mm 0.5

Figure 3-3 Many glasses contain euhedral phenocrysts that grew under conditions of a slower heat or volatile loss than the episode that created the glass. **Vitrophyric** is a combination of porphyritic and glassy fabrics. In this photomicrograph of a vitrophyric rock, phenocrysts of feldspar lie in a matrix of perlitic glass (see Figure 3-4) with flow layers defined by varying concentrations of minute crystallites.

groundmasses. Hence, field and preliminary laboratory classification must rely on phenocrysts alone. A rigorous quantitative breakdown is hardly justified, so the following guidelines for nomenclature may be used:

Rhyolite Phenocrysts of sanidine and quartz; any type of mafic phenocrysts may be present, but they are sparse to absent.

Rhyodacite Phenocrysts of plagioclase, sanidine, and quartz; any type of mafic phenocrysts may be present, but they are sparse to absent.

Dacite Phenocrysts of plagioclase and quartz; mafic phenocrysts of any composition may be present.

Andesite Phenocrysts of plagioclase, with

0 cm 1

Figure 3-4 Nests of concentric curved fractures in glass constitute **perlitic** fabric, as in this rock and in the one shown in Figure 3-3. Internal reflectance of light imparts a lustrous pearly gray color to the hand sample. A small nodule of dense black, uncracked obsidian may be located in the core of a nest of cracks. Up to 10 wt.% of perlitic glass is comprised of H_2O^-, most of which may be driven off by gentle heating, suggesting that it was absorbed into the glass at low temperatures. In contrast, obsidian without cracks generally has less than 1% water, chiefly as H_2O^+, held more tightly in the glass, probably representing original magmatic water. Perlitic fabric develops by slow progressive hydration under atmospheric conditions. Hydration of a thin outer layer of the original obsidian causes expansion and consequent cracking away from the nonhydrated substrate. Repetition of this process causes inward growth of the perlitic cracking.

or without pyroxene, hornblende, or perhaps biotite, rarely olivine.

Latite Sometimes used in lieu of andesite but denotes a more potassic composition so that phenocrysts of plagioclase, biotite, and perhaps sanidine would be appropriate; pyroxene and hornblende, if present, may be subordinate.

0 cm 2

Figure 3-5 In **spherulitic** fabric, spherical to ellipsoidal clusters of radiating fibrous crystals lie in a glassy, or sometimes felsic, aphanitic matrix. Individual spherulites range in size from less than 1 mm up to a meter or so in diameter. A central nucleus such as a phenocryst in the original glass is commonly present and apparently acts as a seed to initiate crystallization. Spherulites are usually comprised of alkali feldspar and some polymorph of SiO_2. Spherulites can form after active flow of the magma, as shown by ghostlike vestiges of the primary flow layering running through them, as in this photo, or before the magma had totally ceased flowing, as shown by spherulites deflecting surrounding layers. Laboratory experiments indicate spherulites form in response to drastic undercooling of viscous silicate melts (see pages 243–245). Spherulites are not phenocrysts.

Figure 3-6 Highly vesicular glass has a **pumiceous** fabric. Gas-charged magma froths upon decompression during extrusion as dissolved gases collect into innumerable small bubbles, or **vesicles**. If the bubbles reside in a crystal-poor or crystal-free melt that subsequently solidifies to a glassy or aphanitic material, the bubbles will be rounded due to the surface tension of the glass. In the hand sample (a) and microscopic view (b) shown here, the individual vesicles are elongate due to stretching of the glass during flow.

(a)

0 cm 1

(b)

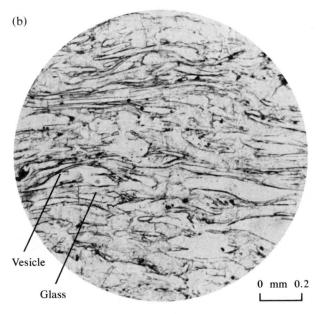

Vesicle

Glass

0 mm 0.2

Figure 3-7 Collection of exsolved gas into irregularly shaped pockets in an already crystalline matrix produces **vuggy** fabric. The vugs are commonly lined with euhedral **vapor-phase** crystals that grew freely in the entrapped vapor. Similar fabric in phaneritic plutonic rocks is called miarolitic fabric.

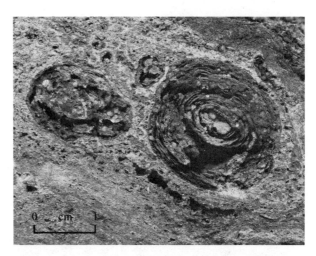

Figure 3-8 **Lithophysal** fabric occurs in glassy or felsic aphanitic rocks and consists of rounded masses generally a few centimeters in diameter with concentric shells of aphanitic material. Lithophysae (singular, lithophysa) means "stone bubbles." Their origin is uncertain, but rhythmic exsolution and expansion of volatiles during crystallization seems to be involved. Lithophysal cavities are sometimes filled with resistant secondary silica to produce geodes. [U.S. Geological Survey photograph courtesy of Robert L. Smith and Susan L. Russell-Robinson.]

Basalt Phenocrysts of olivine, pyroxene, and generally minor plagioclase.

Many aphanitic rocks are nonporphyritic or aphyric, lacking any distinctive phenocrysts, or the feldspar phenocrysts in porphyritic rocks are so altered that they cannot be identified. In such cases the above classification is impractical and we are forced to use color of the rock as a basis for general classification. The term **felsite** can be used for a light-colored aphanitic rock with or without phenocrysts, and the term **basalt** can be used for dark-colored ones.

Various modifying terms can make the rock name more explicit, for example, biotite felsite, porphyritic biotite dacite, strongly porphyritic andesite, and so on. The term **porphyry** (see page 116) should be reserved for strongly porphyritic intrusive bodies.

More detailed or rigorous classification is possible (Streckeisen, 1979) if the modal composition of the aphanitic crystalline component as well as the phenocrysts of the rock can be determined by microscopic petrography or other means. Because of its relative ease of access, a chemical analysis of the rock furnishes one of the best bases for detailed classification, especially in the case of glassy rocks (see references on page 44).

Generalized classification of glassy rocks. Volcanic glasses may be classified in the field or in preliminary laboratory work as follows:

Obsidian A massive, commonly flow-layered glass (see Figure 3-2) occurring in

Figure 3-9 **Flow layering** is expressed by layers of contrasting texture or sometimes mineralogical composition—usually considerably enhanced on weathered surfaces of the rock body. The layers vary in thickness from thin laminae less than 1 mm up to many centimeters or even meters. Existing suspended crystals may somehow be segregated into crystal-rich and crystal-poor zones, or differential shearing flow may influence the crystallization behavior from one layer to the next. Layers may be planar, folded into smooth open to isoclinal forms, or intricately contorted. Planar layers originate by laminar flow within the moving magma. Contorted or folded layers manifest a transition from purely laminar to nonlaminar or turbulent flow as the moving magma experiences local changes in applied stresses, perhaps by encounter with an obstruction. Flow layers are markers of the internal pattern of flow of the viscous magma during its emplacement. This photograph shows flow layering in a rhyolite lava flow that has been enhanced by differential weathering, which has etched-out certain layers. Flow layering on smaller scales is evident in Figures 3-2 and 3-3. [Photograph courtesy of Glenn Embree.]

Figure 3-10 Explosive eruptions, as well as other types of volcanic processes, create fragmental or **clastic** fabric, as seen in this photograph of an outcrop.

colors from black to brown to red, sometimes variegated red to brown within a hand sample because of variable oxidation of iron. Obsidians may be spherulitic, lithophysal, perlitic, or somewhat pumiceous.

Pitchstone Like obsidian but with a waxy, resinous luster rather than vitreous because of absorbed water.

Perlite Pearly gray glass with perlitic fabric (see Figure 3-4).

Pumice Lightweight, highly vesicular, pumiceous fabric (see Figure 3-6).

Vitrophyre Vitrophyric glass.

(a)

(b)

Broken crystal
of feldspar

Glass shards

Pumice

0   cm   2

0   mm   1

Figure 3-11   A common type of clastic fabric is comprised of lumps of pumiceous glass and smaller glass **shards,** which are disaggregated bits of pumice. This **vitroclastic** fabric is created by explosive effervescence of viscous silicic melts, the glassy fragments representing the disintegrated froth. Broken crystals are also usually present.

In the case of vitrophyres, a compositional name can be appended on the basis of the phenocrysts, such as amphibole vitrophyre, dacite vitrophyre, and so on. For all five glassy rocks, more specific rock type names can be appended if chemical data are available, giving names such as rhyolite pumice, dacite obsidian, and so on.

Classification of volcaniclastic rocks.   This section considers the classification of **volcaniclastic** rocks whose clastic fabric owes its origin to volcanic processes (Fisher, 1966), as contrasted with sedimentary or epiclastic processes. **Epiclastic** processes include weathering, erosion, transport, and deposition in the Earth's hydrologic system at its surface.

A sedimentary or epiclastic rock is formed at the surface by consolidation of fragments of preexisting rock; examples are conglomerate and sandstone. It will be apparent in the remainder of this chapter that epiclastic and volcaniclastic processes can sometimes grade into one another, which causes difficulties in classification of deposits.

Explosive volcanic eruptions produce large volumes of volcaniclastic debris, but other volcanic processes also create clastic rocks on a more limited scale (see below).

As with the classification of the common epiclastic sedimentary particles, the *size* of the volcanic fragments is fundamental. The size designations for volcanic and corresponding sedimentary particles are as follows:

	<2 mm	2–64 mm	>64 mm
Volcanic	ash	lapilli	block and bomb
Sedimentary	clay, silt, sand	granule, pebble	cobble, boulder

Blocks are angular clasts produced by fragmentation of solid rock, whereas bombs are fragments that were ejected from the vent as pasty blobs of magma (usually of mafic to intermediate composition) and were streamlined into smooth forms as they solidified during flight through the air.

In addition to size, volcanic fragments may also be classified by *composition* as **vitric, crystal,** or **lithic** (polygranular rock). Ash-size clasts are usually vitric (glass shards) or crystal, whereas blocks are usually lithic, but may be vitric.

Volcanic fragments may be classified in still a third way—according to *heritage.* Thus, clasts formed from the magma involved in the volcanic event itself are **juvenile,** whereas those formed by fragmentation of older, preexisting rock and incorporated into the volcaniclastic deposit are **accidental.** Pieces of older rock torn from the walls of the volcanic conduit or fragmented during explosive cratering are accidental; they are important in providing information on the basement covered by the volcaniclastic deposit.

Clasts of volcanic origin can be cemented together by secondarily precipitated mineral material, as in epiclastic deposits of sandstone, or if still hot at emplacement, sticky glassy fragments may **weld** together. Classification of *consolidated volcaniclastic deposits* is based upon size of clasts (see Figure 3-12). **Tuffs,** comprised wholly of ash-size fragments, are very common, but rocks comprised entirely of blocks, **volcanic breccia,** or of lapilli, **lapilli stone,** are rare because finer particles usually fill in the spaces between the lapilli and blocks. For this reason consolidated mixtures of ash and lapilli, **lapilli tuff,** and of ash and blocks, **tuff-breccia,** are common. **Agglomerate** is a term some-

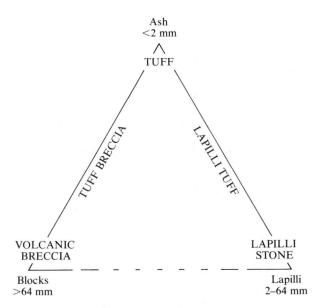

Figure 3-12   Nomenclature for volcaniclastic rocks (in capital letters) based on the size of fragments.

times used for unsorted deposits of bombs accumulated near the volcanic vent.

Where known, compositional terms can be appended to the volcaniclastic rock name; for example, lithic lapilli tuff, vitric tuff-breccia, crystal tuff, dacitic vitric tuff, andesite breccia, and so on.

Alteration

Alteration of volcanic rocks is widespread and locally very intense; it modifies, or even eliminates, original fabrics and mineral constituents, creating new properties more or less in equilibrium with the alteration system. **Deuteric alteration** comes about more or less automatically as the volcanic deposit cools and "stews" in its own gases, or condensed fluids, forming new, more stable minerals and fabrics. Subsequent **hydrothermal alteration** caused by percolation through the deposit of hot mineral-laden aqueous solutions of foreign origin can accomplish the same type of modifications. It is not always easy to distinguish the effects of one type of alteration process from the other, or from the effects of burial metamorphism (see pages 419–421).

Intratelluric crystals of biotite and amphibole formed in subterranean magma systems may decompose upon extrusion, or upon intrusion to shallow crustal depths, as water pressure drops in the magma system (see Figure 8-40), especially if high but submagmatic temperatures are maintained for a time. Fine-grained anhydrous aggregates of feldspar, pyroxene, and Fe–Ti oxides replace the original hydrous mafic silicate, starting as rims around the grain (see Figure 3-1).

If elevated water pressures are maintained in buried systems and temperatures are sufficiently high, primary silicates may recrystallize into fine-grain aggregates of hydrous minerals such as chlorite, amphibole, serpentine, epidote, prehnite, pumpellyite, and Fe–Ti oxides. Many of these same minerals, together with quartz, carbonates, zeolites, pyrite, and sulfates, can be precipitated in and coat the walls of vugs, vesicles, and open fractures. Alteration of this kind is particularly widespread in andesites and common around some ore bodies and central volcanic vents; it is called **propylitic alteration** and produces rocks resembling greenstones (see page 397).

Hot gases escape from recently emplaced volcanic bodies at vents called **fumaroles,** or if the gases are sulfurous, **solfataras.** Yellowstone National Park is famous for these features and associated hot springs, geysers, and boiling mud pots. Escaping gases carry many elements to the surface and, when cooled by contact with the atmosphere, various sublimates including sulfur and hematite are formed around the vent. Hot springs and geysers form mounds of travertine, siliceous sinter, and opal. Associated, usually very porous, volcanic deposits are altered to varicolored but usually yellow-brown mixtures of low-temperature Fe oxides, sulfates, clay minerals, opal, and chalcedony.

All volcanic glasses are unstable under atmospheric conditions and with passing time devitrify to a stable crystalline assemblage. **Devitrification** is delayed crystallization of glass. Hence, glass is progressively more rare the older its time of formation; most glasses are late Cenozoic. Devitrification results in aphanitic or spherulitic fabric (see Figure 3-5). Primary fabrics—such as flow layering and vitroclastic texture—can be surprisingly well preserved in devitrified rocks, making possible accurate interpretations of the history of the glass body (see Figure 3-13). Prior to devitrification, some glasses hydrate by absorbing water from the atmosphere or groundwater, forming perlite and pitchstone.

3.2 VOLCANIC PROCESSES AND FIELD RELATIONS

Our intent here is not to describe volcanic phenomena in detail, as this is covered in a number of books (such as Macdonald, 1972; Francis, 1976; Williams and McBirney, 1979), but rather to summarize some salient geologic processes and field relations so as to provide a better appreciation of how magmas behave as they reach the Earth's surface. The *Atlas of Volcanic Phenomena,* prepared by the U.S. Geological Survey, is a useful pictorial summary; a more comprehensive pictorial atlas is by Green and Short (1971). Pertinent field relations and fabrics are briefly described and nicely illustrated by Compton (1962).

Magma systems near the surface of the Earth may be divided into two broad categories, each having different flows of energy and modes of eruption. First are magmas whose dissolved gas content is low enough that they do not boil, or do so only gently. Second are magmas that boil violently and explosively. Magmas erupted from volcanoes are thus either poured out as coherent fluidal lava flows or blown out as pieces of pyroclastic ejecta. During the history of a volcano the mode of eruption commonly fluctuates between these two extremes.

In another sense, volcanic eruptions are either central or fissure. In a **central eruption,** magma is extruded from a localized vent and builds a conical edifice that the average layman would call a vol-

(a)

0   cm   2

Figure 3-13 Devitrified vitric lapilli tuff. In the hand sample (a) the relict pumice lapilli are comprised of strings of delicate spherulites replacing the filamentlike glassy walls between the original stretched vesicles. In the photomicrograph (b) these spherulites and relict glass shards, now completely crystallized, are obvious.

(b)

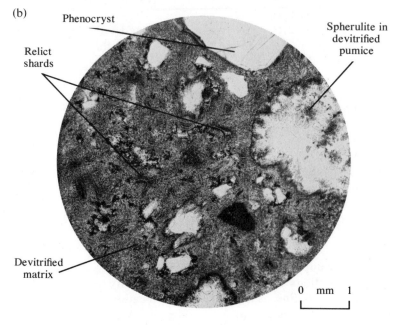

Phenocryst

Relict shards

Spherulite in devitrified pumice

Devitrified matrix

0   mm   1

cano. Other magmas are extruded from long cracks in the crust by **fissure eruption.**

This section briefly reviews the nature of magmatic extrusion—first those poured out of the vent, and then those blown out—with the objective of becoming more familiar with processes and field relations in active volcanoes so that ancient rock bodies can be more accurately interpreted. Then we will describe the nature and evolution of two important types of volcanic edifices: ash-flow–caldera complexes and composite volcanoes.

Extrusions of Magma: Lava Flows and Domes

Although many extrusions of silicic to intermediate-silica magmas are explosive to some degree, coherent effusions commonly occur at some time during the history of the volcano. Many terribly violent explosive eruptions are preceeded, or followed, by harmless extrusion of highly viscous, semisolid magma from the vent.

Extrusions display a wide range of forms, depending upon their mobility or apparent viscosity (see Figure 3-14). Relatively low-viscosity or fluid basaltic flows (see Chapter 5) spread out from the vent as thin sheets with lateral dimensions several orders of magnitude greater than their thickness. Principally as a function of increasing concentration of silica, lava flows become more viscous, or less mobile, and the ratio of lateral dimension/thickness of the flow decreases (see Figure 3-15).

The most silicic magmas thus lie at the opposite end of the spectrum of flow morphology from basalt and are commonly extruded as thick, bulbous **domes,** shaped like a mushroom or artichoke, which pile up, over, and around their vents. The ratio of horizontal diameter to thickness of domes can be near 1. Two types of domes are shown in Figure 3-16. Some grow by expansion from within, pushing slabs of semisolid lava out of and away from the vent in fan fashion. As the dome grows, slender spines may be elevated high above the remainder of the dome before being shattered by minor steam explosions or thermal stresses. Shattering produces a mantle of angular slabs and fragments that covers the surface of the dome and spills down its sides, forming an apron of talus. Sometimes larger, more vigorous avalanches of fragmental debris cascade down the flanks of the domes; these may be relatively cold or, if disintegration of the dome penetrates far inside, the avalanche will be hot and may travel some distance as it rides on a cushion of entrained heated air. Still larger-scale explosions can generate incandescent glowing avalanches that will be discussed later. Some effusions from the vent are so nearly solid that they are simply thrust slowly out of the vent as a spine without much lateral expansion, like a cork withdrawn from a wine bottle. Other domes growing from slightly less viscous magma extrude from the vent as a bulbous mass with more concentric than radial flow layering, as shown in Figure 3-16b.

Planar and linear fabric elements (see Figure 4-6) are widespread in silicic extrusions and constitute markers of the pattern of internal flow in the body during emplacement (Compton, 1962). Conspicuous variations in internal fabric are evident in domes incised by erosion or breached by faulting (see Figure 3-17). Tops and bottoms are of breccia, usually glassy in young domes. Interiors are flow layered, locally folded or contorted due to drag during flowage, and are commonly either a stony felsite due to primary crystallization of an aphanitic matrix during emplacement or spherulitic due to later devitrification of glass. Layered pumice or pumiceous perlite/obsidian may occur near the top of the dome, reflecting mild exsolution of gas from the lava flow during emplacement.

A dome in Japan grew over 300 m vertically in about a year and a half; another in Guatemala started to grow in 1922 and by 1967 its volume was 0.7 km³.

Many domes grow in craters produced by preceding explosive eruption of magma and represent the termination of activity after the magma has lost its propellant gas (see Figure 3-18).

Figure 3-14 Mt. Shasta, a composite volcano in the Cascade Mountains of northern California. View is looking southeast up Whitney Creek. Extruded magmas with differing apparent viscosities have created a range of lava-flow morphologies. The most viscous lava produced Haystack dome in left foreground; it is 150 m high and 900 m in diameter. Less viscous lava formed the flows to the right and closer toward Mt. Shasta; note the lobate tongues at the toe and well-developed lava levees along the flow margins toward the source. Lava flows and volcanoclastic deposits built Mt. Shasta, which rises almost 3 km above Haystack dome and about 16 km away. A flank dome is barely visible just to the right of the summit of Mt. Shasta. [From John S. Shelton, *Geology Illustrated.* San Francisco: W. H. Freeman and Company. Copyright © 1966.]

Extrusions of Magma: Autoclastic

Some extrusions or parts of extrusions of magma are neither coherent lava nor pyroclastic debris. Here we make a brief digression into a special but minor category.

In terms of volcanic processes of fragmentation, most clasts originate in one of two ways: autoclastic or pyroclastic. **Autoclastic** processes are self-inflicted and create essentially monolithologic, poorly sorted deposits of angular juvenile clasts. For example, in a flowing lava, the crusted, more rigid margin breaks up as the more mobile interior continues to move; pieces of this crust spall off on the top, bottom, and sides of the flow, where they may be infiltrated by molten lava or cemented sub-

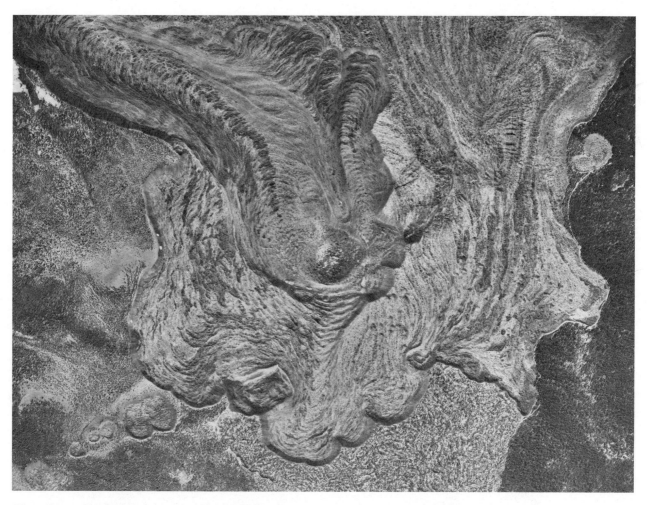

Figure 3-15 Vertical aerial photograph of Big Glass Mountain, a rhyolite obsidian flow complex, east of Mt. Shasta. **Pressure ridges** oriented more or less perpendicular to the direction of flow are conspicuous. Less obvious are the high, steep flow margins, which indicate the lava flows were relatively viscous, but not as much as mushroom-shaped domes. [Photograph courtesy of the U.S. Forest Service.]

sequently by secondary mineral matter. Talus aprons and avalanches around silicic domes are another type of autoclastic deposit.

Autoclastic processes involve relatively small flows of energy, so that fragments are generally of block size. However, as the energy flow in the volcanic process becomes greater in the more explosive eruptions, more mechanical comminution of juvenile and older rock material occurs, yielding finer lapilli- and ash-size particles with greater surface areas relative to their volume. Autoclastic spalling and breakup of silicic domes creates aprons of talus blocks, but more vigorous breakup due to minor steam explosions can generate avalanches of debris that cascade down slope. Still more explosive eruptions are pyroclastic.

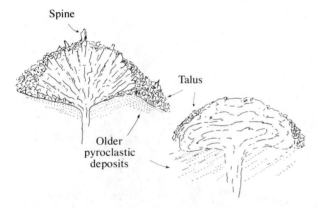

Figure 3-16   Two types of silicic lava domes. [After H. Williams, 1932, The history and character of volcanic domes. University of California Publications in Geological Sciences, Bulletin 21. See also Williams and McBirney, 1979, pp. 188–197.]

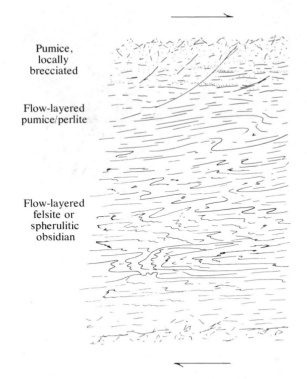

Figure 3-17   Idealized cross section through a silicic lava flow.

Extrusions of Magma: Pyroclastic

Pyroclastic processes involve explosive aerial ejection and dispersal of fragments from a vent. Fragments are called **ejecta, tephra,** or simply pyroclastic material. Explosions are caused by near-surface boiling of viscous magma, sometimes involving failure of older cap rock or more solid magma overlying the gas-charged magma body. Other explosions are caused as near-surface bodies of magma (of any gas content) encounter a mass of water, which expands violently to steam, rupturing proximate rock. In a **phreatic,** or steam blast eruption, ground water heated by an underlying magma body explodes, and the ejected fragments are all accidental pieces of old rock material. In a **phreatomagmatic** explosion, juvenile magmatic material as well as accidental fragments are ejected.

Once the pyroclastic material has been created, three processes of tranport and deposition can take place on land: fall, surge, and flow. Knowledge of these mechanisms of dispersal and emplacement is still incomplete, and understandably so, as geologists usually cannot observe the mechanism at close hand and still live to tell about it! Mechanisms must therefore be deduced from distant observations, combined with knowledge of the properties of the deposits themselves and with theoretical calculations.

Pyroclastic falls.   Very explosive eruptions send tephra of lapilli- and ash-size high into the atmosphere, where it is carried great distances by air currents in the eruptive cloud, or by winds, before finally falling to the surface under the influence of gravity. Accumulations of this tephra constitute **pyroclastic-fall** or **air-fall** deposits. The geographic extent of ejecta from a single eruption can be considerable. Ash from the climactic eruption of Mt. Mazama, which created Crater Lake in Oregon 6,600 years ago, accumulated to a depth of 30 cm some 130 km away; accumulations a few millimeters thick have been found in peat bogs in Canada

Figure 3-18 View toward south over Mono Craters in east central California. Note circular crater almost 1 km in diameter of pyroclastic material occupied by a silicic dome, right foreground. Other viscous flows to the south have spread outward from their vents. Sierra Nevada in distance. [From *Geology Illustrated* by John S. Shelton. W. H. Freeman and Company. Copyright © 1966.]

and in sediments in Great Salt Lake 1,100 km from the vent (see Figure 3-19). Fine volcanic dust from such eruptions can actually get entrained in the high atmosphere so that dispersal is global, producing multihued sunsets for extended periods of time. A single eruption may produce a well-sorted layer (see Figure 3-20) in which the size of particles decreases gradually away from the vent; at a particular locale an upward decrease in clast size may also be evident. But sometimes reverse grad-

ing, with coarser clasts towards the top, occurs in air-fall deposits due to formation of progressively coarser fragments during the course of a progressively more vigorous eruption, or due to fall into standing bodies of water where larger pumiceous clasts sink only as they become waterlogged. Regardless of the sorting, pyroclastic fall deposits show mantle bedding, the tephra falling like snow uniformly over the ground surface (see Figure 3-21). Sequences of size-graded layers manifest

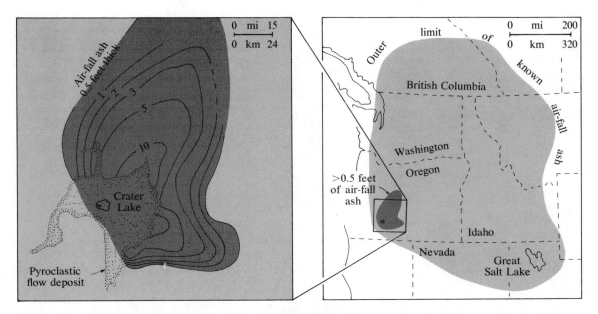

Figure 3-19   Distribution of pyroclastic flow and air-fall deposits from the climactic eruption of Mt. Mazama, which created Crater Lake, Oregon. Thickness of air-fall deposits shown by contours, in feet. Prevailing winds apparently were toward the north-northeast and east. [After H. Williams and A. R. McBirney, 1979, *Volcanology* (San Francisco: Freeman, Cooper and Co.); and H. Williams and G. Goles, 1960, Volume of the Mazama ash-fall and the origin of Crater Lake caldera. *Oregon Department of Geology and Mineral Industries Bulletin* 62.]

Figure 3-20   Well-sorted, evenly stratified beds of air-fall ash and pumice lapilli. Camera lens cap shows scale.

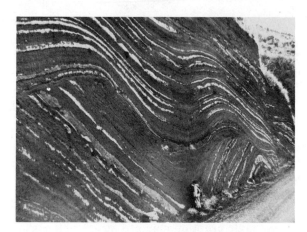

Figure 3-21   Mantle bedding of air-fall ash layers (partly covered by snow), O Shima Volcano, Japan. The beds are not folded; dips are primary and mimic the configuration of the depositional surface on which they rest. Note unconformity. [Photograph courtesy of Jack Green.]

Figure 3-22   Ring-shaped pyroclastic surge in the phreatomagmatic eruption of Capelinhos volcano, Azores. The central black eruption column choked with basaltic ejecta is estimated to be about 400 m high. A white steam cloud lies behind and to the right of the column. [Photograph courtesy of Richard V. Fisher from Othon R. Silveira of Horta, Azores. From A. C. Waters and R. V. Fisher, 1971, Base surges and their deposits: Capelinhos and Taal volcanoes. *Journal of Geophysical Research* 76. American Geophysical Union. Copyright © 1971].

repeated eruptions. The properties of pyroclastic-fall deposits provide a basis for interpreting the nature of the explosive eruption (Walker, 1973).

Because vitric particles are dispersed in the atmosphere for some time before deposition, they are generally not hot enough to weld together and consolidation must occur by secondary cementation. However, air-fall deposits of vitric particles may become welded subsequently where adjacent hot lava flows or intrusions of magma conduct heat into them.

Air-fall layers, if undisturbed, can be as thin as 1 mm or so and can be very persistent over great distances; crystal and vitric particles within a given air-fall layer can have a unique composition. Air-fall layers are hence ideal stratigraphic time horizons in tephra chronology.

Considerable tephra inevitably falls into lakes and the ocean, where it forms **water-laid** deposits interstratified with normal lacustrine and marine sediment.

Beds of ash preserved in older rock sequences are valuable not only stratigraphically but also because they can alter to economically useful deposits of clay and zeolites.

Pyroclastic surges.   During the 1965 phreatomagmatic eruption of Taal volcano in the Philippines, a new type of dispersal/depositional process was recognized that had previously been noted in nuclear bomb explosions (Moore, 1967). A **pyroclastic surge** or **base surge** is a ring-shaped cloud of water droplets and minor solid particles moving horizontally at hurricane velocities away from the base of a vertical eruption column (see Figure 3-22). Surges commonly occur in a series and are initiated by phreatomagmatic eruptions where copious volumes of surface water or ground water

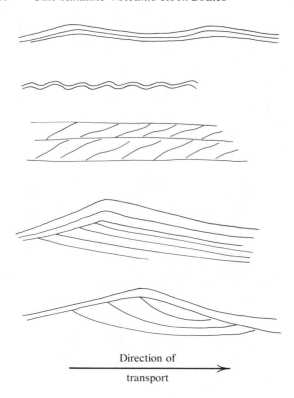

Direction of
transport

Figure 3-23   Schematic sandwave forms found in pyroclastic surge deposits. Structures are on the order of a few centimeters to tens of centimeters in amplitude. Bottom two features are called antidunes. [After K. H. Wohletz and M. F. Sheridan, 1979, A model of pyroclastic surge. Geological Society of America Special Paper 180.]

make contact with the magma. Temperatures in surges can be near ambient.

Observed surges have produced only thin local deposits on the order of a meter thick and extending less than several kilometers from the vent. Deposits are poorly sorted but have distinctive bedding structures such as sandwaves (see Figure 3-23), inversely graded planar beds, and bedding sags beneath heavy large blocks ejected from the vent (Wohletz and Sheridan, 1979; Crowe and Fisher, 1973).

Pyroclastic flows.   Perplexing volcanic deposits were encountered a century ago in the western United States by early government geologists.

Figure 3-24   Ash-flow sheets associated with the Valles Caldera, near Los Alamos, New Mexico. [U.S. Geological Survey photograph courtesy of Robert L. Smith.]

They consist of sheets of silicic to intermediate-silica rock, each covering as much as thousands of square kilometers but with a thickness measured only in meters or a few hundred meters at very most (see Figure 3-24). Hundreds of these sheets of Tertiary age occur in the western United States and Mexico. Their aggregate original volume has never been tallied but easily compares with the great Mesozoic batholiths of the western United States (see Figure 4-15). Some sheets, or portions of sheets, are porous and have an obvious clastic fabric in which lapilli and perhaps blocks rest in an abundant matrix of ash. Most clasts are of glass shards, pumice, and crystals, but accidental lithic fragments are common. Sorting is poor and internally the sheets commonly (but not always—see below) lack stratification (see Figure 3-25). Associated sheets, or portions of them, are hard, firmly indurated, nonporous, generally darker in color, and have lavalike aspects. It seemed incomprehensible that these thin sheets could have been emplaced as flows of viscous silicic lava, yet an alternate origin by air-fall accumulation seemed equally implausible.

Interpretations of natural phenomena are conditioned by past experiences of the observer, his intuition, and available theories. In the case of these enigmatic sheets, no processes were known that could wholly account for their properties. But beginning in 1902 with the tragic eruptions of Mt. Pelée and La Soufrière in the West Indies and continuing with hundreds of similar eruptions in many parts of the world, trained observers have gained much insight into the nature of explosive eruptions. We now have working models, but not complete solutions. The key discovery was a new dispersal mechanism associated with certain modes of explosive eruption.

This newly recognized mechanism is the **pyroclastic flow,** or **ash flow**—a highly mobile, hot mixture of gas and ejecta that moves swiftly and closely along the ground surface away from the vent, all the while retaining an overall flow aspect. Pyroclastic flows were first recognized in the 1902 eruptions of Mt. Pelée as a ground-hugging companion of the

Figure 3-25 Poorly sorted, nonstratified pyroclastic flow deposit overlying sorted and stratified pyroclastic fall beds. Erosional debris (dark boulders) from another rock unit overlies the ash flow.

ash–steam cloud that billowed up in gigantic convolutions overhead (see Figure 3-26). Together, the flow and the overshadowing cloud have been referred to as a **nuée ardente**, a term meaning glowing, or hot, cloud and coined by the French petrologist Alfred Lacroix, who was on the scene of some of the eruptions.

(a)

(b)

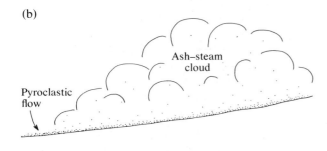

Ash–steam
cloud

Pyroclastic
flow

Figure 3-26 Nuée ardente. The dual nature of this type of eruption is barely discernable in the photograph (a) of a nuée ardente sweeping down the slope of Mt. Pelée on the island of Martinique in the West Indies. The remains of the town of St. Pierre, devastated by the nuée ardente of May 8, 1902, which took 28,000 human lives, lies in the foreground. [Photograph courtesy of The Geological Museum, London.] (b) Conceptual sketch of a nuée ardente showing its dual nature with a ground-hugging pyroclastic flow largely overshadowed by the ash–steam cloud.

Among English-speaking geologists, deposits of pyroclastic flows are referred to as **ash-flow tuffs,** or **ignimbrites.** However, neither term is wholly satisfactory. The latter, coined by Peter Marshall in his 1935 report on the North Island of New Zealand, stems from the Latin *ignis,* meaning "fire," and *nimbus,* meaning "cloud," hence, "fiery cloud rock." But the cloud component of the nuée ardente does not produce the deposits in question— the pyroclastic flow does. Consolidated flow deposits generally include lapilli and sometimes blocks, but most contain at least 50 wt.% ash-size particles (Sheridan, 1979). Ross and Smith (1961) have discussed the terminology and properties of these deposits.

A pyroclastic flow is closely confined to topographic lows and canyons as it moves downslope, diverging and turning according to slope configuration, rather similar in behavior to snow avalanches in mountain canyons. Pyroclastic flows from Mt. Mazama traveled in large part down stream valleys (see Figure 3-19). The accompanying ash–

steam cloud, on the other hand, seems to be undeflected by topographic obstacles as it rolls onward, lagging somewhat behind the faster-moving flow. The racing flow retains its tonguelike identity; the more diffuse cloud, apparently fed by convective updrafts of fine ash and steam from the flow, expands considerably during the travel of the underlying and often hidden flow.

Aside from actual observations of nuée ardentes (for example, Taylor, 1958), several features of pyroclastic flow deposits themselves lend support to emplacement by flowage rather than by air fall. Robert L. Smith, one of the foremost experts on the subject, summarized these features as follows (1960, p. 806): (1) restriction of small deposits to topographic lows; larger deposits are thickest over paleovalleys; (2) nearly level upper surfaces, modified somewhat as compaction and welding occur (see below) in thicker accumulations (see Figure 3-24); (3) general lack of sorting even far from the vent among clasts ranging in size from fine ash to blocks (see Figure 3-25); (4) preservation of high temperatures within the deposited tephra for long periods of time, manifest in long-term fumarolic activity and welding of glass clasts; and (5) incorporation of debris from the ground surface into the deposit.

Observed pyroclastic flows are remarkably mobile and fast moving (Sheridan, 1979). Speeds of up to 200 km/hr have been observed on steep slopes 10 km from the vent, and 25 km/hr on gentle slopes. Some ancient deposits in the rock record extend well over 100 km from their source in terranes where the overall ground slope was nil; large ash flows were not extruded from high conical volcanoes as are most smaller flows (see section on calderas, page 91). The lateral extent, thickness, and depositional slope of many pyroclastic-flow deposits suggest transport mobilities similar to or better than basaltic lavas, which have viscosities some 5 orders of magnitude less than coherent silicic magmas (see pages 233–235).

In some instances a strong motive force is apparently provided by a horizontally directed component in the volcanic explosion that initiates the flow, for example from a cleft at the base of an emerging dome in a crater (see Figure 3-27b). Flows vented from the summit of a steep, lofty volcano undoubtedly can derive much, if not all, of their motive force from gravity. But gravity would also be an important motive factor in the absence of a volcanic edifice if a massive, dense, vertical eruption column were blasted high out of the vent and then collapsed under its own weight (see Figure 3-27c,d). This mechanism could disperse pyroclastic flows radially all around the vent for some distance. The gravitational potential energy (initially derived from the explosion of gas-charged magma out of the vent) of the erupted mass would be transformed to kinetic energy in a fast-moving, far-traveling flow. Flow mobility could thus be viewed as a reflection of dissipating gravitational potential and kinetic energy imparted to the ejecta upon eruption. Collapse of large vertical eruption columns has actually been observed to form flows from major conical volcanoes (Sheridan, 1979). The height of a vertical eruption column and whether it collapses back down around the vent or is dispersed away by wind currents depends upon several factors, including the "muzzle" velocity of the mass of ejecta leaving the vent, which is in turn a function of gas content of the magma, depth of the magma chamber, and radius of the vent (Wilson and others, 1978; Sparks and others, 1978).

Intrinsic properties of the flow itself may enhance its mobility. Some geologists, in an effort to explain how some flows can seemingly override high topographic obstacles, have proposed that flows are greatly inflated by hot gas, which serves as a virtually frictionless cushion between widely dispersed solid particles. A high degree of inflation—that is, very low flow density—would require that deep flow accumulations in valleys would, after compaction to a virtually pore-free rock (see below), have an upper surface drastically lower than the adjacent surfaces of deposits covering hills. However, observed amounts of compaction, together with other considerations, suggest the density of ash flows to be near 1 g/cm³, equivalent to only a little over 50% porosity.

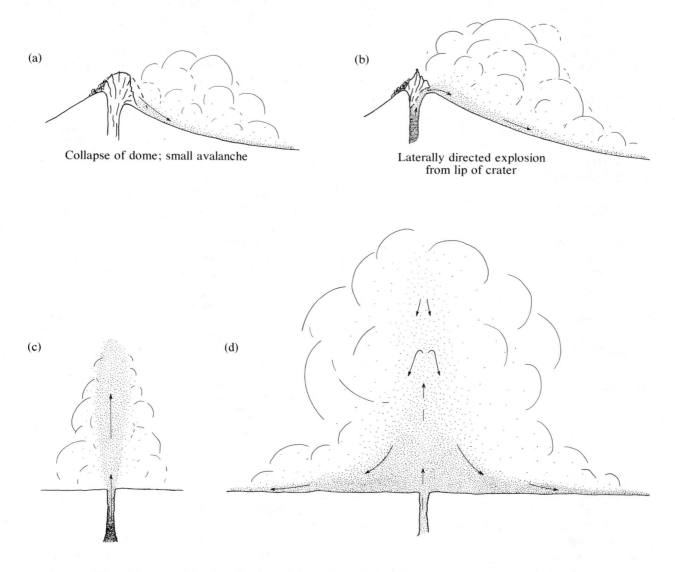

Powerful upward ejection of tephra, followed by gravitational
collapse of vertical eruption column and backfall around vent

Figure 3-27   The motive force for pyroclastic flows might be provided in three ways. (a)
Disintegration and gravitational collapse of side of a dome, sending a small hot avalanche down the
volcano flank. (b) Explosion from the lip of a crater occupied by a dome, directing the ejecta
laterally. (c) Powerful upward "gun-barrel" ejection of tephra from within the conduit, followed by
(d), gravitational collapse of the heavy vertical eruption column and backfall of ejecta around the
vent. Compare Sheridan (1979).

A probable factor in the mobility of pyroclastic flows arises from the phenomenon of **fluidization**—a process widely used in industry for transport of solid particles without recourse to physical conveyors (trucks, belts) or liquids. Sparks (1976, p. 170) summarizes:

> When a bed of loose particles is placed in a tube with a porous base and a gas is passed through the system at increasing gas velocity...it is found that at a certain velocity the bed of particles no longer acts as a coherent mass, but adopts a fluidlike appearance and is said to be fluidized. Fluidization occurs when the gas velocity is sufficiently great to support the weight of individual particles.
>
> For the geologist a very important property of a fluidized system is that it radically alters the mechanical behavior of granular material.

The source of fluidizing gas in a pyroclastic flow can be twofold. First, air from the atmosphere may be entrained in the flow as it moves, a mechanism that can impart fluidized mobility to almost any type of avalanche or debris flow. Second, emission of gas from the glassy particles within the moving flow itself would promote fluidization. Although the initial explosion that initiated the flow from the vent would release pent-up gas in the magma, not all of the dissolved gas would be released at the vent, and continued exsolution is likely. A plausible explanation for the ubiquitous expanding ash–steam cloud associated with ground-hugging pyroclastic flow is that it represents elutriated fines flushed from the interior of the flow by convective updrafts of fluidizing gas.

In fluid–solid mixtures, overall flow mobility decreases in proportion to the concentration of solid particles. In a pyroclastic flow, the finer ash and gas acts as the "fluid" and larger lapilli and blocks as the "solid" component. Elutriation of fines and loss of gas into the overlying cloud would therefore decrease the mobility of the flow. Greater mobility would apparently be possible in flows with fewer larger clasts.

The generally poor sorting in pyroclastic flow deposits has usually been ascribed to turbulent flow. Such a flow regime may well be present, especially at higher velocities. However, poor sorting is also characteristic of fluid–solid mixtures moving in laminar fashion where concentrations of solids are high. If flows are not greatly inflated and flow is laminar, then some degree of sorting of larger lapilli and blocks might be expected. Sparks (1976) illustrates relatively thin deposits where clasts of frothy pumice have apparently floated to the surface of the flow, whereas denser accidental lithic fragments have sunk.

Pyroclastic flows conserve their thermal energy to a remarkable degree, as shown by long-lived fumarolic activity, measured temperatures in modern deposits, intense welding in ancient flows, and experimental determinations of minimum welding temperature for glass particles (Smith, 1960). Indirectly, this retention of thermal energy supports the concept of laminar flow because turbulence would tend to engulf cooler atmospheric air into the flow. But significant variations in the welding of flows from a common magma source implies some cooling mechanism independent of the transport and emplacement of the flow. This could be related to the properties of the vertical eruption column—other factors being the same, higher columns might draw in more cooling atmospheric air and the resulting deposit would be less welded.

These conceptual pyroclastic-flow models have been fruitful in accounting for properties of many deposits. However, it must be quickly admitted that of the hundreds of pyroclastic flows observed since 1902, none have actually produced deposits of the volume so commonly found in the ancient rock record, and only a very few have produced even the slightest degree of welding, which is so characteristic of these ancient deposits. Could our models be wrong? Or, are the models basically accurate, and modern volcanic systems over the past few decades have just not produced the sort of flows erupted over the course of millions of years?

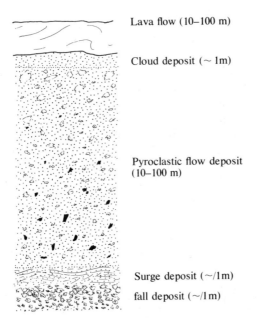

Lava flow (10–100 m)

Cloud deposit (~1m)

Pyroclastic flow deposit (10–100 m)

Surge deposit (~/1m)

fall deposit (~/1m)

Figure 3-28 Volcanic deposits formed in an idealized explosive eruptive cycle showing magnitudes of their thickness in meters. The cycle begins with a **plinian** eruption—a powerful, continuous blast of steam and pumice—that yields an inversely graded pyroclastic fall deposit, rich in lapilli and small blocks of pumice and perhaps lithic fragments. The surge deposit is comprised of thin-bedded, crystal–lithic-rich ash. One or more pyroclastic flow units comprise the bulk of the eruptive deposit. These ash-flow sheets may lack sorting or may show, as in the sketch, subtle sorting of denser lithic and less dense pumice fragments. Fumarolic pipes may be found near the top. The explosive ensemble is capped by fine vitric-rich air-fall ash deposited from the ash–steam cloud; its volume can be a significant proportion of the total pyroclastic deposit, but the air-fall ash is easily eroded away by running water and much of it is deposited beyond the limits of the ash flow. Near the vent a lava flow or dome formed by extrusion of mostly degassed magma may cap the pyroclastic sequence. If the ash flow is welded to any degree, both it and the lava flow are more resistant to erosion, and in old rock sequences they are generally better exposed than the softer surge, fall, and cloud deposits. [After M. F. Sheridan, 1979, Emplacement of pyroclastic flows: A review. Geological Society of America Special Paper 180.]

Pyroclastic sequences. As ash-flow tuffs are almost invariably associated with other types of pyroclastic deposits, it is important that we describe an idealized pyroclastic sequence before considering in more detail the properties of an ash-flow deposit itself.

A cross section through a pyroclastic flow and coeval deposits formed in an idealized eruptive cycle is shown in Figure 3-28. The sequence of fall, surge, flow, and fall is compatible with, but does not prove, the collapsing eruption column model. Pyroclastic surges and ash clouds may move independently of pyroclastic flows and their deposits may therefore be lacking, occur alone, or lie within any part of the eruptive sequence (Fisher, 1979). Large, fast-moving pyroclastic flows, especially near their source, may travel in a turbulent manner, but small flows and large ones at their distal ends may travel in a laminar fashion. The distal portion of a large-volume eruptive sequence may consist of a thin air-fall deposit of fine ash overlaid by an ash flow with several subtle graded layers and capped by more air-fall ash.

Compositional zoning in pyroclastic flow deposits. Ash-flow sheets, especially rhyolitic ones, exhibit vertical variations in chemical and mineralogical composition that are interpreted as representing compositional zoning in the preeruption magma chamber (Smith, 1979). The pyroclastic flow deposit formed by the climactic eruption of Mt. Mazama in the Oregon Cascades circa 6,600 years ago that resulted in caldera collapse and formation of Crater Lake furnishes a good example of such zonation (see Figures 3-29 and 3-30). The initial eruptions deposited crystal-poor ($< 10\%$) rhyolitic flows with 72 wt.% SiO_2, and these were followed by ejecta with 67.5 wt.% SiO_2. Culminating eruptions yielded crystal-rich (50% to 90%) ejecta with 57.5 to 54.5 wt.% SiO_2. Note the compositional gap between 67.5% and 57.5% SiO_2. By way of interpretation, Smith (1979) rules out sampling bias and the presence of two independent magma chambers and prefers a single source

Figure 3-29 Mazama ash-flow tuff along Wheeler Creek, Crater Lake National Park, Oregon. The upper, slightly more welded and erosionally resistant tuff is a crystal-rich hornblende andesite, whereas the underlying lighter-colored tuff into which it grades without any sharp break is a crystal-poor rhyodacite–rhyolite. This sequence is believed to be an inverted representation of the compositional zonation in the source magma chamber; that is, magma from the top of the chamber erupted first, followed by magma from deeper levels (see Figure 3-31). [Photograph courtesy of the Oregon State Highway Department.]

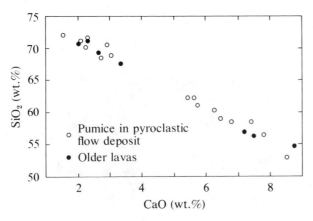

Figure 3-30 Harker variation diagram for two oxides in rocks from Mt. Mazama, Oregon. [After H. Williams, 1942, The geology of Crater Lake National Park, Oregon, with a reconnaissance of the Cascade Range southward to Mt. Shasta. Carnegie Institution of Washington Publication 540.]

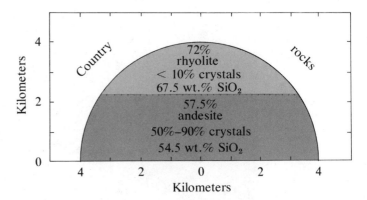

Figure 3-31 Schematic interpretive model of the zoned magma chamber beneath Mt. Mazama, the upper levels of which erupted to form the flow deposit in Figure 3-29. [After R. L. Smith, 1979, Ash-flow magmatism. Geological Society of America Special Paper 180.]

magma chamber that is doubly zoned (see Figure 3-31).

Zoned calc-alkaline magma chambers have long been known from studies of intrusions, but the compositional variation in extruded pyroclastic flow deposits represents a more accurate "quenched" image of this phenomena. This discovery from volcanological studies of a fundamental property of magma bodies merits further investigation. How does zoning develop in a single chamber? To what extent does zonation in one body explain compositional variations in sequences of lava flows erupted from a central vent over the life of a volcano? Can zoned bodies yield the hybrid lavas sometimes found in coherent flows or as ejecta blocks (see Figure 3-32)?

Secondary zonation in pyroclastic flow deposits. After emplacement, an ash-flow deposit is subsequently acted upon by secondary processes that reflect both the high temperature of the deposit and the presence of entrapped gas, mainly steam. These secondary processes, which produce additional zonal variations within the ash flow besides

those due to composition, include the following (see Figure 3-33): (1) welding and compaction; (2) crystallization of solids from the entrapped gas; and (3) crystallization of the glassy material.

Welding is the sticking together of glassy particles due to their softness at elevated temperature. Due to the weight of any overlying material on these soft particles, they are compacted together and entrapped gas is squeezed out, collapsing pore spaces and vesicles within pumice fragments and reducing the bulk porosity of the tuff. The most intense welding and compaction occurs in the lower portion of the flow, but not right at the base, where faster cooling prevents much welding. Welding and compaction operate simultaneously to create the typical **eutaxitic** fabric of welded ash-flow tuffs, in which pumice lapilli and glass shards are flattened into discoidal shape more or less parallel to the depositional plane (see Figure 3-34). Rigid inequant mineral grains such as biotites and feldspars rotate within the soft glassy matrix and so also become aligned in similar orientation. Some intensely welded tuffs with abundant lapilli and crystals resemble foliated schists be-

Figure 3-32 Mixed rhyolite–andesite pumice block from the pyroclastic flow that erupted in 1912 from Novarupta and created the Valley of Ten Thousand Smokes, Alaska.

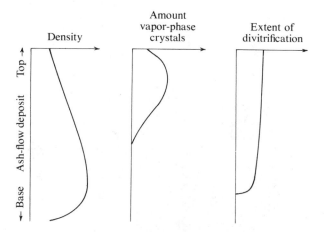

Figure 3-33 Secondary vertical zonal variations in a typical prehistoric ash-flow deposit. The intensity of compaction and accompanying welding of the initially porous material is shown by the density curve, the most dense tuff being the most compacted and welded.

cause of this well-developed planar fabric. Other intensely welded tuffs resemble black vitrophyre (or obsidian if intratelluric crystals are lacking) formed by solidification of coherent lava flows. But the vitroclastic and eutaxitic fabric evident at least in thin section is characteristic of a welded tuff.

Eutaxitic fabric and flow layering such as that shown in Figure 3-9 are sometimes mistakenly confused because they are both planar features. However, their geologic significance differs. Flow layering originates as a primary fabric element during emplacement and reflects the pattern of local flow in the coherent body; it consists of fairly uniform layers of contrasting texture/composition that are continuous along strike for a meter or more. The flattened, discoidal pumice lapilli in eutaxitic fabric show in surfaces oblique to the foliation as discontinuous lenses with flamelike terminations; they are secondary to emplacement and are elongate perpendicular to the direction of compaction.

Two types of crystallization occur simultaneously with welding and compaction in the ash flow. Entrapped gas migrates up through the deposit and escapes into the atmosphere, mainly at fumaroles. These cooling vapors, which contain significant amounts of dissolved silica and alkalies, precipitate en route to the surface various vapor-phase minerals within pore spaces in the weakly welded upper part of the deposit. The main products of such **vapor-phase crystallization** are alkali

(a)

(b)

Figure 3-34   Welded ash-flow tuffs; compare with the nonwelded tuff illustrated in Figure 3-11. Photographs (a) and (b) are the same piece of rock, except (a) is of a fresh surface showing its obsidianlike character, whereas (b) is of the opposite side, a weathered surface. On it can be seen darker pumice lapilli that are collapsed and flattened; note especially their frayed, flamelike terminations. The lighter-colored matrix between the lapilli is composed of ash-size glass shards, which in thin section under the microscope can also be seen to be flattened. In the microscopic view (c) of a more crystal-rich tuff than the hand sample, this **eutaxitic** fabric of flattened shards and pumice lumps is clearly expressed. The flattened glass shards appear white against submicroscopic, gray dust particles. The shards are even more tightly compressed around the large white plagioclase phenocrysts, which behaved as rigid bodies during compaction of the surrounding soft glass fragments. Also note the collapsed pumice lapilli with its typical frayed termination. [Parts (a) and (b) are U.S. Geological Survey photographs courtesy of Robert L. Smith.]

(c)

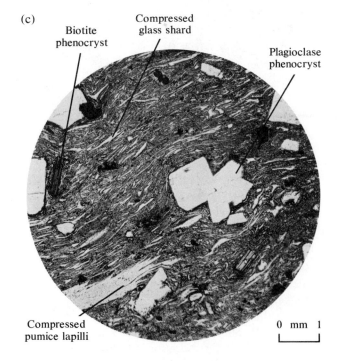

Biotite phenocryst

Compressed glass shard

Plagioclase phenocryst

Compressed pumice lapilli

0 mm 1

feldspar, tridymite, and cristobalite crystals that are attached to walls of pore spaces. The other secondary crystallization that occurs in ash-flow deposits is devitrification of the glassy material. Devitrification affects chiefly the middle to upper parts of an ash flow; its lower bound can be very sharp. In outcrop where this contact occurs within the welded zone, the devitrified tuff is red, pink, brown, or purple and the rock looks stony, in contrast to the black, glassy, underlying nondevitrified tuff. In devitrified tuffs, aphanitic quartz–feldspar intergrowths, locally spherulitic or lithophysal, replace glass, but delicate pumiceous and vitroclastic textures are faithfully preserved, even though the rock may be entirely crystalline (see Figure 3-13). In other instances, devitrification completely erases fragment outlines; vitroclastic and eutaxitic fabrics are obliterated and the tuff assumes a massive, featureless aphanitic fabric similar to that of many lava flows.

The concept of a **cooling unit** was conceived by Smith (1960) to represent the pyroclastic material nearly instantaneously ejected and deposited from a magma chamber. A simple cooling unit would be comprised of one ash flow, or of more than one emplaced in rapid succession so that the sequence cooled as a single sheet. Compound cooling units have some sort of internal interruption so that the cooling was not uniformly continuous (for example, two or more flows emplaced in succession). The implications of this concept in interpreting the evolution of ash-flow–caldera complexes are discussed by Christiansen (1979).

Calderas

A **caldera** may be defined as "a large volcanic collapse depression, more or less circular or cirquelike in form, the diameter of which is many times greater than that of any included vent. A crater may resemble a caldera in form but is almost invariably much smaller and differs genetically in being a constructional form rather than a product of destruction" (Williams and McBirney, 1979, p. 207; compare the terms *cauldron* and *volcano-tectonic depression* in Smith and Bailey, 1968, p. 616).

The cause of collapse is generally withdrawal of magma from an underlying magma chamber, either by extrusion of lavas or by catastrophic eruption of pyroclastic material, and foundering of its roof into the void. Contrary to widespread lay belief, calderas such as the one that holds Crater Lake (the misnomer is unfortunate!) did not form by the volcano "blowing off its top." Had that been the case the ejecta from the caldera-forming explosion would have been chiefly accidental. Instead, it is mostly glassy juvenile material.

Old magma chamber–caldera complexes deeply dissected by erosion show that collapse accompanied only partial withdrawal of magma; Smith (1979) estimates that less than one-tenth of a chamber is evacuated in one pyroclastic eruption.

Volume relations. Smith (1960, 1979) and Smith and Bailey (1968) have emphasized the intimate association—actually cause and effect—between eruption of large volumes of pyroclastic material, chiefly as ash flows, and caldera collapse. Smith finds that the volume of pyroclastic flow deposits in the rock record and formed in historic times ranges over eight orders of magnitude (see Figure 3-35). The smallest deposits, common in historic times, range from 0.001 to 1 km^3 and form from hot avalanches cascading off domes and backfall flows from small central-vent, composite conical volcanoes typical of island arcs. Examples include Mt. Pelée (see Figure 3-26a) and Mt. Lamington (Taylor, 1958). These eruptions do not lead to caldera collapse. Volumes of 1 to 100 km^3 are erupted from larger conical volcanoes such as Krakatoa in the East Indies, Novarupta (source of the 1912 deposit in the Valley of Ten Thousand Smokes, Alaska, Figure 3-32), and Crater Lake. Caldera formation begins in this volume range; the void rapidly created in the magma chamber by explosive eruption is large enough that the roof overlying it cannot support itself and founders catastrophically downward. In sustained pyroclastic eruptions occurring over extended periods of

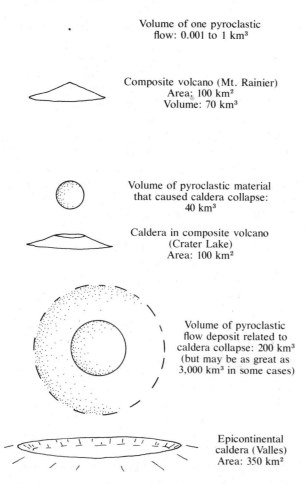

Figure 3-35 Volumes of pyroclastic flows (represented by spheres) in relation to volcanic edifice. Volume of one flow from a conical composite volcano such as Mt. Rainier is represented by a small dot.

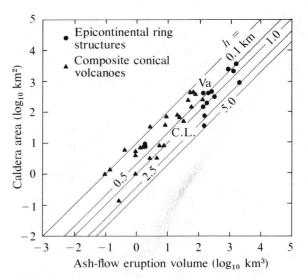

Figure 3-36 Correlation between volume, V, of erupted pyroclastic flow deposits and area, A, of the caldera it produced. Volumes of deposits range from 10^{-1} to 10^{3} km^3. Diagonal lines are the depth of collapse, h, or drawdown of magma chamber, a figure often difficult to determine from field relations, but which equals V/A. Vertical caldera walls and a flat chamber roof are assumed. C.L. is Crater Lake and Va is Valles caldera. [After R. L. Smith, 1979, Ash-flow magmatism. Geological Society of America Special Paper 180.]

time (months or years), collapse is intermittent with eruption as shown by landslide deposits formed by caving of the steep unstable wall of the caldera that interfinger with ash flows within the depression. Volumes of 100 to in excess of 1,000 km^3 are associated with collapse of large epicontinental ring structures such as in Yellowstone Park, the San Juan Mountains of Colorado, and many others in the western United States that are far inland from the continental margin, and in Sumatra and the North Island of New Zealand.

It is notable that eruption of large volumes of ash-flow material correlates closely with large calderas (see Figure 3-36) and that these calderas do not destroy major volcanic edifices. The Valles caldera in New Mexico is an example of a large epicontinental ring structure that was not a large volcano before collapse (Smith and Bailey, 1968).

Valles caldera. The relief model in Figure 3-37 shows the morphology of this Plio-Pleistocene feature, and the generalized geologic map in Figure 3-38 shows (1) distribution of the ash-flow tuff sheets whose eruption prompted caldera collapse; (2) post-caldera clastic material that accumulated in the depression; (3) post-caldera rhyolite domes that were extruded from vents around the ring-fracture zone; and (4) the resurgent uplift of the

Figure 3-37 Relief model of the Valles Caldera and vicinity, New Mexico. [Made by Stephen H. Leedom from U.S. Geological Survey relief maps.]

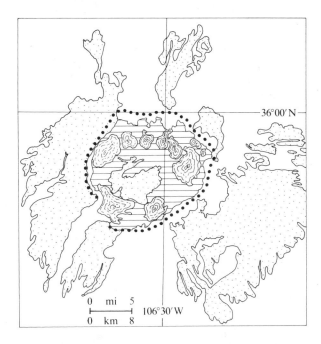

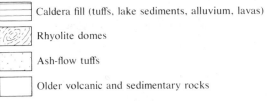

Caldera fill (tuffs, lake sediments, alluvium, lavas)

Rhyolite domes

Ash-flow tuffs

Older volcanic and sedimentary rocks

• • • Topographic rim of caldera

Figure 3-38 Generalized geologic map of the Valles caldera. Faults have been omitted for clarity. The patchy distribution of the caldera-related ash-flow tuff is due to the uneven topography onto which it was deposited as well as to subsequent erosion. [After R. L. Smith and R. A. Bailey, 1968, Resurgent cauldrons, in R. R. Coats, R. L. Hay, and C. A. Anderson (eds.), *Studies in Volcanology.* Geological Society of America Memoir 116.]

floor of the caldera, manifest in the erosional window cut through the clastic fill into the caldera-related ash-flows. The sequence of cross sections in Figure 3-39 show how this caldera is believed to have evolved.

Nests of two or more cross-cutting calderas are common (see Figure 3-40), and indicate recurrent shallow magmatic activity—magma intrusion and boiling, pyroclastic eruption and partial evacuation of the magma chamber, and caldera collapse—above a deeper magma source (compare Figure 4-24).

Deposits of Reworked Material

All of the explosive volcanic processes described so far in this chapter can create deposits of loosely consolidated debris susceptible to subsequent transport and deposition by running water. Pyroclastic-fall and nonwelded tops of pyro-clastic-flow and pyroclastic-surge deposits are all easily transported and redeposited by running water and wind to form **reworked pyroclastic deposits.** Torrential rainfalls occur in the vicinity of the volcanic vent as erupted steam cools and con-

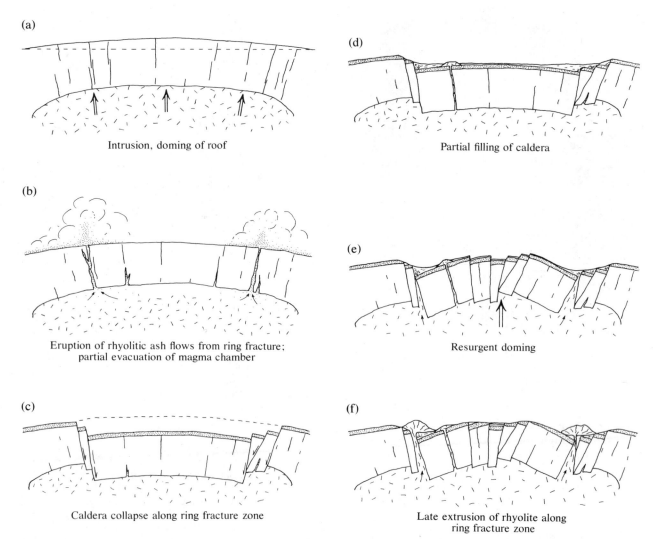

(a)

Intrusion, doming of roof

(b)

Eruption of rhyolitic ash flows from ring fracture;
partial evacuation of magma chamber

(c)

Caldera collapse along ring fracture zone

(d)

Partial filling of caldera

(e)

Resurgent doming

(f)

Late extrusion of rhyolite along
ring fracture zone

Figure 3-39 Schematic cross sections illustrating the evolution of the Plio-Pleistocene Valles caldera. Compare Figure 4-24. (a) Doming of the roof over the intruding magma and formation of ring fracture system. Dashed line is original ground surface. (b) Eruption of rhyolitic pyroclastic material from ring fractures and partial evacuation of magma chamber. (c) Caldera collapse guided by existing ring fractures in roof; central block about 20 km in diameter subsided almost 1 km as an intact "piston." (d) Partial filling of caldera by landslides from unstable wall of depression, by lake sediments, and by post-caldera pyroclastic deposits and lavas. (e) Resurgent uplift, doming, and fracturing of central block due to renewed magmatic activity. (f) Extrusion of viscous rhyolite lava from ring fractures peripheral to central block, forming an arcuate group of domes. Magmatism is now presumed to be dead, but geothermal activity persists today in many hot springs. [After R. L. Smith and R. A. Bailey, 1968, Resurgent cauldrons, in R. R. Coats, R. L. Hay, and C. A. Anderson (eds.), *Studies in Volcanology*, Geological Society of America Memoir 116.]

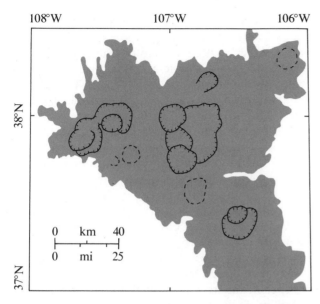

Figure 3-40 Swarm of Oligocene calderas in the San Juan volcanic field (shaded) of southwestern Colorado. Possible buried calderas are shown by dashed lines. Steven and Lipman believe that the calderas formed by partial evacuation of shallow magma bodies that rose out of an underlying larger batholithic mass. [After T. A. Steven and P. W. Lipman, 1976, Calderas of the San Juan volcanic field, southwestern Colorado. U.S. Geological Survey Professional Paper 958.]

Figure 3-41 Water-laid volcanic sandstone comprised of well-sorted, somewhat abraded sand-size particles that are thinly stratified and locally cross bedded.

denses; minute ash particles mingled with the steam facilitate the rainfall process by acting as "seeds" in a nucleating capacity. Raindrops falling through an ash cloud may collect concentric layers of fine ash, forming **accretionary lapilli.** Normal rainfalls subsequent to the eruption also cause reworking. Reworked deposits display features typical of most fluival epiclastic deposits, such as abrasion of clasts, cross bedding, lenticular beds, and so on. Because of these similarities, it is difficult and commonly impossible to distinguish between reworked deposits formed from pyroclastic material that was never consolidated and epiclastic deposits formed from fragments produced by weathering and disintegration of consolidated volcanic rocks. Hence, Fisher (1966 and references

therein) recommends use of nongenetic terms for deposits comprised chiefly of volcanic fragments and based upon grain size, such as volcanic sandstone (see Figure 3-41).

Composite Volcanoes

Composite volcanoes are the lofty, photogenic landmarks that most people picture in their minds as a "volcano." Most active or recently active volcanoes in subduction zones around the margin of the Pacific Ocean, in the Caribbean, and in the Mediterranean are of this type, including famous ones such as Fujiyama in Japan, Vesuvius in Italy, Mayon in the Philippines, and Rainier, Shasta (see Figure 3-14), and Mt. St. Helens in the Cascades of the Pacific Northwest (see Figure 3-42). Many reach great heights because they rest on an already elevated platform of older volcanic deposits. Although imposing topographically, the total volume of any one composite volcano is not as great as might be anticipated (see Figure 3-35).

A **composite volcano,** sometimes called a

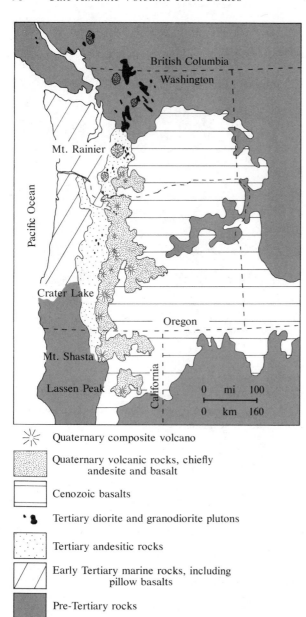

British Columbia

Washington

Mt. Rainier

Pacific Ocean

Crater Lake

Oregon

Mt. Shasta

California

Lassen Peak

0   mi   100

0   km   160

Quaternary composite volcano

Quaternary volcanic rocks, chiefly
andesite and basalt

Cenozoic basalts

Tertiary diorite and granodiorite plutons

Tertiary andesitic rocks

Early Tertiary marine rocks, including
pillow basalts

Pre-Tertiary rocks

Figure 3-42 Generalized geologic map of the Cascade
Range and vicinity. [After Tectonic Map of North
America, U.S. Geological Survey.]

stratovolcano, is a large, more or less symmetric andesitic cone built mostly from a central eruptive vent and consisting of innumerable alternating tongues of lava and volcaniclastic deposits. Until removed by erosion, a small crater lies at the summit. Some magma is extruded from flanking vents. Magma solidified within the central conduit and in flank conduits and fissures as plugs, dikes, and sills forms a reinforcing skeleton for the edifice. Details of this internal structure can be observed in deeply incised canyons in the flanks, or in the wall of the summit crater (see Figures 3-43 and 3-44).

The steep slopes of composite volcanoes reflect the relatively small volume and low rates of extrusion of magma from the summit and the presence of abundant clastic material resting at a steep angle of repose. The extruded usually viscous andesitic to dacitic lavas cool and become immobile near the vent, rather than traveling far downslope as in the more voluminous eruptions of less viscous basaltic lava, which forms subdued shield volcanoes (see pages 157–158).

A major component of the volcaniclastic deposits in any composite volcano is the **volcanic debris flow**, or **lahar** (an Indonesian word) (see Figure 3-45). Volcanic debris flows resemble and behave much like wet concrete and consist of blocks and lapilli suspended in a water-saturated mud (ash) matrix that imparts mobility to the body. A debris flow has rheologic properties (see page 236) similar to a lava flow, its plastic yield strength enabling large blocks to be transported and allowing arrested flows to have fairly steep toes. Debris flows are motivated by gravity and originate as masses of water-soaked, unstable volcanic material on the steep slopes of the volcano. Some are cold, the triggering agent being heavy rainfall or meltwater from winter snows; other flows are hot, formed as active lava, or perhaps a pyroclastic flow, encounters and blends explosively with snow, or perhaps with water in a crater lake, or with a wet mass of rock that is already nearly unstable. Debris flows are usually heterolithologic, poorly sorted tuff breccia (see

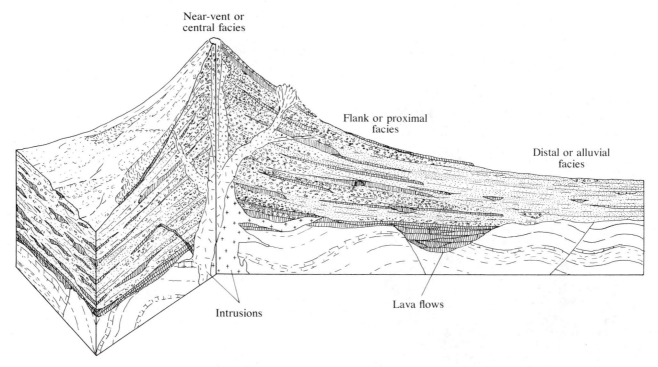

Near-vent or
central facies

Flank or proximal
facies

Distal or alluvial
facies

Intrusions

Lava flows

Figure 3-43 Idealized cross section through a composite volcano showing alternating layers of
lava and volcaniclastic material cut by dikes, sills, and plugs, some of which fed surface lava
extrusions. The platform on which the volcano rests is a hypothetical mass of folded and faulted
sedimentary rocks capped by a sequence of basalt flows. Different facies of the composite volcano
are discussed in the text. [After H. Williams and A. R. McBirney, 1979, *Volcanology* (San
Francisco: Freeman, Cooper and Co.).]

Figure 3-45b). The range of fragmentation processes and modes of mobilization of lahars is considerable (Williams and McBirney, 1979, pp. 171–
178).

It is helpful to subdivide large central vent complexes such as composite volcanoes into vent and
alluvial facies (Parsons, 1969) or into central, proximal, and distal facies (see Figure 3-43) (Williams
and McBirney, 1979, pp. 312–313). The central or
near-vent facies (within 2 km of the central vent) is
a bewildering array of structures and both intrusive and extrusive rock bodies, often with a strong
alteration overprint. Thin lava flows are subordinate to coarse, poorly sorted volcaniclastic deposits with steep initial dips. The proximal or flank

facies (up to roughly 5 to 15 km from the central
vent) comprises thick lava flows, lahars with angular coarse clasts, and some reworked clastic deposits; zones of weathering and soil development
may occur between layers of lava and volcaniclastic deposits. The distal facies comprises layers of
rock with considerable lateral continuity formed of
well-sorted and fairly well-bedded lahars and epiclastic deposits of rounded clasts; interstratified
lake deposits may occur, and lava flows are restricted to more fluid types that flowed down
valleys.

In young island arcs and near-continental-
margin volcanic complexes, **subaqueous** lava
flows and volcaniclastic deposits, deposited in the

(a)

(b)

Figure 3-44   Internal structure within the inner crater
wall of Vesuvius three years after the 1906 eruption.
Most of the wall exposes stratified pyroclastic deposits,
but local lava flows are evident just above the apex of
the talus cone on the left, and many slightly
lighter-colored dikes and sills can be seen cutting the
pyroclastic layers. Outer rim of the volcano is in the
distance. This composite volcano is one of several in the
Mediterranean region built of highly alkaline rocks
containing leucite and is, therefore, unlike the typically
andesitic volcanoes of the circum-Pacific region.
[Photograph obtained through the courtesy of N. M.
Short from glass-plate negative of the Geophysical
Laboratory, Carnegie Institute, Washington, D.C.
From J. Green and N. M. Short, 1971, *Volcanic
Landforms and Surface Features* (New York:
Springer-Verlag).]

Figure 3-45   Volcanic debris flows. (a) Flank of Mt.
Rainier, Washington. The steep dips, up to 30°, are
primary in this 150-foot-high cliff face. Note crude
vertical erosional columns and lenses of partly brecciated
lava alternating with debris flows. (b) Detail of a debris
flow showing heterogeneous, angular, poorly sorted
clasts. [Photograph in (a) courtesy of C. A. Hopson.
From R. S. Fiske, C. A. Hopson, and A. C. Waters, 1963,
Geology of Mt. Rainier National Park, Washington. U.S.
Geological Survey Professional Paper 444.]

ocean, constitute a large portion of the rock record (see, for example, Fiske and others, 1963).

3.3 SPACE–TIME–COMPOSITION RELATIONS IN CALC-ALKALINE VOLCANIC ARCS

The intent of this section is to provide a more integrated view of calc-alkaline volcanic arc bodies, which are typically located along the edge of the overriding plate in a subduction zone (see Figure 1-17). Once we have some insight, questions regarding the evolution of arc volcanoes and the origin of magmas can be considered in the succeeding section.

Cascade Volcanic Arc

Important differences exist between the Quaternary composite volcanoes of the Cascade Range in the northwestern United States, (see Figure 3-42). Mt. Rainier and Mt. Lassen are representative of these differences (McBirney, 1968).

Mt. Rainier has been built, as far as can be determined by sampling at the surface (Fiske and others, 1963), of thin lava flows of monotonously uniform siliceous andesite bordering on dacite (see Table 3-1 and Appendix D). Phenocrysts of augite, hypersthene, and plagioclase are abundant (see, for example, Figure 1-9a). Volcaniclastic deposits are subordinate in this Pleistocene edifice, which rises to over 4,300 m above sea level capping a 2,400-m volcanic plateau of Tertiary andesitic rocks (see Figure 3-42). Unlike Quarternary composite volcanoes to the south, there are no nearby contemporaneous basaltic extrusions around Rainier.

In the Lassen region (Williams, 1932b), the rocks are more varied, ranging through basalt–andesite–dacite–rhyolite (see Table 3-2 and Figures 2-4 and 2-5). No lofty conical edifice like Mt. Rainier exists; instead, on a vast platform of andesites and basalts an andesitic cone rising to perhaps 3,400 m above sea level formed while four shield volcanoes of fluid basalt grew around it (see Figure 3-46). Following this, rhyolite and several

Table 3-1 Range in chemical composition of five rocks from Mt. Rainier (in wt.%)

SiO_2	60.53–63.57
TiO_2	0.20–0.95
Al_2O_3	16.91–17.25
Fe_2O_3	1.12–2.12
FeO	3.32–3.85
MgO	2.28–3.47
CaO	5.11–5.80
Na_2O	4.01–4.21
K_2O	1.58–1.84
P_2O_5	0.24–0.85
H_2O^+	0.09–0.25
H_2O^-	0.04–0.12

Source: After R. S. Fiske, C. A. Hopson, and A. C. Waters, 1963, *Geology of Mt. Rainier National Park*. Washington, D.C.: U.S. Geological Survey Professional Paper 444.

flows of relatively fluid dacite were extruded from scattered vents. Later, apparently due to degassing of the magma chamber(s), more viscous dacite domes pushed out of several vents; the largest of these forms Lassen Peak. Minor pyroclastic showers accompanied these extrusions. A remarkable, though not uncommon, feature of the Lassen dacites is the presence of abundant mafic xenoliths. Sometime during the dacitic activity, the top of the andesitic cone collapsed, forming a small caldera. Basalts with embayed crystals of quartz have been extruded within the last few thousand years, and in 1915 a small dacite extrusion from the summit of Lassen Peak triggered a disastrous debris flow.

Although there are exceptions, dominantly andesitic volcanoes of the Rainier type tend to lie in the central Cascades, where no pre-Tertiary continental rocks occur, or are postulated to occur, beneath a thick cover of Tertiary lavas and mafic volcanic debris; this embayment in the pre-Tertiary margin of the continent is apparent in Figures 3-42 and 4-44. The diverse piles of basalt–andesite–dacite–rhyolite like those at Shasta and Lassen occur where pre-Tertiary continental rocks are present.

Table 3-2 Chemical compositions of representative rocks from the Lassen region of the northern California Cascades (in wt.%).

	Basalt of Cinder Cone	Andesites of Brokeoff Cone		Dacite of Lassen Peak	Rhyolite near Willow Lake
SiO_2	54.56	57.04	63.47	68.32	74.24
TiO_2	0.53	0.47	0.37	0.31	0.20
Al_2O_3	16.04	19.11	16.75	15.26	14.50
Fe_2O_3	0.95	4.37	2.15	1.66	1.27
FeO	6.07	2.48	2.75	1.26	0.67
MnO	0.17	0.12	0.09	0.04	0.06
MgO	8.71	3.94	3.04	1.32	0.25
CaO	8.89	7.34	5.72	3.26	0.11
Na_2O	3.05	3.48	3.94	4.27	3.00
K_2O	1.18	1.16	1.62	2.81	3.66
P_2O_5	0.18	0.08	0.13	0.12	0.07
H_2O^+ H_2O^-	0.28	1.09	0.55	1.37	2.04
Total	100.61	100.68	100.58	100.00	100.07

Source: After H. Williams, 1932, Geology of the Lassen Volcanic National Park, California. *University of California Publications in Geological Sciences, Bulletin* 21.

Tonga–Kermadec–New Zealand Arc

This arc is important because volcanoes comprising the Tonga and Kermadec Islands rest upon oceanic crust and lithosphere, whereas in the New Zealand segment the underlying crust is continental (see Figure 3-47).

Once again diverse but predominantly silicic calc-alkaline magmas are associated with continental crust, whereas more uniformly andesitic to basaltic calc-alkaline to tholeiitic magmas are extruded where the underlying crust is oceanic. These contrasts are shown in Figures 3-48 and 3-49. In the oceanic Tonga–Kermadec segment of the arc, tholeiitic basalts and andesites predominate though dacites are also present. (The volume of extruded dacites is grossly exaggerated in Figure 3-48.) In New Zealand, andesites and basalts are calc-alkaline and are overshadowed about twentyfold by voluminous extrusions of rhyolite, chiefly explosive ash flows. Large-ion-lithophile (LIL) elements (K, Rb, Cs, Ba, U, Th) that are incompatible with respect to mantle peridotite are systematically enriched in these New Zealand rocks; Sr isotopes tend to be more radiogenic (up to 0.706) than the Tonga–Kermadec lavas (<0.704), and there is a positive correlation between $^{87}Sr/^{86}Sr$ ratios and potassium.

3.4 PETROGENESIS OF CALC-ALKALINE ARC VOLCANIC ROCKS

It is not our intent here to consider the full details of the origin and evolution of calc-alkaline volcanic rocks produced along convergent plate boundaries. That is deferred to Chapter 9, after plutonic calc-alkaline rocks and other rock associations that will help put magma genesis and evolution in proper perspective have been discussed. However, some general comments are in order, dealing especially with the sort of interpretations that have been advanced by petrologists.

(a)

(b)

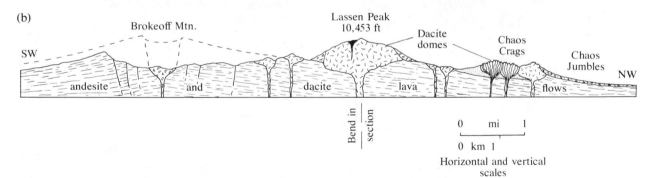

Figure 3-46   Aerial photo (a) and cross section (b) of a part of the Lassen volcanic area, northern California. Brokeoff Mountain is a remnant of a large volcano (dashed lines). Collapse of one flank of Chaos Crags dacite dome, possibly in the nineteenth century, produced the Chaos Jumbles avalanche. Note talus apron around dome complex on either side of flank area that collapsed. A small eruption of dacite lava (black in part b) from the summit of Mt. Lassen occurred in 1915. [Photograph courtesy of John S. Shelton. Cross section after H. Williams, 1932, Geology of the Lassen Volcanic National Park, California. University of California Publications in Geological Sciences, Bulletin 21.]

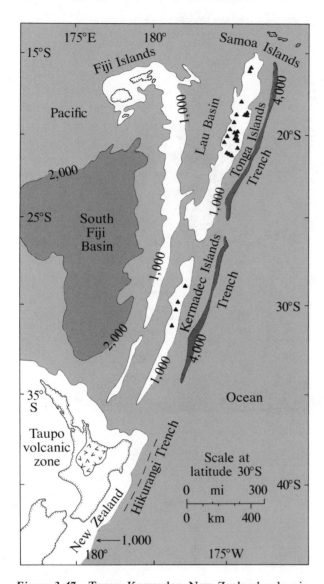

Figure 3-47 Tonga–Kermadec–New Zealand volcanic arc in the southwestern Pacific. Volcanic islands in the arc shown by solid triangles. Bathymetric contours are in fathoms (1 fathom = 1.83 m). [After A. Ewart, R. N. Brothers, and A. Mateen, 1977, An outline of the geology and geochemistry, and the possible petrogenetic evolution of the volcanic rocks of the Tonga–Kermadec–New Zealand island arc. *Journal of Volcanology and Geothermal Research* 2.]

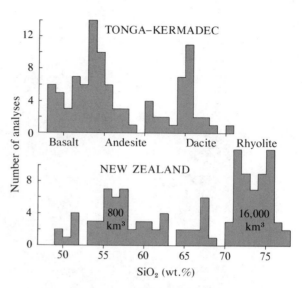

Figure 3-48 Frequency distribution of silica in analyzed samples from the Tonga and Kermadec Islands and from the North Island of New Zealand. The histogram for Tonga–Kermadec is not representative of relative volumes because a far greater proportion of dacites were analyzed; hence, this segment of the arc is predominantly basalt and andesite. In continental New Zealand, on the other hand, rhyolites are 20 times more voluminous than andesites. *Beware of interpreting volumes of rock types from histograms and variation diagrams that only represent what the petrologist has sampled and analyzed.* [After A. Ewart, R. N. Brothers, and A. Mateen, 1977, An outline of the geology and geochemistry, and the possible petrogenetic evolution of the volcanic rocks of the Tonga–Kermadec–New Zealand island arc. *Journal of Volcanology and Geothermal Research* 2.]

Competing Possible Models

The basic evolutionary mechanisms known to cause compositional diversification in magma systems in general have been briefly reviewed in Section 2.3, where it was also indicated that such mechanisms must operate upon already variable magmas rising from lower continental crustal and upper mantle source regions.

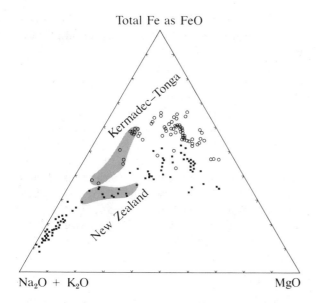

Figure 3-49 AFM diagram for analyzed rocks from the Tonga and Kermadec Islands and the North Island of New Zealand. The oceanic Tonga–Kermadec segment of the arc is comprised of relatively Fe-enriched tholeiitic basalts and andesites with minor dacite, whereas the continental New Zealand rocks are calc-alkaline and include abundant rhyolite. Dacites are shaded. Compare Figure 2-7. [After A. Ewart, R. N. Brothers, and A. Mateen, 1977. An outline of the geology and geochemistry, and the possible petrogenetic evolution of the volcanic rocks of the Tonga–Kermadec–New Zealand island arc. *Journal of Volcanology and Geothermal Research* 2.]

A fact of primary importance regarding arc magmas (see also Chapter 4) is that they are more felsic and diverse where founded on a continental crust but more restricted in composition, to basalt and andesite, in youthful island arcs where the overriding plate is oceanic (Jakeš and White, 1972, Johnson and others, 1978). A thick pile of continental sialic material must somehow be involved in the former situation, whereas in the latter mantle-derived mafic magmas are virtually the only lavas emerging to the surface. Whether the tholeiitic basalt–andesite magmas of the young island arcs such as Tonga–Kermadec are derived by partial melting of basaltic oceanic crust capping the descending lithospheric plate or by partial melting of the overlying wedge of peridotitic mantle due to the fluxing action of water liberated from the descending crust is one of the most controversial problems in petrology today. Perhaps both source regions are viable and mixing of melts from the two is possible. The contribution of marine sediments overlying the oceanic crust is also controversial. Are they subducted and melted, or scraped off the descending plate and piled up in the trench?

Magmatic differentiation of mantle-derived basaltic magmas has in the past enjoyed wide appeal among petrologists to account for the diverse spectrum of lavas found along continental margins such as in the southern Cascades, or in island arcs founded on continental sial settings such as New Zealand. Regretably, in many cases such a mechanism has been invoked with little attention to factual data on the specific rock assemblage at hand and only because it is the most popular idea of the time.

Can thick continental crust somehow provide greater opportunity for long-term residence of mafic mantle-derived magmas in insulated surroundings so that appreciable differentiation occurs, as for example, by crystal–liquid fractionation? Or do the mafic magmas become contaminated by sialic material to produce more felsic rock types? Or do the mafic magmas supply thermal energy to large volumes of continental crust sufficient to produce rhyolitic partial melts? Do these then mix in various proportions to yield intermediate rock types? Ewart and others (1977) conclude that the mantle-derived andesite–basalt magmas of the Tonga–Kermadec Islands locally experienced some crystal–liquid fractionation to yield dacite derivative magmas. They postulate that similar mantle-derived mafic magmas in the continental New Zealand portion of the arc were somewhat contaminated by sialic material, producing more radiogenic and incompatible-element-enriched andesites and basalts. Residence of these mafic magmas in the lower continental crust for sufficient times to contaminate them might also have permitted sufficient heat to be transferred into the sialic wall rocks, creating low-*T* rhyolitic partial melts. Indeed, Sr-isotope and trace-element data point to production of the voluminous rhyolite magmas by partial melting of graywackes and shales whose measured initial Sr-isotope ratios lie in the range of 0.705 to 0.718. Intermediate composition dacites were conceivably created by mixing of mafic and rhyolitic magmas or perhaps by crystal–liquid fractionation of the mafic magmas. The generous volume of mantle-derived magmas that must have been intruded to create 16,000 km³ of rhyolite may be somehow associated with extension of the crust perpendicular to the subduction zone, a process well expressed in the back-arc spreading behind the Kermadec–Tonga island arc to the north, where oceanic basaltic magmatism has been concurrently active.

The spectrum of rock types in the Lassen region may have been produced by differentiation of parental basaltic magmas. Yet it is difficult to account for the relatively large volume of dacites by this model unless large volumes of complimentary, very mafic to ultramafic differentiates are hidden at depth beneath the region. Is there geophysical evidence for this necessary mass? Alternatively, it may be more feasible to postulate some sort of mixing process involving mantle-derived basaltic magma and rhyolitic partial melts that were produced by the thermal energy of these magmas fusing the sialic crust. The occurrence of common mafic inclusions in dacites and partly dissolved quartz grains in basalts might have a significant bearing on this model. Yet, any contributions from old sialic material in the Lassen volcanic rocks seem to be ruled out because of their low initial $^{87}Sr/^{86}Sr$ ratios (0.7032–0.7041), which are very similar to oceanic volcanic rocks (Peterman and others, 1970).

The volcanic rocks of the Crater Lake volcano present still another dilemma because of the clearly expressed gap in the middle of the compositional spectrum (see Figure 3-30). Are there two parental magmas—one silicic and the other mafic—that have mixed slightly but incompletely? Such a compositional gap in the past has been taken as evidence against magmatic differentiation. The discontinuously zoned magma chamber advocated by Smith (1979) might provide an explanation. The question is how it became zoned.

The origin of the compositionally uniform siliceous andesite lavas of Mt. Ranier is also uncertain. Differentiation from a basaltic parent magma seems unlikely because such lavas of similar age are not exposed nearby. A mixing process also fails to account for their origin, for the same reason. Could the lavas represent rather direct extracts from mantle source regions?

Conclusions Regarding Petrogenesis

The foregoing section should make obvious that arc magmas conceivably originate by a variety of mechanisms, such as partial melting of the lower continental crust, subducting oceanic crust, or peridotitic mantle; once the magma has been generated, the full compositional spectrum of the calc-alkaline suite might evolve by additional diversification involving mixing, contamination, and differentiation. But evidence sufficient to support a particular mechanism, or group of mechanisms, is not necessarily compelling and is certainly not proof; other processes not yet conceived could actually be responsible. However, if it is tentatively concluded that the mechanisms listed above explain the origin of arc magmas, then the question is: Which of these mechanisms can best explain the origin of the magmas in a specific volcanic complex? If the magmatic association is truly polygenetic, then each volcano must be interpreted on the basis of its own properties, using the concepts discussed in Chapters 8 and 9. We cannot say that because the Lassen rocks originated in a certain way, the nearby Shasta assemblage did likewise, although that is a possibility.

It can be dangerously misleading to make uncompromising one-to-one correlations between magmatic associations and plate tectonic regimes. The calc-alkaline suite and convergent plate setting is a common partnership, but some subduction zones have tholeiitic associations (for example, Tonga–Kermadec). What are the implications of the occurrence in southern Colorado of a huge volume of calc-alkaline volcanic rocks (Steven and Lipman, 1976) 1,000 km inland from the western margin of the American plate? Although some geologists have postulated a subducting plate beneath western Colorado when this field was active, Gilluly (1971) and other workers are skeptical. Is it possible that calc-alkaline magmatism can in certain regions be somehow triggered independently of plate subduction?

SUMMARY

3.1. Calc-alkaline and minor basaltic tholeiitic volcanic rocks alongside oceanic trenches have a wide variety of fabrics due to complex flows of energy. Compositions are also broad, ranging from basalt to rhyolite, and embody combinations of most major rock-forming minerals. Classification schemes are adapted to the basic fabric types—aphanitic, glassy, and clastic.

3.2. Field relations and fabrics of volcanic bodies are governed in large part by the silica and gas content of the magma. Coherent lava flows form by extrusion of gas-poor magma; in highly silicic magmas the flows are so viscous (immobile) that they pile up over the vent to form mushroom-shaped domes. Some extrusions become broken up by self-inflicting processes, forming usually coarse and monolithologic autoclastic breccias.

Explosive eruption of gas-charged magma propels ejecta or tephra from a volcanic vent. Three disperal and deposition mechanisms operate to produce pyroclastic deposits.

1. Pyroclastic- or air-fall deposits are relatively well-sorted and bedded mantle accumulations of ash and lapilli spread over large areas by fallout from the large steam–ejecta clouds accompanying most explosive eruptions.

2. Pyroclastic- or base-surge deposits are less extensive and have distinctive bedding structures produced as a cloud of water mist and minor solid ejecta moves rapidly away from a phreatic or phreatomagmatic eruption.

3. Pyroclastic flow deposits are generally poorly sorted and stratified sheets of pumice-rich lapilli tuff ranging in volume from 10^{-3} to 10^4 km^3 that are produced by highly mobile, hot mixtures of gas and abundant solid ejecta moving swiftly along the ground surface partly concealed beneath an overlying ash–steam cloud. The dispersal of flows 100 km or more from the vent may be related to gravitational collapse of huge vertical eruption columns, aided by fluidization of the flow itself by entrained and exsolving gas.

Many pyroclastic eruptions produce a sequence of deposits beginning with fall, then surge, and then flow with a capping fall.

Silicic pyroclastic flow deposits show compositional zoning that reflects an inverted record of zonal variations in the source magma chamber. Retention of thermal energy and gas in pyroclastic flow deposits after emplacement facilitates secondary zonal variations in degree of welding, compaction, vapor-phase crystallization, and devitrification of glass particles. A cooling unit consists of flow material nearly instantaneously erupted from a vent.

Eruption of magma from a buried chamber may be sufficient to cause collapse of the overlying roof into the void, producing a caldera. There is a one-to-one correlation between the eruption of a large volume of pyroclastic material and the collapse of a large caldera tens of kilometers in diameter and a kilometer or so in depth. Swarms of calderas are common and reflect recurring shallow-level magma intrusion, boiling, eruption, and collapse.

Many calderas have a late-stage resurgence of activity that domes the earlier-collapsed floor.

Widespread reworking of unconsolidated pyroclastic deposits by running water forms bedded volcanic sands and coarser volcanic breccias and conglomerates.

Composite volcanoes are the large conical edifices that characterize volcanic arcs alongside oceanic trenches. Their steep form around a central vent is related to slow extrusion of small volumes of pasty, generally andesitic lavas and formation of coarse debris flow deposits; lahars, pyroclastic-fall, and pyroclastic-flow deposits are emplaced chiefly in the flank and distal zones of the volcano.

3.3. Knowledge of space–time–composition relations of arc volcanic bodies is crucial to an understanding of their origin. The Cascade composite volcanoes vary from predominantly andesitic simple cones to more complex edifices built of basalt–andesite–dacite–rhyolite that have a thick sialic crust beneath. In the Tonga–Kermadec–New Zealand arc the island arc segment founded on oceanic crust comprises chiefly tholeiitic basalts and andesites, whereas the continental segment is comprised of rhyolites and subordinate dacites, andesites, and basalts of the calc-alkaline suite.

3.4. Continental sialic material is involved in the production of diverse but predominantly felsic rocks of the calc-alkaline suite. More mafic tholeiitic magmas appear to be mantle-derived, conceivably originating in the subducting oceanic crust or in the overlying wedge of peridotitic mantle that is permeated by fluids from the descending crust. Although magmatic differentiation has been in the past a favored model for generation of arc magmas, mixing, contamination, and partial melting of lower crustal sial may be at least as important in specific instances.

Diverse arc magmas appear to be polygenetic; hence the origin of a particular body must be decided on its specific properties. The occurrence of voluminous calc-alkaline bodies far inland from active continental margins is warning that perhaps we don't, after all, understand magma genesis and evolution as well as we think we do.

REFERENCES

Christiansen, R. L. 1979. Cooling units and composite sheets in relation to caldera structures. Geological Society of America Special Paper 180:29–42.

Compton, R. R. 1962. *Manual of Field Geology*. New York: John Wiley and Sons, 378 p.

Crowe, B. M., and Fisher, R. V. 1973. Sedimentary structures in base-surge deposits with special reference to cross-bedding, Ubehebe craters, Death Valley, California. *Geological Society of America Bulletin* 84:663–682.

Ewart, A., Brothers, R. N., and Mateen, A. 1977. An outline of the geology and geochemistry, and the possible petrogenetic evolution of the volcanic rocks of the Tonga-Kermadec-New Zealand island arc. *Journal of Volcanology and Geothermal Research* 2:205–250.

Fisher, R. V. 1966. Rocks composed of volcanic fragments and their classification. *Earth Science Reviews* 1:287–298.

Fisher, R. V. 1979. Models for pyroclastic surges and pyroclastic flows. *Journal of Volcanology and Geothermal Research* 6:305–318.

Fiske, R. S., Hopson, C. A., and Waters, A. C. 1963. Geology of Mt. Rainier National Park, Washington. U.S. Geological Survey Professional Paper 444, 93 p.

Francis, P. 1976. *Volcanoes*. New York: Penguin, 368 p.

Gilluly, J. 1971. Plate tectonics and magmatic evolution. *Geological Society of America Bulletin* 82:2383–2396.

Green, J., and Short, N. M. 1971. *Volcanic Landforms and Surface Features*. New York: Springer-Verlag, 519 p.

Jakeš, P., and White, A. J. R. 1972. Major and trace element abundances in volcanic rocks of orogenic belts. *Geological Society of America Bulletin* 83:29–40.

Johnson, R. W., Mackenzie, D. E., and Smith, I. E. M. 1978. Volcanic rock associations at convergent plate boundaries: Reappraisal of the concept using case histories from Papua New Guinea. *Geological Society of America Bulletin* 89:96–106.

Macdonald, G. A. 1972. *Volcanoes*. Englewood Cliffs, N.J.: Prentice-Hall, 510 p.

McBirney, A. R. 1968. Petrochemistry of the Cascade andesite volcanoes. In H. M. Dole (ed.), Andesite conference guidebook. *Oregon Department of Geology and Mineral Industries Bulletin* 62:101–107.

Moore, J. G. 1967. Base surge in recent volcanic eruptions. *Bulletin Volcanologique* 30:337–363.

Parsons, W. H. 1969. Criteria for the recognition of volcanic breccias: Review. *Geological Society of America Memoir* 115:263–304.

Peterman, Z. E., Carmichael, I. S. E., and Smith, A. L. 1970. $^{87}Sr/^{86}Sr$ ratios of Quaternary lavas of the Cascade Range, Northern California. *Geological Society of America Bulletin* 81:311–318.

Ross, C. S., and Smith, R. L., 1961. Ash flow tuffs: Their origin, geologic relations, and identification. U.S. Geological Survey Professional Paper 366, 81 p.

Sheridan, M. F. 1979. Emplacement of pyroclastic flows: A review. *Geological Society of America Special Paper* 180:125–136.

Smith, R. L. 1960. Ash flows. *Geological Society of America Bulletin* 71:795–842.

Smith, R. L., and Bailey, R. A. 1968. Resurgent cauldrons. In R. R. Coats, R. L. Hay, and C. A. Anderson (eds.), *Studies in Volcanology*. Geological Society of America Memoir 116:613–662.

Smith, R. L. 1979. Ash-flow magmatism. *Geological Society of America Special Paper* 180:5–27.

Sparks, R. S. J. 1976. Grain size variations in ignimbrites and implications for the transport of pyroclastic flows. *Sedimentology* 23:147–188.

Sparks, R. S. J., Wilson, L., and Hulme, G. 1978. Theoretical modeling of the generation, movement, and emplacement of pyroclastic flows by column collapse. *Journal of Geophysical Research* 83:1727–1739.

Steven, T. A., and Lipman, P. W. 1976. Calderas of the San Juan volcanic field, southwestern Colorado. U.S. Geological Survey Professional Paper 958, 35 p.

Streckeisen, A. 1979. Classification and nomenclature of volcanic rocks, lamprophyres, carbonatites and melilitic rocks: Recommendations and suggestions of the IUGS Subcommission on the Systematics of Igneous Rocks. *Geology* 7:331–335.

Taylor, G. A. 1958. The 1951 eruption of Mount Lamington, Papua. *Australia Bureau of Mineral Resources, Geology and Geophysics, Bulletin* 38, 117 p.

Walker, G. P. L. 1973. Explosive volcanic eruptions—A new classification scheme. *Geologische Rundschau* 62:431–446.

Williams, H. 1932a. The history and character of volcanic domes. *University of California Publications in Geological Sciences, Bulletin* 21:51–146.

Williams, H. 1932b. Geology of the Lassen Volcanic National Park, California. *University of California Publications in Geological Sciences, Bulletin* 21:195–385.

Williams, H., and McBirney, A. R. 1979. *Volcanology*. San Francisco: Freeman, Cooper and Co., 397 p.

Wilson, L., Sparks, R. S. J., Huang, T. C., and Watkins, N. D. 1978. The control of volcanic column heights by eruption energetics and dynamics. *Journal of Geophysical Research* 83:1829–1836.

Wohletz, K. H., and Sheridan, M. F. 1979. A model of pyroclastic surge. *Geological Society of America Special Paper* 180:177–194.

4

Calc-Alkaline Plutonic Rock Bodies

Calc-alkaline plutonic rocks are chiefly granitic, comprised mostly of feldspar and quartz, although more mafic rock types, including diorite and gabbro, occur in minor amounts. The phaneritic sequential growth fabric and mineralogical composition of these plutonic rocks indicate slow crystallization of magma under an insulating cover of overlying crustal rocks. The form, size, and internal fabric of plutons vary widely, depending upon such factors as composition and availability of magma, structure of country rocks, depth and mechanism of emplacement, and time of emplacement relative to tectonic processes. Field studies and laboratory experiments suggest that plutons rise from source regions as buoyant bodies of magma that spread out horizontally somewhere near the surface as gravitational equilibrium is approached. Some plutons now exposed at the surface may represent solidified subterranean magma chambers that fed volcanic extrusions.

Enormous volumes of granitic and minor dioritic rocks have developed along continental margins overriding subducting oceanic lithosphere. These belts of batholiths comprise hundreds of in-dividual plutons, most of which are compositionally zoned to some degree. The origin of this zoning remains unsolved but is probably related to diversification processes occurring near or at the site of emplacement. Formation of batholithic volumes of granitic magmas appears to require continental settings. Much of the current research on calc-alkaline plutons is concerned with chemical tracers that provide clues to the nature of the source region of magma genesis, whether crustal igneous or sedimentary rock or mantle material.

4.1 PETROGRAPHY

This chapter deals with phaneritic rocks composed chiefly of feldspar with lesser amounts of quartz, biotite, amphibole, pyroxene, and several accessory minerals. Consult Appendix C for comments regarding these minerals.

Fabric

Generally slow solidification of magma bodies surrounded by insulating older wall rocks yields somewhat more uniform fabrics compared to the more or less compositionally equivalent volcanic rocks described in the previous chapter. The slower loss of heat from intrusive bodies of magma buried beneath the surface allows the magma to crystallize completely so that glass is generally not present. Another contrast between volcanic and plutonic magmatic bodies is the rarity of gas cavities in buried bodies; the confining pressure exerted by the overlying cover of older rocks holds gas in solution in the high-T melt and the crystalline solids, such as biotite, that precipitate from it.

The different types of fabrics commonly found in calc-alkaline plutonic bodies are shown in Figures 4-1 through 4-9. One fabric not shown is **miarolitic,** which may be considered as the plutonic analog of vesicular fabric in volcanic rocks. During crystallization of the intrusive body of magma, especially in relatively shallow settings just beneath the surface of the Earth, small pockets of gas may segregate by boiling processes. Because of the crystalline nature of the walls of the pocket, or **vug,** it is shaped by the outlines of neighboring crystals. In some miarolitic vugs, late-stage euhedral crystals grow freely from the walls into the gas pocket (compare Figure 3-7).

Phaneritic rocks can be designated as **fine-**, **medium-**, or **coarse-grained.** The dimensions of the major grains are arbitrarily set at

$$coarse > 5 \text{ mm}$$
$$medium = 2\text{--}5 \text{ mm}$$
$$fine < 2 \text{ mm}$$

Contrasts in grain size in magmatic rocks of similar chemical composition are related chiefly to contrasts in the concentration of nuclei which can grow into visible grains, and this concentration is a function of how fast the silicate melt falls below its freezing temperature. Small amounts of undercooling generally yield coarser grains.

Porphyritic is a type of inequigranular texture in which there is a distinctly bimodal population of grain sizes—not the more or less continuous range as in **seriate** texture. The large grains are

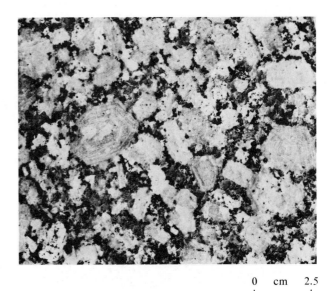

0   cm   2.5

Figure 4-1 The fabric in this polished slab is typical of many granitic rocks. White, generally rectangular grains are plagioclase. Light gray grains that are more **equant** (approximately equal dimensions in all directions) are potassium-rich alkali feldspar; note the oscillatory chemical zoning in the larger grains made evident by delicate alternating layers of whiter and grayer color concentric to their margins. Dark gray grains without apparent crystal faces are quartz; black ones are biotite.

This phaneritic rock is **inequigranular** because of the conspicuous variation in size of the grains. In situations such as this one where the grain-size variation is more or less continuous from smallest to largest, the texture is called **seriate.** Variations in cooling rate during crystallization or in the growth rates and number of nuclei available for growth of the different minerals can be factors in the formation of seriate texture. With regard to grain shape, this rock is **hypidiomorphic-granular** in that some grains are fairly **euhedral,** or idiomorphic, with planar bounding crystal faces (some of the feldspars), whereas other grains are **anhedral,** or xenomorphic, and have no regular crystallographic shape (quartz grains). The hypothetical rock illustrated under the microscope in Figure 1-11 is an idealized example of hypidiomorphic-granular texture. In that example, the texture is denoted by the generic term *sequential* rather than the descriptive term *hypidiomorphic-granular*. This texture, regardless of which name is used, originates by early uninhibited growth of certain crystals, which are euhedral, from the silicate melt, followed by later growth of other grains whose anhedral outlines must conform to surrounding existing grain margins.

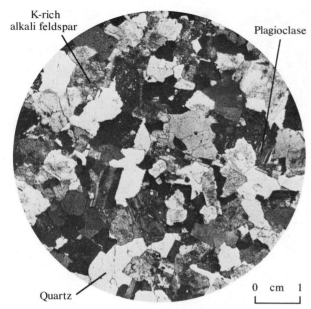

Figure 4-2 labels: K-rich alkali feldspar, Plagioclase, Quartz

0   cm   1

Figure 4-2 **Equigranular** refers to a rock in which grains are all more or less of equal size, in contrast to the seriate texture in Figure 4-1 and porphyritic texture in Figure 4-3. It results from uniform rates of growth and nucleation of the constituent minerals, all of which were growing more or less simultaneously. The fine phaneritic, equigranular fabric shown in this photomicrograph taken with cross-polarized light is sometimes referred to as **aplitic.** In hand sample, it appears sugary, like sandstone, but the grains are somewhat interlocking and pore spaces are nonexistent.

phenocrysts and are surrounded by a finer grained **matrix,** or **groundmass.** Phenocrysts will be euhedral if the magma system was near equilibrium at the time of solidification. Porphyritic fabric can originate in several different ways:

1. An early episode of slight undercooling and growth of large crystals might be followed by an episode of more rapid undercooling with formation of finer crystals as the magma is emplaced at shallower, cooler levels of the crust, or as it loses dissolved gas.

2. Differing rates of nucleation and crystal growth of different mineral species in a magma steadily losing heat and/or gas

(a)

0   cm   1

(b)

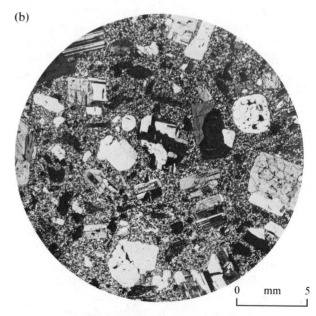

0   mm   5

Figure 4-3 In the hand sample (a) and photomicrograph (b), a porphyritic fabric is defined by phenocrysts of polysynthetically twinned plagioclase and of dark mafic minerals in an aphanitic felsic matrix.

Figure 4-4 **Pegmatitic** fabric applies to rocks of abnormally coarse but usually variable and, therefore, seriate grain size. As an arbitrary lower size limit, to distinguish from coarse phaneritic, we may use 3 cm. Pegmatitic fabric, therefore, generally cannot be discerned in hand sample—outcrops are required. Crystals in some pegmatites attain extraordinary dimensions—over 10 m. Such exceptionally large crystals, enclosed as they usually are in more normal phaneritic rocks, reflect rapid growth in very fluid, aqueous liquids. The photograph shows a quarry face in the Harding pegmatite, Taos County, New Mexico. The box and gallon jug show the colossal size of the spodumene crystals. [Photograph courtesy of Richard H. Jahns. From R. H. Jahns and R. C. Ewing, 1977, The Harding Mine, Taos County, New Mexico, *The Mineralogical Record* 8.]

Figure 4-5 **Graphic** fabric is an intergrowth in which a single crystal of alkali feldspar, usually now unmixed as perthite, encloses numerous small grains of cuneiform-shaped quartz, creating an overall resemblance to ancient hieroglyphic writing. Although physically separate, the quartz grains may be all crystallographically continuous. The size of the grains varies from that illustrated here to much smaller ones visible only under the microscope in **micrographic** texture. The origin of graphic fabric is uncertain; replacement processes and simultaneous growth of the feldspar and quartz from a melt have been suggested. Graphic texture is common in pegmatites.

could produce phenocrysts of one or more minerals different from matrix minerals.

3. Overgrowths may occur on large relict refractory grains derived from the region where magma was produced by partial melting of older rock.

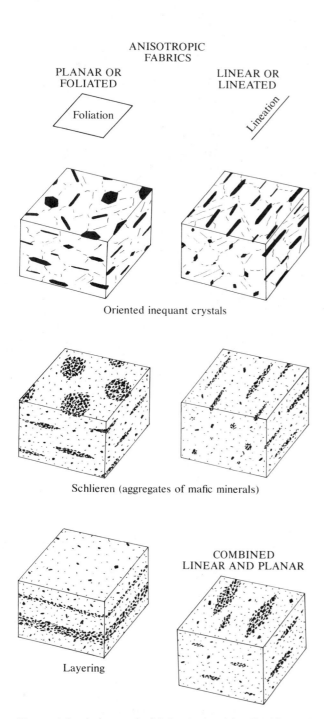

Figure 4-6 Anisotropic fabrics in granitic, dioritic, and some other types of plutonic magmatic rocks.

Figure 4-7 **Schlieren** (schliere, singular) are pencillike, discoidal or bladelike aggregates of mafic minerals more highly concentrated than in the host rock. Their preferred orientation in this outcrop of granitic rock produces anisotropic fabric that indicates flow occurred during emplacement of the magma.

4. Growth in the solid state from alkali-rich vapors may replace magmatic material, producing large crystals, or **megacrysts,** of alkali feldspar in granitic rocks.

It is not unusual that the distinctly different grain sizes of porphyritic fabric are more obvious in thin section under the microscope than in hand sample (see Figure 4-3).

Directional or **anisotropic** fabrics are those which have different aspects in different directions, in contrast to random **isotropic** fabrics where the rock has similar properties in any direction. Anisotropic fabrics are either planar, like a stack of papers, or linear, like a clutched handful of pencils, or combined planar and linear, like a stack of ruled notepaper. Rock bodies with planar fabric are **foliated,** or possess **foliation**; those with linear fabric are **lineated,** or possess **lineation.** If both lineation and foliation are present in a body, the

Figure 4-8 Discontinuous **segregation layers** in granitic rock defined by variable concentrations of mafic minerals.

Figure 4-9 Weak planar fabric expressed by preferred orientation of tabular feldspar grains that may have originated by magmatic flow.

lineation almost always lies within the foliation, like lines of print running across pages of a book. Some of the forms of anisotropic fabrics found in plutonic magmatic rocks, especially of granitic composition, are shown diagrammatically in Figure 4-6; photographs of real rock bodies with planar fabric are Figures 4-7 to 4-9. Note that any surface of exposure oblique to a foliation gives

the *appearance* of a linear fabric; careful scrutiny of exposures are required to discern the true nature of the anisotropic fabric, that is, whether it is planar, linear, or a combination of the two.

Anisotropic fabrics originate in several different ways, such as flow of magma containing suspended inequant crystals, settling of crystals onto a depositional surface, deformation of solid rock

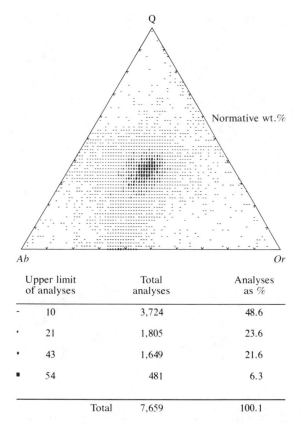

Q

Normative wt.%

Ab   Or

Upper limit of analyses		Total analyses	Analyses as %
-	10	3,724	48.6
·	21	1,805	23.6
✦	43	1,649	21.6
■	54	481	6.3
	Total	7,659	100.1

Figure 4-10 Chemical analyses of over 7,600 granitic rocks from around the world with greater than 80 wt.% normative $Q + Ab + Or$ have been compiled and stored on magnetic tape by Roger W. LeMaitre of the University of Melbourne. In this variation diagram that he created with a computer plotter, each symbol represents a particular number of analyses falling within the range specified under "Upper limit of analyses." About one-half of the analyses plot within only about 5% of the diagram. Compare with a similar diagram in Tuttle and Bowen (1958, fig. 42).

after crystallization, and replacement of originally anisotropic rock by hydrothermal solutions producing an igneouslike appearance (Compton, 1962, pp. 282–290). The latter two processes are metamorphic. Anisotropic fabrics are manifestations of a significant gradient in some form of energy in the magmatic system from which the rock originated. In a flowing magma, there are gradients in velocity, that is, in kinetic energy. The pattern of these flow gradients determines the form of the anisotropic fabrics—hence, they are **flow markers.** Foliation reflects planar, sheetlike laminar flow. Lineation reflects linear streamlined flow, like toothpaste extruded from a tube. Turbulent flow and static conditions yield isotropic fabrics. Inequant mineral grains—prismatic amphiboles, platy micas, tabular feldspars—and inequant mineral aggregates orient during flow with their longer or longest dimension in the direction of the flow, like logs in a river. Field mapping of anisotropic flow markers can thus provide valuable information concerning patterns of flow in the viscous magma during emplacement. In magma where crystals are settling under the influence of gravity, layers of settled crystals reflect a gradient in gravitational potential energy. Such gravity-produced layering would seem to be rare in granitic plutons because of the high viscosity of the magmas and consequent slow rates of crystal settling. Discontinuous, lenticular **segregation layers** are locally associated with flow fabrics and may be produced by some sort of flow process or by crystal settling.

Classification

Both composition and fabric can be used as a basis for classification of calc-alkaline plutonic rocks.

Composition as a basis of classification. Felsic, phaneritic rocks containing conspicuous quartz (say, greater than 10% mode) in addition to large amounts of feldspar can be designated as a whole by the term **granitic rocks.** This nonspecific term is useful where the type of feldspar is not recognizable because of alteration or weathering, for purposes of quick reconnaissance field studies, or for general discussion. Most granitic rocks have subequal proportions of quartz, sodic plagioclase, and K-rich alkali feldspar—either orthoclase or microcline. However, because of extensive solid solution and unmixing into perthite, the proportions of K-Na and Na-Ca end-member feldspars may not be readily apparent; but by recasting the chemical composition of the rock in terms of its normative *Q, Ab,* and *Or* (see page 44), the subequal proportionality is obvious (see Figure 4-10). The petrogenetic significance of this profoundly restricted composition will be considered later.

As with volcanic rocks, classification of granitic rocks based on composition can use either the bulk chemical composition or the normative composition (see Appendix D). Unlike volcanic rocks, however, the modal composition can also be used as a reliable basis. (See Appendix B for techniques for determining the mode.) Because the felsic minerals K-rich alkali feldspar, plagioclase, and quartz are major constituents, specific rock types are defined on the basis of their ratios. The nature of other minerals in a rock may be indicated in the form of a modifier, for example, muscovite granite, biotite diorite.

In an attempt to standardize many different schemes of classification that have appeared during the growth of igneous petrology, an international consortium of petrographers has agreed upon a classification and nomenclature based upon modal composition (see Figure 4-11). This classification overcame a major weakness of many older systems in which most granitic rocks that have subequal proportions of two feldspars and quartz were called quartz monzonite, or adamellite, rather than granite. However, to this writer, there are too many subdivisions in peripheral areas of the triangular diagram in Figure 4-11 where few real rocks exist, and the nomenclature is burdened with too many cumbersome names.

Granite, granodiorite, quartz diorite, and diorite form a common sequence with decreasing K-rich feldspar and quartz but increasing amounts of plagioclase; biotite and hornblende also increase. Diorites typically have about 20% to 40% hornblende; with increasing pyroxene, they grade into hornblende gabbro and gabbro as pyroxene and olivine become the chief mafic minerals. **Trondhjemite** is a leucocratic quartz diorite or tonalite with only minor amounts of mafic minerals, usually biotite; oligoclase–andesine is the plagioclase.

A simple and genetically significant classification proposed by Tuttle and Bowen (1958) recognizes that the two feldspar components—sodic

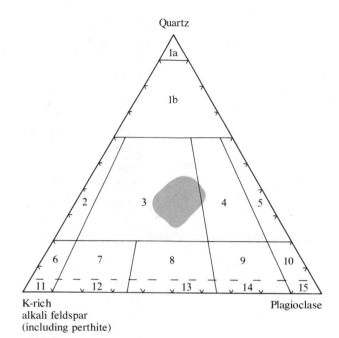

Quartz

K-rich alkali feldspar (including perthite)

Plagioclase

1a Quartzolite (silexite)

1b Quartz-rich granitoids

2 Alkali-feldspar granite

3 **Granite**

4 **Granodiorite**

5 **Tonalite**

6 Alkali-feldspar quartz syenite

7 Quartz syenite

8 **Quartz monzonite**

9 **Quartz monzodiorite**/quartz monzogabbro

10 **Quartz diorite**/quartz gabbro/ quartz anorthosite

11 Alkali-feldspar syenite

12 **Syenite**

13 **Monzonite**

14 Monzodiorite/monzogabbro

15 **Diorite**/gabbro/anorthosite

Figure 4-11 The International Union of Geological Sciences (IUGS) classification of granitic and allied rocks based upon modal composition in volume %. The shaded area represents about one half of the analyzed granitic rocks from around the world with over 80% Q + Ab + Or from Figure 4-10. Rocks in fields labeled 2, 3, 4, and 5 are called **granitoids**. Anorthosite with < 10% mafic minerals and gabbro with pyroxene and olivine as the chief mafic minerals are shown in Figure 5-26. Note that the feldspar end members are reversed relative to those in Figure 4-10. [After A. L. Streckeisen, 1976, To each plutonic rock its proper name. *Earth-Science Reviews* 12.]

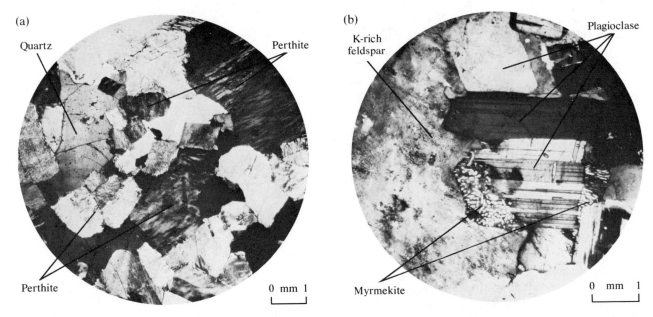

Figure 4-12   Two types of granites have different textural forms of feldspars depending upon water content of the magma. In a **hypersolvus** granite (a), feldspars crystallized as an initially homogeneous crystalline solid solution of K, Na, and Ca components, which almost always has subsequently unmixed to perthite; this would occur in relatively dry magma systems where little water was present. In a **subsolvus** granite (b), feldspar components in the melt precipitated as two separate minerals, plagioclase and K-rich alkali feldspar, each of which may suffer a little subsequent exsolution; such crystallization would occur in more water-rich magmas. In this photomicrograph, taken under cross-polarized light, note the more euhedral plagioclases and later crystallizing K-feldspar growing around them. Secondary alteration of the K-feldspar to clay minerals has made it appear turbid. The origin of the **myrmekite**—a vermicular intergrowth of quartz and sodic plagioclase adjacent to K-feldspar—is uncertain but also appears to be secondary.

plagioclase and K-rich alkali feldspar—in granites occur in two contrasting textural forms (see Figure 4-12) because of differences in the water content of the magma.

Fabric as a basis of classification.   The classifications just discussed are based chiefly upon composition of the phaneritic rock. Rocks with aplitic, strongly porphyritic, and pegmatitic fabric are designated, respectively, as aplite, porphyry, and pegmatite; rocks with these special fabrics have compound names—a compositional term preceding the fabric name—such as granite porphyry, diorite porphyry, monzonite aplite, syenite pegmatite, and so on.

Porphyry is a term restricted to conspicuously porphyritic intrusive felsic rocks with a fine phaneritic or even aphanitic matrix, such as in Figure 4-3. The name is not applied to all porphyritic rocks. Porphyries occur in fairly small, relatively shallow intrusive magma bodies that experienced an extended initial period of relatively uniform crystallization, with growth of abundant large phenocrysts, followed by an episode of rapid crystallization, forming the fine phaneritic or aphanitic matrix. The second event could be related to rapid extrusion into cooler country rocks, or sudden loss of gas, or both. By slight changes in circumstances, such as less crystallization in an early stage before subsequent extrusion, or com-

Figure 4-13 Thin subparallel dikes of leucogranite aplite in a darker granodiorite host; dikes were probably injected as residual melts along fractures in the cooling host body.

plete crystallization of magma in an early stage with no second event transpiring, a rhyolitic rock or an equigranular granite rock, respectively, could originate from the same magma.

Porphyry bodies of Tertiary age are widespread in western North America; some of them are hydrothermally altered and mineralized with disseminated Mo or Cu sulfides, forming so-called copper or molybdenum porphyries.

Aplite has aplitic fabric in which grains are generally subhedral to anhedral (see Figure 4-2). Typically, aplite occurs as thin late dikes within coarser-grained hypidiomorphic-granular granitic plutons (see Figure 4-13) and is commonly intimately associated with pegmatite, even within the same intrusive body. The leucocratic composition of aplites, corresponding closely to the lowest-temperature melts in granitic magma systems and injection into a kindred but slightly more mafic and calcic granitic host rocks, suggest that they are late residual melts differentiated from the magma that formed the host pluton.

Pegmatite has pegmatitic fabric (see Figure 4-4). Although diorite, gabbro, and even ultramafic pegmatites do occur, syenitic and granitic ones are the most common. They are typically found around the margins of relatively large, deep-seated plutons, commonly extending from the pluton itself into the adjacent country rocks. Pegmatite bodies are podlike or lenticular in shape and are relatively small, on the order of meters or tens of meters. Many pegmatites have internal zonation of fabric and mineralogical composition; such **zoned pegmatites** (see Figure 4-14) commonly have cores of massive quartz. Zones may be transected by late, solid-state replacement bodies (see Figure 4-33).

Most pegmatites are mineralogically simple, consisting primarily of quartz and alkali feldspar with lesser amounts of muscovite, tourmaline, and Fe–Mn garnet. These are classed as **simple pegmatites.**

Other pegmatites have relatively high concentrations of gaseous constituents such as H_2O, P, Cl, F, and S, as well as large-ion lithophile and other elements strongly partitioned into the residual melt formed during slow crystallization of a deeply buried magma body. These residual elements, incompatible in quartz, feldspar, and so on because of ionic size, charge, or electronegativity (see pages 53–55) include Li, Be, Zr, Nb, Mo, Sn, Ta, W, Th, and U. High concentrations of these elements cause the appearance, in addition to quartz and feldspar, of such minerals as topaz, beryl, apatite, amblygonite, lepidolite, spodumene, cassiterite, tantalite, columbite, monazite, zircon, and uraninite in **complex pegmatites.** Pegmatites associated with syenitic and highly alkaline plutons are particularly enriched in unusual minerals containing rare-earth elements—La, Ce, and Y.

Pegmatites can be economically very valuable.

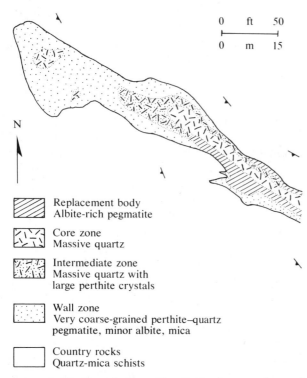

0 ft 50

0 m 15

Replacement body
Albite-rich pegmatite

Core zone
Massive quartz

Intermediate zone
Massive quartz with
large perthite crystals

Wall zone
Very coarse-grained perthite–quartz
pegmatite, minor albite, mica

Country rocks
Quartz-mica schists

Figure 4-14 Zoned pegmatite. Geologic map of a part of the North Star pegmatite, Rio Arriba County, New Mexico, showing simple zonal structure and replacement body. [After E. N. Cameron, R. H. Jahns, A. H. McNair, and L. R. Page, 1949, Internal structure of granitic pegmatites. Economic Geology Monograph 2.]

Simple ones are exploited for large volumes of quartz and feldspar, which are used in the glass and ceramic industries. Complex pegmatites are also mined for this purpose, plus as an important source of the above-named elements.

Pegmatites originate from complex crystal–silicate melt–aqueous fluid systems active in the end-stages of crystallization of magmas (see pages 292–293).

Granophyre is a felsic rock with micrographic texture.

Alteration

Most granitic and dioritic rocks show **alteration**—replacement of the primary high-temperature

magmatic minerals by secondary minerals formed at submagmatic temperatures after the magma completely crystallized. It is similar in many respects to the alteration that occurs in volcanic rocks (see pages 70–71).

The intensity of alteration can vary widely within one body and even on the scale of a single outcrop. Some rocks show no alteration whatsoever; others have biotites partly replaced by chlorite; hornblendes by actinolite, chlorite, epidote, and sphene; alkali feldspars by fine-grained clays and white micas; and plagioclase by epidote and phyllosilicates. Still other rocks will have no vestiges of their primary magmatic minerals. Alteration is commonly more intense along fractures, which are coated with selvages of quartz, epidote, and other moderate-temperature minerals. This localization suggests percolating aqueous fluids were responsible for the alteration.

4.2 FIELD RELATIONS OF INTRUSIVE MAGMATIC ROCKS

Granitic rocks are volumetrically insignificant in oceanic regions of the globe where the crust is thin and basaltic. Dioritic and some granitic rocks occur locally in older island arcs such as the Greater Antilles in the Caribbean, where geophysical data indicate a somewhat thicker, more sialic crust. The largest volume of granitic rocks and local diorite, monzonite, and even syenite bodies occurs along continental margins where oceanic lithosphere has been subducted; for example, calc-alkaline magmatism occurred over several tens of millions of years during the plate convergence that created the Appalachian and Cordilleran orogenic belts in North America (see Figure 4-15). Granitic bodies are also widespread in older, more deeply eroded Precambrian shields, where their tectonic setting of emplacement is uncertain (see Section 15.1). Still other granitic bodies, but with alkaline affinities, occur in extensional intracontinental settings (see Section 6.6).

Folded, faulted, and otherwise deformed sedimentary, igneous, and metamorphic rocks sur-

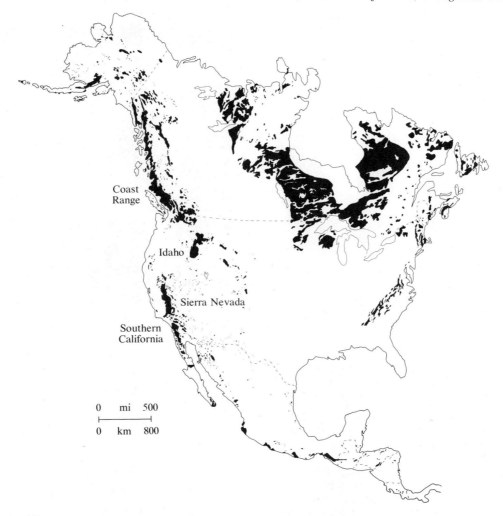

Coast
Range

Idaho

Sierra Nevada

Southern
California

| 0 | mi | 500 |
| 0 | km | 800 |

Figure 4-15 Distribution of exposed granitic rocks in North America. Plutons in the Appalachian orogen along the East Coast are chiefly Paleozoic with some Precambrian; plutons in the Cordilleran orogen along the West Coast are Mesozoic and Tertiary, although some far inland (in Colorado and Wyoming) are Precambrian. Areas of granitic rocks in the Precambrian Canadian shield are generalized and locally include metamorphic rocks (see Chapter 15). Major batholiths in the Cordilleran orogen are labeled. [After the Tectonic Map of North America, U.S. Geological Survey.]

round granitic bodies. In young or still active orogenic belts where topographic relief is substantial, shallow, newly unroofed granitic bodies may show connection to comagmatic extrusive volcanic rocks (see Figure 4-16).

A body of magmatic rock of any composition, size, and shape, emplaced and solidified beneath the surface of the Earth, is an **intrusion,** or **pluton,** though the latter term is generally reserved for large, thick bodies with steep walls.

(a)

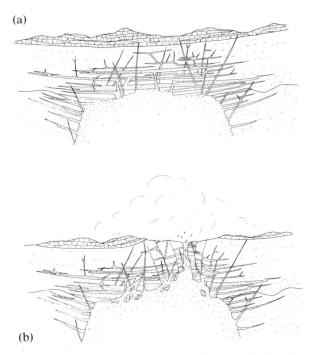

(b)

Figure 4-17 Knife-sharp, discordant, intrusive contact between granodiorite and overlying bedded quartzite.

Figure 4-16 Schematic cross sections illustrating inferred relations between pluton core, overlying dike–sill swarm, and contemporaneous volcanic rocks in the Mio-Pliocene Tatoosh plutonic complex in Mt. Rainier National Park, Washington. In an early wholly plutonic stage (a), a complex of dikes and sills developed in the roof of older volcanic deposits overlying the active magma body. Subsequently (b) the magma rose into this complex and breached explosively through the roof to the surface, showering the landscape with pyroclastic debris. [After R. S. Fiske, C. A. Hopson, and A. C. Waters, 1963, Geology of Mount Rainier National Park, Washington. U.S. Geological Survey Professional Paper 444.]

External Contacts of Intrusions

The existence of an intrusion demands a mass of older, solid or semisolid **country, wall,** or **host rock.** The discontinuity in fabric and composition between the intrusion and its country rock is the **external contact** of the intrusion. Rapid solidification of hot magma against colder country rock in shallow crustal settings may produce a knife-sharp

contact (see Figure 4-17); the thermal gradient was very steep near such a contact at the time of emplacement. Other intrusion boundaries may be blurred and so gradational over several meters or tens or even hundreds of meters that the term *contact* is hardly justified and the term **border zone** is more appropriate. Border zones develop where there was more protracted interaction—chemically, physically, and thermally—between the magma and its wall rock (see Figure 4-18). **Intrusion breccia** or **contact breccia** (see Figure 4-18a) consists of wall rock physically intruded and isolated into free-floating inclusions by innumerable dike offshoots from a main magma body that was emplaced in generally shallow, brittle country rocks. Other border zones are composed of **migmatite,** an intimate mixture of fairly mafic metamorphic rock and granitic rock (see pages 393 and 425) (Compton, 1962, pp. 322–325). Migmatitic border zones are difficult to interpret as to origin because they can form in different ways. Some must develop by intimate injection and per-

(a) Injection

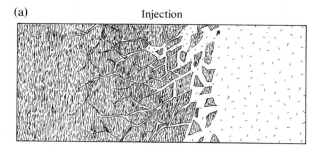

(b) Permeation or
partial melting

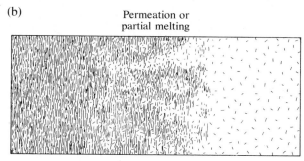

(c) Injection or
partial melting

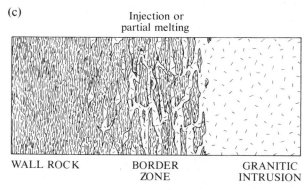

WALL ROCK   BORDER   GRANITIC
               ZONE   INTRUSION

Figure 4-18 Idealized types of border zones between
an intrusion and its wall rock. The width of the zone can
range from a meter to tens or even hundreds of meters.
Any of these border zones can be designated as
migmatite if granitic rock is intimately mingled with
more mafic metamorphic rock.

meation of material from the main intrusion, "inflating" the wall rocks. But under deep crustal conditions already hot wall rocks heated to still higher
temperatures by the adjacent intrusion may partially melt, forming pockets of granitic melt in the
more refractory mafic host.

Contacts and narrow border zones may be

classified according to their geometry as **concordant,** paralleling structures in the country rock
such as bedding or schistosity, or as **discordant,**
cutting across country rock structures. Contacts
may also be classified according to mechanism of
emplacement as **intrusive, tectonic** (or faulted or
sheared), and **passive** (or replacement) (see Figure
4-19).

Interpretation of contact relations can tax even
the experienced geologist. All evidence must be
carefully weighed for each situation. It is not unusual for contact relations to be ambiguous or conflicting in terms of origin and time significance,
especially where magma bodies were intruded
more or less contemporaneously against one another.

Geometric Forms of Intrusions

There are many different forms and sizes of intrusions. A **dike** is a *discordant* tabular or sheetlike
body. It may have any orientation—vertical, horizontal, or inclined. A tabular or sheetlike body
whose intruded rocks have isotropic fabric, such
as most igneous rocks, is also called a dike (see
Figure 4-13).

A **sill** is a *concordant* tabular or sheetlike body of
any orientation. Careful attention to the contacts is
critical in distinguishing a sill from a lava flow,
which may also be a concordant body in a layered
sequence of rocks.

Other intrusive forms—**dike swarm, ring dike,
cone sheet, laccolith, plug, stock,** and **batholith**—
are illustrated in Figures 4-21 to 4-27.

Internal Nature

Careful studies show that very few, if any, plutons
are truly homogeneous in composition. A **zoned
pluton** is one in which changes in fabric and especially composition are gradational within it. The
pegmatite body in Figure 4-14 is zoned. Zoning
could result from wall-rock assimilation, from differentiation processes within the magma body
after emplacement, or from inhomogeneities in the

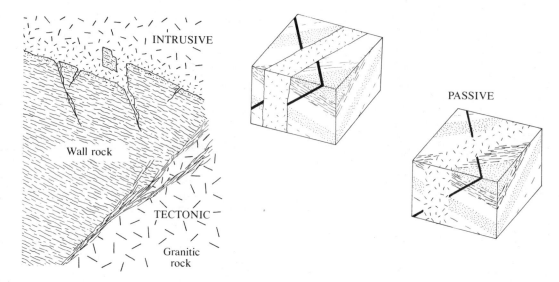

Figure 4-19   Types of contacts between plutons and their wall rocks, based upon the mechanism of emplacement. The scale of the features varies greatly.

(a) **Intrusive contact** (top of left diagram and center block) shows evidence of mobile magma invading more solid host-rock material as follows:
(1)   Magmatic rock is commonly finer-grained near contact, although some truly intrusive contacts against mafic or other types of wall rock have no such fine-grained margin, apparently because this more refractory wall rock was already hot, or perhaps even hotter than the intrusive magma.
(2)   Country rock is recrystallized, or rarely fused at contact.
(3)   Dikes and other fingerlike offshoots of magmatic rock penetrate into wall rocks.
(4)   Pieces of wall rock (inclusions or xenoliths) are embedded in pluton, and contact breccia may be present.

(b) **Tectonic,** or faulted or sheared, **contact** (bottom of left diagram) shows evidence that a solid, or nearly solid, intrusive body was mechanically juxtaposed against its wall rocks, as follows:
(1)   Contact is relatively planar.
(2)   Slickensides, breccia, fault gouge, or plastically deformed rocks are found along contact.
(3)   Grain size in pluton is unchanged toward contact, except for reduction due to deformation.

(c) Bodies with **passive contacts** (right-hand block) show evidence of replacement of preexisting rock without differential movement of wall rocks on either side; all structures regardless of orientation in wall rocks project across contacts without offset. In contrast, an intrusive body dilates the host rocks, forcing them apart so that not all structures project continuously across the contacts.

magma as it left the source region before emplacement. Sometimes the outer margin of the pluton is more mafic and the core more felsic, perhaps because early-forming mafic minerals crystallized preferentially near the cool wall rocks, allowing the felsic components in the magma to concentrate in some manner toward the core. Other plutons show the reverse zonation. A detailed description of zoned plutons is given on pages 136–139.

Most batholiths and larger stocks consist of many plutons and reflect multiple intrusion of

(a)

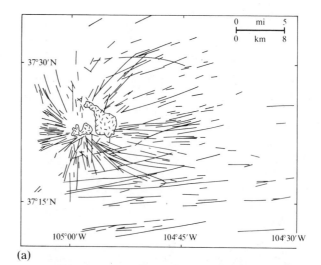

(a)

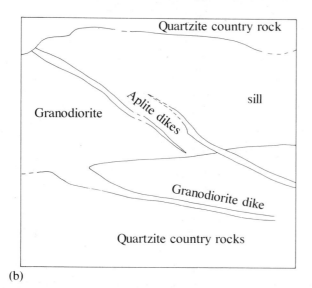

(b)

Figure 4-20   A sill of granodiorite intruded concordantly with bedding in quartzite country rock has a smaller offshoot dike penetrating discordantly across bedding. Late, more leucocratic aplite dikes cut across both country rock and granodiorite. (a) Photograph. (b) Sketch of photograph.

(b)

Figure 4-21   Swarm of subvertical dikes around the Spanish Peaks stocks in southern Colorado. (a) A radial swarm such as seen here can develop around a central intrusion as country rocks are stressed during its emplacement, creating radially oriented tensile fissures into which magma can migrate. Subparallel dike swarms, also evident here, are associated with tensile fissuring imposed by regional tectonic forces. [After R. B. Johnson, 1968, Geology of the igneous rocks of the Spanish Peaks region, Colorado. U.S. Geological Survey Professional Paper 594-G.] (b) Spanish Peaks in distance with resistant dikes standing like ribs above soft subhorizontally bedded sedimentary rocks (partly covered by snow). [Photograph courtesy of John S. Shelton.]

Figure 4-22 Irregular swarm of leucocratic aplite–pegmatite dikes in darker-colored granitic rock. Trees are about 4 m tall.

magma. The Sierra Nevada batholith (see Figures 4-15 and 4-28), for example, is comprised of over 200 (estimated) separate intrusions ranging in age from Triassic to late Cretaceous and in composition from gabbro to granite.

A common feature of granitic and especially granodioritic to dioritic plutons is the presence of enclaves, or **inclusions,** of rock that differ in fabric and/or composition from the pluton itself (see Figure 4-29) (Didier, 1973). Generally they have more calcic feldspar and a higher concentration of mafic minerals, especially biotite and hornblende. They range in size from small aggregates of just a few grains to blocks more than a meter across and even larger **rafts.** Some are equidimensional but others are elongate or discoidal in three dimensions and qualify as **schlieren** (see Figure 4-7). Contacts with their host rock can be sharp or blurred.

Inclusions conceivably originate in different ways. Some may be recognized simply as blocks of wall rock that have been mechanically incorporated into the magma body, a process called **stoping.** Foreign rock inclusions are called **xenoliths.** Some mafic inclusions are aggregates of early-formed crystals precipitated from the magma itself; after segregation along the margin of the pluton, these **autoliths** were then stoped into the intrusion during continued movements of the magma. Still other mafic inclusions might represent clots of refractory residuum left over from partial melting processes in the deep crust where the granitic magma originated, which were brought up with it during ascent. Or, some mafic inclusions might represent clots of solidified mantle-derived magma that had ascended into the granitic source region where, during crystallization and dissipation of thermal energy, the host granitic magma was generated.

Cooler inclusions immersed in a hotter magma will soak up heat. At the very least, there may be some metamorphic recrystallization of the inclusion as it becomes hotter. If it becomes hot enough to partially or even completely melt, the melt fraction will be chemically assimilated into the active magma, changing its chemical composition and producing, ultimately, a hybrid or contaminated igneous rock. However, the consumption of energy by the heating and melting of inclusions cools the host magma, accelerating its crystallization and curtailing further modifications of the inclusions or their assimilation. As wall-rock xenoliths tend to be more concentrated around the margin of the pluton where heat is also flowing into the surrounding country rocks from the intrusion, reconstitution and assimilation processes can be minimal.

If there is sufficient time and availability of thermal energy, as in the interior of a large pluton, any inclusion will tend to equilibrate mineralogically with the P–T–X conditions of the host magma, and hence will tend to possess the same mineralogical composition as the host.

Despite the potentially great significance of inclusions in furnishing insight into the evolution of the magma system, assimilation and internal metamorphic reactions tend to erase most if not all of the important clues within them—their primary fabrics and mineralogical compositions. However, inequidimensional inclusions can be valuable markers of the pattern of magmatic flow during

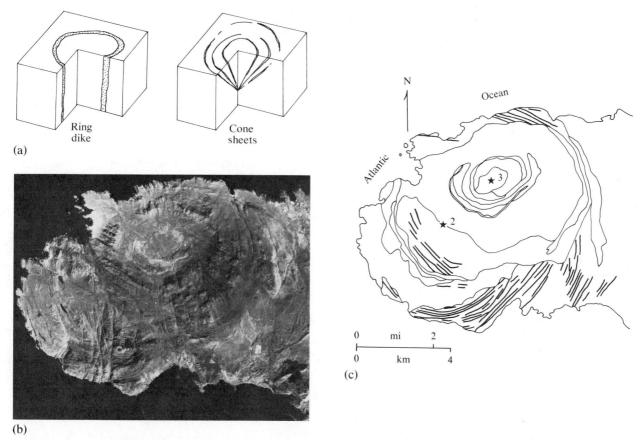

Figure 4-23   A **ring dike** is an arcuate or annular tabular body whose axis is more or less vertical. **Cone sheets** are a family of conical sheetlike bodies converging toward a buried apex. Both are commonly associated together in a **ring complex** and can be created from upsurge of magma and its migration into resulting tensile fractures in the country rock. See also Figure 4-24. (a) Schematic block models. (b) Aerial photo mosaic of Ardnamurchan Peninsula along the western coast of Scotland. (If predominantly north-south stream valleys appear as ridges, turn photo upside down.) [Courtesy of the Institute of Geological Sciences, London.] (c) Generalized geologic map at the same scale showing contacts of ring dikes, chiefly about center 3 (star), and inward dipping, thin cone sheets (heavy lines) about center 2. Faults are omitted. Rocks are Tertiary and mainly of gabbroic composition. [After J. E. Richey, 1934, *Guide to the Geologic Model of Ardnamurchan*, Scotland Geological Survey.]

emplacement, denoting either linear, planar, or linear–planar flow regimes (see Figure 4-6).

Time and Depth of Emplacement

Many plutons are **posttectonic**, their time of emplacement coming after any episodes of deformation affecting the country rocks; deformational features in the country rocks are cut by the pluton. But magmas may also be emplaced during tectonism as **syntectonic** plutons; in such cases, the internal fabric would tend to be anisotropic and would bear the imprint of the deformation, being geometrically concordant with the fabric of the country rocks. Such bodies as well as **pretectonic** plutons are commonly found in relatively deep-

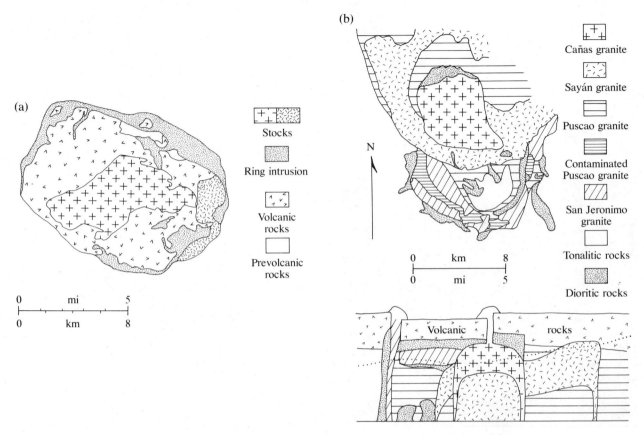

Figure 4-24   Ring complexes. (a) Generalized geologic map of the Liruei ring complex with two central stocks surrounded by a ring intrusion in northern Nigeria. [After R. R. E. Jacobson, W. N. MacLeod, and R. Black, 1958, Ring-complexes in the younger granite province of northern Nigeria. Geological Society of London Memoir 1.] (b) Generalized geologic map and schematic north-south cross section of the composite Huaura ring complex in the Peruvian coastal batholith. Successive emplacement of bodies of granitic magma into tonalitic country rocks about 66 to 61 My ago was followed by presumed pyroclastic eruptions and foundering of caldera blocks, producing a series of ring dikes and cross-cutting, flat-topped, steep-walled stocks. Earlier phases of this activity are obscured by later, as shown by the subsidence of part of the Sayán granite body and emplacement of the Cañas granite. Dotted line in cross section is present topography. [After M. A. Bussell, W. S. Pitcher, and P. A. Wilson, 1976, Ring complexes of the Peruvian Coastal Batholith: A long standing volcanic regime. *Canadian Journal of Earth Sciences* 13.]

seated, high-grade metamorphic terranes; their fabric and mineralogical composition bear some degree of metamorphic imprint. Country rock and pluton structures are more or less continuous (see Figure 4-30).

The depths of emplacement of plutons extend from just below the surface downward through the crust. Shallow plutons such as the Mio-Pliocene Tatoosh pluton in Washington (see Figure 4-16) are exposed by erosion after brief geologic time intervals, whereas deeper ones are exposed only after tens or hundreds of millions of years of erosion.

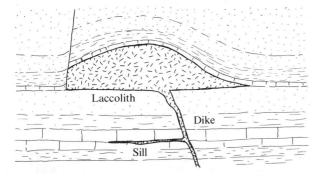

Figure 4-25 A **laccolith** is essentially an inflated sill with a planar floor and an upward arched or partly faulted roof. A feeder dike for a laccolith is postulated. A sill is shown for comparison.

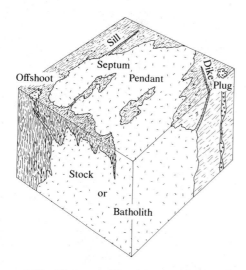

Figure 4-26 Plutons without known floors, that is, of indefinite vertical extent, include plugs, stocks, and batholiths. A **plug** is a steep pipelike body with a relatively small cross-sectional area, less than, say, 100 m in diameter. Many are conduits of volcanic vents. **Batholiths** are the largest-volume intrusions, with surface areas in excess of 100 km²; they are usually composite and consist of many individual plutons. **Stocks** are intermediate in size. Contacts are discordant, concordant, or both. Erosional remnants of country rocks completely surrounded by pluton are **roof pendants,** or if they have peninsulalike connections to the main wall rocks, **septa. Offshoot** is a general, nonspecific term for plugs, dikes, and sills from the main pluton.

Figure 4-27 Split Mountain **roof pendant** in the Sierra Nevada batholith, California. Note somewhat irregular but nearly horizontal intrusive contact between the light-colored granitic rock and overlying darker roof pendant. [Photograph courtesy of John S. Shelton.]

Consideration of rates of erosion, isostatic processes, field relations, hypothesized thickness of the roof rocks over the pluton at the time of emplacement, and mineral geothermometers and geobarometers in the surrounding country rocks permit at least an approximate depth of pluton emplacement to be determined. Depth zonation of emplacement of granitic plutons is summarized in Figure 4-31.

Some continental magmatic arcs have experienced recurring plutonism over a long period of geologic time with emplacement of granitic plutons over a range of depths. An example of this sort of activity can be found in the Colorado Rockies southwest of Denver (see Figure 4-32), where Tertiary epizonal plutons were emplaced into Precambrian mesozonal and catazonal intrusions, and all are now exposed at the surface.

4.3 ORIGIN OF PLUTONS

Decades of debate on the origin of granitic plutons have proved that there are few unequivocal answers but lots of models, some more speculative than others. Even after careful consideration of everything that is known regarding composition, fabric, and field relations, ambiguities in interpre-

tation may still remain. The number of variables tends to be greater than the constraining data.

Basically, the origin of plutons is a fourfold

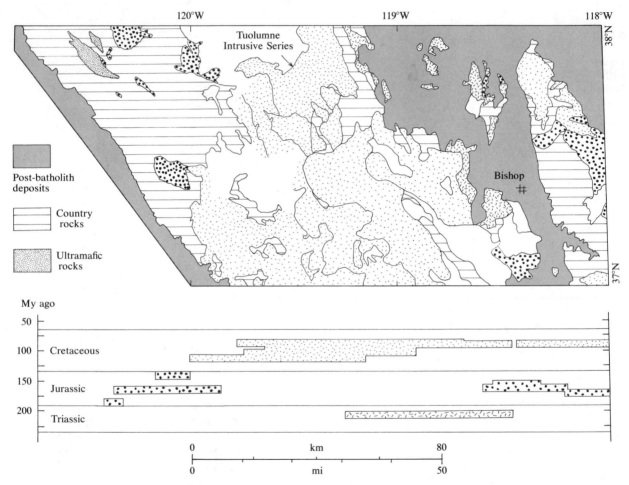

Figure 4-28   Greatly simplified geologic map across the central part of the Sierra Nevada batholith. Contact lines enclose 18 areas that are sequences of several discrete but probably comagmatic plutons interpreted to have evolved from a common parent magma. Age of these sequences and their east-west distribution are shown in graph below map. Mapping or age designation is incomplete in blank areas. To appreciate fully the relationships in a complicated map of this sort, it is recommended that the different units be colored by pencil. [After T. W. Stern, P. C. Bateman, B. A. Morgan, M. F. Newell, and D. L. Peck, 1981, Isotopic U-Pb ages from granitoids of the central Sierra Nevada, California, U.S. Geological Survey Professional Paper 1185.]

problem. First, are they magmatic? Second, how was room made for their volume? Third, what is the specific source of the plutonic rock? If the answer to the first question is affirmative, then the third question asks about the source of the magma. The fourth question concerns the origin of zoned plutons.

Magmas Versus Granitization

Though now mostly of historical interest, the debate (see, for example, Read, 1957) that peaked around 1955 regarding the origin of granitic plutons furnishes us with an opportunity to examine some fundamental data and arguments for their magmatic origin. Before then, many petrologists regarded volcanism and plutonism as fundamentally different and unrelated processes. Plutons of granitic rock were viewed as forming in the continental

Figure 4-29 Mafic inclusions in granodiorite. Note thin subparallel leucocratic aplite dikes above author's wife.

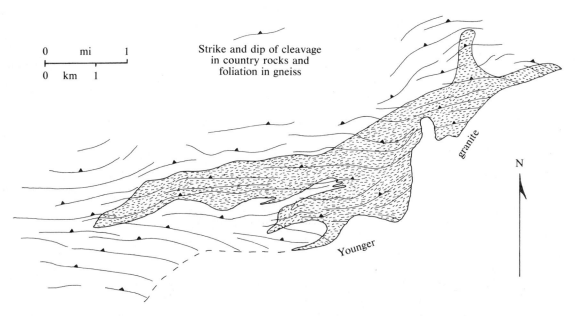

Figure 4-30 Pretectonic St. Jean du Gard granitic pluton north of Paris, France. The original porphyritic rock is now an **augen** (German for "eyes") gneiss in which the metamorphic foliation is continuous with the cleavage in the adjoining slaty country rocks, showing that both country rocks and granitic rocks were deformed and metamorphosed as a single rock mass. [After D. DeWaard, 1950, *Koninklijke Nederlandse Akademic van Wetenschappen Proc.*, 53, nos. 4 and 5.]

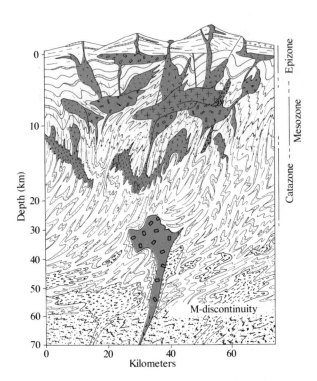

Figure 4-31   Highly idealized and speculative cross section through the continental crust showing relations of plutons (shaded) and their country rocks. Sizes of plutons and structures in country rocks are exaggerated relative to the scale shown along margins of diagram. M-discontinuity is the base of the crust. The depth zones of emplacement described here are from Buddington (1959). **Epizone:** (1) usually posttectonic; (2) sedimentary and volcanic country rocks weakly contact-metamorphosed and locally intensely mineralized and hydrothermally altered; (3) intrusive contacts sharp and dominantly discordant, brecciated and sheared contact common; (4) plutons relatively small with isotropic internal fabric; (5) multiple intrusion common; (6) typical environment of porphyries; (7) close genetic association with volcanic rocks; (8) typical environment of ring dikes, cone sheets, laccoliths, plugs. See Figures 4-16 and 4-24 for examples of epizonal plutons. **Mesozone:** (1) commonly post- and syntectonic; (2) low-grade metamorphic country rocks metamorphosed again and deformed by emplacement of pluton; (3) external contacts sharp to gradational, discordant to concordant; (4) plutons commonly stocks and batholiths built by multiple intrusions; (5) anisotropic internal fabric common but not universal. **Catazone:** (1) commonly pre- to syntectonic; (2) medium- to high-grade metamorphic country rocks; (3) sharp contacts absent, extensive concordant migmatic border zones typical; (4) plutons may consist of gneissic pods and lenses separated by screens of metamorphic rocks in which fabric is concordant with that in granitic gneiss; (5) exact mechanism of emplacement uncertain but involves syntectonic flow; (6) metamorphic effects widespread.

sialic crust, whereas volcanoes formed from basaltic magmas rising from deeper sources, with no connection between them. Although some petrologists held the opinion that plutons originated from magma, others believed that transformation of solid rocks into ones of granitic character without passing through a magmatic stage was the sole means of creating large granitic plutons. This process of **granitization** was considered a type of metamorphism, or, more specifically, **metasomatism,** in which significant amounts of material are introduced into a solid body, other material necessarily being removed, changing its composition, but all in the solid state. The postulated mecha-

nisms of granitization involved diffusion of ions through crystals and/or percolation of thin films of aqueous solutions along grain margins through the solid body.

The proponents of granitization used field relations as evidence against involvement of magma in the formation of granitic rocks. Viewing large plutons as steep-walled bodies extending indefinitely downward created a "room problem." How could magmas displace batholithic volumes of country rock during emplacement? What happened to the rock formerly occupying that volume? Some plutons were alleged to possess "ghost stratigraphies," or internal compositional variations con-

106°00′W

105°30′W

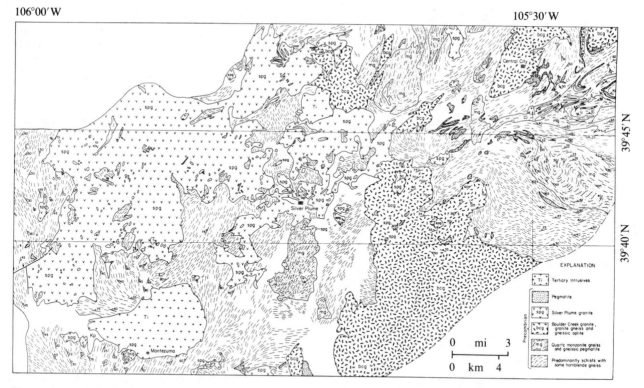

Figure 4-32 Geology of a part of the Front Range, Colorado. Tertiary plutons were emplaced in the epizone, Silver Plume batholith in the mesozone, Boulder Creek batholith in transitional mesozone–catazone, and quartz monzonite gneiss in the catazone. [From A. F. Buddington, 1959, Granite emplacement with special reference to North America. *Geological Society of America Bulletin* 70.]

tinuous with similar variations in the surrounding country rocks. Some small granitic bodies show, on the scale of an outcrop, passive contacts and continuity of anisotropic fabric through them, as if they had been metasomatized in place and were not formed by dilative introduction of the whole body as a magma (see Figures 4-19 and 4-33). Large plutons, usually catazonal and sometimes mesozonal, have a migmatitic border zone, interpreted by granitizers to be a frozen transformation front.

The room problem is obviously a dilemma for the magmatic origin of plutons, but the considerations in the next section ameliorate it. Ghost

stratigraphies and migmatitic border zones may be less of a problem for the magmatist if a magma is not taken to be completely fluid, as defined by extremist advocates of granitization; in deep catazonal levels of the crust, slight partial melting could produce barely mobile crystal-rich magmas that do not migrate far enough to disrupt original source rock heterogeneities and that have gradational zones into unmelted but highly metamorphosed country rock. Small bodies of granitic rock, as shown in Figure 4-33, may well originate by metasomatism.

The contrasting views of magmatists and granitizers in the late 1940s were summarized in

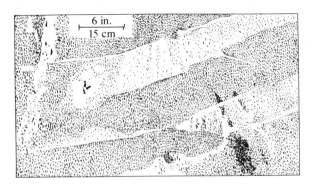

Figure 4-33   Small leucocratic granitic dikes, apparently formed by granitization, in more mafic host. Note continuity of foliation, defined by preferred orientation of mafic grains, from host through dikes and continuity of mafic inclusion from host through dike. Granitizing agent presumably was controlled by a set of fractures, which locally define a sharp contact. [From R. R. Compton, 1962, *Manual of Field Geology* (New York: Wiley).]

Memoir 28 of the Geological Society of America (Gilluly, 1948). Ten years later, in Memoir 74, Tuttle and Bowen (1958) summarized new experimental data pertaining to the chemical and mineralogical composition of granitic rocks. These data, together with careful field studies at about the same time of large granitic plutons in northern Ireland by Read, Pitcher, and colleagues (see Pitcher and Berger, 1972), demonstrated that virtually all plutons are indeed magmatic.

Tuttle and Bowen must be given credit for being the first to place the origin of granitic rocks on a firm experimental base. Subsequent investigations by numerous petrologists using their laboratory techniques have added further support to the magmatic origin of granitic and kindred rocks. We now examine the mineralogical and chemical features, as well as textures, that are compatible with a magmatic origin.

Compositional relations in granitic magmas. Tuttle and Bowen found that the lowest-temperature silicate melts that coexist in equilibrium with quartz or alkali feldspar or both have compositions such that, upon crystallization, they would form subequal proportions of quartz, K-feldspar, and albite (see Figure 4-34). Granitic magmas of this unique composition could conceivably originate in two ways. First, the initial lowest-temperature partial melt formed by heating rocks comprised of quartz and feldspar (plus certain mafic phases as well) will have a composition within the thermal valley. Such a melt, together with possible quartz and feldspar crystals in equilibrium with it, would constitute a mobile, less dense magma capable of being intruded into higher crustal levels. Subsequent crystallization of this magma would yield subequal proportions of quartz, K-feldspar, and albite. Second, magmas comprised of quartz, feldspar (and other minerals), and silicate melt can experience crystal–liquid fractionation (see Section 2.3), separating crystals from melt, and this residual differentiated melt then can crystallize subequal proportions of quartz, K-feldspar, and albite.

The capability of magmatic systems—whether generated by partial melting or by differentiation—of producing rocks identical in composition to those actually observed in granitic plutons (see Figure 4-35 and Table 4-1) constitutes a strong argument for their magmatic origin. Solid-state granitization has not been shown to have this capability.

If plutonic granitic rocks are magmatic, and if the same processes produce plutonic igneous as well as volcanic rocks, then similarities in their composition might be expected. Appendix D does indeed show close similarities between the chemical compositions of average granites and rhyolites. Variation diagrams in Tuttle and Bowen (1958, figs. 41, 42) show the same similarities. Individual plutonic and volcanic associations also show similarities (see Figure 4-36).

Other evidence for magmatic origin of granitic plutons. That bodies of volcanic and plutonic rocks in arc tectonic settings can have similar compositions is not too surprising when direct or implied indirect connections between them are visible in the field (see Figure 4-16). Plutons with sharp contacts and finer-grained borders and offshoots imply penetration of mobile hot magma into cooler country rock. Some only slightly inhomogeneous plutons in the Sierra Nevada batholith (Bateman and others, 1963) are surrounded by diverse wall rocks, suggesting magmatic origin. Strongly zoned plutons have compositional patterns unrelated to diverse wall rocks. Anisotropic internal fabrics in some plutons display patterns related more closely to the external contacts of the pluton than with country rock structures.

Hypidiomorphic-granular fabrics of granitic rocks (see Figure 4-1) are entirely compatible with sequential crystallization of a magma; experimental studies involving crystal–melt systems (see pages 263–264) confirm this sequential precipitation. In contrast, solid-state crystallization in granitization would yield crystalloblastic fabric (see Figure 1-14). Textural distinctions between the feldspars of hypersolvus and subsolvus granitic rocks (see Figure 4-12) correlate with other properties of such bodies that indicate differences in the "wetness" of the magmatic system were responsible; this is additional evidence that such rocks are not the result of solid state granitization.

Proof of the existence of batholithic volumes of granitic magma in the upper crust lies in the occurrence of vast glass–crystal pyroclastic deposits and associated caldera complexes (see pages 91–93). These features indicate the existence of shallow magma chambers thousands of cubic kilometers in volume. Epizonal ring complexes such as those shown in Figure 4-24 provide a genetic link between ash-flow tuff caldera complexes and more deeply eroded granitic plutons.

Altogether, the facts of composition, fabric, and field relations argue strongly for the magmatic origin of granitic plutons. Although no one has ever observed their formation first-hand, we can be confident that if granitization is involved, it must only occur on a small local scale and possibly secondary to larger magmatic systems that are an energy source for solid state metamorphism.

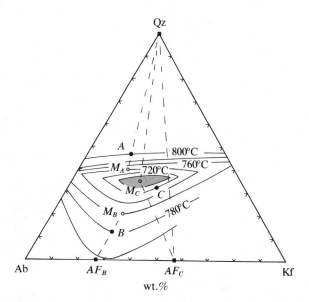

Figure 4-34 Equilibrium crystal–melt compositional relations in mixtures of quartz, K-feldspar, and albite at 0.2 GPa pressure with excess water. Isothermal lines of constant temperature (800°C, 760°C, and so on) in this diagram are analogous to contour lines of equal elevation on a topographic map, except here they show the liquidus surface above which a particular composition is wholly liquid. Mixture A at 740°C consists mostly of silicate melt M_A and of quartz; at temperatures > 800°C, mixture A is completely melted. Mixture B at 760°C consists of alkali feldspar solid solution AF_B, and silicate melt M_B; mixture B is completely melted at and above 780°C. Mixture C at just below 700°C consists of melt M_C plus alkali feldspar AF_C and quartz; this mixture is completely melted at and above 720°C. Any mixture within the 700°C contour—a **thermal valley** (shaded)—is completely melted at or above that temperature. All real granitic magma systems contain Ca, Fe, Mg, and so on, but provided these constituents are only minor, the relations in this diagram are not significantly modified. [After O. F. Tuttle and N. L. Bowen, 1958, Origin of granite in the light of experimental studies in the system $NaAlSi_3O_8$-$KAlSi_3O_8$-SiO_2-H_2O. Geological Society of America Memoir 74.]

Room Problem and Form of Large Plutons

Advocates of granitization had a valid question: How is room made for large intrusions of magma? They chose to reject the magmatic origin of large

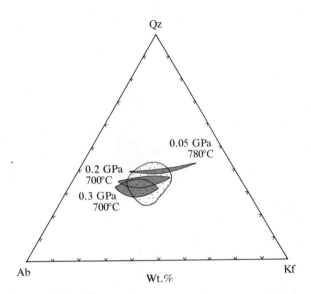

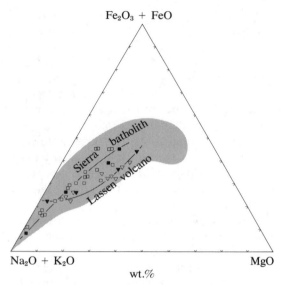

Figure 4-35   There is a strong one-to-one correlation between the compositions of most granitic rocks, expressed in terms of normative felsic minerals within stippled area (see Figure 4-10), and the compositions of thermal valleys (shaded) in experimental quartz–K-feldspar–albite–H$_2$O systems at 0.05, 0.2, and 0.3 GPa pressure, corresponding to depths in the crust of about 2, 7, and 10 km (see Figure 1-15). Temperatures of contour lines denoting thermal valleys at these pressures are indicated. [After O. F. Tuttle and N. L. Bowen, 1958, Origin of granite in the light of experimental studies in the system NaAlSi$_3$O$_8$-KAlSi$_3$O$_8$-SiO$_2$-H$_2$O. Geological Society of America Memoir 74.]

Figure 4-36   The chemical compositions of calc-alkaline plutonic and volcanic rocks are similar, lying within the same region (shaded) of this AFM diagram. Individual plutonic and volcanic suites such as from the Sierra Nevada batholith and the Lassen volcanic complex are also similar. Squares are compositions of granitic rocks from the central Sierra Nevada batholith (Bateman and others, 1963), and filled squares are analyses listed in Table 4-1. Triangles are compositions of volcanic rocks from Mt. Lassen, and filled triangles are analyses listed in Table 3-2.

plutons because the question could not be answered.

The room problem is still not fully resolved, but in view of the compelling evidence for magmatic origins of plutons, we must admit it exists and try to deal with it (see also Section 9.4).

Some intrusions gain space rather passively by mechanically stoping blocks of wall rock. But this is not new space, as the magma and wall rock are merely exchanging places. Other intrusions are emplaced forcibly by upward arching of overlying roof rocks, or by laterally shouldering aside wall rocks that consequently squeeze upward. In the

Sierra Nevada batholith (Bateman and others, 1963), country rock structures are steepened, bowed outward, and locally offset as much as 3 miles around individual plutons, suggesting forcible shouldering aside during magma emplacement. Strain fabrics (see Figure 1-13) are not uncommon around contacts, indicating some energy was expended in mechanical work of crushing, shearing, and solid-state flow.

The traditional view that batholiths and large stocks have no floor and extend indefinitely down into the crust as essentially vertical-walled masses may be faulty (see Figure 4-37). The pattern of

Table 4-1   Chemical compositions of representative granitic rocks from the central part of the Sierra Nevada batholith (in wt.%)

	Quartz diorite	Round Valley Peak granodiorite	Tungsten Hills quartz monzonite	Quartz monzonite of Cathedral Peak type
SiO_2	59.0	63.53	69.60	73.0
TiO_2	0.82	0.63	0.42	0.09
Al_2O_3	17.2	15.61	14.89	15.2
Fe_2O_3	2.0	2.35	1.07	0.04
FeO	4.6	3.25	1.99	0.34
MnO	0.14	0.12	0.07	0.06
MgO	2.8	2.54	0.91	0.16
CaO	6.2	4.58	2.70	1.1
Na_2O	3.3	3.31	3.18	3.8
K_2O	2.1	2.98	4.45	5.0
P_2O_5	0.34	0.23	0.12	0.05
H_2O^+	} 1.0	0.61	0.31	} 0.48
H_2O^-		0.04	0.08	
Total	99.95	99.78	99.79	99.90

Source: After P. C. Bateman, L. D. Clark, N. K. Huber, J. G. Moore, and C. D. Rinehart, 1963, *The Sierra Nevada Batholith: A Synthesis of Recent Work Across the Central Part.* U.S. Geological Survey Professional Paper 414-D.

internal movement of magma during emplacement, as shown by flow markers, is constrained in part by the walls of the intrusion; hence, the internal anisotropic fabric of plutons should provide insight into their shape. The results of a particularly instructive field study of flow fabric in the Rattlesnake Mountain pluton in the southern California batholith by MacCall are shown in Figure 4-38; flow markers reveal that the pluton consists of horizontally extending lobes fanning away from a central funnel-shaped mass. In the deep fiords of east Greenland magnificent three-dimensional exposures, uncluttered by vegetation and soil cover, show horizontally lobate granitic bodies (Haller, 1956) that are similar to the inferred form of the Rattlesnake Mountain pluton. Following this same train of thought, Hamilton and Meyers (1967) have postulated that batholiths are actually laccolithic in form. The relative paucity of granitic plutons in the Paleozoic Appalachian region, as contrasted with their great abundance in the Mesozoic Cordillera of the western Americas, is considered by them to be a reflection of this laccolithic form. The present erosional surface may lie within the most laterally extensive lobes of the plutons in the Cordillera at depths of 5 to 8 km (see Figure 4-31), whereas in the Appalachian belt erosion has cut deeper than, say, 10 km. Hamilton and Meyers further propose that the Boulder batholith in Montana was emplaced as a sheetlike mass beneath a cover of volcanic extrusions of its own making.

Model experiments provide useful insight into the form of granitic plutons at depths (Ramberg, 1970). The physical properties, dimensions, and appropriate time scales of plutonism can be simulated by bodies of soft putty and viscous oils spun in a laboratory centrifuge. The models reveal that buoyant fingerlike protuberances, or **diapirs**, of

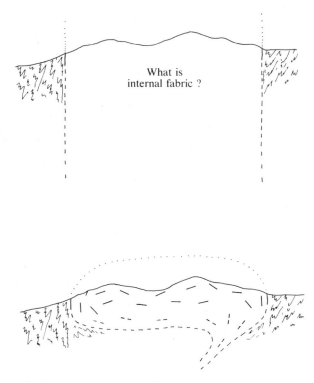

Figure 4-37 Two interpretive cross sections of a pluton whose exposed contacts are vertical. Which one is correct? It is not necessarily true that exposed contacts project with the same orientation indefinitely beneath the surface.

granitic magma can rise into overlying denser material. Continued ascent may cause the diapirs to separate from their source, forming inverted teardrop-shaped blobs; near the surface, as gravitational equilibrium is attained and density contrasts vanish, the diapir spreads horizontally, forming a laccolithic shape with a downward-extending tail (see Figure 4-39).

If granitic magmas are generated in the deep crust by partial melting, as considered below, and then rise as buoyant diapirs, the room problem disappears because there is simply an exchange in position of matter. Satisfactory as this may appear on the large scale, the room problem is not so easily reconciled for the geologist studying a particular pluton in the field, where forcible em-

placement commonly seems to account for only a portion of the room; the remainder is equivocal (Bateman and others, 1963, pp. D42–44).

Zoned Plutons

More or less concentrically zoned plutons are common in magmatic arcs. The Rattlesnake Mountain pluton illustrated schematically in Figure 4-38 is a composite zoned body with significant compositional variations occurring in both the quartz monzonite and quartz diorite intrusions (see Figure 4-40).

The Tuolumne Intrusive Series (Bateman and Chappell, 1979) in the central Sierra Nevada batholith (see Figure 4-28) appears to be a comagmatic group of rocks ranging from quartz diorite on the outer margin of the body through widespread younger granodiorite to youngest granite porphyry in the core (see Figure 4-41). Compositional gradients occur within both the granodiorite and quartz diorite. The higher concentrations of hornblende and biotite, as well as the more calcic plagioclases, occurring in the periphery, contrasted with more abundant K-rich feldspar and quartz in the core, indicate that the pluton crystallized inward with decreasing temperature. During solidification, episodic surges of fresh magma moved into the core of the pluton, inflating the semisolid envelope. Surges broke through this carapace locally, producing some sharp contacts within the pluton and in places invading the country rocks around the pluton. The earliest magma had sufficient water to stabilize hornblende and biotite, but with continued crystallization of anhydrous feldspars and quartz, the water concentration increased in the remaining melt. Eventually, a late residual water-rich magma broke through the roof rocks to the surface, causing a loss of volatiles and rapid crystallization of the final magma and producing the small porphyry body in the center of the intrusion.

Some concentrically zoned plutons have been interpreted to be the consequence of assimilation of wall rocks, but the Tuolemne wall rocks have

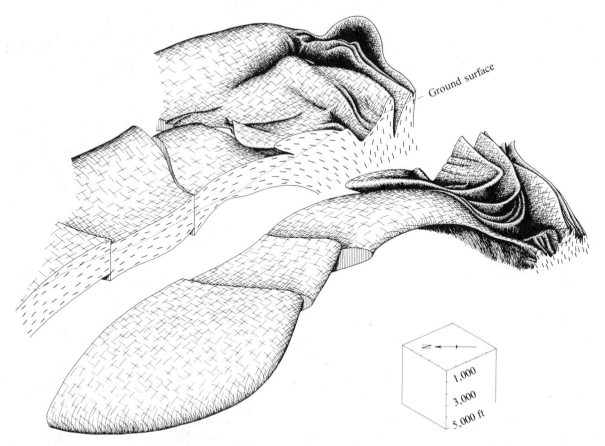

Figure 4-38 Orthographic three-dimensional projection of the Rattlesnake Mountain porphyritic biotite–quartz monzonite pluton shown in map view in Figure 4-40. Screens of hornblende–quartz diorite in the southern part of the complex are represented by wrinkled portion of projection on right. [From R. S. MacColl, 1964, Geochemical and structural studies in batholithic rocks of southern California: Part 1, structural geology of Rattlesnake Mountain pluton. *Geological Society of America Bulletin* 75.]

inappropriate compositions to produce the observed compositional variation in the pluton. Some plutons more simply zoned than the Tuolemne body may have originated from intrusion of a mushy dioritic magma containing abundant calcic, mafic crystalline material—the refractory residuum from a deep crustal source—suspended in a low-temperature granite melt; progressive clearing of the residuum material from the core of the pluton by gravitative settling would leave increas-

ing concentrations of the granite melt fraction. But the Tuolemne pluton grew by multiple surges of fresh magma that inflated, broke through, stoped, and eroded or assimilated the older semisolid envelope material.

It must be admitted that the origin of zoned granitic plutons remains a problem. Models invoking some form of gravitative crystal settling may not be valid and are especially not applicable to zoned magma chambers that erupt voluminous

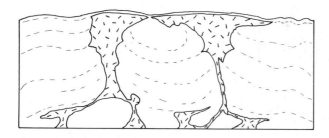

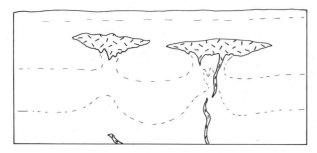

Figure 4-39 Diapiric intrusion of a layer of soft, lower-density putty (patterned) into overlying denser layered media, created in a centrifuged force field. [Drawn from photographs in H. Ramberg, 1970, Model studies in relation to intrusion of plutonic bodies, in G. Newall and N. Rast (eds.), *Mechanism of Igneous Intrusion* (Liverpool: Gallery Press).]

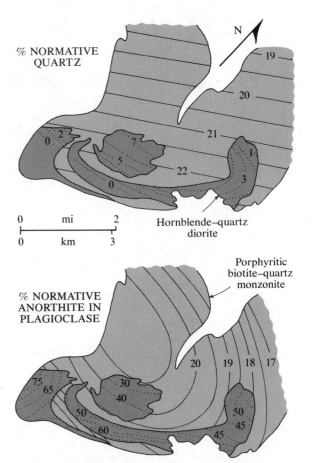

Figure 4-40 Rattlesnake Mountain plutonic complex, part of the southern California batholith (see Figure 4-15). Lines are smoothed, computer-derived compositional parameters based upon 949 shallow drill core samples that were chemically analyzed. Compare Figure 4-38. [After A. K. Baird, D. B. McIntyre, and E. E. Welday, 1967, Geochemical and structural studies in batholithic rocks of southern California: Part II, Sampling of the Rattlesnake Mountain pluton for chemical composition, variability, and trend analysis. *Geological Society of America Bulletin* 78.]

crystal-poor pyroclastic material (see Chapter 3). Other processes must cause such zoning (see pages 323–325).

Origin of Magmas

In view of the apparent genetic link between calc-alkaline plutonism and volcanism, consideration of the origin of magmas for the one regime should also satisfy that for the other. However, plutons provide a somewhat different perspective on the evolution of magmas, and the acquisition of relevant data has followed somewhat different lines. Also, extruded magmas from the tops of shallow magma chambers may not be the same as deeper magmas, now found in deeply eroded old plutons.

For these reasons, and also to make this chapter a self-contained unit, consideration will be given here to genesis of calc-alkaline pluton magmas. Further details are discussed on pages 310–314.

The traditional magmatic viewpoint, before the past two decades or so, of the origin of large gra-

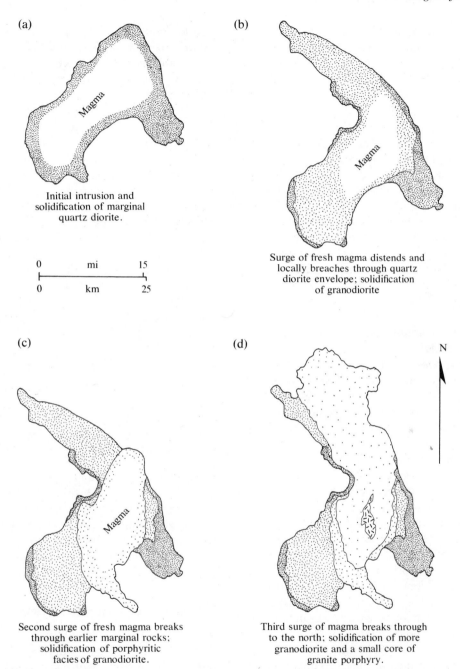

(a)

Magma

Initial intrusion and
solidification of marginal
quartz diorite.

```
0        mi        15
├────────┼────────┤
0        km        25
```

(b)

Magma

Surge of fresh magma distends and
locally breaches through quartz
diorite envelope; solidification
of granodiorite

(c)

Magma

Second surge of fresh magma breaks
through earlier marginal rocks;
solidification of porphyritic
facies of granodiorite.

(d)

N

Third surge of magma breaks through
to the north; solidification of more
granodiorite and a small core of
granite porphyry.

Figure 4-41 Evolution of the Tuolumne Intrusive Series, a zoned pluton in
the Sierra Nevada batholith. [After P. C. Bateman and B. W. Chappell, 1979,
Crystallization, fractionation, and solidification of the Tuolumne Intrusive Series,
Yosemite National Park, California. *Geological Society of America Bulletin* 90.]

nitic batholiths such as the Sierra Nevada involved differentiation of mantle-derived basalt magmas. Separation of early formed olivine, pyroxene, and labradorite plagioclase in the basaltic magma was envisaged to leave a residual granitic melt enriched in elements not highly concentrated in these minerals, such as Si, K, and Na. But even without making a quantitative evaluation of this process, it is obvious that a large volume—at least equal to and probably much greater than that of granitic rock—of complimentary ultramafic differentiates should exist with the granitic rocks. But in the Sierra Nevada region, it is just not present, neither at the surface nor at depth, where these denser minerals ought to have gravitated to the base of the alleged parental basaltic magma chambers. There is no geophysical evidence for a buried mass of dense ultramafic rock within the crust.

Hence, large-scale differentiation of basaltic magmas fails to account for the Sierra Nevada batholith. This is not to say, however, that differentiation cannot operate on a smaller scale within an individual intrusion to produce internal compositional zonation or within a comagmatic sequence of mafic to younger felsic plutons formed from one batch of magma intruded at a high level in the crust.

According to modern concepts of magma genesis in subduction zones, partial melting might occur (1) in the upper mantle peridotite overlying a subducting lithospheric slab, (2) in the basaltic crust in the subducting slab, and (3) in the lower continental crust above the subducting plate. Either or both of the first two mechanisms may well account for the origin of the magmas that formed minor intrusions of dioritic rock occurring in island arcs as well as along continental margins. These dioritic rocks may be viewed as subsurface counterparts of compositionally similar extrusive andesites. But the overwhelming dominance of granitic rocks in continental margin compared to island arc settings would appear to rule out (1) and (2) as viable mechanisms of generation of granitic magmas. Along the Aleutian magmatic arc, for ex-

ample, granitic rocks occur at the eastern end, where there is older sialic crust, but they are virtually absent along the western two-thirds, where the crust overriding the subducting plate is oceanic.

Although older continental sialic crust must be involved in granitic magma production, the exact mechanism is not clear (Brown and Hennessy, 1978). Are mantle-derived mafic magmas, generated by (1) and (2) above, arrested at the base of the continental sialic crust because of their greater density, where they differentiate to produce granitic residual magmas that can then buoy farther upward into the crustal levels now exposed by erosion? (In this model, complimentary ultramafic differentiates become part of the mantle.) Or, do mafic magmas entrapped in the lower continental crust supply sufficient thermal energy to their wall rocks to partially melt them, producing buoyant granitic magmas that rise to shallower levels (Fyfe, 1973)? Or, do both mechanisms possibly operate?

4.4 REGIONAL SPACE–TIME–COMPOSITION PATTERNS IN GRANITIC PLUTONISM

Even though subsequent diversification processes attending ascent and loss of heat will tend to obscure the ancestory of granitic magmas, certain chemical features that provide clues concerning the magma source might survive. These chemical properties are now examined in the context of space and time.

S- and I-Type Granitic Rocks

Aside from the minor gabbroic and dioritic intrusions along continental margins that seem to have been derived from mantle sources, the more felsic granitic plutons of large volume have been found by White and Chappel (1977) to comprise two compositional types presumed to reflect their

sources. These are labeled I-type and S-type granites, derived by partial melting of igneous and
sedimentary materials, respectively.

Most plutons in subduction zone continental
margin settings are **I-type.** They have relatively
high concentrations of Na and Ca, so that
hornblende and commonly sphene are present in
the mode and *Di* in the norm. Common calcic inclusions with abundant hornblende are believed by
some petrologists to represent the residuum of the
partial melting of deep crustal igneous source rock
(likely metamorphosed). Initial $^{87}Sr/^{86}Sr$ ratios are
generally under 0.708 in I-type granitic rocks, suggesting that the igneous source rocks were somewhat poor in Rb, appropriate to earlier derivation
from mantle regions. Mo- and Cu-porphyries are
I-type.

S-type granitic rocks are generally less common.
They occur in some regionally metamorphosed
terranes (see pages 426–427), especially catazonal migmatitic complexes, where they may
have been derived from local partial melting
of the enclosing metasedimentary rocks, and in
lower-grade but thick metamorphic sequences.
Because of their highly aluminous composition,
they characteristically contain no hornblende but
at least biotite and commonly muscovite, garnet,
and cordierite; normative compositions contain
corundum. Quartz is abundant and metasedimentary aluminous to quartzitic inclusions may occur.
Because of the high Rb content of the crustal
source, their initial $^{87}Sr/^{86}Sr$ ratios are >0.710.
Many tin (as cassiterite) deposits are associated
with S-type granites.

In the Mesozoic western United States and
Paleozoic eastern Australia batholith belts, I-type
plutons occur next to the continental margin,
whereas S-type bodies are found farther inland
(see Figure 4-42).

Mesozoic Andean Batholith Belt

Although many details are yet to be examined,
significant patterns have emerged from studies of

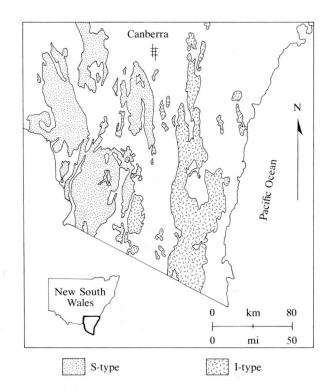

Figure 4-42 Paleozoic plutonic belt in southeastern
New South Wales showing distribution of I- and S-type
granitic plutons. Compare with Miller and Bradfish
(1980) for western United States. [After map distributed
by Australian National University Students' Geological
Society, 1980.]

the batholith belt in the Andes Mountains of western South America. Within the 80-My period of
development of the coastal batholith in Peru
(Pitcher, 1974), there is a clear first-order compositional gradient from early gabbro–diorite intrusions to late granite. Within this sequence, which
must represent some fundamental mechanism of
magma generation at depth, there are second-order
rhythmic compositional variations that may reflect
processes of local diversification near the level of
final emplacement. These local variations are similar to those in comagmatic sequences such as the

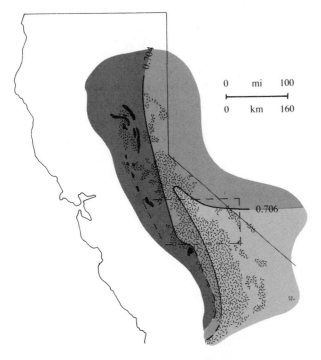

Figure 4-43 Regional variation in the initial $^{87}Sr/^{86}Sr$ in the Sierra Nevada region of California. Batholith is patterned. Mesozoic and Paleozoic(?) alpine ultramafic rocks are black. The area of Figure 4-28 is enclosed by dashed lines. Quartz diorite predominates west of the 0.704 line, quartz diorite and granodiorite between 0.704 and 0.706, and granodiorite and granite in the region where the initial Sr-isotope ratio > 0.706 and where late Precambrian and Paleozoic shelf-type sedimentary deposits occur. [After R. W. Kistler and Z. E. Peterman, 1973, Variations in Sr, Rb, K, Na, and initial $^{87}Sr/^{86}Sr$ in Mesozoic granitic rocks and intruded wall rocks in central California. *Geological Society of America Bulletin* 84.]

Tuolumne in the Sierra Nevada (see Figures 4-28 and 4-41).

In northern Chile (McNutt and others, 1975), plutons generally are younger eastward. This migration of plutonism is believed to correlate in some way with subduction of the oceanic lithosphere beneath the coast. Magma generation began nearest the coast in shallow regions of the descending plate, producing magmas with fairly uniform initial Sr ratios. Subsequent magmas produced farther inland have increasing ratios that are believed to reflect increasing involvement of mantle-derived melts with a more radiogenic component, such as subducted marine sediment or lower crustal sial.

Mesozoic Batholith Belt of the Western United States

Rather similar patterns in composition to those in Chile occur in the western United States. In the Sierra Nevada region, where considerable data are available, initial $^{87}Sr/^{86}Sr$ ratios tend to be lower to the west near the continental margin. Because variations in ratios appear to be independent of age, Kistler and Peterman (1973) suggest that magma composition was mandated by source controls that persisted throughout Mesozoic time. The 0.706 ratio line in Figure 4-43 is the approximate boundary between the region of late Precambrian through Paleozoic shelf sediments (shales, carbonates, and so on) to the southeast and oceanic deposits (mafic volcanics, graywackes, cherts, and so on) to the north and west. Furthermore, the 0.704 ratio line approximately marks the eastern limit of exposures of ultramafic rocks and of diorites and quartz diorites whose trace element compositions resemble oceanic magmatic rocks. Thus, the pattern of initial $^{87}Sr/^{86}Sr$ ratios in the plutons seems to reflect involvement with different source regions of considerable antiquity, ranging from possibly the upper mantle on the west to Precambrian sialic crust on the east.

Throughout the entire length of the North American Cordillera, from Alaska into Mexico, plutonism seems to have been more or less continuous over the past 200 My (Gilluly, 1973). Yet numerous radiometric dates reveal discrete belts of principal chronologic activity in the western United States (see Figure 4-44). In just the central part of the Sierra Nevada, plutonic activity can be viewed as occurring in two eastward migrating se-

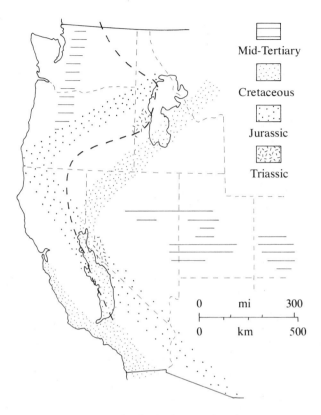

Mid-Tertiary

Cretaceous

Jurassic

Triassic

Figure 4-44 Age of principal plutonic activity in the western United States. Main parts of the Idaho and Sierra Nevada batholiths are outlined. The eastern limit of occurrences of major quartz diorite plutons—the "quartz diorite line"—is marked by a heavy dashed line. [After R. W. Kistler, 1974, Phanerozoic batholiths in western North America. *Annual Review of Earth and Planetary Science* 2.]

quences, the eastern sequence beginning before the western (see Figure 4-28).

These space–time patterns developed during essentially steady subduction of oceanic plate beneath the margin of the continent. The problems of reconciling these patterns, steady subduction, and our simple models of magma production should be obvious. Additional problems concern the cause of episodic plutonism as advocated by Shaw and oth-

ers (1971) in the Sierra Nevada and the occurrence of Cenozoic granitic plutons 1,000 km inland from the coast in western Colorado.

SUMMARY

4.1. Plutonic calc-alkaline rocks are comprised dominantly of phaneritic aggregates of feldspar with lesser amounts of some combination of quartz, biotite, amphibole, pyroxene, and various accessory minerals. Their generally coarse grain size and lack of gas cavities, together with the absence of glass, reflect slow solidification of magma under confining insulating country rock.

Classification uses both fabric and composition as a basis. Most granitic rocks contain subequal proportions of quartz, sodic plagioclase, and K-rich feldspar. They may be classed compositionally on the basis of chemical, modal, or normative composition or into just two simple classes—hypersolvus and subsolvus—depending upon the nature of the feldspars. Intrusive felsic rocks with strong porphyritic fabric and fine-grained matrix are porphyries; those with aplitic texture, usually in thin dikes, are aplites. Pegmatites have very coarse, seriate pegmatitic fabric and are either mineralogically simple or complex with rare minerals; most pegmatite bodies are zoned.

Most granitic and dioritic bodies show some degree of alteration of their primary high-T magmatic minerals to lower-T secondary chlorite, clays, epidote, and so on.

4.2. The most voluminous development of granitic to dioritic plutons is along continental margins overriding subducting oceanic lithosphere.

External contacts of a pluton against its country rocks are geometrically concordant or discordant, sharp or gradational; they are produced by intrusive, tectonic, or replacement processes. Intrusive bodies having a variety of shapes, sizes, and geometric relations to country-rock structures include dikes, sills, dike swarms, ring dikes, cone sheets, laccoliths, plugs, stocks, and batholiths.

Most granitic to dioritic plutons are compositionally zoned because of *in situ* differentiation, assimilation, or variations in composition of the magma rising from deeper source regions.

Mafic inclusions are widespread in granodioritic to dioritic plutons. Primary fabrics and minerals tend to be obliterated during chemical and thermal interaction with the larger mass of the host magma. Clues as to their origin—whether by stoping of mafic wall rocks, fragmentation of early crystallized mafic rinds around the magma body, incorporation of refractory residua from the source region, or incorporation of mantle-derived basaltic material that underplated the sialic source region—are thus obscured.

Emplacement of plutons is pretectonic, syntectonic, or posttectonic relative to the time of any episode of deformation that has affected the country rocks. Plutons are emplaced over a range of depths in the crust and are referred to as epizonal, mesozonal, and catazonal (deepest).

4.3. Questions about the origin of plutons include their mode of emplacement, origin of the magma, and cause of internal compositional variations. The hypothesis of granitization has been discounted except as a possible means of forming small local bodies of granitic rock. The magmatic origin of granitic plutons, on the other hand, is favored by: (1) the coincidence between the composition of most granitic rocks and the lowest-T melts that can form in mixtures of quartz and feldspar, and the striking similarities in chemical composition of volcanic and plutonic felsic rocks; (2) sequential, not crystalloblastic, fabrics; and (3) the existence of batholithic volumes of magma that erupt to form vast pyroclastic deposits.

Room for granitic plutons can conceivably be made by the shouldering aside or doming of country rocks by buoyant magma as it rises in an inverted teardrop shape and then spreads laterally.

Large granitic plutons probably do not originate by differentiation of basaltic magmas at shallow crustal levels because the requisite volume of complimentary ultramafic differentiates are lacking. Minor dioritic bodies, compositionally equivalent to widespread andesites in volcanic arcs, probably originate from mantle-derived magmas. Voluminous granitic magmas generated along continental margins overlying subducting lithosphere may owe their origin to interaction between ascending mantle-derived basalt–andesite magmas and lower crustal sialic material.

4.4. The ancestry and evolution of calc-alkaline magmas can be deduced from their space–time–composition patterns. Relatively calcic I-type magmas are apparently derived by partial melting of igneous source rocks, whereas less common aluminous S-type magmas have sedimentary sources.

Time–space patterns of Mesozoic batholithic activity in the western United States are difficult to reconcile with more or less continuous contemporaneous subduction along the coast. This observation, together with the occurrence of calc-alkaline granitic intrusions far inland from the continental margin, is a warning that overly simplistic models of magma genesis can be misleading.

REFERENCES

Bateman, P. C., Clark, L. D., Huber, N. K., Moore, J. G., and Rinehart, C. D. 1963. The Sierra Nevada batholith: A synthesis of recent work across the central part. U.S. Geological Survey Professional Paper 414-D, 46 p.

Bateman, P. C., and Chappell, B. W. 1979. Crystallization, fractionation, and solidification of the Tuolemne Intrusive Series, Yosemite National Park, California. *Geological Society of America Bulletin* 90:465–482.

Bateman, P. C., and Dodge, F. C. W. 1970. Variations of major chemical constituents across the central Sierra Nevada batholith. *Geological Society of America Bulletin* 81:409–420.

Baird, A. K., McIntyre, D. B., and Welday, E. E. 1967. Geochemical and structural studies in batholithic rocks of southern California: Part II, Sampling of the Rattlesnake Mountain pluton for chemical composition, variability, and trend analysis. *Geological Society of America Bulletin* 78:191–222.

Brown, G. C., and Hennessy, J. 1978. The initiation and thermal diversity of granitic magmatism. *Philosophical Transactions of the Royal Society of London,* Series A, 288:631–643.

Buddington, A. F. 1959. Granite emplacement with special reference to North America. *Geological Society of America Bulletin* 70:671–747.

Bussell, M. A., Pitcher, W. S., and Wilson, P. A. 1976. Ring complexes of the Peruvian Coastal Batholith: A long standing volcanic regime. *Canadian Journal of Earth Sciences* 13:1020–1030.

Cameron, E. N., Jahns, R. H., McNair, A. H., and Page, L. R. 1949. Internal structure of granitic pegmatites. Economic Geology Monograph 2, 115 p.

Compton, R. R. 1962. *Manual of Field Geology.* New York: Wiley, 378 p.

Didier, J. 1973. *Granites and Their Enclaves.* New York: Elsevier, 393 p.

Fiske, R. S., Hopson, C. A., and Waters, A. C. 1963. Geology of Mount Rainier National Park, Washington. U.S. Geological Survey Professional Paper 444, 93 p.

Fyfe, W. S. 1973. The generation of batholiths. *Tectonophysics* 17:273–283.

Gilluly, J. (ed.). 1948. Origin of granite. Geological Society of America Memoir 28, 139 p.

Gilluly, J. 1973. Steady plate motion and episodic orogeny and magmatism. *Geological Society of America Bulletin* 84:499–514.

Haller, J. 1956. Probleme der Tiefen tektonik. Bauformem im Migmatit-Stockwerk der ostgrönländischen Kaledoniden. *Geologische Rundschau* 45:159–167.

Hamilton, W., and Meyers, B. 1967. The nature of batholiths. U.S. Geological Survey Professional Paper 554-C, 30 p.

Jacobson, R. R. E., MacLeod, W. N., and Black, R. 1958. Ring-complexes in the younger granite province of northern Nigeria. Geological Society of London Memoir 1, 72 p.

Kistler, R. W. 1974. Phanerozoic batholiths in western North America. *Annual Review of Earth and Planetary Science* 2:403–418.

Kistler, R. W., and Peterman, Z. E. 1973. Variations in Sr, Rb, K, Na, and initial $^{87}Sr/^{86}Sr$ in Mesozoic granitic rocks and intruded wall rocks in central California. *Geological Society of America Bulletin* 84:3489–3512.

McNutt, R. H., Crocket, J. H., Clark, A. H., Caelles, J. C., Farrar, E., Haynes, S. J., and Zentilli, M. 1975. Initial $^{87}Sr/^{86}Sr$ ratios of plutonic and volcanic rocks of the central Andes between latitudes 26° and 29° south. *Earth and Planetary Science Letters* 27:305–313.

Miller, C. F., and Bradfish, L. J. 1980. An inner Cordilleran belt of muscovite-bearing plutons. *Geology* 8:412–416.

Pitcher, W. S. 1974. The Mesozoic and Cenozoic batholiths of Peru. *Pacific Geology* 8:51–62.

Pitcher, W. S., and Berger, A. R. 1972. *The Geology of Donegal; a Study of Granite Emplacement and Unroofing.* New York: Wiley Interscience, 435 p.

Ramberg, H. 1970. Model studies in relation to intrusion of plutonic bodies. In G. Newall and N. Rast (eds.), *Mechanism of Igneous Intrusion.* Liverpool: Gallery Press, pp. 261–286.

Read, H. H. 1957. *The Granite Controversy.* London: Murby, 430 p.

Shaw, H. R., Kistler, R. W., and Evernden, J. F. 1971. Sierra Nevada plutonic cycle: Part II, Tidal energy and a hypothesis for orogenic-epeirogenic periodicities. *Geological Society of America Bulletin* 82:869–896.

Stern, T. W., Bateman, P. C., Morgan, B. A., Newell, M. F., and Peck, D. L. 1981. Isotopic U-Pb ages from granitoids of the central Sierra Nevada, California. U.S. Geological Survey Professional Paper 1185.

Streckeisen, A. L., 1976. To each plutonic rock its proper name. *Earth-Science Reviews* 12:1–33.

Tuttle, O. F., and Bowen, N. L. 1958. Origin of granite in the light of experimental studies in the system $NaAlSi_3O_8-KAlSi_3O_8-SiO_2H_2O$. Geological Society of America Memoir 74, 153 p.

White, A. J. R., and Chappell, B. W. 1977. Ultrametamorphism and granitoid genesis. *Tectonophysics* 43:7–22.

Williams, H., Turner, F. J., and Gilbert, C. M. 1954. *Petrography.* San Francisco: W. H. Freeman and Company, 406 p.

5

Subalkaline Basaltic and Ultramafic Rock Bodies

Igneous rocks composed principally of some combination of plagioclase, pyroxene, and olivine, and formed from basaltic magmas, are the most widespread rocks on Earth and occur frequently on other planetary bodies in the solar system. They constitute the Earth's oceanic crust, cover parts of every continent and parts of some other planets, and are found as some types of meteorites. The wide-spread creation of basaltic magmas for billions of years reflects the existence of thermal perturbations in peridotitic planetary interiors that cause partial melting, segregation, and ascent of the buoyant liquid fractionate.

Subalkaline basaltic magmas are more voluminous than alkaline. In both oceanic and continental regions of the Earth, extensional fissures in the crust allow mantle-derived magmas to be extruded onto the surface. Upon solidification, magma-filled fissures form dikes. Most basaltic extrusions are of fluidal flows rather than of pyroclastic ejecta because of the low viscosity of the magma. This property no doubt also influences the generation of large differentiated sills and larger stratiform intrusions of anorthosite, pyroxenite, and dunite via mechanisms of diversification *in situ* that are yet to be understood. Among other outstanding prob-

lems are the origin of voluminous but chemically distinct batches of continental flood basalts, the origin of continental bimodal basalt–rhyolite associations, and the character of magmatic systems in the oceanic lithosphere beneath spreading plates and the way their products become obducted onto continental margins as ophiolite.

5.1 PETROGRAPHY OF BASALTIC ROCKS

This section considers aphanitic, glassy, and clastic magmatic rocks in which plagioclase and pyroxene are major constituents and olivine and Fe–Ti oxides are commonly present. Excluded are silica-undersaturated alkaline rocks, which will be considered in the following chapter.

Fabric

Basaltic rocks are aphanitic and sometimes porphyritic. However, in terrestrial settings, basalts are rarely very glassy and vitroclastic because of the lower viscosity of basaltic melts compared to the more silicic ones described in Chapter 3. This contrast means that even in large masses of ex-

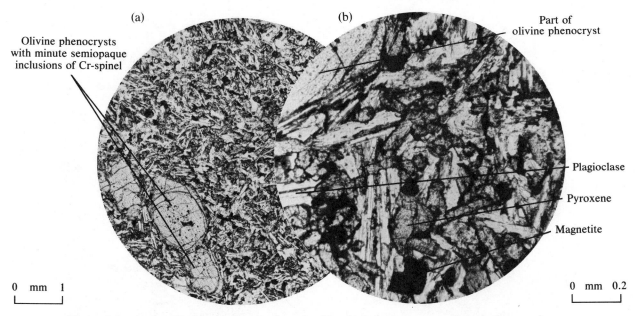

(a)

Olivine phenocrysts
with minute semiopaque
inclusions of Cr-spinel

(b)

Part of
olivine phenocryst

Plagioclase

Pyroxene

Magnetite

0 mm 1

0 mm 0.2

Figure 5-1 An interlocking network of randomly oriented pyroxene and plagioclase grains, with or without olivine and Fe–Ti oxides, is called **intergranular** texture (a, b). It resembles the felty texture of the andesite in Figure 3-1, except that mafic minerals are more abundant. Intergranular texture results from the formation of abundant nuclei of all crystalline phases during substantial undercooling of the basaltic melt, combined with sufficient crystal growth rates to consume all of the melt. No internal flow of the magma occurred during crystallization of this particular sample because the lath-shaped plagioclases show no preferred orientation. In some basalts the angular interstices within an interlocking network of grains are occupied by glass, forming **intersertal** texture. It forms essentially as does intergranular texture, but faster undercooling precludes complete crystallization of all of the melt.

truded and cooled magma, the inherently faster rates of crystallization and nucleation can yield at least semicrystalline rock, rather than massive obsidian. Greater rates of crystallization allow convergence toward a more common crystalline fabric, even when rates of cooling vary. Because of lower viscosity, exsolved gas can escape more readily, so basaltic melts do not explode violently with subsequent generation of pyroclastic flows and vitroclastic ash-flow deposits. The lower viscosity of basaltic melts is directly responsible for highly fluid features such as spatter, Pele's tears and hair, and pahoehoe flows, all of which are unknown in viscous silicic extrusions.

In subaqueous settings, especially on the sea floor, where water can more quickly lower the temperature of basaltic extrusions than air because of its greater heat capacity, glassy rocks are widespread.

In addition to their basic aphanitic fabric, basaltic rocks are either porphyritic or aphyric, and some are also vesicular, vuggy, and trachytic; all of these textures have been previously described and illustrated in Chapter 3. Still others have **intergranular, intersertal, ophitic, scoriaceous,** or **amygdaloidal** fabrics, which are illustrated in Figures 5-1 to 5-3. Note that it is possible for several different fabric elements to occur in a single hand sample; for example, one sample may be aphanitic, porphyritic, vesicular, and intersertal.

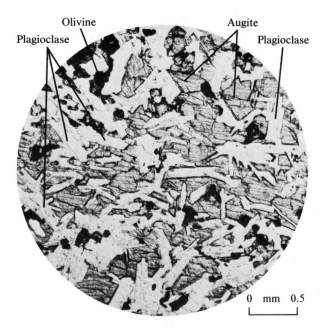

Plagioclase   Olivine   Augite   Plagioclase

0   mm   0.5

Figure 5-2 Several randomly oriented plagioclases are partially to completely surrounded by a single large crystal of pyroxene in **ophitic** texture. Plagioclase began growing from more nucleation centers than did pyroxene, but as crystallization was being completed, both minerals were probably growing simultaneously.

Inclusions occur in some basaltic extrusions. Pieces of crustal rock may be partially melted if they are of appropriate composition (granite, for example) and resided in the hot host magma long enough before it cooled and solidified.

Xenoliths of dunite, peridotite, and pyroxenite having metamorphic fabrics and submagmatic mineralogical compositions generally represent pieces of the upper mantle entrained in their host magma near the region of magma generation. These are discussed further in Section 6.4. **Cognate inclusions,** or **autoliths,** representing clotted accumulations of early precipitating crystals from the enclosing host lava, can be recognized on the basis of interlocking euhedral grains whose compositions are close to those of the isolated phenocrysts in the lava.

Unfilled vesicles

Amygdules

0   cm   2

Figure 5-3 Vesicles are generally more conspicuous in basaltic lava flows than in the silicic to intermediate extrusions described in Chapter 3. Their larger size does not necessarily imply that basaltic magmas contain more dissolved gas that can collect into bubbles as confining pressure is reduced. Rather, their size may be a function of the less viscous nature of the melt, which allows the bubbles to expand more readily before the magma solidifies, and of the higher temperature of the basaltic extrusions, which allows a given amount of gas to occupy a greater volume. Because of the relatively greater ease of flotation of gas bubbles within basaltic magma bodies, the tops of basaltic flows as well as some ejecta are commonly highly vesicular; vesicles as much as a few centimeters in diameter may comprise a large proportion of the rock. The vesicles are thus larger in **scoria** than in rhyolitic pumice and in **scoriaceous** basalt than pumiceous rhyolite. In some lava flows, frozen-in upward-rising gas bubbles form elongate **pipe vesicles.**

Vesicles may become filled with secondary minerals such as quartz, carbonates, and zeolites precipitated from percolating low-T ground water. These fillings are called **amygdules** and the resulting fabric is **amygdaloidal,** as in this photograph.

Upper limit of analyses	Number of analyses	Analyses as %
5	2,391	49.6
10	1,446	30.0
20	919	19.1
25	64	1.3
Total	4,820	100.0

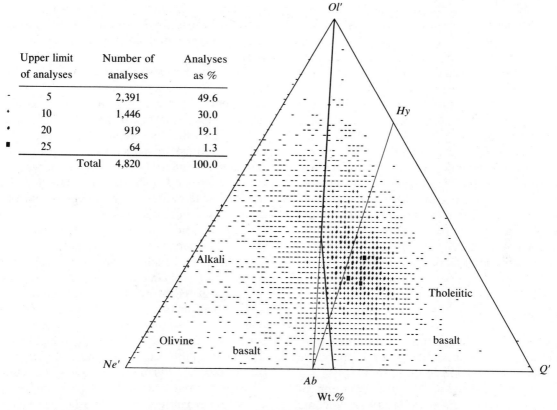

Figure 5-4 A computer plot of over 4,800 basaltic rocks in terms of adjusted major normative minerals, prepared by Roger W. LeMaitre at the University of Melbourne. $Ol' = Ol + [0.714 - (Fe/Fe + Mg)0.067]Hy$; $Ne' = Ne + 0.542Ab$; $Q' = Q + \frac{2}{5}Ab + \frac{1}{4}opx$. The heavy line discriminates between alkali olivine and tholeiitic basalts according to Irvine and Baragar (1971), whose paper should be consulted for details.

Classification

Not all dark-colored aphanitic rocks are basalt. Rigorous distinction between basalt and significantly different rock types must be based on careful analyses of chemical or mineralogical composition (refer to Appendix D, or for quantitative chemical discriminants see Irvine and Baragar, 1971).

Basalts are comprised principally of pyroxene and plagioclase with lesser amounts of some combination of olivine, Fe–Ti oxides, apatite, and glass. Because of extensive solid solution in all of these phases, basalts comprise a spectrum of chemical compositions. A complete representation

of this chemical spectrum would require many variation diagrams or pages of chemical analyses; an alternative first approximation is to show the spectrum in terms of major normative minerals (see Figure 5-4 and page 44). This continuum of composition is split into two basalt types—**alkali olivine basalt** and **tholeiitic basalt**—that belong to the alkaline and subalkaline suites, respectively, as discussed on pages 51–53. Alkali olivine basalts and other rocks of the alkaline suite will be considered in the following chapter.

Without a chemical analysis and a normative composition calculated from it, classification of a basaltic rock into one of these two types can be uncertain. But if orthopyroxene is actually present, the basalt is tholeiitic, and if olivine is abundantly present in the groundmass, the rock probably is an alkali olivine basalt. Additional mineralogical guides can be found in Yoder and Tilley (1962) and Wilkinson (1968).

For basaltic rocks with clastic fabric, the conventional nomenclature for volcaniclastic rocks (see pages 69–70) may be applied; for example, basalt tuff, basalt breccia, and so on.

For rocks more or less transitional in grain size between aphanitic basalt and phaneritic gabbro, usually with an ophitic texture, the name **diabase** is used in the United States. (Among British petrologists the term **dolerite** is used instead because diabase applies to altered basalts and dolerites.) Diabase is common as dikes and sills but does constitute some lava flows.

Alteration

Decomposition of primary high-temperature minerals and glass to more stable lower-temperature minerals is widespread in basaltic rocks.

A common breakdown involves oxidation of iron. Similar effects are seen in other volcanic rocks but are more pronounced in basalts because of their greater concentration of Fe. In basaltic magmas below the Earth's surface, most of the Fe is in the reduced ferrous state, Fe^{2+}, residing either in the melt or in ferrous silicates such as pyroxenes and olivine. As basaltic magmas emerge to the surface and atmosphere, where oxygen is more concentrated, ferrous silicates are unstable and may break down to ferric-ferrous oxide, such as magnetite, and ultimately to ferric oxide, hematite. If sufficient water is available, the ferric hydrate geothite will form. The oxygen for these oxidative reactions comes from two sources. First, the atmosphere has abundant oxygen and undoubtedly promotes oxidation on the exposed surface of the magma body. Second, oxidative effects that occur within extrusive bodies suggest an internal source of oxygen. Dissociation of water originally dissolved in the melt yields oxygen and

hydrogen. The much smaller hydrogen molecules readily diffuse out of the hot body, and the less mobile oxygen molecules participate in oxidative reactions.

Different degrees of oxidation result in the varicolored appearance of basaltic extrusions. Black represents the pristine, still reduced state. Blue-green-yellow iridescent films on olivines and black glass reflect slight oxidation of a thin surface layer and consequent dispersion of white light into its component spectral hues. More pervasive red-brown and red pigmentation indicate deeper, more intense oxidation and development of abundant minute hematite grains.

Deuteric and hydrothermal alteration (see pages 70–71) produce other modifications after emplacement of basaltic bodies. Olivines may be replaced to varying degrees by submicroscopic mixtures of red-brown ferric oxides and clay minerals, sometimes called "iddingsite," by oxides alone, or by aggregates of serpentine, talc, and sometimes carbonate; all of these replacements are usually pseudomorphic, retaining the original form of the olivine. Pyroxenes are less susceptible to breakdown but are replaced by such phases as chlorite, serpentine, clays, and Fe–Ti oxides. Calcic plagioclases are replaced by clays, white micas, and especially hydrated Ca–Al silicates such as epidote minerals, prehnite, pumpellyite, and certain types of zeolites.

Highly unstable basaltic glass, or **sideromelane,** can be replaced by many of the minerals already listed. More massive bodies of glass forming in subaqueous environments, such as rinds on pillows and clasts in tuffs, are replaced by **palagonite,** a yellow to orange-brown poorly crystalline montmorillonite. Palagonite is optically isotropic to weakly birefringent and commonly concentrically banded in colloform fashion. It is significantly depleted in SiO_2, Al_2O_3, CaO, Na_2O, and K_2O but vastly and variably enriched in H_2O (as much as 30 wt.%) compared to the sideromelane from which it formed; most of the Fe in palagonite is in the ferric state. Although many investigators assume the alteration occurs at high temperature immediately following eruption, Hay and Iijima (1968) conclude that most palagonite forms slowly at essentially atmospheric temperatures by the action of percolating ground water, or by the action of sea water in the case of submarine glasses.

In arid climates, vesicular basalts are commonly impregnated with earthy calcium carbonate deposits, or caliche.

5.2 FIELD RELATIONS OF BASALTIC ROCKS

The intent of this section is to summarize fundamental field relations and the geologic processes responsible for them. Further details may be found in Williams and McBirney (1979) and the other references listed in Section 3.2.

Intrusions

Shallow, or **hypabyssal,** intrusions—principally dikes, sills, plugs, and volcanic necks (filled volcanic conduits)—can cool rapidly enough that aphanitic or (rarely) partly glassy textures result. Basalt in such bodies is generally weakly vesicular to nonvesicular; columnar joints (see below) with the columns oriented more or less perpendicular to the external wall are common. Deeper and larger basaltic bodies have increasingly coarser grain size as they grade through diabases into gabbroic rocks and into various mafic differentiates (see Section 5.4). Swarms of dikes and sills occur in the crust underlying plateau basalts (see Section 5.3).

Extrusions

The larger-scale structures and form of basaltic extrusions seen in the field reflect the relatively low viscosity of the magmas and rapid rates of extrusion. This low viscosity is manifest in more fluidal features in flows and in their much subdued, sheetlike profile, which contrasts with the mushroomlike shape of many very silicic extrusions. Low viscosity allows exsolved gas bubbles to escape more readily from the magma body without the tremendous buildup of internal pressures typical of near-surface silicic magmas. Hence, the volume of fluidal basaltic flows greatly exceeds the volume of pyroclastic deposits, which is just the opposite of silicic extrusions.

Lava flows. Coherent, fluidal effusions of basaltic magma from fissure and central vents vary in size, rate of discharge, surface morphology, and internal structure. The volume of one flow, commonly comprised of many gushes of lava extruded in the course of a single eruptive event, generally lies in the range of 0.01 to 1 km^3, although enormous floods of several hundred cubic kilometers are found in the ancient rock record. These volumes are extruded over periods of days to a few years, so that discharge rates range from 1 to 10^5 m^3/s. As will be discussed in Section 7.2, the rate of extrusion is an important factor controlling the length of an individual flow, which is often measured in several tens of kilometers.

Basically, three end-member types of flow can be recognized, with gradations between them. These are aa, pahoehoe, and pillow. The first two derive their name from the Hawaiian native tongue and are pronounced "ah-ah" and "pa-*ho*-e-*ho*-e".

Pahoehoe flows develop from very fluid, low-viscosity magma and consist of thin tongues and lobes, commonly overlapping upon one another, creating a billowy, rolling, or corded surface. A quickly congealed glassy skin insulates the interior and blocks the escape of exsolved gas. Unobstructed downslope movement of the active lava tongue results in a smooth top, whereas restricted flowage causes the skin to wrinkle into corded or ropelike festoons (see Figure 5-5). Momentary slowdowns in rate of flow followed by renewed surge may cause rupture of the distended skin at the toe of the lobe, from which a bud of incandescent lava will emerge and grow into a new tongue. Eruptive temperature and gas content can just about be maintained in the flowing lava beneath the glassy skin, even during flow over several kilometers. Maintenance of such conditions appropriate for low viscosity permits downslope

Figure 5-5 Small pahoehoe tongues a meter or so in width formed during the 1974 eruption of Mauna Ulu, Hawaii. An active lava fountain plays in the left background over the vent; actively moving incandescent flows are somewhat lighter colored. [U.S. Geological Survey photograph courtesy of Robin T. Halcomb.]

drainage of lava from inside the skin if no replenishment occurs, producing an open **lava tube**. Exposure of the interior of a solidified pahoehoe flow, either by faulting or erosion, reveals these tubes. Pahoehoe flow fields commonly have elongate pressure ridges (compare Figure 3-15) caused

Figure 5-6   Aa flow about 20 m thick mantled by jagged scoriaceous slabs.

by buckling of the thickened crust of the flow perpendicular to the local direction of movement.

Some pahoehoe flows grade into aa flows downslope, possibly as the result of increasing viscosity caused by some cooling and loss of dissolved gas.

Aa flows have exceedingly rough surfaces formed by a thick mantle of irregular, clinkerlike scoriaceous fragments (see Figure 5-6) that develop autoclastically by breakup of the solidified crust of the flow during movement. Slabs formed on the steep toe of the active flow fall to the ground, where they form a carpet of rubble over which the flow subsequently advances, like an endless caterpillar tread. **Block flows,** a variant of aa flows, have a mantle of dense, vesicle-poor angular blocks several centimeters across rather than scoriaceous clinker.

Internally, most aa and block flows have well-developed columnar joints formed by shrinkage during cooling of the massive semisolid interior of the flow (see Figure 5-7).

The virtual absence of metamorphic effects on rock material overridden by basaltic flows may seem surprising at first. But most of the heat loss from a lava flow is directed upward into the atmosphere, and the rubble base of the flow serves to insulate the incandescent interior from the substrate, which typically shows only slight discoloration and induration (''baking'').

In addition to ponded lava flows, the sea floor is covered by **pillow lavas.** Pillows are almost always of basaltic composition and generally form wherever hot fluid magma comes into contact with water or water-saturated sediment, even in shallow intrusive situations. Although having the appearance in most planar exposures, such as roadcuts (see Figure 5-8a), of a pile of discrete, independent ellipsoids of pillow or sack shape and size, submarine pillow lavas viewed on the sea floor in three dimensions consist of a tangled mass of elongate, interconnected flow lobes that are circular or elliptical in cross section (see Figure 5-8b). Flattening of the still hot and semisolid pillows produces convex upward tops and bottoms that are cusped and fill in openings between underlying pillows; such forms are useful indicators of stratigraphic ''right-side-up'' direction. Some flattened pillows may resemble pahoehoe toes in cross section, but the lack of open ellipsoidal internal tubes, fewer vesicles, and radial cracks distinguish pillows.

The mechanism of pillow growth was conjectural for many decades but was believed to involve inflation of blobs of magma beneath a stretching viscous glassy skin; the pillow then pinched off and tumbled away so that another could form. However, Moore (1975) actually observed pillows forming off the coast of Hawaii and found that they result from protrusion of elongate lobes, much as toothpaste is squeezed from a tube (see Figure 5-9).

Pillows on the sea floor typically are concentrically zoned in texture because of the drastic

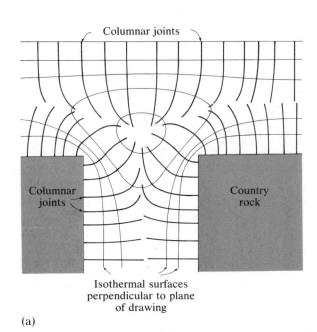

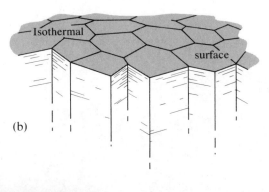

(c)

Figure 5-7   All magmatic rocks are jointed to some degree, because of tectonic processes after consolidation, contraction during cooling from the magmatic state, or both. During thermal contraction of tabular lava flows and intrusive sills and dikes an array of polygonal **columnar joints** may form. As the margin of the body cools and thermally induced stresses exceed the tensile strength, cracks are spontaneously nucleated at many points on this isothermal surface of uniform temperature. Three-pronged cracks approximately 120° apart apparently represent the "least-work" configuration at these nucleation points. As these cracks propagate, they intersect to form irregular polygons of 5 or 6 sides, sometimes 3, 4, or 7. Subsequent secondary cracks may split these polygons. With continued cooling and contraction within the inner parts of the tabular body, the polygonal cracks initiated on the marginal isothermal surface propagate inward to form three-dimensional polygonal joint columns (Ryan and Sammis, 1978).

The geometry of columns relative to isothermal surfaces and shape of the cooling magma body are shown in (a) (compare Figure 7-20), and a perspective view of ideal joint columns is shown in (b). The photograph in (c) shows columnar jointing in a basalt flow about 70 m thick. Above a fairly regular basal "colonnade" of thick vertical columns that grew upward is a haphazard "entablature" of thinner columns that grew in a perturbed thermal regime possibly caused by influx from above of cooling surface water. [Photograph courtesy of W. Kenneth Hamblin.]

(a)

(b)

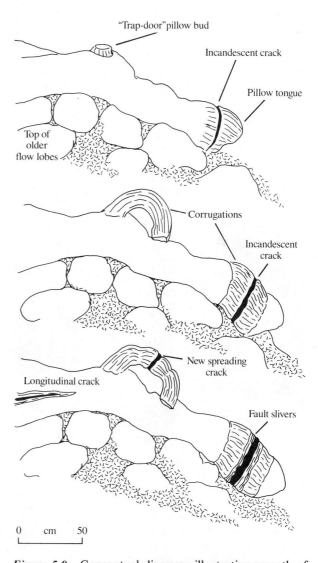

0 cm 50

Figure 5-8 Pillow lavas of basaltic composition. (a) Roadcut at Nicasio Dam in the Mesozoic Franciscan Complex north of San Francisco. [Photograph courtesy of Mary Hill.] (b) Submarine pillows at a water depth of about 2 km along the Puna Ridge east of the Island of Hawaii. [Photograph courtesy of Hank Chezar and Daniel J. Fornari.]

Figure 5-9 Conceptual diagrams illustrating growth of submarine pillow tongues from a larger upslope feeder source. Toothpastelike protrusions of fresh lava squeeze from "trap-door" and girdling cracks in the tongue. Note longitudinal corrugations on protrusions in this drawing and in Figure 5-8b. New tongues may fall off and roll downslope. [After J. G. Moore, 1975, Mechanism of formation of pillow lava, *American Scientist* 63.]

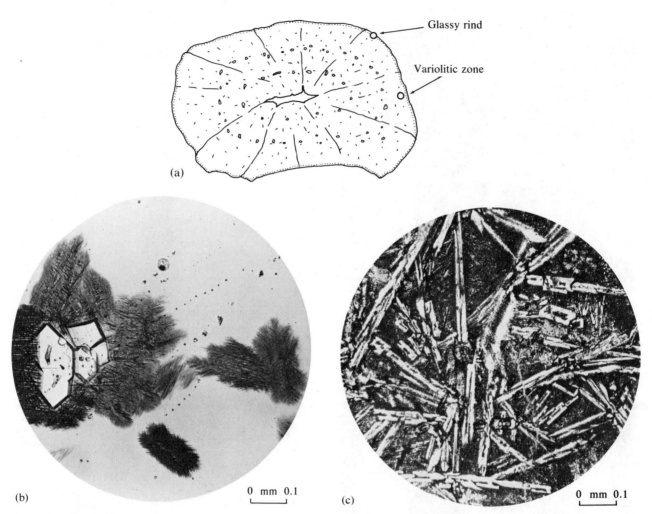

Figure 5-10 Concentric zonal variations in texture of a pillow. (a) Idealized cross section of a pillow with sparse vesicles and radial cracks showing locations of glassy rind and variolitic zone beneath the rind. (b) The glassy outer rind a few millimeters in thickness produced by drastic quenching of the hot magma against cold sea water consists of clear sideromelane with generally < 10% and as little as 0.2% phenocrysts of plagioclase, spinel, and usually, as in this photomicrograph, olivine. Feathery pyroxene aggregates formed during the rapid quenching of the rind have nucleated on the olivines. (c) Inside the glassy rind is the variolitic zone, also a few millimeters thick, of glass, phenocrysts, and abundant feathery skeletal and spherulitic crystals produced by a somewhat extended period of quenching. Intergrown radiating aggregates of feathery pyroxene and plagioclase, known as **varioles,** define the **variolitic fabric** of this zone. The innermost, and most voluminous, zone of fabric development, reflecting the relatively slowest rate of heat loss in the core of a typical pillow, is a partly glassy to holocrystalline intergranular aggregate of plagioclase, olivine, pyroxene, and spinel, but usually exhibiting some quench textures. Photograph in (b) is of a sample collected from the submersible *Alvin* in the FAMOUS area (see Figure 6-1), and (c) is from a Deep Sea Drilling Project core sample. [Both photographs courtesy of A. E. Bence.]

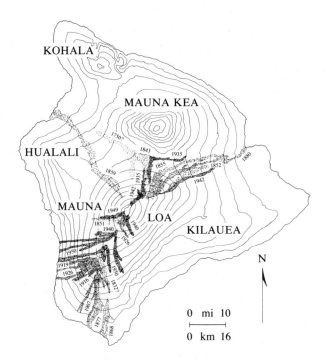

Figure 5-11 The Island of Hawaii is built of five coalescing shield volcanoes, three of which—Hualalai, Mauna Loa, and Kilauea—are still active. Historic lava flows of Mauna Loa (patterned and labeled) were extruded from a north-northeast-trending fissure system. Contours are thousands of feet above sea level. [After H. T. Stearns, 1966, *Geology of the State of Hawaii* (Palo Alto, Calif.: Pacific Books).]

decrease in the rate of cooling from the outside inward (see Figure 5-10). They are also progressively encrusted with Mn–Fe oxides as the glassy material "weathers" to palagonite.

During the drastic quenching of hot magma against cold water accompanying pillow formation, internal stresses in cooling and contracting bodies of glass cause them to spall and shatter into small curved slivers, forming **hyaloclastites,** or **aquagene tuffs** (Carlisle, 1963). These deposits of sideromelane fragments are unlike the strongly cuspate glass shards in the vitroclastic fabric produced by explosive disaggregation of vesiculating magma in subaerial pyroclastic eruptions (see Fig-

ure 3-11). Hyaloclastite material, often altered to palagonite, carbonates, and zeolites, occurs as the matrix between pillows and pillow fragments in **pillow breccias.** Sometimes the matrix is of carbonate mud.

Shield volcanoes. Innumerable extrusions of fluid basaltic flows from a central vent complex, or from a fissure system, can build a subdued but laterally expansive edifice called a **shield volcano,** resembling a warrior's shield. The enormous shields that form the Hawaiian Islands rise as much as 10 km above the sea floor and have diameters of over 100 km; they are the largest volcanoes on Earth (see Figures 5-11 and 5-12).

Volcaniclastic deposits. Fragments of basaltic composition are produced in three ways: (1) boiling of hydrous magma near the Earth's surface; (2) phreatomagmatic eruptions (see page 76); and (3) mild shattering of hot material in contact with water, snow, or ice. As (3) has already been considered in conjunction with hyaloclastites, attention will now be focused on (1) and (2).

Pyroclastic eruptions caused by boiling basaltic magma build local edifices (see Figure 5-12) immediately around the vent; the mostly vertical ejection of material occurs in a much less explosive manner than in the eruptions of more silicic magma described in Chapter 3. Depending upon the force of ejection, size of the ejected pieces of lava, length of their trajectories, eruption temperature, and other factors, the tephra will consist of spatter, ash, cinders, blocks, or bombs.

Spatter is a term for ejected clots of fluid, usually basaltic, magma that, because of their short distance of travel, fall to the ground still hot enough to splash and then to weld together as **welded spatter,** or agglutinate. Where piled around a central vent, the accumulation forms a **spatter cone** (see Figure 5-13) or a spatter rampart along a fissure. Spatter cones may be hundreds of meters in diameter, but most are smaller. If accumulation is rapid enough for the hot spatter to fuse into a large coherent mobile mass, it may become gravi-

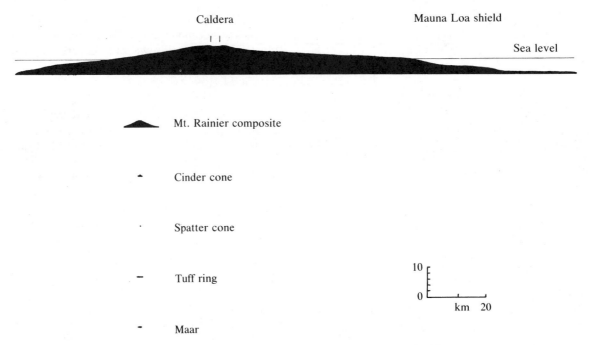

Figure 5-12   Comparative sizes of basaltic volcanoes and a composite andesitic volcano. A composite volcano is an order of magnitude smaller than the large Hawaiian shield volcanoes and an order of magnitude larger than basaltic tephra volcanoes. [After H. Williams and A. R. McBirney, 1979, *Volcanology* (San Francisco: Freeman, Cooper and Co.)]

tationally unstable and move downslope as a rootless lava flow.

Some clots of fluid magma ejected from the vent may solidify during flight and become shaped into streamlined **bombs** (see Figure 5-14a). Smaller droplets of very fluid ejecta congeal into glassy **Pele's tears** or may be drawn out into thin glassy filaments of **Pele's hair** (see Figure 5-14b).

The most common type of explosive basaltic eruption involves a **lava fountain** of ejecta propelled vertically from the vent by expanding exsolved gas (see Figure 5-15). The ejecta consist of fluid spatter as well as cooler and more rigid cinders and ash. **Cinders** are vesicular, dark-colored basaltic to andesitic tephra of lapilli size. Although ash may travel far, especially if windborne, heavier and larger clasts accumulate immediately around

the vent, forming a conical **cinder-and-spatter cone**, or if mostly of cinders, a **cinder cone** (see Figures 5-16 and 5-17). Some degree of cohesion may be produced by slight welding of the ejecta immediately upon deposition, or by subsequent cementation by secondary mineral matter. Obstructions in the vent or prevailing winds cause asymmetric dispersal of ejecta and produce asymmetric cones. Some cones have no summit crater; in older ones this reflects erosion of the rim, but youthful craters may also be filled with lava or with less forcefully ejected material in the closing stage of eruptive activity. Sometimes downwind accumulations from lava fountains form craterless mounds. Some cones may have a mantle of welded spatter, reflecting more moderate terminal fountaining activity as contrasted with the ear-

Figure 5-13 Basaltic spatter discharging during the 1969 eruption of Mauna Ulu, Hawaii, is accumulating in a spatter cone around the small lava fountain. [U.S. National Park Service photograph courtesy of R. E. Fultz.]

(a)

Bombs

0   cm   4

(b)

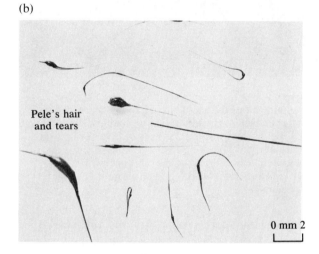

Pele's hair and tears

0 mm 2

Figure 5-14 Examples of fluid basaltic ejecta streamlined during their airborne trajectory from the vent. (a) Volcanic bombs are spindle- or ribbon-shaped tephra, commonly with cracked surfaces resembling the crust on French bread, formed by expansion of gases trapped beneath the crust. (b) Pele's hair and tears formed by drawing out of small, very fluid ejecta.

lier, more explosive emission of solid ejecta. Cinder-and-spatter cones and cinder cones are typically associated with coeval lava flows (see Figure 5-16). Several variations occur: Construction of the cone may represent the termination of activity and cover the vent; an early cone may be broken and partly rafted away by subsequent effusions of lava; lava fountaining may be more or less contemporaneous with extrusion of coherent fluid lava; or lava may spill over the rim of the cone, or issue from openings in the flank, or at the base.

Phreatomagmatic explosions (see pages 76–79) generated by magma encountering water produce relatively small (<1 km^3) accumulations, mostly of ash, which are more subdued topographically but range farther from the vent than pyroclastic basaltic deposits formed by effervescing fountaining magma. These contrasts in the form of the deposits are related to the more or less vertical eruption of boiling magmas, where gas-

driven discharge is guided by the conduit, like shot from a shotgun, and to the shallow depth of focus of phreatomagmatic explosions, which throw ejecta horizontally over a wide area. Volcanic features built by phreatomagmatic eruptions include maars, tuff rings, and tuff cones.

Figure 5-16   S.P. cinder cone and associated basaltic lava flow (covered by less snow) located about 40 km north of Flagstaff, Arizona. Age is ~0.07 My. Note pressure ridges on flow, sharply defined crater at top of cone, and older cinder cones and a tuff ring in the background. [Photograph courtesy of John S. Shelton.]

Figure 5-15   Incandescent 1959 Kilauea Iki lava fountain. Note diffuse cloud of cooler black ash and cinder carried to the left of the fountain by wind currents and, below the fountain, a river of lava feeding the partly crusted lava lake barely visible in the extreme lower left-hand corner of the photograph. [U.S. Geological Survey photograph courtesy of G. A. Macdonald.]

Morphologically, **tuff cones** and **tuff rings** (see Figures 5-18 and 5-19) form a continuum with cinder cones. In cinder cones, the ratio of the height to basal diameter is approximately 1 : 3, and any constructional crater is small relative to the volume of the cone. In tuff rings, at the opposite end of the spectrum, the ratio of edifice height to basal diameter is generally <1 : 5, and the volume of crater space is larger than volume of ejecta. Tuff cones are intermediate. Tuff rings are built chiefly of pyroclastic-surge and some pyroclastic-fall deposits (Waters and Fisher, 1971). Some edifices are of unconsolidated tephra and can be called ash rings or ash cones, but most are fairly well consolidated because of extensive palagonitization of the unstable glassy ash and cementation by mate-

Figure 5-17   Quarry face in cinder cone showing crude stratification defined mostly by higher concentration of blocks.

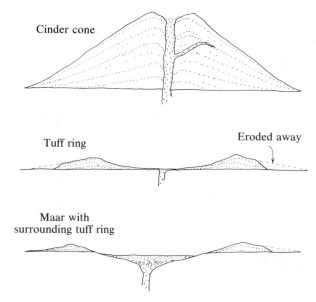

Cinder cone

Tuff ring   Eroded away

Maar with
surrounding tuff ring

Figure 5-18 Idealized cross sections through a cinder cone, tuff ring, and maar with surrounding tuff ring, showing shape and internal stratification of ejecta. All about the same scale. No vertical exaggeration.

Figure 5-19 MacDougal Crater, a maar and surrounding tuff ring in the Pinacate craters area of northwestern Sonora, Mexico. The tuff ring, approximately 1.5 km in diameter, is primarily of juvenile basaltic ejecta with little older material; hence, the maar must have formed by collapse into the partly evacuated shallow magma chamber (see Jahns, 1959). [Photograph courtesy of John S. Shelton.]

rial carried in solution in the associated lake or ground waters.

A **maar** is a broad volcanic crater cut below the preeruption surface into older material, with a shape somewhat between a funnel and a shallow dish (see Figures 5-18 and 5-19). Maars form by the blasting out of older material during a steam-driven explosion, by calderalike collapse into a voided shallow magma chamber, or by a combination of these mechanisms. Ejecta of juvenile or accidental nature carried by surges build a tephra ring around the crater or merely form a broad ejecta blanket. Slump debris and later less-explosive effusions of lava may fill the maar with cinder or spatter deposits, or a lake of lava. In the region around Eifel, Germany, maars and cinder cones are closely associated, even along the same fissure system. But cinder cones tend to form on hills, whereas maars form in the valleys, where there was access to the shallow water table.

5.3 CONTINENTAL BASALTIC ASSOCIATIONS

Basaltic rocks are among the most widespread types of igneous rocks on the continents, and they comprise most of the oceanic crust. In both the continental and oceanic environment, chiefly sub-alkaline basalts are associated with **extensional tectonism**; that is, crustal stretching manifest in normal block faulting and tensional fissuring. Continental subalkaline basaltic rocks occur in basically two different associations: (1) plateau flood basalts and related feeder dike swarms; and (2) cinder-cone lava-flow fields, sometimes with coeval silicic extrusions that produce a bimodal basalt–rhyolite association.

Plateau Basalts

Floods of basalt lava extruded from fissures and central-vent shield systems in regions of continental extension are the most voluminous magmatic effusions on continents, and on the whole Earth

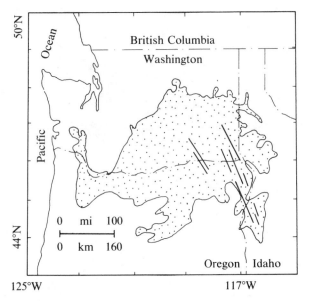

Figure 5-20 Extent of Columbia River Plateau basalts (stippled). Subparallel heavy lines show the approximate location and orientation of known groups of feeder dikes, which number in the thousands. Nearly the entire volume of basalt was extruded in the mid-Miocene over a time span of 2 to 3 My. [After D. A. Swanson, T. L. Wright, and R. T. Helz, 1975, Linear vent systems and estimated rates of magma production and eruption for the Yakima Basalt on the Columbia Plateau, *American Journal of Science* 275.]

Figure 5-21 Columbia River Plateau basalts. Several basalt flows are exposed in the sheer 150-m high east wall of Grand Coulee, Washington. Upper and lower flow margins are marked by erosionally less resistant soil, sediment, or autoclastic rubble. Note variations in columnar jointing between flows. [Photograph courtesy of John S. Shelton.]

are second only to extrusions along oceanic rifts. The lavas are tholeiitic to slightly silica-undersaturated, and only minor sedimentary and volcaniclastic deposits are locally present.

Characteristics. In the Columbia River Plateau (Williams and McBirney, 1979, p. 268), the Roza flow unit, comprised of one or more rapidly emplaced surges of lava, covers at least 40,000 km², averages 30 m thick, and has a volume of more than 1,500 km³, roughly twenty times the entire volume of Mt. Rainier and equivalent to the volume of large epicontinental silicic pyroclastic flow deposits (see Figure 3-35). The entire volume of the Columbia River Plateau basalts (see Figures 5-20 and 5-21), consisting of hundreds of flows and

ponded flows extruded into a subsiding basin, is estimated at about 200,000 km³, over ninety percent of which was erupted between 17 My and 14 My ago, but continuing until about 6 My ago. Though rates of extrusion would appear to be enormous, calculation shows an average rate of about 0.06 km³/y, comparable to discharge rates of Hawaiian shield volcanoes and to oceanic rifts. Withdrawal of this vast volume of magma from subterranean reservoirs was apparently compensated by more or less concurrent sinking of the basalt pile, so that extrusions were emplaced consistently near sea level. Subsequently, the pile was uplifted to form a plateau.

The Proterozoic Keeweenawan basalts of the Lake Superior region had a total volume in excess of 400,000 km³. The Karroo basalts of Jurassic age (Cox, 1970) in southern Africa once covered an area estimated at 2,000,000 km³, and locally the sequence is 9 km thick. The Jurassic-Cretaceous Parana basalts of eastern South America covered

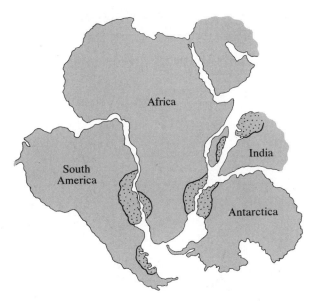

Figure 5-23 Early Mesozoic reconstruction of the southern continents and distribution of Mesozoic basalt plateaus (stippled) possibly associated with their breakup. [After P. Francis, 1976, *Volcanoes* (New York: Penguin Books).]

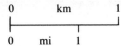

Figure 5-22 Youthful Kings Bowl basaltic fissure eruption in the eastern Snake River Plain of Idaho. Note subparallel fissures on either side of the one that fed the lava flows and the light-colored mantle of ash that covers the east-central part of the flow. Irregular white line crossing photograph from east to west is a road. [Photograph courtesy of the U.S. Department of Agriculture.]

an estimated 1,200,000 km² with a conservatively estimated volume of 350,000 km³ (Herz, 1977).

Vast swarms of more or less vertical basalt and diabase dikes cut older basalt flows and underlying country rocks. These dikes are generally a few meters to tens of meters wide; unusually large ones may be 300 m in width and extend along strike for 100 km. They are interpreted to be the fillings of extensional fissures through which the mantle-derived magmas ascended through the crust and were extruded onto the surface (see Figure 5-22). Basalt and diabase dike swarms of several ages and orientations in the Precambrian Canadian Shield (see Figure 15-10) may have been associated with flood basalts now eroded away.

Many flood basalt extrusions herald the initiation of continental rifting and drift. Pre-Mesozoic reconstruction of the Southern Hemisphere continents shows juxtaposition of the South American, African, Antarctican, and Indian basalt plateaus (see Figure 5-23). Mesozoic-Cenozoic drift of continents in the North Atlantic is marked by remnants of basalt plateaus in the northern British Isles and Greenland. In both instances, basaltic activity started some 20 My before the split of the continent. Remnants of 1.1-Gy-old Keeweenawan plateau basalts around Lake Superior and a buried extension to the southwest in the mid-continent of

Table 5-1 Chemical compositions of selected types of Columbia River basalt.[a]

	Stratigraphic unit			
	Picture Gorge (high-Mg)	Lower Yakima (high-Mg)	Middle Yakima (Frenchman Springs)	Upper Yakima (Ice Harbor)
SiO_2	50.14 wt.%	54.09 wt.%	51.14 wt.%	47.09 wt.%
TiO_2	1.55	1.78	3.00	3.65
Al_2O_3	15.47	14.33	13.71	12.35
FeO[b]	11.20	11.42	14.16	17.65
MnO	0.20	0.20	0.24	0.32
MgO	6.65	4.90	4.19	4.38
CaO	10.62	8.72	8.46	8.84
Na_2O	2.94	2.84	2.86	2.43
K_2O	0.57	1.17	1.33	1.39
P_2O_5	0.22	0.31	0.56	1.58
Total	99.56	99.76	99.65	99.68
Ni	44 ppm		20 ppm	
Ba	440		480	
Y	50		63	
Zr	160		210	

[a] See page 41 for complete analysis of another Yakima basalt, sample BCR.
[b] $FeO = FeO + 0.9 Fe_2O_3$
Source: After T. L. Wright, M. J. Grolier, and D. A. Swanson, 1973, Chemical variation related to stratigraphy of the Columbia River basalt, *Geological Society of America Bulletin* 84; and I. McDougall, 1976, Geochemistry and origin of basalt of the Columbia River Group, Oregon and Washington, *Geological Society of America Bulletin* 87.

North America suggest an aborted continental rift (see Figure 15-15); interestingly, these old basalts are quite similar in composition to Mesozoic and Cenozoic plateau basalts, indicating similar magma-generating conditions and sources for the past billion years. It is debatable whether the Columbia River Plateau represents an aborted rift separation of lithospheric plates, or only localized extensional back-arc spreading behind the western North America subduction zone.

Origin. In oceanic rifts, there can be little doubt that basaltic magmas originate in the mantle, specifically in the already partially molten asthenosphere rising beneath the rift axis. In areas of continental extension where basalt plateaus form, a similar magma-generating system may exist in upwelling mantle. (Crustal sources are compositionally and thermally inadequate to yield the volumes of basaltic magma extruded.) But continental plateau basalts compared to oceanic generally have more evolved compositions, containing higher concentrations of Si, K, Ti, P, and Ba and lower Mg and Ni (compare Tables 5-1 and 5-5). We must decide whether this compositional property results from differentiation as the mantle-derived magmas rise through thicker crust, or from contamination with sialic crustal material, or because of different melting conditions or compositions in the mantle source region. Any one or a combination of these possibililies could conceivably produce the observed compositional contrasts.

Several hundred chemical analyses of flows in the Columbia River Plateau show a significant range of composition (see Table 5-1) (Wright and others, 1973). The Picture Gorge basalts, among the stratigraphically lowest basalt flows in the plateau, are chemically fairly coherent and have the lowest concentration of incompatible (or residual) elements such as K, Ti, Ba, and so on and highest Mg and Ni. The more voluminous and mostly younger Yakima basalts are quite variable, some verging on pyroxene andesite and others having appreciable amounts of Fe, Ti, K, and P, suggesting extensive differentiation. But it has proven impossible to construct satisfactory differentiation models to derive one basalt type from another (Wright and others, 1973), and most petrologists have agreed with Waters (1961) that the compositional types originated independently and also are not derivatives of one hypothetical parent magma.

McDougall (1976) has determined initial Sr-isotope ratios and integrated these data with previous information to constrain the origin of the Columbia River basalts. Picture Gorge flows have uniformly low initial ratios (0.7035 to 0.7039), consistent with an origin by partial melting of upper mantle peridotite; some fractionation, however, likely occurred before these magmas were erupted because of their evolved olivine-poor compositions (compare Baffin Island lavas discussed below). In contrast, the generally younger, more chemically diverse and evolved Yakima flows have more variable and higher initial ratios (0.7043 to 0.7092), which suggests that they were variably contaminated by more radiogenic sialic crustal material. This suggestion is strengthened by the additional observation of an inverse correlation between initial Sr-isotope ratio and Sr concentration, and between initial ratio and age (see Figure 5-24). There may have been progressively greater contamination during plateau building of mantle-derived magma with initial ratios < 0.704 and Sr > 300 ppm by crustal material with initial ratios >

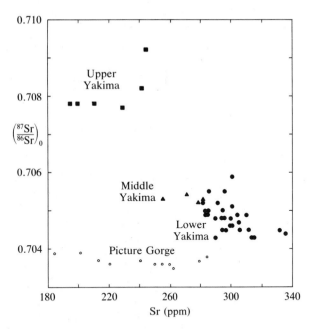

Figure 5-24 Plot of initial Sr-isotope ratios relative to concentration of Sr in Columbia River basalts. [After I. McDougall, 1976, Geochemistry and origin of basalt of the Columbia River Group, Oregon and Washington, *Geological Society of America Bulletin* 87.]

0.709 and Sr < 200 ppm. However, McDougall is bothered by the enormous volumes of extruded magma that would have to have been so contaminated. He suggests as an alternative possibility that the subcontinental mantle in the Columbia River region is inhomogeneous and anomalously radiogenic. But this runs counter to another notion (see Section 15.4) that, during the evolution of the Earth, Rb has been preferentially transferred from the upper mantle into overlying continents, and therefore [87]Sr should be lower in subcontinental mantle than suboceanic. So neither the contamination nor the inhomogeneous mantle-source model is entirely satisfactory as an explanation of the Columbia River Plateau flows.

In a study of Tertiary flood basalts on either side of Baffin Bay on the west coast of Greenland and the east coast of Baffin Island, Clark (1970) also

Table 5-2 Average chemical compositions of three groups of basalts from east coast of Baffin Island.[a]

	Olivine-rich basalts	Olivine basalts	Olivine-poor basalts
SiO_2	44.0 wt.%	45.1 wt.%	47.5 wt.%
TiO_2	0.58	0.76	0.97
Al_2O_3	8.3	10.8	13.8
Cr_2O_3	0.36	0.27	0.13
Fe_2O_3	2.2	2.4	3.3
FeO	8.8	8.1	6.7
MnO	0.2	0.2	0.2
MgO	26.0	19.7	11.8
CaO	7.3	9.2	11.7
Na_2O	0.90	1.04	1.53
K_2O	0.06	0.08	0.10
P_2O_5	0.07	0.09	0.10
H_2O^+	0.97	2.14	2.20
Total	99.7	99.9	100.00
Ni	1,493 ppm	943 ppm	314 ppm
Ba	44	55	56
Y	13	18	21
Zr	45	52	66

[a] High H_2O^+ contents indicates the somewhat altered nature of these basalts.
Source: After D. B. Clarke, 1970, Tertiary basalts of Baffin Bay: Possible primary magma from the mantle, *Contributions to Mineralogy and Petrology* 25.

finds a range of compositions represented. But in contrast to the Columbia River Plateau, the Baffin flows are much poorer in incompatible elements K, Ti, P, Ba, Zr, and Si (compare Tables 5-1 and 5-2) and have much higher concentrations of Mg and Ni, reflecting far greater amounts of olivine in the flows. The three chemical groups of flows in Table 5-2 have correlative stratigraphic and petrographic features. The olivine-rich basalts lie at the base of thick flows or constitute entire flows that are characterized by large olivine phenocrysts; these rocks are believed by Clarke to represent the effects of olivine accumulation in a flow after extrusion or in shallow reservoirs before extrusion. Olivine-poor basalts occur near the tops of thick flows and constitute other particularly feldspathic flows; these are believed by Clarke to represent olivine-depleted magmas.

It is quite apparent that olivine fractionation has played an important role in the evolution of these flows. Variation diagrams clearly show it (see for comparison Figure 9-14). Intermediate composition olivine basalts have a major element composition similar to that predicted from experimental studies to be in equilibrium with mantle peridotite at depths of 100 km (see Figure 9-7). Thus, Clarke believes the parental Baffin magmas originated as partial melts at these depths.

Local Basalt Fields

In some areas of continental extensional tectonism, such as has occurred in the Basin and Range province of the southwestern United States since mid-Miocene time, magmatism has not produced copious floods of basalt. The reason for this contrast with basalt plateaus is unknown. Instead, there are widely scattered fields of basalt flows dotted with young cinder cones (see Figure 5-16). In a few fields, there is a large composite volcano built of intermediate to silicic composition extrusions that possibly developed as a result of diversification within the crust of mantle-derived basaltic magmas. More common associations with these basaltic extrusions are flows and pyroclastic deposits of high-silica, high-alkali rhyolite with little or no intermediate compositions represented. This bimodal basalt–rhyolite association is well expressed in the Snake River Plain–Yellowstone National Park region, where early basaltic extrusions have been succeeded by voluminous rhyolite pyroclastic eruptions that produce large calderas. Still later basaltic lavas bury these silicic deposits. Complete petrogenetic models have yet to be developed, but these chronologic relations are compatible with the following scenerio. Hot mantle-derived magmas ascend into the rifting continental crust, where sufficient magma is arrested at depth to heat crustal sial, producing large volumes of silicic partial melts that buoy to shallow levels and form batholithic masses. While pyroclastic eruptions of rhyolite are being fed from the active magma reservoir, denser basaltic magmas cannot rise through it, but as it cools, solidifies, and becomes rifted, basalt can once again be extruded.

5.4 PETROGRAPHY OF GABBROIC AND ULTRAMAFIC ROCKS

Gabbroic and ultramafic rocks are produced in much the same tectonic settings as basalts; they represent more slowly cooled, intrusive bodies of mafic magma that never reached the surface. Gabbroic rocks are compositional equivalents of basalts but have phaneritic grain size. Ultramafic rocks and anorthosite are sometimes monomineralic, or nearly so, being comprised solely of pyroxene, olivine, or plagioclase. This feature, together with the virtual absence of aphanitic volcanic equivalents, suggests that ultramafic rocks

and anorthosites in general are crystal differentiates of basaltic magmas.

Fabric

There is relatively little variety in the fabric of gabbroic rocks because slow crystallization of low-viscosity basaltic magmas in plutonic environments allows crystallization processes to converge toward similar products. Most intrusive basaltic magmas crystallize by slow sequential growth of minerals, producing hypidiomorphic-granular texture (see Figure 1-11). Some gabbros and ultramafic rocks have **poikilitic** fabric, a type of inequigranular texture in which a single large mineral grain, known as an **oikocryst**, encloses several randomly oriented smaller grains of another phase or phases (see Figure 5-25). Poikilitic relations can develop in at least two ways: (1) sparse nucleation of a mineral from a melt occupying the interstices between many earlier-formed crystals, and envelopment of the preexisting grains by the growing oikocryst; and (2) invasion and dilation of a crystalline aggregate by some sort of fluid, which subsequently crystallizes into large oikocrysts. In many gabbros, flow of the magma, sometimes in concert with gravitative settling or other processes during crystallization, have caused tabular plagioclases to become aligned, creating **igneous lamination** (see Figure 1-11b). The counterpart of this texture in aphanitic felsic rocks is trachytic texture.

Other types of fabrics in ultramafic rocks will be considered later.

Classification

Figure 5-26 shows the IUGS classification of rocks comprised of some combination of pyroxene, plagioclase, and olivine.

Alteration

Like other magmatic rocks, gabbroic and ultramafic rocks are subject to deuteric and hydrothermal alteration. Recrystallization may be

0   cm   3

Figure 5-25 Hand sample of peridotite showing **poikilitic** fabric. Large single pyroxene grain, with light reflecting off its cleavage, encloses numerous smaller grains of olivine.

extensive, but unless there is accompanying deformation of the body the original fabric and mineralogical composition can often be discerned because of the pseudomorphic nature of the replacement. Virtually all alteration involves interaction between the primary high-temperature silicate minerals and introduced water, producing lower-temperature hydrous silicates; except for this introduced water, replacement is more or less isochemical, aiding the determination of original mineralogical compositions.

Olivines and Ca-poor orthopyroxenes commonly break down to serpentine, often with admixed talc and brucite; the Fe in the primary phase that is not tolerated in the hydrous alteration products forms finely disseminated secondary Fe–oxide grains. Decomposition of olivine grains begins on the periphery and along internal fractures, producing anastomosing veinlets of cross-fiber serpentine with islets of residual olivine (see Figure 5-27); complete **serpentinization** results in a mesh-textured aggregate of fine serpentine, with disseminated secondary Fe–oxides, which mimics the shape and, to a degree, the internal cracks of the original olivines. Orthopyroxene grains tend to be replaced by large plates of serpentine. Balanced chemical reactions may be written involving water and olivine (or orthopyroxene) as reactants and serpentine, brucite, and talc as products. If water and silica are introduced, there is an increase in volume (60% to 70%) during formation of the serpentine; this expansion is so great that some petrologists question whether observed fabrics support such a reaction. If, on the other hand, only water is introduced and MgO and SiO_2 are withdrawn from the system, presumably in solution with excess water, then replacement can be volume-for-volume (see pages 398–399).

Orthopyroxenes also may be replaced by Ca-poor amphiboles. Calcic clinopyroxenes are replaced by calcic amphiboles or by fine-grained aggregates of sphene, hydrated Ca–Al silicates, and chlorite.

Calcic plagioclases are replaced by fine aggregates of hydrated Ca–Al silicates, clay minerals,

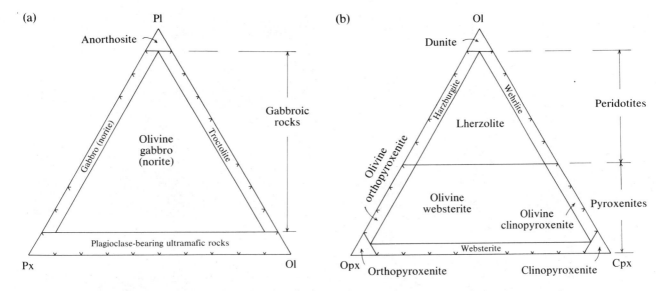

Figure 5-26 Classification of phaneritic rocks comprised of some combination of plagioclase, olivine, and pyroxene. (a) Rocks with major amounts of plagioclase (usually labradorite–bytownite). The field of gabbro is large and modifying prefixes are helpful, such as feldspathic gabbro, leuco-gabbro, olivine-rich gabbro, and so on. Gabbro in which the pyroxene is principally orthopyroxene can be called **norite.** (b) Classification of ultramafic phaneritic rocks comprised of olivine and pyroxenes. [After A. Streckeisen, 1979, Classification and nomenclature of volcanic rocks, lamprophyres, carbonatites, and melilitic rocks: Recommendations and suggestions of the IUGS Subcommission on the Systematics of Igneous Rocks, *Geology* 7.]

calcite, and albite, imparting a dull, pale-green color to the grains in hand sample.

5.5 NATURE OF PLUTONS

Gabbroic rocks transitional into ophitic diabase occur in sills, dikes, and small plugs. As the dimensions of the intrusion increase, grain size generally increases, and the effects of magmatic differentiation may become manifest in fabric and compositional variations. In larger intrusions of basaltic magma with a vertical extent of more than several hundred meters, convection of the relatively fluid magma becomes increasingly likely, and opportunities for crystal–liquid fractionation increase. Layered, virtually monomineralic accumulations of pyroxenite, peridotite, dunite, and anorthosite as well as more felsic, silicic residual differentiates may result.

Although a multitude of geometric, chemical, rheologic, and thermal factors govern the crystallization of intrusions of basaltic magma—and this is somewhat reflected in their variety—certain features are consistent. Two types of intrusion will now be described: first, the internally differentiated but unlayered sill, and second, the large, differentiated stratiform intrusion.

Differentiated Sills

Characteristics.   Many **differentiated sills** from all over the world have been studied (Carmichael and others, 1974, pp. 437–451). The Red Hill intrusion of Jurassic age in Tasmania (McDougall, 1962) provides an example of the sort of internal differentiation resulting from slow cooling of basalt magma under epizonal conditions. The body is a subhorizontal almost concordant sheet about 400 m

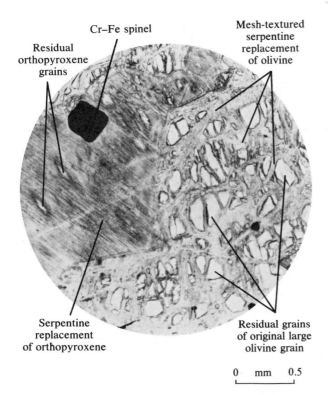

Figure 5-27   Partially serpentinized periodotite, showing serpentine replacements of olivine and orthopyroxene grains.

(a)

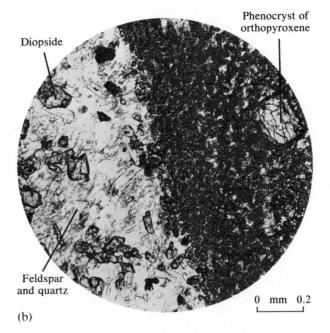

(b)

Figure 5-28   Contact between the Red Hill mafic intrusion and its calcareous sandstone country rock. (a) Outcrop with pen resting on hornfelsed country rocks. (b) Photomicrograph of very fine grained basalt (on right) composed of minute crystals of pyroxene, plagioclase, and Fe–Ti oxides in contact with hornfelsed country rock (on left) comprised of diopside, quartz, and feldspar.

thick with an upward protuberance over 1 km wide and more than 300 m high. The contact between the hornfelsed Permo-Triassic sedimentary rocks and the fine-grained basalt margin is knife sharp (see Figure 5-28). Grain size coarsens away from the contact into the interior of the body, reflecting slower undercooling and lower nucleation density in the melt. We usually refer to the marginal rock of the intrusion that quickly solidified against the cool country rock as the "chilled" margin. Photographs (Figures 5-28 and 5-29) of the chilled marginal basalt, a quartz-bearing diabase 10 m from the contact, and a highly differentiated rock from near the top of the protuberance, show the nature of these changes in fabric. More or less paralleling these variations in fabric

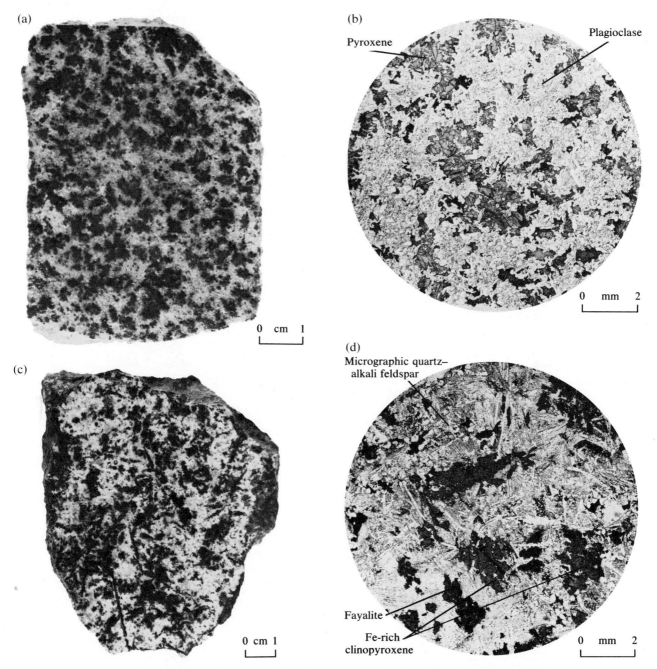

Figure 5-29   Rocks from within the Red Hill intrusion. (a, b) Quartz-bearing diabase 10 m inside body. (c, d) Granophyre near top of intrusion with elongate, somewhat feathery Fe-rich clinopyroxene, fayalite, and micrographic quartz–alkali feldspar.

Table 5-3   Compositional data on representative rocks from the differentiated Red Hill intrusion, Tasmania.

Rock type	Chilled margin basalt	Quartz-bearing diabase		Fayalite granophyre	Granophyre	
Sample number	M172	M212	M210	M12	M162	M19
Modal composition (vol.%)						
Quartz	0.0	10.1	14.6	18.3	29.5	23.8
Alkali feld.	0.0			24.4	41.3	25.9
Plagioclase	3.5	60.4	59.5	41.5	16.1	30.0
Pyroxene	4.0	28.4	21.3	7.7	8.1	7.9
Fayalite	0.0	0.0	0.0	2.3	0.0	0.0
Fe-Ti oxides	0.0	0.7	1.9	2.1	1.4	8.6
Glass	92.5	0.0	0.0	0.0	0.0	0.0
Other	—	0.4	2.7	2.7	3.6	3.8
Mineralogical composition						
Plagioclase	—	—	An_{65}	An_{61}	An_{50}	An_{14}
Pyroxene (atomic Mg/Fe)	—	—	1.54	0.30	0.06	—
Whole rock chemical composition (wt. %)						
SiO_2	54.15	53.05	53.93	60.37	68.94	63.22
TiO_2	0.71	0.56	0.80	1.13	0.79	1.08
Al_2O_3	14.83	15.41	18.88	13.77	11.27	10.92
Fe_2O_3	0.50	0.56	1.38	2.00	3.80	7.89
FeO	8.95	7.93	7.31	8.20	4.03	3.77
MnO	0.12	0.16	0.17	0.17	0.14	0.14
MgO	6.76	7.65	3.09	0.83	0.35	0.66
CaO	10.65	11.71	10.32	6.14	3.20	3.44
Na_2O	1.98	1.44	2.09	2.82	2.50	2.80
K_2O	1.13	0.80	0.95	2.42	3.20	2.88
P_2O_5	0.26	0.14	0.16	0.33	0.18	0.64
H_2O^+	0.26	0.78	1.21	1.26	0.89	1.23
H_2O^-	0.08	0.40	0.20	0.52	1.03	1.0
Total	100.38	100.57	100.49	99.96	100.32	99.75

Source: After I. McDougall, 1962, Differentiation of the Tasmanian dolerites: Red Hill dolerite–granophyre association, *Geological Society of America Bulletin* 73.

are sympathetic variations in chemical and mineralogical composition, shown by data on six samples in Table 5-3. The modal percentage of quartz plus alkali feldspar, occurring in large part as micrographic intergrowths, increases upward in the body at the expense of pyroxene and plagioclase. Pyroxenes show a paralleling increase in Fe/Mg ratio in this sequence, and plagioclases show an

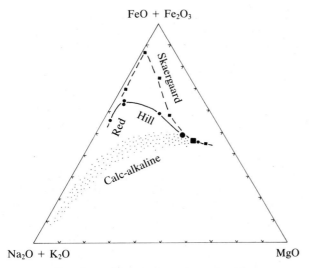

Figure 5-30 Chemical variation in rock suites. Compositions of Red Hill and Skaergaard differentiates from Tables 5-3 and 5-4 are plotted as circles and squares, respectively. The Skaergaard differentiation trend (dashed line) shows an extreme degree of iron enrichment. Typical calc-alkaline trend (strippled area) shows little or no iron-enrichment. Note that most differentiates are more enriched in iron and alkalies than the parental magma, represented by the chilled margin samples (larger symbol), but that a few differentiates are more magnesian.

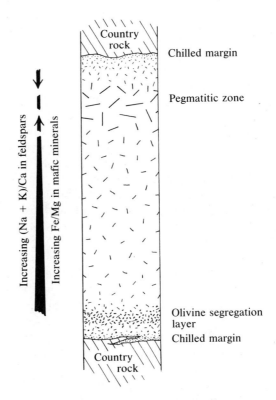

Figure 5-31 Cross section through an idealized hypothetical differentiated mafic sill. Note possible downward differentiation from roof toward pegmatitic zone.

increase in the albitic end member; that is, Na/Ca increases. Fayalitic (Fe-rich) olivine occurs in some granophyres near the top of the intrusion. Chemically, the samples show increasing Si, Fe, Na, and K and decreasing Mg and Ca. These chemical variations are shown graphically in Figure 5-30.

Three features common to thick sills but not observed in the Red Hill body are worth mentioning. First, in the upper, late felsic differentiates, the grain size may be pegmatitic, with large bladed Fe-rich clinopyroxenes attaining lengths of several centimeters. Second, basal, complementary, more mafic, and especially more magnesian differentiates occur in the form of olivine- or pyroxene-

rich rocks. The Palisades sill in New York and New Jersey, for example, has a zone of olivine-rich diabase several meters thick (Walker, 1969). Third, irregularities in the pattern of compositional variations suggest multiple injection of magmas; crystallization of a sheet is underway when another surge of fresh magma enters and mixes with the first. Compositional discontinuities and the location of the olivine-rich zone well above the base of the Palisades sill, for example, indicate reinjection of new magma.

Because of the many factors involved, every sill is somewhat different. Nonetheless, a typical differentiated basaltic sill might appear as in Figure 5-31.

Mechanism of differentiation.   Differentiated sills, and unusually thick basalt flows as well, are important in providing data on chemical variations in crystallizing bodies of magma. They can serve as tests of theoretical models and of small-scale, time-restricted, compositionally simple laboratory crystal–melt systems. If our theoretical and laboratory models cannot adequately account for what is observed in nature, we must revise or even reject the models, no matter how elegant and convincing they may seem. Inconsistencies possibly indicate some unknown magmatic process.

It has been tacitly assumed so far that these sills are differentiated. Could other processes of magmatic diversification have produced the observed compositional variations? Assimilation of country rock can be ruled out because of the sharp contacts and lack of country rock inclusions within typical intrusions. Mixing of magmas appears to have occurred in some sills, but the patterns of compositional variation appear to be developed within individual draughts, and the overall variations observed in many sills are so consistent that exceedingly special mixing conditions would be required to replicate trends in every body.

Of known mechanisms of differentiation, liquid immiscibility and crystal–liquid fractionation should be considered. As will be discussed on page 321, immiscibility has been verified in laboratory studies of compositionally restricted Fe-rich silicate systems not unlike the residual melts that evolve in differentiated sills. So, although no direct evidence of immiscibility can be found in sills, there is the possibility that diverse liquids could develop to produce the observed compositional variations. But some other independent mechanism capable of producing Fe-enrichment appears necessary, and we therefore turn to crystal fractionation. As Figure 5-31 shows, typical variations in fabric and composition within sills are highly asymmetric, with late, coarse granophyric to pegmatitic differentiates lying near, but not at, the top. To account for this, many petrologists have suggested gravitational settling of denser pyroxenes

and olivines in the low-viscosity basaltic melt and attendant crystal–liquid fractionation. This would account for the observed internal asymmetry of the sills and the chemical nature of the variations—increasing Fe, Na, and K and decreasing Ca and Mg in later-solidifying higher parts of the intrusion. During solidification, a few crystals might attach themselves to the roof, allowing it to grow downward with increasingly more K-, Na-, and Fe-rich crystals precipitated from the differentiating melt, essentially duplicating compositional variations upward from the floor but over a shorter distance. The disposition of plagioclase during solidification is uncertain. Theoretical calculations indicate any density contrast with the liquid may be too small in the initial stages to allow segregation; but in the later stages, as plagioclase becomes more sodic and less dense and the contempary melt becomes more Fe-rich and more dense, plagioclase ought to float. No evidence is seen in sills for floating, however. It may be that enough pyroxenes, and in some intrusions olivines, are settling to pull down with them any plagioclases in the melt; clots of crystals may settle together rather than widely separated crystals. This notion would also help to explain why inequant tabular plagioclases and pyroxenes show no preferred orientation, as would be expected if they were settling independently to the floor of the sill and there assuming an orientation of least gravitational potential energy.

Whatever the mechanism involved in their evolution, differentiated sills and thick basalt flows show within a single body what compositional variations can occur in nature; we can, therefore, legitimately postulate that similar observed variations between several coeval bodies in a magmatic center might have evolved in similar manner. For example, several lava flows and shallow intrusions of a volcano may define a compositional pattern paralleling that found in sills; we may suppose that a large hidden subterranean magma body was evolving as sills do, and a succession of extracts from it built the volcano complex.

Layered Intrusions

Few magmatic rock bodies are more intriguing and potentially more capable of providing information regarding crystallization processes and kinetic properties of magmas than **layered** or **stratiform intrusions** of basaltic composition. These layered tholeiitic intrusions possess a number of singular properties not found in other magmatic rock bodies.

Properties of layered intrusions. Many stratiform intrusions are large. The Precambrian Bushveld complex in South Africa is colossal (see page 546); it is exposed over an area of 65,000 km² with a thickness of up to 7 km. The Tertiary Skaergaard intrusion in east Greenland is relatively small, with an area of 170 km² and an estimated volume of 500 km³—a little less than the Roza basalt flow in the Columbia River Plateau. Most layered intrusions appear to be funnel-shaped, though the diameter of the funnel relative to the height of its subvertical axis varies greatly (see Figure 5-32). Layering is discordant with the walls of the funnel.

Layering in stratiform intrusions is as pervasive and distinctive as stratification in sedimentary sequences (see Figures 5-33 to 5-36). Single layers range from millimeters to hundreds of meters in thickness and from meters to tens of kilometers in lateral extent. In the Bushveld complex, the Merensky Reef—a pyroxenitic layer 1 to 5 m thick that is the world's chief repository of platinum—extends along strike for nearly 150 km in the eastern part of the complex and 190 km along strike in the western part, about 300 km distant.

Layering is most conspicuously defined by variations in relative proportions of minerals. Gradational variations within a single layer, from top to bottom, may be obvious. Figure 5-34a shows mineralogical sorting with a greater concentration of mafic grains at the bottom and more plagioclase at the top, whereas 5-34b shows size sorting of diopsides, larger ones lying at the base. Other types of layers show no sorting and are essentially isomodal

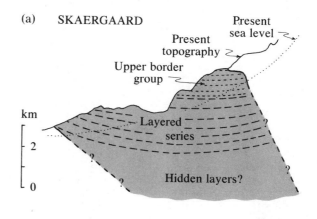

(a) SKAERGAARD

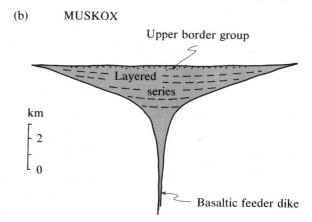

(b) MUSKOX

Figure 5-32 Idealized cross sections through two layered intrusions showing discordance between internal stratification and external contact of funnellike body. Both intrusions were tectonically tilted after emplacement, and these sections have been restored to original horizontality. Along the margin of each is a thin zone of chilled basalt and other nonlayered mafic rocks. Horizontal scales equal vertical. (a) The Skaergaard intrusion in east Greenland was emplaced as an inclined funnel, and the layered rocks are stacked like saucers within it. Granitic differentiates occur within the upper border group. (b) The Muskox intrusion in northern Canada is shaped like the keel of a sailing ship. [After L. R. Wager and G. M. Brown, 1968, *Layered Igneous Rocks* (London: Oliver and Boyd).]

Figure 5-33 Rhythmic layering in the Skaergaard gabbroic intrusion, east Greenland. Lighter layers are richer in plagioclase. Height of cliff face is about 300 m. [Photograph courtesy of G. M. Brown.]

(a)

(b)

throughout, standing out by virtue of their modal contrast with adjacent layers (see Figure 5-36). Rhythmic alternation of sorted or isomodal layers is the most visible aspect of layered intrusions.

A surprising property of some apparently gravity-induced mineral-graded layers (see Figure 5-34a) is the lack of hydraulic sorting of the constituent grains. For example, in the Stillwater complex of Montana, Jackson (1961) found layers comprised of minute chromites near the base grading upward into coarse but less dense olivines. As hydraulic sorting is a function of settling velocity, and as this velocity, according to Stoke's Law (see page 318), is proportional to the first power of the density contrast and to the square of particle size, the chromite–olivine layers are hydraulically upside down.

Locally, layers of contrasting composition thin and then pinch out along strike, merging with unlayered rock (see Figure 5-34a). Other layers are faulted, contorted, or otherwise distorted from planar configuration, implying instability had set in

Figure 5-34 Rhythmic graded layers in stratiform intrusions. (a) Uniform gabbro passes into layers with greater concentration of mafic minerals at the base grading upward into more plagioclase-rich rock. [Courtesy of G. M. Brown.] (b) Size-graded layers in Duke Island intrusion. Base of each layer is dominantly large pyroxene crystals with minor smaller olivines that become more abundant upward; pyroxenes diminish in size upward.

Figure 5-35 Angular unconformity in layering in olivine clinopyroxenite, Duke Island, Alaska.

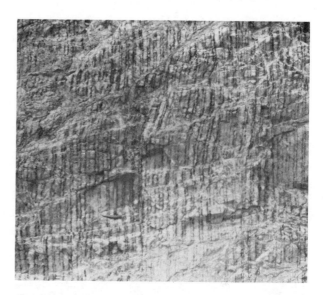

Figure 5-36 Incredibly rhythmic ''inch-scale'' layering in gabbroic rocks, defined by varying proportions of pyroxene and plagioclase, Stillwater complex, Montana. Note hammer for scale. [Photograph courtesy of A. R. McBirney.]

after crystal accumulation but before complete solidification of the layer (see Figure 5-35).

Igneous lamination occurs in many layers.

Another important aspect of layered intrusions implies a parallel between them and stratified clastic sedimentary sequences. In clastic rocks such as sandstones, there is generally a clear textural distinction between the accumulated, gravity-settled detrital grains and the secondary cement surrounding them (see Figure 1-12). In rocks of layered intrusions, especially those comprised of but two or three crystalline phases, a similar distinction may exist between a generation of early formed grains and minerals that formed later around them. These two textural components are interpreted by many petrologists to be the product of magmatic sedimentation. In such rocks, or **cumulates** as they are called (Wager and Brown, 1968), the **cumulus mineral grains** are interpreted to have settled in a melt under the influence of gravity to a site of deposition, other than their place of origin, where the **postcumulus minerals** grew around them.

Cumulus fabric provides a record of processes operative at the site of deposition in the ''mush'' of cumulus grains after their accumulation; basically, three postcumulus processes are significant, as shown in Figure 5-37.

With respect to cumulus phases, there are three features of special interest in layered intrusions: cyclic units, cryptic mineral variation, and phase layering.

Within the lower and usually ultramafic portions of larger intrusions there is commonly a regular repetition of a certain sequence of cumulate layers. Thus, in the Muskox intrusion (Irvine, 1970; also Jackson, 1971) a thick olivine cumulate layer is succeeded by a thinner clinopyroxene–olivine cumulate layer and then by a clinopyroxene–olivine–plagioclase cumulate layer; this triplet of layers, or **cyclic unit,** is repeated many times.

Nearly all layered intrusions display more or less continuous variation in the chemical composition of cumulus phases upward from the base. This

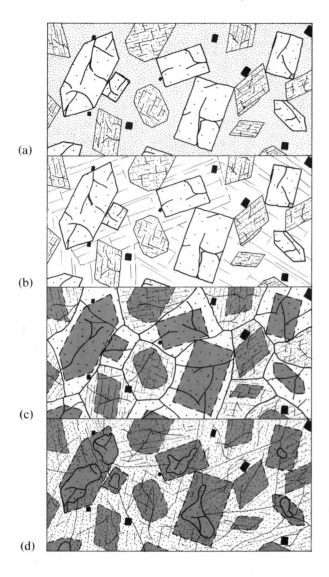

Figure 5-37 Evolution of **cumulus** fabric. The three **postcumulus** processes illustrated are ideal end-members, and real cumulate rocks generally form by some combination of these.

(a) Original cumulus grains of olivine (bold relief), pyroxene (with cleavage), and chromite (small black) in a melt (stippled).

(b) Postcumulus **space filling.** The intercumulus melt surrounding the cumulus grains in (a) crystallized as large plagioclase grains; extraneous ions in the melt—Fe, Mg, Ti, and so on—not required by the growing plagioclase are presumed to have diffused out of this local volume. This fabric is poikilitic.

(c) Postcumulus **secondary enlargement.** The intercumulus melt surrounding the cumulus grains (shaded areas enclosed by dashed outlines) in (a) crystallized as enlargements on the cumulus olivines and pyroxenes until all pore space was eliminated. Note three-grain triple junctions. Plagioclase was either unstable or did not nucleate. Monomineralic dunites, pyroxenites, and anorthosites can originate by secondary enlargement of accumulations of olivines, pyroxenes, and plagioclases, respectively. Unwanted ions in the intercumulus melt must diffuse out of the local volume.

(d) Postcumulus **reaction replacement** and secondary enlargement. The intercumulus melt and the surrounding cumulus grains (shaded areas enclosed by dashed outlines) in (a) reacted together, partly consuming olivines. Simultaneous secondary enlargement of the cumulus pyroxenes eliminated all pore spaces.

gradational variation is independent of local rhythmic layering and of rock type. As it is visible only by analysis of samples in the laboratory, it is called **cryptic mineral variation.** For example, olivines and pyroxenes become more enriched in Fe relative to Mg and plagioclases become more sodic upward (see Figures 5-30 and 5-31).

In addition to continuous variations in composition of solid solution mineral phases, certain phases will disappear at particular positions in the intrusion and others will appear elsewhere, defining what is called **phase layering.** Figure 8-33 shows cryptic variations and phase layering in the Skaergaard.

It may be noted that cryptic mineral variations in cumulus phases require that the bulk chemical composition of whole rocks vary sympathetically; this is born out by the analyses shown in Table 5-4 and plotted in Figure 5-30.

Although far from complete, this description of

Table 5-4   Compositions of representative rocks from the Skaergaard
intrusion (in wt.%).

	1	2	3	4	5	6
SiO_2	48.08	45.08	48.15	43.05	49.12	60.23
TiO_2	1.17	0.94	2.64	2.54	2.36	1.18
Al_2O_3	17.22	16.41	18.02	10.58	9.29	11.29
Fe_2O_3	1.32	2.09	2.52	4.49	3.67	5.52
FeO	8.44	9.29	9.50	18.23	20.84	9.11
MnO	0.16	0.06	0.12	0.22	0.54	0.24
MgO	8.62	11.65	5.25	3.92	0.09	0.51
CaO	11.38	10.46	10.17	10.37	8.98	5.11
Na_2O	2.37	2.06	3.46	3.39	3.08	3.92
K_2O	0.25	0.27	0.14	0.30	0.61	1.94
P_2O_5	0.10	0.05	0.05	2.95	0.54	0.27
H_2O^+	1.01	0.77	0.20	0.11	0.96	0.80
H_2O^-	0.05	0.26	0.02	0.16	0.20	0.10
Total	100.17	99.79	100.24	100.31	100.28	100.12

1. Chilled margin basalt
2. Lower-layered series plagioclase–olivine cumulate (gabbro)
3. Middle-layered series plagioclase–augite–magnetite cumulate (gabbro)
4. Plagioclase–olivine–augite–magnetite–apatite cumulate (ferrodiorite)
5. Plagioclase–iron–wollastonite–augite–olivine–magnetite–apatite cumulate (ferrodiorite)
6. Mafic granophyre
Source: After L. R. Wager and G. M. Brown, 1968, *Layered Igneous Rocks* (London: Oliver and
Boyd).

the properties of layered intrusions is sufficient to
indicate some of the outstanding interpretive
problems.

Origin of layered intrusions.   The basic problem is
to account for all aspects of the rhythmic layering
as well as cryptic mineral variation and phase
layering. Cryptic variations exist in differentiated
sills (see Figure 5-31), but not rhythmic layering,
so mechanisms of formation must differ. The origin of cyclic units will not be considered further
here, though it may be noted that one possible explanation involves multiple injection of batches of
identical magma, which subsequently differentiate,
to produce each cycle.

Emplacement of an independent draught of
magma to form each rhythmic layer was considered possible several decades ago, but this idea
has now been abandoned because of the obvious

difficulty in accounting for the more recently recognized cryptic mineral variation.

The similarities between bedding and grain fabric of clastic sedimentary bodies with the rhythmic
layering and cumulus fabric of stratiform intrusions suggest similar processes of sedimentation
are involved. Basically, magmatic sedimentation
has been considered to involve simultaneous
gravity-induced crystal settling and convective
flow of the magma. Crystal–liquid fractionation associated with the crystal settling is believed to explain cryptic mineral variation, with the more
Mg-Ca-rich crystal fractions accumulating at the
bottom of the pile. Crystal settling would also account for phase layering and the presence of the
latest crystallizing differentiates very near the top
of the intrusion above the thick pile of layered
cumulates (see Figure 5-32).

Wager and Deer, who first described the Skaer-

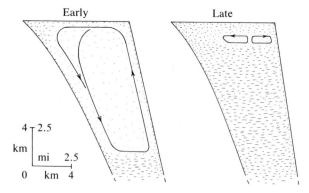

Figure 5-38 Hypothetical pattern of convection currents in the Skaergaard intrusion early and late in the course of its solidification as envisaged by Wager and Brown (1968).

gaard intrusion in 1939, postulated that convection was responsible for the rhythmic layering, igneous lamination, local lenticular and cross bedding, and slump structures. As developed later by Wager and Brown (1968), the convection hypothesis presumed a greater loss of heat from the roof of the intrusion relative to its walls and floor. It was further held that magma near the roof would begin to crystallize first, and being denser and gravitationally unstable, would sink along the walls of the intrusion and spread horizontally along the floor, dropping its suspended load of crystals (see Figure 5-38). Two types of convection were postulated: a slow, steady current producing more or less isomodal layers, and intermittent fast currents scouring into the crystal mush along the floor and, as they came to rest, producing mineral-graded layers and lenticular layers from the abundant suspended crystals.

On theoretical grounds, using reasonable values for the parameters involved, there is little doubt that convection occurs in a large body of basaltic magma (see pages 250–251). However, detailed theoretical modeling of convective flow patterns and velocities in even simply shaped bodies has not been done because of the many complexities involved in compositionally complex silicate magmas. It will be shown later (see pages 318–319) that crystallization is more likely initiated near the

floor of the intrusion. The lack of hydraulic sorting in some layers noted earlier indicates that settling of cumulus grains may not occur as in the familiar processes of clastic sedimentation; instead, more complicated nucleation and crystallization phenomena may be involved in these chemically complex magma systems. An important point to be reconciled in the explanation of the origin of stratiform intrusions is that increasingly more sodic cumulus plagioclases crystallize in increasingly Fe-rich melts during the course of solidification; theoretically these lighter plagioclases ought to float rather than sink, as expected in the cumulus mechanism.

Many tantalizing problems persist regarding the origin of stratiform intrusions. Insights into their solution will be considered later (see pages 318–321).

5.6 OCEANIC SUBALKALINE BASALTIC TO ULTRAMAFIC ASSOCIATIONS

Oceanic regions of the globe harbor two types of basaltic petrotectonic associations: (1) intraplate volcanoes that are distant from junctures of lithospheric plates and in a few instances rise above sea level as islands, and (2) spreading plate junctures comprising major global oceanic rifts, such as the Mid-Atlantic Ridge and the East Pacific Rise, and local dilatant marginal basins behind volcanic island arcs. Spreading junctures, which also have ultramafic rock bodies, are dominated by subalkaline tholeiitic rocks and are the main subject of this section. Tholeiitic basalts are prominant, but not exclusive, in many oceanic islands, and other islands have a varied alkaline suite as the only exposed rocks; hence intraplate oceanic islands are treated in the following chapter (Section 6.2).

Iceland

Brief mention should first be made of this unique island that straddles the Mid-Atlantic Ridge because of the insights it provides on oceanic rifts.

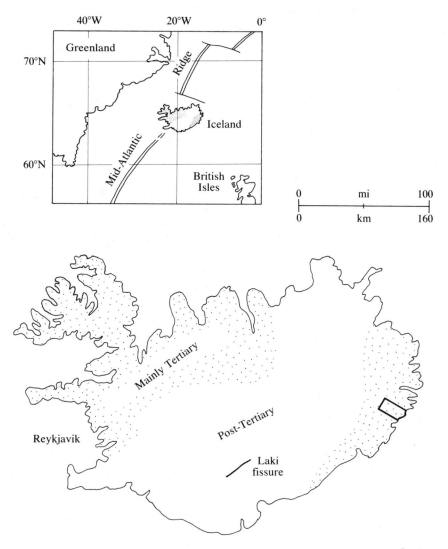

Figure 5-39 Iceland lies on the Mid-Atlantic Ridge and has older lavas flanked by younger. In the rectangular area outlined in the eastern part of the island, dikes have an aggregate width of about 8 km at lowest levels of exposure. [After G. P. L. Walker, 1960, Zeolite zones and dike distribution in relation to the structure of the basalts of eastern Iceland, *Journal of Geology* 68.]

In a sense, the wholly volcanic island of Iceland is a subaerial outcropping of the Mid-Atlantic Ridge (see Figure 5-39). No continental sialic material beneath it has been detected by geophysical studies. The island is built of older lavas on either side of a central axial zone of more recent vol-canism, mimicking age relations across the sub-marine oceanic ridge to the northeast and south-west. The island is also transected by numerous northeast-trending, elongate fault blocks and fis-sures; in 1783, 12 km³ of basaltic lava were ex-truded from the 32-km-long Laki fissure. Innumerable subvertical dikes more or less paralleling this same trend are seen in the topographically lowest exposures, but their concentration diminishes dramatically higher into the overlying pile of sub-horizontal lava flows. The dikes indicate a very significant amount of crustal extension perpendicular to their overall trend.

Most of Iceland is composed of lava flows of oversaturated tholeiitic basalt. Minor amounts of slightly silica-undersaturated basalts are also found together with Fe-rich intermediate silica icelandite and rhyolite. Icelandite may form from tholeiite by fractionation of olivine.

The origin of the vast pile of lava comprising Iceland remains speculative, but a rising and melting mantle plume has been considered possible.

Oceanic Rifts

Oceanic rifts have been the last major unexplored geologic frontier on our planet. Although unrecognized until the past couple of decades, tectono-magmatic systems along oceanic accretionary plate boundaries rank among the most significant on Earth. Rift processes and sea-floor spreading have created, since Jurassic time, the entire present-day oceanic crust, covering 70% of the Earth's surface. The geologic record suggests that similar sea-floor spreading may also have been operative in pre-Mesozoic time, possibly into the Precambrian (see Chapter 15). Petrologic processes, back to at least the Jurassic, in other plate tectonic systems have depended directly or indirectly upon sea-floor spreading and accretionary processes along oceanic rift boundaries. For example, without spreading there would have been no subduction of oceanic plates and all the associated magmatism, metamorphism, tectonism, and sedimentation that have occurred along consumptive plate margins.

Discussion of oceanic rift bodies has been deferred until after that of continental ones because the rift assemblage is rarely exposed to view and very few geologists will ever see even a small part of it in deep water. Nonetheless, its significance on the global scale compels investigation. New discoveries and insights are being published almost monthly, so it is only possible to write of current (1980) data and ideas on this exciting frontier.

General character of oceanic lithosphere. Oceanic rift systems lie at the crest of broad rises or somewhat steeper ridges that stand about 3 km above the average level of adjacent abyssal plains (see Figure 6-1). Within the Mid-Atlantic crestal rift valley, 20 to 30 km wide and 2 km deep, there are broad, extensional fault-controlled terraces flanking a narrow inner valley only a few kilometers wide and 100 to 400 m deep. In contrast, the faster-spreading East Pacific Rise has less relief and no pronounced crestal rift valley.

Although structurally complex on a local scale, seismic investigations show the oceanic crust in general possesses a rather consistent downward seismic velocity gradient, which can be interpreted, for convenience, in terms of layers that grade into one another. These are, from top to bottom (Clague and Straley, 1977):

Layer 1   0 to 1 km thick; sediment with low seismic velocity.

Layer 2   1 to 3 km thick; basalt flows, pillows, fragmental debris, and dike rocks; this layer has a large variation in seismic velocity and appears to hold most of the magnetization of the ocean floor.

Layer 3   4 to 8 km thick; a more uniform, higher seismic velocity that is interpreted to represent less fractured, mafic intrusive rocks.

The thickness of the subcrustal peridotitic lithosphere beneath layer 3 has been found to thicken exponentially with distance and therefore age away from the crest of oceanic ridges (see Figure 5-40). In the vicinity of the crest it is thin, or virtually absent, but at distances where the age is > 100 My its thickness becomes essentially uniform, ~ 100 km. The thickening of the aging lithosphere is intimately related to its thermal evolution.

Measurements of the rate of heat flow from the vicinity of ridges show a wide variation, from less than 1 HFU to as much as 6 HFU (see Appendix A

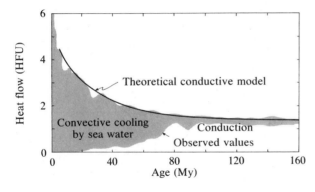

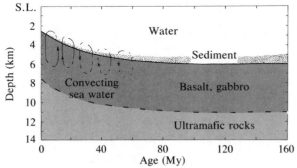

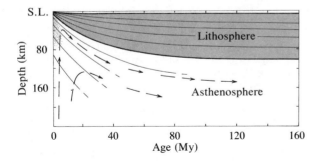

Figure 5-40 Oceanic lithosphere. (a) Heat flow from the sea floor as a function of its age; shaded area approximates distribution of observed values measured in all oceans and solid line is the theoretical heat flow according to a model in which the lithosphere is cooling by conduction only. [After R. N. Anderson, M. F. Langseth, and J. G. Sclater, 1977, The mechanisms of heat transfer through the floor of the Indian Ocean, *Journal of Geophysical Research* 82.] (b) Schematic geologic cross section through the upper lithosphere (ignoring surface topography and structural complications) showing increasing thickness of sediment with age away from crest of ridge, deepening of sea floor away from crest, the uniform thickness of the mafic rock layers 2 and 3 comprising the oceanic crust, and the underlying ultramafic rocks of the uppermost mantle; convective circulation of sea water through the upper part of hot mafic crust diminishes away from rise crest. (c) The lithosphere beneath the mafic crust is a thickening lid of cooling peridotitic mantle covering the hotter asthenosphere; light lines are highly schematic isotherms suggesting that the base of the lithosphere, deepening with respect to age, is essentially an isothermal solidus surface; dashed heavier lines indicate possible directions of flow of the asthenosphere away from axis of the rise; stippled area under the crest is the approximate location of a zone of advanced partial melting, corresponding to high temperatures at shallow depths.

for meaning of HFU), in contrast to a more uniform flow of about 1 HFU at great distances from the ridge, such as in the abyssal plains. The highly variable flow near the axis of the rise has been ascribed to convective circulation of cool sea water through the highly fractured, hot crust, a process that has profound implications for subsea-floor metamorphism (see Section 12.5). If it were not for this convection, the heat flow from the ocean floor would fall off in a regular exponential manner away from the axis of the ridge, obey-

ing a model of conductive cooling of the upper mantle and crust (see Figure 5-40). In such a model, the lithospheric slab thickens with age as it grows from the cooling underlying asthenosphere.

Back-arc or **marginal basins** are equidimensional to linear areas on the opposite side of island arcs from the deep trench side. They are especially common in the western and southwestern Pacific and include the area of the Japan Sea west of Japan and the Lau Basin west of the Tonga island

Table 5-5   Chemical and normative composition of sea-floor basalt.

	Chemical Composition			Normative Composition	
SiO$_2$	50.67 wt.%	V	182–310 ppm	*Or*	1.1 wt.%
TiO$_2$	1.28	Cr	35–510	*Ab*	21.0
Al$_2$O$_3$	15.45	Ni	75–270	*An*	30.6
FeO*	9.67	Rb	0.2–3.9	*Di*	21.5
MgO	9.05	Sr	52–135	*Hy*	16.9
CaO	11.72	Zr	38–130	*Ol*	3.5
Na$_2$O	2.51	Pb	0.4–1.3	*Mt*	3.7
K$_2$O	0.15	Ba	2.7–46	*Il*	2.4
P$_2$O$_5$	0.20	La	1.6–6.7	*Ap*	0.3
Total	99.50	Th	0.7–0.47		
		U	0.05–0.15		

Source: Averages of major, minor, and normative constituents are of 155 glasses from the Atlantic, from W. B. Bryan, G. Thompson, F. A. Frey, and J. S. Dickey, 1976. Inferred geologic settings and differentiation in basalts from the Deep-Sea Drilling Project, *Journal of Geophysical Research* 81; range of trace elements is from many samples in C. J. Allègre and S. R. Hart (eds.), 1978, *Trace Elements in Igneous Petrology* (New York: Elsevier).

arc (see Figures 3-47 and 6-1). Seismic investigations reveal a crust similar to that described above for the general oceanic realm; magnetic signatures, though weak, indicate an origin by slow spreading accretion about a centrally located axis subparallel to the island arc.

Sea-floor basalts.   Basaltic lavas constituting the upper part of the oceanic crust (Liou, 1974; Bryan and Moore, 1977) have been called various names, including abyssal tholeiite, low-K olivine tholeiite, mid-oceanic ridge basalt (MORB), and sea-floor basalt. Similar lavas have been found in marginal basins (Hawkins, 1977).

The fabric of the lavas reflects rapid cooling of fluid, only very slightly crystallized magma extruded into the cold submarine environment. Pillows with glassy rinds are widespread, although ponded flows are not uncommon. Quenched glassy rinds on pillows (see Figure 5-10) have very sparse phenocrysts of bytownitic plagioclase with or without magnesian olivine, Fo$_{80-90}$ (sometimes with minute Cr-rich spinel inclusions). Phenocrysts of augite are rare and usually confined to rocks with fairly abundant olivine and plagioclase phenocrysts. These relationships indicate that olivine ($\pm$ spinel) and calcic plagioclase are the first minerals to crystallize with falling temperature, followed by augite and then Fe–Ti oxides at still lower temperatures. Some phenocrysts are embayed, indicating disequilibrium with the melt just prior to extrusion, or possibly during extrusion. Glomeroporphyritic clots of phenocrysts may represent accumulations of early-formed crystals broken up during the extrusive event.

With respect to subalkaline basalts worldwide, those formed at oceanic rifts are relatively uniform but characterized by low concentrations of incompatible elements including Ti, P, and large-ion lithophile elements such as K, Rb, Ba, and so on (see Table 5-5; compare Table 5-1). They usually have both *Ol* and *Hy* in their normative composition and are thus olivine tholeiites. Olivine-rich (picrite) basalts with accumulative olivine have greater MgO contents. Ferrobasalts with high concentrations of FeO apparently formed as residual magmas from fractionation of early crystallized magnesian olivine, plagioclase, and perhaps magnesian pyroxene (see Figure 5-41).

The presence of corroded phenocrysts of highly magnesian olivines and calcic plagioclases in chemically more evolved groundmasses as well as anomalous melt inclusions in phenocrysts testify to disequilibrium in the crustal magma chamber feeding the submarine extrusions along oceanic rifts. It is apparent that mixing of recurring influxes of mantle-derived basaltic magma with older differentiated magma in the crustal storage chamber (see page 328) moderates the effects of crystal–liquid fractionation; hence, large amounts of strongly accumulative and highly evolved residual magmas seldom develop, and extruded ocean rift lavas are relatively uniform in composition.

The source of sea-floor basalt magmas lies at relatively shallow depths in the upper mantle beneath the oceanic rift (see Figure 5-40), judging from the high heat flow and corresponding geothermal gradient and their subalkaline composition (see Figure 9-3). A startling discovery regarding sea-floor lavas is that their incompatible element concentrations are so low that even with extensive partial melting it appears that the peridotitic source has experienced a prior fusion event, or at least some process to extract the most incompatible elements. The Rb/Sr ratio inferred for the source may also be too low to yield the observed low initial $^{87}Sr/^{86}Sr$ ratios (~ 0.703) of the lavas in 4.6 Gy (the age of the Earth), starting from an acceptable primeval meteoritic composition. The nature and cause of this earlier depletion event is a serious dilemma, as is the means of extracting so much Mesozoic and Cenozoic magma worldwide from an already depleted source (see Section 15.4).

Magmatic processes. Investigation of the FAMOUS area (Figure 6-1) of the Mid-Atlantic Ridge by manned submersibles (Ballard and van Andel, 1977) has shown that volcanic activity is virtually confined to the inner valley of the rift system (see

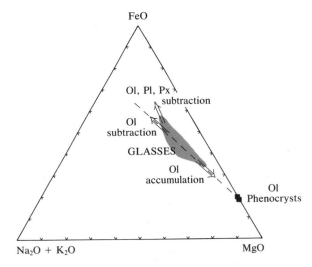

Figure 5-41   Crystal–liquid fractionation in submarine basalts from FAMOUS area. Shaded area represents distribution of analyzed glasses, filled squares are analyzed olivine (Ol) phenocrysts, and dashed line is olivine control line, showing how subtraction or accumulation of olivine phenocrysts in magma reservoir would cause variation in composition of erupted lavas. Subtraction of olivine, pyroxene (Px), and plagioclase (Pl), together in some proportions would cause variation toward the FeO apex of variation diagram. [After W. B. Bryan and J. G. Moore, 1977, Compositional variations of young basalts in the Mid-Atlantic Ridge rift valley near lat 36°49'N., *Geological Society of America Bulletin* 88.]

Figure 5-42). Bryan and Moore (1977) postulate a thermally and chemically zoned magma chamber beneath the inner valley (see Figure 5-43). The "lid" on this chamber is inferred to be only 1 to 2 km thick and is believed to float upon the underlying magma. Cooling along the walls of the chamber induces precipitation of olivine and pyroxene, some of which gravitate to the floor, forming a pile of ultramafic cumulates and creating chemical gradients within the chamber. More primitive lavas erupted along the axis of the valley are fed from the core of the chamber, in contrast to more crystal-fractionated lavas fed from dikes tapping cooler margins. But chemical gradients

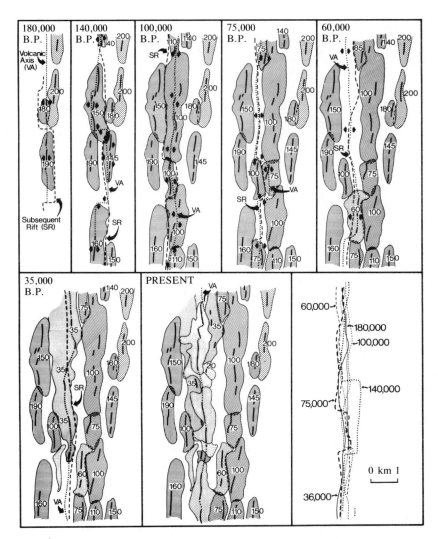

Figure 5-42 Evolution of the inner rift valley in the FAMOUS area during the past 0.2 My. Shaded areas show volcanic edifices built along active volcanic axis (VA) that are later cut by a subsequent rift (SR). Lower right diagram combines all of these rifts. Small numbers give ages of volcanic edifices in thousands of years. [From R. D. Ballard and T. H. van Andel, 1977, Morphology and tectonics of the inner rift valley at lat 36°50′N on the Mid-Atlantic Ridge, *Geological Society of America Bulletin* 88.]

within the magma reservoir are believed to be modified by recurring introduction of new batches of magma from the mantle source that mix with older magma before the chamber completely crystallizes. There is thus a quasi-balance, or steady state, between crystallization and solidification of

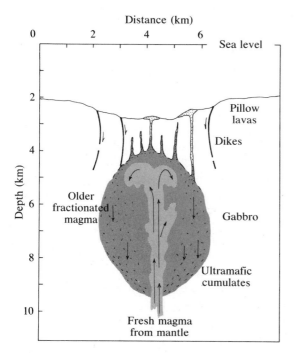

Figure 5-43 Schematic section across an oceanic rift showing a possible magmatic model. [After W. B. Bryan and J. G. Moore, 1977, Compositional variations of young basalts in the Mid-Atlantic Ridge rift valley near lat 36°49′N., *Geological Society of America Bulletin* 88.]

gravity-settled mafic phases would be a consequence. The dimensions of such plutonic magma bodies cannot be decided from basalt petrology, but the bodies may well comprise parts of seismic layer 3.

The presence of cumulate ultramafic bodies is supported by dredge hauls off the sea floor that yield not only basalts but also a varied group of coarse gabbroic rocks and pyroxenites, dunites, and peridotites (Engle and Fisher, 1975). These phaneritic rocks are now exposed on the sea floor, presumably because of extensional rifting and transform faulting. Stratigraphic relations are generally indeterminate in dredge hauls, thwarting any detailed spatial interpretations.

The prospects of obtaining additional information on the evolution of the oceanic rift magma system without expensive and technically difficult deep drilling would appear bleak. However, we now believe that a particular assemblage of rocks exposed along continental margin subduction zones was created in oceanic rift systems. By a stroke of good fortune, geologists have been able to study many of these ancient oceanic lithosphere assemblages—called **ophiolite**—in well-exposed areas on dry land, complementing oceanographic investigations of active rifts on the sea floor.

5.7 OPHIOLITE

Gabbroic and ultramafic rocks on continents occur chiefly in large layered intrusions and in generally smaller **alpine ultramafic complexes** (Jackson and Thayer, 1972; Wyllie, 1967). These on-land alpine complexes and associated rock bodies provide important insights into oceanic rift magmatism.

Historical Development of the Ophiolite Concept

Alpine ultramafic rocks have been recognized for several decades in orogenic belts less than 1200 My old all over the world (see references to work by Benson, Hess, Thayer, and others in Car-

the walls of the magma chamber and injection of new batches of magma into the chamber.

Other petrologists (for example, Hopson and Frano, 1977) have found too many structural complications in ophiolites (see next section) and discontinuities in elemental variation patterns in samples recovered from drill holes near oceanic ridges to accept the model of a single steady-state magma chamber. Instead, they envisage a constantly evolving system of several small magma chambers. Obviously, the nature of the rift magma plumbing system will depend upon rift dynamics, especially the rate of spreading.

One important point stemming from petrologic studies of sea-floor basalts is the widespread action of crystal fractionation. Cumulate bodies of michael and others, 1974, pp. 606–620). Alpine ultramafic bodies comprise two compositional subtypes that seem to grade into each other:

Harzburgitic   Chiefly harzburgite and dunite with minor dikes, veins, and pods of pyroxenite, gabbro, trondjhemite and sodic granite, or "plagiogranite."

Lherzolitic   Chiefly lherzolite with minor pyroxenites, both with prominent clinopyroxene; the main aluminous phase may be either garnet, spinel, or plagioclase. These rocks have the capability of yielding basaltic liquids by partial melting, a property not possessed by the harzburgitic subtype.

The fabric of alpine ultramafic rocks, in general, is decidedly metamorphic and anisotropic as a result of pervasive deformation in the solid state; schistosity and gneissic layers are locally folded. Cumulus fabrics in some rocks are extensively overprinted by metamorphic effects. Alpine bodies are variably, but commonly pervasively, serpentinized and sheared.

The mode of emplacement of alpine bodies has sparked controversy. Although aureoles of thermal contact metamorphism surround some, such effects are generally absent or weak. Emplacement as magmas seems to be ruled out in most instances because of the high temperatures required. Most contacts are tectonic, marked by breccia and slickensides, and emplacement is commonly along fault zones.

Geologists studying alpine bodies and spatially related rocks in the Alps of northern Italy and Switzerland about the turn of the century were impressed by a consistent association of ultramafic rocks, basaltic and spilitic lavas, cherts, argillites, and limestones. (Spilite is a metasomatized basalt in which the primary high-T minerals have been replaced at least in part by albite, hydrated Ca–Al silicates, chlorite, calcite, sphene, and actinolite; primary basaltic textures, however, are preserved.) The presence of marine sedimentary rocks, together with the commonly pillowed character of the lavas, led them to postulate a submarine origin for the entire assemblage. One of these geologists, Steinmann, stressed the intimate association between serpentinized alpine ultramafic rocks, spilite, and chert—which led to the designation "Steinmann Trinity." In the early 1960s, increasing geophysical information on the oceanic crust coupled with the developing concept of sea-floor spreading made it increasingly obvious that this association might well represent a slice of the oceanic upper mantle and crust tectonically emplaced onto the margin of the continent along a subduction zone. The mechanics of **obducting** a slice of dense ultramafic and related rocks from the oceanic lithosphere—**ophiolite**—onto sialic crust have not yet been fully resolved; it may well depend upon local tectonic configurations and processes.

The specific lines of evidence that link on-land ophiolite to oceanic lithosphere are summarized by Williams and Smyth (1973):

1. Transported on-land ophiolite is rooted in oceanic lithosphere at Papua, New Guinea.

2. Similarities between sea-floor basalts and basalts of ophiolite sequences.

3. Sea-floor spreading as a model to account for the sheeted dike complex (see below).

4. High-P mineralogical composition of certain alpine ultramafic bodies that require mantle conditions of formation.

5. Unique metamorphic imprints in ophiolites compatible with the very high thermal gradients prevailing along diverging plates (see Section 12.5).

6. Similarities between the physical properties of ophiolites and those inferred for the oceanic lithosphere from geophysical measurements at sea.

With regard to the last point, Salisbury and Christensen (1978) have made a thorough petrologic study of the Blow-Me-Down ophiolite complex exposed in northwestern Newfoundland and have in addition determined in the laboratory the elastic properties of the rocks as if they were buried under a few kilometers of sea water. They find "remarkable agreement" between the seismic velocity structure of this complex and the oceanic lithosphere in general.

Although there is general agreement that ophiolites are pieces of ancient oceanic lithosphere, there has been considerable debate regarding just where in the oceanic regions of the globe they originated. It would seem likely that the birthplace of ophiolites lies along oceanic rifts marking the divergent junctures of major lithospheric plates, simply because of their considerable linear dimensions. However, Miyashiro (1975) and subsequently many other workers (for example, Upadhyay and Neal, 1979) have noted that some capping lavas and dike rocks in ophiolites possess chemical variation patterns with calc-alkaline affinities such as develop in island arcs and along continental margins. (Metamorphism of rocks can greatly perturb concentrations of major and some minor trace elements. However, certain elements, including Zr, Ti, Y, Cr, Nb, and REE, appear not to be mobilized and reflect the composition of the original rock.) Some ophiolites are covered by thick, nearly contemporaneous deposits of trench-related rock sequences. For these reasons, it appears that at least some ophiolites formed in marginal basins in close proximity to island arcs or to continental margins onto which they were shortly later obducted due to changes in plate motion. Probably no single set of tectonic conditions generates all ophiolites.

Most ophiolites are tectonically dismembered and, in addition, are variably metamorphosed. Ophiolites are nonetheless useful indicators in the ancient rock record of convergence between oceanic and continental plates, large or small.

From a modern viewpoint, a coherent ophiolite sequence consists from bottom to top of the following (Dewey and Kidd, 1977, p. 960; see also Coleman, 1977):

> Harzburgite—minor dunite—rare lherzolite, with metamorphic textures and a complex and pervasive high-temperature deformation history; cumulate dunite below a cumulate transition sequence of dunite, feldspathic dunite, wehrlite, feldspathic wehrlite, clinopyroxenite, and anorthositic gabbro; more isotropic and homogeneous, somewhat leucocratic gabbro with irregular pods and veins of trondhjemite near the top and with diabase dikes increasing in abundance toward the top; a sheeted diabase dike complex with minor trondhjemite bodies near the base; and a pillow lava complex with an increasing percentage of dikes, and sometimes sills, toward the base.

A highly idealized section through ophiolite in its oceanic birthplace is shown in Figure 5-44.

Evolution of Ophiolite

Having established with reasonable certainty that ophiolites are segments of ancient oceanic lithosphere, we should now consider in somewhat greater detail the processes that create them. Basically, an ophiolite sequence embodies three complementary genetic components formed in response to mantle upwelling, partial melting, and magma segregation, intrusion, and extrusion: (1) a *refractory residue* of upper mantle harzburgitic peridotite that has been deformed by solid state flow and drained of its low-melting-temperature basaltic constituents, (2) overlying *"fossil" magma chambers* composed of shallow crustal cumulates that are capped by (3) *solidified fractionated and mixed basaltic magmas* that were tapped out of the magma chamber and occur as a **sheeted dike complex** grading into a carapace of submarine lava flows, in part pillowed.

The pervasively deformed alpine ultramafic rocks at the base of the ophiolite sequence (see Figure 5-44) represent mantle rock that is fertile (lherzolitic subtype) and depleted (harzburgitic subtype) in basaltic constituents. Figure 5-45 illustrates that extraction of basaltic partial melts from fertile lherzolite leaves a sterile harzburgite/dunite residuum. Overlying this refractory residuum is a suite of magmatic cumulus-textured ultramafic and gabbroic rocks, and locally more leucocratic differentiates (such as plagiogranite) formed in floored basaltic magma chambers. Cumulus fabrics may be obscured by deformation and recrystallization during obduction. In the Vourinos ophiolite in Greece, Jackson and others (1975) find

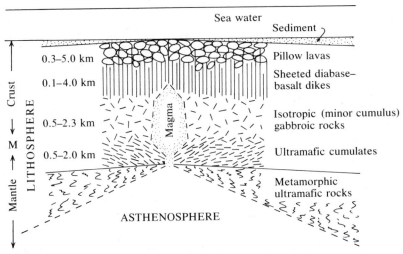

Figure 5-44 Highly idealized section through a typical ophiolite sequence in its oceanic setting—either mid-ocean rift or dilating marginal basin. Range of thickness of units within the ophiolite is given on the left. The M-discontinuity marking the seismic boundary between the crust and the higher-velocity mantle occurs near the top of the ultramafic layer. The lower part of the ophiolite sequence is the alpine ultramafic complex. Overlying marine sediments are usually found with ophiolite. Compare with Figure 5-43. [Range of thickness of units within the ophiolite after D. A. Claque and P. F. Straley, 1977, Petrologic nature of the oceanic Moho, *Geology* 5.]

a remarkably sharp contact between the metamorphic and magmatic ultramafic bodies; their lucid description of this complex is recommended reading. This thick (1,600 m) body displays features typical of most larger stratiform ultramafic-gabbroic complexes described on pages 175–180. Olivines in the Vourinos alpine metamorphic suite are of the same composition as those at the base of the magmatic cumulate sequence, which is to be expected if a partial melt of basaltic composition had equilibrated with the metamorphic suite (representing residual refractory crystals) before moving upward a short distance to begin precipitation of magmatic crystals.

The general lack of intrusive contacts within the gabbroic and ultramafic cumulate components of intact ophiolite sequences is evidence for the exis-

tence of one or more magma chambers beneath an actively dilating crust that is more or less continuously fed by new batches of magma ascending from the mantle source as crystals plate onto their walls and floors (Dewey and Kidd, 1977).

A remarkable component of most intact ophiolite sequences is the sheeted-dike complex composed of diabase and basalt dikes intruded into one another with little, if any, other type of wall rock. Whereas the gabbroic and ultramafic cumulates plate more or less continuously onto the walls and floor of the magma chamber, dikes are intermittently injected into the extensionally stressed overlying crust. For dike widths in the range of 20 to 100 cm (Dewey and Kidd, 1977), the rate of emplacement is one dike every 1 to 50 years, for a half spreading rate of 1 to 10 cm/yr. Anastamosing

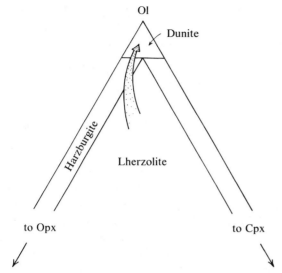

Figure 5-45 During progressive partial melting of lherzolite mantle, the composition of the refractory crystalline residuum changes along the arrow from harzburgite to dunite as first clinopyroxene and then orthopyroxene are consumed into the melt. Compare Figure 5-26b.

zones of brecciation on scales from microscopic to hand sample are widespread in virtually all sheeted dike complexes. Their irregular disposition is incompatible with a tectonic origin and suggests instead some other mechanism, such as shattering due to entry of cold sea water into the hot dike body. Indeed, Dewey and Kidd postulate that the lower reach of penetrative convective circulation of cold sea water into the oceanic crust dictates the depth of the sheeted dike zone. Below this reach, cooling occurs more slowly by simple conduction and the crust accretes by steady state plating of crystals onto the walls of the axial magma chamber(s).

Dikes are usually aphyric, suggesting effective separation of melt from the crystal-bearing magma chamber, or perhaps injection of melt coming almost directly from the mantle source. If the former, then the separation of melt must be very efficient, as an individual dike can seldom, if ever, be traced continuously into massive gabbro via in-

creasing grain size. Subsequent plating of crystals on the roof of the magma chamber(s) might obscure any earlier continuity. The top of the sheeted dike complex is not sharply bounded with the overlying pillow lavas. Local intrusive sills and a gradual reduction in the percentage of dikes relative to pillow lava occurs over a vertical distance of as much as several hundred meters.

One of the deficiencies of the idealized ophiolite model shown in Figure 5-44 is that all of the magmatism is implied to occur immediately beneath the juncture of the diverging plates and everything is symmetric about this axis. Many ophiolites are more complex than this simple picture and appear to have grown in part by off-axis additions of magma. For example, Hopson and Frano (1977) have documented four stages of dike and sill intrusion spanning the entire period of development of the Point Sal, California, ophiolite. Other ophiolites show evidence of multiple contemporaneous magma chambers.

SUMMARY

5.1. Basaltic rocks are composed of major amounts of an intermediate composition plagioclase and pyroxene, with or without olivine, Fe–Ti oxides, glass, and other accessory minerals. Because of their low viscosity and consequently rapid crystallization rates, basaltic melts tend to crystallize to a considerable degree upon extrusion, except in subaqueous environments, where more glass forms. Crustal- and mantle-derived xenoliths and cognate inclusions are common.

Because of extensive solid solution in constituent minerals, basalts comprise an extensive spectrum of chemical composition, which is divided into silica-undersaturated alkali olivine basalt and saturated to oversaturated tholeiitic basalt subtypes. Primary high-T minerals and glass in basalts interact with the oxygen-rich atmosphere and any surrounding waters to form ferric oxides and a variety of hydrated lower-T minerals.

5.2. Basaltic rocks are the most widespread of any igneous rock on Earth; they comprise the oceanic crust and cover parts of every continent. Small, shallow intrusions are generally aphanitic, weakly to nonvesicular, and may have columnar joints subperpendicular to wall-rock contacts. Low magma viscosity and rapid rates of extrusion are responsible for the sheetlike form of basalt lava flows; the volume of pyroclastic deposits is very subordinate to that of fluidal flows.

Basaltic lava flows vary between smooth, ropy pahoehoe, rough, irregular aa, and pillows, depending upon viscosity, rate of extrusion, and extrusive environment (air or water). Hyaloclastites and pillow breccias form in subaqueous environments due to thermal shattering of the cooling magma. Shield volcanoes are built of innumerable tongues of fluid lava extruded from a central vent or from a fissure system. Fountains of effervescing basaltic magma produce ash, cinders, blocks, and semi-solid spatter and bombs that build cinder and spatter cones around the vent. Phreatomagmatic explosions may create maars and tuff-ring accumulations of tephra.

5.3. Subalkaline basaltic extrusions occur in regions of crustal extension.

Plateau flood basalts are the most voluminous magmatic extrusions on the continents. Some plateau-forming events appear to have heralded continental breakup. Plateau basalts have relatively evolved chemical compositions that may have developed by contamination with continental sialic crust. Some flood basalts show evidence of shallow crustal crystal fractionation.

Local cinder-cone lava-flow fields are associated with rhyolitic extrusions; the silicic magmas may have been generated in the lower sialic crust by partial melting driven by heat from stagnated mantle-derived basaltic magmas.

5.4. Subalkaline gabbroic and phaneritic ultramafic rocks occur in the same extensional tectonic settings as do basalts and form from the same sort of magmas. Fabrics formed in slowly cooled intrusive bodies of basaltic magma include hypidiomorphic-granular, poikilitic, and igneous lamination. Phaneritic rocks comprised of some combination of pyroxene, plagioclase, and olivine

are classified according to the diagrams in Figure 5-26. High-T magmatic silicate minerals in gabbroic and ultra-mafic rocks are commonly altered to some degree by lower-T hydrous minerals including chlorite, serpentine, talc, epidote minerals, and clay minerals. Such replacements may be pseudomorphic.

5.5. With increasing size of the intrusion, effects of magmatic differentiation generally become more apparent.

Differentiated but unlayered diabase–gabbro sills are found all over the world. From the base upwards, sills become more felsic and enriched in Fe and (Na + K) relative to Mg and Ca. The origin of these chemical variations is not understood, but crystal fractionation, liquid immiscibility, or perhaps other unknown mechanisms may be responsible.

Stratiform intrusions comprised of layers of gabbro, pyroxenite, peridotite, dunite, and anorthosite provide special insight into the behavior of large bodies of basaltic magma. Rhythmic, graded, and isomodal layers, locally with unconformities and various sorts of disturbance penecontemporaneous with deposition, characterize these intrusions; layers in large intrusions can have lateral extents of tens of kilometers. In cumulates, collected cumulus minerals may react with the intercumulus melt, or the melt may form overgrowths, or simply fill in space. Some stratiform intrusions show cyclic repetition of sequences of cumulus minerals; most of them display cryptic mineral variations, and phase layering. All these features of stratiform intrusions have been explained in terms of crystal–liquid fractionation and crystal settling in a convecting magma; however, the relative densities of alleged cumulus minerals and melt, together with other considerations, cast some degree of doubt on these mechanisms.

5.6. In oceanic regions, basaltic magmatism occurs at isolated centers to produce intraplate volcanoes, some of which form islands, and along spreading oceanic rifts and in dilatant marginal or back-arc basins.

Iceland sits astride the Mid-Atlantic Ridge and

has extensional faults and fissures cutting a thick pile of chiefly tholeiitic basaltic flows that is laced by thousands of subparallel dikes.

All of the present oceanic crust, comprising two-thirds of the Earth's surface, has been formed since the Jurassic by magmatic processes at oceanic rifts where lithospheric plates are moving apart.

Surface morphology of the sea floor near the axial crest of the rift seems to depend upon spreading rate. The peridotitic lithosphere beneath a more or less uniform crust grows thicker with age away from the crest of the rift due to conductive cooling of the underlying asthenosphere. Cooling near the rift is perturbed by convective penetration of cold sea water into the hot nascent crust.

Generally quite glassy sea-floor basalts are fairly uniform *Ol* normative tholeiites whose relatively limited compositional variations can apparently be produced by crystal fractionation and mixing in magma chambers only a kilometer or so below the sea floor. Parental magmas are generated at shallow depths beneath the rift by extensive partial melting of a peridotitic source previously significantly depleted in incompatible elements. Crystal fractionation inferred to take place in magma chambers beneath the axis of the rift must produce significant volumes of mafic to ultramafic cumulate bodies.

5.7. Alpine ultramafic complexes in orogenic belts provide insights into oceanic rift magmatism. Harzburgitic to lherzolitic bodies with metamorphic fabrics comprising alpine complexes are intimately associated with other peridotitic and gabbroic rocks and with overlying basaltic dikes and sills that are capped by flows and pillows. This entire sequence, known as ophiolite, is usually tectonically dismembered and locally metamorphosed during obduction onto continental margins. Ophiolites represent slices of ancient oceanic lithosphere that were created in spreading regimes either in marginal basins or along major plate junctures.

REFERENCES

Anderson, R. N., Langseth, M. G., and Sclater, J. G. 1977. The mechanisms of heat transfer through the floor of the Indian Ocean. *Journal of Geophysical Research* 82:3391–3409.

Ballard, R. D., and van Andel, T. H. 1977. Morphology and tectonics of the inner rift valley at lat 36°50′N on the Mid-Atlantic Ridge. *Geological Society of America Bulletin* 88:507–530.

Bryan, W. B., and Moore, J. G. 1977. Compositional variations of young basalts in the Mid-Atlantic Ridge rift valley near lat 36°49′N. *Geological Society of America Bulletin* 88:556–570.

Carlisle, D. 1963. Pillow breccias and their aquagene tuffs, Quadra Island, British Columbia. *Journal of Geology* 71:48–71.

Carmichael, I. S. E., Turner, F. J., and Verhoogen, J. 1974. *Igneous Petrology*. New York: McGraw-Hill. 739 p.

Clague, D. A., and Straley, P. F. 1977. Petrologic nature of the oceanic Moho. *Geology* 5:133–136.

Clarke, D. B. 1970. Tertiary basalts of Baffin Bay: Possible primary magma from the mantle. *Contributions to Mineralogy and Petrology* 25:203–224.

Coleman, R. G. 1977. *Ophiolites*. New York: Springer-Verlag, 229 p.

Cox, K. G. 1970. Tectonics and volcanism of the Karroo period and their bearing on the postulated fragmentation of Gondwanaland. In T. N. Clifford and I. G. Gass (eds.), *African Magmatism and Tectonics*. Edinburgh: Oliver and Boyd, pp. 211–236.

Dalrymple, G. B., Silver, E. A. and Jackson, E. D. 1973. Origin of the Hawaiian Islands. *American Scientist* 61:294–308.

Dewey, J. F., and Kidd, W. S. F. 1977. Geometry of plate accretion. *Geological Society of America Bulletin* 88:960–968.

Engel, C. G., and Fisher, R. L. 1975. Granitic to ultramafic rock complexes of the Indian Ocean ridge system, western Indian Ocean. *Geological Society of America Bulletin* 86:1553–1578.

Hawkins, J. W., Jr. 1977. Petrologic and geochemical characteristics of marginal basin basalts. In M. Talwani and W. C. Pitman III (eds.), *Island Arcs, Deep Sea Trenches, and Back-Arc Basins.* American Geophysical Union, pp. 355–365.

Hay, R. L., and Iijima, A. 1968. Nature and origin of palagonite tuffs of the Honolulu group of Oahu, Hawaii. Geological Society of America Memoir 116:101–112.

Herz, N., 1977. Timing of spreading in the south Atlantic: Information from Brazilian alkalic rocks. *Geological Society of America Bulletin* 88:101–112.

Hopson, C. A., and Frano, C. J. 1977. Igneous history of the Point Sal ophiolite, Southern California. *Oregon Department of Geology and Mineral Industries Bulletin* 95:161–183.

Irvine, T. N. 1970. Crystallization sequences in the Muskox intrusion and other layered intrusions. Geological Society of South Africa, Special Publication 1:441–476.

Irvine, T. N. 1974. Petrology of the Duke Island Ultramafic Complex, Southeastern Alaska. Geological Society of America Memoir 138, 240 p.

Irvine, T. N., and Baragar, W. R. A. 1971. A guide to the chemical classification of the common volcanic rocks. *Canadian Journal of Earth Sciences* 8:523–548.

Jackson, E. D. 1961. Primary textures and mineral associations in the ultramafic zone of the Stillwater complex, Montana. U.S. Geological Survey Professional Paper 358, 106 p.

Jackson, E. D. 1971. The origin of ultramafic rocks by cumulus processes. *Fortschritte Mineralogie* 48:128–174.

Jackson, E. D., Green, H. W., III, and Moores, E. M. 1975. The Vourinos ophiolite, Greece: Cyclic units of lineated cumulates overlying harzburgite tectonite. *Geological Society of America Bulletin* 86:390–398.

Jackson, E. D., and Thayer, T. P. 1972. Some criteria for distinguishing between stratiform, concentric and alpine peridotite-gabbro complexes. *24th International Geological Congress, Montreal, Proceedings,* section 2:289–296.

Jahns, R. H. 1959. Collapse depressions of the Pinacate volcanic field, Sonora, Mexico. *Arizona Geological Society Southern Arizona Guidebook* 2:165–184.

Liou, J. G. 1974. Mineralogy and chemistry of glassy basalts, Coastal Range ophiolites, Taiwan. *Geological Society of America Bulletin* 85:1–10.

Macdonald, G. A. 1968. Forms and structures of extrusive basaltic rocks. In H. H. Hess and A. Poldervaart (eds.), *Basalts,* Vol. 1. New York: Wiley Interscience, pp. 1–62.

McDougall, I. 1962. Differentiation of the Tasmanian dolerites: Red Hill dolerite-granophyre association. *Geological Society of America Bulletin* 73:279–316.

McDougall, I. 1976. Geochemistry and origin of basalt of the Columbia River Group, Oregon and Washington. *Geological Society of America Bulletin* 87:777–792.

Miyashiro, A. 1975. Classification, characteristics, and origin of ophiolites. *Journal of Geology* 83:249–281.

Moore, J. G. 1975. Mechanism of formation of pillow lava. *American Scientist* 63:269–277.

Ryan, M. P., and Sammis, C. G. 1978. Cyclic fracture mechanisms in cooling basalt. *Geological Society of America Bulletin.* 89:1295–1308.

Salisbury, M. H., and Christensen, N. I. 1978. The seismic velocity structure of a traverse through the Bay of Islands ophiolite complex, Newfoundland, an exposure of oceanic crust and upper mantle. *Journal of Geophysical Research* 83:805–817.

Shaw, H. R., and Jackson, E. D. 1973. Linear island chains in the Pacific: Result of thermal plumes or gravitational anchors. *Journal of Geophysical Research* 78:8634–8652.

Swanson, D. A., Wright, T. L., and Helz, R. T. 1975. Linear vent systems and estimated rates of magma production and eruption for the Yakima Basalt on the Columbia Plateau. *American Journal of Science* 275:877–905.

Upadhyay, H. D., and Neal, E. R. W. 1979. On the tectonic regimes of ophiolite genesis. *Earth and Planetary Science Letters* 43:93–102.

Wager, L. R., and Brown, G. M. 1968. *Layered Igneous Rocks.* London: Oliver and Boyd, 588 p.

Walker, K. R. 1969. The Palisades sill, New Jersey: A reinvestigation. Geological Society of America Special Paper III, 178 p.

Waters, A. C. 1961. Stratigraphic and lithologic variations in the Columbia River basalt. *American Journal of Science* 259:583–611.

Waters, A. C., and Fisher, R. V. 1971. Base surges and their deposits: Capelinhos and Taal volcanoes. *Journal of Geophysical Research* 76:5596–5614.

Wilkinson, J. F. G. 1968. The petrography of basaltic rocks. In H. H. Hess and A. Poldervaart (eds.), *Basalts,* Vol. 1. New York: Wiley Interscience, pp. 163–214.

Williams, H., and McBirney, A. R. 1979. *Volcanology.* San Francisco: Freeman, Cooper and Co., 397 p.

Williams, H., and Smyth, W. R. 1973. Metamorphic aureoles beneath ophiolite suites and alpine peridotites: Tectonic implications with west Newfoundland examples. *American Journal of Science* 273:594–621.

Wright, T. L., Grolier, M. J., and Swanson, D. A. 1973. Chemical variation related to stratigraphy of the Columbia River basalt. *Geological Society of America Bulletin* 84:371–386.

Wyllie, P. J. 1967. *Ultramafic and Related Rocks.* New York: John Wiley and Sons. 464 p.

Yoder, H. S., Jr., and Tilley, C. E. 1962. Origin of basalt magmas: An experimental study of natural and synthetic rock systems. *Journal of Petrology* 3:342–532.

6

Alkaline Rock Bodies

Silica-undersaturated alkaline rocks are relatively rare but constitute the sole products of magmatism in many intraplate regions that are tectonically stable or rifting apart only very slowly. Alkaline magmas are virtually nonexistent in tectonically active regions of plate convergence and vigorous spreading where geothermal gradients are high.

Considerable interest and controversy surround both the nature and origin of alkaline associations, especially because of their diverse and, in some instances, bizarre compositions. They range from ultramafic to wholly felsic, include virtually silica-free carbonatites, and many have such unusually high concentrations of normally minor or trace elements—for example, C, P, Ti, Nb, Zr—as to cause the appearance of unique minerals, such as perovskite and priderite, and to cause ordinarily accessory minerals, such as calcite, ilmenite, or apatite, to be major constituents.

Inclusions of mantle-derived rocks and minerals, including diamond, occur almost solely in mafic to ultramafic alkaline igneous rocks; they provide fascinating insights not only into what comprises this inaccessible part of our planet to depths of at least 200 km but also into the processes occurring in these magma source regions. The recent realization that source regions of alkaline magmas have experienced widespread and locally intense metasomatism by aqueous fluids and melts laden with incompatible elements opens the door to new models of alkaline magma genesis and evolution.

6.1 PETROGRAPHY OF ALKALINE ROCKS

As defined on page 51, alkaline rocks have high concentrations of alkalies (Na and K) relative to Si so that *Ne* appears in the norm. Actual major modal minerals that may be present in these silica-undersaturated rocks include alkali feldspar, plagioclase, feldspathoids such as nepheline and leucite, analcime, biotite–phlogopite, Na–Ca pyroxene, alkalic amphiboles, olivine, and Fe–Ti oxides, plus a host of unusual minerals or ones generally occurring only in accessory amounts. Correspondingly, the spectrum of chemical composition represented in the alkaline suite is broad, ranging from ultramafic to felsic and from virtually no SiO_2 into the intermediate silica range. Additional dimensions to the compositional spectrum stem from the unusually high concentrations of certain elements (Zr, Nb, Rb, Ti, P) normally present in only trace or minor amounts in subalkaline rocks. These compositional features have consequently spawned a complex nomenclature. Indeed, one-half of all formal igneous rock names apply to alkaline rocks, in spite of the fact that their volume is less than one percent of all magmatic rocks combined.

The fabric of alkaline rocks encompasses virtually all of those previously described in Chapters 3–5. They are both aphanitic and phaneritic and, in addition, some are clastic and a few are glassy. In plutonic bodies, variations in grain size and fabric can be conspicuous and patchy on the scale of an outcrop. In some highly alkaline plutonic rocks, alkalic amphiboles and Na–Ca clinopyroxenes tend to form very elongate and even feathery and radiating habits, rather than the more robust prisms of subalkaline rocks.

Classification

There is no justification for going into the voluminous and intricate nomenclature of alkaline rocks here. Instead we will introduce a few new names for the more common alkaline rocks and for ones of special petrologic significance (see Williams and others, 1954; Appendix D).

Aphanitic rock types. Some aphanitic alkaline rocks in volcanic settings comprise suites or lineages of several rock types closely associated in space and time. A common lineage of rather sodic nature embodies the following rock types (Coombs and Wilkinson, 1969) arranged in order of increasing differentiation index (DI) (see page 50):

Alkali olivine basalt Contains abundant olivine (including in the groundmass), labradorite-andesine, and augite with substantial Ti and Al. *Ol* and minor *Ne* in the norm.

Basanite Same as alkali olivine basalt but more *Ne* and modal nepheline (or analcime, leucite) actually present. This rock may have a lesser DI than alkali olivine basalt or take its place in the lineage.

Hawaiite Modal olivine, augite, andesine, with minor anorthoclase and in the norm minor *Ne*. **Mugearite** is similar but has oligoclase instead of andesine.

Trachyte Approximately silica-saturated felsic rock with sanidine or anorthoclase and small amounts of quartz or a feldspathoid in rocks transitional to rhyolite and phonolite, respectively. Minor amounts of fairly Fe-rich biotite, amphibole, pyroxene, or olivine.

Phonolite Undersaturated felsic rock with conspicuous (> 10%) modal feldspathoid plus sanidine or anorthoclase. Mafic silicate is generally Na–Ca clinopyroxene.

This lineage may be incomplete to varying degrees in any specific area, and as few as only a couple of the rock types may be associated together. Also, certain of these rock types can occur with other types—a common association, for example, is tholeiitic and alkali olivine basalts.

Other aphanitic rocks include the following:

Shoshonite Marginally saturated rock with a relatively high K_2O/Na_2O ratio (commonly > 1) and with olivine, Ca-rich and Ca-poor pyroxene, plagioclase, and sanidine. It commonly occurs with latites and andesites, into which it grades with decreasing K.

Nephelinite A mafic rock more undersaturated and with greater *Ne* and modal nepheline than basanite. It has Na–Ti–Al-rich clinopyroxene, olivine, and accessory amounts of Fe–Ti oxides, feldspar, and possibly leucite.

Kimberlite This important rock type is defined on page 209.

Lamprophyre A group of mafic dike rocks. Though encompassing a range of composition, they possess in common a strongly porphyritic fabric with euhedral mafic minerals as both phenocrysts and groundmass constituents. Feldspars and feldspathoids are confined to the aphanitic matrix. They are barely silica-saturated to undersaturated and have abundant volatiles—mainly H_2O and CO_2. Many are compositionally similar to kimberlite, nephelinite, and alkali olivine basalt. Lamprophyres tend to occur in dike swarms and they commonly cut granitic plutons as if they were in some way genetically related to the granitic magmatism. Yet their origin remains a puzzle (Carmichael and others, 1974, p. 508).

Phaneritic rock types. Most of the foregoing aphanitic rock types have phaneritic counterparts that occur within intrusive bodies. These include the following:

Syenite (see Figure 4-12) and **feldspathoidal syenite** These are the phaneritic equivalents of trachyte and phonolite, respectively. Some complex pegmatites (see page 117) associated with nepheline syenite intrusions are important repositories of rare elements—Be, Cs, Th, U, Nb, Zr, and rare earth elements—and hence a paradise of unusual minerals. For example, in the Lovozero district on the Kola Peninsula in the northwestern USSR (Sørensen, 1974, p. 45, 217), no less than 150 different mineral species have been recorded.

Ijolite Essentially a phaneritic counterpart of nephelinite.

Theralite. Essentially a phaneritic counterpart of basanite.

Carbonatite A magmatic rock composed chiefly of carbonates of Ca, Mg, and Na with subordinate feldspar, pyroxene, olivine, biotite–phlogopite, apatite, perovskite, barite, pyrochlore, and other unusual minerals. Although usually intrusive and phaneritic, carbonatite also occurs as youthful lava flows.

6.2 ALKALINE ROCKS OF OCEANIC REGIONS

Hundreds of volcanoes dot the ocean floors (see Figure 6-1), and they have mostly grown by magmatic activity far removed from the junctures of lithospheric plates. Only some of these intraplate volcanoes resting on oceanic lithosphere emerge above sea level as islands where they can be sampled and studied. Rocks exposed on volcanic islands range in composition from tholeiitic to highly alkaline and from ultramafic to felsic. Some representative islands will now be described. It should be kept in mind that only the minor cap rock of any island is accessible for close study; the greater part of the volcanic edifice is unknown.

Tholeiitic to Alkaline Associations: Galapagos and Hawaiian Islands

The Galapagos Islands, lying near the equator, 1,000 km off the west coast of South America (see Figure 6-1), are built of still active, coalescing shield volcanoes. The central and eastern islands expose principally alkali olivine basalts, whereas the western two islands are comprised of Fe-rich tholeiitic basalts, icelandites (see page 182), and siliceous trachytes that are believed to represent parental and derivative differentiated magmas (McBirney and Williams, 1969).

The Hawaiian Islands (Carmichael and others, 1974, p. 406) are a chain of shield volcanoes that are younger toward the southeast and terminate at the big island of Hawaii, which has experienced historic activity from three of its five coalescing shield volcanoes (see Figure 5-11). Large as they are (see Figure 5-12), an entire shield can apparently be constructed in something like one million years, according to radiometric dates on accessible

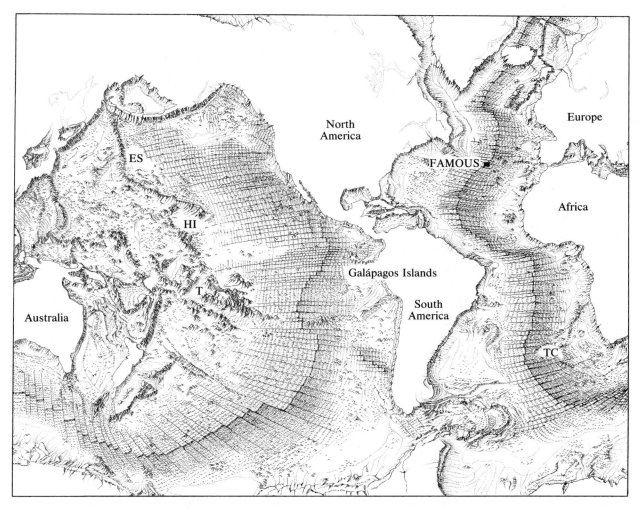

Figure 6-1   The Pacific and Atlantic ocean basins and nearby continents
(unpatterned). ES, Emperor seamounts; HI, Hawaiian Islands; T, Tahiti; TC,
Tristan da Cunha; FAMOUS, joint French–American Mid-Ocean Undersea Study
area. [Drawn by Robert T. Pack; from W. Kenneth Hamblin, 1981, *The Earth's
Dynamic Systems* (Minneapolis: Burgess Publishing Co.).]

lavas and historic rates of extrusion. The shield
volcanoes appear to be built of basaltic lavas that
range in composition from oversaturated (Q normative) to saturated (Hy) tholeiite to minor alkali
olivine basalt (Ne normative). On the average
(see Table 6-1), shield lavas are slightly oversaturated tholeiite with slightly more incompatible
elements than sea-floor basalt (see Table 5-5) but
less than continental flood basalts (see Table 5-1).
During or shortly after the waning of shield-
building activity, a very small volume of slightly
more alkaline and felsic lavas commonly have
formed; these include hawaiites, mugearites, and
trachytes. On some islands, Oahu, for example,

Table 6-1 Chemical and normative compositions of volcanic rocks from intraplate oceanic islands.

	Hawaiian Islands[a]	Tristan da Cunha[b]	Tahiti[c]
SiO_2	49.36 wt.%	46.7 wt.%	48.50 wt.%
TiO_2	2.50	3.6	2.68
Al_2O_3	13.94	17.3	17.53
Fe_2O_3	3.03	3.8	4.43
FeO	8.53	7.1	3.73
MnO	0.16	—	0.12
MgO	8.44	4.7	4.32
CaO	10.30	9.7	6.21
Na_2O	2.13	4.1	6.79
K_2O	0.38	3.0	3.54
P_2O_5	0.26	—	0.62
Total	99.03	100.0	98.47
Q	2.28 wt.%	—	—
Or	2.22	17.7 wt.%	20.71 wt.%
Ab	17.82	11.2	20.87
An	27.52	20.0	6.84
Ne	—	12.7	23.70
Di	17.49	22.6	15.72
Hy	21.90	—	—
Ol	—	3.5	3.10
Mt	4.41	5.5	3.04
Il	4.71	6.8	3.70
Ap	0.67	—	1.29
Rb	—	110–300 ppm	—
Sr	250–500 ppm	800–1600	—
Nb	—	85–130	—
Pb	1–2	8–35	—
Ba	45–180	800–1200	—
La	1–30	160–250	—
Th	0.3–1.7	—	—
U	0.2–1.0	—	—

[a] Average of 181 shield-forming tholeiitic lavas.
[b] Average of 10 trachybasalts.
[c] "Tahitite"; includes H_2O^+ 0.91, H_2O^- 0.47, and *Hm* 1.03.
Source: Average composition for Hawaiian Islands after G. A. Macdonald and T. Katsura, 1964, Chemical compositions of Hawaiian lavas, *Journal of Petrology* 5; approximate range of trace elements for Hawaiian Islands after C. J. Allègre and S. R. Hart (eds.), 1978, *Trace Elements in Igneous Petrology* (New York: Elsevier), and W. P. Leeman, J. R. Budahn, D. C. Gerlach, D. R. Smith, and B. N. Powell, 1980, Origin of Hawaiian tholeiites: Trace element constraints, *American Journal of Science* 280-A; composition for Tristan da Cunha after P. E. Baker, I. G. Gass, P. G. Harris, and R. W. Le-Maitre, 1964, The volcanological report of the Royal Society expedition to Tristan da Cunha, 1962, *Philosophical Transactions of the Royal Society of London,* Series A, 256; composition for Tahiti after A. R. McBirney and K. Aoki, 1968, Petrology of the Island of Tahiti, Geological Society of America Memoir 116.

deep erosion of the shield for a million years or so has been followed by highly alkaline magmatism that produced basanites, nephelinites, and melilite-bearing volcanic rocks.

Alkaline Associations: Tristan da Cunha and Tahiti

Tristan da Cunha is a group of three islands in the South Atlantic about 500 km east of the mid-oceanic rift (Baker and others, 1964). The mostly *Ne* normative mafic to intermediate silica lavas comprising these islands are relatively rich in K, Rb, Sr, Ba, and Nb and consist chiefly of trachybasalt with minor trachyandesite and trachyte (see Table 6-1). Pyroxene-rich, presumably accumulative basalts called ankaramite are uncommon. Many of the trachybasalts contain a small amount of leucite, one of the rare occurrences of this mineral in the oceanic environment. If there is a single parental magma for the volcanic sequence, it has not been found; all the exposed lavas have evolved compositions with high contents of incompatible elements and low Ni and Cr, which suggest at least some fractionation of olivine has occurred.

Tahiti is a mid-Pacific island comprised of alkaline lavas with a suite of chemically similar plutonic rocks exposed in the cores of its two volcanoes. Most of the lavas are strongly porphyritic alkali olivine basalts and basanites; included in this group are ankaramites, in which phenocrysts of olivine and augite comprise about one-half the rock. Trachytes and phonolites are subordinate, as are tahitites—basic lavas with zoned Ti–Na–Al-rich clinopyroxene, kaersutite amphibole, glass, alkali feldspar, and the feldspathoids haüyne, sodalite, and perhaps leucite (see Table 6-1).

Phaneritic rocks exposed in the deeply eroded cores of the volcanoes include theralites (containing zoned labradorite, Ti–augite, olivine, Fe–Ti oxides, and minor analcime, nepheline, kaersutite amphibole, and apatite), monzonites, syenites, and nepheline syenites.

Processes responsible for the diverse alkaline rocks of Tahiti are unknown (McBirney and Aoki, 1968), although crystal fractionation at shallow levels involving major minerals occurring in the rocks is quite unlikely.

Chronology and Origin of Linear Island Chains in the Pacific

Some oceanic volcanoes, particularly in the Pacific, form long chains across the ocean basin, rest on older crust, and may have little regard for sea-floor structures such as transform faults. The Hawaiian Islands constitute one such linear volcanic chain; it has been recognized for over a century that the chain is apparently younger toward the southeast, terminating with the active Kileaua volcano on the big island. Recent oceanographic investigations have disclosed that the Hawaiian Islands are actually only the exposed southeast end of the much longer Emperor seamount–Hawaiian Island chain of 80 volcanoes stretching 6,000 km all the way to the junction of the Kurile–Aleutian trench (see Figure 6-1). Radiometric ages of the volcanoes increase regularly to the northwest, giving a propagation rate of about 8 to 9 cm/yr. A conspicuous bend occurs midway along the chain, corresponding to a time of about 42 My before the present.

Other linear island chains south of the Hawaiian chain show similar age patterns (Duncan and McDougall, 1976). The younger segments of all these Pacific chains lie subparallel to one another and subperpendicular to the East Pacific Rise—the present axis of sea-floor spreading in the Pacific.

J. Tuzo Wilson was the first to postulate that linear volcanic chains may be created by episodic pulses of magmatism at a fixed melting "hot spot" beneath a moving oceanic plate. Currently active volcanoes at the southeastern extremity of the island chains in the Pacific are believed to overlie the melting spot at present. If the melting spot is fixed relative to a stationary lower mantle, then the propagation rate and linear trend of the chain gives the vector of absolute motion of the overriding plate. The bend in the longer Pacific chains is considered to reflect a change in direction of plate motion about 42 My ago.

The true nature and origin of the melting spot is controversial, as discussed by Dalrymple and others (1973). It is mind-boggling to consider how magma generation could persist in one locale for some 60 My, as postulated for the Emperor–Hawaiian chain. One suggestion is that the melting anomaly corresponds to a long-standing convective plume of hot mantle material rising from the vicinity of the mantle–core boundary. An alternate suggestion by Shaw and Jackson (1973) is that melting is occurring over a descending "anchor" of dense refractory mantle. After initiation of melting, by a cause unknown, the sterile crystalline residuum sinks, sucking fertile asthenosphere into the melting focus; localized shear melting perpetuates volcanism and the sinking anchor. However, residual refractory peridotite that is enriched in Mg may actually be less dense than fertile mantle and would be expected to buoy upward.

One difficulty with both the plume and the anchor models is the question of why, in the Hawaiian Islands, the cyclic magmatism that builds each island begins with apparently large volumes of tholeiitic magmas and terminates with minor volumes of diverse alkaline magmas before starting another cycle.

6.3 CONTINENTAL ALKALINE ROCKS

The minor volumes and compositional diversity of alkaline rocks in oceanic regions are properties duplicated in continental sectors, where, however, compositional variations are even more extreme. Most continental alkaline rocks are broadly associated in time and space with continental rifting—the extensional spreading apart of plates that are capped by sialic crust. Other alkaline rocks have formed in areas that are apparently tectonically stable, except perhaps for vertical uplift.

Breakup of the South American and African plates during the early Mesozoic was accompanied by alkaline magmatism as well as extrusion of plateau-forming flood basalts. In eastern Brazil,

basaltic magmatism began ∼ 147 My ago in the Jurassic and persisted until about 100 My ago in the mid-Cretaceous; alkaline magmatism began roughly 138 My ago, continuing to ∼ 50 My ago (Herz, 1977). Highly alkaline rocks of Quaternary age occur on islands off the coast of Brazil (Carmichael and others, 1974, p. 381–383).

In east Africa (Malawi), highly alkaline magmatism occurred soon after breakup of the continent in the late Mesozoic.

Intracontinental rifts with associated alkaline magmatism have spreading rates as much as two orders of magnitude slower than across oceanic rifts (1 to 10 cm/y), where relatively uniform tholeiitic basalt magmas are generated. The Rhine graben system in western Germany, active since the late Cretaceous with diverse alkaline and some tholeiitic rocks, has been spreading at about 0.01 cm/y. In Kenya, across the east African rift, the rate of spreading is about 0.05 cm/y. A tenfold faster rate, about 0.5 cm/y, characterizes the Basin and Range Province in the western United States, which since the Miocene has evolved into a very broad intracontinental region of extensional block faulting with bimodal basalt–rhyolite volcanism.

In this section we briefly review the nature of magmatism in the east African rifts, then examine the carbonatite–nephelinite association, ultrapotassic rocks, and, finally, kimberlites.

Magmatism in the East African Rifts

Broadly concurrent magmatism, crustal uplift, and extensional faulting are well expressed in the east African rift system, which has long been a classic region for this type of activity. The general geologic framework has been documented by Gass (in Clifford and Gass, 1970), Baker and others (1972), and Bailey (in Sørensen, 1974).

The east African rift system extends some 3,700 km from Mozambique in central east Africa northward through Ethiopia, where it splits into the Gulf of Aden and Red Sea oceanic rifts (see Figure 6-2). The Red Sea rift continues northward 2,400 km through the Dead Sea area into Turkey, and the Gulf of Aden rift extends eastward as the Carlsberg ridge system in the Indian Ocean. The east African rift system is generally 80 km wide but flares to 480 km in the Ethiopia triple junction region. In its central part, the system bifurcates into eastern and western branches. Grabens within these continental rifts have dropped as much as 3 km. Magmatism during the 70-My history of the rift system has been quite varied.

Major crustal upwarp occurred in the early Tertiary, forming the broad Ethiopia and Red Sea domes (see Figure 6-2). During this time, a vast flood of alkaline basaltic lavas covering nearly 10^6 km², locally to depths of more than 1 km, was erupted in Ethiopia and southwest Arabia. After a period of relative quiescence, separation of the continental crust formed the Gulf of Aden in the Miocene and the Red Sea in Pliocene time; these have evolved as true oceanic rifts with typical low-K, sea-floor olivine tholeiites. In Ethiopia, continental extension (but not separation) beginning in the mid-Miocene was accompanied by extrusion of basaltic lavas transitional between alkaline and tholeiitic whose volume was only perhaps 0.2 percent of the earlier Tertiary flood. Magmatism continued into the Plio-Pliestocene with some basaltic extrusions but with more voluminous (3×10^5 km³) peralkaline rhyolite and trachyte, in large part as ash-flow eruptions.

Southward into Kenya, continental attenuation began slightly later and has been weaker. Basaltic activity has been widespread in time and space, but overshadowed by eruption of nephelinite and intrusion of carbonatite complexes in the mid-Miocene, fissure eruption of phonolite lavas over a wide area in the late Miocene, and extrusion of trachyte flood lavas and ash flows in the Pliocene.

The western rift is noted for its Pliocene to Recent highly potassic volcanism; where it merges with the eastern rift, the volcanism ranges from alkalic basalt to trachyte to phonolite.

The southern part of the African rift system in Tanzania and Malawi and including parts of the western rift is well known for its occurrences of carbonatite. Though minor in volume, these un-

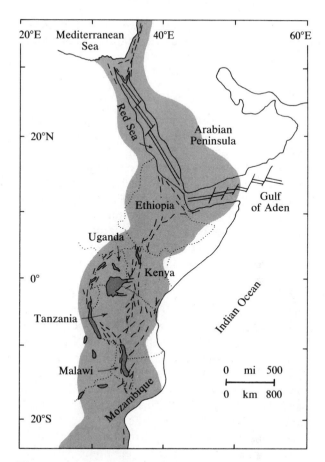

Figure 6-2 East African continental rift system and its union in northern Ethiopia with the Red Sea and Gulf of Aden oceanic rifts. These rifts lie within a chain of coalescing broad domical upwarps (*shaded*). High-angle normal faults are shown by dashed lines. Double lines denote oceanic rifts cut by transform faults. Dotted lines outline countries mentioned in text. Note several lakes (*darkly shaded*). [After I. G. Gass, 1972, Proposals concerning the variation of volcanic products and processes within the oceanic environment, *Philosophical Transactions of the Royal Society of London,* Series A, 271.]

usual magmatic bodies are both economically valuable and a great petrologic curiosity. Some are prerift whereas others are quite young, a notable example being the active volcano Oldoinyo Lengai in northern Tanzania, which has recently erupted carbonatite and highly undersaturated nephelinitic lavas and pyroclastics.

Carbonatite–Nephelinite Association

Most carbonatite is intimately associated with olivine-poor nephelinite, or its phaneritic plutonic counterpart, ijolite. Less common is an association of carbonatite and kimberlite or of syenitic rocks and carbonatite. These magmatic associations occur in some continental rifts, as in the east African system, and in apparently stable tectonic regions such as southern Africa (Mathias, in Sørensen, 1974). Carbonatite associations are far more abundant in Africa than in any other continent and are virtually unknown in oceanic regions.

Although carbonatite associations are extremely small in areal extent (a few hundred square kilometers altogether, worldwide), petrologists have been attracted to them because of their unusual composition, economic value (apatite, niobium, rare earth elements; see Table 6-2), and controversial origin. Three full-size books (Heinrich, 1966; Tuttle and Gittins, 1966; LeBas, 1977) have been written on the topic.

Although very young carbonatite–nephelinite associations comprise conical composite volcanoes of lavas and volcaniclastic rocks, most older ones are subvolcanic, shallow plutonic complexes (see Figure 6-3). These are typically circular or elliptical in outline, and the constituent rock units form plugs, arcuate ring dikes, and cone sheets. Brecciation is widespread. Rock types include, in addition to carbonatite and ijolite, alkaline pyroxenite, phonolite, nepheline syenite, and, most characteristic of all, **fenite**. This is a metasomatic rock produced by solid state transformation of older wall rocks by infiltration of hot reactive gases or hydrothermal solutions rich in Na and K derived from the nephelinitic or carbonatitic magmas. The combination of brecciation, multiple intrusion, assimilation, and fenitization produces complicated overprinted fabrics.

Table 6-2   Comparison of average abundance of trace elements in carbonatite, all igneous rocks, and limestone (in ppm).

	Carbonatite	Igneous rocks	Limestone
Sc	10	13	1
Co	17	18	0.1
Y	96	20	30
Zr	1120	170	19
Nb	1951	20	0.3
Mo	42	2	0.4
La	516	40	<1
Ce	1505	40	12

Source: After E. W. Heinrich, 1966, *The Geology of Carbonatites* (Chicago: Rand McNally).

Activity in carbonatite–nephelinite complexes typically begins with silicate magma and ends with carbonatite magma (LeBas, 1977) (see Figure 6-3). Early nephelinitic magma is emplaced as a shallow intrusive plug, part of which may erupt explosively, forming a volcano. Accumulation of pyroxenes in the intrusion yields alkaline pyroxenites. Expulsion of hot, highly reactive volatiles into the wall rocks fenitizes them to crystalline phases compatible with the P–T–X conditions of the magma system, namely nepheline, alkali feldspar, sodic amphiboles and pyroxenes, biotite, and carbonates. This metasomatism erases any possible chilled borders on the magma body and permeates the country rocks for up to several tens of meters from the contact (see Figure 6-4). Late carbonatite magma penetrates through the earlier ijolite body and its fenite envelope. Commonly, there are three or four pulses of carbonatite emplacement, each accompanied or preceded by a wave of related brecciation and fenitization of intruded rocks.

Petrogenesis of the carbonatite–nephelinite association has basically three facets. First (and this is now mostly of historical interest), what is the origin of the carbonate rocks and their mode of

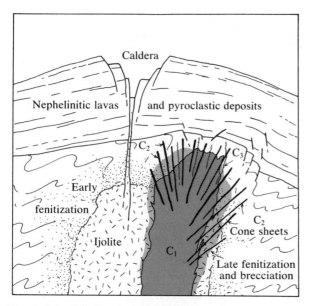

Figure 6-3 Idealized spatial relations of a carbonatite–nephelinite association as seen in east Africa. Intrusion of ijolite magma is accompanied by early fenitization of wall rocks; eruption of magma to surface builds a volcanic cone of nephelinite, which subsequently experiences caldera collapse. Carbonatite magma C_1 (dark shaded) intruding into the ijolite and older country rocks and doming the roof rocks is accompanied or preceded by a wave of fenitization and brecciation (light shaded). Resurgence of the magma locally breaches its fenite envelope. A swarm of later carbonatite cone sheets (C_2) is then emplaced, followed by still later carbonatite dikes (C_3). [After M. H. LeBas, 1977, *Carbonatite-Nephelinite Volcanism: An African Case History* (New York: Wiley Interscience).]

Figure 6-4 Thin dike of carbonatite with alkali amphibole emplaced into granite has produced fenite borders of similar amphibole and biotite.

emplacement? Second, how does this paired rock association originate? Third, granting that this silicate–carbonate assemblage represents a unified magmatic system, how and where did the parent magma form and how is it related to the stable tectonic or continental rift setting?

The magmatic origin of carbonatites has been one of most elusive concepts in petrology and only came to be accepted within the past couple of decades after a half century of controversy, fueled by conflicting field interpretations, misleading labora-

tory experiments, and no doubt blind prejudice. Mobilized limestone (or marble), hydrothermal carbonate solutions, and carbonate gases were all considered at different times as agents of emplacement. However, field relations strongly indicated a magmatic origin for some perceptive petrologists. But not until 1958 did O. F. Tuttle and P. J. Wyllie experimentally confirm that melts of carbonate could exist at realistically low temperatures (600°C to 700°C) in the presence of a dense H_2O-CO_2 vapor. And then, in 1960, geologists found unassailable proof of carbonatite magma by witnessing extrusion of a sodium carbonate lava in the crater of the nephelinitic volcano Oldoinyo Lengai, Tanzania. Die-hard objections that this lava was nothing more than trona lake beds melted by the nephelinitic magma were laid to rest by the demonstration that the Sr-isotope composition was that of primitive mantle-derived material, not sediment. Trace element concentrations in carbonatite and sedimentary carbonate rocks are entirely different (see Table 6-2).

Experimental investigations, summarized by LeBas (1977, pp. 285–287), indicate two possible mechanisms of producing a carbonatite–nephelinite magmatic association: (1) Low-T (< 800°C) crystal–liquid fractionation of a nephelinitic parental magma (see Table 6-3); or (2) a higher-T (> 800°C) immiscibility between nephelinitic and carbonatitic melts. The latter mechanism is particularly intriguing. In experimental systems with high-CO_2 vapor pressures, two liquids and a dense vapor coexist in equilibrium. Al, Fe, Mg, and K partition together into a silicate melt with some dissolved carbonate; a coexisting carbonate melt is enriched in Na and Ca, and the vapor phase carries CO_2, H_2O, Na, and some Si. This three-phase experimental system is a remarkably good model of the association of nephelinite and carbonatite melts and fenitizing solutions.

The third question posed above regarding the origin of the parental magma of the carbonatite–nephelinite association will be deferred to Section 6.5, after discussion of other alkaline associations.

Highly Potassic Rock Bodies

In contrast to the sodic nephelinite–carbonatite association just described, we now consider highly potassic rocks in which $K_2O/Na_2O > 3$. Such rocks (Carmichael and others, 1974, p. 252–262,

Table 6-3 Chemical compositions of carbonatites and a hypothetical parent magma for the carbonatite–nephelinite association (in wt.%).

	Average carbonatite	Average alkali carbonatite, east Africa	Nephelinite parent magma, east Africa
SiO_2	10.29	0.58	42.03
TiO_2	0.73	0.10	2.41
Al_2O_3	3.29	0.10	12.42
Fe_2O_3	3.46	0.29	6.77
FeO	3.60	0.00	4.96
MnO	0.68	0.14	0.31
MgO	5.79	1.17	5.75
CaO	36.10	15.54	12.22
Na_2O	0.42	29.56	5.64
K_2O	1.36	7.14	2.69
P_2O_5	2.09	0.95	0.67
CO_2	28.52	31.72	1.41
H_2O	1.44	5.15	2.76
SrO	0.46	—	0.03
BaO	0.40	—	0.03
F	0.81	—	0.07
Cl	—	—	0.09
SO_3	1.40	—	0.08

Source: After E. W. Heinrich, 1966, *The Geology of Carbonatites* (Chicago: Rand McNally); and M. H. LeBas, 1977, *Carbonatite–Nephelinite Volcanism: An African Case History* (New York: Wiley Interscience).

511–517), in which leucite is generally a major constituent, seem to be confined to continental regions. Unique volcanic rocks and shallow intrusive plugs comprised chiefly of phlogopite, leucite and diopside are found in the Leucite Hills of Wyoming, in the West Kimberly province of western Australia, and in southern Spain. Basically, two unusual rock types comprise the Tertiary bodies in Wyoming:

Wyomingite Phenocrysts of phlogopite set in a matrix of leucite, diopside, apatite, and locally glass; **orendite** is a variant with sanidine in the matrix.

Madupite Phenocrysts of diopside poikilitically enclosed by phlogopite in a groundmass of diopside, leucite, and apatite.

Other minerals in these rocks include olivine, potassic amphibole, perovskite, $(Ca, Na, Fe, Ce, Ti, Nb)O_3$, wadeite, $ZrK_4Si_6O_{16}$, and priderite, $(K,Ba)_{1.5}$ $(Ti,Fe^{2+},Mg)_{8.25}O_{16}$. There are no Fe–Ti oxides, and the leucites and sanidines are rich in Fe^{3+} (substituting for Al^{3+}). Chemically these peralkaline rocks are oversaturated to highly undersaturated in silica (see Table 6-4).

Highly potassic rocks of somewhat different character, made famous through the work of Arthur Holmes, are found in southwest Uganda in the western rift of east Africa (see Figure 6-2). Among these are mafic, ultrabasic, highly undersaturated volcanic rocks with major amounts of olivine, melilite, and the feldspathoids leucite, nepheline, and kalsilite, $KAlSiO_4$, together with perovskite and biotite. Three principal rock types are recognized:

Katungite Olivine, melilite, and K-rich glass
Mafurite Olivine, augite, and kalsilite
Ugandite Olivine, augite, and leucite

Inclusions of alkaline peridotites and biotite pyroxenites are abundant in these rocks.

The trace element and Sr-isotopic composition of highly potassic rocks are as unusual as their major element concentrations. Table 6-5 shows high concentrations of Rb, Sr, Zr, and Nb, as are found in some very felsic rocks, yet Cr and Ni are also high (see Table 6-4), as in most ultramafic rocks; this feature is also typical of kimberlites (to be discussed next).

In a study of Sr-isotopic ratios by Bell and Powell (1969), highly potassic African rocks have a pronounced positive correlation between initial $^{87}Sr/^{86}Sr$ and Rb/Sr and negative correlation with Sr, Nb, and Zr. They use these data to test numerous petrogenetic hypotheses advanced by many

Table 6-4 Chemical and normative compositions of highly potassic rocks in Wyoming and Uganda (in wt.%).

	Wyomingite	Madupite	Katungite
SiO_2	55.43	43.56	35.37
TiO_2	2.64	2.31	3.87
ZrO_2	0.28	0.27	—
Al_2O_3	9.73	7.85	6.50
Cr_2O_3	0.02	0.04	0.01
Fe_2O_3	2.12	5.57	7.23
FeO	1.48	0.85	5.00
MnO	0.08	0.15	0.24
NiO	—	—	0.19
MgO	6.11	11.03	14.08
CaO	2.69	11.89	16.79
SrO	0.27	0.40	0.04
BaO	0.64	0.66	0.25
Na_2O	0.94	0.74	1.32
K_2O	12.66	7.19	4.09
P_2O_5	1.52	1.50	0.74
H_2O^+	2.07	2.89	2.78
H_2O^-	0.61	2.09	1.15
SO_3	0.46	0.52	—
Total	99.75	99.51	99.65
Q	6.03	—	—
Z	0.42	0.40	—
Or	53.12	1.57	—
Lc	—	32.09	18.75
Ne	—	0.19	5.96
Th	0.82	0.92	—
Ac	4.35	2.21	—
Ks	6.01	—	—
Wo	0.88	19.72	—
Hy	15.21	17.04	—
Ol	—	7.31	40.89
Cm	0.03	0.06	—
Il	3.28	2.08	7.30
Hm	0.62	4.81	2.72
In	2.24	—	—
Pf	—	2.07	—
Ap	3.60	3.55	1.55
Mt	—	—	6.50

Source: After I. S. E. Carmichael, F. J. Turner, and J. Verhoogen, 1974, *Igneous Petrology* (New York: McGraw-Hill).

investigators to explain the origin of potassic rocks. Once popular hypotheses of crystal fractionation and assimilation of limestone are rejected in favor of three models (between which they can-

Table 6-5 Trace element (in ppm) and Sr-isotope composition of highly potassic rocks listed in Table 6-4.[a]

	Wyomingite	Madupite	Katungite
Cr	(140)	(270)	(70)
Ni	—	—	1500
Rb	—	205	118
Sr	(2300)	3400	2560
Zr	(2000)	(2000)	580
Nb	—	—	223
Rb/Sr	—	0.060	0.046
$^{87}Sr/^{86}Sr$ (initial)	0.7070	0.7066	0.7051

[a] Values in parentheses recalculated from Table 6-4.

Source: After I. S. E. Carmichael, F. J. Turner, and J. Verhoogen, 1974, *Igneous Petrology* (New York: McGraw-Hill), p. 515.

not choose): two involve contamination of nephelinitic or carbonatitic magma rich in Nb, Sr, and Zr with old sialic crustal material rich in ^{87}Sr, and the third involves partial melting of old K–Rb-rich ultramafic source rocks.

It is difficult, however, to account for the origin of highly potassic rocks in Wyoming and western Australia by the first two favored models. No carbonatitic or nephelinitic rocks occur in these areas, and initial $^{87}Sr/^{86}Sr$ and Rb/Sr show no positive correlation.

Kimberlite

Kimberlites share certain chemical properties with highly potassic rocks yet have several distinctive aspects worthy of special attention. Because they are the only primary source of natural diamonds, their economic importance is considerable. Because of the presence of other high-P crystalline inclusions derived from depths in the mantle where diamond is stable, kimberlites provide an unparalleled opportunity to investigate the actual

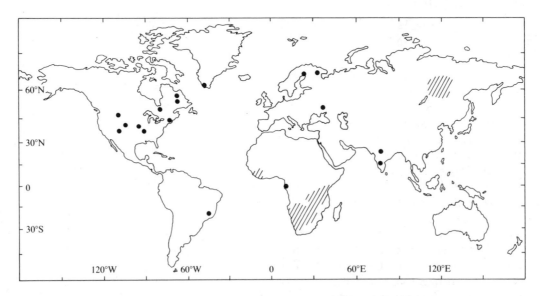

Figure 6-5 Distribution of kimberlite. Major occurrences in Africa and Russia denoted by diagonally ruled areas. Minor occurrences shown by filled circles. [After J. B. Dawson, 1971, Advances in kimberlite geology, *Earth Science Reviews* 7.]

composition and fabric of upper mantle rocks and, in themselves, furnish a sample of the sort of melts that are generated at these depths. For these reasons, much effort has been devoted to study of kimberlites (Boyd and Meyer, 1979a,b; Ahrens and others, 1975; Nixon, 1973).

Kimberlites have been sporadically emplaced into stable, nonorogenic continental platforms, especially in Africa and Siberia, since Proterozoic time (see Figure 6-5). Generally, the significance of any structural control on the occurrence of kimberlite bodies is obscure at the surface; hidden deep crustal or upper mantle structures could govern their site of emplacement.

Characteristics. **Kimberlite** may be defined as a potassic ultrabasic hybrid igneous rock containing large crystals (megacrysts) of olivine, enstatite, Cr-rich diopside, phlogopite, pyrope–almandine, and Mg-rich ilmenite in a fine-grained matrix of serpentine, phlogopite, carbonates, perovskite, and other minerals. Not all kimberlites contain diamond, and in those that do, it is a very widely dispersed mineral. The minute concentrations of diamond are attested by the fact that in the famous Kimberly mine, 24 million tons of kimberlite yielded only 3 tons of diamond, or one part in 8 million.

Most kimberlite near the surface occurs in pipelike **diatremes** (see below). Diatremes of kimberlite are small, generally less than one km² in horizontal area. They tend to occur in clusters, and some can be demonstrated in deep mine workings to coalesce at depth with dikes of nonfragmental kimberlite (see Figure 6-7). These dikes are thin, less than 10 m, but may be as much as 14 km long. Dikes and diatremes may be mutually crosscutting. Flow and cumulate fabrics have been observed in thin sills formed from very fluid, carbonate-rich kimberlite melt.

The hybrid nature of kimberlite is manifest in the rock and mineral inclusions that are commonly present. Because of unquestioned chemical interaction between these inclusions and the matrix, it is difficult to be sure of the composition of the kimberlite melt, now represented essentially by

the matrix, before inclusions were entrained. Many of the megacrysts, typically with anhedral and even embayed shapes, are obviously derived from fragmentation of garnet peridotite and eclogite xenoliths (see pages 214–215) that are locally abundant in some kimberlite bodies (see Figure 6-8). Other xenoliths are of crustal derivation and include: (1) angular pieces, sometimes very large, of stratigraphically higher sedimentary rocks that apparently subsided into the kimberlite diatreme; (2) fragments of adjacent sedimentary country rocks; (3) metamorphic xenoliths from the deeper crust.

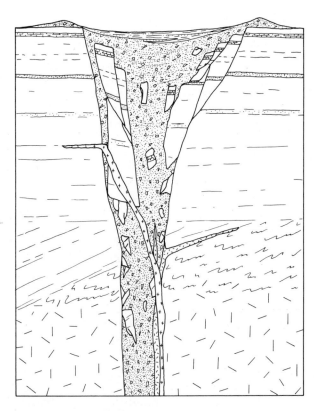

Figure 6-6 Idealized diatreme terminating in a maar at the surface. The maar is surrounded by a tuff ring and is partly filled with lake sediments and alluvium. Note inward-dipping stratification near the top of the diatreme, caved fragments of wall rock, and late magmatic dikes. The horizontal scale is greatly exaggerated relative to the vertical.

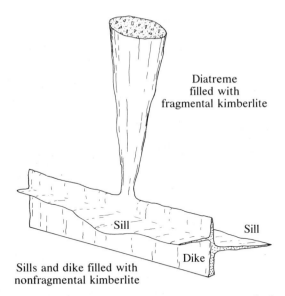

Figure 6-7 Schematic relationship between explosive kimberlite diatreme and deep-seated sills and feeder dike filled with nonfragmental kimberlite magma that solidified *in situ*. [After J. B. Dawson, 1971, Advances in kimberlite geology, *Earth Science Reviews* 7.]

Figure 6-8 Xenolithic kimberlite. The abundant lighter-colored and rounded inclusions (note especially the large one on which the hammer rests) are mantle-derived and olivine-rich. Smaller, angular dark sedimentary xenoliths are barely visible in the kimberlite block behind and to the left of the hammer. [Photograph courtesy of Keith G. Cox.]

The hybrid character of kimberlite is also reflected in the chemical composition of clean, inclusion-free kimberlite (see Table 6-6). Some elements, such as Cr and Ni, have concentrations similar to typical mantle peridotite (see Section 6.4); other constituents—including H_2O, CO_2, P_2O, K, Ti, Rb, Sr, and Nb—have concentrations similar to those in carbonatites and other highly alkaline rocks (see Tables 6-2 and 6-5). Possible reasons for these hybrid compositional features will be considered in Section 6.5.

Emplacement of diatremes. **Diatremes** are pipes or narrow, funnel-shaped bodies filled with accidental and gas-charged juvenile magmatic material (see Figure 6-6). Characteristically, (1) the juvenile material is kimberlite or chemically related volatile-rich alkaline mafic to ultramafic rock; (2) blocks of now eroded, stratigraphically higher rocks, some many meters across and highly fri-

able, are found concentrated around the margin of the pipe, whereas well-rounded and apparently abraded xenoliths from deeper crustal or even upper mantle regions tend to be concentrated in the center; (3) thermal contact metamorphic effects on xenoliths and wall rocks are weak or absent, even on highly susceptible coal and even though kimberlite diatremes in South Africa connect at depth to dikes emplaced as hot magma; (4) where connections to the original surface are observable, diatremes terminate in maars and tuff rings; (5) fragmental material, especially near the original top of the diatreme, may be bedded with inward dips; (6) ring faults and subsidence of the diatreme filling may be apparent; (7) many diatremes are composite, showing multiple emplacement of fragmental material and of latter intrusion by fluid magma; (8)

Table 6-6 Average composition of eight garnet peridotite xenoliths in alkaline volcanic rocks, Lashaine volcano, Tanzania; South West Africa micaceous kimberlite; and eclogite from the Roberts Victor Mine, Orange Free State.

	Garnet lherzolite[a]	Kimberlite	Eclogite
SiO_2	43.7 wt.%	31.1 wt.%	45.67 wt.%
TiO_2	0.07	2.03	0.42
Al_2O_3	1.63	4.9	17.85
Fe_2O_3	0.88	5.9	2.88
FeO	6.00	10.5	8.46
MnO	0.13	0.1	0.17
MgO	44.5	23.9	11.90
CaO	1.09	10.6	7.35
Na_2O	0.12	0.3	2.01
K_2O	0.09	2.1	0.39
CO_2	—	7.1	—
P_2O_5	0.06	0.66	0.04
Total	99.1	99.19	100.29[b]
Cr	3,400 ppm	1,000 ppm	
Ni	2,800	1,200	
Rb	3.6	250	
Sr	15	600	
Zr	4.6	190	
Nb	1.4	200	

[a] Modal composition calculated from the chemical analysis is olivine 72%, enstatite 18%, diopside 2%, and garnet 6%.
[b] Includes H_2O^- 1.07 and H_2O^+ 2.01.
Source: Garnet lherzolite after J. M. Rhodes and J. B. Dawson in L. H. Ahrens, J. B. Dawson, A. R. Duncan, and A. J. Erlank (eds.), 1975, *Physics and Chemistry of the Earth,* Vol. 9 (New York: Pergamon); kimberlite after J. B. Dawson in P. J. Wyllie (ed.), 1967, *Ultramafic and Related Rocks* (New York: John Wiley and Sons); eclogite after I. Kushiro and K. Aoki, 1968, Origin of some eclogite inclusions in kimberlite, *American Mineralogist* 53.

swarms of Tertiary diatremes and maars in Europe and the Colorado Plateaus in western North America are associated with lake sediments, testifying to abundant surface water during magmatic activity.

It is uncertain whether phreatomagmatic explosions producing the maar in some way trigger formation of an underlying diatreme due to propagation of a decompression wave down the magma column, or whether the maar is secondary to the diatreme. Current thinking favors rapid rise of highly gas-charged (CO_2 + H_2O) kimberlitic magma into the crust. Somewhere en route, sufficient gas boils out of the decompressing magma to form an expanded fluidized column (see page 85) that may cause upwarping of the overlying lid of rock and eventual rupture, allowing the fluidized mixture of gas and particulate material to erupt. Velocities of particles within the fluidized column may not be great, but sufficient internal abrasion might occur to produce the observed rounding of fragments; a convecting and pulsating system with surges of pressure and relaxation would allow intermittent spalling and enlargement of the walls of the pipe and subsidence of wall-rock fragments. The fluidizing gas might be relatively dense, depending upon the confining pressure, and might not be very hot due to the cooling effects of its exsolution from the melt and consequent adiabatic expansion and the work of transporting solids up the diatreme.

Laboratory model experiments by Woolsey and others (in Ahrens and others, 1975) have shown that fluidized intrusive systems can form features strikingly similar to those seen in natural diatremes.

6.4 MANTLE-DERIVED MAFIC AND ULTRAMAFIC XENOLITHS IN ALKALINE MAGMAS

Study of the Earth's mantle has been one of the frontier efforts in geological sciences during the past couple of decades, besides the more visible explorations of the ocean floors and other planetary bodies. Once the dream of drilling into the mantle—the "Mohole" project—was abandoned, petrologists were obliged to find other means of investigating the mantle. To complement geophysical measurements and laboratory experiments, tangible, real pieces of mantle rock were needed. Ultramafic rocks with metamorphic fabric in alpine complexes and ophiolite are such mantle samples, but the usually strong overprinting of deformation and recrystallization on material obducted onto continents obscures their original character.

The value of mantle-derived xenoliths in volcanic rocks lies in their generally fresh and unmodified condition. Although some have visibly reacted with the enclosing host magma during ascent—as shown by penetrating veinlets of melt (now glass), compositional zoning in xenolith crystals next to the host, and rounding and partial dissaggregation of the xenolith—other inclusions are sharply angular and pristine in appearance. Xenoliths of several centimeters in diameter, locally even of football size, must have been carried rapidly to the surface in a matter of days within the entraining magma. Otherwise, during slow magma ascent their greater density ($\sim$ 3.3 g/cm^3) than the host magma ($\sim$ 3.0) would have caused them to settle and to be assimilated.

The obvious disadvantage of xenoliths is that the "stratigraphy" of the mantle region represented in them is scrambled during the fragmentation and entrainment processes. Despite such limitations, xenoliths provide a wealth of information on the actual fabrics and chemical and mineralogical compositions of the mantle, and from these data various processes of evolution can be inferred. Certain mineralogical compositions serve as geothermometers and geobarometers to constrain the location in the mantle from which the xenolith sample was taken.

Pieces of mantle rock are found in igneous rocks at hundreds of localities all over the world in both oceanic and continental regions. In any given host body, these fragments will range from very sparse small clots comprised of but a few grains to occurrences in which football-size fragments choke over half the body. In many occurrences, the inclusions are ejected explosively from a volcanic vent and are found as free, clean pieces with juvenile magmatic tephra or as coated or cored bombs (see Figure 6-9).

Host rocks of mantle-derived inclusions are virtually always alkaline with at least normative *Ol*

and usually *Ne*. They also range from basic (basaltic) to ultrabasic (kimberlite and allied highly undersaturated rock types). The reason subalkaline magmas, including mantle-derived tholeiitic types such as those extruded from oceanic rifts, are not host to mantle-derived inclusions is not entirely clear. Perhaps subalkaline magmas with sources in the mantle drop any entrained fragments during slow or interrupted ascent through the crust. Perhaps the intrinsically more gas-rich alkaline magmas ascend more vigorously and even explosively through the crust. There may be other reasons.

Use of the term xenolith implies the rock inclusion is foreign to the host rock in which it occurs. Evidence of this exotic character is supported by the presence of metamorphic rather than magmatic fabrics and isotopic compositions that are out of equilibrium with the host rock (see, for example, Barrett in Ahrens and others, 1975).

Textural and Compositional Types of Inclusions

Fragments whose mineralogical composition bespeaks crystallization at upper mantle *P–T* conditions include both rocks and single, isolated crystals. If foreign to their host, they are xenoliths and xenocrysts; if this character is uncertain, they can simply be referred to in a nongenetic way as inclusions or nodules and, in the case of the single crystals, which can be quite large, as megacrysts.

Inclusions in basaltic host rocks. Xenoliths of peridotite are overwhelmingly predominant in most basaltic hosts (see Figure 6-9). Generally, modal proportions are such that olivine $\gg$ enstatite $>$ diopside, so that most are lherzolite, less are harzburgite, and dunite is rare (see Figure 5-26). Olivine and enstatite typically have about 90 mol % of the magnesian end-member and the emerald-green (in hand sample) chromian diopside has less than 5 mol % of the ferrosilite end-member. Minute spinels (deep red-brown in thin section but black in hand specimen) present in most of these xenoliths are compositionally variable solid solutions

Figure 6-9 Cored bomb. (a) A rounded, mantle-derived fragment of spinel-bearing chromian diopside lherzolite is surrounded by black vesicular basaltic rock that adhered to the fragment as magma when it was ejected from an explosively erupting volcano. Abundant gray grains are olivine ($\sim$Fo$_{90}$), black grains are enstatite ($\sim$En$_{90}$) and chromian diopside. (b) Photomicrograph under cross-polarized light showing metamorphic texture produced by grain growth in solid state, with a late overprint of deformation that produced the bands of variable optical extinction in an olivine grain.

$(Mg,Fe)(Cr,Al)_2O_4$ in which usually Al > Cr. Textures are very diverse but are all metamorphic.

Many rock bodies containing chromian diopside peridotite xenoliths also have a distinctively different but much less abundant and, therefore, frequently overlooked suite of inclusions characterized by black Fe–Ti–Al–rich augite. Megacrysts of this mineral, which can be several centimeters in diameter, typically show good conchoidal fracture but no cleavage. At some localities, megacrysts of black kaersutitic amphibole are also found together with generally less common crystals of hypersthene, phlogopite, spinel, olivine, feldspar, and apatite. Most megacrysts are anhedral and show evidence of corrosion, presumably in the host magma. Rock fragments composed of the

same sort of minerals are chiefly various pyroxenites (see Figure 5-26), but peridotites and gabbros also occur; the latter type may be of crustal derivation in view of pressure limitations on feldspar stability. Textures are variable but not uncommonly magmatic; rare inclusions have poikilitic texture with either amphibole, augite, or phlogopite serving as oikocrysts.

Composite rock inclusions with both the green chromian diopside and black Ti–Al–rich augite assemblages present in the same fragment are rare but petrologically very significant (Irving, 1980). The black augite assemblage is found as thin veins cutting the green diopside peridotite (see Figure 6-10), or fragments of the latter are included in the former. Both relations indicate invasion of the

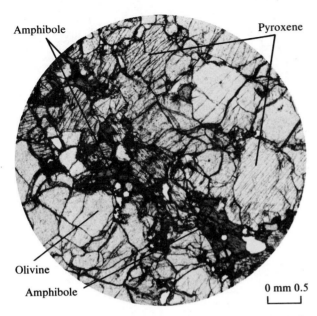

Figure 6-10   Thin, discontinuous vein of amphibole in chromian diopside lherzolite.

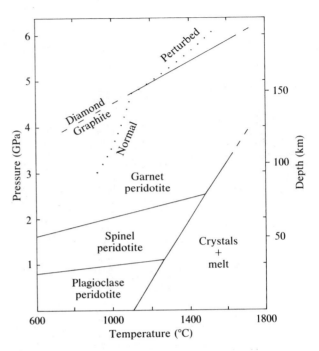

Figure 6-11   Mineralogical relations at high pressures relevant to inclusions in kimberlite. Dotted lines show range of *P* and *T* conditions under which garnet lherzolite xenoliths in northern Lesotho kimberlites crystallized. Lherzolites crystallizing along the normal geothermal gradient might contain graphite, whereas those crystallizing along the upper part of the perturbed gradient might contain diamond. [Lines delineating the stability fields of garnet, spinel, and plagioclase peridotite (olivine + enstatite + diopside) and the field of partial melted peridotite (crystals + melt) are after P. J. Wyllie, 1970, Ultramafic rocks and the upper mantle, Mineralogical Society of America Special Paper 3; boundary between stability fields of diamond and graphite after C. S. Kennedy and G. C. Kennedy, 1976, *Journal of Geophysical Research* 81; dotted lines after Boyd and Nixon in L. H. Ahrens, J. B. Dawson, A. R. Duncan, and A. J. Erlank (eds.), *Physics and Chemistry of the Earth*, Vol. 9 (New York: Pergamon).]

green diopside peridotite by fluid material that crystallized into the black augite assemblage before fragmentation and entrainment into the host magma as xenoliths. Strong chemical zoning occurs in green diopside minerals within a millimeter or so of their contact with the chemically disparate vein material.

Megacrysts may be derived by fragmentation of veins with pegmatitic fabric.

Inclusions in kimberlitic and other highly alkaline host rocks.   These fragments have been described by Ahrens and others (1975), Sobolev (1977), and Boyd and Meyer (1979b). The dominant mantle rock type is again a green chromian diopside peridotite, mostly lherzolite but including harzburgite; dunite and pyroxenites are still less common. However, in kimberlitic host rocks spinel is rare or absent and instead the highly aluminous mineral is a pyropic garnet, sometimes with a substantial Cr content. According to laboratory studies, these garnetiferous peridotites crys-

tallized at greater pressures (corresponding to depths of 100 to 200 km) than the more or less chemically similar spinel peridotites typically occurring in basaltic hosts (see Figure 6-11).

There are significant variations in the metamorphic fabric of the garnet lherzolite xenoliths in kimberlite from northern Lesotho that correlate with composition and *P–T* conditions of crystallization. Those formed at lowest *P* and *T* are coarsely (2 to 4 mm) granular, whereas ones crystallized at progressively higher *P* and *T* show evidence of progressively greater solid-state shear deformation, producing a finer-grained matrix around surviving larger grains. Boyd and Nixon (in Ahrens and others, 1975) believe that the granular lherzolites crystallized along a normal geothermal gradient in the upper mantle, whereas deeper lherzolite was perturbed off the normal gradient by some unknown process that produced the sheared fabric and the inflection in the *P–T* range of lherzolite crystallization. However, these interpretations of model geothermal gradients have been challenged on a number of grounds (see, for example, Harte, 1978; Mercier, 1979). Some of the granular lherzolites contain accessory phlogopite, chromite, and rare graphite, whereas the sheared rocks lack these minerals. The granular lherzolites are depleted in Na, Ti, Fe, Ca, and Al (but not K) relative to the sheared ones. These constituents are among those in basaltic melts produced by partial fusion; hence, Boyd and Nixon refer to the granular lherzolite as sterile and the sheared as more fertile for basalt magma production. These designations are relative only.

Generally less common than the garnet peridotites, but in some kimberlite bodies more abundant, are a variety of other inclusions. Among these are eclogite and megacrysts of minerals similar to those that constitute xenoliths. **Eclogite** is essentially a bimineralic rock comprised of pyrope–almandine garnet and a clinopyroxene rich in jadeite and diopside end-members; minor accessories include kyanite, rutile, corundum, and phlogopite. Some eclogite xenoliths have mineralogical compositions indicative of crystallization in the deep continental crust and are quite similar to lenses of eclogite found in some deeply eroded Precambrian metamorphic terranes. Other eclogites, sometimes called "griquaites," occur as discrete xenoliths and as lenses in garnet peridotite

and have higher-*P* mineralogical compositions, including the very rare presence of diamond (Nixon, 1973, p. 309). The similarity in bulk chemical composition between eclogite and basaltic rocks (compare Table 6-6 and tables in Chapter 5) suggests that eclogite xenoliths are samples of basaltic magma that crystallized in pockets within the peridotitic mantle.

Major Implications of Mantle-Derived Inclusions

The foregoing discussion should make amply clear that mantle-derived inclusions in kimberlitic and basaltic rocks are significantly heterogenous in fabric and composition. It is impossible to say whether this mantle sample is truly representative of the upper 100 km or so of the mantle worldwide or only represents its character in regions of magma genesis. However, even if only the latter is true, we have gained some significant insights into processes of magma genesis and the nature of their mantle sources.

Mantle source regions. To a first approximation, the uppermost mantle sample in magma source regions is an olivine-rich peridotite of lherzolite composition (see Figure 6-9 and Table 6-6). It contains spinel or garnet as the Al-rich mineral, depending upon depth (and pressure) in the mantle. In detail, however, this universal uppermost mantle rock is variable in texture and composition, undoubtedly because of any number of solid-state metamorphic processes of deformation and recrystallization and magmatic processes of partial melting, magma segregation, and intrusion. Although it may be tempting to tabulate the chemical composition of the uppermost mantle based on such xenoliths, this could be quite misleading. A major pitfall lies in the uncertainty about how much mafic melt might have been previously extracted from the mantle sample. While it is possible to say that a certain xenolith has a more fertile composition than another, it is questionable whether any mantle-derived xenolith represents pristine mantle.

Metasomatism in the uppermost mantle.   One of the significant discoveries made during investigations of mantle-derived inclusions is that much of the sampled mantle has experienced variable degrees of metasomatism (Lloyd and Bailey, 1975). In this process, sterile, refractory Mg-rich peridotite more or less depleted in constituents that have been extracted in basaltic partial melts has been infiltrated by solutions that modify its bulk chemical and mineralogical composition. The nature of the solutions is uncertain; they may contain substantial CO_2, almost certainly are rich in water, and could be either metamorphic-type aqueous fluids at submagmatic temperatures or silicate melts. The solutions most notably carry incompatible elements (such as P, Ti, and large ion lithophile elements K, Rb, Cs, Ba, Pb, Th, U, and so on) that are deposited in the refractory peridotite. In a sense, this type of metasomatism in the mantle is analogous to the fenitization that occurs in the crust around carbonatites.

The most obvious evidence for this metasomatism lies in veins and secondary grains of phlogopite and amphibole and the black augite-bearing rock assemblages that have invaded chromian diopside peridotite xenoliths (see page 213; Lloyd and Bailey, 1975; Best, 1974; and many other references in Boettcher and O'Neil, 1980). Less conspicuous but nonetheless compelling evidence for infiltration metasomatism was adduced by Frey and Green (1974) from detailed chemical data on some chromian diopside lherzolite xenoliths. They found that they are a mix of two chemically and mineralogically distinctive components: (1) major minerals and chemical elements together with compatible trace elements (Cr, Ni, heavy rare earth elements) that have the characteristics of a refractory sterile residuum left over after extraction of a basaltic partial melt; and (2) accessory minerals including apatite, phlogopite, and amphibole and incompatible elements (K, Ti, P, Th, U, light rare earth elements) that have the characteristics of a silica-undersaturated alkaline nephelinite or some sort of highly potassic rock. It may be noted that the phlogopites and kaersutitic amphiboles (Appendix C, Tables C-6 and C-7) of concern here have silica-undersaturated alkaline compositions, as may be verified by calculating their norms, and they also have high concentrations of incompatible elements.

It appears that at least part of the uppermost mantle represented in chromian spinel lherzolite xenoliths has experienced a two-stage chemical evolution. Subsequent to an early melting event that left it depleted to some degree in basaltic constituents, it has been infiltrated, perhaps repeatedly, by metasomatic fluids forming veins and isolated grains enriched in much the same constituents as first extracted. In the second metasomatic stage(s), portions of the mantle were exceedingly enriched in the same suite of incompatible elements that are so characteristic of the alkaline magmas discussed earlier in this chapter (see Table 6-7).

The petrogenetic ramifications of a metasomatized mantle will be briefly considered in the next section and in more detail in Section 15.4.

6.5 ORIGIN OF ALKALINE MAGMAS

Once-popular notions that alkaline magmas originate by contamination of mantle-derived basaltic magmas with granitic crustal rocks and by assimilation of limestone into magmas must be generally abandoned. Although very small volumes of alkaline rocks demonstrably occur at some basalt–limestone contacts, a wealth of data on trace element concentrations and isotopic composition and the occurrence of alkaline rocks in oceanic islands indicates that these notions are untenable. Moreover, the presence of dense xenoliths with high-pressure mantle mineralogical compositions in kimberlite and other alkaline rocks indicates rapid ascent of these highly gas-charged host magmas from mantle source regions.

The relatively small volumes of alkaline rocks exposed in intraplate settings that are tectonically stable or only very slowly rifting apart contrast strikingly with the copious volumes of subalkaline magmas produced along more vigorously spread-

Table 6-7 Comparative concentrations of certain elements in alkaline volcanic rocks and associated xenoliths and in average nonalkaline ultramafic rock.

	Ultramafic rock	Ugandan volcanics[a]	Concentration factor	Phlogopite clinopyroxenite xenolith	Amphibole clinopyroxenite xenolith
Na	3,647 ppm	14,600 ppm	4	5,300 ppm	17,400 ppm
P	195	4,300	22	800	3,100
K	180	40,700	226	28,500	10,400
Ti	300	31,600	105	22,200	17,800
Rb	2	126	74	84	—
Sr	10	2,127	207	519	878
Zr	37	441	12	175	188
Nb	8	246	31	31	50
Ba	<1	1,781	2,968	1,146	481
La	2	162	81	—	60
Cr	2,000	364	0.2	21	105
Ni	1,833	142	0.1	34	28

[a] Average of 23.
Source: After F. E. Lloyd and D. K. Bailey, 1975, Light element metasomatism of the continental mantle: The evidence and the consequences, *Physics and Chemistry of the Earth* 9.

ing oceanic and continental rifts and converging plate margins. Alkaline magma generation typifies regions of thicker lithosphere that have lower thermal gradients than vigorously spreading or converging plate regimes. Where the lithosphere is thick, melting conditions are reached only at considerable depths; kimberlite magmas, as an extreme example, must have been derived from depths of at least 100 to 200 km in order to incorporate diamonds and their other inclusions (see Figure 6-11).

The small volume of alkaline magmas and their high concentrations of residual or incompatible elements implies derivation as residual, highly differentiated melts or, alternatively, as very small-degree partial melts. Some magmatic differentiation may well occur in particular areas and could account for recurring alkaline lineages, such as basanite–hawaiite–trachyte–phonolite. However, in other situations, such as the highly potassic rocks of Wyoming, there are no associated, less-evolved

rock types exposed that might be representative of parent magmas. It will be shown later (see pages 308–310) that small degrees of partial melting of peridotite in the presence of abundant CO_2 at high pressures (corresponding to depths of 100 to 200 km) can produce alkaline melts. But the composition of such melts does depend greatly on the concentrations of incompatible elements in the peridotite, that is, on whether it contains accessory apatite, phlogopite, amphibole, and so on. Highly alkaline volcanic rocks can have greater than two-hundred-fold enrichments in some elements such as K and Ba relative to possible source rocks (see the concentration factors in Table 6-7); this would require less than 0.5% partial melting, assuming all of the incompatible element entered the melt. Such small volumes of melt would be difficult to segregate from the crystalline residuum.

Another promising model for local generation of alkaline magmas follows from the demonstrated incompatible-element metasomatism in mantle rocks sampled as xenoliths in these magmas. As argued by Lloyd and Bailey (1975), Boettcher and O'Neil (1980), and many others, metasomatic infiltration over an extended period of time by incompatible-element-enriched fluids could build up high concentrations of these elements in local regions of the mantle, as shown by the black augite suite of xenoliths that carry amphibole and phlogopite (see Table 6-7). Partial melting of such metasomatized mantle could then yield alkaline magmas. Variably metasomatized mantle could conceivably yield, upon partial melting, a spectrum of alkaline magmas ranging from alkali olivine basalt to highly potassic wyomingites, katungites, nephelinites, and kimberlites.

6.6 PERALKALINE ASSOCIATIONS

Brief mention should be made of the rare peralkaline rocks in which molecular (Na_2O + K_2O) > H_2O so that *Ac, Ns,* and rarely *Ks* appear in the norm; this chemical property is manifest in the presence of alkali-rich pyroxenes and amphiboles, such as aegirine and riebeckite, rather than the more common augite, hornblende, and biotite of calc-alkaline rocks. Peralkaline rocks can be silica-undersaturated or oversaturated, such as the wyomingite and madupite listed in Table 6-4. Some of the more common peralkaline volcanic rocks are rhyolitic comendites and pantellerites with > 70 wt.% SiO_2 and > 25% *Q* (Carmichael and others, 1974, p. 236). In addition to major element and mineralogical contrasts with ordinary rhyolites (Appendix D), peralkaline rhyolites have much higher concentrations of F, Cl, Nb, Ta, Zr, Mo, and Cd.

Silicic peralkaline rocks are found in oceanic and continental regions of crustal extension, commonly in association with basaltic rocks that are transitional between tholeiites and alkali olivine basalts. Ascension Island on the Mid-Atlantic Ridge just south of the equator consists chiefly of such transitional basalts plus hawaiites and mugearites; trachytes and aegirine peralkaline rhyolites are minor.

More voluminous peralkaline volcanic rocks are found on the continents. Thousands of cubic kilometers of late Cenozoic peralkaline ash-flow tuffs and minor rhyolite flows in the Nevada portion of the Basin and Range Province of the western United States are associated with basalts. In eastern Australia, the Miocene Nandewar shield volcano consists predominantly of basalts and trachyandesite flows capped by peralkaline trachytes and rhyolites. Vast volumes of peralkaline rhyolite and trachyte pyroclastic deposits in the Kenya and Ethiopia rift valleys were noted previously in this chapter. In the Oslo, Norway graben area, Permian lavas include mildly alkaline basalts and peralkaline trachytes, the so-called rhomb porphyries.

The common association of transitional basalts (hawaiite, mugearite), peralkaline trachyte, and rhyolite suggests a recurring petrogenetic process. Many investigators (Carmichael and others, 1974, p. 496) have appealed to crystal fractionation of mafic parent magmas even though in some instances—Kenya and the Basin and Range Province—the volume of exposed felsic rock types is relatively excessive. However, trace element variation patterns appear to support the crystal fractionation model. Certainly, fractionation of Al-rich calcic plagioclases and Al-bearing pyroxenes is one way of depleting the residual melts in Al while enhancing their Na and K contents.

It should be noted that some peralkaline rocks comprise intrusive bodies. Among these are the White Mountain magma series of New Hampshire and the younger granites of Northern Nigeria (Black and Girod in Clifford and Gass, 1970). Both comprise Jurassic epizonal ring complexes. The Nigerian province lies within a northeast-trending graben system extending inland from the coastal reentrant and may constitute a failed arm of the three-pronged rift that grew into the Atlantic Ocean as Africa and South America separated. The Nigerian granites, like many other **anorogenic** granites of similar character, have high concentrations of F, Sn, Ta, Nb, Zr, Mo, Y, Th, U, and rare earth elements. Some of these elements in particular intrusive complexes are of sufficient concentrations to be of economic value.

SUMMARY

6.1. Alkaline rocks comprise a broad compositional spectrum typified by normative and modal feldspathoids together with unusually high concentrations of elements—K, Zr, Nb, Rb, Ba, F, Ti, P, and so on—incompatible in mantle minerals. Alkaline rocks are encumbered by more names than all other igneous rocks even though their aggregate worldwide volume is less than one percent of the total.

6.2. The ocean floors are dotted by hundreds of volcanoes that have formed by intraplate magmatism. Some island groups, such as the Galapagos and Hawaii, are built of tholeiitic shields capped by minor subalkaline differentiates and alkaline rocks. Other islands expose only alkaline rocks that contain prominant feldspathoids. Linear chains of volcanoes resting on older oceanic crust show a regular pattern of ages that suggests the moving oceanic plate has passed over a fixed, long-lived magma source in the mantle.

6.3. Relatively small volumes of alkaline rocks with extreme compositional variations are emplaced in continental regions experiencing uplift and very slow crustal extension or in tectonically stable cratons.

The east African rift system evolved during the Cenozoic by broadly concurrent crystal uplift, extensional faulting, and volcanism that has included voluminous outpourings of basaltic flood lavas and peralkaline trachytes and rhyolites together with minor highly alkaline rocks.

Associations of carbonatite with nephelinite and ijolite occur in small continental central volcanic and intrusive ring complexes. The origin of these small complexes, which have unusually high concentrations of apatite, Nb, and rare earth elements, is possibly linked to separation of an immiscible carbonate magma from a parental nephelinitic magma; alkali-rich fluids fenitize country rocks.

Small volumes of highly potassic but diverse rocks with leucite, kalsilite, and sanidine are petrogenetic puzzles.

Kimberlites are fine-grained, potassic, ultrabasic rocks containing megacrysts of olivine, enstatite, chromian diopside, phlogopite, pyropic garnet, Mg-ilmenite, and very rare diamond in a carbonated and serpentinized matrix. The high-P megacrysts are at least in part disaggregated mantle-derived xenoliths that together with common crustal xenoliths contribute to the hybrid character of kimberlite. Chemically, this character is manifest in high concentrations of elements that both compatible (Cr, Ni) and incompatible (K, Rb, Ti, C, Nb, and so on) in mantle minerals. Kimberlite pipes are emplaced into upper crustal levels as highly gas-charged (CO_2 and H_2O) fluidized systems.

6.4. Relatively fresh inclusions of rocks and minerals from the upper mantle are brought rapidly to the surface only in alkaline magmas. Most inclusions fall into one of two compositional categories. The most abundant suite comprises lherzolites with green chromian diopside and either Mg–Fe–Al–Cr spinel (basaltic host rocks) or Mg–Fe–Cr garnet (kimberlitic hosts from more deeply derived magmas). Less abundant, chiefly pyroxenitic xenoliths and megacrysts are characterized by black Fe–Ti–Al rich augite, kaersutitic amphibole, and Ti-rich phlogopite. In rare composite xenoliths, the black augite assemblage occurs as veins cutting chromian diopside lherzolite.

At least the upper 200 km of the mantle in source regions of alkaline magmas is quite inhomogeneous, and even the apparently uniform chromian diopside lherzolites are variably depleted in basaltic constituents and incompatible elements that could have been extracted during one or more episodes of partial melting. The black augite–amphibole–phlogopite assemblages that locally demonstrably vein chromian diopside lherzolite are evidence for infiltrating, metasomatizing hydrous fluids laden with incompatible elements.

6.5. Generation of small volumes of alkaline magmas typifies regions of thicker lithosphere and lower geothermal gradients, in contrast to the copious volumes of subalkaline magmas produced in regions of converging and rapidly diverging plates, where geothermal gradients are higher. The unusually high enrichment in alkaline rocks of incompatible elements, locally to factors of 200 or more relative to subalkaline rocks, is not wholly satisfied either by differentiation mechanisms or by processes involving very small degrees of partial melting of mantle peridotite. But the recently recognized phenomenon of mantle metasomatism opens the door to some promising models of alkaline magma genesis.

6.6. Peralkaline silicic rocks with molecular $(Na_2O + K_2O) > Al_2O_3$ that are commonly associated with and may be derived by differentiation of marginally saturated to undersaturated basaltic rocks occur in much the same tectonic environments as alkaline rocks.

REFERENCES

Ahrens, L. H., Dawson, J. B., Duncan, A. R., and Erlank, A. J. (eds.). 1975. *Physics and Chemistry of the Earth,* Vol. 9 (Papers presented at the First International Conference on Kimberlites, Capetown, South Africa, September, 1973). New York: Pergamon, 936 p.

Baker, P. E., Gass, I. G., Harris, P. G., and LeMaitre, R. W. 1964. The volcanological report of the Royal Society expedition to Tristan da Cunha, 1962. *Philosophical Transactions of the Royal Society of London,* Series A, 256:439–575.

Baker, B. H., Mohr, P. A., and Williams, L. A. J. 1972. Geology of the eastern rift system of Africa. Geological Society of America Special Paper 136, 67 p.

Bell, K., and Powell, J. L. 1969. Strontium isotopic studies of alkalic rocks: The potassium-rich lavas of the Birunga and Toro-Ankole regions, east and central equatorial Africa. *Journal of Petrology* 10:536–572.

Best, M. G. 1974. Mantle-derived amphibole within inclusions in alkalic-basaltic lavas. *Journal of Geophysical Research* 79:2107–2113.

Boettcher, A. L., and O'Neil, J. R. 1980. Stable isotope, chemical and petrographic studies of high-pressure amphiboles and micas: Evidence for metasomatism in the mantle source regions of alkali basalts and kimberlites. *American Journal of Science* 280-A:594–621.

Boyd, F. R., and Meyer, H. O. A. (eds.). 1979a. *Kimberlites, Diatremes and Diamonds: Their Geology, Petrology and Geochemistry.* American Geophysical Union, Proceedings Second International Kimberlite Conference, Vol. I.

Boyd, F. R., and Meyer, H. O. A. (eds.). 1979b. *The Mantle Sample: Inclusions in Kimberlite and Other Volcanics.* American Geophysical Union, Proceedings Second International Kimberlite Conference, Vol. II.

Carmichael, I. S. E., Turner, F. J., and Verhoogen, J. 1974. *Igneous Petrology:* New York: McGraw-Hill, 739 p.

Clifford, T. M., and Gass, I. G. (eds.). 1970. *African Magmatism and Tectonics.* Edinburgh: Oliver and Boyd, 461 p.

Coombs, D. S., and Wilkinson, J. F. G. 1969. Lineages and fractionation trends in undersaturated volcanic rocks from the east Otago volcanic province (New Zealand) and related rocks. *Journal of Petrology* 10:440–501.

Dalrymple, G. B., Silver, E. A., and Jackson, E. D. 1973. Origin of the Hawaiian Islands. *American Scientist* 61:294–308.

Duncan, R. A., and McDougall, I. 1976. Linear volcanism in French Polynesia. *Journal of Volcanology and Geothermal Research* 1:197–227.

Frey, F. A., and Green, D. H. 1974. The mineralogy, geochemistry and origin of lherzolite inclusions in Victorian basanites. *Geochimica et Cosmochimica Acta* 38:1023–1059.

Gass, I. G. 1972. Proposals concerning the variation of volcanic products and processes within the oceanic environment. *Philosophical Transactions of the Royal Society of London,* Series A, 271:131–140.

Harte, B. 1978. Kimberlite modules, upper mantle petrology and geotherms. *Philosophical Transactions of the Royal Society of London,* Series A, 288:487–500.

Heinrich, E. W. 1966. *The Geology of Carbonatites.* Chicago: Rand McNally, 555 p.

Herz, N. 1977. Timing of spreading in the south Atlantic: Information from Brazilian alkalic rocks. *Geological Society of America Bulletin* 88:101–112.

Irving, A. J. 1980. Petrology and geochemistry of composite ultramafic xenoliths in alkalic basalts and implications for magmatic processes within the mantle. *American Journal of Science* 280-A: 389–426.

LeBas, M. H. 1977. *Carbonatite-Nephelinite Volcanism: An African Case history.* New York: Wiley Interscience, 347 p.

Lloyd, F. E., and Bailey, D. K. 1975. Light element metasomatism of the continental mantle: The evidence and the consequences. *Physics and Chemistry of the Earth* 9:389–416.

Macdonald, G. A., and Katsura, T. 1964. Chemical composition of Hawaiian lavas. *Journal of Petrology* 5:82–133.

McBirney, A. R., and Aoki, K. 1968. Petrology of the Island of Tahiti. Geological Society of America Memoir 116:523–556.

McBirney, A. R. and Williams, H. 1969. Geology of the Galápagos Islands. Geological Society of America Memoir 118, 197 p.

Mercier, J-C. C. 1979. Peridotite xenoliths and the dynamics of kimberlite intrusion. In F. R. Boyd and H. O. A. Meyer (eds.), *The Mantle Sample: Inclusions in Kimberlites and Other Volcanics.* Washington, D.C.: American Geophysical Union, pp. 197–212.

Nixon, P. H. (ed.). 1973. *Lesotho Kimberlites.* Maseru: Lesotho National Development Corporation, 350 p.

Sobolev, N. V. 1977. *Deep-Seated Inclusions in Kimberlite and the Problem of the Composition of the Upper Mantle.* (Translated from the Russian by D. A. Brown.) Washington, D. C.: American Geophysical Union, 279 p.

Shaw, H. R., and Jackson, E. D. 1973. Linear island chains in the Pacific: Result of thermal plumes or gravitational anchors. *Journal of Geophysical Research* 78:8634–8652.

Sørensen, H. (ed.). 1974. *The Alkaline Rocks.* New York: John Wiley and Sons, 622 p.

Tuttle, O. F., and Gittins, J. 1966. *Carbonatites.* New York: Wiley Interscience, 591 p.

Williams, H., Turner, F. J., and Gilbert, C. M. 1954. *Petrography.* San Francisco: W. H. Freeman and Company, 406 p.

Wyllie, P. J. (ed.). 1967. *Ultramafic and Related Rocks.* New York: John Wiley and Sons, 464 p.

Part II
MAGMATIC SYSTEMS

Unlike sedimentary systems, in which most flows of energy and matter can be readily observed and studied in modern environments, natural magmatic systems are usually not amenable to direct investigation. Only a few extruded magmas of nonexplosive nature can be investigated at close range.

In view of these obstacles, igneous petrologists have turned to controlled laboratory experiments and theoretical studies in an effort to learn more about the behavior of magmatic systems. Although such investigations have been under way for several decades, many questions remain unanswered. These investigations can never exactly mimic the large dimensions and very long time scales of natural systems, but they do have the advantage of controlling many influencing variables; for example, pressure and all possible compositional parameters can be held fixed while temperature is allowed to vary in a known manner. Within the past two to three decades, laboratory techniques and equipment have been devised that enable experimental petrologists to simulate conditions throughout the crust and upper mantle, where magmatic phenomena are important. Experimental and "paper-and-pencil" systems provide insight into what might happen and what cannot happen in magmatic systems.

The objective of this part of the book is to review salient information on magmatic behavior, obtained mostly from the laboratory, compare it to the properties of real rock bodies described in the previous chapters, and demonstrate how it can be used to formulate interpretive models of natural magmatic systems. It must be kept in mind at all times that natural magmatic systems and the rock bodies created in them are almost always more complicated than the models conceived to explain them. Any model that is adopted and believed can be a serious impediment to further understanding; established popular ideas, regretfully, do not die easily, no matter how fallacious they may be.

This part of the book builds upon the principles outlined in Chapter 1, a review of which would be beneficial at this point.

7

Composition and Kinetic Aspects of Magmas

Magma is a liquid solution or melt that generally contains suspended crystals and dissolved gaseous constituents, or volatiles. Magmas are initially undersaturated in volatiles, but as they ascend into regions of lower pressure in the crust and begin to crystallize, they may also become water-saturated and boil. Exsolution and subsequent expansion of gas in magmatic systems can cause explosive volcanic eruption.

Melts are comprised of polymers of interconnected but distorted Si–O tetrahedra, with other major ions such as Ca, Mg, Fe, and Na occurring in looser coordination with the oxygens. High-silica, water-poor melts are more polymerized and hence more viscous, flow is more sluggish, and nucleation and growth of crystals are slower. Silica-poor, wet melts behave in an opposite fashion. Rates of nucleation and crystal growth increase with undercooling below the freezing temperature of the crystal, come to a maximum, and then decrease. Grain size in magmatic rocks is dictated by the integrated number of nuclei forming per unit volume relative to the crystal growth rate; abundant nuclei yield finer grain size.

The thermal history of a magmatic body and its surrounding wall rocks is dictated by many factors, including size and geometry of the body and its composition, particularly the volatile content. A convecting body of less viscous magma contrasts in many ways with respect to one that is static and only conducting heat into the wall rocks.

7.1 CONSTITUTION OF MAGMAS

Magma is hot, mobile, molten rock material. Measurements on active lava flows together with evaluations of mineral geothermometers in magmatic rocks indicate that temperatures of magmas near the Earth's surface range from about 1200°C to 700°C, with the higher values for mafic compositions, the lower for silicic. Overall densities of bodies of magma range from about 2.2 to 3.0 g/cm³ and are generally somewhat lower than values for solid rocks of the crust.

At sufficiently high temperatures, any crystalline material or any mixture of crystals melts to form a homogeneous liquid solution, or **melt.** (Certain uncommon magma systems consist of not one but two separate and distinct melts; this phenomenon is known as liquid immiscibility and will be discussed on pages 275 and 321.) In geologic systems, melts contain ions of O, Si, Al, Ca, H, Na, and so on. Under some conditions, however, certain ions may reside predominantly in a separate phase—that is, in separate gas or crystals that are in equilibrium with the melt.

A **phase** may be defined as a homogeneous, mechanically separable part of a system that is bounded by an interface with adjacent phases. A phase may be in a gaseous (vapor), liquid, or solid (crystalline or amorphous) state. A glass of water containing ice cubes consists of two phases. A bottle of soda pop with the cap just removed consists of two phases—liquid pop solution and bubbles of carbon dioxide.

According to thermodynamic principles, if hydrogen, for example, exists chiefly in a separate gas phase in a magma system, a small amount of hydrogen can still be found in the melt and in any crystals that might coexist. In contrast, Ca in magmatic systems is usually present in relatively very small concentrations in the gas phase and is found principally in the melt and crystals.

Depending upon the pressure, temperature, and chemical composition, magma may consist solely of a crystal-free melt with dissolved gas, or of melt plus crystals, or of melt plus bubbles of gas, or of melt plus crystals plus gas bubbles. Thus, only in special instances are the terms *melt* and *magma* synonymous; in most situations the melt is only part of the whole magma system. But the existence and to some extent the properties of this melt fraction largely govern the overall behavior of the whole magma.

Composition

There are many ways of denoting the amount of any constituent in a melt; two common ones are by its concentration and by the pressure of its vapor in equilibrium with the melt, as this pressure is a function of the concentration. For a **condensed constituent**, i, whose equilibrium vapor pressure is rel-

atively small, the amount in a melt or in a crystal is usually denoted by its concentration in terms of the mole fraction, X_i. Actual concentrations of condensed constituents in magmas vary greatly, as shown by their range of concentration in corresponding igneous rocks (see Table 2-1). The numerous factors that govern their concentration are related to processes at site of magma generation and to subsequent processes of diversification.

Volatile constituents at low pressures reside chiefly in a gaseous or vaporous state, and their amount in the system may be specified in terms of either concentration or equilibrium vapor pressure. Because melts of petrologic importance are comprised of many constituents, more than one of which may be volatile, it is necessary to define rigorously what is meant by the term **equilibrium vapor pressure** (Burnham, 1967, p. 36). For water, the equilibrium vapor pressure, P_{H_2O}, is that pressure of pure water, at any specified temperature, that would be in equilibrium with the melt through a membrane permeable only to water; P_{CO_2}, P_{O_2}, and so on may be similarly defined (see Figure 7-1). The **fluid pressure**, P_f, may be defined as the sum of all the equilibrium vapor pressures of the different volatile constituents, $P_f = P_{H_2O} + P_{CO_2} + \ldots$. If water is the only volatile, then $P_f = P_{H_2O}$. The confining pressure, P, is conceptually different from the fluid and equilibrium vapor pressure (see box on page 25). However, the magnitude of P_f, measured in pascals (see Appendix A), may be less than, equal to, or more than P, depending upon circumstances. The magnitude of P_f thus varies over a wide range of values.

Vapor pressures of volatile constituents in melts depend upon the temperature and often on the confining pressure on the magma system and upon the amount of the constituent in the melt, which is determined by processes at the site of magma generation and subsequent diversification.

In gases, there is a wide dispersion of constituent atoms or molecules relative to the spacing between them in liquid or crystalline states (see Table 7-1). This is clearly manifest in properties

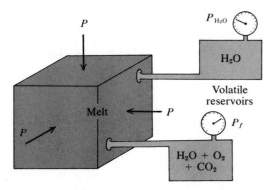

Figure 7-1 A container on the left holds a high-temperature melt (with or without suspended crystals) with dissolved volatiles H_2O, CO_2, and O_2. A confining pressure P constrains the melt with equal magnitude on all sides. On the side of the container are two small membranes. To one of these membranes, permeable only to H_2O molecules, is attached a reservoir of pure H_2O at the same temperature as the melt and in equilibrium with it. A pressure gauge on the reservoir shows P_{H_2O}. A second reservoir containing all three volatiles is attached to the melt container at the other membrane, which is permeable to all three volatiles. A pressure gauge here shows $P_f = P_{H_2O} + P_{CO_2} + P_{O_2}$.

such as compressibility, density, and the reciprocal of density, the **specific molar volume**. In a tenuous gas at high T or low P whose molecules or atoms exert little or no interactive forces on each other, the relation between molar volume, V, and P and T is expressed in the perfect gas law:

$$V = \frac{nRT}{P} \qquad (7.1)$$

where n is the amount of the gas, R is the gas constant, and T is in K. The significance of T and P in determining the molar volume of water is shown in Figure 7-2. The volume at 1000°C and one atm (0.1 MPa) is ~ 6000 times the volume under atmospheric conditions. With increasing P, water is compressed into smaller volumes until at $P = 100$ MPa and $T = 1000$°C its volume is only six times that at atmospheric conditions. Liquid and solid water (ice), in contrast, have volumes close to 1

Enthalpy

The **molar heat**, or **enthalpy, of fusion** is the quantity of thermal energy, at constant pressure, that must be put into one mole of a crystalline solid at its melting temperature to produce the change in state to a liquid. The heat of vaporization is similarly defined for the change in state from liquid to vapor. The change in state from liquid to crystal is accompanied by the release of the **latent heat of crystallization**, which is equivalent to the heat of fusion.

cm^3/g over a wide range of pressures and are relatively incompressible compared to high-temperature steam.

The volume of a hydrous melt at some P and T is not equal to the sum of the separate volumes of the dry melt and the water because of interactive atomic forces between them. The **partial molar volume** of water in the melt is defined as the change in volume of a large amount of melt when a small amount of water is dissolved in it at constant P and T. The partial molar volume of water dissolved in a silicate melt diminishes with increasing confining pressure on the melt, but for the same P and T, this volume is generally less than the volume of pure water. This inequality suggests that P and T would affect the amount of water that can be dissolved in melts. Consider a generalized reaction

more hydrous melt =
less hydrous melt + water (7.2)

and examine it in terms of LeChatelier's Principle. Because the partial molar volume of water in the melt is less than the volume of pure water, the right-hand side of the equation will have greater volume. Disturbing the equilibrium represented in this equation by increasing the confining pressure on the system drives the reaction toward a state of lower volume, the more hydrous melt on the left.

Table 7-1   Some comparative properties of crystalline, liquid, and gaseous states.[a]

	Crystalline	Liquid	Gaseous
Primary Properties			
1. Bonding cohesion between atoms, molecules, or ions	Strong, essentially permanent bonds	Strong but not permanent bonds	Very weak bonds and no permanence
2. Degree of geometric order among atoms, molecules, or ions	Highly ordered (but never perfect in real crystals); both long- and short-range order	Some short-range order	No geometric order
Secondary Properties Related to Atomic Structure			
3. External form and volume	Has a definite shape (bounded by planar crystal faces where free to grow) and occupies a definite volume	Shape is governed by the container and the viscosity of liquid; occupies a definite volume	Shape and volume governed entirely by the container
4. Rigidity and shear strength	High	Low to very small	Zero
5. Incompressibility	High	Moderate	Low
6. Thermal expansion	Low	Moderate	High
7. Density	High	Somewhat smaller than for crystalline state[b]	Small

[a] Amorphous solids have some properties like the liquid state.
[b] Our most familiar liquid—water—is a striking exception to this, since ice floats on water.

If water is available, increasing the confining pressure forces more of it into the melt. Increasing temperature favors a state of greater volume; hence, the equilibrium shifts to the right, diminishing the concentration of dissolved water in the melt.

The solubility of water in melts as a function of P and T, just shown qualitatively, can be measured quantitatively using the experimental apparatus and technique described in Figure 7-3. By making a number of experimental runs at different pressures and water contents, all at the same temperature, it is possible to define a region on a pressure–water concentration diagram where a melt will accomodate all the water, as opposed to a region where the melt accomodates only part of the total water in the system. The boundary between these two regions is a curve representing the solubility of water in the melt (see Figure 7-4). The solubility of water in silicate melts increases significantly at higher confining pressures.

In a review of data, Burnham (1979) finds that on a molar basis, rather than on a weight basis as shown in Figure 7-4, water solubilities in melts are very similar, regardless of composition. Experimentally determined solubilities of carbon dioxide in silicate melts are lower than those of water and depend in a complex way on the composition of the melt (Burnham, 1979).

There is no justification for assuming that melts in nature must contain the water concentrations indicated in Figure 7-4. These solubilities simply indicate the maximum possible concentration and not what actually has to occur in natural magmas at depth. Actual H_2O or CO_2 concentrations in natural magmas are seldom known, and direct

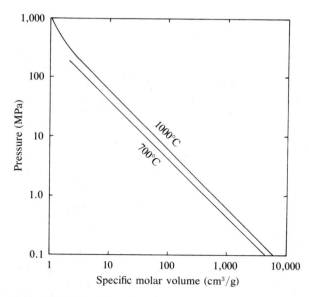

Figure 7-2 Specific molar volume of water as a function of pressure at two temperatures. [Data from *Handbook of physical constants*, Geological Society of America Memoir 97.]

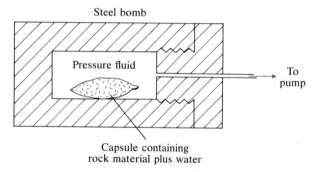

Figure 7-3 To determine the solubility of water in a melt as a function of P and T, rock powder and water in known proportions are placed in an inert metal foil capsule, which is welded shut and placed in a hydrothermal pressure vessel. This vessel or "bomb" consists of a hollow cylinder of high-strength alloy steel into which an end plug is screwed. The plug is fitted with steel tubing to carry a fluid (usually inert gas) under pressure from a pump. The bomb is placed in a furnace, at the desired temperature, and then the fluid pressure is raised to the desired value. Inside the bomb, the fluid bears against the flexible walls of the impervious foil capsule to yield the desired confining pressure on the material inside. After sufficient time has passed for a state of equilibrium to be attained within the capsule, it is very rapidly cooled, or **quenched,** to room temperature in less than a minute or so by dropping the bomb in a bucket of water or by directing a blast of compressed air at it. Any melt present in the capsule is quenched to glass. If the concentration of water in the capsule system exceeds the solubility in the melt at the particular P and T of the experiment, bubbles of water exist in the melt, and upon rapid quenching, they are frozen into a glass as vesicles. If the concentration is less than the solubility, no vesicles are present.

measurement is beset with many difficulties. In erupting lavas, the drastic reduction of confining pressure allows most of the water in the melt to escape into the atmosphere (see Section 7-3). In addition, atmospheric gases may contaminate juvenile volatiles in the lava. Solidified lava flows commonly contain absorbed meteoric water. Intrusive magmas usually behave as open systems, gaining or losing water to the surrounding country rocks during the protracted period of slow crystallization.

Pillow basalts extruded on the ocean floor offer a means of evaluating dissolved volatile concentrations of submarine lavas because they are quickly quenched and appear to have neither gained nor lost volatiles. In addition, the water depth at which vesicles appear is dictated by the volatile content or vapor pressure of the lava (see Section 7-3). Moore (1970) and co-workers found volatile concentrations in submarine basalts to be quite low, only a fraction of experimentally determined solubilities.

Volatiles other than H_2O and CO_2 occur in natural magmas. One way to determine what types is to sample the gases emitted from active volcanic vents and lava flows. The main elements found are O, H, C, and S with minor B, N, Ar, Cl, and F. The exact nature of the molecular gas species in equilibrium with the magma, or dissolved in the melt, depends upon the P and T and relative element concentrations. However, H_2O, SO_2, and CO_2 usually appear to be dominant, at least in basaltic magmas. There is reason for doubting all

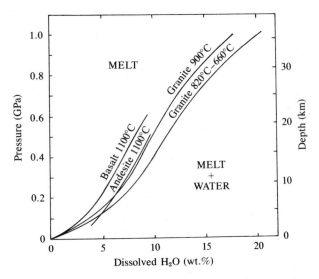

Figure 7-4 Solubility of water in selected silicate melts. To the right of a particular curve at the specified T, the system contains more water than can be dissolved in the melt and the system consists of hydrous melt plus a separate water phase whose density depends upon P (see Figure 7-2). To the left of a curve, the solubility limit is not exceeded and all of the water is dissolved in the melt. [After I. S. E. Carmichael, F. J. Turner, and J. Verhoogen, 1974, *Igneous Petrology* (New York: McGraw Hill).]

published volcanic gas analyses, however, because of the uncertainty as to how much atmospheric or wall-rock contamination has occurred prior to sampling of the gas (Gerlach and Nordlie, 1975). In other words, we have no assurance that the sampled gases are wholly juvenile. In summary, H_2O, CO_2, and minor SO_2 appear to be the chief volatiles in magmas, but their concentrations at depth are usually significantly less than their experimentally determined solubility in melts; that is, natural melts well below the surface are usually undersaturated in volatiles.

Atomic Structure

The ease with which a magma flows, the rate of growth of crystals, the formation of glass upon

rapid cooling, and the explosiveness of erupting lava are some of the consequences of properties that can be related to the atomic structure of silicate melts.

In pictorial representations of crystalline, liquid, and gaseous states (see Figure 7-5), the individual particles have to be drawn as fixed in position relative to one another, but these positions are only average, or instantaneous. Even in crystals, the individual atoms have motion. As Lonsdale (1949) puts it: "A crystal is like a class of children arranged for drill, but standing at ease, so that while the class as a whole has regularity both in time and space, each individual child is a little fidgety."

Many of the properties of liquids are intermediate between those of the crystalline and gaseous states (see Table 7-1). The changes in a given property between the three states reflect their thermal energy. The greater the energy, the greater the motion of the particles constituting the material. In the transition (melting) from the crystalline state to the liquid state, there is some loss in cohesion and complete loss in long-range order among atoms; loss of cohesion becomes virtually complete in the transition (vaporization) to the gaseous state. Secondary properties in Table 7-1 reflect these differences in atomic structure, or lack of it.

Several important clues are available regarding liquid structure near the melting temperature: (1) the modest increase of molar volume (decrease of density) of 5% to 15% upon melting of a silicate suggests only a slightly looser atomic arrangement; (2) the molar heat of fusion (see box on p. 227) is always much smaller than the molar heat of vaporization, suggesting that cohesive forces between atoms decreases only slightly upon melting; (3) the molar entropy change for melting of silicates is much less than for vaporization, suggesting that atomic structural order is not greatly different in the liquid and crystalline states; (4) X-ray diffraction analysis of silicate liquids reveals a lack of **long-range structural order,** on the scale of more

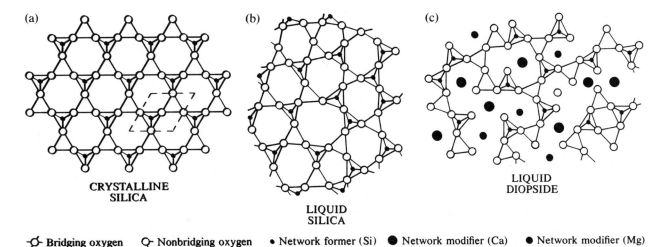

CRYSTALLINE SILICA

LIQUID SILICA

LIQUID DIOPSIDE

⊸⊶ Bridging oxygen   ⊶ Nonbridging oxygen   • Network former (Si)   ● Network modifier (Ca)   ● Network modifier (Mg)

Figure 7-5 Conceptual models of atomic structures of silicate melts (see Carmichael and others, 1974, p. 133) with a model of a crystalline solid for comparison. (a) Crystalline silica (high tridymite). Layers of hexagonal rings of Si–O tetrahedra with alternating apices pointing up and down are stacked on top of one another, creating a three-dimensional structure in which each oxygen is shared by two silicons. Tetrahedra with apices pointing up have the upper apical oxygen left out of drawing so as to reveal underlying silicon. Dashed line indicates outline of one unit cell. (b) Model of liquid silica. Si–O tetrahedra are only very slightly distorted. Long-range order is absent. Structure is very highly polymerized, with all tetrahedra interconnected. (c) In this model of liquid diopside, there is less polymerization than in liquid silica. Note presence of network modifiers.

than one lattice unit cell, but the presence of **short-range structural order**—that is, neighboring atoms of like kind are more or less regularly spaced; (5) many silicate melts, unlike most liquids, including water, can be readily quenched to form **glass,** an amorphous metastable solid with a liquid atomic structure, and within the time of cooling the cohesion between atoms in the liquid silicate is too great to allow them to reorganize into crystalline entities; and (6) the relatively high electrical conductivity of silicate melts indicates the presence of some loosely bound ions in the structure.

A conceptual model of the atomic structure of silicate liquids can be constructed on the basis of these observations. Basically, a silicate melt can be thought of as an array of ions, not unlike the corresponding crystalline structure, but possessing

no long-range order; melt structures are essentially linkages of Si^{4+} and Al^{3+} ions in tetrahedral coordination with O^{2-} ions. These groupings are more interconnected, or **polymerized,** in the more silicon–aluminum-rich melts.

Conceptual models of the atomic structure of liquid silica and diopside are shown in Figure 7-5. Liquid silica (SiO_2) consists, according to the model, of a complex network of Si–O chains, or **polymers,** which are entangled in a highly polymerized structure. The Si–O tetrahedra are only slightly distorted, so the liquid possesses the short-range order of the crystalline state but lacks long-range order—the structure cannot be generated by repetition of a unit lattice cell, for no repeated unit can be identified. Each oxygen ion forms a bridge between two silicon ions. Liquid diopside also has the short-range order and lacks

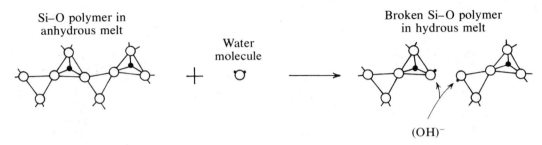

Figure 7-6 Roughly speaking, dissolved water in a silicate melt breaks polymers by replacing a bridging oxygen with two nonbridging hydroxyls. For a more precise description, see Burnham (1979, p. 443).

the long-range order of crystalline diopside. But, in contrast, liquid diopside is not as polymerized as silica. Many oxygens do not bridge between two silicons; the Si–O polymers are not as long and continuous as in molten silica.

A second contrast between the models of diopside and silica liquid structures is the presence in the diopside of **network-modifying ions,** Ca and Mg, besides the **network-forming** Si cations. A number of different cations function as network modifiers, including Ca, Mg, Fe^+, Mn, and Al (in octahedral coordination); variations in their concentration in a melt seem to have little effect on the degree of polymerization (Kushiro, 1975). Al seems to have a dual function, as it enhances polymerization when in tetrahedral coordination. Small, highly charged cations such as Ti^{4+} and P^{5+} are also network formers and enhance polymerization.

Dissolved water in silicate melts has a profound effect on melt structure (see Figure 7-6). Solubility experiments indicate that the concentration of dissolved water increases in proportion to the square root of the vapor pressure of water in the system,

suggesting a reaction of the type

$$H_2O + \underset{\substack{\text{dissolved} \\ \text{in melt}}}{O^{2-}} = \underset{\substack{\text{dissolved} \\ \text{in melt}}}{2(OH)^-} \qquad (7.3)$$

(but see Burnham, 1979). In other words, water is not held in solution as neutral water molecules, but combines with oxygen to form nonbridging hydroxyl (OH) ions. Hence, dissolved water breaks Si–O chains and reduces the degree of polymerization.

Kushiro (1975) finds experimental evidence that addition of K_2O and Na_2O to silicate melts plays a role similar to H_2O. Dissolved CO_2 acts in the opposite manner; by forming CO_3^{2-} linkages with nonbridging oxygens, it increases the polymerization of the melt.

7.2 KINETICS OF MAGMAS

Kinetics is the study of the motion of matter and the forces causing it. In the context of igneous petrology, kinetics involves movement of atoms, crystals, and vapor bubbles in a melt, and flow of magma as it intrudes the crust and perhaps is extruded onto the surface. *Time is a significant variable in kinetics.*

Various forces drive matter toward a more stable state. For example, there may be chemical transport due to disturbed chemical equilibrium. Or a melt may have higher free energy than coexisting crystals due to withdrawal of heat from the system, resulting in movement of atoms through the melt to growing crystals. Or melting may occur due to an elevation in temperature, resulting in a deep-seated magma body that is less dense than its surroundings and buoys upward due to gravitational forces. It may be noticed, however, that transport of matter must involve differential motion of atoms or of groups of atoms bonded together by chemical forces; in other words, any mass transport in the magma system depends upon the mobility of atoms in the melt. This mobility should be profoundly influenced by the atomic structure, especially by the degree of polymerization of the melt and indirectly by its chemical composition.

The validity of any model is determined by its success in accurately predicting phenomena whose behavior was not an input or constraint in initially formulating the model. We now evaluate our model of atomic structure by examining some of these time-dependent kinetic transport phenomena.

Viscosity

All liquids flow, some more readily than others. **Viscosity**, the inverse of mobility or fluidity, is a measure of the internal resistance of a liquid body to flow. Water is highly fluid and has a very low viscosity. Honey or heavy machine oil is much more viscous, and when poured onto a table flows more slowly outward and in a thicker sheet than does water. Asphalt (tar) on a cool day (15°C) is very viscous and its rate of flow is imperceptible;

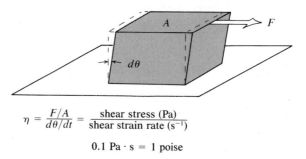

$$\eta = \frac{F/A}{d\theta/dt} = \frac{\text{shear stress (Pa)}}{\text{shear strain rate (s}^{-1}\text{)}}$$

$$0.1 \text{ Pa} \cdot \text{s} = 1 \text{ poise}$$

Figure 7-7 Definition of the coefficient of viscosity, η, in a fluid body of surface area A subjected to a tangential force F. The shear stress is F/A and the time rate of shear strain is $d\theta/dt$. The poise ($=0.1$ Pa $\cdot$ s) is the conventional unit for the coefficient of viscosity.

but if left unsupported under the influence of gravity, a body of cold tar will eventually, over a period of many days, assume a somewhat flattened shape.

Newtonian viscosity. Formally, viscosity is expressed in terms of the **coefficient of viscosity**, η, which is the ratio of the applied shear stress to the time rate of shear strain (see Figure 7-7). If the coefficient is constant regardless of the magnitude of the shear stress and strain, the viscosity is said to be **Newtonian**. A simple way to measure the coefficient of viscosity is by timing the fall of a heavy sphere through a liquid (see Stoke's Law on page 318). Measured coefficients on silicate melts without crystals or gas bubbles as a function of temperature and composition are shown in Figure 7-8.

In order for a melt to flow, there must be rearrangement of atoms, either by distortion of the Si–O polymer network or by fragmentation of the network and reuniting of broken polymers into a new configuration; network-modifying ions only shuffle about in response to the changing polymer network. In highly polymerized silicic melts with entangled Si–O chains, the strong bonds of the network must be ruptured to permit flow; this calls

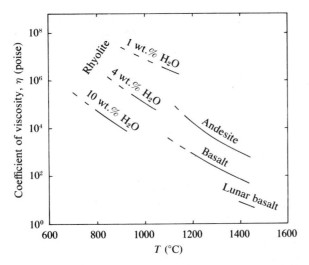

Figure 7-8 Newtonian viscosities of some crystal-free melts as a function of temperature. The curves for rhyolite are for hydrous melts containing the indicated concentrations of dissolved water. Andesite and basalt melts are anhydrous. Curves are dashed at temperatures where crystallization should occur under equilibrium conditions. [After H. R. Shaw, 1965, Comments on viscosity, crystal settling, and convection in granitic magmas, *American Journal of Science* 263; and T. Murase and A. R. McBirney, 1973, Properties of some common igneous rocks and their melts at high temperatures, *Geological Society of America Bulletin* 84.]

Table 7-2   Comparison of dissolved water concentrations in silicate melts on a weight and a molecular basis.

Melt	Weight %	Molecular %
SiO_2	5	13.5
$NaAlSi_3O_8$ (albite)	5	43.4

The melt structure loosens up at higher temperatures with greater motion of the atoms, which are able to free themselves more easily from near neighbors. Hence, viscosity is reduced.

Bottinga and Weill (1972) have devised an elegant model for direct calculation of the viscosity of a melt from its chemical composition.

Non-Newtonian apparent viscosity.   So far we have considered only Newtonian viscosity for the simplest possible type of magmas, namely, ones comprised only of a melt above its crystallization temperature and without bubbles of separate gas. As silicate melts crystallize or vesiculate, the foregoing concepts regarding viscosity and measured viscosity coefficients become invalid for the overall magma body. However, they may still hold true for the melt fraction, or for melts supercooled metastably below their freezing temperature so that no crystals form.

Magma bodies containing crystals and gas bubbles are more viscous than a melt of equivalent chemical composition. Suspended crystals impede the shearing flow of the magma body because they are rigid. Suspended bubbles also retard shear flow, due to the effects of gas pressure inside the bubbles and surface tension at the gas–liquid interface. An example of the effect of bubbles is shown by well-beaten, frothy egg white, which is more viscous than unbeaten egg white. Froths of highly viscous silicic melts may, however, have lower apparent viscosities than the bubble-free melt if the shear stress required to disrupt a bubble is less than that needed to shear a corresponding volume of liquid.

If the effects of suspended crystals and bubbles

for large applied stresses, so viscosities will be large. In contrast, basaltic melts are less polymerized, and flow can occur by breaking of the weaker, nonbridging oxygen bonds and only shuffling the shorter Si–O polymers; hence lower stresses are needed and viscosities are lower. Figure 7-8 shows that basaltic melts are up to 10^4 times less viscous than silicic rhyolite melts.

Dissolved water reduces viscosity because it depolymerizes the melt. Water-rich silicic melts can be as fluid as basaltic melts. The reason small weight percentages of dissolved water can be so effective in reducing viscosity is that water has a much lower molecular weight than silicate melts, so the molecular percentage of dissolved water is a significantly larger number (see Table 7-2).

in a magma body are recognized, we should then denote the resistance to flow as the apparent viscosity. Taking only crystals into account, the apparent viscosity shows a discontinuity at the temperature of beginning of crystallization and rises steeply at lower temperatures (see Figure 7-9). Presence of bubbles might increase this apparent viscosity still more.

Another reason for using the concept of apparent viscosity is the yield strength of magma, a property that can be illustrated by a simple experiment. A small weight placed on the surface of a magma body merely depresses the surface by an amount proportional to the total mass of the weight. When the weight is removed, the surface rebounds back to its original configuration. This type of nonpermanent or reversible behavior is called **elastic**; the distortional strain of the body is proportional to the applied stress and strain is eliminated when the stress is zero. However, a larger weight of some critical mass placed on the surface of the magma will penetrate and fall through the body at a certain rate. The applied stress (the weight) is now sufficiently large to cause a continuous, irreversible flow without rupture. The applied stress has exceeded a threshold, or **yield strength,** characteristic of the magma. Bodies that possess a yield strength are said to exhibit **plastic** behavior.

Real magmas are plastic as well as viscous; they are plastico-viscous. If viscosity were the only factor governing the capacity of an extruded lava to flow, then given sufficient time without cooling on any flat surface, it should spread almost indefinitely until very thin.

Experimental measurements (Williams and McBirney, 1979, p. 24) indicate that the yield strengths of basalt and andesite melts at temperatures above their crystallization range are negligibly small, whereas the strengths of partially crystallized magmas of these compositions can be nearly 100 times greater. No measurements have been reported for vesiculated melts, but they may also have substantial yield strengths. Preliminary data (Hulme, 1974) indicate that yield strengths for

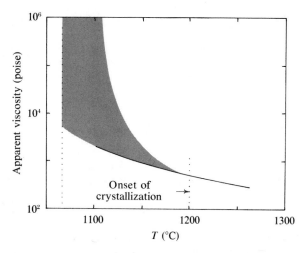

Figure 7-9 Apparent viscosity increases steeply but variably (shaded area) for a particular Hawaiian basalt melt within its range of crystallization because of the presence of suspended crystals. Crystallization under equilibrium conditions begins at 1200°C and is 50% complete at 1070°C. Solid line is viscosity for a melt without crystals. [After H. R. Shaw, T. L. Wright, D. L. Peck, and R. Okamura, 1968, The viscosity of basaltic magma: An analysis of field measurements in Makaopuhi lava lake, Hawaii, *American Journal of Science* 266.]

silicic lava flows are about 100 times greater than for basaltic flows.

For comparing the flow behavior of whole bodies of magmas, the **apparent viscosity** should be defined as the ratio of the total shear stress required both to initiate and to sustain flow, relative to shear strain rate (Shaw and others, 1968).

Still another complexity in the flow of magmas, which causes additional departure from an ideal Newtonian behavior, is the rate at which shear occurs. Fast rates of shearing motion can increase the thermal energy of the viscous material faster than heat can be dissipated. Consequently, the temperature rises, which tends to reduce apparent viscosity, not only because the viscosity of the melt fraction is lowered but also because crystals may go into solution, reducing their concentration.

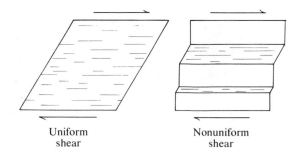

Uniform   Nonuniform
shear   shear

Figure 7-10 Localized zones of thermal feedback during flow of a plastico-viscous body may cause nonuniform shear flow.

This thermal feedback phenomenon (Shaw, 1969) may be localized along discrete internal flow surfaces so that most, if not all, of the subsequent shear becomes concentrated into narrow zones. Continued application of stress then causes flow in these zones rather than throughout the body (see Figure 7-10).

Magmas flowing through confined conduits or channels sometimes move as a more or less stagnant "plug" in the center with most of the relative displacement between it and the stationary walls of the body occurring in only a thin shear zone near the walls.

Morphology of Lava Flows and Volcaniclastic Deposits

The shape, especially the lateral extent, of a lava flow depends not only upon the apparent viscosity of the extruded lava, but also upon the topography, angle of slope of the surface, the total volume of lava extruded, and the rate of extrusion.

Even if the slope angle is high and the initial apparent viscosity of the magma is low, extruded lava will tend to pile up near the vent if its effusion rate is low; the amount of cooling relative to flow volume will be sufficient to increase the apparent viscosity drastically. Greatest flow lengths would be realized in large volumes of rapidly erupted fluid magma, such as some extrusions of basalt on Hawaiian shield volcanoes (Walker, 1973).

Volcanic debris flows probably behave in a plastico-viscous manner if the proportion of coarse fragments to matrix mud is large; the rigid clasts impede flow in the same way as crystals in magma. In pyroclastic flows, fluidizing hot gases reduce or even eliminate the yield strength and consequently even small shear stresses will induce flow; that is, the apparent viscosity is very low.

Diffusion

Diffusion governs the rate at which changes in the chemical state of a system, such as growth of crystals, or chemical reactions between different phases can take place. Diffusion of atoms or molecules is a spontaneous process that occurs within a gaseous, liquid, or solid phase because of uneven concentration of thermal energy or of chemical constituents. Examples include the diffusion of dye molecules through water, coloring it uniformly, and the diffusion of H_2S, "rotten egg gas," through the air in a chemistry laboratory.

The time rate of diffusion of an atom or molecule, J, is directly proportional to the concentration gradient, dc/dx

$$J = -D(dc/dx) \qquad (7.4)$$

where the proportionality constant, D, is the **diffusion coefficient**. Viscosity and temperature govern D, as might be anticipated:

$$D \propto T/\eta \qquad (7.5)$$

Because viscosity varies through several orders of magnitude in liquids and is far less in gases and much greater in solids, values of D must also vary greatly in these same materials. The size, charge, and polarizability of the diffusing atomic species also dictate D, values being smaller for larger size and charge. Unfortunately there is little data on D values for magmas, but Table 7-3 shows some representative numbers.

To obtain some feeling for the actual distance that something will diffuse, we may define a

Table 7-3 Representative values of the diffusion coefficient, D.

D (cm^2/s)	Chemical species	Medium	P	$T(°C)$
2.5×10^{-6}	Na	Basalt melt	1 atm	1300
4.5×10^{-8}	O	Basalt melt	1 atm	1300
5.0×10^{-8}	V	Basalt melt	1 atm	1300
2×10^{-10}	Na	Orthoclase	0.2 GPa	800
2×10^{-14}	Na	Orthoclase	0.2 GPa	500
6×10^{-14}	Rb	Orthoclase	0.2 GPa	800
9.5×10^{-10}	H_2O	Silica glass	1 atm	1050
1×10^{-12}	O_2	Silica glass	1 atm	1050
2.4×10^{-6}	H_2	Silica glass	1 atm	1050
1×10^{-8}	H_2O	Obsidian	0.2 GPa (P_{H_2O})	800
$\sim 1 \times 10^{-11}$	Sr	Obsidian	1 atm	700
$\sim 1 \times 10^{-9}$	Sr	Obsidian	1 atm	950
3×10^{-12}	Ba	Obsidian	1 atm	700
1×10^{-6}	Na	Obsidian	1 atm	800
$\sim 1 \times 10^{-9}$	Na	Obsidian	1 atm	400
$\sim 3 \times 10^{-8}$	K	Obsidian	1 atm	800

Source: After I. S. W. Carmichael, F. J. Turner, and J. Verhoogen, 1974, *Igneous Petrology* (New York: McGraw-Hill); H. R. Shaw, Diffusion of H_2O in granitic liquids, Parts I and II, and K. A. Foland, Alkali diffusion in orthoclase, both in A. W. Hofmann, B. J. Giletti, H. S. Yoder, and R. A. Yund (eds.), 1974, *Geochemical Transport and Kinetics*, Carnegie Institute of Washington Publication 634; A. W. Hofmann and M. Margaritz, 1977, Equilibrium and mixing in a partially molten mantle, *Oregon Department of Geology and Mineral Industries Bulletin* 96; and M. Margaritz and A. W. Hofmann, 1978, Diffusion of Sr, Ba, and Na in obsidian, *Geochimica et Cosmochimica Acta* 42.

characteristic penetration distance as

$$x = \sqrt{Dt} \qquad (7.6)$$

where t is time in seconds and x is distance in centimeters. In solids, even at high temperatures, D values are so small that penetrations are only on the order of centimeters in one million years. In a fairly viscous obsidian at 800°C, even over a million years, H_2O moves only 10 m. Data is scanty, but apparently H_2 diffuses much faster than H_2O and O_2 in melts or glasses.

As the rate of cooling is several orders of magnitude faster than the rate of diffusion, it is obvious that there can be no large-scale diffusion of water in a static magma body before it solidifies (see Section 7.4). However, if the body is convecting, then large-scale transport is possible. Or, if the body boils, rapid transport via rising bubbles is possible for volatile constituents in the separate vapor phase.

Formation of Crystals in a Liquid

Crystallization of salt crystals such as alum or halite from undercooled, supersaturated aqueous solutions at room temperature is a familiar experiment (Holden and Singer, 1960). Such an undercooled and supersaturated solution can persist metastably for a long period of time without crystallizing. However, once seeds form or are artificially introduced into the metastable brine, growth of crystals occurs rapidly. Crystallization is a two-step process involving nucleation, or the formation of a seed or crystal nucleus, and crystal growth, or accretion of atoms onto the nucleus.

Nucleation and crystal growth must operate hand in hand; without either one, there would be no crystals. Nucleation within liquids, including chemically complex silicate melts, depends upon the probability of transient fluctuations from the normal atomic structure of the melt, whereas crystal growth is chiefly a diffusional process. Nucleation phenomena strongly influence grain size in magmatic rocks, whereas crystal growth phenomena influence both grain size and the morphology of freely growing crystals.

Homogeneous nucleation, the subject of our discussion here, involves formation of embryonic seed crystals spontaneously within a melt. In contrast, **heterogeneous nucleation** occurs on some sort of preexisting surface in contact with the melt, such as the walls of the container or on surfaces of preexisting crystals suspended in the melt.

A note on surface energy.   Nucleation phenomena depend critically upon energy relations in the crystallizing system. It was previously shown (see page 13) that the Gibbs free energy, G, of two states in equilibrium is equal; that is, the difference between their energies is zero. Accordingly, in a system composed of liquid plus crystals of the same composition at the equilibrium freezing or melting temperature, T_e, the free energies of liquid and crystals are equal: $G_l = G_c$ and $\Delta G = G_l - G_c = 0$. But if $T < T_e$, then $G_l < G_c$, and crystals are more stable because their free energy is less (see Figure 7-11).

These concepts regarding energy are accurate when the phases involved are large. But as their size diminishes there is proportionately more and more surface area encompassing a given volume of the phase. [This may be easily verified by calculating the ratios of surface area ($6a^2$) to volume (a^3) for progressively smaller cubes. As the edge length (a) of the cubes diminishes to submillimeter di-

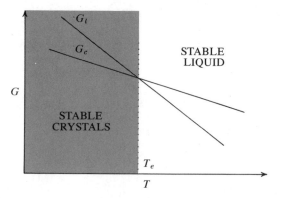

Figure 7-11   Free energies of both liquid and crystal decrease with increasing temperature, but on different slopes, so that a crossover occurs at the freezing (melting) temperature, T_e. From Equations (1.5) and (1.9), at constant pressure

$$\left(\frac{dG}{dT} \right)_P = -S$$

Therefore the G–T function has a negative slope because S is always positive and because $S_l > S_c$, the slope of the free energy function for the liquid is greater.

mensions, the ratio $6a^2/a^3 = 6/a$ increases very rapidly.] The energy distribution at the surface of any phase will be different from that in the interior because of unsatisfied, or at least distorted, chemical bonds that tend to pull the surface particles inward. In a liquid, this attractive force, called the **surface tension,** makes droplets spherical, which is the shape with the smallest surface/volume ratio. Very small spherical droplets in a mist tend to unite into more stable larger spherical drops, as in rain, further reducing the surface/volume ratio. The unit of surface tension is J/cm², or energy divided by area, so that surface tension times area is energy, which nature minimizes by forming spherical shapes of smallest possible surface/volume ratio. Other forces, such as gravity, that distort raindrop spheres into elongate raindrops, counteract this effort to minimize the surface-related energy.

Solids also have more energetic particles at their surface, which leads to what may be called surface energy, E_s. This term is actually a misnomer, as its units are J/cm², like surface tension; but because we generally multiply E_s by its associated area, the quantity E_sA has units of energy. One measure of E_sA is the work required to create a new surface in a solid, such as a cleavage face.

Nucleation. We are now in a position to analyze the stability of minute embryonic crystals that form in a simple liquid of the same composition. The change in free energy accompanying formation of the embryo, assumed to be spherical, from the liquid at $T < T_e$ is

$$\Delta G = G_c - G_l = \frac{\frac{4}{3}\pi r^3}{V} (g_c - g_l) + 4\pi r^2 E_s \quad (7.7)$$

where r is the radius of the embryo, V is the volume of an atom in the embryo, and g_l and g_c are the free energies per atom in the liquid and crystalline states, respectively. The $4\pi r^2 E_s$ term is the surface energy contribution of the embryo. For the embryo to be stable, $\Delta G < 0$. The term involving $(g_c - g_l)$ is negative at $T < T_e$, but the $4\pi r^2 E_s$ term is positive, and ΔG can only be negative when the latter is smaller than the former, as shown in Figure 7-12.

Recalling our earlier description of the liquid state (under Atomic Structure in Section 7.1), it should be obvious that minute, local, transient fluctuations in the atomic structure might just happen to possess the long-range order of the crystalline state. These local transients constitute an embryo. They will be extremely small, perhaps several times the size of a unit cell of the crystalline lattice. At the equilibrium temperature $T = T_e$, where the degree of undercooling, ΔT, is zero, these embryos are unstable because in Equation (7.7) $g_c = g_l$ and so $\Delta G > 0$; the larger their radius, r, the more positive is ΔG (see Figure 7-12). Hence no embryo has a chance of living and no crystals can permanently form. For small ΔT, on the order of a degree Celsius or so, many small embryos are again unstable because, as shown in Figure 7-12,

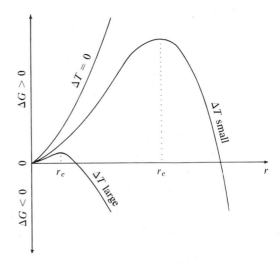

Figure 7-12 Schematic plots of Equation (7.7) at three different temperatures ($T = T_e$, $T < T_e$, and $T \ll T_e$) so that the degree of undercooling, $\Delta T = T_e - T$, is zero, small, and large. The critical radius for the particular degree of undercooling is r_c.

$\Delta G > 0$, and no growth occurs because increasing r would cause ΔG to become even more positive; such crystallites are again eliminated. However, a very few fluctuations might occur whose radius exceeds a critical radius, r_c; growth of these decreases ΔG, perhaps enough for it to become negative. These large nuclei are now stable and can continue to grow. For greater degrees of undercooling, ΔT large, the $(g_c - g_l)$ term in Equation (7.7) becomes larger negative and r_c decreases. Now the more probable small fluctuations can grow into nuclei as well as the larger fluctuations. Therefore, the rate of formation of nuclei increases for increasing degrees of undercooling. But as T continues to fall below T_e, the liquid structure becomes less flexible—manifest in an increasing viscosity—and so the probability of any fluctuations occurring diminishes.

There are almost no controlled experimental data on nucleation rates in chemically complex silicate melts relevant to geology; some prelimi-

Table 7-4   Entropy of melting (ΔS_m) and temperature of melting (T_e, °C) of some rock-forming minerals.

	$\Delta S_m{}^a$	T_e (°C)		$\Delta S_m{}^a$	T_e (°C)
Cristobalite	0.32	1713	Enstatite	2.03	1557
Albite	0.75	1118	Fayalite	2.11	—
Sanidine	0.77	—	Fluorophlogopite	2.21	1397
Anorthite	1.22	1553	Sphene	2.22	1397
Diopside	1.88	1392	Magnetite	2.52	—
Forsterite	1.94	1890	Ilmenite	2.64	—

^{a} In entropy units per atom in gram formula-weight.
Source: Mostly after I. S. W. Carmichael, F. J. Turner, and J. Verhoogen, 1974, *Igneous Petrology* (New York: McGraw-Hill).

nary data on the alkali feldspar–water system has been presented by W. Luth and colleagues (in Bailey and Macdonald, 1976). Carmichael and others (1974, p. 167) have concluded on theoretical grounds that the nucleation rate for a particular crystal is proportional to the square of its entropy of melting, ΔS_m. Some values are listed in Table 7-4. Feldspars with smaller ΔS_m values thus ought to nucleate at a slower rate than pyroxenes and olivines; large ΔS_m values for Fe–Ti oxides suggest they should have the fastest nucleation rates at some particular ΔT.

There are too many unknowns in real magma systems and the rock bodies that form from them to evaluate absolute values of nucleation rates. The amount of undercooling is unknown in an absolute sense even in simple small dikes. The total number of separate mineral grains in a particular rock volume is the integrated, cumulative effect of nucleation over the range of temperature and time during which crystallization occurred, that is, the effect of the total nuclei population per unit volume. There is an infinite number of rate versus undercooling curves that, when integrated over time, would give the same total nuclei population. Still another complexity in natural magmas is that most of the precipitating minerals are solid solutions whose compositions change significantly

with falling T while the chemical composition of the melt is also changing.

Crystal growth.   There is considerable information regarding growth of a crystal from a melt of the same composition. The phenomenon is quite complicated, but in simplest terms the rate of growth depends upon how fast particles can attach to the growing crystal face or upon how fast the latent heat of crystallization can be dissipated from the face. Slow dissipation may cause T to rise, approaching T_e and hence arresting growth altogether. Whichever rate—heat dissipation or particle attachment—is slower determines the overall rate of crystal growth. Increasing undercooling might allow faster heat dissipation but also causes increasing viscosity, so that particle mobility is reduced and attachment to the crystal is slowed. Hence, rates of crystal growth increase for increasing undercooling up to a maximum value and then diminish.

Again, conclusions must be tempered by the fact that, in natural magmas, the crystals growing from the melt are solid solutions whose compositions change with falling T.

Crystallinity and grain size in magmatic rocks.   It requires little experience with magmatic rocks to recognize their enormous range of grain size, from submicroscopic (< 0.001 mm) to the giant crystals of pegmatites (several meters), a range of well over one million. Why is this range so great? What governs grain size in magmatic rocks? Why are some minerals, such as magnetite, generally small regardless of their environment of formation? What factors dictate whether the texture is glassy, aphanitic, porphyritic, phaneritic, or pegmatitic?

Some insights, but few firm answers, for complex natural magma systems can be gained by using the models of nucleation and crystal growth just described for chemically simple systems. Figure 7-13 indicates that for very small degrees of undercooling, $\Delta T = T_e - T_A$, for example, the probability of forming viable nuclei is very small. The few that form subsequently grow to large size because of the fairly high rate of crystal growth; the resulting grain size could be relatively coarsely phaneritic. This would correspond to deep plutonic conditions in which the rate of heat loss from the magma body is very slow. At a moderate degree of undercooling, $\Delta T = T_e - T_B$, the nucleation rate is about an order of magnitude greater, whereas the growth rate has increased only threefold. Many growing crystals compete for solute particles, and the final grain size is less than in the first example. At $\Delta T = T_e - T_C$, the nucleation rate is still greater and the growth rate lower, so that a finer, perhaps aphanitic grain size develops. This might correspond to conditions along the margin of a thin dike emplaced near the surface where rapid undercooling was produced by loss of heat to the cool wall rocks. Still more drastic undercooling sweeping rapidly down to T_D and below might occur in an extruded lava flow rapidly losing heat to the atmosphere; only a small number of nuclei form, but as growth rates are nil, no crystals appear, and the melt solidifies to glass.

An amorphous liquid, sensibly solid but in reality only very viscous, that because of drastic undercooling has not crystallized is called **glass** (see Carmichael and others, 1974, p. 128 for a qualifying distinction between glass and supercooled

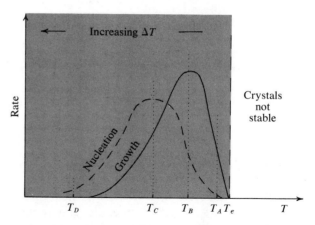

Figure 7-13 Wholly hypothetical rates of nucleation and crystal growth as a function of temperature, T, or degree of undercooling, ΔT, for a mineral that is not a solid solution and that is growing from a chemically simple melt.

liquid). All glasses are metastable and eventually crystallize, or **devitrify.** Most natural glasses are Cenozoic; progressively older glasses are progressively more rare, although lunar glasses billions of years old seem common.

Grain size in magmatic rocks is governed by the ratio of integrated crystal growth rate to the integrated nuclei population. High nuclei populations yield fine-grained rocks.

Porphyritic fabrics are widespread in magmatic rocks. They are conventionally interpreted in terms of a two-stage cooling history: an initial episode of slow cooling with a small degree of undercooling, such as $\Delta T = T_e - T_A$ in Figure 7-13, producing a few large crystals, followed by a second episode of more rapid cooling with greater ΔT ($T_e - T_C$, for example) producing a finer-grained matrix. This two-stage history is easily accomplished in lavas that began to crystallize slowly in an insulated subterranean magma chamber and then were extruded.

Porphyritic intrusive rocks might form in two ways. First, the magma may have begun to crystallize very slowly at depth and then ascended to a shallower level and stopped, there cooling more rapidly—yielding a bimodal grain size. Second, a slowly crystallizing intrusion might lose some of its dissolved water, raising its freezing temperature (see pages 258–260). This produces a greater undercooling at the same prevailing temperature because T_e increases, causing ΔT to increase.

Two other possible explanations of porphyritic fabric can be conceived from the nucleation–growth rate model shown in Figure 7-13: (1) crystallization occurs at a fixed degree of undercooling, but different minerals have different nucleation and growth rate curves; and (2) crystallization occurs at a uniformly increasing degree of undercooling. Consider first a fixed degree of undercooling, say at T_B, with simultaneous crystallization of minerals M, N, O, and P, all of whose freezing temperature is T_e. Suppose the nucleation and growth rates for mineral M are as shown by the two curves. Suppose further that minerals N, O, and P have rate curves shifted strongly toward T_e and compressed between T_B and T_e such that at T_B these phases nucleate rapidly and grow slowly, as M would do at T_C. The resultant fabric at constant $\Delta T = T_e - T_B$ would be prophyritic with phenocrysts of M in a fine-grained matrix of N, O, and P.

Consider next a uniformly increasing ΔT, due to a steady rate of heat loss, for only one mineral whose rate curves are like those shown in Figure 7-13. As T falls below T_e, the first few nuclei to form would grow to large size. At lower temperatures, T_B to T_C, progressively more abundant nuclei form but are subject to falling growth rates. From T_C to T_D, even though the nucleation rate is diminishing, it is increasingly greater relative to growth rate; hence crystals are abundant and small. Although this particular model of crystallization might yield a seriate fabric in which grain size varies uniformly from large to small rather than being bimodal, a more porphyritic fabric could develop if the rate curves had fewer overlapping limbs.

Obviously, controlled experimental information on nucleation and crystal growth in complex silicate melts is necessary to evaluate these models.

It has been convenient in the foregoing discussion to regard undercooling as the principal parameter controlling nucleation and growth phenomena. The essentially synonymous term supercooling could have been used. And we could have used supersaturation in lieu of undercooling because it is sympathetic with undercooling; as one increases so does the other, though not necessarily in direct linear proportion. The point to be made here is that undercooling or supersaturation in silicate melts can be caused not only by reduction in temperature, but also by change in composition, particularly the concentration of dissolved water, and by change in pressure (see Section 8.1).

Constitutional undercooling.   (For a full appreciation of this concept, pages 276–279 should be studied first.) Up to this point our model of crystal growth has been based more or less upon formation of crystals from a melt of the same chemical composition; in such simple systems growth rate depends on the rate of dissipation of the latent heat of crystallization from the crystal–melt interface and on the rate of attraction of particles to the interface.

In natural silicate melts where the crystallizing phases are solid solutions that differ in composition from the melt, crystal growth phenomena are governed additionally by diffusion rates of the different constituents, by the composition of the melt, and by the degree of supersaturation. Compositional zoning in crystals, their shapes, and unusual rock fabrics not accounted for by previous simple models might be related to the disequilibrium phenomenon known as **constitutional undercooling**. This is a local type of supersaturation in which chemical constituents other than those forming the growing crystal become highly concentrated in the melt around the crystals because their growth rate is faster than the diffusion rates of the different chemical species in the melt.

Crystallization of plagioclase illustrates constitutional undercooling. From any melt ΔT below its liquidus temperature, crystals form that are more calcic than the melt. If the rate of crystal

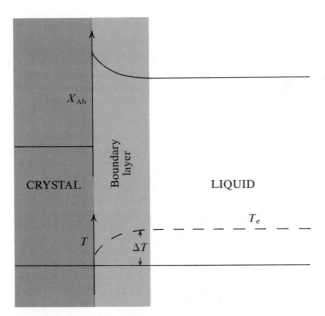

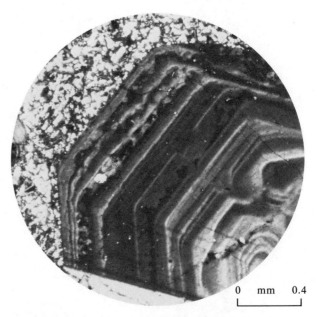

Figure 7-14 Constitutional undercooling occurs in a melt boundary layer near the interface of the melt and growing crystal. The temperature of the entire system is assumed constant, although the latent heat of crystallization would actually increase the temperature and, therefore, decrease ΔT even more than shown at the interface. X_{Ab} is the concentration of the albite component and is greater in the melt than in the crystal.

Figure 7-15 Oscillatory zoning in a plagioclase phenocryst in a diorite porphyry.

growth and consumption of nutrient ions from the melt exceeds the rates of diffusion of unwanted Na and Si ions, then their concentrations in the melt will increase progressively toward the crystal–melt interface within a thin (several micrometers thick) boundary layer. The increasing concentration of the albitic component in this boundary layer lowers the local liquidus temperature, T_e, and therefore ΔT as well, toward the interface (see Figure 7-14).

Compositionally inhomogeneous or zoned plagioclases are virtually ubiquitous in magmatic rocks. In addition to the common **normal** and more rare **reverse zoning,** in which the rim is more and less albitic, respectively, than the core, many plagioclases have **oscillatory zoning** (see Figures 7-15 and 7-16). Such zoning is especially typical of

plagioclases in intermediate silica volcanic rocks and shallow granitic intrusions, such as porphyries. Sibley and others (1976) interpret major abrupt zones correlating throughout plagioclases in an entire magmatic body to be caused by external factors, such as cyclic buildup and release of water pressure in the magmatic system. They believe that finer-scale, more rhythmic zones not correlative between crystals reflect an interplay of crystal growth and diffusion rates at the expanding planar crystal face.

Growth morphologies of plagioclase and other silicates have been found to be strongly dependent upon the initial degree of undercooling of the melt. Experiments by Lofgren (1974) show that as the degree of undercooling increases, crystals form with greater departure from an equilibrium tabular habit (see Figure 7-17). A similar pattern has been documented in silicate minerals in quenched submarine pillow lavas (see Figure 5-10). Lofgren believes the variable crystal morphologies may be

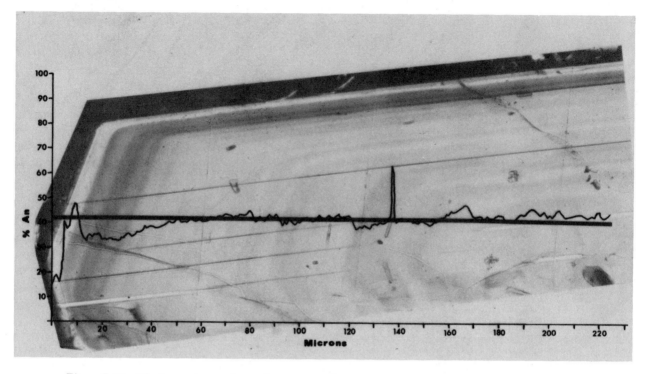

Figure 7-16 Electron microprobe analyzer trace of the anorthite concentration (percent An) in an oscillatory zoned plagioclase phenocryst in andesite. Sharp spike at about 140 micrometers is the response to an inclusion (apatite?). Note fine-scale variation in anorthite content of several percent along trace. [National Aeronautics and Space Administration photograph courtesy of Gary Lofgren and Robert Smith.]

caused by progressive changes in the relative rates of crystal growth and diffusion. Just below T_e, the ratio of these two rates is assumed to be near unity, and tabular crystals with planar faces develop. At lower temperatures for increasing ΔT, the rate of crystal growth increases and diffusion decreases; there is an increasing departure from a stable equilibrium. Perturbations on the growing crystal faces, in the form of protrusions of accreting particles extending beyond the constitutionally undercooled boundary layer, become more frequent and closely spaced. These protrusions grow faster than the adjacent face because they lie in a more strongly undercooled region where growth rate is greater. Skeletal crystals represent growth of perturbations whose wavelengths are on the

order of the length of the crystal. Dendritic crystals have more abundant and closely spaced perturbations, and spherulitic clusters of crystals represent many small, very closely spaced perturbations that penetrated the boundary layer surrounding the initial nucleus.

As ΔT is specific to each phase, different minerals in the same undercooled system could show different morphologies. In less viscous melts, more extreme growth morphologies may not be capable of developing because diffusion rates are relatively high; hence spherulites are rare in basaltic systems, in which drastic quenching commonly produces only skeletal or perhaps dendritic forms, whereas spherulites are common in silicic glasses.

Comb layering, spinefex, and harrisitic fabrics

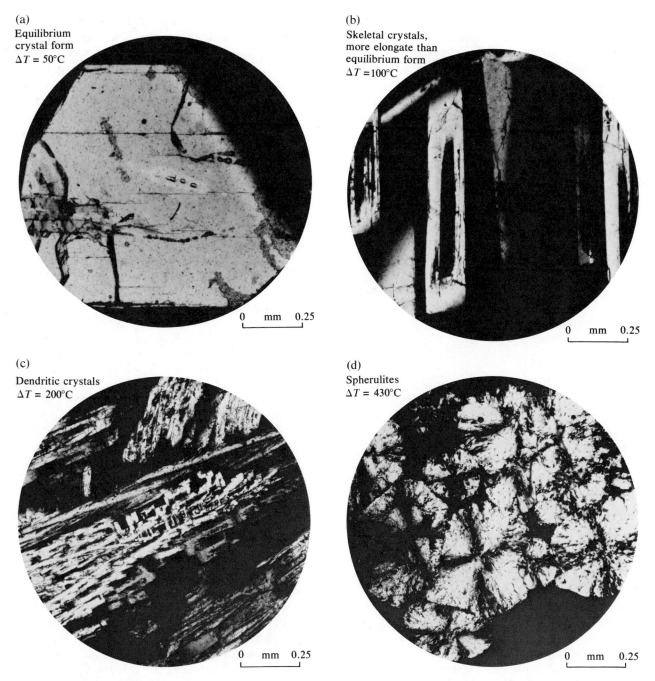

(a)
Equilibrium
crystal form
$\Delta T = 50°C$

0   mm   0.25

(b)
Skeletal crystals,
more elongate than
equilibrium form
$\Delta T = 100°C$

0   mm   0.25

(c)
Dendritic crystals
$\Delta T = 200°C$

0   mm   0.25

(d)
Spherulites
$\Delta T = 430°C$

0   mm   0.25

Figure 7-17   Experimentally produced spectrum of freely growing plagioclase crystal forms as a function of the degree of undercooling, ΔT. All shown under cross-polarized light. [From G. Lofgren, *American Journal of Science* 1974, vol. 274; courtesy of NASA.]

all involve development of elongate and commonly feathery or skeletal crystals oriented more or less perpendicular to the isothermal cooling surface of the magma body (Lofgren and Donaldson, 1975; Drever and Johnson, 1972). Possibly following heterogeneous nucleation on the surface, crystals begin growing into the supersaturated melt, producing a compositionally zoned boundary layer next to the surface. Farther into the boundary layer away from the surface, ΔT and hence crystal growth rates are greater, so the growing crystals expand into the layer rather than sideways. Growth consumes nutrient constituents and maintains the constitutionally undercooled boundary layer so that directional growth continues into the magma body approximately perpendicular to the cooling surface. The final result is an aggregate of elongate crystals that grew rapidly under fairly high degrees of undercooling in the direction of most available nutrient constituents.

7.3 VESICULATION

If a silicate melt becomes supersaturated with respect to some dissolved mineral constituent, crystals of that mineral will nucleate and grow. If a high-temperature melt becomes supersaturated with water or some other volatile—that is, the volatile concentration exceeds its solubility (see Figure 7-4)—the volatile **exsolves** from the melt by nucleating and growing bubbles. The process of **exsolution** is also called **vesiculation** and **boiling.**

Causes of Boiling

There are two ways that a melt initially undersaturated in gas can become supersaturated so that it boils. Under more or less constant temperature, or **isothermal** conditions, the confining pressure can be reduced—that is, the melt can be decompressed—so that the volatile concentration exceeds the solubility (see Figure 7-4). Boiling occurs when the volatile vapor pressure equals the confining pressure.

Boiling can also be caused under more or less constant pressure, or **isobaric** conditions, by falling temperature. This may sound like a contradiction because, as Figure 7-4 shows, water solubility increases with falling T. However, if falling T promotes crystallization of anhydrous minerals such as olivine, feldspar, and so on, the water concentration in the residual melt may increase sufficiently to counteract the T effect. For example, if a granitic magma initially all melt at $P = 0.1$ GPa, $T = 900°C$, and a water content of 3 wt.% were to crystallize as a closed isobaric system during a T drop of $\sim 200°C$ to 50% feldspars plus quartz, the residual melt ought to contain 6% H_2O, but would boil before reaching that concentration because the solubility is only 4% to 5%. This type of boiling promoted by crystallization is called **retrograde** boiling because it occurs by a process that is thermally the reverse of the way most boiling happens.

Boiling produces the familiar vesicular fabric of very shallow intrusions and surface lava flows. In a magma with only a few suspended crystals, the vesicles, if not moving, will tend to be spherical or ellipsoid because of the surface tension of the surrounding melt. In crystal-rich magmas, exsolution of volatiles will create gas cavities whose shape is conditioned by the nearby crystals, forming angular vugs and vuggy fabric.

Controls on Explosive Volcanic Eruptions

The significance of boiling magmas goes far beyond the creation of vesicular fabrics. In certain types of systems, exsolution of gas from a magma provides the driving force for explosive volcanic eruption. This phenomenon (Burnham, 1979; Williams and McBirney, 1979, chap. 4) depends upon the interplay of several factors.

Generally speaking, intermediate to silicic magmas (andesite, dacite, rhyolite) but sometimes also alkaline mafic magmas with very low SiO_2 (such as nephelinites) are most explosive, whereas basaltic magmas erupt as relatively tame lava fountains. Hence, silica content and therefore apparent viscosity cannot be the only factor governing explosivity. Neither can initial volatile content, since some extensive rhyolitic pyroclastic deposits have mineralogical compositions indicative of low water contents; mafic phenocrysts are anhydrous pyroxenes rather than biotite and hornblende and there is no evidence for appreciable CO_2 in the magma. However, both initial volatile content and apparent viscosity are indirectly involved in another factor—the degree of gas expansion during vesiculation. This factor dictates whether the energy is released steadily or explosively and all at once.

Consider two vertical columns of rising magma containing similar volatile contents that begin to boil at a similar depth. In the first, because of lower viscosity or initially slower rate of ascent, exsolved gas in bubbles has expanded to a considerable degree by the time the magma column appears at the surface; this expansion accelerates the rate of ascent, but because there is little residual gas pressure remaining in the bubbles once the magma reaches the vent, there is little explosive tendency. In the second magma column, on the other hand, suppose there has been little gas expansion en route to the vent because of a greater apparent viscosity that has retarded stretching of bubble walls. In this system, there is considerable residual energy still stored in the magma as it reaches the vent; ultimately the stiff magma disintegrates violently in an explosion that can blast ejecta (fragmented magma) over a broad area.

Formation and expansion of gas bubbles is analogous to nucleation and growth of crystals. Experiments show that bubble nucleation is very rapid, even in highly silicic and very viscous melts (Williams and McBirney, 1979, p. 82). Diffusion of water (as $(OH)^-$ ions?) through melts, however, is slow, especially in silicic ones; hence many small bubbles are anticipated. Indeed, in blocks and lapilli of silicic pumice in pyroclastic flows the vesicles are commonly only a fraction of a millimeter in diameter.

Expansion of gas in the bubbles is governed by the confining pressure on the magma and the apparent viscosity of bubble walls, which retards their stretching. Figure 7-2 shows that, for water,

expansions of several hundred times might be possible in shallow magmatic systems that experience a reduction in confining pressure from, say, 50 MPa to near atmospheric pressure. Even allowing for the fact that water constitutes only a fraction of a melt and may not exsolve completely out of the melt, the expected volumetric expansion is still enormous.

Exsolution of water from the melt increases its apparent viscosity. This will reduce diffusion rates for bubble growth and will also make stretching of the bubble walls more difficult. The tensile yield strength of the melt and diminished diffusion rates may be sufficient to impede bubble expansion significantly. Sparks (1978) argues that bubble growth may effectively cease in silicic melts with coefficients of viscosity $> 10^8$ poise, primarily because of the strength of their walls. But slow diffusion of water from these walls into the bubbles causes the pressure in the bubble to increase and approach the equilibrium vapor pressure. Ultimately, the pressure inside the bubble may overcome the tensile yield strength of its enclosing walls, causing the bubble to burst explosively. Disintegration of the melt starting near the top of the magma body progresses downward while the rapidly expanding gas from the disrupted bubbles propels fragments upward out of the vent as pyroclastic ejecta (see Figure 7-18).

Figure 7-4 might be misconstrued to mean that actual water content increases with increasing pressure or depth in magma bodies. If this were true, volcanic eruptions, once started, would become progressively more explosive as deeper, more volatile-rich magma boiled out of the vent until all of the magma was expelled from the chamber. Observations of eruptions and study of the rock record indicates that this is not the case. Rather, only part of the magma from a particular chamber is expelled during one eruption, and a sequence of episodic eruptions can occur from one magma body. A possible reason for this is a downward diminution of the amount of water in the preeruption chamber. Implications of this concept will be considered on pages 322–325.

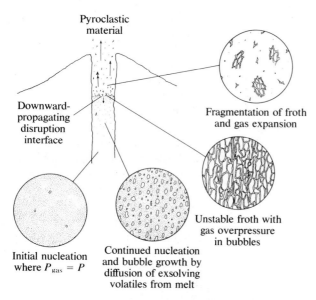

Figure 7-18 Nucleation, growth, and explosive expansion of gas bubbles disintegrate magma and eject pyroclastic material from volcanic vent.

It is often difficult to pinpoint the exact mechanism that triggers an explosive volcanic eruption of a shallow body of magma (Williams and McBirney, 1979, p. 75). One possibility is that the rock capping the magma chamber suddenly fails due to rupturing induced by the expanding magma body. As the confining pressure exerted by the caprock is reduced or even eliminated, further expansion of enormous explosive proportions might occur. Burnham (1979) calculates that P_{H_2O} in boiling magma chambers can exceed the tensile strength of rocks; water escaping from the magma can penetrate into and further open fractures in the cap rock, much like the mechanism of hydrofracting to promote greater recovery of oil or gas from wells. He further notes that the mechanical energy (PV) from retrograde boiling released almost instantaneously upon failure of the cap rock can theoretically equal the estimated kinetic energy released in real pyroclastic eruptions.

7.4 COOLING OF MAGMATIC BODIES

The following brief discussion of the cooling of magmatic bodies is intended to place the thermally related, time-dependent kinetic processes just considered in better perspective.

Cooling lava flows can form a crust sufficiently rigid to walk on in just a few hours. However, complete cooling of the entire flow to ambient atmospheric temperatures takes several years to tens of years, depending upon thickness. Radiative and convective heat transport into the atmosphere allow extrusions to cool much faster than intrusive bodies, whose slower rates of heat loss are dictated by slower conduction and sometimes penetrative convection of heat through rock.

Many factors govern the rate of cooling and the temperature distribution within a magmatic body, including the temperature of the body at emplacement, volatile content, latent heat of crystallization, apparent viscosity, thermal conductivity, density, specific heat, dimensions, and shape, as well as the temperature, conductivity, specific heat, and volatile content of the country rocks. It is impossible to evaluate all of these factors in any but the most ideal, simplified cases. Two general cases can be recognized: (1) where the magma cools statically, without internal movement, by conduction; and (2) cooling of a convecting magma.

Static Cooling by Conduction

The two most important aspects of static cooling are the internal temperature distribution in the magma body at any particular moment and the time rate of change in temperature, or the thermal history of the body.

The approximate relative temperature distribution within two-dimensional bodies of simple geometry is easily represented by isothermal lines, or **isotherms**, which are lines of equal temperature. In three-dimensional bodies representation would involve isothermal surfaces. In Figure 7-19, a series of isotherms drawn at different periods of

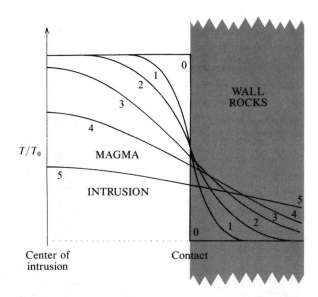

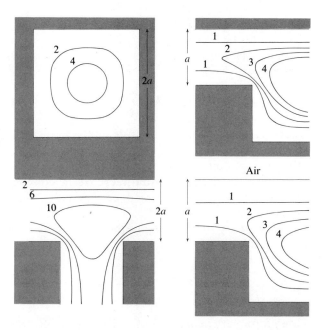

Figure 7-19 Schematic variation in temperature across the contact of a body that extends indefinitely upward and downward. Isotherms 1, 2, 3, 4, and 5 show T from center of intrusion into country rocks at five intervals of time after emplacement of magma at a uniform initial temperature T_0. Steplike line labeled 0 is the T distribution at the instant of emplacement.

Figure 7-20 Position of the $0.8T_0$ isotherm at time intervals 1, 2, 3, and so on after emplacement of magma initially at a uniform temperature T_0. Bodies extend indefinitely at open ends and are infinite in third dimension perpendicular to section. Country rocks (shaded) are initially at a uniform temperature. (a), (b), and (c) are intrusions, and (d) is an extrusion. Note different sizes of bodies. [After J. C. Jaeger, 1968, Cooling and solidification of igneous rocks, in H. H. Hess and A. Poldervaart (eds.), *Basalts* (New York: Wiley Interscience).]

time show that the thermal gradient near the contact immediately after emplacement is very steep, but that it smooths out as time advances and heat conducts outward into the wall rocks. The exact shapes of these isotherms and the absolute time period involved depend upon many factors (Jaeger, 1968). Figure 7-20 shows the inward progress of the $0.8T_0$ isotherm with time in some simple two-dimensional bodies, T_0 being the assumed initial uniform temperature of the magma body at the time of emplacement. If a family of isotherms at different temperatures were to be drawn for these same bodies at an instant of time, their pattern would be similar.

The shape of the body has an obvious influence on the shape of the isotherms. Heat is conducted away from corners faster than along planar sides because of the larger mass of neighboring country rock into which heat can sink. Isotherms, there-

fore, are located farther into the body near corners. Reentrants into the body behave in an opposite manner; heat from the two adjacent sides is conducting into the same mass of country rock, which thermally "saturates," thus impeding cooling; isotherms are crowded together and thermal gradients are higher. In the extrusive body, heat loss into the atmosphere—mostly by radiation and convection as well as by conduction—is more efficient and the hottest part of the body is displaced toward its base (compare (b) and (d) in Figure 7-20).

Although the foregoing qualitative descriptions of cooling bodies are helpful, the following semiquantitative descriptions might also be valuable. For simple bodies, the distance x in meters of a particular isotherm within the body from a planar contact is

$$x = A \sqrt{t} \qquad (7.8)$$

where A is a number that depends upon the specific situation and t is time in years. Hence, other things being equal, doubling the dimension of a body requires a fourfold increase in cooling time; if the dimension is tripled, cooling takes nine times as long; and so on.

The velocity of a moving isotherm away from a contact is

$$v = \frac{dx}{dt} = \frac{A}{2 \sqrt{t}} \qquad (7.9)$$

Thus, for a very small time after emplacement, the velocity is high but then slows down. Margins of magma bodies chill rapidly, but the interior cools at a much slower rate, a conclusion in agreement with observed variations in grain size.

Large intrusions, say on the order of 20 km in diameter, would require about one million years to crystallize, but even at that time the temperature in the core would still be several hundred degrees. Silicic lava flows < 1 My old are commonly used as initial target areas for geothermal power exploration; the rationale is that a sizable associated intrusion might occur in the upper crust beneath the flow and is there transferring heat to the groundwater system.

Rates of conductive heat transfer are several orders of magnitude faster than rates of diffusion in liquids and solids, so little net diffusion (on the order of meters or less) can occur within the time of conductive cooling of a static magma body (a million years or less; see page 237).

Cooling of a Convecting Body

In this case, heat is still dissipated into the country rocks by conduction, or into the atmosphere by a combination of transfer mechanisms, but movement of matter occurs within the body; this can have a strong influence on its thermal history.

The **natural convection** considered here is a consequence of gravitational instability in the magma body related to density contrasts established by uneven cooling and contraction (but see also pages 319–321). In a chemically homogeneous liquid body at equilibrium the density, ρ, at any point is dependent upon P and T. For a body of vertical extent h under the influence of gravity, the confining pressure, P, and density are related by the equation $P = \rho g h$, where g is the acceleration of gravity. Thus, the lower regions of the body, being under greater confining or hydrostatic pressure, will be denser. But this self-compression causes the temperature to rise in an insulated or adiabatic system—work done on the system is manifest as an increase in thermal energy. This adiabatic situation applies approximately to a magma body beneath the Earth's surface. The resulting adiabatic temperature gradient is just a few tenths of a degree per kilometer but does cause the lower regions of the body to expand very slightly, counteracting the self-compression. Therefore, gravitational equilibrium obtains if the actual T gradient in the body is equal to or less than the adiabatic gradient.

Conduction of heat out of the top of a body will produce a T gradient that is larger than the adiabatic gradient (see Figure 7-21b). The cooler and more dense upper part (a') overlying less dense magma (b') will be gravitationally unstable. The cooler magma will sink, displacing the hotter, less dense magma, which takes its place at the top of the body. The T gradient is now less than the adiabatic gradient, and the body is gravitationally stable (see Figure 7-21c). Further heat loss through the roof rocks over the magma body could induce subsequent overturn.

Obviously, other factors besides the difference between the actual T gradient and the adiabatic gradient determine whether convection will occur. A fluid with a large coefficient of thermal expansion, α, would be expected to convect more read-

(a)   Emplacement

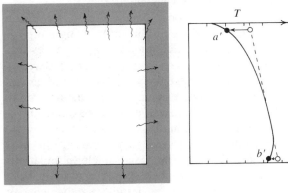

(b)   Conductive heat loss, chiefly through roof;
        new thermal gradient

(c)   Convective overturn

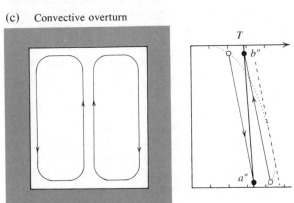

Figure 7-21 Schematic vertical sections through a
cooling magma body showing thermal processes of
conduction and convection on the left and temperature
distribution on the right.

ily, as would a body with larger vertical dimension, h. However, a high thermal diffusivity, K,
would tend to retard development of a large vertical thermal gradient, thus inhibiting convection.
High viscosity, η, would impede convective flow.
The effectiveness of the buoyancy forces in the
liquid body acting against the combined resistance
of diffusivity and viscosity is shown in the dimensionless Raleigh number, R.

$$R = \frac{g\alpha T' h^3 \rho}{K\eta} \qquad (7.10)$$

where g is the acceleration of gravity and T' is the
difference in temperature between the top and the
bottom of the body (Shaw, 1974). Convection is
assured if $R > 10^3$, depending upon certain boundary conditions (Bartlett, 1969). For natural magma
systems the greatest variations occur in h and η,
each of which can range over several orders of
magnitude. Substitution of realistic appropriate
values and with T' only 1°C and $h = 1$ km into
Equation (7.10) shows that a body of basaltic
magma ought to convect for any viscosity less than
10^{13} poise. With T' and h larger, the limiting viscosity increases. As measured viscosities of basaltic
melts are only $\sim 10^3$ poise, convection would appear to be certain in such a body.

Theoretical analysis of convection in viscous
systems indicates higher flow rates in a boundary
layer along the margin of the body, with diminishing velocity into a stagnant core and toward the
outer contact (Shaw, 1974).

Convection and the consequent stirring of the
magma permit a greater opportunity for migration
of chemical constituents than static cooling by
conduction. A convecting body will cool faster
than one not convecting because of the transfer of
heat from the base to the walls and especially to
the top, where it can be more quickly dissipated.

SUMMARY

7.1. Depending upon the pressure, temperature,
and chemical composition, magma may consist

solely of a crystal-free melt with dissolved volatiles, or of melt plus crystals, or of melt plus bubbles of volatile constituents, or of melt plus crystals plus volatile bubbles.

The concentration of a condensed chemical constituent, i, in a melt or crystal is conveniently denoted in terms of its mole fraction, X_i. The amount of a volatile constituent, v, in a magma system is usually denoted by its equilibrium vapor pressure, P_v. The specific molar volume of water ranges over four orders of magnitude in magmatic systems depending upon the confining pressure, P. The solubility of water in silicate melts increases significantly at higher confining pressures. H_2O, CO_2, and minor SO_2 appear to be the chief volatiles in magmas, but their concentrations are usually well below their experimentally determined solubility in melts; that is, natural melts well below the surface are usually undersaturated in volatiles.

A silicate melt can be thought of as an array of ions, not unlike the corresponding crystalline structure, but possessing no long-range order; the melt structure is viewed as linkages of Si^{4+} and Al^{3+} ions in tetrahedral coordination with O^{2-} ions. These groupings are more interconnected or polymerized in the more silicon–aluminum-rich melts. Dissolved water breaks Si–O chains and reduces the degree of polymerization.

7.2. The kinetic properties of magmas—how they flow and crystallize with respect to time, for example—must depend upon atomic mobility in variably polymerized silicate melts.

Viscosity is a measure of the resistance to flow of a liquid; more viscous liquids are less mobile. More polymerized silicic melts are more viscous; dissolved water depolymerizes a melt and makes it less viscous. For a given composition, melts are less viscous at higher temperatures. Most real magmas exhibit non-Newtonian plastic behavior in their crystallization range; their apparent viscosities, which can depend upon shear rate and magnitude of shear stress, are orders of magnitude greater than compositionally equivalent metastable melts free of crystals and gas bubbles. Highly silicic melts may behave similarly.

The shape and especially the lateral extent of a lava flow depends upon its apparent viscosity, the slope angle, the rate of extrusion, and the total volume extruded.

Diffusion occurs in response to unevenness in concentration of thermal energy or chemical constituents and smooths out these gradients in time at a rate governed by temperature, viscosity, and the nature of the diffusing species.

Crystallization is a two-step process involving nucleation, or the formation of a seed or crystal nucleus, and crystal growth, or accretion of atoms onto the nucleus.

All surfaces of liquids and solid bodies have an associated surface tension and surface energy that become a significant part of the overall energy of the body when it is very small.

Nucleation can only occur after a significant overstepping of the equilibrium freezing temperature. The rate at which nuclei form increases to a maximum at some particular amount of undercooling and then decreases to zero for very large ΔT. Rates of growth increase for increasing undercooling up to a maximum value and then diminish.

Grain size in magmatic rocks is governed by the ratio of integrated crystal growth rate to the integrated nuclei population. High nuclei populations yield fine-grained rocks. Glasses are amorphous solids or highly viscous liquids that because of drastic cooling could not crystallize. Porphyritic fabrics do not necessarily indicate a simple two-stage cooling history.

Constitutional undercooling can occur in chemically complex melts as solid solution minerals grow at a faster rate than rates of diffusion. This phenomenon generates a chemically zoned melt boundary layer at the crystal–melt interface that may be involved in the growth of oscillitory zoned silicate minerals, in the formation of growth morphologies that depend upon degree of undercooling, and in the development of comb layering.

7.3. In evolving magma systems, boiling, exsolution, or vesiculation occurs when the concentration of a volatile exceeds its solubility in the melt.

Boiling can be caused by isothermal decompression to the point where $P_{volatile} = P$ and by isobaric crystallization of volatile free minerals so that the content of volatiles in the residual melt exceeds their solubility, even though T has fallen, increasing solubility. In real evolving magma systems, there must be a complex interplay between these processes.

Explosive volcanic eruptions are driven by rapid expansion of gas bubbles that cause disintegration of the uppermost volatile-rich part of the magma body. Violently explosive eruptions of intermediate to highly silicic magmas may be related to the high apparent viscosities of their melts, which retard gas expansion until something triggers disintegration.

7.4. The thermal history of a magma body depends upon its emplacement temperature, volatile content, latent heat of crystallization, apparent viscosity, thermal conductivity, density, specific heat, dimensions, and shape, as well as the temperature, conductivity, specific heat, and volatile content of the country rocks.

Rates of conductive heat transfer are so slow that large static magma bodies (tens of kilometers across) would take a million years or so to crystallize. Diffusion rates, however, are orders of magnitude slower, so that little chemical transport is possible in static bodies.

Convecting magma bodies can cool faster than static ones, produce anisotropic fabrics, and permit significant transfer of chemical constituents.

STUDY QUESTIONS

1. Look up the atomic structure of forsterite and propose a model of its liquid structure. Compare it with the liquid structures of silica and diopside. Comment, qualitatively, on the viscosity of liquid forsterite.

2. List and briefly describe aspects of magmatic behavior that depend upon viscosity.

3. A basaltic magma begins to convect immediately after intrusion. Would it be expected to convect throughout its range of crystallization? Discuss.

4. Newtonian viscosities of extruded lavas have often been calculated from an equation using flow velocity, thickness, density, and angle of slope. Discuss the validity of such calculations.

5. What effect would dissolved water have on the nucleation rates of silicate melts?

6. Would you intuitively expect the rate of nucleation to be any different in a flowing melt in contrast to a static melt, other conditions being the same? Discuss. In what types of melts—silicic or mafic—might there be a difference? Why?

7. Feldspars in granitic rocks are commonly several centimeters or more in diameter, whereas magnetites in igneous rocks generally are usually only a few millimeters in diameter, at most. Why?

8. Draw and justify a diagram like Figure 7-11 for liquid and vapor.

9. Under what conditions are the terms *melt* and *magma* precisely synonymous?

10. What is the effect of dissolved CO_2 on melt viscosity?

11. Discuss the evolution of magma systems that are volatile undersaturated at depth.

12. Discuss the possible effects of dissolved F^- on melt viscosity.

13. Some highly alkaline magmas contain several tenths of one weight percent ZrO_2. How would this constituent affect melt viscosity? Justify your answer.

14. Contrast the development of the characteristic shapes of shield and composite volcanoes.

15. Some silicic to intermediate magma chambers appear to have higher water concentrations near the top. Considering diffusion rates, could this develop in a static body? Justify your answer.

16. Discuss different ways that porphyritic fabric could originate.

17. Subaerial extrusions of rhyolitic lava are commonly quite glassy, whereas basaltic ones are not. Explain.

REFERENCES

Bailey, D. K., and Macdonald, R. (eds.). 1976. *The Evolution of the Crystalline Rocks*. New York: Academic Press, 484 p.

Bartlett, R. W. 1969. Magma convection, temperature distribution, and differentiation. *American Journal of Science* 267:1067–1082.

Bottinga, Y., and Weill, D. F. 1972. The viscosity of magmatic silicate liquids: A model for calculation. *American Journal of Science* 272:438–475.

Burnham, C. W. 1967. Hydrothermal fluids at the magmatic stage. In H. L. Barnes (ed.), *Geochemistry of Hydrothermal Ore Deposits*. New York: Holt, 670 p.

Burnham, C. W. 1979. The importance of volatile constituents. In H. S. Yoder, Jr. (ed.), *The Evolution of the Igneous Rocks: Fiftieth Anniversary Perspectives*. Princeton: Princeton University Press, pp. 439–482.

Carmichael, I. S. E., Turner, F. J., and Verhoogen, J. 1974. *Igneous Petrology*. New York: McGraw-Hill, 739 p.

Drever, H. I., and Johnson, R. 1972. Metastable growth patterns in some terrestrial and lunar rocks. *Meteoritics* 7:327–340.

Gerlach, T. M., and Nordlie, B. E. 1975. The C-O-H-S gaseous system: Part I. Composition limits and trends in basaltic cases. *American Journal of Science* 275:353–376.

Hofmann, A. W., and Margaritz, M. 1977. Equilibrium and mixing in a partially molten mantle. *Oregon Department of Geology and Mineral Industries Bulletin* 96:37–41.

Holden, A., and Singer, P. 1960. *Crystals and Crystal Growing*. New York: Anchor, 320 p.

Hulme, G. 1974. The interpretation of lava flow morphology. *Royal Astronomical Society Geophysical Journal* 39:361–383.

Jaeger, J. C. 1968. Cooling and solidification of igneous rocks. In H. H. Hess and A. Poldervaart (eds.), *Basalts*. New York: Wiley Interscience, pp. 503–536.

Kushiro, I. 1975. On the nature of silicate melt and its significance in magma genesis: Regularities in the shift of the liquidus boundaries involving olivine, pyroxene, and silica minerals. *American Journal of Science* 275:411–431.

Lofgren, G. E. 1974. An experimental study of plagioclase crystal morphology: Isothermal crystallization. *American Journal of Science* 274:243–273.

Lofgren, G. E., and Donaldson, C. H. 1975. Curved branching crystals and differentiation in comb-layered rocks. *Contributions to Mineralogy and Petrology* 49:309–319.

Lonsdale, K. 1949. *Crystals and X-rays*. Princeton: Van Nostrand, 99 p.

Moore, J. G. 1970. Water content of basalt erupted on the ocean floor. *Contributions to Mineralogy and Petrology* 28:272–279.

Shaw, H. R. 1965. Comments on viscosity, crystal settling, and convection in granitic magmas. *American Journal of Science* 263:120–152.

Shaw, H. R. 1969. Rheology of basalt in the melting range. *Journal of Petrology* 10:510–535.

Shaw, H. R. 1974. Diffusion of H_2O in granitic liquids. Part I, Experimental data; Part II, Mass transfer in magma chambers. In A. W. Hofmann, B. J. Giletti, H. S. Yoder, and R. A. Yund (eds.), *Geochemical Transport and Kinetics*. Carnegie Institute of Washington Publication 634, pp. 139–170.

Shaw, H. R., Wright, T. L., Peck, D. L., and Okamura, R. 1968. The viscosity of basaltic magma: An analysis of field measurements in Makaopuhi lava lake, Hawaii. *American Journal of Science* 266:225–264.

Sibley, D. F., Vogel, T. A., Walker, B. M., and Byerly, G. 1976. The origin of oscillatory zoning in plagioclase: A diffusion and growth controlled model. *American Journal of Science* 276:275–284.

Sparks, R. S. J. 1978. The dynamics of bubble formation and growth in magmas: A review and analysis. *Journal of Volcanology and Geothermal Research* 3:1–38.

Walker, G. P. L. 1973. Lengths of lava flows. *Philosophical Transactions of the Royal Society of London*, Series A, 274:107–118.

Williams, H., and McBirney, A. R. 1979. *Volcanology*. San Francisco: Freeman, Cooper and Co., 397 p.

Wright, T. L., Peck, D. L., and Shaw, H. R. 1976. Kilauea lava lakes: Natural laboratories for study of cooling, crystallization, and differentiation of basaltic magma. American Geophysical Union Monograph 19:375–392.

8

Crystal–Liquid–Vapor Equilibria in Magmatic Systems

Magmas crystallize because of significant changes in state variables—*P, T, X*—such as falling temperature or decreasing water concentration. During crystallization, minerals precipitate sequentially from the melt over a range of temperature and perhaps pressure. The exact sequence and the particular minerals formed depend upon their relative solubilities and the chemical composition of the melt and its confining pressure. As most igneous rock-forming minerals are solid solutions, crystal–melt equilibria are dominated by incongruent behavior; subalkaline residual melts are enriched in Na, K, and Fe relative to precipitated crystals, which are enriched in Ca and Mg. Incomplete reaction relations occurring between crystals and melt during precipitation create crystalline reaction series, manifest in compositionally zoned crystals, reaction rims, and regular patterns of compositional variation in the magmatic rock body.

Near-surface magma bodies commonly boil, particularly if the initial concentration of volatiles in the melt fraction is high. Loss of water from the magma causes decomposition of primary biotites

and hornblendes. In confined plutonic magma systems, the separation of a volatile phase is responsible for production of chemically differentiated, giant-textured pegmatites and of hydrothermal solutions, which form ore deposits and alter preexisting rocks. Heated meteoric ground water may also become involved in plutonic magma systems.

The aim of this chapter is to investigate compositional relations between crystals, melt, and a separate volatile phase under equilibrium conditions. Changes in *P, T*, and bulk composition of the system modify the equilibria between the constituent phases. As these equilibria in natural systems can be quite complex, we approach them through simplified model systems, looking for general principles that will hold up in more complex situations. We seek insight into the nature of states of stable equilibrium in magma systems, whereas the previous chapter considered time-dependent rates of change between equilibrium states.

It is very important to bear in mind the distinction between (1) an equilibrium state and (2) the kinetics, or rates of processes, involved in moving to that state from some initial state. This distinction can be illustrated by means of a football stadium. In one equilibrium state it is filled with thousands of frenzied fans rooting wildly for their respective teams. In a second equilibrium state at a later time the stadium is empty and quiet. Kinetics has to do with the rates of changing states, or how fast and in what way the sports fans leave the stadium at the end of the game, whether they swarmed out over fences, through gates, or whatever. Kinetics of changing magmatic states considers such things as viscous flow rates, diffusion rates, how fast heat is dissipated, how many nuclei form in the melt with respect to time, and so on.

In this chapter our concern is basically with equilibrium states—that is, by analogy, with the condition of the stadium at game time as compared to its state during the rest of the week when empty. A review of the principles outlined in Chapter 1 would be useful before reading further.

8.1 CRYSTALLIZATION OF MAGMA

It is a familiar fact that magmas crystallize as their thermal energy diminishes. Crystallization can also be caused by changes in composition and pressure. The aim of this section is to consider crystallization of magmas in the light of some simple thermodynamic relationships and to examine in a simplified manner the sequential precipitation of minerals that occurs in magmas.

Crystallization and Melting as Changes in State

The state of any homogeneous system at equilibrium is uniquely determined by the intensive state properties T, P, and X, where X represents the concentration (specifically the mole fraction) of all of the chemical constituents comprising the system. An appropriate change in any one of the variables T, P, or X can change a state from wholly liquid to one in which crystals exist.

Such a change in state is shown schematically in

Figure 8-1. The change might involve only T with P and X remaining constant (isobaric and isochemical melting or crystallization), or a change only in X (isobaric and isothermal melting or crystallization), or a change only in P (isochemical and isothermal melting or crystallization). Most magma systems crystallize, or are formed through melting, by combinations of these changes.

Changes in systems of constant composition. With composition constant, a P–T section through the crystal and liquid stability volumes of Figure 8-1 appears as in Figure 8-2. Over the range of T and P where crystals are more stable, the Gibbs free energy of crystals is less than that of liquid, $G_c < G_l$, whereas in the liquid region the reverse is true, $G_c > G_l$. Along the boundary line or melting curve between the two regions where the two states are in equilibrium, their free energies are equal, $G_c = G_l$.

Suppose we have a system in equilibrium with both crystals and liquid present. Qualitatively, we can predict how the equilibrium will shift if P or T change by applying LeChatelier's principle and noting the inequalities in molar volumes and entropies:

$$V_c < V_l$$

$$S_c < S_l$$

States of small molar volume are more stable at higher pressure and the more disordered states of high entropy are more stable at higher temperature. Hence in Figure 8-2 the higher-entropy liquid is stable at higher temperatures and the lower-volume crystals are stable at higher pressures.

Quantitative expressions for equilibrium can be derived from the mathematical formulations of thermodynamics. In systems of constant chemical composition, an important master equation derived from Equations (1.5) and (1.9) (see also Atkins, 1978, pp. 151–154) relates infinitesimally small changes in P, T, and G:

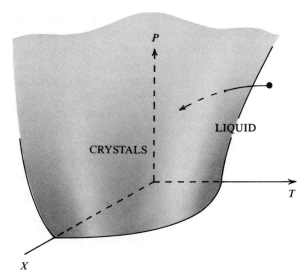

Figure 8-1 In *P–T–X* space, the field volume where liquid is stable is separated from the stability field of crystals by a curved boundary surface. Change in the state properties can cause an initially liquid system to crystallize along the indicated path.

$$dG = VdP - SdT \qquad (8.1)$$

If a change from an equilibrium state in which $G_c = G_l$ on the melting curve is prompted by an infinitesmally small change in temperature only, dT, with P constant so that $dP = 0$, then from Equation (8.1)

$$\left(\frac{dG}{dT}\right)_P = -S \qquad (8.2)$$

The subscript P reminds us that P is fixed. This expression indicates that the state of higher entropy, liquid, is made stable for a positive increment in dT, increasing temperature, because its free energy is made more negative, $G_c > G_l$. Hence, in changing from a state in which crystals and liquid are in equilibrium, at a point on the melting curve in Figure 8-2, an increase in temperature shifts the system into the stability field of liquid. In contrast, a negative dT shifting to lower

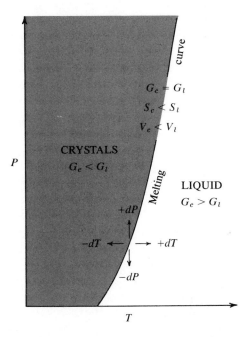

Figure 8-2 At constant composition, in a *P–T* section through the *P–T–X* volume of Figure 8-1, the melting curve separates the stability fields of the liquid and crystalline states.

temperatures favors the lower entropy state and so crystals are stable.

If T is constant and P changes, we have from equation (8.1)

$$\left(\frac{dG}{dP}\right)_T = V \qquad (8.3)$$

Increases in P (dP positive) hence favor the state of least volume, as this will minimize G. Starting from a system of crystals plus liquid, increasing pressure therefore shifts the system into the stability field of crystals.

These changes in P and T accord with our experience. Melting curves, with only a few exceptions, have positive slopes in a P–T diagram. An expression for the exact slope of the melting curve

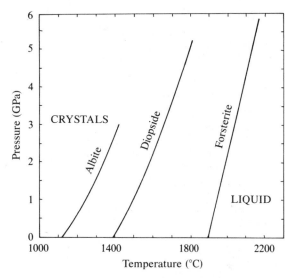

Figure 8-3 Melting curves for some pure end-member minerals. [After F. R. Boyd and J. L. England, 1961, Melting of silicates at high pressures, *Annual Report, Director Geophysical Laboratory, Carnegie Institution of Washington, Yearbook* 60; B. T. C. Davis and J. L. England, 1963, Melting of forsterite, Mg_2SiO_4, at pressures up to 47 kilobars, *Annual Report, Director Geophysical Laboratory, Carnegie Institution of Washington, Yearbook* 62.]

at a point in P–T space can be found by noting again that for crystals in equilibrium with liquid, $G_c = G_l$. In a new equilibrium state on the melting curve with these phases at a different P and T, the change in free energies must be equal:

$$dG_c = dG_l$$

or from (8.1),

$$V_c dP - S_c dT = V_l dP - S_l dT$$

Collecting terms,

$$(V_c - V_l)dP = (S_c - S_l)dT$$

$$\frac{dP}{dT} = \frac{(S_c - S_l)}{(V_c - V_l)} = \frac{\Delta S}{\Delta V} \qquad (8.4)$$

This relation, known as the **Clapeyron equation**, gives the slope of the melting curve of a pure substance, such as quartz, on a *P–T* diagram. More generally, it gives the slope of the boundary line separating any two phases, or two assemblages of phases, of identical composition. Some actual melting curves determined experimentally in the laboratory are shown in Figure 8-3. At constant chemical composition, the melting temperature of a volatile-free mineral, or mineral assemblage, shifts to higher temperatures at higher pressures.

An important implication of this pressure dependence of the melting temperature is that decompression, or release of pressure, on hot, deep-seated solid rock might induce isothermal melting.

Systems of variable composition: effect of water on melting. The freezing temperature of water can be depressed by mixing it with certain types of chemical compounds. An example is the addition of antifreeze to water in a car radiator. In general, any dissolved constituent that does not enter into precipitating crystals, or does so only in relatively small concentrations, dilutes the concentrations in the solution of those that do, necessitating lower temperatures to saturate the solution.

For solid or liquid solutions whose composition can vary, the overall free energy depends upon the chemical potential of the different components in the solution (see box on page 259). Increasing the concentration of water in a silicate melt dilutes other components, reducing their chemical potentials and stabilizing the melt relative to the crystalline state.

Another way to look at the depression of the freezing temperature of a melt by dissolved water is to note that at equilibrium, the reaction

water + crystals = hydrous melt

has $V_m < (V_w + V_c)$; therefore the hydrous melt is the more stable at higher pressures. The equilibrium boundary between these two states—water + crystals and water-saturated melt—has a negative slope in *P–T* space, and there is a drastic lowering

Chemical Potential

As G is an extensive state function (see page 9), it should depend not only on P and T, as shown in Equation (8.1), but also on composition. Indeed, in systems of variable chemical composition, such as silicate melts and solid crystalline solutions like feldspars, G depends upon the concentrations (mole fractions) of all the chemical species in the system $X_a, X_b, X_c, \ldots$. It is possible to define a new state function that plays the same role for variations in composition that V plays for variations in pressure and S plays for variations in temperature. We call it the **chemical potential**, μ, and define it formally by

$$\mu_a = \left(\frac{dG}{dX_a} \right)_{P,T,X_b,X_c,X_d,\ldots} \tag{8.5}$$

These symbols mean that the chemical potential of chemical species a equals the infinitesmally small change in G that accompanies the addition of a very small amount of species a to the system, holding P, T, and the amounts of all other chemical species constant. Note that the chemical potential handles compositional aspects of the Gibbs function like V and S handle P and T aspects [see Equations (8.2) and (8.3)]. The total change in G is then represented by the master equation of chemical thermodynamics:

$$dG = VdP - SdT + \Sigma_i \mu_i dX_i \tag{8.6}$$

where Σ_i means the sum for all chemical components, $i = a, b, c, \ldots$.

The reason for the name "chemical potential" is that μ in chemical systems is analogous to gravitational potential energy in gravitational systems; the most stable state is one of lowest potential, and at equilibrium potentials of neighboring states are equal. For example, in a hydrous magma system at equilibrium the chemical potential of water in the compositionally complex silicate melt must equal the chemical potential of water in the associated vapor, which must equal the chemical potential of water in biotite crystals suspended in the melt, and so on. Symbolically,

$$\mu_{H_2O}^{melt} = \mu_{H_2O}^{vapor} = \mu_{H_2O}^{biotite} = \ldots$$

This **principle of uniform chemical potential** applies regardless of the number of phases in equilibrium in the system. If $\mu_{H_2O}^{melt} > \mu_{H_2O}^{vapor}$, the vapor is more stable and some water from the melt should move into it.

of the melting curve relative to anhydrous conditions at higher pressures (see Figure 8-4a).

If only a limited amount of water is available in the system, the depression of the melting curve with increasing P will progress only until the available H_2O equals the solubility; with further increase in pressure, the silicate melt cannot saturate any more and the melting curve follows a positive slope as if the system were anhydrous (see Figure 8-4b).

Where do we go from here?   Up to this point, crystallization of silicate melts has generally been considered as if a pure mineral freezes from a liquid of its own composition at a unique temperature point, or crystals precipitate from solution at a particular temperature. The freezing of ice from water at 0°C, or precipitation of halite from a saturated brine at a particular T and concentration of NaCl, come to mind as examples of this sort of simple behavior. However, crystallization of magmas is not nearly so simple because of the large number of dissolved chemical components in naturally occurring silicate melts and because of extensive solid solution between the principal rock-forming minerals. Most melts ultimately will precipitate from two to six or more minerals, all of which, except for quartz, are solid solutions.

(a)

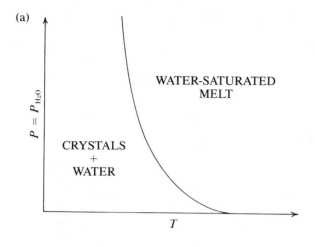

(b)

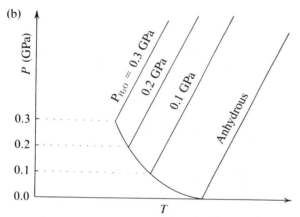

Figure 8-4 Comparison of melting curves in hydrous and anhydrous melts. (a) **Water-excess** conditions in which the melt is water-saturated ($P = P_{H_2O}$) at any P and T. The water concentration of the melt increases upward along the curve. (b) Schematic anhydrous melting curve and family of melting curves for systems of limited water content, $P_{H_2O} = 0.1, 0.2,$ and 0.3 GPa. Above $P = 0.2$ GPa, the melt with $P_{H_2O} = 0.2$ GPa is undersaturated in water and follows a positively sloping melting curve. Compare page 289.

In igneous petrology, there are no real differences in principle between melt and solution, between melting and dissolution, and between freezing and precipitation. The only differences are in usage. For example, halite dissolving in water and

pyroxene melting at high temperature in a basaltic melt are similar processes. Likewise, in discussions of silicate melts, we do not bother with any distinctions between solvent and solute because of the multitude of chemical components; expressed as elements, these would include O, Si, Al, Fe, Ti, Mg, Ca, K, Na, H, and so on.

The fact that naturally occurring silicate melts and rock-forming minerals are complex solutions has some very important implications regarding the origin of magmatic rocks, as follows:

1. Magmas do not completely crystallize under equilibrium conditions at one unique P and T, but rather over a significant range of T and P.

2. The several mainly solid-solution mineral phases—olivines, feldspars, and so on—that crystallize over a range of conditions do so sequentially, one after another, but with considerable overlaps.

3. During crystallization, each solid solution mineral, as well as the coexisting silicate melt, progressively change composition.

4. The minerals that crystallize from a particular melt, and their sequence, depend entirely upon the intensive variables P, T, and especially X and how these vary as the system changes from the liquid to the crystalline state.

It is easy to state these important concepts but quite another thing to demonstrate them convincingly and to apply them in detail to particular magma systems. Basically, five approaches can be used: (1) analysis of rock fabric, (2) sampling of semimolten lava in active volcanoes, (3) laboratory determination of melting relations in rocks, (4) laboratory determination of crystal–melt equilibria in simplified model systems, and (5) thermodynamic calculations of mineral solubility. The first three approaches are discussed in the next section; the latter two are considered in the section after the next.

Sequential Precipitation During Crystallization

Historically, study of rock textures in thin section with the petrographic microscope has played an important role in the determination of the sequence of precipitation of magmatic minerals. However, many textures allow only ambiguous interpretations, so caution is advised. Two textural relationships that indicate sequential chronologic development are the habits of minerals in sequential fabric and overgrowths. Examination of the sequential fabric in Figure 1-11 discloses that certain minerals are more **euhedral**—possess better-developed crystal faces—than others. In that figure, oxide grains are more euhedral than pyroxenes, which appear to have grown around olivines and plagioclases. Early precipitating crystals in melts grow freely and develop characteristic crystal faces (although characteristic crystal faces do not develop in drastically undercooled melts; see Figure 7-17). However, later-precipitating minerals growing from the residual melt and constrained between these early-formed crystals must assume the shape of the intercrystal space and cannot create their own characteristic crystal faces; they are **anhedral.** It should not be assumed that the euhedral grains necessarily stopped growing before the anhedral grains crystallized, only that much of their growth preceded that of the anhedral grains. As postmagmatic recrystallization may produce euhedral grains by metamorphism or alteration, caution is urged in using relative development of crystal faces to determine sequence of precipitation.

Figure 8-5 illustrates an **overgrowth** of amphibole on pyroxene. The anhedral, corroded and embayed outline of the pyroxene suggests that after growth it had become unstable in the system at the time the amphibole started to grow around it. This textural relation must be used with considerable caution, however, because exsolution, secondary replacement, pseudomorphism, and other processes may confuse the relative age relations.

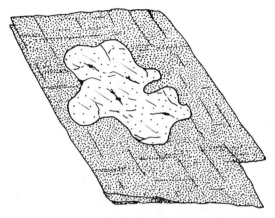

Figure 8-5 Overgrowth of euhedral hornblende around partly resorbed and deeply embayed pyroxene that became unstable in the melt subsequent to its growth.

Another way of determining sequence of crystallization is to sample a cooling lava, as has been done in a recent Hawaiian lava lake. After the lava, which partly filled a preexisting crater, had developed a hard crust, U.S. Geological Survey petrologists drilled into the still molten interior of the lake, collecting samples on the ends of probes and simultaneously measuring temperatures by a thermocouple. Photomicrographs of samples collected from the Makaopuhi lava lake (Wright and Okamura, 1977) at various temperatures clearly show the sequence of precipitation in this basaltic magma at near atmospheric pressure (see the photomicrographs reproduced in Plates I through IV). Each sample is presumed to represent an equilibrium assemblage of crystals and melt. The temperature at which crystallization begins—that is, the temperature above which the system is entirely liquid—is known as the **liquidus;** the

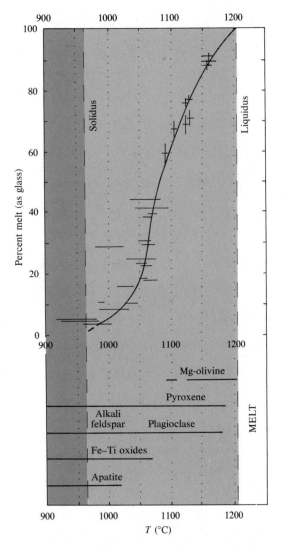

Figure 8-6 Crystallization of the Makaopuhi, Hawaii, basalt. Temperatures (measured in drill hole) and amounts of melt (determined in thin sections as glass) are indicated by horizontal line segments and crosses; length of line segments indicates uncertainty in measured values. Heavy curve is best-fit line to data showing how the amount of melt varies with temperature. Temperature range over which each mineral precipitated from the melt is shown at bottom of diagram. [After T. L. Wright and R. T. Okamura, 1977, Cooling and crystallization of theoleiitic basalt, 1965, Makaopuhi lava lake, Hawaii, U.S. Geological Survey Professional Paper 1004.]

temperature below which the system is entirely crystalline is the **solidus.** Just below the liquidus and at $T = 1170°C$ a few, almost colorless olivines are large and euhedral (see the photomicrograph in Plate I). Dark pyroxenes show a greater nucleation population than rare colorless plagioclase and a smaller grain size. Grains of these three minerals were suspended in a melt that is seen now as a brown glass. With decreasing temperature, more crystals precipitate. Although not obvious in the photomicrographs, there is a progressive change in the equilibrium composition of the minerals with falling temperature. Plagioclases become more sodic, and pyroxenes and olivines become more enriched in Fe relative to Mg. Interestingly, olivines below $\sim 1100°C$ (not shown in Plates III and IV) are corroded and embayed. They were apparently unstable in the melt, and had there been more time in the quickly cooling crust of the lava lake their resorption would have been complete.

It is obvious from the changes in the color of the glass that the melt was also changing composition as its temperature changed. From 1170°C to $\sim 1070°C$ the glass becomes darker brown due to enrichment in Fe and Ti, the concentrations of these elements being less in the crystallizing olivines, plagioclases, and pyroxenes than in the overall magma system. An accompanying enrichment of Na, K, and P also occurs because these elements are also entering into the crystalline phases in lower concentrations than they occur in the whole magma system. Ultimately, at $\sim 1070°C$ the concentrations of Fe and Ti build up sufficiently in the melt to meet the solubility limit of ilmenite; as this phase precipitates, the concentration of these elements in the melt drops, causing its color to change from red-brown to gray-green. At 1030°C the melt becomes supersaturated with respect to P-bearing apatite, which then precipitates. As the last residual dregs of melt enriched in Na and K are consumed at the solidus temperature, the precipitating feldspar growing as thin rims on earlier-formed plagioclases is an alkali feldspar.

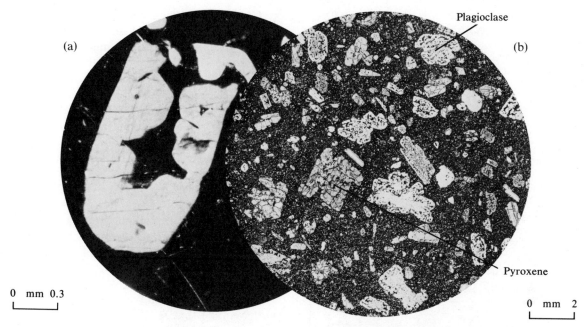

(a)

(b)

Plagioclase

Pyroxene

0 mm 0.3

0 mm 2

Figure 8-7 Corroded phenocrysts that have been partly resorbed due to instability in the melt. (a) Plagioclase in extrusive vitrophyre. (b) Pyroxene and plagioclase in andesite lava.

In this particular basaltic magma, crystallization occurred over a temperature range in excess of 200°C by sequential and considerably overlapping precipitation of solid solution minerals whose equilibrium compositions varied with respect to temperature (see Figure 8-6). After initial stable growth near liquidus temperatures, one of the phases—olivine—was subsequently **resorbed** at lower temperature. Thus, corroded mineral grains, especially of a single phase, in an igneous rock do not necessarily indicate temperature has increased to cause melting. Resorption can occur by shifting stability relations as T falls (retrograde solubility) or as P or X changes (see Figure 8-7).

Another method of determining sequence of crystallization is by laboratory study of a natural rock. The rock is finely pulverized and small amounts are placed either in a furnace, for determinations at 1 atm, or in a hydrothermal pressure vessel (see Figure 7-3) at a predetermined P and T until equilibrium is attained. Many experimental runs are required to delineate the P–T regions where the different minerals precipitate stably.

Melting relations in a granodiorite from the Sierra Nevada batholith have been determined by Piwinskii (1968) at 0.2 GPa—corresponding to a depth of ~ 7.5 km (see Figure 8-8 and Table 8-1). Crystallization occurs over a range of ~ 300°C, beginning with mafic minerals and plagioclase and finally terminating with quartz and alkali feldspar, which together constitute 45% of the rock, within 50°C of the solidus. This order of precipitation is compatible with the sequential or hypidiomorphic-granular texture typical of granitic rocks in which early crystallizing Fe–Ti oxides, biotites, hornblendes and calcic plagioclases are more or less euhedral whereas late anhedral quartz and alkali feldspar fill in available remaining space (see

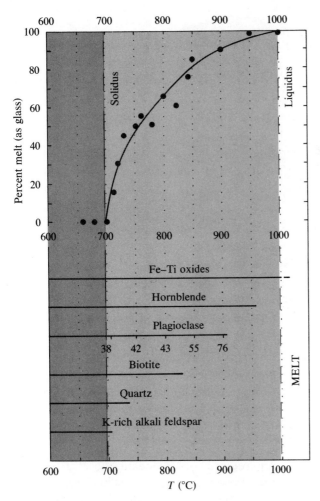

Figure 8-8 Melting relations of the Mt. Givens granodiorite, Sierra Nevada batholith, at 0.2 GPa pressure and under water-excess conditions, that is, $P_{H_2O} = P$. Circles are estimated amount of melt at the indicated temperature and heavy curve is best-fit line to these data points. Numbers beneath plagioclase indicate its composition at the indicated temperature in weight percent anorthite. [After A. J. Piwinskii, 1968, Experimental studies of igneous rock series: Central Sierra Nevada batholith, California, *Journal of Geology* 76.]

Plate V and Figures 4-1 and 4-12b). Note once again that equilibrium plagioclase compositions become more sodic at lower temperatures.

Confining pressure can exert an important control on the nature and sequence of crystalline phases precipitating from a melt. In a basalt from the Snake River Plain, Thompson (1972) found, at atmospheric P with decreasing T, that the sequence is olivine, plagioclase, and then clinopyroxene (see Figure 8-9 and Table 8-1). But plagioclase replaces olivine as the liquidus phase between 0.5 and 1 GPa, and above 3.2 GPa garnet is the liquidus phase. (The **liquidus phase** is the first crystalline phase to precipitate from a crystal-free melt with falling T.) Above ~ 2.5 GPa there is considerable solid solution in a pyropic (Mg–Fe–Ca) garnet and in a jadeitic (Na–Ca–Al) clinopyroxene so that these two phases constitute most of the rock, which is called an **eclogite**. Olivine is not stable above 1.3 GPa and plagioclase is not stable above 2.9 GPa.

From these data on two basalts and granodiorite we conclude that:

1. Magmas crystallize over a range of P and T; the chemical compositions of the melt and the precipitating solid solutions change over this range. Figures 8-1, 8-2, and 8-4, which show only a single freezing boundary, must be amended to recognize that crystallization in chemically complex silicate magmas occurs over a region of P, T, and X (see Figure 8-10).

2. Minerals precipitate from melts sequentially, overlapping one another. Once precipitated, a mineral may subsequently be resorbed back into the melt due to retrograde solubility.

3. The mineralogical composition of a magmatic rock depends upon the chemical composition of the magma and the pressure at which it crystallizes. Hence, Mg olivine will never precipitate from a granitic melt because the concentration of Mg is much too low and that of silica is too high. Olivine and plagioclase do not crystallize at high pressures from a basaltic melt because these phases are less sta-

Table 8-1 Compositions of rocks whose melting relations are discussed on pages 261–265.

	Makaopuhi, basalt	Mt. Givens, Sierra Nevada, granodiorite	King Hill, Snake River Plain, basalt
SiO_2	50.24 wt.%	67.36 wt.%	47.17 wt.%
Al_2O_3	13.32	14.72	13.34
Fe_2O_3	1.41	1.42	2.54
FeO	9.85	2.61	12.93
MgO	8.39	1.74	4.42
CaO	10.84	3.90	7.68
Na_2O	2.32	3.07	3.40
K_2O	0.54	3.45	2.00
H_2O^+	—	0.40	0.33
TiO_2	2.65	0.57	4.15
P_2O_5	0.27	0.13	1.39
MnO	0.17	0.09	0.23
Total	100.00	99.44	99.73
Modes			
Olivine	5 vol.%		
Pyroxene	51		
Plagioclase	30	40 vol.%	
Fe–Ti oxides	9		
Glass	5		
Quartz		25	
Alkali feldspar	trace	20	
Mafic minerals		15	

ble than jadeitic pyroxene plus pyropic garnet. Amphibole cannot crystallize from a water-free melt. Not all basaltic magmas precipitate olivine first, nor does alkali feldspar precipitate last from all granitic magmas.

Crystal–Melt Equilibria in Simple Model Systems

Much can be learned about sequential crystallization of magmas using the techniques described in the preceding section. A drawback in working with natural magmatic systems, however, is their chemical complexity, or the large number of constituents in the melt that govern the crystalline sequence. It is difficult to assess unambiguously the importance of any one constituent in the presence of so many. An alternate approach to the determination of crystal–melt equilibria is to study simple systems with fewer constituents that can serve as models of more complex natural magmas. We might, for example, investigate a simple combination of an end-member feldspar and an end-member pyroxene, such as albite and diopside, as a model of basalt magma, or of quartz and albite, as a model of granite magma, and so on. With

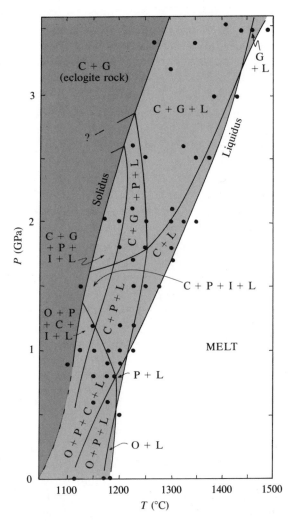

Figure 8-9 Melting relations of the King Hill basalt, Snake River Plain, as a function of *P* and *T*. Filled circles are experimental data points. C, clinopyroxene; G, garnet; I, ilmenite; O, olivine; P, plagioclase; L, liquid. Note that the lines delineating *P–T* stability fields are interpolated and extrapolated from only a few data points. This is the way that virtually all phase diagrams, such as those in this chapter, are constructed; but usually the constraining data points are omitted for clarity. [After R. N. Thompson, 1972, Melting behavior of two Snake River lavas at pressures up to 35 kb, *Annual Report, Director Geophysical Laboratory, Carnegie Institution of Washington, Yearbook* 71.]

standard laboratory equipment *P* and *T* can be precisely controlled on systems of known chemical composition, *X*. With only a few chemical species in a simple model system, *T* and *X* can be represented in a single isobaric phase diagram. **Phase diagrams** are graphic devices illustrating equilibria between different phases and their compositions. From them we can determine solubilities and sequences of precipitation.

The acknowledged pioneer in the experimental determination of phase diagrams of petrologically significant systems was N. L. Bowen, who began his fruitful career in the early 1900s at the Geophysical Laboratory of the Carnegie Institution of Washington, D.C. Since then, this laboratory and its hundreds of investigators have been chiefly responsible for the elucidation of the hundreds of phase diagrams that form an important part of the foundation of igneous petrology. Some of this effort is summarized in Ehlers (1972), Morse (1980), Ernst (1977), and Muan (1979).

Phase rule. Study of phase diagrams and crystal–melt equilibria is facilitated by use of the **phase rule,** a formulation by J. W. Gibbs that tells us what must be known about a system in order to predict its properties (Atkins, 1978, pp. 181, 281).

As systems are composed of phases and components, we need to be sure what these are before describing the phase rule.

A **phase** is a chemically and physically homogeneous part of a system that is bounded by an interface with adjacent phases. Several phases can coexist in equilibrium in a system; for example, at 1100°C and 1 atm in the Hawaiian basalt system represented in Figure 8-6, phases include melt, olivine, pyroxene, plagioclase, and minor steam. A phase may be gaseous, liquid, or solid (crystalline or amorphous). A phase may be large or small in volume; some minute perthitic intergrowths of two feldspar phases can only be recognized by X-ray diffraction.

Up to now we have used the terms constituent and species rather loosely for the different chemical entities, such as Ca, Al, FeO, CO_2, or Cl, that make up magmatic systems. This practice will be

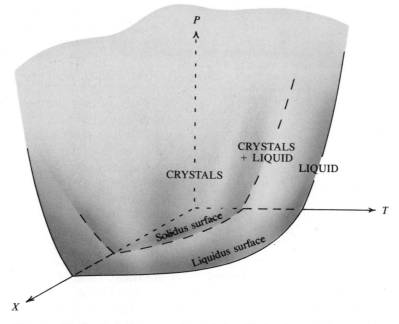

Figure 8-10   In *P–T–X* space, the stability field volumes of crystals, crystals plus liquid, and liquid are separated by the solidus and liquidus surfaces, respectively.

continued wherever a degree of informality exists. But, with regard to the phase rule and phase equilibria in silicate systems, the more restricted term component will be used. The number of **components**, C, comprising a system is the *minimum* number of chemical constituents necessary to define completely the composition of every phase in the system.

Consider a system composed of the elements H, O, Na, Al, and Si. What are the components? Some possible choices include:

1. H, O, Na, Al, Si
2. H_2O, O_2, Na, Al, Si
3. H_2O, Na_2O, Al_2O_3, SiO_2
4. H_2O, $NaAlSi_3O_8$
5. OH, $NaAlSi_2O_6$
6. H_2O, $NaAlSi_2O_6$, SiO_2

If the only phases occurring in the system under a certain P and T are water and albite crystals, the appropriate choice of components would be number 4. In another system comprised of water, albite crystals, and jadeite crystals, the appropriate choice would be number 6, as the composition of albite can be represented by combination of $NaAlSi_2O_6$ and SiO_2. In a melt of dissolved water and albite, number 4 would again be the appropriate choice, as their quantities may be varied to express the composition of the liquid system.

Under certain $P–T$ conditions, mixtures of K-feldspar and albite react together to form stable leucite crystals—$(K,Na)AlSi_2O_6$—and a silicate melt more silica-rich than these feldspars. Obviously no combination of $KAlSi_3O_8$ and $NaAlSi_3O_8$ can represent the compositions of the phases in this system, and therefore they cannot be its components. We could take as components O, Na, Al, Si, and K, but the total number 5 is excessive. An appropriate choice would be SiO_2, $NaAlSi_2O_6$, and $KAlSi_2O_6$.

In formulating the phase rule, we can start by

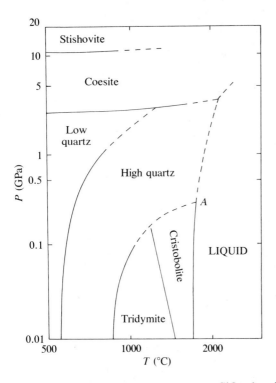

Figure 8-11 The one-component system SiO_2 showing *P–T* stability fields of the various polymorphs and of liquid. Note the logarithmic *P* and *T* scales. Point *A* is a triple point where three phases—high quartz, liquid, and cristobalite—are in state of stable invariant equilibrium. There are no degrees of freedom ($F = 2 + C - \varphi = 2 + 1 - 3 = 0$) because, given these three coexistant phases, there is one, and only one, set of *P* and *T* values possible. [After I. A. Ostrovsky, 1966, PT-diagram of the system SiO_2-H_2O, *Geological Journal* 5.]

asking some questions. How much freedom is possible in changing the *P*, *T*, and *X* of a system without disturbing its properties and equilibrium? If *T* changes, for example, do the stable phases change or remain the same? To what extent are the number of stable phases related to the number of components and to *T* and *P*? How is the composition of a phase related to change in *T*? How much must be known to specify all properties of a system at equilibrium?

We can easily formulate a phase rule for a system of only one component ($C = 1$). Consider a system comprised only of albite. In *P–T* space there is a simple boundary line—the melting or freezing curve—separating the stability field of crystalline albite from liquid (Figures 8-2 and 8-3). In a system of only one phase, either the liquid or the albite crystals, both *P* and *T* can be varied independently to a large extent without affecting the state of equilibrium. We say that this one-phase system is divariant, or that it has two degrees of freedom, or that the **variance** $F = 2$. But when two phases—liquid and crystals—in this one-component albite system are in equilibrium, only one variable, *P* or *T*, may be independently changed. This is the implication of the melting (or freezing) boundary line in Figure 8-2. If we select an arbitrary value for *P*, there is no freedom in what *T* must be so that liquid and crystals coexist in equilibrium together. For our choice of a particular *P* the *T* is uniquely fixed. Hence the system is univariant, with one degree of freedom, and $F = 1$. When three phases coexist in equilibrium in a one-component system (see Figure 8-11), the system is invariant with no degrees of freedom, $F = 0$; there is no freedom in either *P* or *T*; they are unique to the three-phase equilibrium. In summary, it appears the phase rule for a one component system is that the variance is three minus the number of phases, φ, present, $F = 3 - \varphi$.

In formulating a phase rule for multicomponent systems at equilibrium, we first take an inventory of the number of variables possible. These are *P*, *T*, and *X*. If a magnetic field or other factors are significant, they must be added to our list. The composition of each equilibrium phase is denoted by the mole fractions of its components, X_a, X_b, X_c...; but as the sum of these mole fractions is 1, we actually only have $C - 1$ independent mole fractions as variables, the last one being fixed by the others. So if the total number of phases is φ, the total number of compositional variables is $\varphi(C - 1)$ and the overall total number of variables in the equilibrium system is $2 + \varphi(C - 1)$, the 2 standing for *P* and *T*.

For a complex system of several phases and components, the number of variables, or its freedom, appears awesome. However, actual degrees of freedom are considerably fewer because of the equilibrium relations existing within the system. For example, we saw in the one-component system that the equilibrium coexistence of two phases, liquid and crystals, reduced the variance to only one. In the box on page 259 we noted that, for equilibrium, the chemical potential of a particular component must be the same in every phase of the system; that is,

$$\mu_a^\alpha = \mu_a^\beta = \mu_a^\gamma = \ldots = \mu_a^\varphi$$

and

$$\mu_b^\alpha = \mu_b^\beta = \mu_b^\gamma = \ldots = \mu_b^\varphi$$

and so on, for each component, where the Greek letter superscripts denote phases and the subscripts components. Hence, for each component there are $\varphi - 1$ equations, and for the entire system, $C(\varphi - 1)$ equations.

These $C(\varphi - 1)$ equations reduce the number of total variables so that the variance of the system is $F = [2 + \varphi (C - 1)] - [C(\varphi - 1)]$ or

$$F = 2 + C - \varphi$$
the Gibbs phase rule

For a one-component system the phase rule becomes $F = 2 + 1 - \varphi = 3 - \varphi$, as stated in our previous example.

There is the temptation to extract false information from the phase rule. As Weill and Fyfe (1964, p. 569) emphasize, the phase rule

does no more (or less) than to tell us the maximum number (F) of intensive thermodynamic variables we may independently vary without changing the number of phases...in a system of (C) components at equilibrium....It cannot be used to predict the number of phases occurring in a system at equilibrium. Indeed, it is impossible to speak of a system of C components over a range of condi-

tions unless we know the equilibrium phases occurring throughout the range. A knowledge of the phases present is *a priori* necessary to apply the phase rule.

The application of the phase rule to petrologic systems simply allows us to detect possible divariant equilibrium ($F = 2$), possible univariant equilibrium ($F = 1$), possible invariant equibrium ($F = 0$), and above all, disequilibrium ($F < 0$). In every such case the phases in the rock must be known before the phase rule can be applied since otherwise it is impossible to determine C.

Binary eutectic systems and solubility relations in T–X diagrams. In many phase diagrams representing crystal–liquid equilibria, P is held constant (isobaric conditions) and T and X are variables. Hence one degree of freedom is eliminated and the phase rule becomes $F = 1 + C - \varphi$. The reason P is held constant stems partly from the experimental technique; phase equilibria at 1 atm pressure are readily determined in a simple inexpensive furnace that is essentially open to the atmosphere. In natural magma systems, of the three intensive variables influencing equilibria, P is perhaps the least subject to variation when the magma has stagnated at some level in, or on, the Earth. Clearly, however, holding any variable constant can be misleading in our efforts to try to understand the behavior of an evolving natural system.

A one-component isobaric T–X phase diagram is simply a line in temperature space (see Figure 8-12). At the equilibrium freezing (melting) temperature, T_e, the system consists of two coexisting phases, liquid and crystals, and is therefore invariant:

$$F = 1 + C - \varphi = 1 + 1 - 2 = 0.$$

The T–X phase diagram of a two-component, or binary, system is an extension of the one-component line diagram. In Figure 8-13, with the addition of a second component, a liquidus line now delineates the region of crystal-free liquid in T–X space. The sympathetic variations in mole fractions, X_A and X_B, of the two components are shown along the bottom of the diagram. Re-

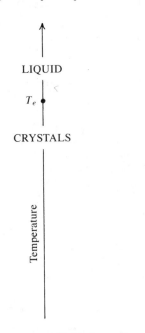

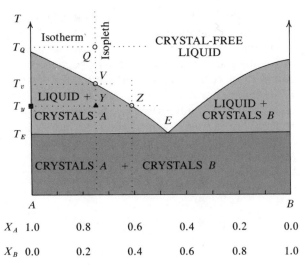

X_A 1.0   0.8   0.6   0.4   0.2   0.0

X_B 0.0   0.2   0.4   0.6   0.8   1.0

Figure 8-13   Phase relations at a fixed pressure in a binary system, A–B. Lines of constant temperature are isotherms and lines of constant composition are isopleths.

Figure 8-12   One-component T–X phase diagram at constant P.

member that $X_A + X_B = 1.0$. With increasing concentration of component B in the system, the liquidus is depressed to lower temperatures. Any point in the crystal-free liquid field, such as Q, represents a divariant equilibrium, as the variance $F = 1 + 2 - 1 = 2$. System Q is completely described if the temperature, T_Q, and the composition, $X_A^L = 0.75$ or $X_B^L = 0.25$, of the liquid phase are specified. Divariant liquid states have two degrees of freedom; the liquid phase can freely vary (within certain limits) in T and X and still remain a liquid. In contrast, at some point on the liquidus, V for example, the system is univariant because it now consists of two phases—crystals and liquid; by fixing one variable, X_A^L, or X_B^L, or T, the other two are uniquely determined.

The liquidus has a threefold significance. It is: (1) a line bounding the lower thermal stability field of crystal-free liquid in T–X space, as we have just seen; (2) the locus of points describing compositions of liquids at various temperatures coexisting in equilibrium with crystals; and (3) a solubility line showing the concentrations of components in the liquid.

The second of these functions of the liquidus is also illustrated in Figure 8-13. From the melt V, as the temperature drops from T_v to T_y, progressively more crystals of pure A precipitate, depleting the liquid in this component. The liquidus line between points V and Z defines this change in liquid composition as crystallization of pure crystals A progresses. A horizontal isothermal line, or **isotherm,** drawn across the diagram from T_y through Y intersects the liquidus line at point Z, which represents the unique liquid in the system at T_y in equilibrium with crystals of pure A. An **isopleth,** or line of constant composition, drawn vertically through Z to the base of the diagram indicates this liquid has a composition $X_A^L = 0.61$ (or $X_B^L = 0.39$) compared to 0.75 at T_v. Hence, once T is known in this univariant crystal–liquid system, the composition of the liquid can be determined from the diagram—it is represented by a point on the liquidus—and the crystals are seen to be pure

A. Note that if we had initially specified the composition of the liquid ($X_A^L = 0.61$), then the temperature, T_y, of the univariant system would have been fixed.

The third role of the liquidus is in showing the solubility of a component in a liquid coexisting with crystals. In Figure 8-13, component *A* decreases in concentration in the liquid from *V* to *Z* with decreasing *T*. A steeper liquidus curve would indicate that the concentration of component *A* remains high over a large *T* range and precipitation of pure *A* crystals tends to be retarded because of this greater solubility. A flatter liquidus indicates the reverse case of low solubility and more immediate precipitation. Assuming ideal solution behavior, the solubility of a component *A* in a binary liquid, *L*, coexisting with crystals of *A* as a function of temperature (Carmichael and others, 1974, p. 173) is

$$\log X_A^L = \frac{-\Delta S_m(T_e - T)}{2.303 \, RT} \qquad (8.7)$$

where ΔS_m is the change in entropy upon fusion of the mineral component, T_e is the equilibrium or freezing temperature of pure *A*, and *R* is the gas constant. The smaller the ΔS_m, the steeper the liquidus, and vice versa. Table 7-4 shows that alkali feldspars and especially crystalline silica have small ΔS_m values and hence have steep liquidi, high solubilities in a melt, and delayed crystallization. In contrast, pyroxenes, olivines, and especially Fe–Ti oxides have large ΔS_m values, and we would anticipate flat liquidi, low solubilities, and early crystallization.

These contrasting solubility relations have a significant effect on the sequence of precipitation of crystalline phases from a melt. To appreciate this concept, consider additional aspects of Figure 8-13. Two liquidus curves meet at a common **eutectic point**, *E*, through which runs the horizontal isothermal **solidus line**. Anywhere below the solidus, the system at equilibrium is comprised of crystals of pure *A* plus crystals of pure *B*. As a melt system of composition $X_A = 0.75$ crystal-

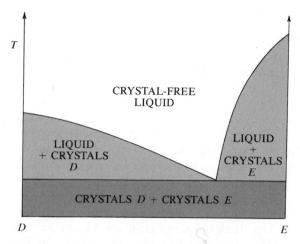

Figure 8-14 Binary eutectic diagram in which T_e and ΔS_m differ significantly for pure end-member phases.

lizes, the residual melt migrates in composition down the liquidus line toward the eutectic. At a temperature just above T_E, the system is composed of pure crystals *A* and of a liquid with a composition about $X_A^L = 0.48$. Upon reaching T_E, additional crystals of *A* plus crystals of pure *B* coprecipitate. Because three phases coexist in equilibrium at the eutectic temperature, the system is invariant ($F = 1 + 2 - 3 = 0$) and has no freedom to change until one of the phases—the eutectic liquid $X_A^L = 0.47$—is consumed by formation of crystals *A* and *B*. After elimination of the liquid phase, the system becomes univariant ($F = 1 + 2 - 2 = 1$) and is free to drop further in temperature.

The sequence of precipitation for this particular model magma system is, therefore, crystals of pure *A* followed by crystals of pure *B*. For any system whose bulk composition lies to the right of the eutectic point *E*, that is, $X_A < 0.47$, crystals *B* precipitate first and are then joined at the eutectic temperature by crystals *A*. In Figure 8-14, another hypothetical eutectic binary diagram shows how crystals of pure *D*, with a large ΔS_m and hence low solubility, can precipitate first over a broad range of melt compositions even through pure crystals of *D* melt at a lower T_e than crystals of pure *E*.

Hence, the relative temperature and the position in the crystallization sequence at which a phase precipitates from a melt depend upon confining pressure, the chemical composition of the melt, the melting temperature of the pure phase, and its entropy of melting. Phases with low ΔS_m, such as quartz and alkali feldspar, are more soluble in common subalkaline melts and tend to crystallize last, whereas olivines, pyroxenes, and Fe–Ti oxides are less soluble and generally precipitate early.

A phase diagram also furnishes information regarding the modal proportions of phases, in addition to their mineralogical compositions. The technique used to find this information involves the **lever rule**. In Figure 8-13, an isothermal tie line $T_y YZ$ passing through the bulk composition point for the system, Y, has end points T_y and Z, representing the two coexisting phases—pure crystals A and liquid Z. Imagine this tie line to be a mechanical lever resting on a fulcrum at Y, as shown in Figure 8-15. At equilibrium, the fractional proportion of crystals, $\mathbf{C}$, multiplied by the lever arm, x, must equal the fractional amount of liquid, L, multiplied by its lever arm, y, or

$$\mathbf{C}x = Ly$$

But as $\mathbf{C} + L = 1$, we have for the relative amount of $\mathbf{C}$

$$\mathbf{C}x = (1 - \mathbf{C})y = y - \mathbf{C}y$$

$$\mathbf{C} = \frac{y}{x + y} \qquad (8.8)$$

The closer the fulcrum, or the bulk composition, lies to the liquidus, the greater is the proportion of liquid. In Figure 8-13, 36% of the system Y is crystals and 64% is liquid. Application of the lever rule to any crystallizing system can show quantitatively the increasing proportions of crystals that form with falling T (see Figure 8-16).

The phase diagram for the familiar binary system water–antifreeze is shown in Figure 8-17.

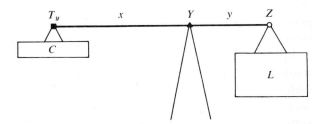

Figure 8-15 Lever rule for system Y at $T = T_y$ in Figure 8-13.

Two final comments should be made before discussing diagrams for model systems involving petrologically significant end-member components. Boundary lines enclosing stability fields are drawn on the basis of a few experimental control points; hence, such lines are interpretive, like contacts between a few exposed rock units on a geologic map (see Figure 8-9). Most phase diagrams are expressed in terms of weight percent (wt.%) or molecular percent (mol%) of the end-member components, rather than mole fraction as previously shown.

The system forsterite–silica. The phase diagram of this simple system demonstrates phase incompatibility, incongruent melting, reaction relation, and liquid immiscibility.

Countless observations of natural rocks indicate that certain minerals virtually never occur together, perhaps only in rare accidental circumstances. There may be two reasons for this. Because of the way natural magmas originate and evolve, minerals that typically precipitate at highest temperatures from the least-evolved systems are seldom, if ever, found to coexist with minerals that precipitate at lowest temperatures from highly evolved magmas. For this reason, forsterite and albite are unlikely associates, even though they are not a disequilibrium pair. In contrast, nepheline–quartz and forsterite–quartz are two mineral pairs that are thermodynamically unstable; rare magmatic rocks might contain both magnesian olivine and quartz, but one or both will be corroded, indi-

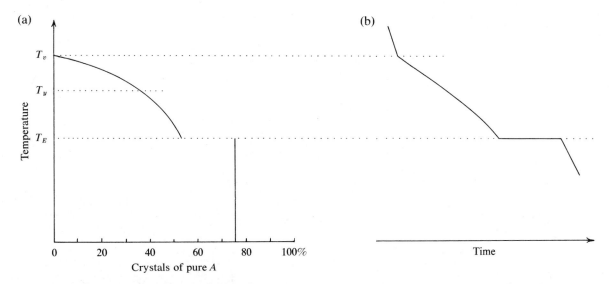

Figure 8-16 (a) Proportion of crystals of pure A that form in the system $X_A = 0.75$ (in Figure 8-13) as a function of temperature. Immediately before reaching the eutectic temperature, T_E, the system is 53% crystals A. At T_E, before the system can cool any more, the coexisting liquid of eutectic composition must crystallize into about equal proportions of crystals A and B, giving finally 75% crystals A and 25% crystals B, which is equivalent to the bulk composition of the system. (b) Schematic plot of temperature versus time for the system $X_A = 0.75$ as it loses heat to its surroundings at what is assumed to be at a constant rate. The crystal-free melt falls rapidly in temperature above the liquidus ($T > T_v$) because only its specific heat must be dissipated. During its excursion through the two-phase region (T_v to T_E), the crystallizing melt cools much slower because of the latent heat of crystallization that must flow out of the system. At the eutectic temperature, T_E, the invariant system remains fixed in temperature for a period of time as the latent heat is dissipated sufficiently so the system can completely crystallize. It may be noted that in a well-insulated system, the latent heat would be dissipated very slowly; rapid crystallization could not occur because it would generate enough heat to raise the temperature of the system above T_E, arresting the eutectic crystallization. Finally, below T_E, the system cools fairly rapidly because only the specific heat of the crystalline solids has to be lost.

cating a state of disequilibrium that was frozen into the rock. The binary system forsterite–silica at $P = 1$ atm, shown in Figure 8-18, confirms that forsterite and quartz cannot coexist in equilibrium but react to form stable enstatite:

$$Mg_2SiO_4 + SiO_2 = 2MgSiO_3$$
forsterite   quartz   enstatite

Most pure minerals with a fixed composition such as albite, diopside, forsterite, and quartz melt **congruently** to a liquid of exactly the same com-

position. Ice melts to liquid water congruently. However, at its melting point (1557°C, 1 atm) enstatite melts to a slightly more siliceous liquid, L_R, represented by point R, plus forsterite crystals (see Figure 8-18).

thermal energy + enstatite =
$$\text{forsterite} + L_R \quad (8.9)$$

The liquid has about 61 wt.% SiO_2 compared to just under 60% in enstatite, and the lever rule discloses a little more than 5% forsterite crystals

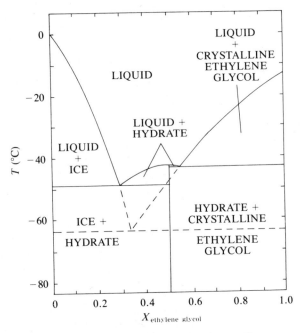

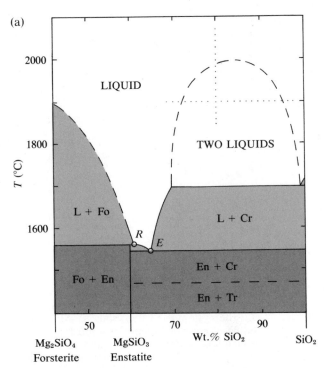

Figure 8-17 The phase diagram for the antifreeze system water–ethylene glycol serves as an illustration of a mixture familiar to anyone who has dealt with water-cooled engines in subfreezing climates. The diagram shows two interesting features. Optimum depression of the freezing temperature occurs for mixtures with approximately equal mole fractions of water and ethylene glycol. Under some circumstances, such intermediate mixtures can be metastably supercooled (dashed boundary lines) without formation of the 1–1 ethylene glycol hydrate; apparently it nucleates with considerable difficulty. [After J. B. Ott, J. R. Goates, and J. D. Lamb, 1972, Solid-liquid phase equilibria in water + ethylene glycol, *Journal of Chemical Thermodynamics* 4.]

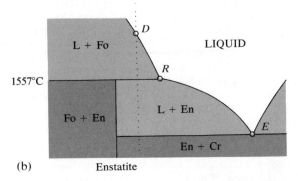

Figure 8-18 (a) The system forsterite–silica at $P = 1$ atm. Fo, forsterite; En, enstatite; Cr, cristobalite; Tr, tridymite; L, liquid. (b) Enlarged and slightly distorted region of the phase diagram in the vicinity of the reaction point, R. [After E. C. Ehlers, 1972, *The Interpretation of Geological Phase Diagrams* (San Francisco: W. H. Freeman and Company).]

forms in the melting reaction. The **incongruent melting** exemplified by enstatite involves a dissociation reaction of the crystalline phase to another solid plus a different liquid.

As will be shown later, crystalline solid solutions intermediate between their pure end members, such as any plagioclase, olivine, or pyroxene, melt incongruently. The significance of this behavior lies in the potential it offers for crystal–liquid fractionation to split an initial starting crystalline material into liquid and crystals of different composition.

The incongruent melting of enstatite can be viewed from another perspective. Consider in Figure 8-18b a cooling liquid whose bulk composition is only slightly more silica-rich than enstatite. It will begin to precipitate forsterite at D where the isopleth intersects the liquidus. At 1557°C there will be a **reaction relation**, Equation (8.9), in which all of the forsterite crystals are consumed by reaction with liquid R simultaneously with precipitation of enstatite crystals. Under equilibrium conditions, all of the forsterite must be consumed before the temperature can fall below 1557°C. The isobaric invariant point R on the diagram is called a **peritectic**, or **reaction point**. Liquids moving down the liquidus from D to E must pause isothermally, under equilibrium conditions, at R until all forsterite is consumed by reaction. Once all the olivine has been converted to enstatite, the temperature can continue to fall; more enstatite crystallizes until the liquid reaches the eutectic point E, where the last drop of liquid crystallizes to more enstatite plus cristobalite.

This type of reaction relation models the behavior of the Makaopuhi basalt magma (see the photomicrographs reproduced in Plates I through IV and Figure 8-6), in which early-precipitated magnesian olivine was resorbed at lower temperature into a relatively silica-rich melt, with enhanced crystallization of pyroxene. The presence of substantial Fe, Ca, Na, Al, and so on in this natural basaltic system changed the details of the reaction relation, but not its fundamental character.

Incomplete reaction relations in natural magmas will be manifest by anhedral, corroded grains of the unstable high-T phase, and sometimes these may be surrounded by a **reaction rim** of the lower-T crystalline phase. Reaction rims in magmatic rocks will have the appearance of overgrowths such as that shown in Figure 8-5. But before an overgrowth texture can be interpreted as originating by an incomplete reaction relation, pertinent crystal equilibria must be verified experimentally.

Liquid immiscibility is still another phenomenon illustrated by the system forsterite–silica. There is a significant T–X range on the silica-rich side of the diagram above 1695°C where two liquids coexist in equilibrium. A liquid $X^L_{SiO_2} = 0.8$ at 2000°C, for example, upon cooling splits into two immiscible liquids; at 1900°C these two melts have $X^L_{SiO_2}$ 0.72 and 0.96.

Binary solid solutions: K-feldspar–anorthite. Most major minerals in magmatic rocks are solid solutions; only quartz is not. Thus any realistic model of magmatic systems must involve crystal–melt equilibria in which both the solid and the liquid phases are multicomponent solutions. In the binary systems considered thus far, the solids have fixed, pure compositions with no solid solution between the component end members; for example, no solid solution occurs between silica and forsterite in Figure 8-18. However, in systems with solid solution, a **solidus curve** reflects the solid solubility of end-member components in the crystalline phases that make up the system. Its slope is a function of the extent of solid solution with respect to T. In the system K-feldspar–anorthite, shown in Figure 8-19, the solidus line on the right side shows how much K-feldspar is "dissolved" in anorthite solid solution crystals, An_{ss}, coexisting with melt, as a function of temperature. The solid solubility of crystals coexisting with liquid increases with decreasing T; at 1100°C it is ~ 2 wt.%, and at 850°C ~ 4 wt.%. Similarly, the solid solubility of An in K-feldspar-rich solid solutions, Kf_{ss}, on the left side of the diagram increases at decreasing T. Below each sloping solidus only a single phase is stable, K-feldspar solid solution (Kf_{ss}) on the left and anorthite solid solution (An_{ss}) on the right. In the stability field of either one of these single phase solid solutions, the variance is $F = 2 + 1 - 1 = 2$. Thus a complete description must specify T and X, which is the same as in the single-phase liquid field.

The isothermal tie line through a bulk composition point W at 950°C intersects the liquidus at An_{20}—that is, $X^L_{An} = 0.20$, and the solidus at An_{97}. Such a system consists of 65% crystals $An_{97}Kf_3$

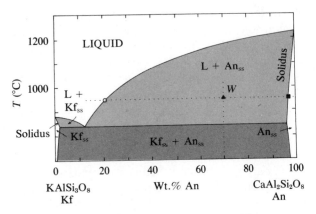

Figure 8-19 The system K-feldspar–anorthite at $P = 0.5$ GPa under water-excess conditions ($P = P_{H_2O}$). The presence of a separate water-rich phase makes this system ternary, but for our purposes here its presence can be ignored. The subscript "ss" on the crystalline phases (An_{ss} and Kf_{ss}) denotes solid solutions. The significance of the two steeply sloping solvus line segments extending below the ends of the isotherm through the eutectic will be discussed later. The symbol Or is more conventionally used to denote the component $KAlSi_3O_8$ but Kf, for K-feldspar, is used here to remind the reader that different polymorphic forms of potassium feldspar occur, and orthoclase is only one of these. [After E. C. Ehlers, 1972, *The Interpretation of Geological Phase Diagrams* (San Francisco: W. H. Freeman and Company).]

and 35% liquid $An_{20}Kf_{80}$. In the same sense that the liquidus is the locus of points describing compositions of liquids that coexist with crystals, the solidus is the locus of points giving the composition of crystalline solid solutions coexisting with liquid.

Binary solid solutions: Plagioclase. Of the model silicate systems considered here, the plagioclase system is the most important, for two reasons. First, plagioclases are virtually ubiquitous in magmatic rocks, and second, this system provides insight into the behavior of other systems involving extensive or even complete solid solution between end-member components. Although crystalline K-feldspar and anorthite display only limited solid solution, to the extent of only a few percent,

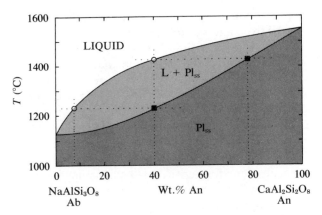

Figure 8-20 The system albite–anorthite at $P = 1$ atm. [After N. L. Bowen, 1928, *The Evolution of the Igneous Rocks* (New York: Dover).]

complete solid solution occurs in the plagioclases between the albite and anorthite end members (see Figure 8-20). The liquidus and the solidus form a double loop running continuously with little change in slope from the melting point of pure anorthite to the one for pure albite. For systems within the univariant two-phase region of crystals plus liquid, progressively more albitic plagioclase crystals coexist with more albitic liquids at lower temperatures. For these univariant equilibria, specification of the temperature uniquely fixes the compositions of both liquid and crystals. Or, specification of the composition of one phase fixes the composition of the other as well as T. Note, however, that the modal proportion of crystals to liquid at any particular T depends entirely on the bulk composition of the system.

Any plagioclase of intermediate composition between pure Ab and pure An will melt incongruently with increasing T, forming liquids more albitic and crystals more anorthitic than the original. For example, initial melting of crystals An_{40} yields liquid with a composition of An_8; as T increases the melting progresses, creating a larger proportion of liquid to crystals, both of which become more enriched in anorthite, as may be seen by drawing a series of isothermal tie lines at successively higher temperatures. At 1425°C the last minute crystal to be consumed, An_{78}, is in equilibrium with a melt of the original bulk composition.

Crystallization of liquid by decreasing the temperature produces crystals more anorthitic than the liquid. For example, the first crystals precipitating from a liquid An_{40} have a composition of An_{78}; as T decreases, crystallization advances, creating more crystals; but both crystals and liquid are more albitic, as may be seen by drawing some representative tie lines.

It may seem surprising that progressive crystallization makes both crystals and liquid more albitic, and the reverse for progressive melting. This is possible because the bulk composition of the system lies between the compositions of the liquid and crystals comprising it; both phases change sympathetically in composition with concomitant changes in their modal proportions.

In order for these sympathetic changes in composition of coexisting solid and liquid solutions to occur within the univariant two-phase region, there must be continuous reaction between them as changing intensive variables, such as T, modify the state of the system. These **continuous reaction relations** occurring over a range of T may be contrasted with the peritectic reaction relation discussed earlier with regard to pure enstatite and forsterite and melt occurring at a unique T.

Adjustment of liquid and crystalline compositions in the course of these reactions in the plagioclase system demands exchange of tetrahedrally coordinated Si^{4+} for Al^{3+}, coupled with exchange of octahedrally coordinated Na^+ for Ca^{2+}, or $(NaSi)^{5+} \rightarrow (CaAl)^{5+}$. Exchange must occur by diffusion of these ions in the melt and crystals across their interface. Obviously, the larger the crystals, or the more viscous the melt, or the faster the change in state variables, the less chance there is that equilibrium will be maintained in the system.

As changing intensive variables disturb the state of equilibrium in a magma system, diffusion-controlled reaction relations will tend to restore equilibrium. The outcome of this contest lies

somewhere between two extremes: (1) perfect equilibrium crystallization or melting in which $P-T-X$ conditions change slowly enough that reactions are complete, and (2) perfect fractional crystallization or melting in which no reaction relations occur between crystals and liquid.

In **perfect equilibrium crystallization,** or its melting counterpart, P, T, and X change reversibly in infinitesimally small steps at a rate slower than the diffusion-controlled reactions that must occur between crystals and liquid. Equilibrium is constantly maintained. This ideal may be approached in deep plutonic systems where T, P, or water concentrations change only very slowly. The total range of transient liquid and crystal compositions from onset to completion of perfect equilibrium melting or crystallization is relatively restricted (see Figure 8-21). The final liquid or crystalline product of the equilibrium melting or crystallization process is chemically homogeneous.

The virtually universal inhomogeneity in plagioclases of magmatic rocks, manifest in their zoning (see Figures 7-15 and 7-16), together with compositional variations throughout a particular magmatic body indicate a general lack of perfect equilibrium during crystallization. Hence most systems crystallize to some extent in the fractional manner. In the extreme case of **perfect fractional crystallization** or melting, no reaction whatsoever occurs between liquid and crystals with changing states of the system. There is a complete separation, or fractionation, of liquid and crystals, either by mechanical processes or by chemical effects associated with rapidly changing states.

During perfect fractional crystallization of a liquid, such as An_{40} (see Figure 8-22), each liquid fraction, isolated from all previously precipitated crystals, is effectively a new system with no knowledge of its prior history. Because of the lack of reaction, NaSi is conserved and CaAl suffers the maximum depletion within the liquid. When crystals An_{40} are precipitating, all of the more calcic crystals, up to An_{78}, are isolated and hence there is still much liquid remaining enriched in NaSi. The end result of perfect fractional crystalli-

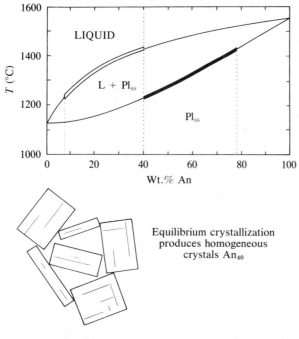

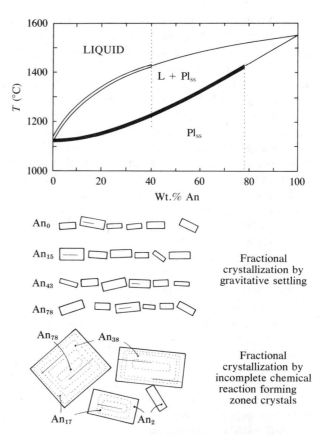

Figure 8-21 Equilibrium behavior in a plagioclase system whose composition is An_{40}. During equilibrium crystallization, or melting, the range of transient liquid and crystal compositions is indicated by the heavy lines along the liquidus and solidus. The final products of equilibrium crystallization are homogeneous crystals An_{40}, and the final product of equilibrium melting is homogeneous liquid An_{40}.

zation is that the last liquid to crystallize is An_0, and its precipitate is An_0—pure albite. Hence the crystalline products in this example range from An_{78} to An_0.

The process of perfect fractional melting is not simply the reverse of fractional crystallization, as it is under equilibrium conditions. Starting with crystals An_{40}, liquid An_8 first appears (see Figure 8-23). Isolation of this liquid demands that the residual crystals, say An_{41}, are a new independent system, and upon melting they form liquid An_9, and so on. The last crystals to melt will be An_{100}, and the liquid will approach An_{100}. The temperature range over which the fractionation process operates is large.

Figure 8-22 Perfect fractional crystallization in a plagioclase system whose composition is An_{40}. The range of transient liquid and crystal compositions is indicated by the heavy lines along the liquidus and solidus. Gravitative settling produces a stack of compositionally variable crystals that are more albitic toward the top. The proportions and compositions of the crystals shown here are schematic only; no quantitative properties are implied. Incomplete reaction produces zoned crystals more albitic toward their margins. All An values are arbitrary except for initial crystals, which are An_{78}.

Fractional melting and crystallization extend the range of transient liquid and crystal compositions and of operational temperatures relative to the equilibrium type of behavior.

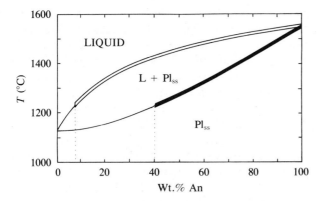

Figure 8-23 Perfect fractional melting of plagioclase An$_{40}$. Range of transient liquids and crystals is indicated by heavy lines.

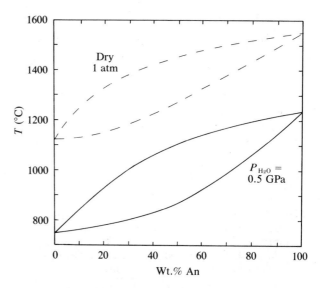

Figure 8-24 The plagioclase system at $P = P_{H_2O} = 0.5$ GPa, water-excess conditions, compared to the dry system at 1 atm. [After E. C. Ehlers, 1972, *The Interpretation of Geological Phase Diagrams* (San Francisco: W. H. Freeman and Company).]

Fractional crystallization can occur in at least four ways: incomplete reaction, gravitative separation, filter pressing, and flowage segregation.

Incomplete reaction occurs during rapid growth of crystals associated with rapidly changing P, T, or X. Additional crystalline material precipitates around previously formed crystals before reaction with the melt can occur by diffusional processes. Then before the melt can react with that newly accreted plagioclase, conditions change once again, inhibiting reaction again. As the process continues, the melt never has a chance to equilibrate with precipitated crystals. In effect, the melt at any stage of the fractional crystallization process is a completely new system in which crystallization is just being initiated because already formed crystals are always sensibly nonexistent. The importance of this process on a small scale is undeniable, for it is manifest in zoned plagioclases in nearly all magmatic rocks.

In the early stages of magmatic crystallization, large, freely suspended crystals might be expected to rise upward or sink if their density is less or greater, respectively, than the enclosing melt. Crystals would become segregated within the magma body and, for the most part, would be isolated from further reaction with the remaining body of melt. Upon further change in the system,

additional precipitation produces a new crop of more albitic crystals, which then separate in turn. By this means an accumulative sequence of crystals would form. It must be realized that this process, called **gravitative separation,** is an imperfect fractionation mechanism; the mass of segregated crystals must lie within a significant volume of interstitial melt, which would ultimately be expected to react to some degree with the adjacent crystals.

In a magma body comprised largely of crystals, the interstitial liquid might be squeezed, or **filter pressed,** out of the interstitial spaces by movement of wall rocks, somewhat like squeezing water from slushy snow. Suspended crystals in magma moving through restricted channels such as dikes may migrate toward the center of the channel by **flowage segregation.**

If water is dissolved in the melt, a substantial lowering of the liquidus and solidus occurs because of the stabilizing effect on the melt relative to crystals (see Figure 8-24). This depression of

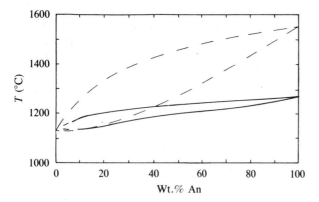

Figure 8-25 Locus of coexisting plagioclase liquid and crystal compositions at 1 atm where diopside is also precipitating in the system Ab–An–Di. The binary system (dashed lines) is shown for comparison. Pure albite does not crystallize from melts containing Ca-bearing components. [After S. A. Morse, 1980, *Basalts and Phase Diagrams* (New York: Springer-Verlag).]

crystallization temperatures due to dissolved water was previously shown in Figure 8-4.

In systems of plagioclase plus additional components in which common chemical elements go into the precipitating phases, not only are crystallization temperatures depressed, but the plagioclase solidus–liquidus loop is also distorted. For example, in the system Ab–An–Di (for a thorough description see Morse, 1980), the loop is depressed more at the An end (see Figure 8-25), where liquids are crystallizing diopside as well as plagioclase. Compared to relations in the binary plagioclase system, crystallization temperatures are depressed between 200°C and 300°C, well into the range at which relatively dry basaltic magmas crystallize (see Figure 8-6). It is also of interest to note that for a given change in T, there is a greater change in compositions of liquid and coexisting crystals due to the small slope of the liquidus and solidus.

It may be anticipated that in still more complex systems with additional components further modification of the liquidus–solidus loop might occur, but the basic behavior of plagioclase crystallization and melting does not change.

Binary solid solutions: Alkali feldspar–water. The system Kf–Ab is important in understanding crystallization of feldspars in general and granitic magmas in particular. However, when dry or with small concentrations of water, the system is complicated by a large stability field of leucite, $KAlSi_2O_6$, whose composition cannot be expressed in terms of Ab and Kf. Hence the system is not binary. However, at elevated water pressures the field of leucite shrinks and is eventually eliminated, and a **temperature minimum** appears on the liquidus and solidus curves (see Figure 8-26a). This minimum is significant in fractional crystallization because, like a eutectic, residual liquids move to it and there precipitate an alkali feldspar solid solution (Af_{ss}) with slightly more of the Ab than the Kf component. Except at this unique minimum composition, melting is incongruent throughout the system; initial liquids are always nearer the minimum composition than the original crystals.

Below the solidus is the **solvus** (or unmixing, or exsolution) **curve,** which separates a field of two stable feldspar solid solutions from an overlying higher-T field of one alkali feldspar solid solution, Af_{ss}. In Figure 8-26a, once below the solvus, initially homogeneous alkali feldspar solid solutions exsolve, or unmix, under equilibrium conditions into two discrete phases. For example, at 600°C an initially homogeneous feldspar $Ab_{40}Kf_{60}$ exsolves into a Kf-rich feldspar solid solution $Ab_{32}Kf_{68}$ and an Ab-rich feldspar solid solution whose composition is $Ab_{79}Kf_{21}$. The lever rule indicates 83% and 17% of each. These two feldspars produced by the exsolution process usually segregate within the original crystal as thin subparallel lamellae, forming the intergrowth known as **perthite.** At a lower T, 500°C, for example, $Ab_{93}Kf_7$ and $Ab_{25}Kf_{75}$ should coexist at equilibrium. Use of perthite as a geothermometer to indicate the temperature of exsolution is hindered by the very sluggish rates of diffusion within the crystal, which generally prevent attainment of equilibrium, and by different solvi for different structural states of the feldspars.

At high water pressures, about 0.5 GPa, the

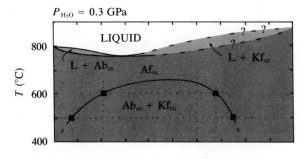

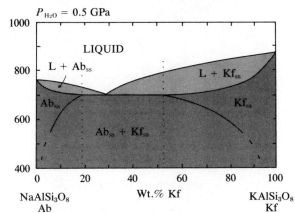

Figure 8-26 The system Ab–Kf–H₂O. Water is present everywhere in addition to the indicated phases. [After E. C. Ehlers, 1972, *The Interpretation of Geological Phase Diagrams* (San Francisco: W. H. Freeman and Company).]

liquidus–solidus loop is depressed so far that it intercepts the solvus (see Figure 8-26b). The Kf–Ab–H₂O diagram now resembles the Kf–An diagram shown in Figure 8-19, except that the extent of solid solution between the two end members is greater in the alkali feldspars. Liquids crystallizing in the equilibrium manner and lying between about $Ab_{81}Kf_{19}$ and $Ab_{48}Kf_{52}$ (and liquids of any composition undergoing extreme fractional crystallization) will solidify directly into two feldspars. Each of these phases may subsequently experience exsolution upon cooling below the solvus, but this is generally only conspicuous for the Kf-rich feldspar solid solution.

Ternary solid solutions: Feldspars. As feldspars occur prominently in most magmatic rocks, it is worthwhile to examine phase relations in the complete ternary feldspar system Ab–An–Kf. For realism and to promote reaction rates, laboratory studies of feldspars have involved water, making the system quaternary, Ab–An–Kf–H₂O. However, to keep the discussion simple, P_{H_2O} will be assumed constant and can be ignored in the diagram.

To represent the system at constant P_{H_2O}, four intensive variables $T–X_{Kf}–X_{Ab}–X_{An}$ are shown in Figure 8-27. A liquid such as X upon cooling down the isopleth intersects the liquidus surface at X', where plagioclase begins to precipitate. As the system cools, the melt follows a curved path $X'X''$ down the liquidus, precipitating progressively more plagioclase of increasing albite concentration. X'' lies in the bottom of the liquidus valley on the **cotectic boundary line** between the composition field where plagioclase first precipitates as a liquidus crystalline phase and the compositional field where Kf solid solutions are liquidus crystals. When the melt reaches X'', it coprecipitates plagioclase and Kf-rich alkali feldspar and continues to do so as it moves down the univariant cotectic boundary line. A wholly liquid system Y upon cooling will first crystallize K-rich feldspar as it impinges upon the liquidus at Y'. When the system cools to the temperature where the melt is at Y'', there is coprecipitation of plagioclase and alkali feldspar solid solutions. Although the boundary line is shown to run continuously to the Ab–Kf–T face of the prism, this is an oversimplification of complex behavior near the Ab–Kf minimum on the liquidus (Steward and Roseboom, 1962).

In Figure 8-28 the feldspar–temperature prism is viewed from another perspective and with the igloo-shaped ternary solvus emphasized in heavy lines. The smooth line loop within the prism, from Q to P around the curve to O and then to R, encompasses the *surface* common to both the ternary solvus and the ternary solidus. $QPOR$ is the line of intersection of the ternary solvus and solidus and

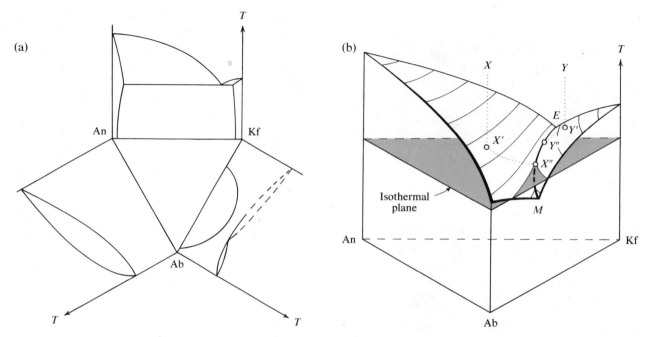

Figure 8-27 The system Ab–An–Kf–H₂O at moderately high pressures appropriate to shallow plutonic conditions. (a) Three binary systems shown in previous figures are linked around a compositional equilateral triangle in which proportions of the three condensed components are represented. The binary Kf–An system has been slightly distorted for clarity. (b) In this perspective view, the three binary systems are folded up to form a triangular prism whose axis represents temperature. The upper surface of the prism is the liquidus—now an isobaric *T–X* surface rather than a line—on which curved isothermal contour lines are drawn parallel to the compositional base of the prism. One isothermal plane (shaded) parallel to the base is shown cutting the liquidus surface along an isotherm. The underlying solidus and solvus surfaces have been omitted for clarity.

is the locus of compositions of two feldspar crystals coexistent with liquid on the cotectic boundary line. *ME* is the field boundary line on which the melt *X″* lies in Figure 8-27b. An isothermal plane (shaded) passing through *X″* intersects the solvus–solidus common surface along a straight line *PO*. This tie line *PO* connects the two feldspars—*P*, plagioclase, and *O*, Kf-rich alkali feldspar solid solution—that coexist in equilibrium with liquid *X″*. The **three-phase triangle** *X″PO* is defined by three isothermal tie lines connecting the three coexisting phases. The bulk composition

of the system is represented by a point located somewhere within the three-phase triangle. Progressively higher-*T* isothermal planes cut the solvus along the indicated two feldspar tie lines, up to tie line *QR*. The higher-*T* melts in equilibrium with these feldspar pairs are represented (but not shown in the figure) by points lying along *ME* to the right of *X″* and to the left of the associated feldspar tie line. The diagram from *X″* to *M* is grossly oversimplified, as previously indicated.

Because of the difficulty in working with a three-dimensional prism in only two dimensions,

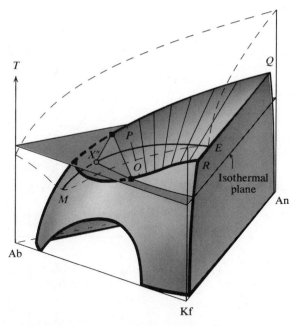

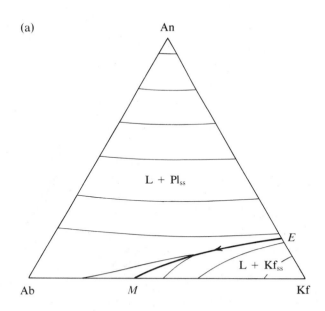

(a)

(b)

Figure 8-28 Another schematic view of the system Ab–An–Kf–H$_2$O at moderately high pressures appropriate to shallow plutonic conditions. The prism has been rotated relative to the view in Figure 8-27; the liquidus lines in the three bounding binary systems are dashed, and the igloo-shaped solvus volume within the prism is heavily outlined.

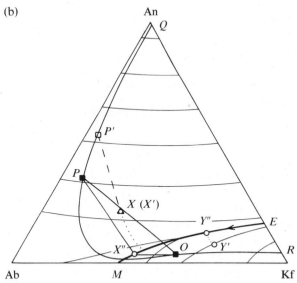

we can use the same procedure employed in making a topographic map of the land surface; that is, features within the prism can be projected down the *T* axis onto its triangular composition base. In this isobaric *T–X* projection, shown in Figure 8-29b, the point *X* (triangle), representing the bulk composition of our sample system, is superposed upon *X'*, the point on the liquidus surface representing melt from which the first plagioclase *P'* precipitated. The compositions of the two feldspars in equilibrium with the last drop of residual liquid in a system whose bulk composition is *X* can be found by drawing a tie line through *X* to intersect *QPOR*. These feldspars are just slightly more albitic than *P* and *O*. Additional details on

Figure 8-29 (a) *T–X* projection of isotherms and cotectic boundary line *ME* from the triangular prism shown in Figure 8-27b onto its base. (b) *T–X* projection of isotherms, boundary line *ME*, solidus–solvus intersection, and crystal–liquid tie lines shown in Figure 8-28.

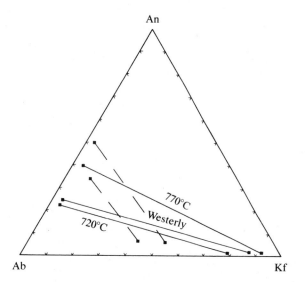

Figure 8-30 Compositions of equilibrium pairs of plagioclase and alkali feldspar solid solutions. Solid tie lines (compare tie line *PO* in Figure 8-29b) connect coexisting feldspars under high water pressure plutonic conditions; 770°C and 720°C are temperatures at P_{H_2O} = 0.5 GPa for experimentally determined equilibria. Westerly is a natural granite from Rhode Island. Dashed tie lines connect coexisting feldspars in a basaltic lava flow from southeastern California. [After E. C. Ehlers, 1972, *The Interpretation of Geological Phase Diagrams* (San Francisco: W. H. Freeman and Company); dashed tie lines after A. L. Smith and I. S. E. Carmichael, 1969, Quaternary trachybasalts from southeastern California, *American Mineralogist* 54.]

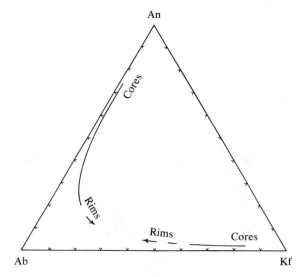

Figure 8-31 Fractionation paths showing how feldspars become zoned during fractional crystallization.

reading ternary diagrams can be found in Ehlers (1972), Muan (1979), and Cox and others (1979).

Compositions of equilibrium feldspar pairs in magmatic rocks resemble those in the experimental ternary system (see Figure 8-30). During fractional crystallization, the ternary phase diagram predicts plagioclases and alkali feldspars should be zoned toward more albitic rims (see Figure 8-31).

Other ternary solid solutions: Pyroxenes. There is extensive solid solution in the pyroxenes of igneous rocks. In alkaline rocks, pyroxene compositions extend from more or less pure diopside to-

ward aegerine ($NaFe^{3+}Si_2O_6$), commonly with significant amounts of Al and Ti. In the more widespread subalkaline rocks, pyroxene compositions are well represented in terms of the Ca, Mg, and Fe^{2+} end-members (see Figure 8-32), and Al, Ti, Mn, Na, and so on are generally present in only minor amounts. Actual phase relations are complicated by polymorphism, subsolidus unmixing, and incongruent melting relations (to give Mg-olivine as the primary liquidus phase), but residual melts are enriched in Fe relative to initial higher-*T* crystalline precipitates.

Reaction Series in Magmatic Systems

We have now briefly examined crystallization of the important rock-forming minerals, except olivines, amphiboles, and micas. Olivines constitute a complete solid-solution series, and the phase diagram is similar to that of plagioclase; magnesian olivine is the high-temperature component. Hydrous amphiboles and biotites are complex solid solutions with variable oxidation of Fe. Representation of pertinent intensive variables P, T, X_{O_2},

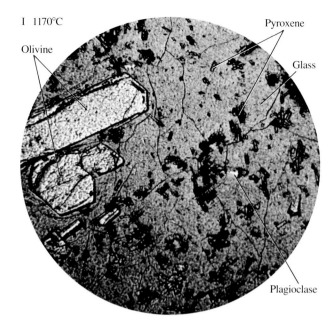

I 1170°C

Olivine

Pyroxene

Glass

Plagioclase

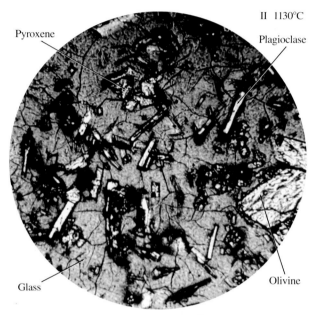

II 1130°C

Pyroxene

Plagioclase

Glass

Olivine

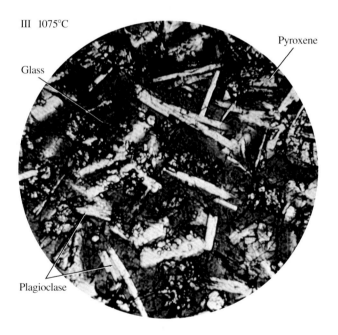

III 1075°C

Glass

Pyroxene

Plagioclase

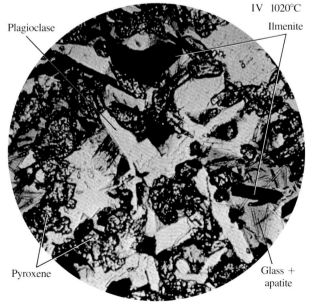

IV 1020°C

Plagioclase

Ilmenite

Pyroxene

Glass + apatite

This sequence of photomicrographs shows progressive crystallization of Makaopuhi basalt. The diameter of each view is about 1 mm. Plate I shows the initial stage of crystallization, at $T = 1170°C$, in this sequence. Compare with Figure 8-6. (Courtesy of Thomas L. Wright, U.S. Geological Survey.)

V

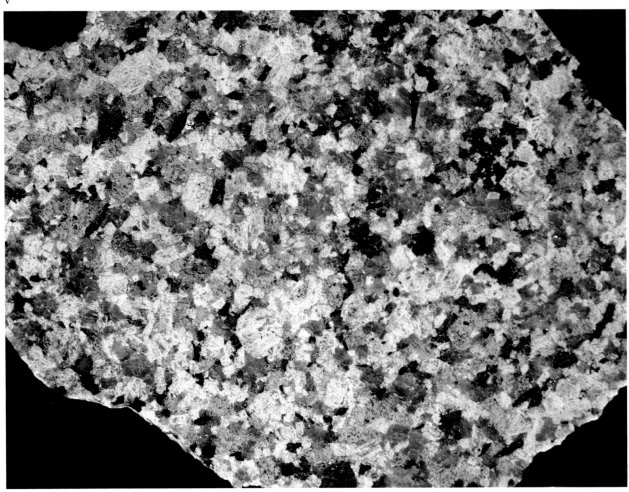

Polished, etched, and stained slab of granite showing K-rich alkali feldspars stained
by yellow potassium cobaltinitrite, gray quartz, and black biotite and hornblende.
Note hypidiomorphic-granular texture.

and so on in a phase diagram is difficult but is summarized at the beginning of Section 8.2. Amphiboles and biotites crystallizing at lower temperatures are commonly, but not invariably, more Fe-rich than higher-T precipitates, as is the case for the anhydrous mafic silicates olivine and pyroxene.

During crystallization of common subalkaline magmas, residual liquids and late-precipitating crystals become enriched in K, Na, and Fe relative to Ca and Mg.

It should be obvious from the foregoing discussion of phase diagrams of pertinent model systems that reaction relations are a fundamental feature of any crystallizing body of magma. They are a consequence of changes in equilibrium melt and crystal compositions with shifting T and P.

There are two types of reaction relations that produce two types of crystalline **reaction series.** The plagioclase system exemplifies a **continuous reaction series,** which develops by uninterrupted reactions between previously formed solid-solution crystals and the melt as T or P change over a wide range.

$$\text{plagioclase}_A + \text{melt}_M =$$
$$\text{plagioclase}_B + \text{melt}_N \quad (8.10)$$

In most natural magmas, the reaction relations are incomplete to some degree; that is, fractional crystallization prevails, with the result that inhomogeneous crystals form as a manifestation of the continuous reaction series. A **discontinuous reaction series** is created by a reaction relation between previously formed crystals and melt, causing them to dissolve with attendant precipitation of a new mineral species. An example is

$$\text{olivine} + \text{melt}_M = \text{pyroxene} + \text{melt}_N \quad (8.11)$$

In chemically complex natural magmas involving olivine and pyroxene solid solutions, this reaction relation can ensue over a limited range of T and P. And, once again, in the normal mode of fractional crystallization both reactant and product phases

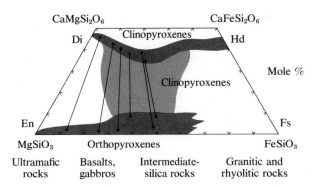

Figure 8-32 Compositions of pyroxenes in the common subalkaline igneous rocks. Heavy shade indicates compositions of pyroxenes occurring in plutonic rocks and as phenocrysts in volcanic rocks. Light shade indicates approximate range of (metastable) groundmass clinopyroxenes in lavas. Representative tie lines connect coexisting equilibrium pairs of pyroxenes. Temperature of precipitation decreases for more Fe-rich pyroxenes. Variable but usually minor amounts of Al, Ti, Mn, and Na are present. Rock types in which pyroxenes of a particular Mg/Fe ratio generally occur are indicated below the trapezohedron. [After I. S. E. Carmichael, F. J. Turner, and J. Verhoogen, 1974, *Igneous Petrology* (New York: McGraw-Hill).]

can be frozen into the rock, rarely as a disequilibrium reaction rim of Ca-poor pyroxene around olivine.

Reaction relations producing crystalline reaction series are ubiquitous. Commonly two or more series are simultaneously developed.

Discussion of ternary and quaternary phase diagrams that demonstrate paralleling reaction series is beyond the scope of this text (but see Osborn, 1979). Their existence, however, is clearly shown in the crystallization behavior of Makaopuhi basalt and Mt. Givens granodiorite discussed previously. In the series shown in Figure 8-6, the basalt was shown to have formed by coprecipitation over a wide range of T of four solid-solution series; in Figure 8-8 the crystallization of the granodiorite was also shown to be accomplished by simultaneously operating reaction relations that produced several parallel reaction series.

No discussion of this topic would be complete without at least brief reference to the parallel reaction series in the Skaergaard intrusion (see pages 173–180). Although the actual mechanism of crystal–liquid fraction in this classic stratiform intrusion is still controversial, the reaction series are unusually well defined and show broad cryptic compositional variations with respect to height in the intrusion (see Figure 8-33).

Many different reaction series are possible in magmas, depending primarily upon their bulk composition but also on the pressure at which they crystallize (see box on p. 287).

8.2 ROLE OF VOLATILES IN THE PHASE EQUILIBRIA OF MAGMATIC SYSTEMS

All of the phase equilibria discussed so far involve essentially only condensed constituents. However, volatiles play an important role not only in stabilizing certain crystalline phases in which they are significant components, but also in promoting explosive volcanic behavior and formation of pegmatites and hydrothermal veins.

Stability of Anhydrous Fe-Bearing Minerals

The stability of Fe-bearing phases is dictated by the amount of oxygen in the system, manifest by its vapor pressure, P_{O_2}, because oxidation-reduction reactions will promote formation of ferrous (Fe^{2+}) or ferric (Fe^{3+}) valence states.

We present here only a simple example of how P_{O_2} influences phase equilibria. Further discussion is deferred to pages 505–506. Figure 8-34 shows phase relations in the system Fe–Si–O. Although the amounts of oxygen that govern phase stability are extremely minute in an absolute sense, their relative variations involve many orders of magnitude. In most natural magmas, the amount of oxygen in the system is such as to stabilize magnetite and fayalite (Fe_2SiO_4). In reality, each of these

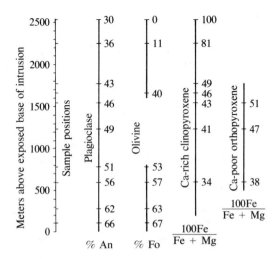

Figure 8-33 Contemporaneous continuous reaction relations have produced systematic cryptic chemical variations in major cumulus minerals in gabbroic rocks of the Skaergaard intrusion. As crystallization of an initially basaltic magma progressed, layers of crystals accumulated from the base of the intrusion upward (see pages 173–180). Olivine did not precipitate in the middle of the intrusion. [After L. R. Wager and G. M. Brown, 1968, *Layered Igneous Rocks* (London: Oliver and Boyd).]

will generally exist as end-members in Fe–Ti oxide and olivine solid solutions. Only in lunar magmas (see pages 572–573) and in very rare terrestrial basaltic lavas that have been reduced by interaction with coal does iron occur in the metallic state or as wüstite (FeO). As terrestrial magmas make contact with near-surface meteoric waters or with the oxygen-rich atmosphere, the predominantly ferrous iron, Fe^{2+}, in the melt and in mafic silicates can be oxidized to the ferric state, Fe^{3+}. Under these more oxygen-rich conditions, hematite is stable. Hence, iron-rich basaltic lavas extruded onto the surface are commonly red because of the presence of very finely divided particles of this phase. The occurrence of abundant Fe^{3+} within basaltic bodies emplaced near the surface may indicate that exsolved water has dissociated into O_2 plus H_2; the faster-diffusing hydrogen could prefer-

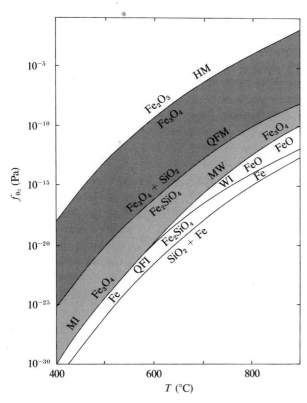

Figure 8-34 Stability fields and reaction boundaries for crystalline phases in the system Fe–Si–O. The fugacity of oxygen, f_{O_2}, is very roughly equivalent to P_{O_2} at high temperatures. Boundary lines represent f_{O_2}–T conditions for reaction between the indicated solid phases and oxygen. For example, the lowest curve (QFI) represents the univariant equilibrium between fayalite (F), quartz (Q), and iron (I)

$$Fe_2^{2+}SiO_4 = SiO_2 + 2Fe + O_2$$

Note that the stability field (shaded) of magnetite (M) depends upon whether SiO_2 is available in the system. Thus, at $f_{O_2} = 10^{-25}$ Pa magnetite is unstable above 400°C in the presence of SiO_2, but without SiO_2 it only becomes unstable for $T \gtrsim 500$°C. The stability of metallic iron (I) also depends upon availability of SiO_2. W, wüstite; H, hematite. [After W. G. Ernst, 1976, *Petrologic Phase Equilibria* (San Francisco: W. H. Freeman and Company).]

Bowen's Reaction Series

A very famous reaction series known to almost every geologist was first formulated by Bowen (1928) and now bears his name—**Bowen's reaction series.** It consists of two parallel series: a discontinuous series of (from higher to lower temperatures) olivine, pyroxene, amphibole, and biotite; and a continuous reaction series of calcic to sodic plagioclase. Unfortunately, however, some textbook writers and petrologists since 1928 have taken his series without reservation, implying it has universal application to all magma systems. Careful reading of Bowen (1928, p. 60) is enlightening: "An attempt is made ... to arrange the minerals of the *ordinary subalkaline* rocks as reaction series. The matter is really *too complex to be presented in such simple form.* Nevertheless *the simplicity, while somewhat misleading,* may prove of service in presenting the subject in concrete form" (Italics added). Obviously, Bowen intended his series to be an overly simplified but instructional viewpoint of how one type of magma can crystallize. We still recognize, with Bowen, that magmatic crystallization is complex and that few generalizations about compositional patterns can be made. Now, fifty years later, with a much greater body of available factual data and with greater insight into magmatic phenomena, we should carry on in Bowen's inquisitive spirit and not be constrained by the notion that Bowen's reaction series tells us how all magmas crystallize under all conditions.

entially escape, leaving the less mobile oxygen to participate in oxidation reactions.

Role of Water in Magmatic Behavior

We have previously shown that dissolved water in silicate melts has a profound influence on their behavior. Specifically, dissolved water

1. Depolymerizes the melt (page 232) and therefore

2. Reduces melt viscosity (page 234) and

3. Increases diffusion rates (pages 236–237);

4. Depresses crystallization temperatures (pages 260 and 279);

5. Exsolves and expands in shallow magma bodies, causing explosive volcanism (see pages 246–248).

The manner in which the very low concentrations, commonly less than 1 to 2 wt.%, of water dissolved in melts can be so significant is more readily appreciated from its much greater concentration on a molecular basis (see Table 7-2).

This section explores the following additional roles of water in magmatic behavior:

6. Water stabilizes hydrous crystalline phases such as micas and amphiboles.

7. Water in subsolidus systems promotes alteration replacement of unstable higher temperature minerals.

8. Liberation of water from volcanic and shallow intrusive bodies causes breakdown of micas and amphiboles.

9. Retrograde boiling with separation of an aqueous solution in confined plutonic systems produces pegmatites and quartz-sulfide veins.

Stability of hydrous minerals.   The stability of minerals containing water is dependent upon X_{H_2O} or P_{H_2O} in addition to other components and the P and T of the system. If the bulk composition of the condensed constituents (X) in the system is held constant, a **breakdown** or **decomposition reaction** of an isolated hydrous phase can be written as

thermal energy + hydrous phase →
  anhydrous or less hydrous phase(s)
                    + water   (8.12)

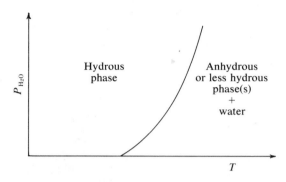

Figure 8-35   Phase diagram for reaction shown in Equation (8.12) under water-excess conditions ($P_{H_2O} = P$).

This reaction defines the **upper thermal stability limit** of the hydrous phase with respect to P_{H_2O} in Figure 8-35. It may be noted that the decomposition products have a greater volume than the hydrous phase and hence are stable at lower pressure, so under water-excess conditions the breakdown curve has a positive slope.

Phase relations in magma systems with hydrous minerals.   These relations are far more variable and complex than for wholly anhydrous systems because of the possible variations in the P_{H_2O} that cause significant shifts in the stability fields of crystalline phases. Two extreme cases limit the range of phase relations, namely, water-excess and water-deficient conditions.

Consider first **water-excess conditions;** this implies that the melt is water-oversaturated, a separate water-rich phase is present, and if the concentration of other volatile components in it is small, then $P_{H_2O} = P$. Melting relations of the Mt. Givens granodiorite with excess water as a function of P are shown in Figure 8-36. Anhydrous crystalline phases—quartz and feldspars—in the system have upper thermal stability limits in the presence of silicate melt that slope negatively in P_{H_2O}–T space; that is, their saturation limits lie essentially parallel to the negatively sloping solidus curve for the whole rock. The hydrous biotite and hornblende, on the other hand, are totally dissolved

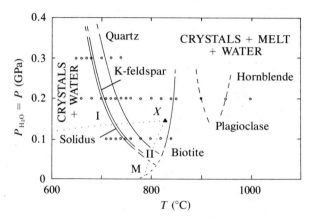

Figure 8-36 Phase boundaries for the Mt. Givens granodiorite with excess water as a function of P and T. The boundary lines labeled by a crystalline phase, such as biotite, indicate the P and T at which it first precipitates when T is lowered, or disappears when T is raised. Circles represent experimental control points. Compare Figure 8-8. See text for discussion of composition X and solidification paths I and II. [After A. J. Piwinskii, 1968, Experimental studies of igneous rock series: Central Sierra Nevada batholith, California, *Journal of Geology* 76.]

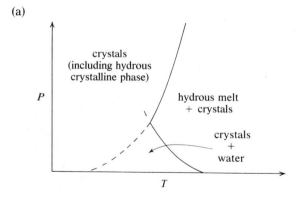

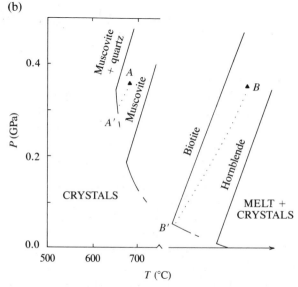

Figure 8-37 Melting relations in water-deficient granitic systems. (a) Conceptual diagram shows a negatively sloping solidus at temperatures greater than the breakdown curve of the hydrous phase, where its water is liberated into the rock. At temperatures below the breakdown, no free water is present in the rock and its melting behavior is as if it were anhydrous with a positively sloping solidus. (b) Solidus curves for granitic systems of appropriate composition such that muscovite plus quartz, muscovite alone, biotite, or hornblende are stable. Positions of solidus curves governed by biotite and hornblende are schematic because of the numerous variables controlling them. [Muscovite-bearing systems after I. S. E. Carmichael, F. J. Turner, and J. Verhoogen, 1974, *Igneous Petrology* (New York: McGraw-Hill).]

in the melt at temperatures above the positively sloping curves; their stability field is limited at high temperatures by curves like that in Figure 8-35.

Under what may be called **water-deficient conditions,** the only water in the system is structurally bound within crystalline phases such as micas and amphiboles. Hence, at temperatures below the upper thermal stability limit of the least refractory hydrous phase, which is biotite in Figure 8-36, there is no water available to depress the melting temperatures of the quartz and feldspars. However, at temperatures above the stability limit of the hydrous phase, it decomposes into anhydrous crystals plus water. This water serves to depress the melting curve, along a negatively sloping trajectory, of the felsic phases in the rock (see Figure 8-37a). Note the similarity between this solidus and those for the limited water contents in Figure 8-4. The solidus curve for a rock under water-deficient conditions thus changes from a negative slope at temperatures above the upper thermal stability boundary of an included hydrous phase to positive above the point where this sol-

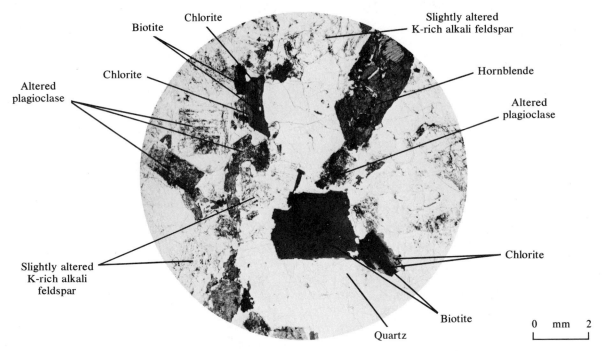

Biotite   Chlorite

**Slightly altered
K-rich alkali feldspar**

Chlorite

**Altered
plagioclase**

Hornblende

**Altered
plagioclase**

Chlorite

**Slightly altered
K-rich alkali
feldspar**

Biotite

0   mm   2

Quartz

Figure 8-38   Altered granite in plane-polarized light showing partly decomposed primary high-temperature magmatic minerals. Biotite is partly replaced by slightly lighter-colored chlorite. Plagioclases are extensively replaced by a turbid, fine-grained aggregate of phyllosilicates and epidote minerals. K-rich alkali feldspar is less extensively altered to phyllosilicates. Quartz is fresh.

idus segment intersects the upper thermal stability boundary.

Magma *A* in Figure 8-37b, upon slight cooling and ascent toward the surface, would solidify at *A'*. Magma *B*, following a similar dP/dT trajectory, would solidify at much shallower depth *B'*; somewhere en route toward *B'* this magma would likely boil.

If a granitic magma is saturated with both silica and water and is sufficiently rich in Al (that is, peraluminous), then muscovite can exist with the melt under a restricted *T* range only for $P \geq 0.35$ GPa. Primary magmatic muscovite never occurs in rhyolites crystallized near or on the Earth's surface because under such low pressures the stability field of muscovite + quartz lies several hundred degrees below the solidus. However, if the

composition of the system is appropriate, secondary muscovite might form as a secondary alteration product in the solid rhyolite at subsolidus temperatures.

The stability fields of Fe-bearing biotites and amphiboles are impossible to represent fully in a single two-dimensional diagram such as Figure 8-37b because of extensive solid solution and their dependence upon the P_{O_2}. However, generally speaking, biotites can precipitate from silicic melts at lower pressures than muscovites, and indeed this phase is a widespread phenocryst in rhyolites.

Consider a hypothetical body of granitic magma with excess water represented by point *X* in Figure 8-36 at equilibrium several kilometers beneath the surface. Assume the bulk composition of the magma is such that stable plagioclase, biotite, and hornblende crystals are present. Let us examine two contrasting hypothetical cooling paths. In the first—a plutonic course denoted by line I—the system cools and decompresses very slowly as the region is uplifted and eroded; eventually the pluton is exposed at the surface after many millions of years. As the crystalline assemblage cools, it passes into the stability fields of muscovite and various lower temperature hydrous phases such as epidote minerals and other hydrated Ca–Al silicates, chlorites, and clay minerals. As it does so, a sequence of very complex coupled reactions requiring additional water consumes the initial high-T magmatic assemblage of plagioclase, alkali feldspar, biotite, and hornblende, forming a lower-T subsolidus assemblage of more hydrated secondary minerals. Although it is impossible to write stoichiometrically balanced reactions, a highly idealized coupled reaction would be of the form

biotite + feldspar + hornblende + water →
    chlorite + muscovite + epidote + sphene
                            + sodic plagioclase

Biotite, for example, decomposes to chlorite with Ti, K, and Si left over; the Ti and Si, together with Ca liberated by coupled breakdown of plagioclase and hornblende, forms sphene; the K goes into muscovite.

These alteration reactions progressing down temperature all require water, much more than is initially present in biotite and hornblende or possibly held in minute pore spaces in the rock. Hence these reactions occur only to the extent that there is water available from the surroundings and the rock body is permeable to their entrance. In a dry or water-deficient plutonic body, the high-T magmatic assemblage will persist metastably; the rock sampled at the outcrop will be a fresh granite. However, in the presence of a small amount of water, insufficient to allow complete reaction, some decomposition will occur; a typical partial breakdown is shown in Figure 8-38. Outcrops of granitic rocks commonly show thin veins of epi-

Figure 8-39 Hydrothermal vein of epidote (dark-colored) in granodiorite. A light-colored border zone of altered wall rock has chalky feldspars replaced by phyllosilicates and chloritized biotites.

dote deposited along fractures by hydrothermal solutions that had dissolved the Ca–Al–Si elsewhere from plagioclase and amphiboles, as shown in Figure 8-39.

Path II in Figure 8-36 represents a route in which P_{H_2O} drops much faster than T and the system falls below the solidus before leaving the stability field of biotite at point M. Unstable biotite crystals cannot be resorbed into any melt but must decompose in a solid-state system into a very fine-grained aggregate of anhydrous phases such as Fe–Ti oxides, magnesian pyroxene, and K-feldspar. Decomposition (see box on p. 292) begins on the rim and progresses inward, but is commonly

Subsolidus Mineralogical Modifications

A brief comment regarding terminology may be helpful at this point. Mineralogical and fabric changes occurring locally in a magmatic rock body below the solidus—that is, secondary to magmatic processes—are collectively called **alteration** (see pages 70–71). If this alteration proceeds more or less automatically during cooling of the body in the presence of its own aqueous fluids, it may be denoted as **deuteric alteration,** or **autometamorphism.** Alteration associated with percolation of hot hydrothermal solutions through the body, especially along fractures, may be called **hydrothermal alteration** (see Figure 8-39). These solutions originate either from a cooling magmatic intrusion or from meteoric ground waters heated by the intrusion. The distinction between alteration and metamorphism is that alteration is less pervasive and more localized—along veins or otherwise irregularly distributed through the body; but in many cases the difference is a semantic one. **Weathering** includes changes in mineralogical and chemical composition and fabric near the Earth's surface due to interaction with surface waters of the hydrosphere and atmosphere. As one may grade into, or be superposed upon, the other, it will obviously not always be easy to distinguish between the effects of weathering and alteration. **Replacement** or **decomposition** are general terms that may be used within the context of weathering, metamorphism, or alteration to indicate that a single mineral grain or a fine-grained secondary aggregate of one or more minerals has taken the place of a primary mineral grain formed under different conditions, usually higher T. Many such replacements are pseudomorphic, preserving the original shape of the mineral grain (see Figures 8-38, 8-40, and 5-27).

tion, possibly by reduction of P_{H_2O} in the magma system, are illustrated in Figure 8-40.

Retrograde boiling: Late stages of confined magmatic systems.

We have just seen how alteration in the closing stages of hydrous magmatic systems can influence rock fabric and mineralogical composition. Innumerable studies of magmatic intrusions, especially those of granitic composition, reveal the occurrence not only of alteration but also of pegmatite and aplite bodies and of quartz–sulfide veins within and around the granitic pluton. Field and fabric relations indicate all of these formed late in the evolution of the magmatic system, apparently from different types of solutions.

The origin and diverse composition of these solutions are a consequence of retrograde boiling and the way in which chemical constituents are partitioned between crystals, melt, and volatile phases during the cooling of the magma. Crystallization tends to enrich common subalkaline residual melts in Na, K, and Fe relative to Ca and Mg. Because of ionic size, charge, and other bonding factors, additional elements initially in minor or trace concentrations in the magma are not incorporated into precipitating crystals, but instead become more highly concentrated in the residual melt. The list of residual elements is long, but includes Li, Be, B, C, P, F, S, Cl, Cu, Zn, Nb, Mo, Ag, Sn, W, Au, and Pb (see page 54).

Precipitation of anhydrous phases (quartz, feldspars) in a cooling magma may increase the relative concentration of water in the residual melt so that at some late stage, somewhere in the system, with an impervious cap rock, P_{H_2O} becomes equal to the confining pressure and the melt experiences retrograde boiling (see pages 246–248). As the water-rich phase that boils off has a density dependent upon the confining pressure, its nomenclature becomes complicated; vapor, gas, steam, aqueous fluid, and hydrothermal solution are all possible names. More significant than the name, however, is the fact that this separate water-rich phase contains substantial concentrations (several percent by weight) of dissolved Na, K, Si, and of course the more volatile elements

arrested as the rock body cools. Grain size in these decomposition aggregates is invariably small, usually submicroscopic. Hornblende phenocrysts that suffered a similar sort of subsolidus decomposi-

(a)

(b)

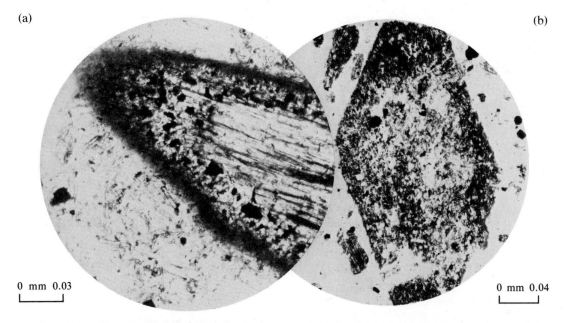

0 mm 0.03

0 mm 0.04

Figure 8-40   Decomposed hornblende phenocrysts. (a) Magnified view of andesite in Figure 3-1 showing rim of fine-grained Fe–Ti oxides, pyroxenes, and feldspars around hornblende. (b) Hornblende phenocryst in a diorite porphyry has been completely replaced by the same anhydrous mineral assemblage as in (a), but its original shape is still evident.

such as C, P, F, S, Cl, plus some of the other residual elements listed above (Burnham, 1967).

Atoms can diffuse much more rapidly in a water-rich than in a condensed silicate melt phase; consequently rates of crystal growth are greatly enhanced. This enhanced rate of growth might have an important bearing on the origin of giant crystals in pegmatites found generally in fairly deep-seated plutonic systems. Jahns and Burnham (1969) have experimentally modeled the origin of pegmatite–aplite bodies. Finely crushed alkali feldspar and quartz mixed with a little water were sealed in gold capsules that were placed in a high P–T bomb (see Figure 7-3). After melting, the mixture was allowed to cool very slowly for a period of time in the bomb before quenching. Subsequent examination of the run revealed that fine-grained feldspar and quartz had crystallized on the walls of the capsule, whereas the core was comprised of much larger crystals. Jahns and Burnham con-

cluded that the system became saturated with water after initial crystallization had created the fine-grained aplite margin; retrograde boiling then created a separate aqueous phase in which the large crystals readily grew.

While the last residual melt in the magma system is crystallizing and long afterwards, any separate aqueous phase will be very active in promoting metasomatic replacement of pegmatite bodies and altering higher-temperature minerals in the wall rocks along its channels of migration. Boiling phenomena and volumetric expansion of late magmatic–hydrothermal systems in epizonal settings can cause fracturing of wall rocks and, due to the rapid loss of volatiles, quenching of any remaining melt in the magma, producing a shallow intrusive porphyry (Phillips, 1973; Cunningham, 1978). Hydrothermal solutions migrating into fractured wall and porphyry rocks can precipitate economically important amounts of Mo, Cu, Ag, Au,

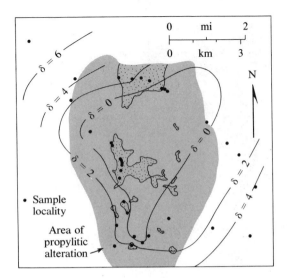

Figure 8-41   Generalized geologic map of the Bohemia mining district, Oregon, showing enrichment in ^{16}O, where $\delta = \delta\ ^{18}O/^{16}O$ [from Equation (2.1)], toward dioritic intrusions (stippled), which is interpreted to have resulted by exchange reactions between hot igneous rocks ($\delta \sim +6$) and heated meteoric waters ($\delta \sim -9$). Country rocks (unpatterned) are volcanic. [After H. P. Taylor, Jr., 1974, The application of oxygen and hydrogen isotope studies to problems of hydrothermal alteration and ore deposition, *Economic Geology* 69.]

Zn, Pb, and U as sulfides and oxides together with large amounts of quartz.

Convective Meteoric Water Systems Around Magmatic Intrusions

Taylor (1978) has shown that rocks in a significant proportion of investigated plutonic systems (intrusion plus country rocks), particularly epizonal granitic ones, have oxygen and hydrogen isotopic ratios that are enriched in the light isotope. This enrichment commonly produces a roughly concentric pattern of δ values around the magmatic body with rocks richest in ^{16}O, for example, occurring in the intrusive complex itself (see Figure 8-41). Normal fresh granitic rocks have δ values of $+7$ to $+10$ whereas meteoric ground waters have en-

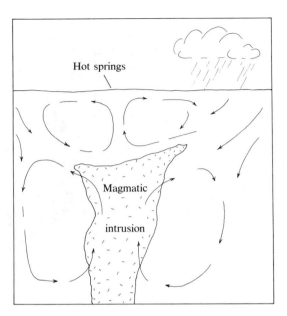

Figure 8-42   Schematic cross section of convective circulation of heated ground water in country rocks around an intrusion. Movement occurs through pervasive fractures.

riched ^{16}O values of approximately -10 (see pages 56–57). Hence, if a hot magmatic intrusion can drive convective circulation of ground water over a sustained period of time, the heated ^{16}O-enriched water will participate in isotopic exchange reactions with the more ^{18}O-enriched rocks of the intrusion and sedimentary ($\delta = +10$ to $+20$) or igneous country rocks. The nature of such convective systems, shown schematically in Figure 8-42, depends upon the existence of pervasively fractured rocks that have high permeabilities. Such fracturing is expected, however, around epizonal plutons forcefully emplaced into brittle country rocks that are stretched above the intrusion and because of the boiling phenomena in such systems alluded to earlier (Phillips, 1973).

This discovery of truly gigantic convective circulation systems involving ground water that extend tens of kilometers into the country rocks has profound implications for the origin of hydrothermal ore deposits associated with magmatic intrusions and for the cooling of them and their country rocks. Purely conductive models (see Section 7.3) may be a limiting case realized only in the case of dry, impermeable country rocks.

SUMMARY

8.1. An appropriate change in any of the variables *T, P,* or *X* can change a state from wholly liquid to one in which crystals exist.

States of small molar volume are more stable at high pressure and the more disordered states of high entropy are more stable at high temperature. At constant chemical composition, in the absence of volatiles, the melting temperature shifts to higher temperatures at higher pressures. Dissolved water depresses melting temperatures, however.

Magmas crystallize over a range of *T* and *P*. Most rock-forming minerals are crystalline solid solutions that crystallize sequentially but with significant overlaps over a broad range of *T* and *P*. Crystalline solid solutions and the coexisting silicate melt progressively change composition during crystallization. The minerals that crystallize from a particular melt, and their sequence, depend entirely upon the intensive variables *P, T,* and especially *X* and the way these vary as the system changes from the liquid to the crystalline state.

Once precipitated, a mineral may subsequently be resorbed back into the melt due to changing stability conditions.

Phase diagrams are graphic devices illustrating equilibrium between different phases and their compositions. A phase is a chemically and physically homogeneous part of a system that is bounded by an interface with adjacent phases. The number of components comprising a system is the minimum number of chemical constituents necessary to define the composition of every phase in the system completely. The Gibbs phase rule for systems of variable *P* and *T* is $F = 2 + C - \varphi$.

The liquidus represents the lower-*T* boundary of crystal-free liquid in *P–T–X* space and is the locus of points that indicate the composition of liquid coexisting in equilibrium with crystals. Phases with low ΔS_m such as quartz and alkali feldspar are more soluble in common subalkaline melts and tend to crystallize last, whereas olivines, pyroxenes, and Fe–Ti oxides are less soluble and generally precipitate early.

All crystalline solid solutions and some pure phases such as enstatite melt incongruently; that is, the crystalline phase dissociates into another solid plus a different liquid. A peritectic reaction relation with falling *T* is exemplified by forsterite + $L_R \rightarrow$ enstatite + thermal energy. In some silicate systems two crystal-free melts can coexist in equilibrium (liquid immiscibility).

The solidus is the locus of points giving the composition of crystalline solid solutions coexisting with liquid.

As changing intensive variables disturb the state of equilibrium in a crystal plus melt system, diffusion-controlled reaction relations will tend to restore equilibrium. The outcome of this contest lies somewhere between two extremes: (1) perfect equilibrium crystallization or melting in which *P–T–X* conditions change slowly enough that reactions are complete at every step of the way, and (2) perfect fractional crystallization or melting in which no reaction relations occur between crystals and liquid. Fractional melting and crystallization extend the range of transient liquid and crystal compositions and of operational temperatures relative to the equilibrium type of behavior. Addition of other components to a melt from which a particular crystalline phase crystallizes can thermally depress and modify the shape of liquidus and solidus but will not change the fundamental behavior of the crystal–liquid equilibria.

During crystallization of common subalkaline magmas, residual liquids and late-precipitating crystals become enriched in K, Na, and Fe relative to Ca and Mg. Both continuous and discontinuous reaction relations between melt and precipitating crystalline solid solutions can occur simultaneously to produce paralleling reaction series. The exact nature of such series in igneous rocks depends upon the *P* and *X* of the magma system.

8.2. Volatile constituents in magmatic systems exert a powerful influence on their behavior.

The stability of Fe-bearing phases is dictated by the amount of oxygen in the system, manifest by its vapor pressure, P_{O_2}, because oxidation-reduction reactions will promote formation of ferrous (Fe^{2+}) or ferric (Fe^{3+}) valence states.

H_2O and CO_2 in magmatic systems can stabilize crystalline phases such as micas, amphiboles, and carbonates provided P, T, and other compositional properties are appropriate. In volatile-saturated systems, upper thermal breakdown curves of volatile-bearing phases have positive slopes in P_v–T space.

Water-deficient magma systems have a "dog-leg" solidus that is governed by breakdown of a hydrous crystalline phase such as mica or amphibole. Primary magmatic hydrous minerals such as biotites and amphiboles crystallized in water-rich magmas can decompose into fine aggregates of feldspars, pyroxenes, and Fe–Ti oxides under subsolidus conditions near the Earth's surface due to reduction in P_{H_2O}. In confined plutonic systems, primary minerals (except quartz) can break down to fine-grained, more hydrous phases under subsolidus conditions to the extent that water is available.

Retrograde boiling and separation of a compositionally distinct aqueous phase whose density depends upon P and T may occur in the terminal stages of the crystallization of magma bodies. In confined systems, this phase is instrumental in creating giant textured pegmatites and, due to its solvent properties, is capable of transporting metals and silica that form sulfide and quartz veins, while simultaneously altering wall rocks.

Although some igneous intrusions may be the sole or at least the principal source of hydrothermal solutions, investigations of oxygen and hydrogen isotopes indicate that other plutons, especially in epizonal settings, served as thermal energy sources to heat meteoric ground waters in country rocks and drive gigantic convective circulation systems.

STUDY QUESTIONS

1. Discuss factors that determine the sequence of precipitation of phases from silicate melts.

2. Outline the role played by water in magmatic behavior.

3. Using simple model silicate systems as examples, discuss which is more important in evaluating P, T, and X in a magmatic system, the modal or the mineralogical composition of the rock body.

4. Why should the country rocks around a pluton be included as part of the magmatic system? In your discussion utilize the concepts of open, closed, and isolated systems and sinks and sources.

5. Were the melting relations shown in Figure 8-9 determined under water-free or water-saturated conditions? Why?

6. Sketch the phase diagram for the binary system forsterite–fayalite. Solid solution is complete; T_e for Fo is 1890°C, and for Fa 1205°C.

7. Explain why natural granites never contain pure Kf or Ab.

8. Explain from basic thermodynamics why the reaction shown in Equation 8.12 has a positive slope on a T–P_{H_2O} diagram.

9. Why does biotite not crystallize from basalt melts?

10. Chlorite and epidote are never observed to be primary precipitates of any magma. Why?

11. Compare and contrast the processes and the consequences of a melt boiling under confined plutonic conditions and under near-surface volcanic conditions.

12. Show why late crystals are generally enriched in Na, K, and Fe relative to Ca and Mg during fractional crystallization of subalkaline magmas.

13. A rock body has three compositional aspects (see page 17). Identify analogous composi-

Table 8-2   Properties of ten systems (all at $P = 1$ atm) composed of mixtures of forsterite and silica.

System			
T (°C)	SiO_2 (wt.%)	Modal composition (wt.%)	Mineralogical composition
1. 1700	50	—	—
2. 1800	—	60% L 40% Fo	—
3. 1750	60	—	—
4. —	—	50% L 50% En	—
5. —	—	—	$X^L_{64.5}$ En, Cr
6. —	50	—	Fo, En
7. 1800	—	50% L_1 50% L_2	—
8. 1900	90	—	—
9. —	50	—	$X^L_{60.5}$

tional aspects of simple model silicate systems as represented in binary phase diagrams and indicate what factors control these aspects.

14. Why do plagioclases in volcanic rocks have more K than in deep plutonic rocks (see Figure 8-30)?

15. Properties of ten systems composed of mixtures of forsterite and silica are listed in Table 8-2. Fill in the blanks.

16. Texturally, how do grains subjected to subsolidus decomposition differ from ones partly resorbed into a melt?

17. Fe liberated by subsolidus decomposition of biotites and amphiboles may form as magnetite or hematite. Discuss factors that dictate which of these two Fe-oxides should be stable.

18. Describe the variance of the system shown in Figure 8-29b which consists of melt X'' and feldspars P and O.

19. Justify the following statement: The composition of the first liquid to form with increasing T

at the solidus changes as a function of P. Use Figure 8-9 as a basis.

20. For water and common ice $V_w < V_i$. Discuss melting–crystallization relations with respect to P and T in the system H_2O.

REFERENCES

Atkins, P. W. 1978. *Physical Chemistry*. San Francisco: W. H. Freeman and Company, 1018 p.

Bailey, D. K., and Macdonald, R. (eds.). 1976. *The Evolution of the Crystalline Rocks*. New York: Academic Press, 484 p.

Bowen, N. L. 1928. *The Evolution of the Igneous Rocks*. New York: Dover, 332 p.

Burnham, C. W. 1967. Hydrothermal fluids at the magmatic stage. In H. L. Barnes (ed.). *Geochemistry of Hydrothermal Ore Deposits*. New York: Holt, pp. 34–76.

Carmichael, I. S. E., Turner, F. J., and Verhoogen, J. 1974. *Igneous Petrology*. New York: McGraw-Hill, 739 p.

Cox, K. G., Bell, J. D., and Pankhurst, R. J. 1979. *The Interpretation of Igneous Rocks*. London: George Allen and Unwin, 450 p.

Cunningham, C. G. 1978. Pressure gradients and boiling as mechanisms for localizing ore in porphyry systems. *U.S. Geological Survey Journal of Research* 6:745–754.

Ehlers, E. C. 1972. *The Interpretation of Geological Phase Diagrams*. San Francisco: W. H. Freeman and Company, 280 p.

Ernst, W. G. 1976. *Petrologic Phase Equilibria*. San Francisco: W. H. Freeman and Company, 327 p.

Fyfe, W. S., Turner, F. J., and Verhoogen, J. 1958. Metamorphic reactions and metamorphic facies. Geological Society of America Memoir 73, 259 p.

Jahns, R. H., and Burnham, C. W. 1969. Experimental studies of pegmatite genesis: I. A model for the derivation and crystallization of granitic pegmatites. *Economic Geology* 64:843–864.

Morse, S. A. 1980. *Basalts and Phase Diagrams*. New York: Springer Verlag, 493 p.

Muan, A. 1979. Crystallization in silicate systems. In H. S. Yoder, Jr. (ed.), *The Evolution of the Igneous Rocks: Fiftieth Anniversary Perspectives*. Princeton: Princeton University Press, pp. 77–132.

Osborn, E. F. 1979. The reaction principle. In H. S. Yoder, Jr. (ed.), *The Evolution of the Igneous Rocks: Fiftieth Anniversary Perspectives*. Princeton: Princeton University Press, pp. 133–169.

Phillips, W. J. 1973. Mechanical effects of retrograde boiling and its probable importance in the formation of some porphyry ore deposits. *Institute of Mining and Metallurgy Transactions* 82B:B90–B98.

Piwinskii, A. J. 1968. Experimental studies of igneous rock series: Central Sierra Nevada batholith, California. *Journal of Geology* 76:548–570.

Stewart, D. B., and Roseboom, E. H., Jr. 1962. Lower temperature terminations of the three-phase region plagioclase-alkali feldspar-liquid. *Journal of Petrology* 3:280–315.

Taylor, H. P. 1978. Oxygen and hydrogen isotope studies of plutonic granitic rocks. *Earth and Planetary Science Letters* 38:177–210.

Thompson, R. N. 1972. Melting behavior of two Snake River lavas at pressures up to 35 kb. *Carnegie Institution of Washington, Geophysical Laboratory Yearbook* 71:406–410.

Weill, D. F., and Fyfe, W. S. 1964. A discussion of the Korzhinski and Thompson treatment of thermodynamic equilibrium in open systems. *Geochimica et Cosmochimica Acta* 28:565–576.

Wright, T. L., and Okamura, R. T. 1977. Cooling and crystallization of tholeiitic basalt, 1965. Makaopuhi lava lake, Hawaii. U.S. Geological Survey Professional Paper 1004, 78 p.

Wood, B. J., and Fraser, D. C. 1976. *Elementary Thermodynamics for Geologists*. Oxford: Oxford University Press, 303 p.

9

Magma Generation, Diversification, and Ascent

In the essentially solid silicate mantle and crust, magmas can only be created where there are significant local perturbations in *P, T,* or *X*. The exact cause and nature of such perturbations in source regions are often controversial, but they are concentrated along converging and diverging lithospheric plate boundaries. The upper peridotitic mantle plays a prominent role in magma generation. It is the source of the world's volumnous basaltic magmas, and at least some of the lesser alkaline magmas as well. Such magmas penetrating upward into the continental crust carry thermal energy from the mantle that might trigger partial melting of already hot, deep sialic material, particularly if large volumes of mafic magma are arrested in their ascent. Because of inhomogeneous source rocks and variable partial melting processes and depths of melting, the possible compositional range of primary partial melts is considerable.

Regardless of whether magma has formed by liquid segregated from a crystalline residuum or exists as a mobile crystal-rich mush, the motivating force for ascent out of the source region is gravity. Less dense magma bodies are gravitationally unstable with respect to the surrounding solid rocks and will tend to buoy upward to heights governed by hydrostatic equilibrium. Other factors, such as magma composition and apparent viscosity, tectonic setting, magma volume, and geometry of any existing conduits, also influence magma ascent, chiefly its velocity.

There are ample opportunities for diversification mechanisms, including differentiation, mixing and contamination, to operate upon the already compositionally diverse magmas rising out of source regions. The fundamental problem in igneous petrology is to decide which of many possible processes of generation, ascent, and diversification actually operated in the history of a particular magma body or in a recurring worldwide magmatic association. [For a brief but lucid review of this topic, see Harris (1974). A more lengthy treatise on the current state of the art, compiled to celebrate the fiftieth anniversary of the publication of N. L. Bowen's classic work *The Evolution of the Igneous Rocks* (Bowen, 1928; reference at the end of Chapter 8) is *The Evolution of the Igneous Rocks: Fiftieth Anniversary Perspectives* (Yoder, 1979).]

9.1 THE GENERATION OF MAGMAS IN THEIR SOURCE REGIONS

Explanations of the generation of magmas deep in the Earth are still largely speculative, and many fundamental uncertainties remain to be solved. However, remarkable advances in our understanding of this complex phenomenon have been made in the past couple of decades. Information comes from a union of both geophysical and petrological data that is based upon observations, experiments, and theory. An iterative procedure is used to create models that account for the data. Thus, a given set of geophysical data can be explained by several models, but petrologic data rule out some of these; additional geophysical data and arguments refine the remaining favored models, new petrologic information further narrows the possibilities, and so on.

Boundary Conditions

Geophysical properties of the Earth indicate that the only wholly liquid region is the outer core, which is comprised essentially of Fe. This is unacceptable as a source of silicate magmas. Hence, magmas must originate in the mantle or crust, both of which, however, are essentially solid because seismic shear waves are transmitted through them. By "essentially solid" is meant a rock body with no more than very small concentrations of liquid, so that crystalline grains can be in mutual contact. The **magma source region** is the first place where magma comes into existence, not a secondary staging area or subterranean magma chamber fed by magma rising from the source.

A magma source region is a volume of essentially solid rock where P, T, or X is sufficiently different from the surroundings so that melting can produce a body of magma potentially able to rise toward the surface. The boundary between the magma source and surrounding rock is gradational because there are no long-term discontinuities in the gradients in P, T, and X.

Where are the most geophysically favorable sites for local perturbations in P, T, or X that could create magma? Temperatures of lava erupted from volcanoes place only a minimum limit on the depth of these perturbed source regions because the magma may have thermally equilibrated in a shallow staging area; or the magma might have cooled adiabatically from its deep source region because of expansion during ascent. Thus, Hawaiian basaltic lavas—with eruption temperatures of $\sim$ 1200°C—probably have a source deeper than 40 km, assuming a linear geothermal gradient of 30°C/km. Earthquake swarms preceding their eruption have epicenters 50 to 60 km beneath the surface. But again it is uncertain whether such epicenters mark the source or simply a region where magma is gathering and moving upward with greatest vigor.

The seismic low-velocity zone is virtually a worldwide layer within the uppermost mantle that, according to the most reasonable interpretations of its seismic and electrical properties, consists of partially melted peridotite. Seismic shear waves are transmitted, but with significant attenuation, through the layer. The geophysical data indicate that only a very small proportion of melt exists in the peridotite, perhaps 1% to 3%, as thin intergranular films and "pipes" (Yoder, 1976, p. 162). Generally, on a global scale, this very slightly melted, low-velocity zone would not appear to constitute a major source region because in the vast global intraplate areas far removed from plate junctures magmatic activity is relatively sparse. It might be argued, however, that the wholly solid lithosphere in intraplate areas is simply impervious to the rise of magmas from an underlying source; that is, no tectonic fractures or pathways are available for its ascent. In any case, there can be no doubt that the most volumetrically productive magma source regions lie along converging and diverging lithospheric plate boundaries.

Plate tectonic processes along convergent and divergent boundaries driven by interacting thermal and gravitational energy perturb P, T, and X in potentially viable source rock systems to cause local partial melting and magma production.

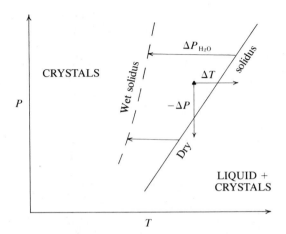

Figure 9-1 A potential solid source rock (solid triangle) that is already at a *P* and *T* near the anhydrous solids can be perturbed by changes in *P*, *T*, or *X* so that melting occurs. The source might decompress ($-\Delta P$), become hotter (ΔT), or change in composition, perhaps by influx of a small amount of water (ΔP_{H_2O}), depressing the solidus below its existing temperature. The wet solidus is not a water-saturated one.

Mechanisms of Melting

Altogether, some two dozen mechanisms of magma generation have been proposed (see Yoder, 1976, chap. 4, for references and an extended discussion of basaltic source mechanisms). But every one is simply a change in *P*, *T*, or *X* (see Figure 9-1). Let us examine some possible ways of changing these state variables in geologic systems so that melting can occur.

Temperature increase. The most obvious way to melt essentially solid rock in a potential source region is to raise its temperature. In all considerations of melting phenomena due to changes in temperature, it is important to keep in mind the significant difference between the heat of fusion (equivalent to the latent heat of crystallization; see box on page 227) and the specific heat of silicate materials (see page 6). Specific heats are about 0.8 to 1.3 J/g degree, whereas heats of fusion are approximately 300 times greater, 270 to 420 J/g (Williams and McBirney, 1979, p. 40). Hence, 100

Table 9-1 Order of magnitude of rate of heat production due to radioactive decay of K, Th, and U.

Type of Rock	Rate of Heat Production (J/g · y)
Granite	3.4×10^{-5}
Basalt	5.0×10^{-6}
Peridotite	3.8×10^{-8}

J of thermal energy can raise the temperature of one gram of solid rock by approximately 100 degrees, but the same energy produces only a few tenths of a gram of silicate liquid from solid rock already just about at its melting temperature (compare Figure 8-16b). Melting consumes relatively large amounts of thermal energy, moderating temperature variations within a rock system in its melting range.

Decay of radioactive isotopes of K, Th, U, and other less abundant isotopes in rocks generates thermal energy (see Table 9-1). However, the rate of production is extremely slow; a body of granite initially within 300°C of its solidus could partially melt only after 10 to 20 My and a body of peridotite only after 10 to 20 Gy, assuming no thermal energy produced leaked out of the body in that time. Melting by such heating is obviously unrealistic and it may be concluded that radioactive decay alone probably cannot generate sufficient perturbations in *T* to cause melting. Even if it could, the first melting episode would extract much of the K, Th, and U from the source rock, because these elements partition more strongly into a liquid phase than most crystals, so that subsequent heat generation would be greatly retarded. However, heat production in metasomatized mantle could be significantly enhanced by radioactive decay (see pages 216–218).

Some of the mechanical work of rock deformation could conceivably be transformed into sufficient thermal energy to induce melting (see page 486). Simple compressional squeezing of rock at depth due to the weight of overlying material produces a temperature increment of only a few tenths

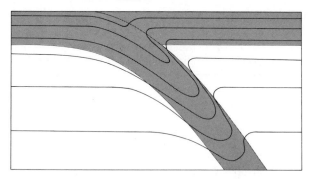

Figure 9-2 Highly schematic isotherms in a subduction zone. Surface topography is ignored. Lithosphere is shaded. Isotherms are calculated from theoretical models and depend greatly upon the relative importance given to such factors as rate of descent of the slab and its thickness; heating produced by adiabatic compression, radioactivity, and deformation; latent heat of high-P phase transitions; cooling effects of endothermic dehydration reactions; migration of fluids, including magma; and so on. For a variety of temperature distribution models, see Wyllie (1979).

of a degree per kilometer (the adiabatic gradient). Some geologists have proposed that sliding of the lithospheric slab into the mantle in subduction zones would generate ''frictional'' heating; however, no correlation has been found between rate of subduction and rate of magma production, as might be manifest in volcanic activity.

Shearing flow of highly viscous mantle or lower crustal rock might lead to melting by means of thermal feedback (see page 236; Shaw, 1973; Anderson and Perkins, 1975). Initially the shear deformation might be distributed over a broad zone and any related heating would be uniform. But due to heat conduction from the perimeter of the sheared zone, the interior remains hotter; any further shear becomes concentrated there because of its lower viscosity (recall $\eta \propto 1/T$), causing further more localized heating and reduction of viscosity and so on. This self-accelerating instability or ''thermal runaway'' can theoretically lead to local melting that effectively arrests the runaway, as shear motion is easily accommodated in the liquid. Migration of the liquid out of the deforming system

might allow the whole process to start over again. In this regard, Shaw (1970) notes that a significant proportion of the tidal energy (see Figure 1-2) due to the gravitational attraction of the Sun and Moon is dissipated within the solid Earth; he proposes that shear melting may be the chief dissipating mechanism.

Mass transfer of rock material and consequent differential heating can also perturb temperatures. Two possibilities may be considered. First, the descending cool oceanic lithosphere in subduction zones is an enormous heat sink and is heated by the surrounding mantle in the manner shown in Figure 9-2. Phase transitions that occur with rising pressure and uncertainties regarding the actual T distribution in the slab cloud the situation, but it is possible that the thin basaltic crustal layer is eventually heated above its solidus; its extensive hydration due to sub-sea-floor metamorphism could dictate a negatively sloping liquidus in P–T space, enhancing the chance of melting. Second, sialic crustal rocks might be heated above their solidus as large bodies of basaltic mantle-derived magma are intruded into them. Deep continental granitic rocks, especially the most hydrous systems, have solidus temperatures well below basalt; the latent heat of crystallization of basalt magma plus some specific heat could be used to melt already warm sialic crustal rock.

Decompression. Because of the positive slope of the solidus curve of dry silicate systems in P–T space, decompression of hot solid rock can induce melting if its temperature is already close to the solidus.

The most effective, and geologically most plausible, means of decompressing a mass of solid rock is by allowing it to rise in the gravitational field of the Earth, as in the upwelling mantle beneath an oceanic or possibly continental rift. Melting in an upwelling peridotitic mantle can by analyzed from the simplified diagram in Figure 9-3a. The system is assumed to behave adiabatically; that is, it is thermally insulated from its surroundings and the only change in T of the system is related to the

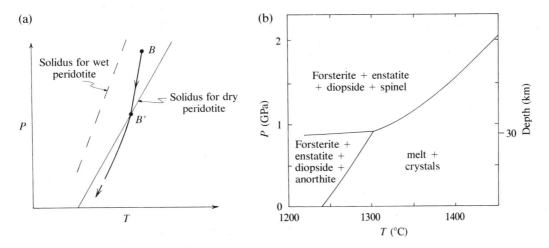

Figure 9-3 Melting relations in mantle peridotite. (a) Schematic solidus curves and geotherm (heavy line) to illustrate concepts. Solidus for wet peridotite refers to a system with less water than that required to saturate a melt at the indicated pressure. (b) The solidus curve for model lherzolite in the volatile-free system $CaO-MgO-Al_2O_3-SiO_2$ shows a pronounced cusp or reentrant at 0.9 GPa and 1300°C where the transition from plagioclase- to spinel-bearing lherzolite intersects the solidus and forms a low-melting-temperature invariant point. It is believed that a similar cusp serves to localize melting in the real upwelling mantle beneath oceanic rifts, where extensive partial melting (up to 35%) of constantly renewed peridotite by equilibrium fusion could produce large amounts of compositionally uniform primary melts that are later modified by crystal–liquid fractionation and mixing in crustal reservoirs (see page 328). [After D. C. Presnall, J. R. Dixon, T. H. O'Donnell, and S. A. Dixon, 1979, Generation of mid-ocean ridge tholeiites, *Journal of Petrology* 20.]

pressure decrement, which for essentially solid mantle rock is a few tenths of a degree per kilometer. Solid mantle initially at B is assumed to well upward, decompressing and cooling along the adiabatic path BB', whose slope will be much steeper than the solidus. At B' melting begins and the thermal energy required for the latent heat of fusion must be drawn from the specific heat stored in the body. Hence, during continued decompression and melting, the upwelling mantle cools at a greater rate at pressures below B' than above and its path may closely follow the solidus. If the solidus has a slope ($dP/dT = \Delta S/\Delta V$ in Figure 8-3) of ~15 MPa per degree, then a pressure fall of 15 MPa over a depth interval of about half of a kilometer yields about one joule of heat per gram

(Williams and McBirney, 1979, p. 41). To create about 30% melting, as postulated for oceanic rift source regions, the mantle would have to rise roughly 60 km after intersecting the solidus.

Changes in composition: Water.   Among the possible changes in X that could induce melting, variations in P_{H_2O} are perhaps the most significant. Even small water pressures can significantly depress the solidus of silicate systems. Hence, in Figure 9-3a, if water were to migrate into initially anhydrous and solid mantle peridotite at B the solidus temperature could fall sufficiently to the wet position so that B would become partially melted. A plausible site for influx of water in mantle peridotite is in the wedge of peridotitic mantle overlying the descending oceanic plate in subduction zones. Widespread sub-sea-floor metamorphism of the oceanic crust (see Section 12.5) produces an assemblage of hydrous phases, including epidote minerals, micas, amphiboles, serpentines, talc, and chlorites. As this assemblage is heated in the subducting plate, a series of endothermic dehydration reactions ensues, liberating water that might move upward into the overlying mantle, depressing the solidus and causing isothermal melting.

9.2 GENERATION OF MAGMA FROM SOLID ROCK

Most of Chapter 8 was devoted to a discussion of the phenomena of sequential crystallization of chemically complex silicate melts and the concomitant reaction relations that take place. Phase diagrams served as a tool for that discussion. We now want to look at the phenomena from the opposite direction. Given a solid multiphase crystalline rock, how does it melt? Once again, phase diagrams will serve as a vehicle for discussion. The guiding principle can be stated as follows: Incongruent melting occurs over a range of T, and reaction relations assure that partial melts are enriched in the more soluble, lower-melting-T components of the system, leaving a more refractory crystalline residue.

In this section we briefly discuss different types of melting and how different liquid compositions can be generated, beginning with the production of basaltic melts from solid peridotite and concluding with generation of granitic melts from solid crustal sources.

Basalt–Peridotite Systems

The principles governing partial melting of complex silicate systems with respect to their major elements have been formalized by Presnall (1969). In parallel with the two ideal modes of crystallization discussed on pages 277–279, Presnall defines **equilibrium fusion** as production of a liquid that continually reacts and equilibrates with the crystalline residue, and **fractional fusion** as production

of liquid that is immediately isolated from the system as soon as it forms so no reaction with the crystalline residue is possible. Equilibrium fusion is a single melting episode and is essentially the reverse of equilibrium crystallization. However, fractional fusion and fractional crystallization, being irreversible processes, are not simply the opposite of one another. As with crystallization, the actual mode of fusion in natural geologic systems would be expected to lie somewhere between the two ideal end members.

Partial fusion, anatexis, or **partial melting** are general terms for any process that produces a melt of some proportion less than the whole.

System Di–Fo–Qz. Consider partial melting in the diopside–forsterite–quartz system at $P_{H_2O} = 2$ GPa (Yoder, 1973). This is only a model system and resemblance to any natural system might be remote. Disregarding extensive real solid solution, systems below the solidus in Figure 9-4a will consist of either the crystalline assemblage Fo + En + Di or En + Di + Qz, depending upon their bulk chemical composition. Figure 9-4b shows the projected liquidus surface and cotectic boundary lines between fields that first precipitate Fo, Di, En, or Qz upon cooling. In Figure 9-5a, the point C represents a mixture of diopside, enstatite, and quartz. If feldspar were present in addition, rock C would model a dacite or granodiorite. With increasing T, holding P_{H_2O} constant, melting begins at 960°C, forming liquid M. This liquid, formed at the invariant point M, is the only one that can coexist with the two pyroxenes plus quartz. It is also the beginning-of-melting liquid with the lowest T in the system and has formed by melting and mutual solution of a little diopside and enstatite together with a larger portion of quartz. If alkali feldspar was also dissolved, M would model a rhyolite melt.

As more thermal energy is introduced into the system, and as long as crystals of the two pyroxenes plus quartz are present—that is, under conditions of univariant equilibrium—the T remains unchanged. The energy consumed in the latent heat of fusion produces more liquid M and drives the crystalline residuum toward C'; note

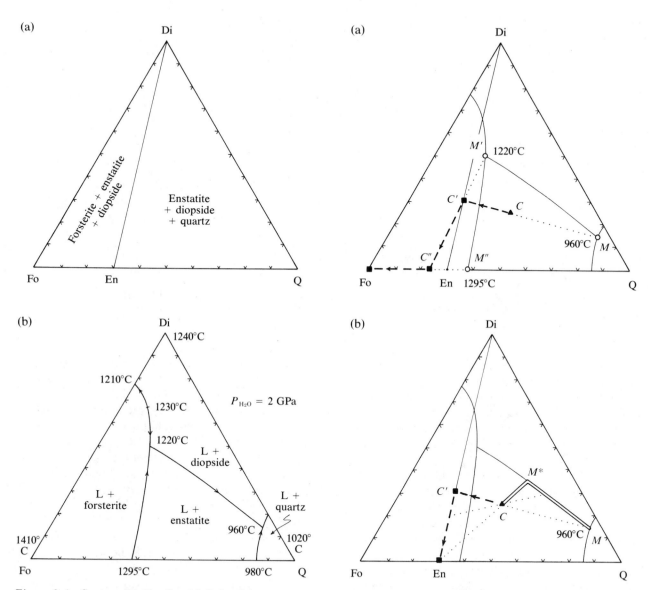

Figure 9-4 System Di–Fo–Qz. (a) Subsolidus diagram indicating that systems composed of the components Fo, Di, and Q will consist of combinations of forsterite + enstatite + diopside or enstatite + diopside + quartz. Solid solution between the two pyroxenes is assumed to be nil. (b) Projected liquidus features at P_{H_2O} = 2 GPa (Kushiro, 1969) indicating temperatures of invariant points and cotectic boundaries between first crystalline phases to precipitate from the liquid.

Figure 9-5 Analysis of melting of rock C in the system Di–Fo–Qz at P_{H_2O} = 2 GPa. (a) Fractional fusion produces discrete melts M, M', M'', and Fo at T = 960°C, 1220°C, 1295°C, and 1890°C. The crystalline residua change continuously from C to C' to C'' to Fo. (b) Equilibrium fusion produces a continuous spectrum of melts ranging from M to M^* to C; crystalline residua, linked to the coexisting melt by a tie line through C, change continuously from C to C' to En.

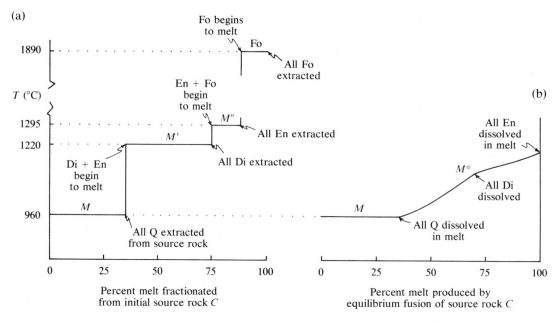

Figure 9-6   Partial melts and crystalline residue produced by (a) fractional and (b) equilibrium fusion as a function of temperature of rock C in Figure 9-5.

that the tie line $C'M$ passes through C. Ultimately, the crystalline residuum reaches C', where it now consists of only the two pyroxenes, modeling a pyroxenite. All of the quartz is now dissolved in the melt M. According to the lever rule, on tie line $C'CM$ the system consists of 35% liquid M and 65% crystalline residuum C'.

What happens now when additional thermal energy is fed into the system? From here on there can be two different modes of melting. Let us first pursue *fractional fusion* and assume that all of the liquid M has been removed from contact with crystals C'. Conceivably, with increased thermal energy going into the system, either the T would increase, or there would be melting of C', or both. Melting might produce a small amount of liquid displaced a little from M toward M' along the cotectic boundary line where Di + En + L coexist in equilibrium. This, however, would force the composition of the crystalline residuum off the En–Di join into the Fo–En–Di triangle as liquid is

withdrawn from the system, which means that the crystalline assemblage would be comprised of forsterite and two pyroxenes. But the only liquid that can coexist with these three crystalline phases is M'. Hence, what must happen as more thermal energy is put into the system C' is that it simply gets hotter, and no further melting can occur until T reaches 1220°C, where liquid M' can form. M' models the composition of an oversaturated basalt or an andesite. Point M' is an invariant point because there are four phases in equilibrium with three components at constant pressure. Fractional fusion of C' continues to yield liquids M'—every drop of which is removed from the system as soon as it is produced—and crystalline residua that move toward C''. With the crystalline residuum at C'', a harzburgite peridotite, no further melting occurs until added energy raises the temperature to 1295°C, where liquid M'' appears. The final crystalline residuum of this fractional melting process is pure forsterite crystals, which melt at 1890°C. The liquid products of the fractional fusion process are therefore four distinct homogeneous compositions *M, M', M"*, and finally pure liquid forsterite.

We now follow progressive *equilibrium fusion* of the system *C*, with the crystalline part already at *C'*, and remembering that a tie line connecting the coexisting liquid and crystals must pass through the bulk composition point *C* (see Figure 9-5b). Input of energy raises the *T* of the system somewhat above 960°C, causing a little diopside and slightly less enstatite to dissolve, thus moving the crystalline residuum from *C'* toward En. Continued heating and melting cause the liquid composition to move to *M** while the crystalline residuum moves to pure enstatite. From *M** the liquid moves to *C* as pure enstatite crystals melt, and finally the system is comprised wholly of liquid *C*.

The entire process of equilibrium fusion is just the reverse of equilibrium crystallization; the liquids produced form a continuous but relatively limited compositional spectrum (see Figure 9-6). In contrast, fractional fusion produces discrete melts spanning a greater compositional and temperature range. Because of the incongruent melting of enstatite, removal of melt *M'*, which is more silica-rich than enstatite, leaves forsterite in the residuum. Subsequent melting gives forsterite liquid. Hence, melts range from oversaturated to saturated.

The composition of the source rock, the type of melting—whether fractional or equilibrium—and the extent of melting all govern the composition of a partial melt. More realistic melting models should take into account the inevitable solid solution in the crystalline phases; this feature has been ignored in the foregoing illustration.

Influence of pressure. A change in pressure significantly shifts the position of cotectic boundary lines and invariant beginning-of-melting points. In Figure 9-7, partial melting of solid peridotite at 1 atm pressure produces a basaltic liquid M_0 in equilibrium with two pyroxenes, olivine, and plagioclase. At 1.5 GPa, the basaltic liquid $M_{1.5}$ has

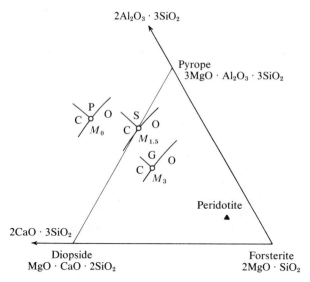

Figure 9-7 Schematic polythermal, polybaric diagram for a model basalt–peridotite system, showing shift in composition of basaltic initial partial melts *M* as a function of pressure (M_0 = 1 atm, $M_{1.5}$ = 1.5 GPa, M_3 = 3 GPa). Heavy lines delineate areas where plagioclase (P), clinopyroxene (C), spinel (S), garnet (G) and olivine (O) are in equilibrium with orthopyroxene plus liquid. Note that the nature of the major aluminous phase changes with increasing pressure from plagioclase to spinel to garnet (compare Figure 6-11). [After M. H. O'Hara, 1968, The bearing of phase equilibria studies in synthetic and natural systems on the origin and evolution of basic and ultrabasic rocks, *Earth Science Reviews* 4.]

much less silica and more dissolved olivine. At 3 GPa (~100 km depth) basaltic liquid M_3 in equilibrium with pyropic garnet, two pyroxenes, and olivine is even poorer in silica and has still greater concentrations of dissolved olivine. These compositional changes can be appreciated by careful attention to the oxide ratios of the end-member components in the diagram.

Partial melts of mantle peridotite are basaltic. With increasing pressure, the stability field of magnesian olivine shrinks in anhydrous peridotitic systems so that more olivine is dissolved into basaltic partial melts at higher pressure. With

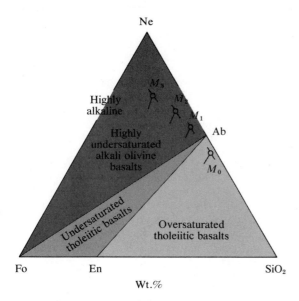

Figure 9-8 Shift of beginning-of-melting invariant points M_0, M_1, M_2, and M_3 at 1 atm, 1, 2, and 3 GPa under volatile-free conditions in the system nepheline–fosterite–silica. Basaltic melts modeled by system are indicated. [After I. Kushiro, 1968, Compositions of magmas formed by partial zone melting of the Earth's upper mantle, *Journal of Geophysical Research* 73.]

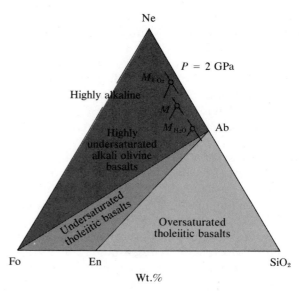

Figure 9-9 Shift of beginning-of-melting invariant points, all at 2 GPa, as a function of volatile composition in the system nepheline–forsterite–silica. M_{CO_2} is CO_2-saturated condition, M_{H_2O} is water-saturated condition, and M is the dry system. [After D. H. Eggler and J. R. Holloway, 1977, Partial melting of peridotite in the presence of H_2O and CO_2: Principles and review, *Oregon Department of Geology and Mineral Industries Bulletin* 96.]

progressively higher degrees of melting at a given pressure, partial melts dissolve more of the crystalline phases in the parent peridotite. Mantle peridotites are mostly olivine with lesser amounts of orthopyroxene and still less clinopyroxene and an aluminous phase; the latter two phases are generally the first consumed in progressive partial melting, leaving an orthopyroxene–olivine (harzburgite) residuum. Still greater melting may consume the orthopyroxene, leaving a dunite residuum (see Figure 5-45). The general pattern is

peridotite source rock = *basaltic partial melt +*
olivine + orthopyroxene   *crystalline residuum*
+ clinopyroxene +   olivine ±
(garnet or spinel)   orthopyroxene

At pressures near 3 GPa basaltic partial melts can crystallize to **eclogite,** composed of pyropic garnet plus a jadeitic pyroxene (see Figure 8-9).

Figure 9-8 illustrates part of the model system forsterite–nepheline–silica at high pressures. Beginning-of-melting invariant points shift markedly with increasing pressure toward more silica-undersaturated compositions and toward slightly more olivine-rich compositions.

Volatiles in basalt–peridotite systems. The concentrations of CO_2 and H_2O in peridotitic systems profoundly affect melting temperatures and phase relations (Wyllie, 1979) and hence must have a strong influence on the compositions of the basaltic partial melts produced (see Figure 9-9) (Eggler and Holloway, 1977). Dissolved H_2O tends to depolymerize silicate melts, thereby stabilizing olivine, which has independent SiO_4 tetrahedra. Stabilization of olivine shifts partial melts toward more silica-rich compositions, counteracting the effect of increasing P shown in Figures 9-7 and 9-8. In contrast, dissolved CO_2 polymerizes silicate melts, causing the stability field of olivine to contract at the expense of the more stable enstatite with its chain structure; the shifting of phase boundaries and invariant points toward more silica-poor compositions caused by increasing P in Figures 9-7 and 9-8 is thus enhanced, and initial high-P partial melts of CO_2-bearing peridotite are predicted to be highly alkaline, extremely undersaturated kimberlites and nephelinites.

Low confining pressures and relatively high water concentrations favor generation of the more silica-rich, tholeiitic basalt or even andesite partial melts of peridotite. High pressures and CO_2-rich peridotite favor generation of silica-undersaturated melts. Alkaline magmas are not only silica-undersaturated but are also greatly enriched in incompatible elements. Enrichment factors of these elements relative to possible peridotitic source rocks (see Table 6-7) are commonly so high that very small degrees—a few percent, at most—of partial melting would be required. More extensive melting would dilute them by dissolution of major silicate components. Prolonged metasomatic activity in the upper mantle furnishes a means of enriching source rocks in incompatible elements, H_2O, and CO_2 (see pages 216–218); subsequent partial melting could then produce silica-undersaturated and incompatible-element-enriched alkaline liquids.

Mantle-derived primary melts. Extensive laboratory experiments on peridotitic–basaltic systems under mantle conditions, beginning with the pioneer work of Yoder and Tilley (1962), leaves no doubt as to the very wide range of silicate liquids that can be generated by partial melting of peridotitic source rocks. Variations in the composition of the source, in confining pressure, and in the degree and mechanism of melting conspire to yield melts ranging from carbonate-rich highly alkaline kimberlite and nephelinite to tholeiitic basalt and perhaps andesite. The popular viewpoint, held a few decades ago, that fractional crystallization promoted this compositional spectrum remains a possibility, but now has to be tested against the equally viable models of partial melting. At least under low confining pressures, it is impossible via fractional crystallization to derive tholeiitic basalt magmas from alkaline ones, or vice versa (Yoder and Tilley, 1962).

Probably the most persistent uncertainty regarding a particular mafic magma is whether it represents an unmodified **primary melt** segregated from peridotitic mantle source rock, or whether it is a **derivative magma** that has been modified by one or more diversification processes after leaving the mantle source region (see Section 9.3). The term **mantle-derived magma**, frequently used in this text, is a general term encompassing both of these genetic categories. (It is conceivable that a partial melt generated at a certain depth in the mantle could rise to a shallower level and reequilibrate with the peridotite at that lower P; further melting might also occur there. In view of this, it may be more accurate to speak of an equilibrated depth rather than a depth of first source.) There are certain criteria that allow petrologists to test whether a particular mafic rock solidified from an unmodified primary melt (Yoder, 1976, chap. 8). However, none are wholly infallible and most constitute necessary but not sufficient evidence to compel a primary character.

Alkaline mafic rocks containing large, relatively dense peridotitic xenoliths of mantle rock are presumed to have ascended rapidly at least from the region where the xenoliths were picked up; the required rapid ascent to lift the fragments [see Stokes' Law, Equation (9.2)] leaves little opportunity for diversification processes to operate, except possibly mixing with other melts or flowage segregation. If the rock represents a segregated primary liquid, then its experimentally determined liquidus in *P–T* space should lie close to the solidus of the source peridotite. However, the peridotite solidus is significantly influenced by minor compositional variations, especially with respect to CO_2 and H_2O; discrepancies may be difficult to interpret unambiguously.

More stringent criteria relate to the composition of a basaltic rock with respect to near-liquidus, high-pressure crystal–melt equilibria. Primary melts of mantle peridotite formed at high pressures should have compositions near beginning-of-melting invariant points, such as M_3 in Figure 9-7. Such a melt at 3 GPa should be in equilibrium with olivine, orthopyroxene, clinopyroxene, and garnet —a garnet peridotite source rock—and would begin to precipitate these phases almost at once on the liquidus with falling temperature. Many basaltic rocks do not precipitate olivine near the liquidus at high pressures—a property that indicates substantial olivine has been fractionated out of the system in their ancestry and precludes their having solidified directly from primary melts. Another facet of this argument is that primary melts, being partial melts of peridotite with abundant olivine $Fo_{90\pm}$, should have Mg(/Mg + Fe) ratios of ~0.72 to 0.74 according to experimental data (Green, 1971). Moreover, their Ni and Cr contents should be high, as these elements are concentrated in olivine.

Since a primary melt represents a liquid that was in equilibrium with a specific crystalline residuum at the *P–T* conditions of partial melting before being segregated from that source, then the crystalline phases first appearing at the liquidus upon cooling of the primary melt should correspond to those minerals remaining in the depleted source at the *P* and *T* of generation.

Granitic Systems

The impossibility of generating granitic liquids by partial melting of mantle peridotite or subducted oceanic basalt crust is demonstrated by the absence of significant volumes of granitic rocks in oceanic regions. (This absence also precludes the possibility of production of large volumes of granitic magma by differentiation of parental basalt magma.) Experimental studies (Wyllie and others, 1976) have shown that quartz or coesite at $P > 3$ GPa occurs prominently near the liquidus of granitic magmas, providing further proof that such liquids could never have been partial melts of peridotite or basalt at such pressures. Also, for increasing $P > 1$ GPa, the compositions of beginning-of-melting liquids in equilibrium with quartz and alkali feldspar are progressively removed from the compositions of most granites (see Figure 9-10), so granitic rocks cannot be unmodified primary melts of subducted arc-trench quartzo-feldspathic sediments. However, subsequent differentiation of these melts during ascent into the crust might convert them into more normal granitic compositions.

Compositional patterns of granitic plutons in western North and South American cordillera seem to suggest that the ancestry of granitic magmas is somehow linked to old continental sialic crust (see pages 142–143). Indeed, Brown (1973), Fyfe (1973), and Presnall and Bateman (1973), among others, have argued that melting processes in the lower continental crust are involved. With some minor puzzling exceptions, the great bulk of Phanerozoic granitic plutons are concentrated along known or suspected subduction zones; hence, processes localized there must be responsible for triggering melting. Geothermal gradients within continents distant from converging plate margins must have been too low to induce melting during Phanerozoic time.

Evidence for water undersaturation in granitic magmas. Much of the experimental work on granitic systems has been done at water-saturated conditions ($P = P_{H_2O}$) in order to hasten the attainment of equilibrium between melt and crystals (for example, Figures 9-10, 8-36, and 4-34). However, this convenience cannot, in any way, be taken as

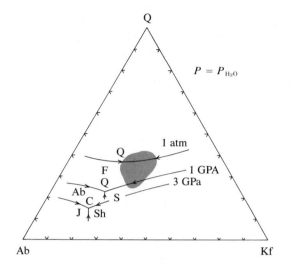

Figure 9-10 Effect of pressure on water-saturated ($P = P_{H_2O}$) liquidus cotectic boundaries in the system Ab–Kf–Q–H₂O. Area enclosing the composition of most granitic rocks from Figure 4-10 is shaded, J, jadeite; Sh, sanidine hydrate; C, coesite; Ab, albite; S, sanidine; Q quartz; F, feldspar. [After P. J. Wyllie, W. Huang, C. R. Stern, and S. Maaløe, 1976, Granitic magmas: Possible and impossible sources, water contents, and crystallization sequences, *Canadian Journal of Earth Sciences* 13.]

evidence that natural granitic magmas generally are water-saturated. If they were, then their solidus would have a negative slope in *P–T* space (see Figure 8-4), so that decompression, abetted by some cooling, during ascent through the crust would cause them to crystallize. As such immobilization would occur not far above their deep crustal sources (still assuming water saturation from their birth), shallow crustal granites and extrusive rhyolites would be precluded. This is certainly not the case! In many batholithic complexes, such as the Sierra Nevada, features suggestive of excess water—pegmatites and miarolitic cavities—are scarce.

One of the most common mineral assemblages in granitic and rhyolitic rocks consists of biotite, alkali feldspar, and Fe–Ti oxide. Simplifying to end-member components, this assemblage suggests an equilibrium reaction of the sort

$$\tfrac{1}{2}O_2 + KFe_3^{2+}AlSi_3O_{10}(OH)_2 =$$
$$\text{in biotite}$$

$$KAlSi_3O_8 + Fe_3O_4 + H_2O$$
$$\text{in alkali} \quad \text{in Fe–Ti}$$
$$\text{feldspar} \quad \text{oxide} \qquad (9.1)$$

If magmas had excess water, this reaction should generally proceed to the left, except at very special T–P_{H_2O} conditions. The widespread coexistence of these three crystalline phases suggests instead that there is limited H₂O in the most granitic magma systems.

Hence, granitic magmas are generally not water-saturated and ascend through the crust governed by positively sloping solidus curves.

Partial melting in mica–amphibole-bearing sialic rocks. All of the above arguments indicate the possibility that granitic magmas might originate by partial melting of water-deficient lower crustal rocks. Various metamorphic rocks containing mica, amphibole, feldspar, and quartz are likely candidates for such source rocks. These will contain less than approximately 1% to 2% water, which is nonetheless sufficient to depress melting temperatures somewhat but not nearly enough to yield saturated melts with 10% to 15% H₂O (see Figure 7-4). Near-solidus melting relations in such water-deficient systems are exemplified by those in Figure 8-37b. Experimental data summarized by Fyfe (1973) show that the bulk chemical composition of the source rock has little influence over the earliest small-degree partial melts but that more extensive melting produces liquids closer toward the bulk composition. Partial melts formed at lowest *T* along reasonable crustal geotherms accompany breakdown of muscovite and are of potassic granite composition (see Figure 9-11). Deeper, hotter melts accompanying biotite and hornblende breakdown are granodioritic to Na–Ca-richer quartz diorites.

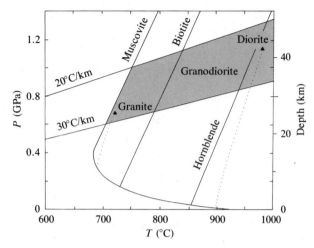

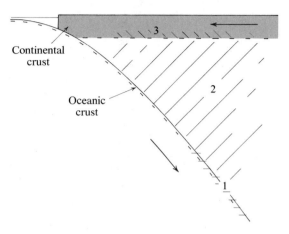

Figure 9-11 Solidus curves for quartzo-feldspathic crustal source rocks containing micas and hornblende are intersected by two geothermal gradients. Plausible *P*–*T* range for partial melting is shaded and approximate composition of melts produced is indicated. Positively sloping solidus line segments are only hypothetical; extensive solid solution actually smears them over a wide *P*–*T* range. Granitic magmas formed at lowest *P* and *T* cannot rise far before intersecting the negatively sloping solidus. Dioritic magmas could ascend all the way to the surface. Compare Wyllie (1979) and Figure 8-37. [After W. S. Fyfe, 1972, The generation of batholiths, *Tectonophysics* 17.]

Figure 9-12 Conceptual diagram showing geologically plausible sites where primary melts may be generated in subduction zones. Any one or a combination of these might be productive; resulting buoyant magmas might mix en route to the surface. (1) Melting of oceanic basaltic crust that has been transformed into a high-*P* eclogitic phase assemblage. (2) Partial melting of mantle peridotite that has been infiltrated by hydrous solutions emanating from the dehydrating subducting oceanic crust. (3) Partial melting of sialic lower crust due to the heat from high-*T* mafic magmas ascending from the underlying mantle. Compare Figures 1-17 and 12-21b.

In subduction zones, a plausible thermal energy source for partial melting of sialic crustal rocks is the mafic magmas rising from the mantle. Due to adverse densities (see Section 9-4), they may be blocked from farther ascent in the lower crust and there cool and crystallize. Sialic country rocks already within a hundred degrees or so of their solidus *T* could be heated to above their solidus by the mantle-derived magmas. First, these magmas would supply their latent heat of crystallization, so that approximately equal volumes of sialic magma could be fused as mafic magma crystallized. Second, with the solidus of mantle magmas no less than ~ 1200°C, they would impart their specific heat down to the solidus of the sialic crustal source rocks, 700°C to 1000°C (see Figure 9-11). Presnall

and Bateman (1973, p. 3181) conclude that "upward transport of andesitic and basaltic magmas generated along a Mesozoic subduction zone dipping beneath the Sierra Nevada would have provided sufficient additional heat to make fusion of the lower crust unavoidable."

Origin of Chiefly Calc-Alkaline Magmas in Subduction Zones

A brief summary statement is in order regarding the origin of magmatic rocks of island arcs and continental margins overlying descending lithospheric plates. Although it is impossible to review here in any detail the vast literature that has accumulated in just the past few years, it is plainly

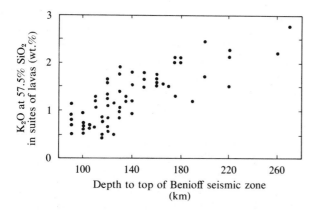

Figure 9-13 Sixty-four suites of lavas from active island arc and continental margin subduction zones around the world show increasing K$_2$O, for a specific SiO$_2$ content, away from the boundary of the plates. Volcanoes are absent for depths to Benioff seismic zone less than 80 km and greater than 280 km. [After W. R. Dickinson, 1975, Potash-depth (K-h) relations in continental margins and intra-oceanic magmatic arcs, *Geology* 3.]

evident that subduction zone magmas are polygenetic. Apparently incontestable evidence has been documented for just about every conceivable mechanism in some body of rock somewhere. Among the countless papers, recent reviews by Wyllie and others (1976), Ringwood (1977), Frey (1979), and Hawkesworth and others (1979) furnish a starting base.

Island arc magmas.   Theoretical and experimental considerations indicate that mafic primary melts can reasonably be expected to form within the oceanic crust capping the descending lithospheric plate and in the wedge-shaped region of the overlying peridotite mantle (see Figure 9-12).

Vast quantities of water are incorporated into the oceanic crust not only in extensive pore spaces but also as structural water in a host of low-temperature hydrous silicates—chlorite, serpentine, actinolite, epidote, and so on—produced by sub-sea-floor metamorphism. As this hydrated crust heats up during descent of the lithospheric plate, pore water is first driven off and then, with increasing T (see Figure 9-2), the hydrous silicates undergo a complex series of endothermic dehydration reactions (Equation 8.12) that liberate structurally bound water. These reactions consume a significant amount of the thermal energy otherwise available for melting, a process that is therefore depressed to deeper levels (Oxburgh and Turcotte, 1976). Water liberated by these dehydration reactions will move in such a direction as to equalize its chemical potential. Gravitational and thermal fields govern this movement. In most subsurface geologic settings these two fields act together, causing upward concentration of hot water, but in the upper region of a subducting plate the geothermal gradient is negative, T decreasing with depth. Hence, if the thermal field overrides the gravitational, water will migrate into the plate and would then be carried to still deeper depths where the gradient becomes more normal and water can be released into the overlying mantle. The liberated water must carry appreciable dissolved silica and alkalies that metasomatize the overlying mantle peridotites as it penetrates upward at experimentally measured rates of millimeters per hour. Entrance of water into the hot peridotite system depresses the solidus sufficiently to promote partial melting. The composition of the primary melts produced is controversial but is probably more or less saturated tholeiitic basalt due to the water-rich nature of the system and relatively extensive degree of melting. These primary melts could experience subsequent diversification to produce more evolved daughter magmas such as andesite.

The correlation between increasing potassium content (at a given SiO$_2$ concentration) and increasing depth to the underlying Benioff seismic zone found by Dickinson (1975) in many (but not all) subduction zone volcanic sequences has defied satisfactory explanation (see Figure 9-13). But perhaps it is a reflection of the distance that ascending magmas can flush out K (and other incompatible elements) from the metasomatized peridotite in the wedge of mantle overlying the descending plate (see Figure 9-12).

Ultimately, during continued descent and heating, the low-grade metamorphosed basaltic crust is dehydrated and transformed to a high *P–T* eclogitic mineral assemblage of pyropic garnet and jadeitic pyroxene. This eclogitic oceanic crust may be brought to its solidus temperature in the descending plate. The primary melts thus generated are believed by some investigators to be calc-alkaline andesites, but experimental data of others indicate a non-calc-alkaline but intermediate silica composition that could evolve into the typical andesite of arc settings. Any magmas derived from eclogitic crust might be expected to blend locally with magmas derived from the melting of the overlying hydrated and metamomatized peridotite.

Isotopic investigations are equivocal regarding incorporation of any subducted deep marine or trench sediment into the partial melts of the subducted crust. Some studies preclude any contribution, whereas in others it is allowable but not proven.

Continental calc-alkaline magmas. Formation of voluminous granitic magmas appears to require a combination of a thick sialic slab plus subduction-related thermal energy sources.

Phanerozoic granitic magmas have not been produced in tectonically stable continental areas, presumably because of inadequate temperatures along low geothermal gradients. However, along continental margins large volumes of ascending high-*T* mafic magma rising from the underlying mantle can apparently furnish the requisite thermal energy to promote partial melting. These mafic magmas must be produced with equal facility regardless of whether the overriding plate is oceanic, in the case of island arc settings, or continental. Hawkesworth and others (1979) observe similar Nd and Sr isotope signatures in magmatic rocks of both settings. Mantle-derived mafic magmas not only can supply thermal energy, but they might also mix with the granitic magmas so produced, yielding intermediate tonalitic and granodioritic magmas.

Old lower crustal source rocks could have quite variable isotopic ratios due to their variable composition and long history of metamorphism and accompanying metasomatic migration of Rb out of lower, hotter regions. Hence, intepretations of radiogenic tracers in terms of granitic magma sources must be made with caution.

Water-rich S-type granite magmas produced at lowest *T* accompanying muscovite breakdown generally cannot rise far from their source. Higher-temperature, drier I-type granodioritic and quartz dioritic magmas associated with hornblende breakdown in igneous source rocks can theoretically rise all the way to the surface and have the capacity to assimilate wall rocks en route.

9.3 DIVERSIFICATION OF MAGMAS

Effects of magmatic diversification are overprinted onto an already very broad range of magmas rising from source regions. A body of magma leaving its source is potentially subjected to **diversification processes** that will change its overall bulk chemical composition before it finally solidifies into solid rock; these processes include:

1. **Magmatic differentiation** Separation of initially homogeneous parent magma into two or more daughter magmas of contrasting chemical composition.

2. **Assimilation or contamination** Chemical incorporation of solid wall rock into the magma, producing a hybrid rock.

3. **Magma mixing.**

Diversification of magma carries the implication of a parent–daughter relation in the magmatic rocks produced (Yoder, 1976, p. 13). The **parent magma** will have the highest liquidus temperature, and in common subalkaline magmas this will be manifest by early precipitation of calcic feldspars and magnesian mafic minerals. The **daughter, derived, differentiated,** or **evolved magma** will accordingly be enriched in constituents found in a liquid fraction, usually Na, K, Fe, and residual elements.

Although it might be supposed that the parent should also be relatively large in volume and appear chronologically earliest in the magmatic activity, circumstances may prevent this. It is not impossible that in some magma systems the parent will be completely eliminated in the diversification processes. Inadequate exposure may not reveal the parent in other situations. Obviously, two parents are required in magma mixing.

Because of the variety of potential mechanisms of diversification and what must generally be a slow and interrupted upward migration from source regions, most magmatic rock bodies are likely to have solidified from evolved magma systems. In other words, rock bodies solidifying directly from unmodified primary melts (see pages 309–310) are probably the exception rather than the rule. It should be realized that the concept of parent magma is relative and depends upon the evolutionary development of a system. Hence, it is quite possible that a primary melt will serve as a parent magma to diversify into daughter magmas; then, one or more of the evolved daughters may serve in a parental role to be further diversified, and so on. (It all begins to sound like biological families!)

The purpose of this section is to illustrate the various mechanisms of diversification, where possible by specific data from particular well-studied rock bodies.

Magmatic Differentiation

Four potentially significant mechanisms of differentiation may be envisaged: crystal–liquid fractionation, liquid immiscibility, vapor transport, and thermogravitational diffusion. These mechanisms involve, respectively, separation of crystals from liquid, liquid from liquid, liquid (and crystals) from vapor, and ions from other ions.

Crystal–liquid fractionation. In partly crystallized magmas, separation of crystals from melt, though never perfect, is a powerful means of magmatic differentiation because of the contrasts in chemical composition. The evidence is most clearly found in chemical variation diagrams that show a regular pattern of compositional variation in a suite of magmatic rocks closely related in space and time, or within a single body. The suite of basalt lavas erupted in 1959 from Kilauea volcano, Hawaii, described by Murata and Richter (1966), is a good illustration. The lavas form a more or less continuous spectrum from strongly porphyritic, olivine-rich basalts, or **picrites,** to olivine-poor basalts; the picrite lavas were hotter and erupted during periods of high discharge rate, presumably reflecting mobilization of a crystal-rich mush near the base of magma chamber where gravitational accumulation had occurred.

In the composite two-element variation diagram of Figure 9-14, the concentration of MgO in the lavas provides an effective monitor for the fractionation of the Mg-rich olivines (for an elegant analysis see Irvine, 1979). Each of the other major and minor oxide constituents in the suite of lavas is plotted against MgO. Clearly, each set of weight percent values—MgO vs. Al_2O_3, MgO vs. SiO_2, and so on—follow a regular linear pattern. Also shown in Figure 9-14 are the weight percent values for olivine phenocrysts occurring in the lavas and for the most Ca–Mg-rich glass that was erupted (as vesicular cinders); this glass is assumed to represent the parent melt because it has no accumulated crystals and has the least evolved composition. Two-element **control lines** drawn through the data points of the lavas also neatly pass through the phenocryst and glass composition points. We can therefore hypothesize that accumulation of the indicated olivine crystals created the lavas more Mg-rich than the glass, while their extraction (after precipitation) from the parent generated the few Mg-poorer lavas.

Techniques for interpreting more complex systems involving more than one fractionating crystalline phase or crystalline solid solutions are outlined by Cox and others (1979, chap. 6). Crystal–liquid fractionation is but one type of process that can be analyzed by graphic or computer-based **mixing calculations;** other obvious processes in-

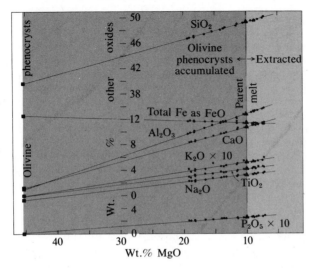

Figure 9-14 Harker variation diagram for the lavas erupted in the 1959 eruption of Kilauea volcano, Hawaii. Triangular points are the presumed parent melt composition represented by the most calcic, mafic basaltic glass erupted. Squares represent the average composition, $Fo_{87.5}$, of analyzed olivine phenocrysts in the lavas. Dots are analyzed whole rock lavas with control lines drawn through them. These lines show that accumulation (heavy shaded) and extraction (light shaded) of olivine from the parent melt can account for the entire spectrum of chemical variation in these lavas. [After K. J. Murata and D. H. Richter, 1966, The settling of olivine in Kilauean magma as shown by lavas of the 1959 eruption, *American Journal of Science* 264.]

volving mixing of two end members to give a spectrum of intermediate compositions are magma mixing, contamination, and partial melting.

Many other documented magmatic bodies appear to have been derived as residual melts from parental magmas subjected to crystal–liquid fractionation, but the evidence is less direct. The composition of the parent magma and the fractionated crystals must sometimes be inferred from experimentally determined phase equilibria. For example, melting experiments (Yoder and Tilley, 1962) on many basalts have shown that olivine, pyroxene, and plagioclase all begin to precipitate within a very limited temperature range near the liquidus at essentially atmospheric pressure (for

example, the Makaopuhi basalt in Figure 8-6). Moreover, these same basalts have bulk chemical compositions that fall near the 1-atm invariant point M_0 in the model system shown in Figure 9-7. These features preclude such basalts having solidified from *primary* melts because it is *geologically* impossible for a peridotite at 1 atm to be partially melted to produce a melt M_0. O'Hara (1968; see also Cox and others, 1979, p. 244) believes extruded basaltic magmas with compositions near M_0 have evolved by polybaric olivine fractionation of parental magmas during their ascent toward the surface from mantle source regions. Primary melts are postulated to have been produced in the mantle with compositions somewhere near or between points $M_{1.5}$ and M_3 in Figure 9-7. As these melts rose toward the surface and decompressed, the stability field of olivine expanded, allowing copious precipitation of olivine. If ascent was slow enough, the olivine crystals could be extracted, say, by gravity settling, causing the residual melts to move along an olivine control line from the forsterite apex of Figure 9-7 through points $M_{1.5}$ or M_3 toward M_0. Some of the basaltic lavas on Baffin Island discussed on pages 165–166 have been interpreted to represent more or less unmodified primary melts using the type of diagram shown in Figure 9-7.

At least three mechanisms to produce crystal–liquid fractionation have been proposed: (1) flowage segregation; (2) filter pressing; and (3) gravitational segregation. Crystal–liquid fractionation is not necessarily equivalent to fractional crystallization (see pages 277–278). The latter process is made possible by filter pressing and flowage and gravitational segregation—the three mechanisms of crystal–liquid fractionation—plus incomplete chemical reaction between crystals and melt where these two phases are not mechanically isolated from one another. Incomplete reaction between nonsegregated crystals and melt results in compositionally zoned crystals.

Crystal–liquid fractionation in flowing magma, or **flowage segregation,** has been documented in many dikes (see Figure 9-15) and sills and has also

(a)

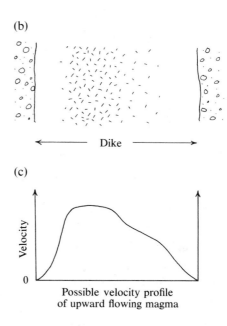

Figure 9-15   Flowage segregation of megacrysts in mafic intrusive dike located 15 km south of Hoover Dam, Arizona. Anhedral black amphibole and white plagioclase crystals did not grow *in situ* after emplacement of the dike; together with anhedral olivines and rare spinel peridotite xenoliths, they were brought up from deep sources and segregated into the center of the dike during flow.

been confirmed as a viable mechanism in theoretical and experimental studies (Komar, 1976). In a magma with more than a few percent suspended crystals, laminar flow parallel to the walls of a restricted channel concentrates crystals in the center of the channel, with regard to both their number and their size. Flow velocity ranges from zero at the wall, where drag resistance is greatest, to a maximum in the center of the conduit, with the greatest gradient in velocity near the walls (as shown in Figure 9-15c). In this gradient region, laminae of magma shear parallel to the walls and suspended crystals bump into one another, which gives them a component of motion perpendicular to the walls. Consequently, the crystals migrate

toward the center of the conduit, where the velocity is more uniform and grains are not colliding and therefore tend to remain in place without transverse displacement.

Greater concentrations of larger size euhedral crystals commonly occur in the cores of dikes and sills simply by crystallization *in situ*; this is less likely at the margins, which cool faster. Hence caution must be exercised in interpreting cases of potential flowage segregation.

It is theoretically possible for the melt in mushlike, crystal-rich magmas to separate from the crystals, either by draining or by being pressed out. Squeezing water from slushy snow is an example. However, the geologic reality of this **filter**

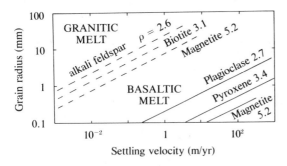

Figure 9-16 Settling velocities as a function of melt viscosity and grain size, assuming Newtonian behavior and a spherical grain shape. Growth of grains during settling is assumed to be zero. Settling in a granitic melt with $\eta = 10^8$ poise and $\rho = 2.3$ g/cm³ is shown by dashed lines. Settling in a basaltic melt with $\eta = 10^3$ poise and $\rho = 2.65$ g/cm³ is shown by solid lines.

pressing remains to be demonstrated. It might occur on the floor of mafic intrusions, where the weight of accumulating crystals presses some of the intercumulus melt out of the underlying crystal mush.

Gravity-driven crystal–liquid fractionation, or **gravitational segregation,** depends upon the difference in densities of crystal and melt, $\Delta\rho$, viscosity of the melt, η, and size of the crystal (assumed to be a sphere of radius r). According to **Stokes' Law**

$$v = \frac{2r^2 g \Delta\rho}{9\eta} \qquad (9.2)$$

where v is velocity of the settling or rising crystal and g is the acceleration of gravity. Some representative values of r, v, and ρ are graphed in Figure 9-16, which shows that settling velocities could permit effective crystal fractionation on the time scale on which large deep intrusions must cool (10^4 to 10^6 years). It must be recalled, however, that silicic and partly crystallized basaltic melts have significant plastic yield strengths that invalidate Stokes' Law, which holds only for Newtonian viscosity. Yield strengths of some basaltic

melts are sufficient to inhibit settling of pyroxenes < 5 mm in radius.

Gravitational settling and crystal–liquid fractionation in stratiform mafic intrusions: Fact or fiction? The Skaergaard and other large stratiform mafic to ultramafic intrusions have served for decades as examples par excellence of gravitational settling and crystal–liquid fractionation (see pages 175–180). Nonetheless, recent work (Irvine, 1979, 1980; McBirney and Noyes, 1979) has pointed out some serious deficiencies in the model of an origin by gravitational settling.

The classic interpretive model for the formation of thick layered intrusions is that cyclic convection currents transport crystals growing near the cooler roof of the magma chamber down along the floor where, because of their greater density, they settle to the base of the melt and are sorted according to hydraulic parameters of size and density (Stokes' Law). Cumulus fabrics within layers are believed to form in an analogous manner to settling of sand grains in water. However, the following arguments and data cast doubt on aspects of this model:

1. Although most of the heat lost from the convecting magma body is into the roof and upper wall rocks, crystals must actually precipitate along the lower walls and floors (see Figure 9-17). Depending upon current velocity and other factors, the convecting melt might be supercooled in the lower reaches of its circuit.

2. Careful analysis of mineral grain densities and sizes in some layers of stratiform intrusions indicates the grains were not sorted according to hydraulic processes (Stokes' Law); for example, coexisting cumulus olivine and chromite grains in a single layer of the Stillwater complex are of such size and density that the much larger olivines should have settled as much as 200 times faster (see page 180).

3. The densities of cumulus plagioclases in

some intrusions are less than that calculated for presumed coexisting melts; plagioclases should theoretically have floated.

4. Sedimentation of sand grains in water is probably a misleading analogy to origin of layering and cumulus fabrics in mafic intrusions because in the latter densities of solids and liquids are not so different and, especially, the silicate magma has a greater apparent viscosity. Indeed, the plastic yield strength of the melt might preclude gravity segregation of most normal-sized (a few millimeters) grains of rock-forming minerals.

5. Layers occur along the walls and even the roof of the Skaergaard, apparently in defiance of gravity.

In view of these and other inconsistencies, a new explanation—or explanations, for there may be more than one genetic type of layering—is needed.

Irvine (1979, p. 294) and McBirney and Noyes (1979) suggest two possibilities: deposition from density currents and *in situ* bottom crystallization. In Irvine's current deposition model, surging density currents of crystal-rich magma are believed to sweep across the floor of an intrusion, similar to the well-known turbidity currents in sedimentary basins. Irvine has performed model experiments in which magma density and viscosity, grain size, and geometry of the chamber are accurately scaled down in a relative sense. Layers of grains showing size and density sorting are found to form by dynamic flow processes in the head of the density current. In viscous magma bodies, a flowing density current might intrude itself between the main stagnant body and the crystalline mush on the floor. Due to the yield strength of a stagnant melt, less dense plagioclase crystals could not buoy out of the current deposit once it came to rest. Evidence of density currents in plutons includes cross bedding, scour-and-fill structures, and transported lithic fragments within rhythmic layers.

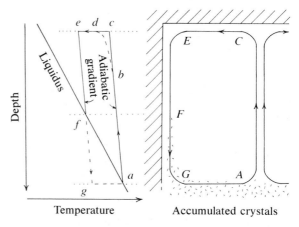

Figure 9-17 Schematic diagram of a convection cell in a thick magma body with possible temperature–depth–liquidus relations shown on the left. Path of magma movement is shown on right; capital letters indicate position of magma corresponding to lowercase letters in depth–*T* space on left. The liquidus temperature at which one or more crystalline phases should precipitate increases with depth at the rate of a few degrees per kilometer and shifts as a whole to lower temperatures with overall cooling of the magma. The actual temperature in the magma body follows an adiabatic gradient of only a few tenths of a degree per kilometer. Path *bde* reflects cooling of the magma near the roof. Path *fa* reflects negligible undercooling of the melt so that during crystal growth the latent heat more or less maintains the liquidus temperature. A supercooled melt that does not crystallize until it reaches the floor follows path *fga*. (Compare Figure 7-21). [After T. N. Irvine, 1970, Heat transfer during solidification of layered intrusions. I. Sheets and sills, *Canadian Journal of Earth Sciences* 7.]

Other layers, especially rhythmic inch-scale layers (see Figure 5-36), may owe their origin to *in situ* crystallization along the margins and especially the floor of a mafic intrusion. Actually, two processes might be involved, nucleation-diffusion and stratified double-diffusive convection. The nucleation model follows from the concept of chemically zoned boundary layers discussed on pages 242–246, only the scale is much greater, so

(a)

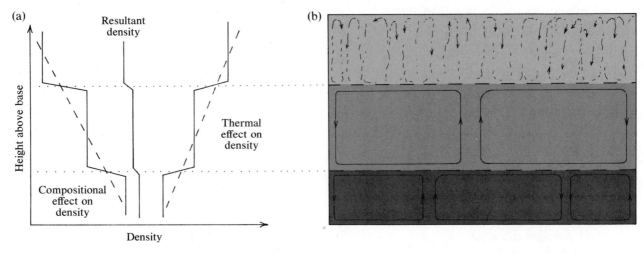

(b)

Figure 9-18 Schematic diagrams illustrating the origin and nature of stratified double-diffusive convective layers in a liquid solution whose composition differs from that of the precipitating crystals. (a) Vertical thermal and composition gradients in a body of melt above a crystallizing layer exert opposing effects (dashed lines) on the vertical density gradient. A positively sloping gradient is dictated by the upward decrease in temperature due to loss of heat through the roof of the magma chamber (see Figure 9-17), whereas a negatively sloping density gradient is favored by the formation just above the layer of growing crystals of a residual melt enriched in Fe. The density gradient due to compositional variations is gravitationally stable, whereas the one associated with the thermal gradient may not be. Depending upon melt viscosity and other factors it may override the former so that convection results. However, instead of a body-wide convective system, as might occur in a compositionally homogeneous melt (see Figures 9-17 and 5-38), the combination of compositional and thermal gradients promotes formation of stratified layers (solid line). (b) In each of these layers (shaded), the melt is well mixed by convection and is sharply separated above and below by interfaces where both heat and chemical constituents diffuse across. Because of this continuous double diffusion, the layers can thin or thicken with time. Two different styles of convection can occur—fingerlike penetrations and larger circuitous cells; these two styles are only schematically assigned to the layers indicated. [After A. R. McBirney and R. M. Noyes, 1979, Crystallization and layering of the Skaergaard intrusion, *Journal of Petrology* 20; and T. N. Irvine, 1979, Rocks whose composition is determined by crystal accumulation and sorting, in H. S. Yoder, Jr. (ed.), *The Evolution of the Igneous Rocks: Fiftieth Anniversary Perspectives* (Princeton: Princeton University Press).]

that instead of a growing single crystal the phenomenon is associated with layers of growing crystals in an intrusion. It was previously indicated (page 240) that the sequence Fe–Ti oxides, olivine, pyroxenes, and plagioclase might have decreasing rates of nucleation. As a zone of melt in the magma chamber potentially able to coprecipitate all four crystalline phases undercools, Fe–Ti oxides would nucleate first. Diffusion of Fe and Ti ions to the growing crystals form a boundary layer depleted in these constituents and eventually, due to slowness of diffusion, this depleted layer prohibits further growth. Olivines later nucleate, grow, and form a boundary layer depleted in olivine components. Finally, plagioclases nucleate, consuming remaining chemical components. In this manner, and because of overlaps in nucleation of the different phases, a mineralogically graded (and apparently gravity-sorted) layer similar to those commonly occurring in stratiform intrusions might develop (see Figure 5-34a). Eventually, a subsequent zone next to the first will undercool and the process is repeated; many such cycles might produce rhythmic layers.

Various other aspects of layering, on both large and small scales, might reflect the transient existence of stratified convective layers. As described and illustrated by McBirney and Noyes (1979) on the basis of model experiments, these double-diffusive layers occur in a liquid in which (1) two (or more) constituents, say Si^{4+} and Fe^{2+}, with different rates of diffusion have different concentrations in the vertical direction, and (2) the two constituents have opposing effects on the density in the vertical direction (see Figure 9-18). The term double-diffusive layers arises from the fact that the layers are separated by a compositionally discontinuous interface across which there is transfer (diffusion) of both matter and heat via convective circulation within each layer.

Liquid immiscibility. The acceptance of liquid immiscibility as a viable petrogenetic process has swung like a pendulum since it was first suggested in the late nineteenth century. At that time it was used to explain the origin of juxtaposed but compositionally contrasting rock types such as basalt and rhyolite. Later, in the 1920s, experiments at the Geophysical Laboratory on simple model silicate systems disclosed that addition of alkalies or alumina to binary systems displaying a significant *T–X* field of liquid immiscibility, such as MgO–SiO_2 (see Figure 8-18), eliminated the immiscibility. This discovery, together with the growing prominence of N. L. Bowen's ideas on crystal–liquid fractionation and lack of documented evidence in natural rock bodies, spelled the doom of this model as a viable mechanism of magmatic diversification. However, in 1951 Roedder reported experiments in the model system K_2O–FeO–Al_2O_3–SiO_2 that showed immiscibility for geologically reasonable temperatures and compositions. Some 20 years later additional experimental evi-

dence was found in model carbonate–silicate systems, and finally petrologists began recognizing evidence in natural rocks. In an extensive and thorough review of the petrogenetic significance of liquid immiscibility, Roedder (1979) was able to list 40 references in the past decade proposing immiscibility in various natural rock bodies. The process has once again gained a degree of acceptance and legitimacy.

Immiscibility appears to exist in basically two silicate compositions—high-Fe basaltic melts (see Figure 9-19) and highly alkaline melts. Experimental melting of synthetic compositions similar to postulated late residual Fe-rich melts in the Skaergaard yielded coexisting conjugate liquids—one rich in Fe and P and low in Si and the other rich in Si and poor in Fe and P (see Figure 9-19b). Evidence for this immiscibility lies in the presence of two glasses of different composition, manifest most obviously by contrasts in color and refractive index, that are sharply separated along a meniscus in quench experimental runs (see Figure 9-20). Similar evidence for conjugate Fe-rich and Si-rich melts are found in microscopic interstices between mineral grains in lunar ferrobasalts. A substantial field of two liquids exists in simple experimental alkali silicate–carbonate systems and this discovery has prompted investigators of nephelinite (ijolite)–carbonatite complexes to invoke liquid immiscibility to explain their formation. Minute inclusions containing immiscible liquids have actually been documented in mineral grains in ijolite.

Petrographic and experimental investigations have unequivocally established the immiscibility of sulfide and silicate melts. Concentrations of only a few hundredths of one percent by weight of S are sufficient to saturate basaltic melts, with the resultant separation of a sulfide melt that is chiefly Fe and S with minor Cu, Ni, and O; it can ultimately crystallize to pyrrhotite, chalcopyrite, and magnetite. Although trace amounts of Ni are strongly partitioned into crystallizing olivine in basaltic melts the uptake into an immiscible sulfide liquid is ten times greater; for Cu it is 100 to 1000 times greater. Pb and Zn apparently tend to remain

(a)

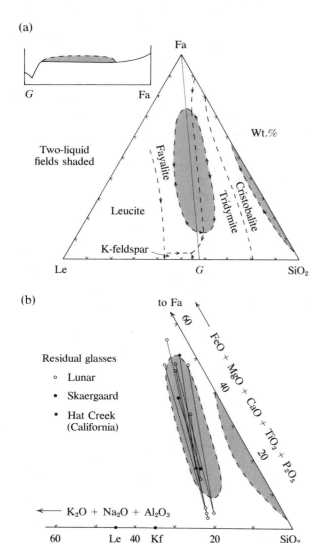

(b)

Figure 9-19 Compositional relations in systems with liquid immiscibility. (a) Preliminary ternary diagram of the system leucite–fayalite–SiO$_2$; the inset in upper left is a *T–X* section along Fa–G of the ternary diagram. (b) Pseudoternary diagram of the system in (a) but with added components so that compositions of some naturally occurring residual conjugate glass pairs can be plotted. Note close relation to experimentally determined two-liquid field. [After E. Roedder, 1979, Silicate liquid immiscibility in magmas, in H. S. Yoder, Jr. (ed.), *The Evolution of the Igneous Rocks: Fiftieth Anniversary Perspectives* (Princeton: Princeton University Press).]

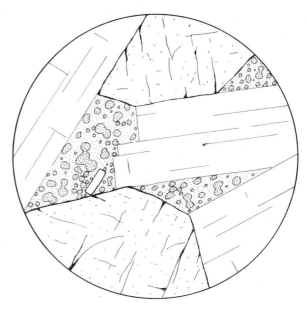

Figure 9-20 Schematic drawing of immiscible globules of brown mafic glass in a lighter-colored siliceous glass, both occupying interstitial volumes between early formed mineral grains. Textures such as this have been found in a number of Fe-rich terrestrial and lunar basaltic rocks. The globules range in diameter from a few micrometers to perhaps 100 μm.

in the silicate melt. These data have significant implications for the genesis of some magmatic ore deposits.

Vapor transport. It is a well-known fact that certain constituents—especially Na and to some extent K, Fe, and Si—normally considered condensed constituents can be concentrated to a significant extent in any separate volatile phase, especially water at high *T*. The existence of fenitized aureoles in the country rocks surrounding alkaline complexes is a manifestation of such transport. But no concrete evidence for a natural magma body enriched by volatile transported constituents has been reported.

Thermogravitational diffusion. In a paper that is already a classic, Hildreth (1979) pointed out the complete inadequacy of the standard mechanisms of diversification in producing the compositional zonation in the magma chamber that was partially erupted to form the Bishop ash-flow tuff in eastern California. More or less continuous compositional zonation in this 0.7-My-old rhyolite tuff represents in inverted form the zonation in the top of the preeruption chamber (see Figure 9-21). Although Si varies by only ~ 2 wt.%, other elements, especially minor and trace elements, have a tenfold variation between the earliest and latest magma erupted. The significant upward enrichment in H_2O in the chamber is especially significant. Although phenocrysts were more abundant in the hotter lower part of the erupted cap of the magma chamber, crystal–liquid fraction is precluded as a mechanism of differentiation. Among many lines of evidence for this conclusion is the fact that crystals of allanite (a rare earth element epidote) with a density of > 4 g/cm^3 and a combined La and Ce content of 16 wt.% occurred only in the topmost part of the chamber whereas whole rock La and Ce values increased threefold downward. Clearly, the concentration of La and Ce was not governed by crystal settling. Extensive quantitative chemical and mineralogical data together with field relations and petrographic observations too numerous to detail here also preclude the operation of large-scale assimilation, liquid immiscibility, and magma mixing.

It is obvious that some other mechanism of diversification must have prevailed in the Bishop magma chamber and, by implication, in other chambers that were partially erupted to produce very similarly zoned rhyolitic pyroclastic deposits (see pages 86–88). Apparently, the zonation is created in an essentially liquid system, yet it was emphasized in Chapter 7 that chemical diffusion rates in static bodies of melt are too slow to achieve transport over distances of more than a few meters on a time scale of a million years. Repeated production of zoned pyroclastic deposits from large magma chambers underlying epicontinental ring fracture caldera systems occurs on the order of a million years, whereas smaller compo-

site volcano systems have times of regeneration of the zoned chamber on the order of a thousand years; obviously, a more rapid mechanism of chemical transport is involved.

The double-diffusive convective model briefly discussed above has considerable promise in explaining compositional zonation in calc-alkaline magma chambers. As envisaged by H. R. Shaw, R. L. Smith, and W. Hildreth (see Hildreth, 1979), the combined influence of a vertical thermal gradient and the gravitational field on a body of silicate melt is to produce a stratified convective system that significantly enhances chemical transport by diffusion. In their model, a cap of cooler, less dense, more siliceous and water-rich magma develops at the top of the magma chamber (see Figure 9-22). The formation of this cap likely depends upon several interrelated but poorly understood phenomena. Water may diffuse a short distance into the magma from the roof and wall rocks; low ^{18}O contents of early-erupted tuffs reflect this influx of meteoric water. Along the walls a thin boundary layer is made less dense by this inbibed water and by the possible effects of crystallization along the contact that leaves a silica–H_2O-enriched melt. Though cooler, this boundary layer is nonetheless less dense, and buoys upward into the cap of the magma chamber. (In contrast to crystallization in tholeiitic basaltic magmas that produce a more dense Fe-rich residual liquid, calc-alkaline residual magmas tend to be enriched in Si, Na, and K and are therefore less dense; see Figure 5-30. Figure 9-23 makes it clear that compositional variations in crystallizing magma bodies can exert a strong influence on densities.) The cap grows by diffusional interchange with the wall rocks, partial crystallization, and double-diffusion (of heat and matter) across the horizontal interfaces of stratified convecting layers in the top of the chamber. Compositional gradients (zonation) develop for each chemical element as a function of the thermal gradient, gravitational field, and gradients in atomic structure of the melt. Uppermost layers of melt are less polymerized, due chiefly to their greater water content, and this must exert an

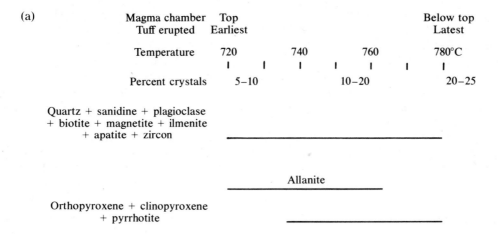

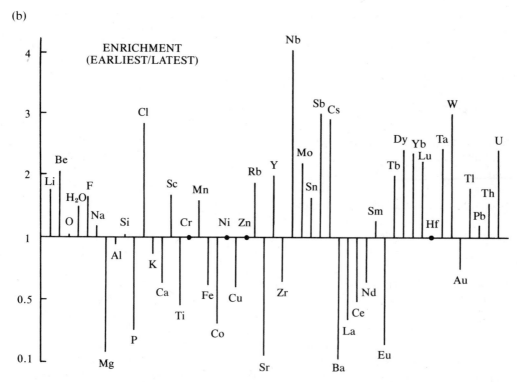

Figure 9-21 Compositional zonation in the Bishop tuff, which is an inverted picture of the variations in the upper part of the preeruption magma chamber. Compare Figure 3-29. (a) Phenocryst assemblage and magma temperatures based on the Fe–Ti oxide geothermometer. (b) Enrichment factors for elements arranged by atomic number; the factor for each element is its concentration in the earliest-erupted tuff divided by the concentration in the latest. Cr, Ni, Zn, and Hf do not exhibit systematic trends. [After W. Hildreth, 1979, The Bishop tuff: Evidence for the origin of compositional zonation in silicic magma chambers, Geological Society of America, Special Paper 180.]

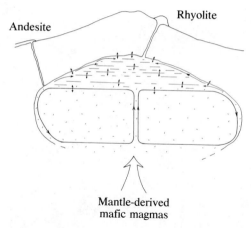

Andesite

Rhyolite

Mantle-derived
mafic magmas

Figure 9-22 Schematic relations in a stratified double-diffusive convection system within a shallow calc-alkaline magma chamber. A buoyant water- and silica-rich cap overlies a more crystal-rich intermediate silica core. Chemical gradients in many elements are strongest in the cap, which is shown to be highly stratified. In some magma chambers these gradients continue uninterrupted into the core, whereas in others there is a significant compositional gap (see Figure 3-31). There may be no zonation in the crystal-rich core magma. Double-headed arrows indicate double diffusive transfer of heat and matter. The calc-alkaline chamber may be underplated by mantle-derived mafic magmas that supply thermal energy and dissolved gases to the convecting system. Tapping the zoned chamber at different levels yields extrusive lavas of variable composition. [After W. Hildreth, 1979, The Bishop tuff: Evidence for the origin of compositional zonation in silicic magma chambers, Geological Society of America, Special Paper 180; and H. Williams and A. R. McBirney, 1979, *Volcanology* (San Francisco: Freeman, Cooper and Company).]

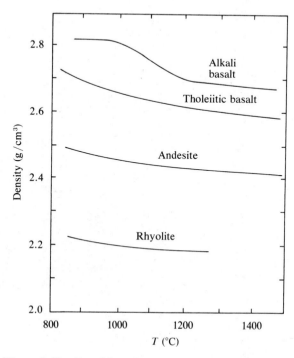

Figure 9-23 Densities of some anhydrous silicate melts. The composition of a melt is more important in dictating its density than temperature. Dissolved water lowers the densities somewhat (see Table 9-2). [After H. Williams and A. R. McBirney, 1979, *Volcanology* (San Francisco: Freeman, Cooper and Company).]

influence on the bonding characteristics of different cations in the melt (Hildreth, 1979).

Interestingly, the elements concentrated in the high-silica rhyolitic roof of the magma chamber—Li, Be, Na, Rb, Cs, Nb, Ta, Mo, W, Th, U, and so on in Figure 9-21—are the same ones concentrated in pegmatites and in tin granites and Mo-porphyry bodies of economic significance. Also of interest is the downward enrichment of Cu and Au into the more crystal-rich, intermediate silica magma; such magmas may give rise to the equally significant Cu-porphyry deposits.

Magma Mixing

In long-lived magmatic centers, two or more unlike magmas may blend together, forming a hybrid daughter product compositionally intermediate between them. Unequivocal evidence for mixing in magmatic bodies is elusive and commonly only permissive, rather than compelling. Lavas with corroded xenocrysts are sometimes interpreted to have originated by mixing of two phenocrystic magmas, but xenocrysts could also reflect assimilative disaggregation of solid rock (see next

(a)

Figure 9-24   Examples of mixed magmatic rocks. (a) Rhyolite–basalt lava from Gardiner River, Yellowstone National Park. Note intimate mingling; thin sections show euhedral calcic plagioclase and mafic silicate phenocrysts from the basaltic component side by side with corroded quartz and sanidine phenocrysts from the rhyolite. Compare Figure 3-32. [R. E. Wilcox, personal communication.]

section). The so-called **mixed lavas** and net-veined agmatites, usually of calc-alkaline affinities, are more compelling evidence for mingling of dissimilar magmas (see Figure 9-24). Such features presumably would represent the initial stage of physical mixing. To achieve more complete chemical blending and homogenization would require considerable time and especially thermal and/or mechanical energy. Just how this might occur is uncertain; magma blending is not analogous to the action of a kitchen appliance! Mixing would tend to cool and solidify the more mafic component while superheating the felsic one above its liquidus T. Blending might be more readily achieved if there are large proportions of the more mafic magma member. Many petrologists currently view mafic inclusions in silicic host rocks as evidence for at least physical mixing, perhaps as mantle-derived mafic magmas were intruded into overlying more silicic magma bodies that they might themselves have created. Other petrologists (for example, Eichelberger and Cooley, 1977) contend

(b)

(b) Net-veined syenite–gabbro agmatite. Note that the pieces of gabbro have curved pillowlike boundaries, not straight and intersecting at sharp angles as would be the case if the syenite had veined between broken rigid fragments. Rather, the pillow form is compatible with globules of perhaps semisolid gabbro magma making contact with cooler syenite magma. Very thin syenite veins possibly reflect blobs of gabbro squeezing tightly together. One might wonder if this agmatite could be a large-scale version of the liquid immiscibility fabric shown in Figure 9-20. [U.S. Geological Survey photographs courtesy of R. E. Wilcox and Louise Hedricks.]

that most if not all calc-alkaline magmas intermediate between basalt and rhyolite are the products of mixing of these two end members.

If the two parental magmas can be identified and chemically analyzed, any blended daughter products should have intermediate compositions falling on a straight line between the parents on a Harker variation diagram (Cox and others, 1979, chap. 6). But a similar linear variation can develop during assimilative processes. Glasses representing quenched melts in volcanic rocks furnish a means of testing for magma mixing in their ancestry (Anderson, 1976). The evidence lies in the presence of a glass whose composition lies outside the range between the bulk rock composition and any residual glass within interstices between grains. This first glass could occur as inclusions within grains.

Sea-floor basalts display only a limited range of compositional variation, which has been widely interpreted to reflect crystal–liquid fractionation in the crustal magma chamber lying beneath the axis

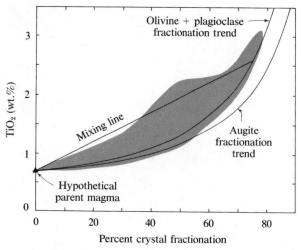

Figure 9-25 Effects of crystal–liquid fractionation and magma mixing on the TiO_2 content of sea-floor basalts. A hypothetical parent magma (SiO_2, 49.9; TiO_2, 0.7; Al_2O_3, 16.1; total Fe as FeO, 7.7; MgO, 10.1; CaO, 13.1; Na_2O, 1.9; and K_2O, 0.09) from which olivine, plagioclase, and augite of realistic compositions are fractionated will produce residual melts along the curved fractionation trend lines. Intermittent mixing of such residual melts with additional draughts of parental magma would produce magmas whose compositions lie within the lens-shaped area bounded by the mixing line and the curved fractionation trend lines. The shaded area represents over 600 chemically analyzed sea-floor glasses. Although there is a close correspondence between the predicted and observed TiO_2 compositional variations that supports the proposed model, other constituent oxides would have to be tested as well. [After J. M. Rhodes, M. A. Dungan, D. P. Blanchard, and P. E. Long, 1979, Magma mixing at mid-ocean ridges: Evidence from basalts drilled near 22°N on the Mid-Atlantic Ridge, *Tectonophysics* 55; shaded area after W. G. Melson, T. L. Vallier, T. L. Wright, G. Byerly, and J. Nelen, 1976, Chemical diversity of abyssal volcanic glass erupted along Pacific, Atlantic and Indian ocean seafloor spreading centers, in *The Geophysics of the Pacific Ocean Basin*, Geophysical Monograph 19 (Washington, D.C.: American Geophysical Union).]

of the oceanic rift (see Figure 5-43). However, an inconsistency with this model is that the major element pattern is characterized by sympathetic variation in Mg/(Mg + Fe) and CaO/Al_2O_3 ratios (Rhodes and others, 1979). Fractionation of olivine, usually the first major mineral to precipitate in sea-floor magmas, cannot alone produce such a sympathetic variation. If plagioclase, the next phase to appear, also fractionates it would cause the CaO/Al_2O_3 ratio to increase from its ratio of ~ 0.9 in lavas with the highest Mg/(Mg + Fe). Substantial augite fractionation would yield the observed elemental variations, yet this phase is generally the last to begin crystallization.

It seems unlikely that the magma chamber beneath the rift, once filled, would remain as a completely closed system throughout its entire history of solidification. Instead, one would expect that fresh draughts of undifferentiated parent magma would be periodically injected into the chamber from underlying mantle source regions. Many of the larger olivines and especially plagioclases in sea-floor basalts are strongly corroded and fritted cores are surrounded by clear crystalline material. These highly magnesian (Fo_{85-90}) and calcic (An_{83-86}) xenocrysts were obviously unstable in their host melts, suggesting the possibility of mixing of phenocryst-bearing magma with more chemically evolved magma. The composition of this less-evolved, porphyritic parental magma can be inferred from the composition of glass inclusions within the xenocrysts, as well as from the nature of the xenocrysts themselves, and is a hypothetical magma more or less equivalent to some naturally occurring, glassy sea-floor lavas with the highest Mg/(Mg + Fe) and CaO/Al_2O_3 ratios. Rhodes and others (1979) conclude that sea-floor basalts worldwide show only a relatively limited compositional variation because fresh draughts of mantle-derived magma are episodically injected into and mixed with magma that is undergoing crystal–liquid fractionation in shallow crustal chambers (see Figure 9-25).

Assimilation

Experimental studies of silicate melts at high temperature invariably use an inert platinum or gold container to avoid contamination problems. But the silicate or locally carbonate wall rocks surrounding natural magma bodies are by no means inert and may be expected to interact with and contaminate the magma. **Assimilation** is the incorporation of matter from wall rocks, or from physically ingested inclusions, into the magma to form a **contaminated** or **hybrid** igneous rock.

As N. L. Bowen emphasized several decades ago, chemical assimilation requires thermal energy, first to heat the solid country rock to temperatures where reactions or fusion can occur and second to furnish latent heat of melting or heat for reactions. The source of this energy can only be the magma itself. If the specific heat of rock is 1.0 J/g degree and the latent heat of melting (or crystallization) is 300 J/g, then to raise the temperature of solid rock in an epizonal setting from 200 to 800°C would require 600 J/g. Two grams of magma would have to freeze to provide this energy; twice as much magma must crystallize than rock heated. Then melting the rock would require that an equivalent amount of magma crystallize.

The amount of assimilation of cool wall rock into a magma is limited by the thermal energy of the magma itself. Transfer of heat from magma to cooler rock leads to solidification of the magma in the vicinity of the transfer, building an armoring barrier of solid rock around the still active part of the magma system, so that further chemical interaction via diffusion is inhibited.

To the extent that thermal energy might be available in a particular system, such as in ascending dry or water-deficient magmas governed by positively sloping solidus curves, individual minerals can be assimilated in two ways (McBirney, 1979). First, the crystalline phase may dissolve if the silicate melt is not already saturated with respect to that phase; for example, quartz can dissolve in basaltic melts that are not already saturated in quartz (see Figure 9-26). Second, the phase may react with the melt so that liberated constituents combine with melt constituents to form another crystalline phase, or phases, with which the melt is in equilibrium; foreign pyroxene crystals may react with a granodioritic melt precipitating hornblende, forming by ionic diffusion a

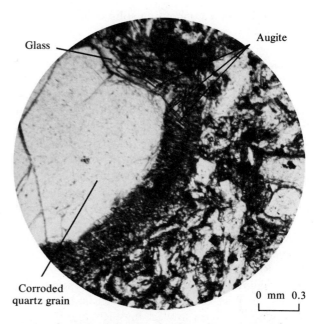

Glass   **Augite**

Corroded quartz grain   0   mm   0.3

Figure 9-26 Partially dissolved quartz xenocryst in a basalt is surrounded by a jacket of glass and minute prismatic augites. Such embayed quartz grains are generally interpreted to have originated by assimilation of quartz-bearing wall rock into the magma, but the possibility that a particular corroded mineral might have formed stably earlier in a higher-*P* regime of the system should always be kept in mind.

reaction rim of hornblende surrounding and possibly eventually replacing the unstable pyroxene.

Small-scale assimilation effects resulting in contaminated magmatic rock can be perceived in many magmatic systems on the basis of criteria such as the following:

1. The contaminated rock is located between, or is at least near, the interface between the parent magmatic material and foreign rock.

2. The presence of physically disaggregated foreign rock xenoliths and embayed, corroded xenocrysts in magmatic material.

3. The contaminated rock is chemically intermediate; in a Harker variation diagram, the contaminated rock plots on a

straight line connecting solid contaminant and parent magmatic material.

These criteria are permissive, but insufficient to prove assimilation. In the older petrologic literature, assimilation of limestone into granitic magma was invoked to account for the origin of feldspathoidal alkaline rocks. Although such desilication effects have been found on a very local scale at limestone–granite contacts, experiments show that prohibitively large amounts of limestone must be digested to produce nepheline and, in the process, the solidus T is raised significantly so as to arrest further solution.

A possible test for contamination of mantle-derived mafic magmas by old continental crust lies in Sr–Rb data (see pages 57–58 and 164–165). If these magmas, having relatively high Sr and initial ratios $^{87}Sr/^{86}Sr$ of 0.704–0.706, assimilate sialic crust with relatively low Sr, high Rb, and $^{87}Sr/^{86}Sr$ > 0.706, there will be a negative correlation between the initial ratio and Sr and a positive correlation between the ratio and Rb/Sr. See Cox and others (1979) for further discussion. But the importance of assimilation of crustal rocks in the evolution of calc-alkaline magmas continues to be debated. On one side of the argument is Taylor (1978, p. 206), who claims that oxygen and hydrogen isotope data show that "many magmas may be strongly affected by widespread melting or assimilation of hydrothermally altered roof rocks above a magma chamber." Roof rocks in long-lived igneous centers are commonly pervasively fractured and intensely hydrothermally altered to water-rich mineral assemblages. This is viewed by Taylor as a favorable setting for block stoping (see next section), assimilation, and partial melting as water is absorbed in the magma.

9.4 ASCENT AND EMPLACEMENT OF MAGMAS

The mechanics of the rise of magmas from deep source regions toward the surface of the Earth depend upon principles of hydrostatic and hydro-dynamic equilibria. A simple overview is presented here; for more details see Williams and McBirney (1979, pp. 43–63).

With partial melting of source rocks of only a few percent, the liquid will exist as barely interconnected films between grains; in this condition the liquid can hardly be drawn away from the grains because of capillary attraction. However, with more advanced melting, larger volumes of melt exist between neighboring grains, and the melt might segregate into a dendritic network of increasingly larger channels, forming a crystal–liquid slush, like partly melted snow. Whether melts can segregate from such mixtures and rise independently toward the surface is difficult to say. One possibility is that they do not segregate but rather that the whole crystal–liquid mass rises as a unit in the form of a finger-like plume or **diapir**. Continued rise and decompression of this diapir could lead to further melting.

Magma Buoyancy

The fundamental driving force causing a crystal–liquid diapir or of a body of crystal-free primary melt to rise upward from its source region is the buoyancy of the less dense magma. Except for ice and a few other geologically unimportant systems, the liquid state is always less dense and occupies a greater volume than the crystalline state. In addition, the incongruent melting relations of silicate systems are such that light elements such as H, C, O, Na, K, Al, and Si tend to be more highly concentrated in the melt than in the crystalline residuum. Heavier Fe, however, goes preferentially into the melt, and the lighter more refractory Mg stays in the residuum. The net result is that melts are less dense by a few tenths of a gram per cubic centimeter than the crystalline residuum (see Table 9-2).

In the gravitational field of the Earth, a lower-density body of magma surrounded and overlaid by higher-density rock is hydrostatically unstable and is acted upon by a buoyant force that causes it to rise. For any body immersed in a fluid, Archimedes' principle states that the fluid exerts an up-

Table 9-2 Some representative densities of rocks and melts.[a]

	$T(°C)$	Density (g/cm^3)
Rocks		
Peridotite		3.2
Basalt		2.9
Diorite		2.8
Sedimentary rocks near Earth's surface		2.1–2.8
Melts		
Andesite	1100	2.4
Basalt	1300	2.60
	1100	2.63
Basalt		
Dry	1313	2.68
0.8 wt.% H_2O	1313	2.66
Fe-rich basalt		
Dry	1120	2.78
6.9 wt.% H_2O	1120	2.64

[a] See also Figure 9-23.
Source: Handbook of physical constants, Geological Society of America Memoir 97; and Y. Bottinga and D. F. Weill, 1970, Densities of liquid silicate systems calculated from partial molar volumes of oxide components, *American Journal of Science* 269.

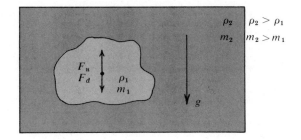

Figure 9-27 Gravitationally unstable system of two viscous fluids.

ward force on the body equal to the weight of the fluid displaced by the body. This upward force on the body (see Figure 9-27), $F_u = m_2 g$, is counteracted by the downward weight of the body itself, $F_d = m_1 g$. The resultant imbalanced positive buoyant force $F_b = F_u - F_d$ is directed upward if $m_2 > m_1$.

Williams and McBirney (1979, p. 44) note that the pressure resulting from volumetric expansion on melting provides a driving force for magma ascent to lower-P regions but that its magnitude is insignificant compared to the buoyancy.

The creation of liquid in a source region and its subsequent ascent constitute a classic example of a pair of geologic processes that are driven by the two fundamental but opposing energy sources within the Earth—thermal and gravitational energy (see Chapter 1). A positive thermal pertur-

bation causes melting, and this would proceed without bound if no gravity field were present until the causitive mechanism (decompression, radioactive heating, volatile influx, and so on) stopped. But in a gravitational field the melting process is relaxed as the liquid body buoys upward from the perturbed source region. The gravity-driven ascent is eventually arrested as the body loses thermal energy and becomes immobile. Renewed growth of the thermal perturbation at the source can initiate the magmatic cycle over again.

Once a less dense body begins to move upward in a hydrodynamic system, the buoyant force is counteracted by the frictional and viscous resistance of the surroundings. These opposing forces are embodied in Stokes' Law, which was previously used to find the velocity of sinking denser crystals. A less dense body of magma may not rise at all if resistive forces exceed buoyancy.

Preceding paragraphs imply that the essentially solid upper mantle and crust above magma source regions behave as viscous fluids. On a geologic time scale of thousands to millions of years this is indeed the case, especially if the rocks are within a couple of hundred degrees or so of their solidus temperature (see pages 466–467). The apparent viscosity is considerable but depends upon T and X of the body. On smaller time scales the upper mantle and crust transmit elastic seismic waves. Hence, rocks may be viewed as elasticoviscous bodies, and their exact behavior depends upon time factors as well as T and X.

Two idealized environments of magma ascent

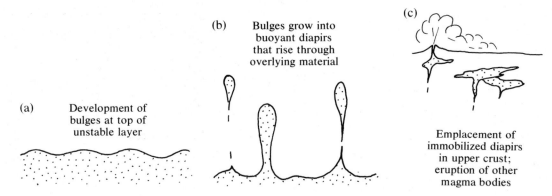

Figure 9-28   Rise of magma diapirs from a gravitationally unstable less dense layer overlaid by denser material.

may be recognized: first, through the hotter viscous upper mantle and lower crust, and second, through the cooler, more brittle upper crust.

Mechanisms of Magma Ascent

Experimental studies and theoretical analyses (Ramberg, 1970; Elder, 1976) show that sinusoidal bulges (Raleigh–Taylor instabilities) will appear in the interface between a low-density layer and an overlying more dense layer (see Figure 9-28). As the bulges extend upward, smaller ones cool, losing their density contrast, and die. Larger bulges grow into fingers that may separate from the mother layer, forming inverted teardrops or diapirs that continue buoying toward the surface. For a diapir of granitic magma of, say, 8 km in radius rising through crust of viscosity 10^{20} poise with $\Delta\rho = 0.1$ g/cm^3, Stokes' Law [Equation (9.2)] predicts an ascent velocity of almost 5 cm/yr, or a total rise of 50 km in one million years. Note that the ascent velocity varies as the square of the radius of the body; a body half as large will rise only one-fourth as far. Decorative "lava lamps" are good examples of this type of viscous behavior.

The wavelength of the bulges in the low-density layer depend upon its thickness and the viscosity contrast with the overlying denser layer. Fyfe (1973) calculates wavelengths on the order of tens of kilometers for a granitic model system. In the chain of andesitic composite volcanoes extending from the Aleutians through to the Cascades, Marsh (1979) finds their spacing is near 70 km. A rising blob of magma will leave in its wake a trail of thermally perturbed and volumetrically expanded rocks that may favorably prejudice the location of subsequent magma ascent and activity, explaining why many magmatic centers show a prolonged period of activity (several million years).

An important process that could theoretically facilitate ascent of a deep magma body is solution stoping or **zone melting.** If a body of magma with a limited water content has an appreciable vertical extent, it cannot be at more or less uniform temperature and still be in equilibrium with the same crystals at its top and bottom. This follows because of the substantial increase in the temperature of the solidus with depth (compare Figure 9-17). The column of magma would be expected to crystallize at its base while melting an equivalent amount of roof rock, provided the temperature of the wall rocks is not far removed from the solidus, allowing the column to melt its way upward. Zone melting has been advocated by Harris (see 1974) as an effective means of enhancing the concentration of incompatible elements in partial melts of mantle peridotite; longer paths of vertical zone melting will tend to create greater concentrations.

Ascending diapiric plutons sometimes appear to

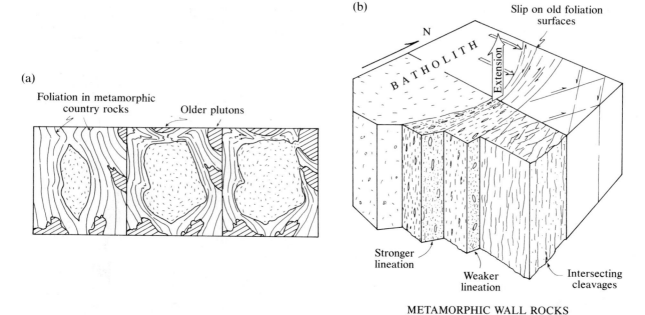

Figure 9-29 The Bald Rock batholith in the northern Sierra Nevada is believed by Compton (1955) to have been emplaced by forcefully shouldering aside its ductile metamorphic wall rocks. (a) Three hypothetical maps showing, at one particular horizon, outward growth of pluton and disposition of foliation in metamorphic wall rocks. The section on the right suggests that late in the emplacement process about one-fourth of the area of the pluton was created by stoping and assimilation of wall rock, with resulting discordance between the pluton and foliated wall rocks. (b) Accommodation of pluton by metamorphic wall rocks. Subhorizontal compression and vertical extension in the metamorphic rocks created stretched pebbles and strong lineation and foliation nearest the contact. [After R. R. Compton, 1955, Trondhjemite batholith near Bidwell Bar, California, *Geological Society of America Bulletin* 66.]

have made room for themselves at a particular horizon in the crust by forcefully shouldering aside country rocks (see Figure 9-29). In the shallower crust, rocks are more rigid and tend to deform in a more brittle manner. Ascent and emplacement of plutons in such environments may be accomplished by engulfment of blocks of country rock into the magma, known as **stoping,** or uplift of the roof rocks, making use, wherever possible, of surfaces of mechanical weakness in the country rocks such as faults and bedding planes (see Figures 9-30 and 9-31). Stoping of wall rocks is facilitated by the thermal effects of the intruding magma. Heated

ground water in cracks expands and pushes free blocks into the magma; steep thermal gradients in wall rocks around near-surface intrusions produce stresses that cause spalling. Stoping and assimilation create no new room for a magmatic intrusion and only allow an exchange in position with country rock.

Highly gas-charged magmas ascend through the crust to form diatremes (see pages 209–211).

Swarms of subparallel vertical dikes of basaltic rock in regions of tectonic rifting (see Figures 5-20 and 15-10) and radial dike systems emanating from central high-level intrusions (see Figure 4-21a) fol-

(a) Vertical
cross sections

Horizontal
plan views

Ground
surface

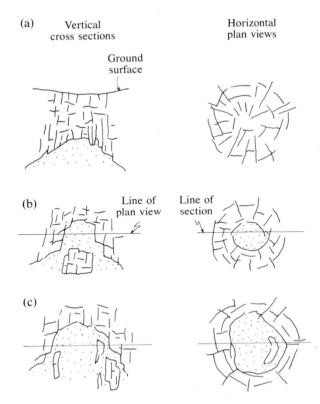

(b)

Line of
plan view

Line of
section

(c)

Figure 9-30 Schematic views of emplacement of
magma at Mt. Monadnock, Vermont, by stoping of
fractured blocks of denser roof rock. [After R. W.
Chapman, 1954, Criteria for the mode of emplacement
of the alkaline stock at Mount Monadnock, Vermont,
Geological Society of America Bulletin 65.]

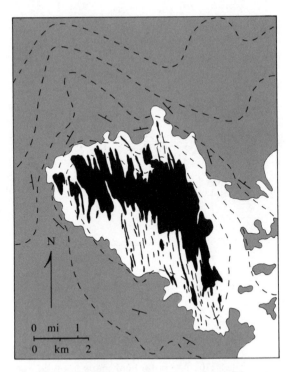

Figure 9-31 Forcible emplacement of granitic magma
into strongly foliated schists in the Black Hills of South
Dakota was like pushing upward on the bottom of an
upright deck of cards. Magma (black) penetrated along
the northerly striking, steeply dipping foliation surfaces,
forming innumerable fingerlike offshoots from the main
body. Doming of the overlying sedimentary strata
(shaded) is indicated by the strike and dip symbols and
by structure contours (dashed lines), which are lines of
equal elevation, spaced 500 feet apart, on the contact
surface between the schists and sedimentary strata.
Many faults are omitted. [After J. A. Noble, 1952,
Evaluation of criteria for the forcible intrusion of
magma, *Journal of Geology* 60.]

low dilatant tensile cracks in the crust (see Figure
9-32). In the first instance, horizontal elastic
stretching of the crust by tectonic processes is followed by rupture along a vertical crack oriented
perpendicular to the direction of the tensional
stress. In the case of radial dikes, the hoop of wall
rocks surrounding the intrusive plug is in a state of
compression (Johnson, 1970) and dilatant cracks
form perpendicular to the direction of least stress.

In real anisotropic crustal rocks, mechanically
significant bedding and persistent fractures are always present, and those with favorable orienta-

tions may become dilatant, so that magma has
easy ingress.

Height of Magma Ascent

The height to which an intrusive body of magma
will rise and whether it ever reaches the surface as
an extrusion depends upon many factors. For the

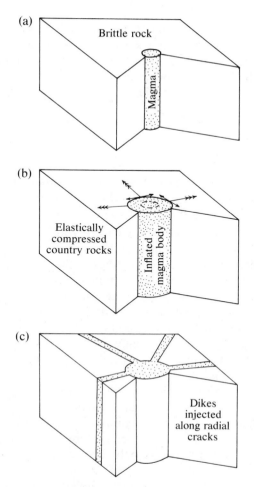

Figure 9-32 Idealized model of formation of radial dikes around an intrusion. Inflation of the magma body (b) compresses the wall rocks so that the maximum principal stress (triple-barbed arrow) is oriented radially to the body and the least principal stress, actually a tension (single-barbed arrow), is oriented tangentially. The stretched wall rocks ultimately break perpendicular to the least stress direction (c), forming radially oriented tensional cracks, into which the magma penetrates. Compare Figure 4-21a.

magma body itself, temperature, composition (especially gas content), viscosity, volume, shape, and depth of source must be involved; for country rocks relevant factors might include volatile content, especially ground water, density, and me-

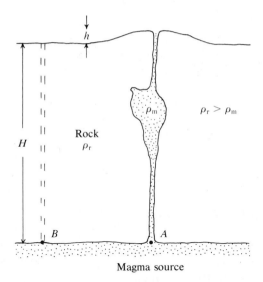

Figure 9-33 Hydostatics of columns of basaltic magma; ρ_m and ρ_r are the average densities of the magma and rock columns, respectively.

chanical properties such as strength and degree of anisotropy furnishing potential avenues for insinuation of magma, as reflected in bedding, foliation, and tensile fractures. As previously discussed (pages 246–248), magmas that heat ground water near the surface or that boil internally and experience bubble expansion can generate sufficient gas pressures to cause explosive volcanic eruption. But not all volcanic eruptions are explosive; basaltic lavas are commonly extruded rather calmly, and other factors must dictate why they are so common as extrusive bodies. The presence of dilatant fissures in tectonic rift settings, where most basalts are extruded, provides easy avenues of egress to the surface of the Earth. Yet, basaltic magmas are the most dense of the common extrusions.

In a simple hydrostatic model, the height to which magma can rise depends upon the depth of the source and the density relations of the magma and the rock overlying the source. For equilibrium, the pressure at the base of the magma column at point A in Figure 9-33a must equal the pressure at the bottom of any nearby vertical rock column of the same cross-sectional area at the same level, such as at point B. Symbolically (see box on page 25),

$$\rho_m g(h + H) = \rho_r g H \qquad (9.3)$$

or upon rearranging terms,

$$h = \frac{H(\rho_r - \rho_m)}{\rho_m} \qquad (9.4)$$

showing that the excess hydraulic head for a volcano, h, or the height to which lava can rise above the ground, is the product of the depth to the level of magma segregation times the ratio of the density contrast to the magma density.

This model is obviously oversimplified because it assumes uniform densities for rock and magma and, more importantly, the existence of a continuous static magma column several tens of kilometers high. Whether such a column can actually exist is doubtful. Nonetheless, some geologists have noted that summits of mature volcanoes in a particular region have similar summit heights irrespective of the topographic elevation of their base (Williams and McBirney, 1979, p. 57). If low density rock lies in the path of the conduit, $(\rho_r - \rho_m)$ and hence h may approach zero, or $(\rho_r - \rho_m)$ may be made negative; in either case, buoyant rise of magma is arrested. For example, during the life of a silicic magma chamber (erupting explosive ash flows), basaltic magma eruptions are found to occur only outside its periphery; but once solidified and fissured, basaltic lavas erupt through it.

The development of sills and laccoliths in horizontally layered rocks is a companion to, or an alternate expression of, volcanic eruption. Diapirs of granitic magma with limited vertical extent likely possess little excess head at their tops. They are easily blocked and solidified, especially if they are rich in water. Another possibility is that the gas pressure becomes sufficient to lift the cap rock, allowing horizontal insinuation of the magma to form a sill. An especially wet layer of sediment might localize sill emplacement by providing the opportunity for some of the thermal energy of the magma to heat the water; the expanding water then facilitates lateral penetration of magma along the bed. Other models of sill emplacement have been proposed (Williams and McBirney, 1979, p. 60).

SUMMARY

Figure 9-34 presents a schematic summary of the processes of magma generation and diversification described in this chapter.

9.1. A magma source region is a volume of essentially solid rock where P, T, or X is sufficiently different from the surroundings so that melting can produce a body of magma potentially able to rise toward the surface. Plate tectonic processes along convergent and divergent boundaries driven by interacting thermal and gravitational energy perturb P, T, and X in potentially viable magma source systems to cause local partial melting and magma production.

Melting consumes relatively large amounts of thermal energy, moderating temperatures of the rock system in its melting range. Geological mechanisms of raising the temperature within a rock body above its solidus include radioactive heating, conversion of mechanical work into thermal energy, and mass transport coupled with differential heating of a body by a hotter one. Because of the positive slope of the solidus curve of dry silicate systems in P–T space, decompression of solid rock can induce melting if its temperature is already close to the solidus. Increase in the water concentration of a system lowers the solidus and can also induce melting.

9.2. Incongruent melting occurs over a range of T, and reaction relations assure that partial melts are enriched in the more soluble, lower-melting-T components of the system, leaving a more refractory crystalline residue.

Partial melting, or anatexis, produces a melt of some proportion less than the whole. The composition of the source rock and its depth, the type of melting—whether fractional or equilibrium—and

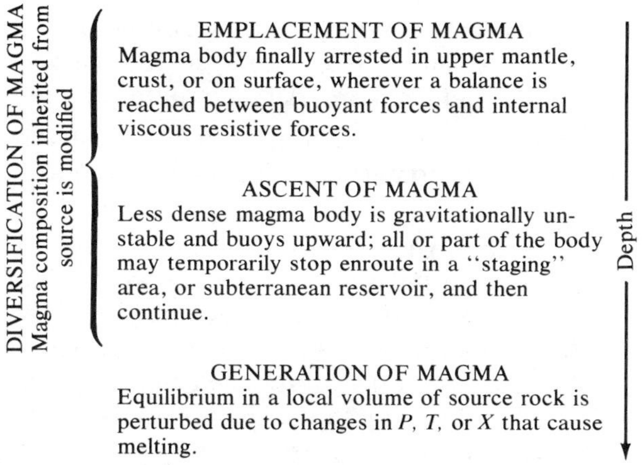

DIVERSIFICATION OF MAGMA
Magma composition inherited from source is modified

Depth

EMPLACEMENT OF MAGMA
Magma body finally arrested in upper mantle, crust, or on surface, wherever a balance is reached between buoyant forces and internal viscous resistive forces.

ASCENT OF MAGMA
Less dense magma body is gravitationally unstable and buoys upward; all or part of the body may temporarily stop enroute in a "staging" area, or subterranean reservoir, and then continue.

GENERATION OF MAGMA
Equilibrium in a local volume of source rock is perturbed due to changes in P, T, or X that cause melting.

Figure 9-34 Schematic summary of magma generation and diversification.

the extent of melting all govern the composition of a partial melt.

Partial melts of mantle peridotite are basaltic. With increasing pressure, the stability field of magnesian olivine shrinks in anhydrous peridotitic systems so that more olivine is dissolved into basaltic partial melts at higher pressure. Low confining pressures and relatively high water concentrations favor generation of the more silica-rich, tholeiitic basalt or even andesite partial melts of peridotite. High pressures and CO_2-rich peridotite favor generation of silica-undersaturated melts. Since a primary melt represents a liquid that was in equilibrium with a specific crystalline residuum at the P–T conditions of partial melting before being segregated from the source, the crystalline phases first appearing at the liquidus of the primary melt should correspond to those minerals remaining in the depleted source at the P and T of generation.

Granitic magmas are generally not water-saturated and ascend through the crust governed by positively sloping solidus curves. Partial melting of water-deficient quartzo-feldspathic crustal source rocks along reasonable geotherms can be expected to produce a spectrum of granitic–dioritic melts. Early melts formed at lowest T accompanying muscovite breakdown are water-rich, granitic, and cannot rise far above their amphibo-litic residua before solidifying. Hotter melts accompanying hornblende breakdown are Na–Ca-rich quartz dioritic liquids that can rise all the way to the surface, leaving anhydrous residua of calcic plagioclase and pyroxene.

Subduction zone magmas are polygenetic. Theoretical and experimental considerations indicate that mafic primary melts can reasonably be expected to form within the oceanic crust capping the descending lithospheric plate and in the wedge-shaped region of the overlying peridotite mantle. Formation of most voluminous Phanerozoic granitic magmas apears to have required a combination of a thick sialic slab plus subduction-related thermal energy sources.

9.3. Effects of magmatic diversification are overprinted onto an already very broad range of magmas rising from source regions. Demonstrable diversification processes include magmatic differentiation, assimilation of country rock, and magma mixing.

Known mechanisms of magmatic differentiation, in which an initially homogeneous parent magma splits into two or more daughter magmas of contrasting chemical composition, include crystal–liquid fractionation, liquid immiscibility, vapor transport, and thermogravitational diffusion.

In partly crystallized magmas, separation of crystals from melt, though never perfect, is a powerful means of magmatic differentiation because of their contrasts in chemical composition. Crystal–liquid fractionation can occur by gravity settling or floating, flowage segregation, and filter pressing.

Certain properties of stratiform mafic–ultramafic intrusions that were once believed to have formed by simple gravity settling associated with body-wide convection actually preclude such an origin. Instead, in cooling bodies of chemically complex silicate melts that have compositionally contrastive liquid and solid phases, stratified double-diffusive convective layers form as a consequence of vertical gradients in temperature and composition.

The essence of immiscibility is that an initially homogeneous melt is no longer stable and splits into two homogeneous melts as magmatic conditions change. This process appears to be petrogenetically significant in the evolution of some alkaline associations, but its importance in other magma systems is uncertain.

A stratified double-diffusive convective mechanism may account for compositionally zoned calc-alkaline magma chambers in which a water-rich rhyolitic cap enriched in Be, Rb, Nb, Mo, Sn, Cs, Ta, W, Th, and U overlies a crystal-rich intermediate composition core.

In long-lived magmatic centers, two or more unlike magmas may blend together, forming a hybrid daughter product compositionally intermediate between them. This mechanism has been advocated for calc-alkaline associations along continental margin subduction zones and for oceanic rifts, where evidence indicates it operates in conjunction with crystal–liquid fractionation.

The amount of assimilation of cool wall rock into a magma is limited by the thermal energy of the magma itself.

9.4. The fundamental driving force causing a crystal–liquid diapir or a body of crystal-free primary melt to rise upward from its source region is the buoyancy of the less dense magma. Rise of magma in shallow regions of the brittle crust is permitted by stoping, assimilation, diatreme formation, and uplift of roof rocks, in contrast to viscous rise of magma diapirs perhaps aided by zone melting in the lower crust and especially the upper mantle. Rise and emplacement of magma in brittle rocks is facilitated by dilatant tensile cracks.

In a simple hydrostatic model, the height to which magma can rise depends upon the depth of the source and the density relations of the magma and the rock overlying the source.

STUDY QUESTIONS

1. Melt crystalline aggregates (a) 50Di:20Fo:30Q; (b) 30Di:45Fo:25Q; and (c) 40Di:50Fo:10Q by the fractional and equilibrium processes, describing crystalline residues and liquid paths.

2. Discuss the effects of allowing liquids M_3 and $M_{1.5}$ in Figure 9-7 to ascend to the surface, cooling and decompressing en route.

3. Under what conditions might relatively aluminous primary basaltic melts be generated?

4. Suppose a batch of basaltic melt is produced in a source region at a depth of 100 km and rises to about 50 km, where it equilibrates with the mantle peridotite and then segregates at that level from the crystalline assemblage before rising rapidly to the surface. What determines the major-element composition of the extruded magma? Discuss.

5. Explain how chemical variation diagrams may help in ascertaining which rock type in a rock suite represents the parent magma.

6. Examine quantitatively the result of fractionating (a) 25% amphibole, (b) 15% olivine, and (c) 40% plagioclase of appropriate compositions (refer to Appendix C) from a particular basaltic magma. Express your answers as chemical analyses of the three daughter magmas.

7. Discuss factors that might control recurrent volcanic eruption. Contrast these factors with controls on recurrent granitic plutonism.

8. Specify the boundary conditions under which large volumes of homogeneous magma can be produced by fusion of upper mantle.

9. Sometimes superheated magmas whose temperatures lie well above the liquidus are postulated to assimilate large volumes of wall rock. What textural evidence would indicate superheated magma? How could superheated magmas realistically form?

10. How fast would a basaltic magma have to rise to the surface in order to carry 10-cm chunks of mantle peridotite? Indicate assumptions and show calculations.

11. Discuss the possibility that the King Hill basalt is a primary melt of spinel lherzolite at 1.6 GPa or of garnet lherzolite at 3 GPa (see

Figure 8-9 and Table 8-1). How might it have originated?

12. Using Figure 9-10 , propose a mechanism that would allow normal granitic magmas to evolve from an albite-rich invariant melt at 3 GPa.

13. Consider a model magma system whose composition is An_{50} at 1350°C (see Figure 8-20). Discuss the behavior of the system as it assimilates an equal mass of crystals (a) An_{64} at 1350°C; (b) An_{64} at 200°C; (c) An_{80} at 1350°C; and (d) An_{20} at 1100°C.

REFERENCES

Anderson, A. T. 1976. Magma mixing: Petrological process and volcanological tool. *Journal of Volcanology and Geothermal Research* 1:3–33.

Anderson, O. L., and Perkins, P. C. 1975. A plate tectonics model involving non-laminar asthenospheric flow to account for irregular patterns of magmatism in the southwestern United States. *Physics and Chemistry of the Earth* 9:113–122.

Bowen, N. L. 1928. *The Evolution of the Igneous Rocks.* New York: Dover, 332 p.

Brown, G. C. 1973. Evolution of granite magmas at destructive plate margins. *Nature,* Physical Science, 241:26–28.

Carmichael, I. S. E., Turner, F. J., and Verhoogen, J. 1974. *Igneous Petrology.* New York: McGraw-Hill, 739 p.

Chapman, R. W. 1954. Criteria for the mode of emplacement of the alkaline stock at Mount Monadnock, Vermont. *Geological Society of America Bulletin* 65:97–114.

Compton, R. R. 1955. Trondhjemite batholith near Bidwell Bar, California. *Geological Society of America Bulletin* 66:9–44.

Cox, K. G., Bell, J. D., and Pankhurst, R. J. 1979. *The Interpretation of Igneous Rocks.* London: George Allen and Unwin, 450 p.

Dickinson, W. R. 1975. Potash-depth (K-h) relations in continental margins and intra-oceanic magmatic arcs. *Geology* 3:53–56.

Eggler, D. H., and Holloway, J. R. 1977. Partial melting of peridotite in the presence of H_2O and CO_2: Principles and review. *Oregon Department of Geology and Mineral Industries Bulletin* 96:15–36.

Eichelberger, J. C., and Cooley, R. 1977. Evolution of silicic magma chambers and their relationship to basaltic volcanism. *American Geophysical Union Monograph* 20:57–77.

Elder, J. 1976. *The Bowels of the Earth.* London: Oxford, 222 p.

Frey, F. A. 1979. Trace element geochemistry: Applications to the igneous petrogenesis of terrestrial rocks. *Reviews of Geophysics and Space Physics* 17:803–823.

Fyfe, W. S. 1973. The generation of batholiths. *Tectonophysics* 17:273–283.

Green, D. H. 1971. Composition of basaltic magmas as indicators of conditions of origin: Application to oceanic volcanism. *Philosophical Transactions of the Royal Society of London,* series A, 268:707–725.

Harris, P. G. 1974. Volcanic liquids, their origin and nature. *Science Progress,* Oxford, 61:515–533.

Hawkesworth, C. J., Norry, M. J., Roddick, J. C., Baker, P. E., Francis, P. W., and Thorpe, R. S. 1979. $^{143}Nd/^{144}Nd$, $^{87}Sr/^{86}Sr$, and incompatible element variations in calc-alkaline andesites and plateau lavas from South America. *Earth and Planetary Science Letters* 42:45–57.

Hildreth, W. 1979. The Bishop tuff: Evidence for the origin of compositional zonation in silicic magma chambers. Geological Society of America, Special Paper 180:43–75.

Irvine, T. N. 1979. Rocks whose composition is determined by crystal accumulation and sorting. In H. S. Yoder, Jr. (ed.), *The Evolution of the Igneous Rocks: Fiftieth Anniversary Perspectives.* Princeton: Princeton University Press, p. 244–306.

Irvine, T. N. 1980. Magmatic density currents and cumulus processes. *American Journal of Science* 280-A:1–58.

Johnson, A. M. 1970. *Physical Processes in Geology.* San Francisco: Freeman, Cooper and Co., 577 p.

Komar, P. D. 1976. Phenocryst interactions and the velocity profile of magma flowing through dikes or sills. *Geological Society of America Bulletin* 87:1336–1342.

Kushiro, I. 1967. Compositions of magma formed by partial zone melting of the earth's upper mantle. *Journal of Geophysical Research* 73:619–634.

Kushiro, I. 1969. The system forsterite-diopside-silica with and without water at high pressures. *American Journal of Science* 267-A:269–294.

McBirney, A. R. 1979. Effects of assimilation. In H. S. Yoder, Jr. (ed.), *The Evolution of the Igneous Rocks: Fiftieth Anniversary Perspectives,* Princeton: Princeton University Press, pp. 307–338.

McBirney, A. R., and Noyes, R. M. 1979. Crystallization and layering of the Skaergaard intrusion. *Journal of Petrology* 20:487–554.

Marsh, B. D. 1979. Island arc volcanism. *American Scientist* 67:161–172.

Murata, K. J., and Richter, D. H. 1966. The settling of olivine in Kilauean magma as shown by lavas of the 1959 eruption. *American Journal of Science* 264:194–203.

Noble, J. A. 1952. Evaluation of criteria for the forcible intrusion of magma. *Journal of Geology* 60:34–57.

O'Hara, M. H. 1968. The bearing of phase equilibria studies in synthetic and natural systems on the origin and evolution of basic and ultrabasic rocks. *Earth Science Reviews* 4:69–133.

Oxburgh, E. R., and Turcotte, D. G. 1976. The physico-chemical behavior of the descending lithosphere. *Tectonophysics* 32:107–128.

Presnall, D. C. 1969. The geometrical analysis of partial fusion. *American Journal of Science* 267:1178–1194.

Presnall, D. C., and Bateman, P. C. 1973. Fusion relations in the system $NaAlSi_3O_8$-$CaAl_2Si_2O_8$-$KAlSi_3O_8$-SiO_2-H_2O and generation of granitic magmas in the Sierra Nevada batholith. *Geological Society of American Bulletin* 84:3181–3202.

Presnall, D. C., Dixon, J. R., O'Donnell, T. H., and Dixon, S. A. 1979. Generation of mid-ocean ridge tholeiites. *Journal of Petrology* 20:3–35.

Ramberg, H. 1970. Model studies in relation to intrusion of plutonic bodies. In G. Newall and N. Rast (eds.), *Mechanism of Igneous Intrusion.* Liverpool: Gallery Press, pp. 261–286.

Rhodes, J. M., Dungan, M. A., Blanchard, D. P., and Long, P. E. 1979. Magma mixing at mid-ocean ridges: Evidence from basalts drilled near 22°N on the Mid-Atlantic Ridge. *Tectonophysics* 55:35–61.

Ringwood, A. E. 1977. Petrogenesis in island arc systems. In M. Talwani and W. C. Pitmann III (eds.), *Island Arcs, Deep-Sea Trenches, and Back-Arc Basins.* Washington, D.C.: American Geophysical Union, pp. 311–324.

Roedder, E. 1979. Silicate liquid immiscibility in magmas. In H. S. Yoder, Jr. (ed.), *The Evolution of the Igneous Rocks: Fiftieth Anniversary Perspectives.* Princeton: Princeton University Press, pp. 483–520.

Shaw, H. R. 1970. Earth tides, global heat flow, and tectonics. *Science* 168:1084–1087.

Shaw, H. R. 1973. Mantle convection and volcanic periodicity in the Pacific: Evidence from Hawaii. *Geological Society of America Bulletin* 84:1505–1526.

Taylor, H. P., Jr. 1978. Oxygen and hydrogen isotope studies of plutonic granitic rocks. *Earth and Planetary Science Letters* 38:177–210.

Williams, H., and McBirney, A. R. 1979. *Volcanology.* San Francisco: Freeman, Cooper and Co., 397 p.

Wyllie, P. J. 1979. Petrogenesis and the physics of the Earth. In H. S. Yoder, Jr. (ed.), *The Evolution of the Igneous Rocks: Fiftieth Anniversary Perspectives,* Princeton: Princeton University Press, pp. 483–520.

Wyllie, P. J., Huang, W., Stern, C. R., and Maaløe, S. 1976. Granitic magmas: Possible and impossible sources, water contents, and crystallization sequences. *Canadian Journal of Earth Sciences* 13:1007–1019.

Yoder, H. S., Jr. 1973. Contemporaneous basaltic and rhyolitic magmas. *American Mineralogist* 58:153–171.

Yoder, H. S., Jr. 1976. *Generation of Basaltic Magma.* Washington, D.C.: National Academy of Sciences, 165 p.

Yoder, H. S., Jr. 1979. *The Evolution of the Igneous Rocks: Fiftieth Anniversary Perspectives.* Princeton: Princeton University Press, 588 p.

Yoder, H. S., Jr., and Tilley, C. E. 1962. Origin of basalt magmas: An experimental study of natural and synthetic rock systems. *Journal of Petrology* 3:342–532.

Part III

METAMORPHIC BODIES AND SYSTEMS

In contrast to sedimentary and magmatic systems, in which at least some generative processes are directly observable, metamorphism is completely concealed beneath the surface of the Earth. Although metamorphic and igneous petrology share many common concepts and approaches to the study of rocks, there are important contrasts in the nature of magmatic and metamorphic systems. Magmatic systems are dominated by a viscous silicate melt, but crystalline and often gaseous phases are also present; interactions between these three states of matter under highly variable pressures and temperatures produce a wide range of compositions and fabrics. Displacement of magma bodies from source regions involves large-scale flows of energy.

In contrast, metamorphic systems consist of, at most, crystalline phases and a volatile fluid. Although equilibrium states are commonly approached, metamorphic reactions are very slow, hence difficult to study experimentally. Thus, metamorphic petrology relies heavily on theoretical thermodynamic considerations. Solid metamorphic rocks suffer tectonic deformation, which strongly influences their fabric.

10

The Nature of Metamorphism

Metamorphism means change. Significant change in any of the intensive variables P, T, and X characterizing the state of a rock system disturbs the existing equilibrium; restoration to a new state of more stable equilibrium—the process of metamorphism—involves reconstitutive changes in fabric and composition of the rock body. Reconstitutive metamorphic changes occur virtually in the solid state, in the presence of usually minor amounts of aqueous or carbonic fluids, and in the $P–T$ realm between magmatic and sedimentary systems.

Solid rock bodies respond to changes of state in an almost endless variety of ways, depending on the composition and fabric of the initial body, the metamorphic processes, or the path of flow of energy and matter, and the P, T, and X of the final state of the system. The nature of the original body may partly survive the metamorphic imprint and be manifest in relict minerals and fabrics and in the bulk chemical composition. The metamorphic path is reflected in the newly imposed fabric and field relations of the body. The new mineralogical composition of the body will be dictated solely by the P, T, and X of that final state if stable equilibrium was attained.

Metamorphic systems have gradients in P, T, and sometimes X that produce mineralogical zonation in the resulting body. Prograde metamorphism usually involves a series of dehydration and decarbonation reactions, liberating H_2O and CO_2 from the rock.

As metamorphism is a response to flow of energy and matter, it, like magmatism, is concentrated in but not restricted to areas near plate boundaries. Highly deformed mylonitic rocks occur along transform faults. Sub-sea-floor metamorphism occurs near oceanic rifts. Subduction zone orogenic belts along continental margins harbor (1) deep, widespread regional metamorphism where rocks are concurrently deformed and recrystallized, (2) contact metamorphism around magmatic intrusions, and (3) mylonitization along deep shear zones.

The early history of the Earth is revealed in the superimposed fabrics and compositional features of ancient metamorphic bodies.

10.1 INTRODUCTION

In a dynamic planet such as the Earth, changes in geologic systems due to flow of energy and matter are continually occurring, chiefly along plate margins. Metamorphism encompasses all solid-state changes in the fabric and composition of a rock body occurring well beneath the surface of the Earth, but without the intervention of a silicate melt. Although metamorphic systems are solid and lack a silicate melt, aqueous and carbonic fluids are almost universally present and play a very important role in the reconstitutive processes.

The P–T realm of metamorphism is thus bracketed between diagenetic sedimentary and weathering systems and the solidus T for the particular rock body at hand, which of course, can vary greatly with P and P_{H_2O} (see Figure 1-15). However, there are no clear-cut distinctions agreed upon between low-grade metamorphism and diagenesis with regard to processes, P–T conditions, fabric, or mineralogical composition. Also, the effects of magmatism, involving a silicate melt, and solid-state metamorphism may be impossible to distinguish in some high-T rock bodies where field relations, fabrics, and mineralogical compositions merge and overlap.

Bodies of sedimentary and volcanic rock that are heated, buried, and deformed in orogenic subduction zones tend toward new equilibrium states by modifying their fabric and composition. Later erosion exposes these rock sequences, which show a gradual transition from unquestioned sedimentary or volcanic character into fabrics and compositions that bear a different imprint—that of metamorphism. The transition is especially obvious in aluminous shales whose original fine clay minerals are replaced by somewhat coarser micas and chlorites, which are more stable under the new metamorphic conditions. The first appearance of micas in aluminous rocks is sometimes used to denote the onset of metamorphism. Original bedding features, angular outlines of detrital quartz grains, and perhaps some fossils are still preserved. Weakly metamorphosed shale, limestone, and sandstone are recognizable as slate, marble, and quartzite. Farther into the metamorphic sequence and in response to different P and T, slates give way to coarser-grained schists composed of an assemblage of muscovite + biotite + quartz and perhaps kyanite and garnet. Fewer vestiges of the original sedimentary fabric are conspicuous and layers of contrasting modal composition formed by metamorphic processes may be intricately folded and contorted. At still deeper crustal levels, micas may be absent in aluminous rocks, their place being taken by a still higher-T, anhydrous assemblage of K-feldspar + sillimanite + pyroxene. The rock is here a coarse-grained gneiss whose overall fabric is wholly metamorphic; only its bulk chemical composition might be suggestive of the original shale from which it was derived.

In a magmatic rock such as gabbro, the assemblage of olivine + pyroxene + calcic plagioclase is unstable at low temperatures in the presence of water; the original olivines are replaced by a fine-grained aggregate of serpentine, primary pyroxenes are replaced by chlorite + actinolite + sphene, and plagioclase by hydrated Ca–Al silicates such as epidote, together with sericite + albite. In a granitic rock within a deep-seated fault zone, coarse grains of quartz and feldspar may be deformed into smaller grains, absorbing some of the mechanical energy of displaced rock masses. Limestone equilibrates to the changed compositional constraints and temperatures imposed by percolating hydrothermal solutions emanating from a nearby cooling magmatic intrusion, recrystallizing to aggregates such as grossular + epidote + idocrase.

All metamorphic processes imprint new fabrics over old, commonly obliterating them completely. New fabrics have new grain sizes and commonly a parallel arrangement, or preferred orientation, of inequant grains. New mineralogical compositions develop. All such changes in fabric and composition of the rock body are an adjustment of the system to a new, more stable equilibrium state.

In metamorphic petrology we are concerned not only with the final state and its new fabric and composition, but also with the processes of transition to that final state. But because of limitations of exposure, or structural complications such as faulting, such transitions are not always exposed in a particular metamorphic terrane, as in the example given above of progressive metamorphism of shale. The entire observable metamorphic body may have been subjected to metamorphism so intense and thorough as to eliminate all vestiges of the original state. But even without directly seeing the transition in states, we can usually be sure that solid-state reconstitution was involved because of distinctive compositional properties and fabrics. For example, the history of the Earth's mantle is not fully understood; yet the mineralogical composition and fabrics perceived in small, mantle-derived xenoliths in basaltic and kimberlitic rocks testify to a strong metamorphic imprint.

As a further illustration of the processes of metamorphic transition, consider a rock comprised of albite + actinolite + chlorite + epidote + sphene. This mineralogical state could be reached by deuteric alteration of a gabbro pluton, metamorphism of a buried basalt flow heated by a granitic intrusion, recrystallization of a basaltic tuff in a subduction zone, recrystallization of a diabase dike in a hydrothermal system above a cooling magma body (Muffler and White, 1969), and so on. The same final state of mineralogical equilibrium can be reached via different paths of energy flow (see page 9). The paths from the different initial states may be manifest in the field relations and possible remaining vestiges of the initial fabric.

In contrast to magmatic systems, which tend to give up internal energy in forming rock bodies, metamorphism is mostly, though not always, an endothermic process, absorbing heat and also mechanical energy. Differential movement of rock bodies transfers kinetic and gravitational potential energy into the metamorphic rock in the form of increased internal energy and mechanical work of reforming grains. Heat is absorbed around cooling magmatic intrusions or in the deep crustal roots of orogenic belts, promoting growth of new and larger grains and driving endothermic chemical reactions.

The response of a solid rock body to changing conditions is almost infinitely variable, but the basic pattern depends upon the following factors:

1. The nature of the original rock, as manifest in any relict fabrics or minerals that might be preserved and in its bulk chemical composition if there has been no substantial change in this during metamorphism.
2. The nature of the metamorphic path.
   a. The change in P, T, or X, or the metamorphic process, as reflected chiefly in the imposed metamorphic fabric of the body.
   b. The reason for the flow of energy and matter in the change of state, as reflected in field relations of the body.
3. The final equilibrium state governed by P, T, and X, as seen in the mineralogical composition of the metamorphic body.

The next three sections consider these three factors at greater length.

10.2 METAMORPHIC RECORD OF THE ORIGINAL ROCK

Metamorphic bodies were originally sedimentary or magmatic rock, or perhaps earlier metamorphic rock. Vestiges of the original fabric and composition of this **protolith** are generally preserved to some degree.

Inherited Fabric

As fabric connotes the flow of energy and matter, or the path of metamorphic processes, it is important to distinguish vestigal, **inherited**, or **relict fabric** elements from the newly imposed metamorphic fabric features.

Inherited fabrics are best preserved in weakly metamorphosed bodies, where the metamorphism was relatively brief, of low grade, and with little or no deformation. (Grade refers basically to the relative temperature of metamorphism; low-grade metamorphism occurs at low *T.*) The larger and sharper the discontinuities defining the original igneous or sedimentary fabrics, the more likely will be their preservation through the one or more episodes of metamorphism experienced by the rock body. Even though none of the original minerals may have survived, their original outlines, or outlines of characteristic mineral aggregates, may be discernible. In low-grade metamorphosed igneous rocks, original grains tend to be replaced more or less pseudomorphically by finer-grained aggregates that do not disturb original grain boundaries. For example, ophitic grain relations may be clearly preserved in a metamorphosed diabase where laboradorite has been replaced by aggregates of epidote + sericite + albite and pyroxene by sheaves of actinolite with sphene + chlorite (see Figure 10-18a). Relict porphyritic fabric in a metamorphosed andesite is obvious in Figure 1-9b. Relict graphic texture can be observed in Figure 10-4. In metamorphosed sandstones, relict epiclastic fabric may be quite conspicuous even where some deformation of the rock body has occurred (see Figure 10-20b). Relict bedding is commonly inherited in low-grade metamorphosed sediments (see Figure 10-1). Because layers of contrasting mineralogical composition and fabric can apparently develop during metamorphism, caution must be exercised in interpreting layering as relict bedding. Preserved epiclastic grain outlines as well as graded and cross bedding support recognition of relict bedding. Generally, as the grade of metamorphism increases, fewer and less conspicuous relict fabrics can be discerned because the growth of new metamorphic mineral grains obliterates original grain outlines. In such cases, however, larger-scale contrastive fabric features may survive, such as cobbles in a metamorphosed conglomerate (see Figure 10-2) or pillows in a metamorphosed submarine lava (see Figure 10-3). Severe pervasive deformation can erase even the largest-scale features of the original body, such as its overall shape. Deformation of a basaltic dike,

for example, could make its original tabular form obscure (see Figure 11-8).

Chemical Composition

The bulk chemical composition of a metamorphic body sets definite constraints on the nature of the protolith, barring substantial changes in X during metamorphism. Most metamorphic rocks have chemical compositions close to their protolith, except for possible losses or gains of volatile constituents such as H_2O, CO_2, O_2, and S. Such **isochemical metamorphism** occurs where nonvolatile constituents have only moved on the scale of individual grains, no doubt reflecting slow rates of diffusion or low concentration gradients. Though often difficult to prove, these conditions are implicit in any attempt to correlate the chemical composition of the metamorphic rock with the original. However, in the type of metamorphism called **metasomatism**, transport of select chemical constituents occurs over substantially larger distances, commonly on the order of meters or more, causing significant change in X.

The chemical compositions of metamorphic rock bodies span a very broad spectrum, encompassing not only magmatic and sedimentary compositions but also metasomatic modifications.

Pelitic metamorphic rocks derived from clay-rich shales are highly aluminous and contain abundant micas, chlorite, and other aluminous minerals. **Calc-silicate** rocks represent metamorphosed shaly or quartz-bearing carbonate rocks as well as metasomatized carbonate rocks. Mafic metamorphic rocks are derived from igneous and uncommon sedimentary rocks containing appreciable Si, Mg, Fe, and Ca.

10.3 THE RECORD OF METAMORPHIC PROCESSES

Metamorphic processes accompany changes in state of the system. *P, T,* and *X* change as energy and matter flow within the system and between the system and its surroundings. In this section we

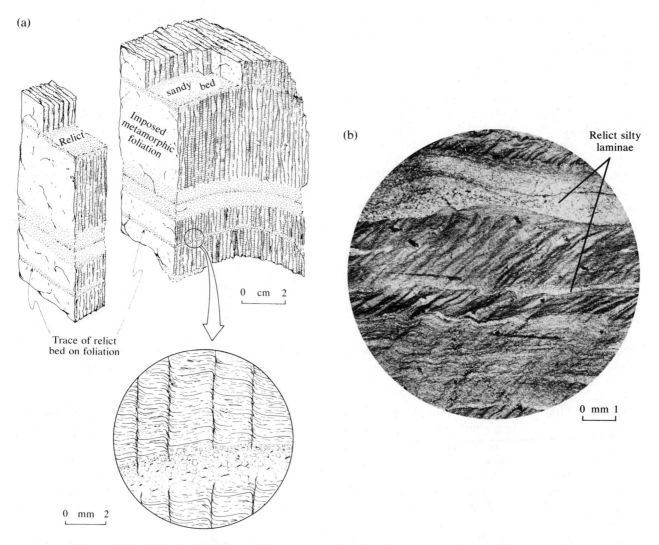

Figure 10-1   Relict bedding in weakly metamorphosed sedimentary rock. (a) Relict beds of alternating metamorphosed sandstone and shale cut by an imposed metamorphic foliation. [After J. Gilluly, A. C. Waters, and A. O. Woodford, 1975, *Principles of Geology,* 4th Ed. (San Francisco: W. H. Freeman and Company.] (b) Photomicrograph of metamorphosed shale with thin laminae of metamorphosed siltstone showing grading of grain size. Orientation of flakes of micas and chlorites, formed during metamorphism, has mimicked the original orientation of clay particles parallel to bedding in the shale. Subparallel limbs of microfolds define a weaker second metamorphic foliation oblique to relict bedding. Note absence of microfolds in quartz-rich silt layers.

examine these flow paths and types of changes in *P, T,* and *X.* Field relations are especially important in suggesting why the state of the system changed. Several illustrations of metamorphism are then given, dealing mostly with the character of imposed metamorphic fabrics.

(a)

(b)

Figure 10-2 (a) Bed of metaconglomerate lying between strongly foliated and lineated quartz schists (originally sandstones). (b) Another view of the same metaconglomerate looking down onto the bed, showing intensely stretched cobbles. Note pocketknife for scale.

(a)

(b)

Figure 10-3   Metamorphosed pillow basalts of Archean age from the Yellow Knife area of the Northwest Territories, Canada. (a) Undeformed pillows. (b) Highly deformed lighter-colored pillows have been stretched and folded. [Geological Survey of Canada photographs courtesy of Maurice Lambert and W. R. A. Baragar.]

Types of Metamorphic Processes

Thermal metamorphism is related to change in T and produces a new mineralogical composition and, because of the accompanying grain growth, a new fabric as well.

Metasomatism is caused by a significant change in chemical composition of the system. Metasomatism involves major gains or losses of matter, usually made possible by flow of large quantities of fluid through an "open" rock body along significant temperature or composition gradients.

Flow of fluid laden with dissolved constituents must be driven by some form of energy; a common energy source is a hot magmatic intrusion emplaced into cooler country rocks, causing convective flow of fluid through fractures. More widespread metasomatic systems exist just below the ocean floor, where sea water penetratively convects through highly fractured hot basaltic rocks freshly emplaced along the oceanic rift. Conse-

quently, changes in T commonly accompany changes in X.

The metasomatism known as fenitization is widespread around carbonatite intrusions (see pages 204–206). In Figure 10-4, quartz in graphic granite wall rock next to carbonatite has been replaced by fine-grained sodic amphibole, faithfully preserving the characteristic outlines of the original quartz. Alkali feldspar was stable but quartz unstable in the presence of the fenitizing fluid emanating from the carbonatite. This **constant volume replacement,** which preserves the form and the size of the original grains, is not always so easy to demonstrate in metasomatic processes.

Changes in chemical composition occurring on the restricted scale of grains and involving only a redistribution of constituents already a part of the rock body is not referred to as metasomatism. Segregation of Mg, Fe, Ca, Al, Si, and O from the homogeneous matrix of a shale to form a large garnet grain is not considered to be metasomatism,

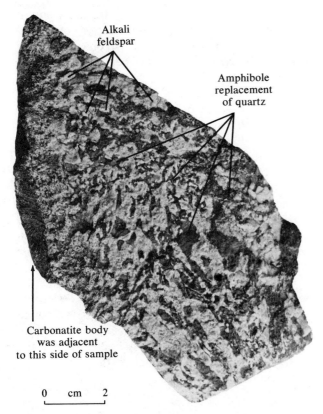

Alkali feldspar

Amphibole replacement of quartz

Carbonatite body was adjacent to this side of sample

0   cm   2

Figure 10-4 Metasomatism in graphic granite wall rock adjacent to magmatic carbonatite. Quartz in the graphic granite has been replaced by fine-grained aggregates of sodic amphibole. Compare Figure 4-5.

but rather is small-scale **metamorphic differentiation.**

Metamorphism related to changes in pressure has different facets. Confining pressure P acts more or less equally in magnitude in all directions, as if the rock body were a fluid and hydrostatic conditions prevailed. Change in P may stabilize new minerals, and their growth can modify fabric. But changes in P are only likely to occur at different depths beneath the surface where temperatures are also different. P and T generally increase sympathetically, but at different rates, with depth (see Figures 1-15 and 10-22).

During orogeny, forces that differ in magnitude in different directions cause folding and thrusting of upper crustal layers. At deeper levels and on a smaller, more pervasive scale of individual grains, associated **nonhydrostatic stresses** deform rock bodies by cohesive, solid-state flow below the melting T of the body. This solid-state flow produces **anisotropic fabrics** such as slaty cleavage and the foliation and lineation commonly seen in schists and other metamorphic rocks. Original fabrics are strongly modified or obliterated.

In some instances, this solid-state flow is not accompanied by any significant thermal or metasomatic effects that cause changes in mineralogical or chemical composition of the rock body. Such metamorphism is sometimes referred to as dynamic metamorphism, although mechanical or deformational metamorphism might be a better term, as all metamorphic changes, whether due to variations in stress, temperature, or composition, are dynamic. More commonly, nonhydrostatic stresses act in concert with thermal activity, simultaneously deforming and recrystallizing the rock. This widespread type of metamorphism has been called dynamo-thermal, but **syntectonic metamorphism** is more appropriate.

Metamorphic Field Relations

Field relations of metamorphic bodies provide insight into the cause of flow of energy and matter responsible for the metamorphism.

Contact metamorphism occurs in the wall rocks surrounding a magmatic intrusion. The principal changes are in T and locally X related to flow of thermal energy and matter. Significant thermal gradients around hot intrusions are limited to epizonal and some mesozonal levels of the crust; at greater depths, the rocks are already too hot to show much thermal imprint from an intrusion. In limestones and dolomites around granitic intrusions, outwardly expelled silica-rich hydrothermal solutions cause gradients in X and T, metasomatizing the initially relatively pure calcite or dolomite rock to calc-silicate assemblages of garnet, epidote, pyroxene, idocrase, and so on. In other country rocks, metamorphism may be essentially isochemical and thermal.

Regional metamorphism is the most widespread of any type of metamorphism, occurring over broad expanses of deeper levels of the crust in orogenic belts where lithospheric plates are converging. More or less simultaneous changes in *P*, *T*, and the stress environment extend over tens of millions of years; episodic, multiple thermal and deformational events are typical in long-lived subduction zones. Products of such regionally extensive, syntectonic metamorphism are the familiar schists and gneisses possessing anisotropic fabrics. In regional metamorphism, the rates of change in *P* and *T* tend to be slower than rates of mineralogical reactions that involve nucleation and growth by diffusion. Thus reconstitutive changes in these bodies are believed to progress more or less continuously as *P* and *T* change. In contact metamorphism, on the other hand, change in *T* and especially *X* conceivably might be more rapid than reaction rates so that the final metamorphic product may have jumped, rather than progressed continuously, to its final state.

Sub-sea-floor metamorphism occurs near oceanic rifts where hot, newly-formed magmatic crust interacts with cold sea water by metasomatic and thermal metamorphic processes that involve flow of thermal energy and matter. Due to relentless spreading of the sea floor, the products of such processes are widespread, but they are usually hidden by the ocean. Some petrologists prefer to call such processes hydrothermal alteration rather than metamorphism because the reconstitution is localized along fractures where heated sea water carrying dissolved ions percolated through the rock body. Whatever it may be called, the process and its products play an integral part in the plate tectonic cycle.

Burial metamorphism occurs in thick sequences of volcanic and sedimentary rock due to changes in *P* and *T* unrelated to orogeny or magmatism. Penetrative deformation is absent. Thermal energy flows into the buried body. Equilibrium mineral assemblages and fabrics are rarely attained, so relict features of the original rock are apparent. It is basically a deepened continuation of low-*T* diagenesis, and many geologists so consider it, but it grades imperceptibly into regional metamorphism.

Shear metamorphism occurs in deep-seated shear zones where nonhydrostatic stresses cause localized cohesive solid-state flow. Mechanical energy of displaced rock masses is transformed into strain and excess surface energy in the finer-grained, deformed rock. Mylonite is a typical rock formed in this way.

Shock or **impact metamorphism** occurs in bodies of rock subjected to meteorite impact (French and Short, 1968). The flow of energy is complex. Transient stresses may locally produce a fabric like that in cohesive flow. New mineral assemblages may be stabilized, such as coesite and stishovite, the dense high-pressure polymorphs of silica (see Figure 8-11). Temperatures may be high enough to cause melting. Terrestrial impact rocks are rare, but on highly cratered planetary bodies such as the Moon the products of impact metamorphism are widespread.

Imposed Fabrics Resulting from Metamorphism

Original inherited fabrics are modified or obliterated by pervasive destructional and constructional metamorphic processes, forming the **imposed metamorphic fabric**. Deformation changes the shape and reduces the size of original grains and modifies boundaries of grain aggregates in response to nonhydrostatic stress. Part of the mechanical energy of the deformation process is stored in the rock body in the form of elastic strain and surface energy of smaller grains that have larger areas relative to their volume. Solid-state crystallization creates new grains or enlarges existing grains. This constructional process tends to consume the stored energy of any prior deformation, bringing the system to a more stable state of lower energy.

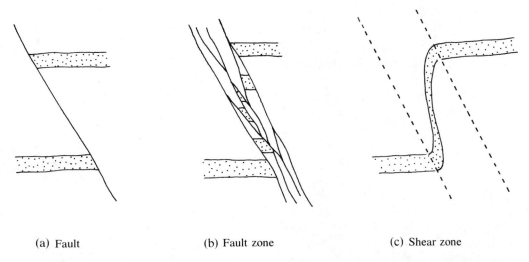

(a) Fault   (b) Fault zone   (c) Shear zone

Figure 10-5 Contrasting expression of a fault, fault zone, and shear zone. Faulting along a discrete, localized fracture surface is a brittle phenomenon. Pervasively distributed displacement in a shear zone is a type of cohesive ductile flow. In all three situations, there is displacement, or dislocation, of rock masses. [After B. E. Hobbs, W. D. Means, and P. F. Williams, 1976, *An Outline of Structural Geology* (New York: John Wiley and Sons).]

Deformation fabrics.   Along faults or fault zones near the surface of the Earth, brittle rocks are broken, crushed, and pulverized to form a loose fault breccia or gouge (see Figure 10-5); subsequent percolation of mineral-laden ground water along the fault may cement the loose particles together. The clasts in such randomized, **cataclastic** material are commonly polygranular, and individual fractures cut several grains. Folds in layered rocks near the surface have wavelengths and amplitudes measured in at least tens of meters, commonly kilometers.

Metamorphic bodies in deeper regions of the crust respond to nonhydrostatic stress chiefly by solution and redeposition and by cohesive deformation of individual grains or small aggregates of grains rather than by crushing and breaking into large-scale faults or by warping of layers into large folds. This pervasive grain-scale deformation commonly produces structures that appear to have formed in a partially molten, and therefore magmatic, state (see Figure 10-6a), but mineral as-semblages in such deformed rocks indicate that subsolidus temperatures prevailed. Over large volumes of rock, these small-scale effects can be quite considerable when integrated together. Deformation can also be remarkably uniform as to orientation and style on all scales of observation because of its penetrative nature (see Figure 10-6b).

Experimental deformation of rocks and minerals under metamorphic conditions in the laboratory confirm the phenomenon of pervasive, grain-scale deformation inferred from field observations. Though superficially resembling cataclastic fabric, the effects of this cohesive ductile flow are different (see Figure 10-7). **Ductile flow** involves bending of grains, crystallographically controlled plastic slip within grains, and thermally activated recrystallization processes that change grain shape and reduce grain size. Ductile flow features do not originate by breaking, crushing, and pulverizing grains as in cataclasis of brittle rocks near the surface of the Earth. Progressive grain destruction by

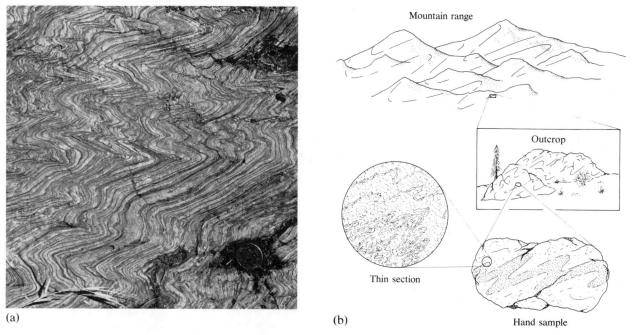

(a)   (b)   Hand sample

Figure 10-6   Folds in metamorphic bodies. (a) Folds produced by solid flow in metamorphosed thin-bedded chert, Sutter Creek, California. [Photograph courtesy of L. E. Weiss, from L. E. Weiss, 1972, *Minor Structures of Deformed Rocks* (New York: Springer-Verlag).] (b) Some metamorphic bodies have been homogeneously deformed on all scales of observation—from mountain range down to thin section. In this hypothetical illustration, looking down the axis of folding, the orientation and style of folds are the same regardless of the size of the area observed.

ductile flow of grains in granitic rock can be seen in the sequence of samples collected across a mylonite zone shown in Figure 10-8. Note that a strongly foliated rock can be created from an initially isotropic one during ductile flow.

Existing layers in a rock body can be transposed into new ones of different orientation during deformation (Hobbs and others, 1976, p. 252–264). This process of **transposition** is complex and likely accomplished by a combination of mechanical rotation and ductile flow. A purely mechanical model is shown in Figure 10-9a, and a more realistic model of transposition related to ductile folding of layers is displayed in Figure 10-9b. Note the drastic shortening perpendicular to the new metamorphic foliation that develops during transposition and the concomitant elongation parallel to it. Figure 10-9c is a photograph of transposed layering in

a calcareous rock; similar transposition occurs on smaller scales of thin sections as well as on much larger scales.

Crystal growth in the solid state.   More or less simultaneous growth of new mineral grains in the solid state (recrystallization), chiefly because of increase in temperature in the system, involves adjustment of boundaries of grains already present, or formation of entirely new mineral grains by chemical reaction, nucleation, and growth. The fabric of quartzites furnishes an example of simple boundary adjustment uncomplicated by chemical reactions because quartz is a stable mineral under most metamorphic conditions. Quartz grains may be made unstable because of high strain imparted by deformation or because of high surface-related energy due to small grain size. To minimize

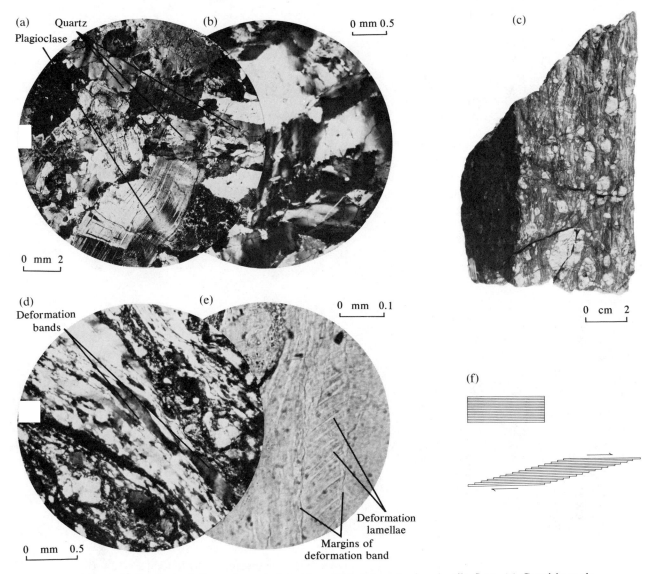

Figure 10-7 Grain deformation produced by solid-state cohesive ductile flow. (a) Granitic rock with no unusual appearance in hand sample discloses in thin section under cross-polarized light conspicuous deformation of virtually every grain. Twin lamellae in plagioclase are bent. Most grains also show **undulatory optical extinction,** different parts of the grain under cross-polarized light extinguishing at slightly different orientations due to distortion of the crystal lattice. (b) Enlarged view of (a) showing strong undulatory extinction, mainly in quartz grains. (c) Hand sample of conspicuously deformed granitic rock with a very fine-grained streaked matrix surrounding large anhedral feldspars. (d) Thin section of rock in (c) viewed under cross-polarized light shows that lenticular to ribbon-shaped **deformation bands** of quartz and feldspar comprise most of the intensely deformed fine-grained matrix. (e) Enlarged view of (d) showing smaller oblique **deformation lamellae** within the deformation bands. Ductile flow of the rocks illustrated here and in the next figure has been accomplished chiefly by plastic slip along preferred crystallographic directions *within* grains, producing deformation bands and the smaller deformation lamellae within them. The slip process is analogous to distorting a pack of cards, as in (f).

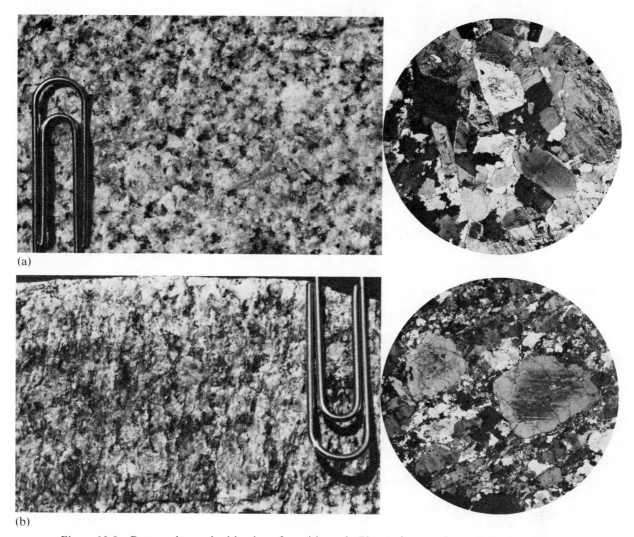

(a)

(b)

Figure 10-8   Progressive mylonitization of granitic rock. Photomicrographs on right are 10× each hand-sample photograph. (a) Original granitic rock. Photomicrograph is under cross-polarized light. (b) Hand sample has a subtle streaked appearance. Photomicrograph under cross-polarized light discloses effects of deformation in the finer-grained matrix that surrounds larger relict grains.

energy, unstrained larger grains will form. (see Figure 10-10).

Perfect uniformity of surface-related energy is manifest in spherical soap and other kinds of bubbles. In aggregates of bubbles, each individual bubble is no longer a sphere but a multisided polyhedron whose shape is such as to enable the entire aggregate to possess minimum surface area and hence minimum energy. The energy of some crystals in rocks varies little with respect to crystallographic orientation of the surface; that is, the crystal is sensibly energy-isotropic.

(c) More advanced deformation has produced a conspicuously foliated and fine-grained mylonite with numerous larger relict feldspar grains. (d) Uniformly very fine-grained mylonite resembles slate in hand sample; only a few small relict grains seen in the thin section testify to the pervasive ductile flow that has occurred. [From *Geology Illustrated* by J. S. Shelton. W. H. Freeman and Company. Copyright © 1966.]

Figure 10-11 illustrates schematically how growth of new grains to mutual contact produces a bubblelike, stable grain boundary configuration— the **crystalloblastic fabric.** Each grain has the shape of a truncated octahedron. Dihedral angles between faces of this form are 120°; but in random slices through an aggregate, as in thin sections cut through a rock, the interfacial angles at triple grain junctions only approximate 120° (see Figures 10-10b, 10-11, and 10-12).

Monomineralic rocks in which well-developed crystalloblastic fabric occurs include marble,

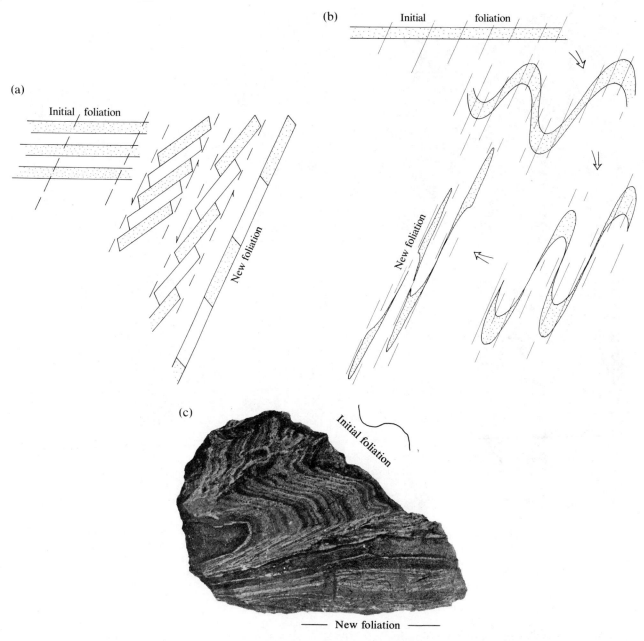

Figure 10-9 Transposed foliation. (a) Purely mechanical rotation of rigid slices of an initial layering produces a new foliation orientation. This model is unrealistic because of the assumed rigidity of the slices and the voids created at the ends of the slices. (b) Progressively tighter ductile folding transposes layer into a new foliation oblique to the initial one. (c) Hand sample showing transposition of a folded layering into an oblique, planar, new layering. Note that this new foliation, as well as the one in (b), is parallel to the axial plane of related folds.

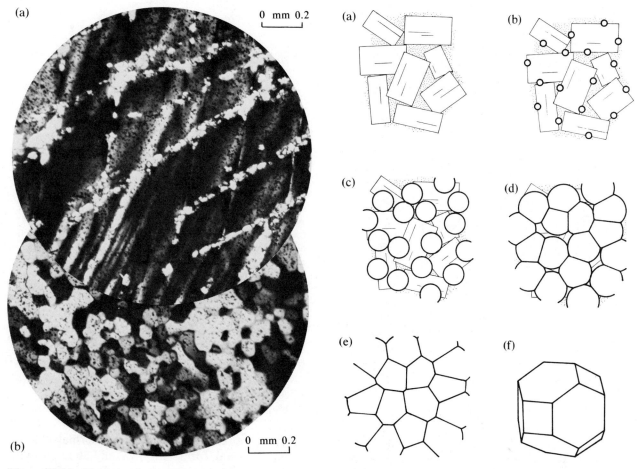

Figure 10-10 Experimental recrystallization of strained quartz. (a) A single crystal of quartz was shortened 25% at 650°C in a laboratory squeezer, producing deformation bands. Numerous small, strain-free grains have nucleated preferentially along the boundaries of the bands. (b) A mosaic of essentially strain-free quartz grains has formed, completely replacing the deformation bands in (a). Abundant specks are impurities in thin-section cement. [Photomicrographs made from thin sections loaned by John M. Christie via Bruce E. Hobbs.]

Figure 10-11 Highly schematic model of solid-state crystal growth producing a monomineralic crystalloblastic aggregate. The initial rock (a) consists of two phases with an igneous fabric. Nucleation and initial growth of a new, more stable phase has begun in (b). The circular form of the new grains is for illustrative convenience only and has no relevance to real crystals. Continued growth of grains of this new phase consumes the original grains and creates a situation (d) where interference between adjacent growing grains occurs. The final crystalloblastic fabric, (e), shows that the growing grains have compromised in their competition for space and have formed polyhedral grains with triple-grain junctions meeting at about 120° angles. (f) In three dimensions a single grain has the shape of a truncated octahedron.

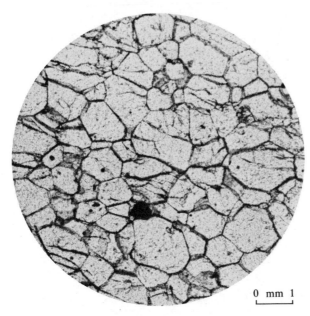

0 mm 1

Figure 10-12   Crystalloblastic fabric in an aggregate of olivine (dunite). Note common triple junctions with approximately 120° angles between grain boundaries.

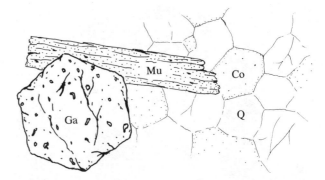

Figure 10-13   Four common metamorphic minerals showing varying degrees of euhedralism as a manifestation of the crystalloblastic series. Quartz (Q) and cordierite (Co) are of equal rank in the series and are equally anhedral against each other. Muscovite (Mu) is more euhedral against these two phases but is anhedral against garnet (Ga), which ranks high in the series.

quartzite, metamorphic dunite, and glacier ice (see box on page 359).

Polymineralic metamorphic rocks whose fabric originated by solid-state growth also have the characteristics described above, but with certain modifications related to grain-energy anisotropy and to differing rates of nucleation. Many minerals have significantly different surface-related energy in different crystallographic directions. Micas, for example, have a strong tendency for growth of platy crystals because the {001} form has a lower overall surface energy than any other form. Quartz, on the other hand, has less surface energy anisotropy (that is, variations in surface energy for different crystallographic directions; Spry, 1969, p. 148) and so commonly occurs as equant, anhedral grains in metamorphic rocks. In an aggregate in which solid-state grain growth tends toward a state of lowest energy, grains with strong surface energy anisotropy will tend to create characteristic euhedral crystal forms at the expense of neighboring grains that are less anisotropic. Irregular or anhedral growth forms of the latter grains impart less additional energy to the aggregate than if the former were anhedral. Hence, micas will tend to be euhedral at the expense of quartz grains. But another mineral such as garnet may dominate in the competition for space over mica (see Figure 10-13).

Hence there is a heirarchy, or a **crystalloblastic series,** among minerals indicating their relative tendency to the development of normal characteristic crystal forms during solid-state growth. Denser minerals with tighter packing of ions tend to be more euhedral.

Crystalloblastic series

MOST EUHEDRAL

↑ Sphene, rutile, pyrite
  Garnet, sillimanite, staurolite, tourmaline
  Epidote, magnetite, ilmenite
  Andalusite, pyroxene, amphibole
  Micas, chlorite, dolomite, kyanite
  Calcite, idocrase, scapolite
↓ Plagioclase, quartz, cordierite
LEAST EUHEDRAL

Glaciers as Metamorphic Bodies

Ice is a mineral, and a solid aggregate of ice crystals is a rock, large bodies of which are called glaciers. Although requiring cold climatic conditions to form, glaciers otherwise satisfy the criteria of a metamorphic rock. They deform during downslope movement under their own weight by solid-state ductile flow. To permit flow, individual ice crystals experience internal plastic slip and solid state growth. Thin sections of glacier ice show well-developed crystalloblastic fabric characteristic of metamorphic rocks. Deformation by ductile flow produces folds in dirty layers within a glacier; these folds are strikingly similar to folds in deep crustal metamorphic bodies (see Figure 10-6).

As in other metamorphic rocks, melting occurs in glacier ice if temperatures become sufficiently high.

Photomicrograph under cross-polarized light of glacier ice. Each grid measures 1.0 cm. Small circular spots are bubbles of air. [Photograph courtesy of A. J. Gow, U.S. Army Cold Regions Research and Engineering Laboratory.]

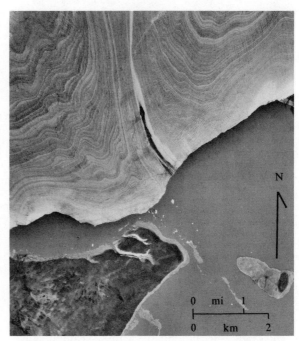

Aerial photograph of the southern margin of the Barnes Ice Cap, Baffin Island, Canada, showing ductile flow folds in dirty layers. [Courtesy of Canadian National Air Photo Library.]

Euhedral development during solid-state growth depends upon relative surface-related energy, whereas in magmatic systems order of crystallization from the melt is the dominant control. This contrast is especially evident in the case of plagioclases, which are widespread in both metamorphic and magmatic rocks; not only are crystalloblastic plagioclases less euhedral, but

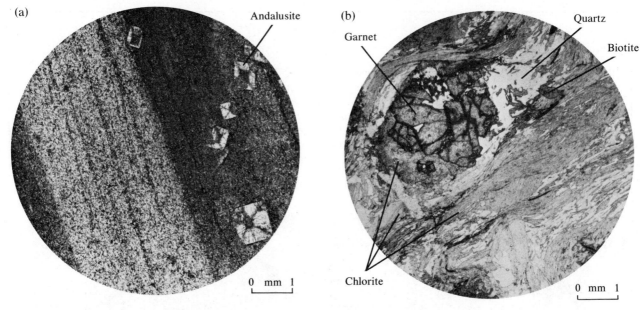

Figure 10-14 Metacrysts, or porphyroblasts, in pelitic rocks. (a) Andalusite with minute graphite inclusions in the form of a cross is sometimes called chiastolite; these poikiloblastic metacrysts lie in a very fine matrix of biotite + muscovite + chlorite + quartz. Note their absence in the coarser, lighter-colored relict silty bed, which lacked sufficient aluminum to stabilize them. (b) Metacryst of garnet in pelitic schist with quartz, biotite, and chlorite. One side of the garnet has been replaced by a fine-grained aggregate of chlorite.

they also tend to be unzoned and have less polysynthetic twinning than magmatic plagioclases.

Different metamorphic minerals must nucleate at different rates because some occur in a particular rock only as large grains, called **porphyroblasts,** or **metacrysts,** surrounded by a finer-grained matrix of other minerals. Almandine-rich garnet (see Figures 10-14b and 10-21) and staurolite invariably occur as metacrysts in metamorphosed shales, as does some andalusite (see Figure 10-14a), presumably because of a very limited population of nuclei relative to micas, chlorites, and other metamorphic minerals of the matrix.

Metacrysts are commonly filled with many small inclusions of other minerals that may or may not be present in the surrounding matrix. These spongelike, inclusion-filled **poikiloblasts** likely

originate in different ways, but one model is shown schematically in Figure 10-15. Poikiloblasts have a large surface area and are, therefore, less stable than an inclusion-free grain of the same volume. Growth of isolated metacrysts may deprive the immediate surroundings of particular constituents, forming a halo that differs in composition from both metacryst and overall matrix (see Figures 10-15 and 10-16. This is an example of **metamorphic differentiation.** In some instances, nucleation of a new phase takes place on a substrate of roughly similar atomic structure, forming **epitaxial overgrowths.** Secondary overgrowth of amphiboles on a seed of pyroxene is an example (see Figure 10-17).

In conclusion, it should be realized that solid-state grain growth cannot always produce the most stable, lowest-energy grain boundary configura-

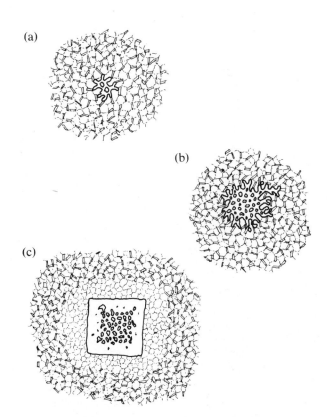

(a)

(b)

(c)

0   cm   2

Figure 10-15 Schematic growth of an isolated metacryst, *M*, by chemical reaction between two phases $K + L \rightarrow M$. (a) Growth from a single nucleus has progressed outward along grain margins of phases *K* and *L* that serve not only as a localized source of nutrient ions from their breakdown but also as easy channels for diffusion of chemical constituents toward the growing grain. Because there is more of phase *L* (anhedral grains) in the rock than *K* (flaky grains), some of *L* is left over after the reaction and grains are incorporated into the metacryst as inclusions. (b) Continued growth along margins of grains of *L* and *K* and incorporation of leftover grains of *L* has produced a poikiloblastic metacryst. If other phases not involved in the reaction producing the metacryst were present in the rock, they too might be incorporated as inclusions. (c) Growth conditions might change from those in (b) so that breakdown of flaky *K* grains occurs some distance from the metacrysts in sufficient amounts to equal the amount of *L* grains consumed at the metacryst interface. The metacryst now develops crystal faces and has no inclusions of *L* because they are entirely consumed. A halo free of *K* has developed.

Figure 10-16 Metacrysts of corundum surrounded by halos of potassic feldspar in a biotite-rich matrix.

Figure 10-17 Epitaxial overgrowth and replacement of pyroxene by a bundle of amphibole needles, a process sometimes called **uralitization**.

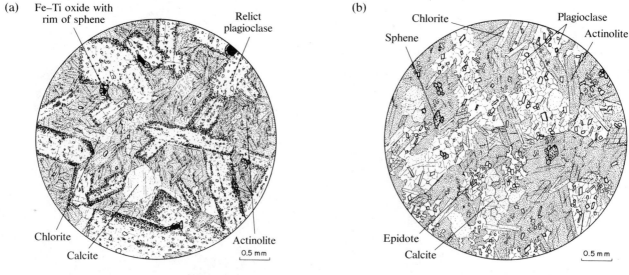

Figure 10-18 Prograde thermal metamorphism of diabase under essentially hydrostatic stress conditions. (a) Weakly metamorphosed diabase or metadiabase. Relict magmatic fabrics and mineral phases (see Figure 5-2) are still evident because solid-state grain growth has not advanced far enough to obliterate magmatic grain boundaries. The original ophitic pyroxenes are represented by randomly oriented, somewhat fibrous, chemically zoned actinolites. Embedded within the actinolite aggregate are patches of fine-grained, felty textured chlorite and tiny epidotes. Incomplete rims of sphene occur around Fe–Ti oxides. Original lath-shaped calcic plagioclases are partly replaced by myriad minute epidote and white mica grains; the remaining

tions because of insufficient time and temperatures that are too low. Grain growth can be inhibited, allowing persistence of relict unstable fabrics. Poikiloblastic fabrics are widespread, not only in metacrysts, but in any relict grain being replaced by finer secondary minerals.

Illustration of crystal growth resulting from thermal metamorphism. Progressively higher-temperature (higher-grade) metamorphism of diabase illustrates some of the concepts just presented (see Figure 10-18). Nonhydrostatic stresses were absent, so the imposed crystalloblastic fabric is isotropic, and the metamorphism was essentially isochemical, only water being mobile. Overall changes in fabric are striking: (1) increase in size of new grains produced by crystal growth at higher grade; (2) relict magmatic fabric is destroyed at the expense of newly formed, imposed crystalloblastic

fabric; (3) newly formed grains attain a more uniform size with increasing grade in response to the demands of minimum energy.

Combined deformation and solid-state crystal growth. Crystal growth may precede deformation of the rock body; the resulting **pretectonic fabrics** are characterized by bent, kinked, and plastically strained grains (see Figures 10-7 and 10-8). **Posttectonic fabrics** developed by crystal growth following deformation display some of the features shown in Figure 10-19. Many granitic intrusions along continental margin subduction zones are posttectonic, and the contact metamorphism around these passively emplaced plutons is posttectonic and modifies or obliterates any earlier deformational fabrics.

In many systems, especially the volumetrically most important regional metamorphic bodies,

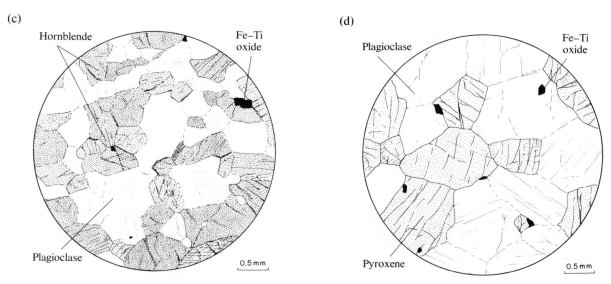

plagioclase is more albitic. Local calcite grains represent relict amygdules. (b) Greenstone. An isotropic aggregate of the same minerals comprising (a), except here grains formed by solid-state growth are larger with simpler, cleaner outlines. The original magmatic fabric is virtually obliterated except for vague suggestions of lath-shaped magmatic plagioclases now represented by crystalloblastic aggregates of untwinned albite + epidote. (c) Amphibolite. Coarser, isotropic and well-developed crystalloblastic fabric. All vestiges of the original magmatic fabric have been erased. (d) Plagioclase–pyroxene granofels. Clean, fairly planar grain margins and coarse size typify this high-grade rock.

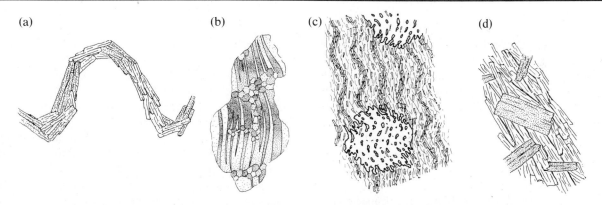

Figure 10-19 Posttectonic fabrics illustrating effects of solid-state crystallization after deformation, as seen under the microscope. (a) Folia of mica define a fold but individual flakes are not bent or strained and each has uniform optical extinction. These mica grains grew after the fold formed, mimicking its form. (b) Relict deformed plagioclase grain, viewed under cross-polarized light, shows undulatory extinction and has been partly recrystallized to small polygonal strain-free grains; note approximate 120° triple junctions. Compare Figure 10-10a. (c) Undeformed garnet poikiloblast shows trains of internal inclusions in continuity with an earlier crenulated folia that it has partially enveloped. (d) Folia of oriented micas is transected by posttectonic, randomly oriented micas.

(a)

(b)

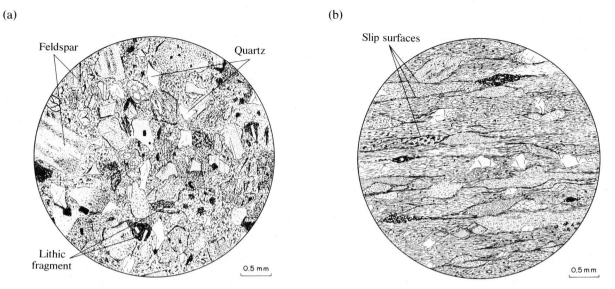

Figure 10-20   Progressive syntectonic metamorphism of volcanic graywacke from New Zealand. (a) Original volcanic graywacke in textural zone 1 is composed of clasts of quartz, feldspar, ferromagnesian minerals, and volcanic rock fragments in an abundant clay matrix; fabric is isotropic. Feldspars—both plagioclase and K-rich alkali feldspar—are extensively altered to sericite and secondary Ca–Al silicates. (b) Foliated metagraywacke or phyllite of textural zone 2. Relicit clasts, especially of quartz, are still evident even though the rock has been pervasively deformed, reducing the size of coarser grains and promoting development of through-going cleavage surfaces, commonly marked by films of iron oxide. Recrystallization of some of the original clasts and clay matrix has produced new but very small grains of quartz + epidote + albite + chlorite + white mica—the latter two forming platy grains, which, because of a strong

crystal growth and deformation have been more or less concurrent, or syntectonic. Ductile flow can progress by more or less simultaneous grain deformation and crystal growth. The characteristic feature of **syntectonic fabrics** is preferred orientation of inequant mineral grains (micas, amphiboles). The pattern of cohesive ductile flow under nonhydrostatic stress influences the patterns of grain growth, producing anisotropic planar and linear fabric geometries.

At high temperatures, where diffusion-controlled recrystallization proceeds rapidly, syntectonic grain growth may produce a fabric apparently of posttectonic origin. Distinctions are ambiguous. Some geologists prefer the suffix *-kinematic,* as in **pre-, syn-,** and **postkinematic,** rather than *-tectonic,* to emphasize movement of rock masses to produce the fabric, instead of making correlation with tec-

tonic episodes. Conceivably some grain deformation might occur during isostatic uplift, gravity-induced sliding, and other nontectonic phenomena.

A sequence of regionally metamorphosed rocks from the South Island of New Zealand illustrates syntectonic fabric development (see Figure 10-20). Segregation of constituents to form compositional layering in the schist from the initially homogeneous and isotropic volcanic sandstone is another type of **metamorphic differentiation.** *The mere presence of layers of contrasting mineralogical composition in a metamorphic body is no assurance that it was originally so layered, or that the layers are relict beds in a metasedimentary rock.* The nature of the differentiation process that produces compositional layering and allied segregations in many intermediate and high-grade rock bodies has been a long-standing problem in petrol-

(c)

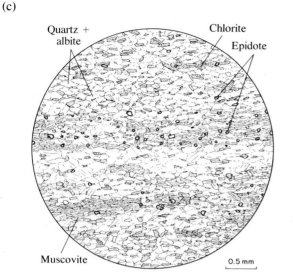

(d)

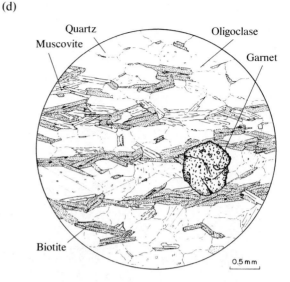

preferred orientation, enhance the foliation of the rock. (c) Fine-grained schist of textural zone 3. Continued growth of new mineral grains under nonhydrostatic stress has produced a well-foliated schist in which all vestiges of the original graywacke fabric and mineral composition have been obliterated. Weak segregation of minerals into alternating quartz–albite-rich and chlorite–epidote-rich layers is evident. (d) Schist of textural zone 4. Continued crystal growth under directed stress has produced still coarser grains, platy ones possessing a strong preferred orientation. Metamorphic differentiation has produced segregated layers of quartz + oligoclase and muscovite + biotite. Complex prograde reactions have eliminated the chlorite + epidote + albite in (c) to give biotite + garnet + oligoclase in this rock.

ogy. The ordering of constituents seems to defy the law of increasing entropy.

Evidence that the fabrics illustrated in Figure 10-20 are syntectonic is indirect. All anisotropic fabric elements in regional bodies—preferred mineral grain orientations, other planar and linear fabric features, folds—commonly display a unity in geometry that is most easily explained by synchronous ductile flow and crystal growth under nonhydrostatic stress. Posttectonic crystal growth mimicking early deformation patterns might not be so regular.

Some regionally metamorphosed bodies with anisotropic fabrics have garnet or other metacrysts with helical trains of inclusions that indicate the metacryst was rolled by ductile flow during growth (see Figure 10-21) (Spry, 1969, p. 255). This is direct evidence of syntectonic mineral growth.

Any rock that possesses an imposed anisotropic fabric reflecting the operation of cohesive deformational flow can be called a **tectonite.**

10.4 MINERALOGICAL COMPOSITION

Regardless of prior states, or the path from them, the final state of thermodynamic equilibrium of a system as dictated by P, T, and X is manifest in the mineralogical composition of the rock body.

Metamorphic Grade

In any body of metamorphic rock, variations in mineral assemblage from place to place are readily apparent. (The terms **mineral assemblage** or **mineral paragenesis** are sometimes used by pet-

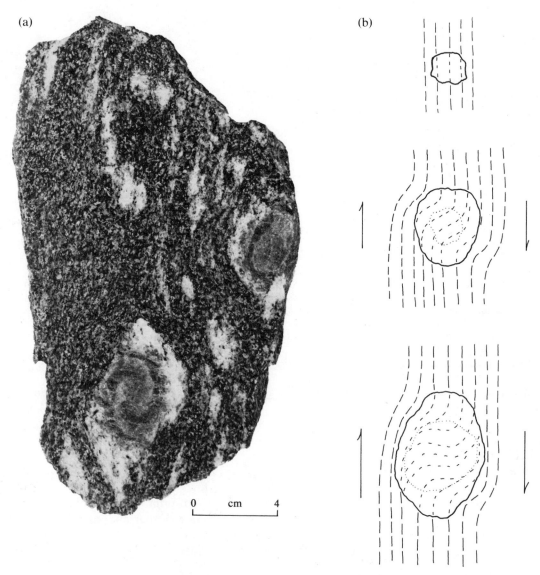

(a)

(b)

0   cm   4

Figure 10-21   (a) Intermediate grade pelitic rock containing metacrysts of almandine-rich garnet in a finer-grained matrix of quartz + biotite + oligoclase. Note segregations of quartz + oligoclase, especially around garnets, on this sawn and polished surface. The poikiloblastic garnet metacryst just above the scale mark shows a helical array of inclusions, indicating it was rolled, like a snowball, by differential shear during growth. (b) Schematic model showing the origin of snowball garnet. Rotation is only 90°, much less than the one in (a).

rologists in lieu of the term **mineralogical composition**.) For example, the following four mineral assemblages might be found:

1. Chlorite + epidote + actinolite + albite
2. Muscovite + biotite + quartz
3. Diopside + calcite + tremolite
4. Plagioclase + garnet + hornblende

Clearly, these different metamorphic mineral assemblages formed under different P, or T, or X, or combinations of these. Chemical analyses of the rocks in the laboratory might show that assemblages 1, 2, and 3 have different bulk chemical compositions (X) but that assemblages 1 and 4 have equivalent chemical compositions, except for less H_2O in 4. Experimentation in the laboratory has revealed that assemblage 4 is stable at relatively high T and that assemblage 1 is stable at lower T. These two different assemblages of similar bulk chemical composition, therefore, reflect different T conditions, or **metamorphic grades.**

Generally, low-grade metamorphic mineral assemblages are characterized by hydrated or carbonated phases (containing H_2O and CO_2), whereas higher-grade assemblages are more anhydrous and CO_2-poor. **Prograde metamorphism** liberates these volatiles from minerals by dehydration or decarbonation reactions.

As an example, muscovite decomposes by a prograde reaction with increasing temperature according to

thermal energy + $KAl_2AlSi_3O_{10}(OH)_2 \rightarrow$
$$\text{muscovite}$$

$$KAlSi_3O_8 + Al_2O_3 + H_2O$$
K-feldspar   corundum

Retrograde metamorphism, going down grade, generally with decreasing T, generally involves hydration or carbonation reactions. For example, there could be a reaction between water, corundum, and K-feldspar to form muscovite.

Metamorphic grade is typically used in a rela-

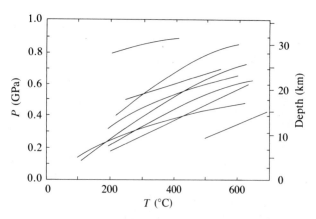

Figure 10-22 Inferred P–T metamorphic gradients in nine regional metamorphic terranes. [After F. J. Turner, 1981, *Metamorphic Petrology: Mineralogical, Field and Tectonic Aspects,* 2nd Ed. (New York: McGraw-Hill).]

tive sense. The mineral assemblages shown in Figure 10-18 developed by a complex sequence of dehydration reactions. The low-grade assemblage of chlorite + epidote + albite + actinolite in (a) and (b) reacted to form hornblende + oligoclase in (c), and these two phases reacted to form the anhydrous mineral assemblage of pyroxene + andesine in (d). Though we usually correlate grade with T, Figure 10-22 discloses that in most metamorphic systems, P changes more or less sympathetically with T as a function of depth in the crust. In some contact metamorphic aureoles surrounding magmatic intrusions, however, there may only be a gradient in T and insignificant variation in P.

Winkler (1976) has proposed some specific guidelines for classifying metamorphic rocks according to grade.

Although higher-grade rocks tend to be coarser (see Figure 10-18) because higher temperatures accelerate diffusion-controlled grain boundary adjustments, grade is only one factor influencing grain size. Original grain size, deformation, and mineral composition are also relevant. For example, the presence of finely disseminated graphite or certain other minerals among quartz grains appar-

(a) (b)

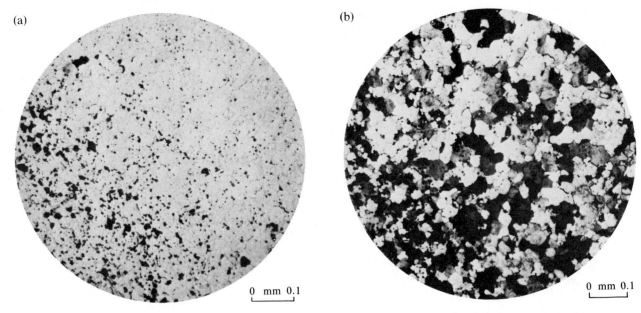

Figure 10-23 Metamorphosed ironstone composed of magnetite + quartz. (a) Plane-polarized light. (b) Cross-polarized light of same area as in (a). Where there is a greater concentration of magnetite grains, the quartz grains are smaller.

ently inhibits growth of larger quartz (see Figure 10-23). Ascribing metamorphic grain size just to $P-T$ conditions can be too simplistic and dangerously misleading. Equally invalid is any correlation between metamorphic grade and the intensity of rock deformation—as expressed, for example, in folds, imposed metamorphic foliation, and other ductile flow features. To avoid misleading correlations, *metamorphic grade should refer specifically to mineral assemblage* rather than to some other attribute of the rock. Grain size, or intensity of deformation, or other fabric aspects should be designated independently of grade.

Metamorphic Zones

Most large contact and regional metamorphic bodies show variations in mineralogical composition. As previously discussed, such variations are dictated by variation in P, T, or X. Within a particular chemical rock type, with X fixed, any variation in mineral assemblage reflects variation only in $P-T$ conditions, or grade.

A **metamorphic zone** is a mappable part of a metamorphic body in which rocks are of similar grade. A metamorphic zone can be identified by the occurrence of a similar mineral assemblage throughout rocks of similar bulk chemical composition and exposed over a particular area of the metamorphic terrane. Adjacent metamorphic zones on a map are separated by **isograds,** or lines of equal grade. An isograd is delineated by the first appearance of a critical **index mineral** or mineral assemblage, after which the higher zone next to it may be named. For example, the first appearance of sillimanite in pelitic rocks delineates the lower-grade boundary of the sillimanite zone. This first appearance of sillimanite may be due to its production by the reaction muscovite + quartz → K-feldspar + sillimanite + water or by kyanite → sillimanite.

In the simple situation of the first appearance of

a critical index mineral by change in T at constant P, the isograd is the line of intersection between the ground surface and the paleo-isothermal surface. But because of the usually complex crystalline and fluid solutions involved in progressive metamorphism, together with inevitable variations in bulk chemical composition of the rocks and variations in pressure as well as possible kinetic factors, the significance of an isograd is more difficult to describe. It usually does not represent simply a paleo-isotherm.

Some rock types are more sensitive to changes in P–T conditions because their constituent minerals have limited stability ranges. Pelitic rocks are usually good indicators of grade and of metamorphic zones.

Metamorphic zones and their bounding isograds can be independent of the conventional stratigraphic and structural contacts usually shown on geologic maps (see Figure 10-24). Zones and isograds reflect conditions of metamorphism, not patterns of deposition, composition, or tectonism.

Toward the end of the nineteenth century, George Barrow mapped a series of regional metamorphic zones in Scotland using mineral assemblages in sensitive pelitic rocks. Isograds were drawn at the first appearance of biotite, garnet, staurolite, kyanite, and sillimanite (see Figure 10-25). These critical index minerals appeared in the course of complex prograde reactions, chiefly in response to increasing T. Such zones are now referred to as **Barrovian metamorphic zones.**

Metamorphic zones have also been defined in the field on the basis of imposed metamorphic fabric rather than mineral assemblage. For example, several decades ago in southern New Zealand, F. J. Turner (see 1981) distinguished four textural subzones within more or less mineralogically uniform chloritic rocks. Representatives of his subzones are shown in Figure 10-20.

Whether based on fabric or mineral assemblage, metamorphic zones reflect the spatial variations in the P–T conditions of metamorphism and flow of energy in the system, independent of the original rock. Metamorphic mineral zones in a contact au-

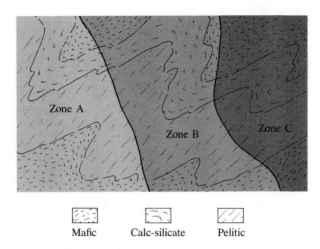

Figure 10-24 Metamorphic isograds (heavy lines) and metamorphic zones (different shades) can be independent of stratigraphic or depositional contacts, as in this schematic sequence of different rock types.

reole around a magmatic intrusion reflect outward flow of heat from the magma; zonal variations in mineral assemblages manifest the resulting thermal gradient. Metamorphic zones of regional extent occurring in orogenic belts near convergent plate boundaries represent regionally extensive gradients in P and T correlated with depth and proximity to heat sources; postmetamorphic tectonic disturbances and isostatic adjustments tilt the sequence of zones so that isograd surfaces obliquely intersect the subsequent erosion surface.

Metamorphic Facies

Definition of metamorphic grade and recognition of metamorphic zones in the field may be hampered or impossible if the particular rock type used for reference is not present everywhere. Consideration of additional mineral assemblages in other rock types may then be useful in disclosing variations in P–T conditions. The fact that a stable mineral assemblage is dictated by P, T, and X underlies the concept of metamorphic facies, as it did the concept of metamorphic zones.

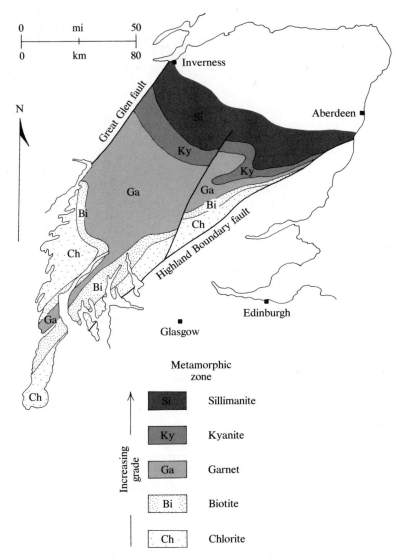

Figure 10-25   Regional Barrovian metamorphic zones in the Scottish Highlands. [After W. Q. Kennedy, 1948, On the significance of thermal structure in the Scottish Highlands, *Geological Magazine* 85.]

One **metamorphic facies** is defined by the whole set of mineral assemblages occurring in spatially associated rock types of diverse chemical composition. If a state of stable equilibrium is represented in this set of mineral assemblages, then we may interpret the one facies as having been formed under a restricted *P–T* condition.

The contrast between the concepts of metamorphic facies and metamorphic zones should be evident. In defining metamorphic zones (plural) we focus attention on only one chemical composition (*X* is fixed) and interpret the occurrence of different mineral assemblages in different areas to be due to varying *P–T* conditions during metamor-

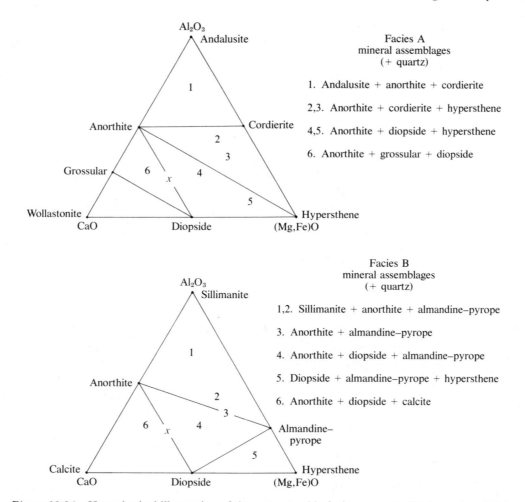

Figure 10-26 Hypothetical illustration of the metamorphic facies concept. Each number within the triangle represents the bulk chemical composition of a rock; the mineral assemblage comprising each rock is shown at the side.

phism. In defining one metamorphic facies (singular) we relate the different mineral assemblages in spatially associated rocks to their diverse chemical composition and interpret the set of assemblages as forming under restricted P–T conditions.

To illustrate the concept of metamorphic facies, let us limit the range of possible chemical compositions to just four components: Al_2O_3–CaO–$(Fe, Mg)O$–SiO_2. (Under other conditions FeO and MgO should be considered as two separate and distinct components.) It is assumed that the com-

ponent SiO_2 is present in excess so that quartz occurs in all mineral assemblages. The other three components can be represented at the apices of a triangle, as shown in Figure 10-26. In this hypothetical world of only these four chemical components, some closely spaced outcrops or layers of rock within one outcrop might consist of the four mineral assemblages shown in Figure 10-26a. Those four assemblages would comprise one metamorphic facies and can be presumed to have formed under some particular P–T condition.

Table 10-1   Standard metamorphic facies.

Facies name	Mineralogical characteristics
Zeolite	Zeolites, especially laumontite and heulandite; also analcime; the assemblage quartz + laumontite + chlorite is diagnostic
Prehnite–pumpellyite	Prehnite ± pumpellyite + quartz is typical (without zeolites or glaucophane or lawsonite)
Glaucophane–lawsonite schist or Blueschist	Glaucophane + lawsonite; also jadeite + quartz + aragonite
Greenschist	Albite + epidote + actinolite ± chlorite ± calcite in mafic rocks and pyrophyllite in pelitic rocks
Amphibolite	Hornblende + plagioclase (An > 20) in mafic rocks and kyanite in pelitic
Granulite	Augite + orthopyroxene + plagioclase; also Mg–Fe garnet
Eclogite	Feldspar-free assemblages typified by jadeite-rich clinopyroxene and pyrope-rich garnet in mafic rocks
Contact facies	Similar to greenschist and amphibolite facies but occur in contact aureoles around igneous intrusions; andalusite occurs in pelitic rocks

Elsewhere, rocks of the very same chemical composition, again closely associated in space, might consist of the five assemblages in Figure 10-26b. They would be in a second metamorphic facies that formed under different $P–T$ conditions from the first. Note that a particular mineral assemblage, such as anorthite + diopside comprising rock X in Figure 10-26, is not diagnostic of either facies, as it occurs in both.

Several different facies are formally recognized (Turner, 1981; Vernon, 1976; Winkler, 1974; Chinner in Nockolds and others, 1978) and have names referring to distinctive aspects of the facies; the greenschist facies, for example, is characterized by schists in which green chlorite, talc, serpentine, epidote, and actinolite are predominant. A more or less standard list of facies is given in Table 10-1.

Individual zeolite minerals can be formed at low temperatures in sedimentary rocks and as cavity-fillings in volcanic rocks. Some petrologists prefer the view that the zeolite facies is diagenetic rather than metamorphic. However, the distinction between these two realms of geologic processes is very difficult to define; one person's diagenesis is another's metamorphism.

The zeolite and the prehnite–pumpellyite facies assemblages cannot form where CO_2 is highly concentrated, and assemblages of chlorite + calcite + clay minerals take their place. The granulite facies appears to require very low P_{H_2O} and, if this condition is not satisfied, then assemblages of the amphibolite facies encroach into its $P–T$ range.

The $P–T$ range over which the several facies are interpreted to have formed is shown in Figure 10-27. Boundaries between facies are not sharp because most characterizing mineral assemblages form by continuous, rather than discrete, reactions as a consequence of the fact that most minerals are solid solutions and most metamorphic fluids are compositionally complex solutions. In addition, some experimental and theoretical data on $P–T$ stabilities of mineral assemblages are still ambiguous, so different petrologists place facies boundaries in different positions. Some petrologists, for example, show a substantial realm of prehnite–pumpellyite facies between the zeolite and blueschist facies.

Metamorphic Pressure–Temperature Regimes

Miyashiro (see 1973) recognized in the 1960s that certain facies are commonly associated to the exclusion of others in metamorphic terranes. These

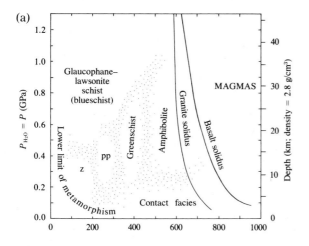

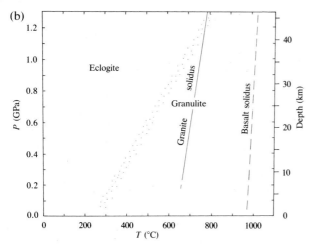

Figure 10-27 Ranges of *P–T* conditions over which metamorphic facies are interpreted to form. For a number of reasons the boundaries between facies are not definable by a single line and hence are shown by stippled bands. z, zeolite facies; pp, prehnite–pumpellyite facies. (a) Water-excess condition. Compare solidi for water-deficient conditions in Figure 8-37. (b) Water-free conditions for mafic systems showing transition zone (stippled) between eclogite and granulite facies. [After G. A. Chinner, in S. R. Nockolds, R. W. O'B. Knox, and G. A. Chinner, 1978, *Petrology for Students* (London: Cambridge University Press); part (a) also after W. S. Fyfe, N. J. Price, and A. B. Thompson, 1978, *Fluids in the Earth's Crust* (New York: Elsevier); water-saturated solidus in part (a) is from O. F. Tuttle and N. L. Bowen, 1958, Origin of granite in the light of experimental studies, Geological Society of America Memoir 74).]

associations, or **metamorphic facies series,** develop under different geothermal gradients in the crust. As can be seen by comparing Figures 10-27 and 10-28, a high *P/T* facies series forms where temperatures increase only slowly with depth (a low geothermal gradient, $\Delta T/\Delta Z$; see box on page 25); that is, the ratio *P/T* is high. Metamorphic rocks in such settings have mineral assemblages appropriate to the zeolite, prehnite–pumpellyite, and blueschist facies. An intermediate *P/T* facies series expressed in typical Barrovian metamorphic zones embodies the greenschist and amphibolite facies; pelitic rocks contain sillimanite and kyanite. Finally, a low *P/T* facies series develops in regions of high geothermal gradients, and andalusite forms in pelitic rocks.

The geothermal gradients in Figure 10-28a are drawn in straight lines for convenience and illustrative purposes only. Real geotherms in the Earth must be markedly concave upward with temperatures increasing at a slower rate with depth; otherwise, with downward extrapolation of the surface gradient, wholesale melting would occur in regions where geophysical evidence indicates there is essentially solid rock. Thus, in some parts of the Earth, there may be a near-surface gradient of, say, 50°C/km, but this must diminish with depth, otherwise considerable melting would take place at depths of only 20 km. But it should be noted that the **metamorphic gradients** shown in Figure 10-22, based upon mineral assemblage geothermometers and geobarometers, are generally convex upward; if extrapolated to higher *P* and *T*, they too would indicate geophysically impossible melting situations, even worse than for the illustrative linear geotherms in Figure 10-28a. Hence, these regional metamorphic *P–T* gradients generally cannot represent real Earth geotherms. Their origin and geologic significance is not always easily assessed, as discussed by England and Richardson (1977) and Turner (1981, pp. 421–455). But some basic concepts can be discerned in Figure 10-29. In subduction zone settings where regional metamorphism occurs, rocks become more deeply buried by tectonic and depositional processes that lead any particular volume along con-

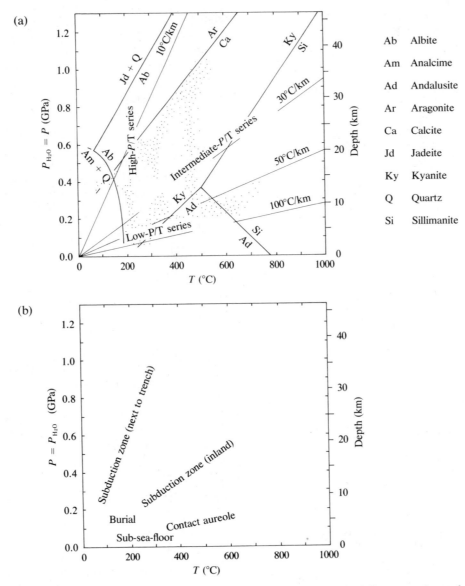

(a)

(b)

Ab	Albite
Am	Analcime
Ad	Andalusite
Ar	Aragonite
Ca	Calcite
Jd	Jadeite
Ky	Kyanite
Q	Quartz
Si	Sillimanite

Figure 10-28 (a) Metamorphic facies series related to hypothetical linear geothermal gradients, reaction curves, and interpreted *P–T* ranges from Figure 10-27. (b) Pressure–temperature regimes of metamorphism occurring in different geologic settings.

cave upward geotherms; lower geotherms (steeper curves on the *P–T* diagram) are associated with more rapid rates of burial, as in trench settings. In contrast to the more or less instantaneous changes in *P* due to burial, changes in *T* at depth occur at a very slow rate, on the order of tens of millions of years. In Figure 10-29 it is assumed that some volume of rock, in which *P* and *T* increased during

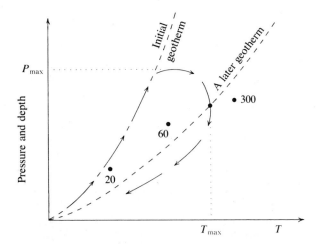

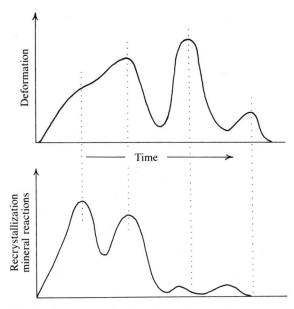

Figure 10-29 Qualitative and highly schematic model of burial and heating of a volume of rock in a regional (subduction zone) setting. Arrows indicate P–T path of a particular volume during burial and uplift. Filled circles represent T_{max} for a series of arbitrary rock volumes that were buried to deeper levels and together define in a very approximate manner a metamorphic gradient (compare Figure 10-22). The numbers indicate the sort of relative times (in My) required for the particular volume of rock to be eventually exposed at the surface after uplift and erosion; more deeply buried volumes require much greater times. [After P. C. England and S. W. Richardson, 1977, The influence of erosion upon the mineral facies of rocks from different metamorphic environments, *Journal of the Geological Society* (London) 134.]

Figure 10-30 Hypothetical time relations of recrystallization (metamorphic reactions) and deformation. Episodic waxing and waning during one orogeny is on the order of tens of millions of years. Multiple orogenies may be separated by hundreds of millions of years. Peaks of recrystallization and metamorphic reaction events need not coincide with episodes of deformation. Compare these hypothetical relations with respect to the concepts in the previous figure.

burial along a steep geotherm, reached its maximum depth where $P = P_{max}$. Thereafter, the thickened crust was essentially stabilized for some time and then uplifted and subjected to erosion. The reduction in P is again virtually instantaneous, but T has continued to increase due to slow conduction and perhaps convection of heat into the volume of rock. T_{max} is attained at less than P_{max}. Volumes of rock more deeply buried will tend to have longer durations of heating, so their T_{max} lie at slightly higher values. For a sequence of rock volumes buried to increasingly greater depths, the maximum T they experience defines a convex upward pattern, or metamorphic gradient. This assumes the mineral assemblage in

the rock reflects its peak temperature, which may be open to some doubt because of complexly interacting factors (see pages 522–524). The actual slope of the metamorphic gradient is clearly dictated by many variables, among which is the erosional history of that sector of the Earth's crust.

10.5 POLYMETAMORPHISM

Repeated episodes of heating and deformation cause **polymetamorphism**. The final metamorphic event maybe of insufficient intensity or duration to erase earlier metamorphic fabrics and mineralogical compositions completely, making it possi-

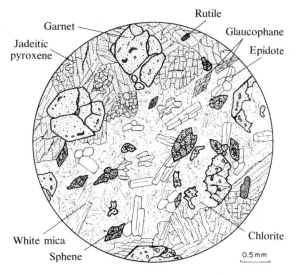

Figure 10-31   Partial hydration of eclogite. Only embayed remnants of the dry, high-pressure eclogite assemblage pyroxene + garnet + rutile survive in the hydrous vein cutting diagonally across the sample.

ble to unravel at least a part of the sequence of metamorphic episodes. Sequences of episodes (see Figure 10-30) developed over broad areas may reflect changes in direction or rates of convergence of lithospheric plates and other tectonic and thermal phenomena.

Many examples of polymetamorphism can be found in rocks. Polymetamorphic effects are found in contact aureoles around magmatic intrusions emplaced after the peak regional metamorphic event. The posttectonic fabrics in Figure 10-19 illustrate the effects of solid-state crystal growth postdating deformation. Transposition of slaty cleavage or schistosity, formed in an early metamorphic episode, by later deformation and possibly concurrent recrystallization has already been discussed (see Figure 10-9). The relict bedding in Figure 10-1 has a parallel, early metamorphic foliation in micaceous layers that is cut obliquely by a later imposed foliation locally transposing the oriented micas. One of the most

common examples of polymetamorphism is retrogressive metamorphism. This is usually seen as an incomplete and localized replacement of anhydrous silicates such as garnet and staurolite, which are stable at higher temperatures, by hydrous phases such as chlorite and micas that are stable at lower temperatures (see Figure 10-14b). Another example of hydration reactions involves a high-P dry, eclogite assemblage of jadeitic pyroxene + almandine–pyrope garnet + rutile that has been replaced by the more hydrous assemblage glaucophane + chlorite + epidote + white mica + sphene (see Figure 10-31). Hydration was prompted by ingress of water into the rock, as manifest by the confinement of the hydrous assemblage to a cross-cutting vein.

On a worldwide scale, exposed metamorphic rocks of any particular chemical composition present a wide spectrum of mineralogical compositions and generally show little textural evidence of retrograde effects. These facts suggest that most observed assemblages reflect the highest, or peak, grade conditions. Why is retrograde metamorphism not generally expressed to the same degree as prograde? Surely whatever original rock went up in P and T must have come down to atmospheric conditions at the outcrop.

Prograde reactions liberate H_2O and CO_2 from the metamorphic body. As T falls after the peak climatic metamorphic event, no volatiles are available and little can enter into the impermeable body to permit the formation of more stable, lower-T, volatile-bearing phases. A volatile fluid also serves a catalytic role in that it is a favorable medium for easy diffusion of nutrient ions to fertile crystal nuclei. Consequently, high-grade minerals usually persist metastably to low P and T.

The observed nature of retrograde effects is compatible with this explanation. Retrograde reactions are generally incomplete (see Figures 10-14b and 10-31), suggesting that there was an insufficient amount of one of the reactants—namely H_2O. Retrogressive metamorphism is commonly localized along fault or shear zones where grains have been deformed. This not only produces a higher energy state in the body but improves the chance for influx of volatiles into the body.

10.6 THE SPECIAL SIGNIFICANCE OF METAMORPHIC ROCKS

The composition, fabric, and field relations of a metamorphic rock body are important records of changing *P, T, X* and stress conditions through time in a segment of the crust. Reading this record provides valuable insight into the nature of inactive or "fossil" orogenic and magmatic belts and zones of dislocation. Elucidation of the early history of the Earth and how the continents have evolved is partly an exercise in interpretive metamorphic petrology (see Chapter 15).

SUMMARY

10.1. Metamorphism is a solid-state reconstitution caused by significant changes in the intensive variables *P, T,* and *X* and in the state of stress; as a result, the system shifts toward a new, more stable state of equilibrium. Some sort of aqueous or carbonic fluid is nearly always present during metamorphism, albeit in usually small quantities. Metamorphism occurs at higher *P* and *T* than weathering and diagenetic processes. More stable compositions and fabrics, commonly characterized by parallel alignment of inequant mineral grains, are overprinted on the original rock as it adjusts to the new *P–T–X* conditions.

10.2. The identity of the protolith of the metamorphic body is recorded in (1) relict sedimentary, magmatic, and even metamorphic fabrics, which are best preserved in weakly metamorphosed, lower-grade rocks; and (2) chemical composition, which can be approximated rather well from knowledge of the mineralogical composition; this compositional correlation to the protolith assumes that the metamorphism was essentially isochemical.

10.3. The record of metamorphic processes by which the system changed states with flow of energy and matter is found in the imposed metamorphic fabric and the field relations of the body. Metamorphic processes can be classified according to whether the dominant change in state was in *P, T, X,* or in the state of stress.

High-pressure metamorphism favors growth of dense minerals. Thermal metamorphism produces new minerals and fabrics. Metasomatism involves major changes in chemical composition of the rock due to transport of matter in moving fluids; this also produces new minerals and fabrics. Metasomatism may take place by volume-for-volume replacement. Nonhydrostatic stresses in orogenic belts produce anisotropic fabrics of aligned inequant mineral grains.

Field relations of metamorphic bodies furnish insight into the reason for changes in state and flow of energy and matter. Contact metamorphism occurs in aureoles surrounding magmatic intrusions, which may serve as a source for aqueous fluids and always as a source of thermal energy. Regional metamorphism extends broadly through the deeper parts of orogenic subduction zones at convergent plate margins where complex flows of energy and matter occur; more or less synchronous thermal and deformational events can recur episodically causing polymetamorphism. Sub-sea-floor metamorphism (alteration) occurs near oceanic rifts where sea water penetratively convects through freshly emplaced and highly fractured hot submarine basalts. Burial metamorphism can be considered a deeper extension of diagenetic processes where increasing confining pressures and temperatures are a consequence of burial. Shear metamorphism, typically producing fine-grained, streaked mylonites, takes place where deep-seated rock bodies are differentially displaced. Shock metamorphism is produced by meteorite impact.

Imposed metamorphic fabrics erasing relict fabrics develop through deformational processes and solid-state grain growth. Deformation in solid-state metamorphic bodies is cohesive and pervasive down to the scale of individual grains. This ductile flow is accomplished by bending of grains, by crystallographically controlled plastic slip within grains, as well as by diffusion-controlled recrystallization. Crystalloblastic fabric is produced by the more or less simultaneous growth of mineral grains in the solid aggregate. New grains form by nucleation and growth following chemical reactions between old grains. Existing grains may enlarge or otherwise adjust their boundaries.

Crystal growth leads toward lower energy states in the grain aggregate by consuming any stored strain energy produced by deformation and by minimizing the surface-related energy of the grains. The stable grain boundary configuration in some aggregates (carbonates, quartz, feldspar) is characterized by approximately 120° triple junctions. A crystalloblastic series is evident where grains with relatively greater surface energy anisotropy grow competitively into more euhedral shapes. Metacrysts, or porphyroblasts, probably nucleate at slower rates and therefore form larger grains than the surrounding matrix. Progressive recrystallization creates coarser, more uniform grain size.

Thermally related crystal growth may precede or follow tectonic deformation of grains, creating pre- and posttectonic fabrics, respectively. However, most imposed metamorphic fabrics are syntectonic and the pattern of ductile flow in the deforming body produced by nonhydrostatic stress is preserved in the pattern of oriented inequant grains or anisotropic fabric that formed during flow. Segregation layers of contrasting modal composition can apparently develop by syntectonic recrystallization from originally isotropic rocks. This is an example of metamorphic differentiation.

10.4. In the case of stable equilibrium, the mineralogical composition of a metamorphic rock is dictated solely by the P, T, and X of its final state. For rocks of a particular fixed chemical composition, any variation in mineralogical composition reflects variation in P and T. As prograde reactions liberate volatiles—H_2O and CO_2—higher-grade rocks are composed of less hydroxyl- or carbonate-rich minerals.

A metamorphic zone is a mappable area of rocks on the ground surface of similar grade. An isograd is a mappable line on the ground surface that marks the first appearance of a critical index mineral or mineral assemblage, after which the higher-grade zone may be named. Zones chiefly reflect the pattern of gradients in P, T, and volatile concentrations during metamorphism.

A metamorphic facies is the entire set of mineral assemblages occurring in closely associated rocks of diverse chemical compositions. A facies is interpreted to have formed under restricted P–T conditions. Different metamorphic facies series develop in response to different geothermal gradients in the crust. High-P/T series develop in trench settings, whereas intermediate- to low-P/T series typify inland regions of subduction zones.

10.5. Polymetamorphic bodies show evidence of recurring episodes of recrystallization and deformation. Retrograde metamorphism generally does not occur more extensively because of lack of volatiles to react with the peak highest-grade minerals.

10.6. Metamorphic bodies are especially important in providing clues regarding continental evolution by recurring episodes of deformation and heating related to ancient plate motions.

STUDY QUESTIONS

1. Compare and contrast metamorphic and magmatic systems with regard to: (a) states of matter involved (solid, liquid, gas); (b) mass transport (movement of matter within the system); (c) fabric; (d) changes in state that create rock bodies; (e) chemical composition; and (f) techniques of investigation by the petrologist.

2. Some migmatites are interpreted as deep crustal rocks that have experienced partial melting, producing mixed parts of granite and of metamorphic rock comprised of hornblende +

plagioclase (An > 20) + almandine-rich garnet. Under what P and T might these migmatites form (see Figure 10-27)?

3. Metamorphosed mafic rocks are fairly widespread in the Scottish Highlands. Would the chlorite, biotite, and garnet isograds in these mafic rocks coincide with those shown in Figure 10-25 for pelitic rocks? Discuss.

4. Write generalized prograde reactions showing progressive dehydration of characterizing minerals in the intermediate-P facies series.

5. What is the significance of preferred orientation of inequant platy and columnar grains in a metamorphic rock?

6. Compare the controls on grain size in magmatic and metamorphic rocks.

7. Describe and sketch the fabric of the rock illustrated in Figure 10-18c if it had been metamorphosed under nonhydrostatic stress.

8. Show three areas in Figure 10-31 that were originally garnet grains.

9. In a rock of calc-silicate composition, what minerals might be found at a low grade and at a high grade of metamorphism? (Begin by listing common minerals comprised chiefly of Ca, Mg, Si, and O).

10. On what basis might metamorphic zones be delineated in a rock body deformed in a deep-seated shear zone? Discuss.

11. In the illustrations of prograde metamorphism, it was assumed that the rate of mineral reactions exceeded the rate of changing state; thus low-grade minerals and assemblages were replaced sequentially by higher. Discuss what might be observed in a metamorphic body if the rates were reversed.

12. Metamorphism is sometimes defined as "the result of heat, pressure, and chemically active fluids." Discuss this critically.

13. If a metasomatic rock has no relict fabric, how can the nature of the original rock be determined?

14. Discuss criteria for distinguishing between plagioclase grains formed by growth from a melt and in the solid state on the basis of habit, twinning, and zoning.

15. Discuss criteria for distinguishing between (a) phenocrysts in magmatic rocks, (b) relict phenocrysts in a metamorphic rock, (c) metacrysts, and (d) relict grains surviving ductile deformation in a mylonite.

16. Discuss the different ways in which compositional layering may originate in metamorphic bodies.

17. Sketch idealized grain boundary relations in a rock formed by solid-state crystal growth and composed of sphene, epidote, chlorite, and feldspar.

18. Refer to Figure 10-16. What was the protolith of this rock? Propose an explanation for the halos around the corundum metacrysts.

19. Why are migmatites (see question 2) not found in rocks of the high-P/T metamorphic series?

20. Contrast the appearance and origin of subsolidus overgrowths of amphibole on pyroxene and magmatic reaction rims of amphibole on pyroxene (see Figures 8-5 and 10-17).

21. How might the rate of uplift and erosion account for the virtual absence of blueschist facies rocks in pre-Mesozoic terranes?

REFERENCES

Compton, R. R. 1962. *Manual of Field Geology* New York: John Wiley and Sons, 378 p.

England, P. C., and Richardson, S. W. 1977. The influence of erosion upon the mineral facies of rocks from different metamorphic environments. *Journal of the Geological Society* (London) 134:201–213.

French, B. M., and Short, N. M. 1968. *Shock Metamorphism of Natural Materials*. Baltimore: Mono Book, 644 p.

Fyfe, W. S., Price, N. J., and Thompson, A. B. 1978. *Fluids in the Earth's Crust*. New York: Elsevier, 383 p.

Higgins, M. W. 1971. Cataclastic rocks. U.S. Geological Survey Professional Paper 687, 97 p.

Hobbs, B. E., Means, W. D., and Williams, P. F. 1976. *An Outline of Structural Geology* New York: John Wiley and Sons, 571 p.

Miyashiro, A. 1973. *Metamorphism and Metamorphic Belts.* New York: Halsted Press, 492 p.

Muffler, L. J. P., and White, D. E. 1969. Active metamorphism of Upper Cenozoic sediments in the Salton Sea geothermal field and the Salton trough, southeastern California. *Geological Society of America Bulletin* 80:157–182.

Nockolds, S. R., Knox, R. W. O'B, and Chinner, G. A. 1978. *Petrology for Students.* London: Cambridge University Press. 435 p. (The chapters on metamorphism by Chinner are an outstanding introduction to metamorphism as illustrated by thin section petrography.)

Short, N. M. 1966. Shock processes in geology. *Journal of Geological Education* 14:149–166.

Spry, A. 1969. *Metamorphic Textures.* New York: Pergamon, 350 p.

Turner, F. J. 1981. *Metamorphic Petrology: Mineralogical, Field and Tectonic Aspects,* 2nd Ed. New York: McGraw-Hill, 524 p.

Turner, F. J., and Weiss, L. E. 1963. *Structural Analysis of Metamorphic Tectonites.* New York: McGraw-Hill, 545 p.

Verhoogen, J., Turner, F. J., Weiss, L. E., Wahrhaftig, C., and Fyfe, W. S. 1970. *The Earth.* New York: Holt, Rinehart and Winston, 748 p.

Vernon, R. H. 1976. *Metamorphic Processes.* London: George Allen and Unwin, 247 p.

Winkler, H. F. G. 1976. *Petrogenesis of Metamorphic Rocks,* 4th Ed. New York: Springer-Verlag, 254 p.

11

Petrography and Composition of Metamorphic Bodies

The chemical compositions of metamorphic rocks, as reflected in their mineral assemblages, virtually span the entire spectrum of magmatic and sedimentary compositions and include metasomatic modifications of these rocks. Simplifying assumptions can be made that allow multicomponent metamorphic rocks to be plotted on triangular composition diagrams. Such plots show the relation between the bulk chemical composition of a rock and its constituent minerals and how this mineral assemblage can vary with grade (P and T).

The fabric of a metamorphic rock is determined by the original fabric, solid-state crystal growth and deformational processes during metamorphism and their chronologic relations, state of stress (hydrostatic or nonhydrostatic), and mineralogical composition of the rock. Imposed planar metamorphic fabrics are common and some have associated linear features; both of these anisotropic fabrics reflect cohesive solid-state ductile flow under nonhydrostatic stress.

Classification of metamorphic rocks is based primarily on the expression of imposed planar fabric and to a lesser extent on composition.

11.1 COMPOSITION

The enormous range of chemical composition of metamorphic rocks reflects their derivation from magmatic and sedimentary rocks (see Figure 11-1), as well as metasomatic modifications of these. Metamorphic equivalents of rare alkaline and silica-undersaturated magmatic rocks of intraplate tectonic settings and of sedimentary evaporite deposits are very seldom found.

This broad compositional spectrum can be subdivided into just a few chemical rock types, each comprised of a limited list of minerals. The mineralogical composition of a metamorphic rock reflects its chemical composition. If the chemical compositions of constituent minerals are known, even approximately, then the chemical composition of the whole rock can be roughly evaluated from its mode. Once ascertained, the chemical composition of the rock provides insight into its protolith, whether basalt, shale, and so on (see Figure 11-1). The box on page 384 provides a list of eight general chemical rock types and their major

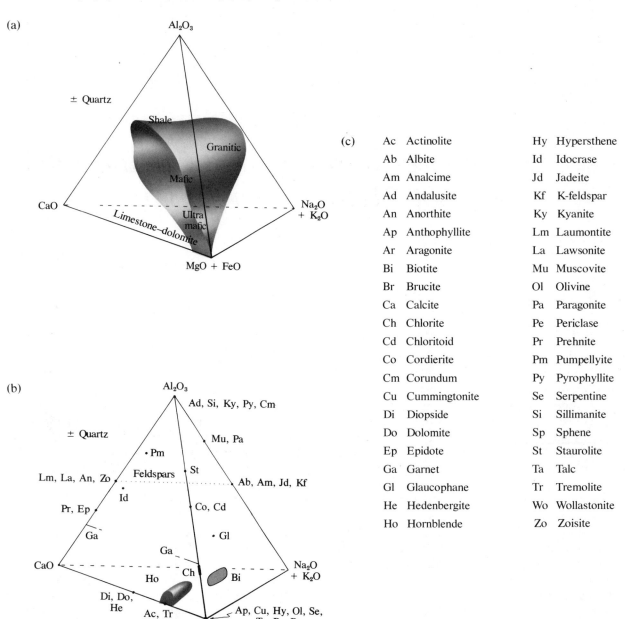

(a)

Al₂O₃

± Quartz

Shale

Granitic

Mafic

CaO

Limestone–dolomite

Ultra mafic

Na₂O + K₂O

MgO + FeO

(c)

Ac	Actinolite	Hy	Hypersthene
Ab	Albite	Id	Idocrase
Am	Analcime	Jd	Jadeite
Ad	Andalusite	Kf	K-feldspar
An	Anorthite	Ky	Kyanite
Ap	Anthophyllite	Lm	Laumontite
Ar	Aragonite	La	Lawsonite
Bi	Biotite	Mu	Muscovite
Br	Brucite	Ol	Olivine
Ca	Calcite	Pa	Paragonite
Ch	Chlorite	Pe	Periclase
Cd	Chloritoid	Pr	Prehnite
Co	Cordierite	Pm	Pumpellyite
Cm	Corundum	Py	Pyrophyllite
Cu	Cummingtonite	Se	Serpentine
Di	Diopside	Si	Sillimanite
Do	Dolomite	Sp	Sphene
Ep	Epidote	St	Staurolite
Ga	Garnet	Ta	Talc
Gl	Glaucophane	Tr	Tremolite
He	Hedenbergite	Wo	Wollastonite
Ho	Hornblende	Zo	Zoisite

(b)

Al₂O₃

Ad, Si, Ky, Py, Cm

± Quartz

• Pm

Mu, Pa

Lm, La, An, Zo

Feldspars   • St

Ab, Am, Jd, Kf

Id

Pr, Ep

Co, Cd

Ga

• Gl

Ga

CaO

Ch

Na₂O + K₂O

Ho

Bi

Di, Do, He

Ac, Tr

Ap, Cu, Hy, Ol, Se, Ta, Pe, Br

Mole %

MgO + FeO

Figure 11-1 Composition of metamorphic rocks. (a) In terms of major chemical constituents, most igneous rocks (granitic, mafic, ultramafic) lie in a pod-shaped volume (shaded) within the tetrahedron. By processes of sedimentary differentiation, these primary rocks become sedimentary limestone, dolomite, shale, and quartz–feldspar sandstone (approximately in same region as granitic rocks). (b) Mineral constituents of metamorphic rocks. Feldspar (Ab, Kf, An), garnet (Ga), and chlorite (Ch) solid solutions plot as lines; biotite solid solutions (Bi) plot as an elliptical patch on the right-hand face and hornblende solid solutions (Ho) plot as a pluglike volume (shaded) extending into the tetrahedron.

mineral constituents (for additional information on metamorphic minerals, consult Appendix C).

The chemical compositions of some magmatic and sedimentary rocks overlap. However, many sedimentary rocks have more extreme concentrations of Si, Al, or Ca. In metamorphosed sedimentary rocks, extremes in Si or Al can be reflected in abundant quartz or muscovite, either one comprising more than half the rock. Highly aluminous minerals, such as Al_2SiO_5 polymorphs and staurolite, also usually indicate a pelitic sedimentary protolith. Calcareous sedimentary parents yield carbonates and calcium-rich silicates such as idocrase and wollastonite. Except for protoliths of intensely weathered rocks, metamorphosed igneous rocks have only minute concentrations of calcite.

11.2 FABRIC

Some aspects of fabric were introduced in the previous chapter, and the origin of imposed fabric will be treated in Chapter 13. The aim of the present section is to discuss petrographic aspects of fabric—the description and classification of metamorphic rocks in hand sample and outcrop.

Geometric Aspects of Fabric

The three-dimensional geometric array of grain boundaries and grain aggregate boundaries in a metamorphic body can be classified into two major and three subsidiary fabric groups:

1. Isotropic, or random boundaries
2. Anisotropic, or boundaries having different aspects in different directions
   a. Planar
   b. Linear
   c. Planar–linear

The geometry of the imposed metamorphic fabric reflects the pattern of cohesive deformational flow, or lack of flow, during the metamorphic event, and indirectly the nature of the stresses applied to the system, whether nonhydrostatic or hydrostatic (uniform in all directions).

Isotropic fabric. Isotropic fabrics have random aspect and appear the same in any direction, as in a mass of soap bubbles or a bag of marbles. Most magmatic rock bodies, thick-bedded sandstones, and limestones are isotropic. Nonhydrostatic stresses were insignificant during their formation. Sometimes the term **massive** is applied to isotropic rocks on the scale of a hand sample or outcrop. **Granoblastic** fabric is another term that refers to a mosaic of equidimensional anhedral grains (see Figures 10-12, 10-18c,d, and 10-20d). Inequant grains, such as micas, if present, are randomly arrayed.

Isotropic fabrics typically develop around passively emplaced magmatic intrusions where the aureole rocks recrystallized under hydrostatic conditions. Some large dry bodies of magmatic rock in regional environments may resist deformation and the resulting imposed fabric may mimic the original isotropic magmatic one.

Many rocks with an isotropic imposed fabric possess an anisotropic inherited fabric, which makes the overall fabric anisotropic. An example is a granoblastic rock from a contact metamorphic aureole with relict bedding. Such rocks emphasize the value of accurately distinguishing between imposed and inherited fabrics if one is seeking to elucidate the nature of stresses in the metamorphic system.

Anisotropic fabrics. Imposed anisotropic fabrics reflect pervasive solid-state flow of the rock body in response to nonhydrostatic stresses. Anisotropic fabrics look different from different directions, like a stack of plywood or a stack of papers (see Figures 10-1, 10-2, 10-8c,d, and 10-20b,c,d). Rocks with such fabric are the familiar typically metamorphic slates, schists, and gneisses encountered in a first course in geology. Elongate mineral grains display a preferred orientation; the rock may also have layers of contrasting composition, as in gneisses. That anisotropic fabrics in tectonites are produced by deformation is shown by intimate geometric relationships to associated folds, deformed pebbles, oolites, fossils, and so on within the same body.

Anisotropic fabrics may be classified into pla-

General Chemical Types for Metamorphic Rocks

The minerals marked with an asterisk (*) in the following lists are particularly diagnostic of the chemical type.

Pelitic Aluminous rocks derived from shales, mudstones, and other sediments rich in clay minerals. Highly calcic minerals such as diopside, wollastonite, grossularite, and epidote are absent but quartz and micas are generally present. Typical minerals include:

Quartz	SiO_2
*Corundum	Al_2O_3 (never with quartz)
Chloritoid	$(Fe^{2+},Mg,Mn)_2(Al,Fe^{3+})$– $Al_3O_2(SiO_4)_2(OH)_4$
Graphite	C
*Andalusite, *Sillimanite, *Kyanite	Al_2SiO_5
Almandine-rich garnet	$(Fe,Mg,Mn)_3Al_2Si_3O_{12}$
*Staurolite	$(Mg,Fe)_2(Al,Fe^{3+})_9$– $O_6(SiO_4)_4(O,OH)_2$
White mica	$(K,Na)Al_2AlSi_3O_{10}(OH)_2$
Biotite	$K(Mg,Fe)_3(Al,Si)_4O_{10}$– $(OH,F,Cl)_2$
Chlorite	$(Mg,Fe,Al)_6(Si,Al)_4O_{10}$– $(OH)_8$
Cordierite	$Al_3(Mg,Fe)_2AlSi_5O_{18}$
Alkali feldspar	$(K,Na)AlSi_3O_8$
Fe–Ti oxides	

Calcareous Ca and CO_3 are dominant chemical constituents, but Mg may also be present in these rocks, which are derived from relatively pure limestones and dolomites. Dominant minerals are calcite or dolomite.

Calc-silicate Ca and Si are dominant constituents in these rocks, which are derived from shaly or quartz-bearing dolomites and limestones, or from relatively pure carbonate rock metasomatized by siliceous solutions from contiguous granitic intrusions. Typical minerals include:

Calcite, dolomite	$CaCO_3$, $CaMg(CO_3)_2$
Quartz	SiO_2
Calcic plagioclase	$(Ca,Na)(Al,Si)_4O_8$
Epidote	$Ca_2Fe^{3+}Al_2O(SiO_4)(Si_2O_7)$– (OH)
*Grossular–andradite garnet	$Ca_3(Al,Fe^{3+})_2Si_3O_{12}$
*Idocrase (vesuvianite)	$Ca_{10}(Mg,Fe)_2Al_4(SiO_4)_5$– $(Si_2O_7)_2(OH)_4$
*Diopside–hedenbergite	$Ca(Mg,Fe)Si_2O_6$
Sphene	$CaTiSiO_5$
*Wollastonite	$CaSiO_3$
Tremolite–actinolite	$Ca_2(Mg,Fe)_5Si_8O_{22}(OH)_2$

nar, linear, and planar–linear. Each reflects a more or less corresponding pattern of deformational flow under applied nonhydrostatic stress. Planar fabric is expressed by a set of closely spaced, subparallel surfaces within the rock body (see Figure 11-2). One set of such pervasive subparallel surfaces is referred to as a **foliation**. The foliation may be irregular, curved, or even folded if deformed. Relict bedding, the **slaty cleavage** of slates, the schistosity of schists, and the compositional layering of gneisses are different types of foliation (see Figure 11-2; see also Hobbs and others, 1976, pp. 215–266). In some rocks, foliation is mechanically significant; the rock readily breaks along it when struck with a hammer, as in slates and schists. Other foliation is expressed in a more subtle, passive manner and the rock may break across it.

Most bodies possess only one foliation, but

Forsterite	Mg_2SiO_4
Talc	$Mg_3Si_4O_{10}(OH)_2$

Mafic Rocks with high concentrations of Mg and Fe, derived from gabbroic and basaltic rocks and less common siliceous shaly dolomites. Important minerals include:

Epidote	$Ca_2Fe^{3+}Al_2O(SiO_4)-$
	$(Si_2O_7)(OH)$
Sphene	$CaTiSiO_5$
Calcite, dolomite	$CaCO_3$, $CaMg(CO_3)_2$
Chlorite	$(Mg,Fe,Al)_6(Si,Al)_4O_{10}-$
	$(OH)_8$
Fe–Ti oxides	
Amphibole	$NaCa_2(Mg,Fe)_5(Si,Al)_8-$
	$O_{22}(OH)_2$
Plagioclase	$(Na,Ca)(Si,Al)_4O_8$
Prehnite	$Ca_2Al[AlSi_3O_{10}](OH)_2$
Pumpellyite	$Ca_4(Mg,Fe^{2+},Mn)-$
	$(Al,Fe^{3+},Ti)_5O(OH)_3-$
	$[Si_2O_7]_2[SiO_4]_2 \cdot$
	$2H_2O$
Pyroxene	$(Ca,Mg,Fe,Ti,Na)_2-$
	$(Al,Si)_2O_6$
Almandine–pyrope garnet	$(Fe,Mg)_3Al_2Si_3O_{12}$
Serpentine	$Mg_3Si_2O_5(OH)_4$
Talc	$Mg_3Si_4O_{10}(OH)_2$

Ultramafic Rocks especially rich in Mg and Fe derived from peridotites, dunites, and pyroxenites, lacking feldspar, and containing the following:

Phlogopite	$K(Mg,Fe)_3AlSi_3O_{10}(OH)_2$
Talc	$Mg_3Si_4O_{10}(OH)_2$
Serpentine	$Mg_3Si_2O_5(OH)_4$
Anthophyllite	$(Mg,Fe)_7Si_8O_{22}(OH)_2$
Forsterite	$(Mg,Fe)_2SiO_4$
Enstatite	$(Mg,Fe)SiO_3$
Magnesite, dolomite	$MgCO_3$, $CaMg(CO_3)_2$
Fe–Ti oxides	
Cordierite	$Al_3(Mg,Fe)_2AlSi_5O_{18}$
Brucite	$Mg(OH)_2$

Ferruginous Fe is the dominant constituent in these rocks, derived from ironstones ("iron formations"). Typical minerals include:

Fe–Ti oxides	
Siderite	$FeCO_3$
Cummingtonite	$Fe_7Si_8O_{22}(OH)_2$
Fe-serpentine	$Fe_3Si_2O_5(OH)_4$
Fayalite	Fe_2SiO_4
Quartz	SiO_2

Siliceous Rocks with high concentrations of Si containing abundant quartz derived from quartz-rich sandstones and chert. Common associates in metasandstones are alkali feldspar and mica and spessartine in metacherts.

Quartzo-feldspathic These assemblages are derived from graywackes, granitic, rhyolitic, and arkosic rocks and consist of abundant feldspars and quartz with minor micas, amphiboles, and other accessory minerals.

some low-grade metasedimentary and polymetamorphic rocks have two or even more sets of nonparallel penetrative surfaces. A common situation is for an imposed foliation expressed by aligned mica flakes to be oblique to relict bedding or to compositionally and texturally contrasting layers (see Figure 11-3, bottom).

Lineation is a set of subparallel pervasive linear features in a rock body, like a clenched handful of pencils (Hobbs and others, 1976, p. 267–288). Lineation is generally straight (rectilinear), but may be curved, and is usually imposed rather than inherited. A pure lineation without associated foliation may be expressed by preferred orientation of columnar or acicular mineral grains, such as amphibole, or by linear segregations of contrasting grain aggregates.

But rock bodies with only a linear fabric are rare

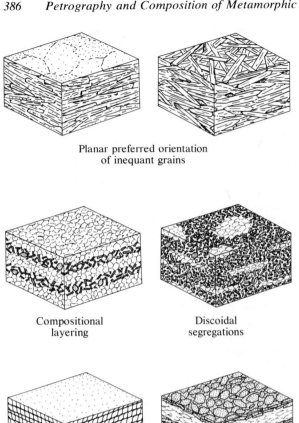

Figure 11-2 Expressions of foliation in metamorphic rocks. Schistosity of schists and cleavage of slates are chiefly represented by the upper left-hand block. Middle blocks typify some granoblastic gneisses. Lower left block can be compared to Figure 10-1, in which a late planar foliation has transected an earlier preferred orientation of platy grains. Lower right-hand block is the most common expression of foliation.

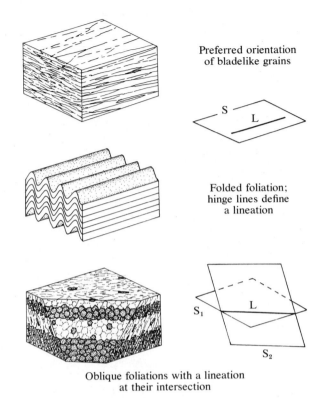

Figure 11-3 Combined expressions of foliation (S) and lineation (L). Preferred orientation of bladelike grains is geometrically similar to the aligned bladelike deformed pebbles in the metaconglomerate shown in Figure 10-2b.

and most lineated rocks are also foliated. These two directional fabric elements are seldom independent of one another (see Figures 11-3, 11-4, and 11-5). Linear streaks of micaceous aggregates, for example, also define a foliation because the basal planes of the mica grains are more or less parallel. Wrinkled or folded foliation defines a lineation. Two oblique foliations create a lineation where their surfaces intersect. Rocks with combined planar–linear fabric will ordinarily break along the foliation, and it is on these surfaces that the lineation is seen.

In natural anisotropic fabrics there is never a perfect preferred orientation of inequant grains or grain aggregates. Not every one of the amphiboles in Figure 11-5, for example, is exactly parallel. The degree to which the preferred orientation is developed varies between the ideal end-member

(a)

(b)

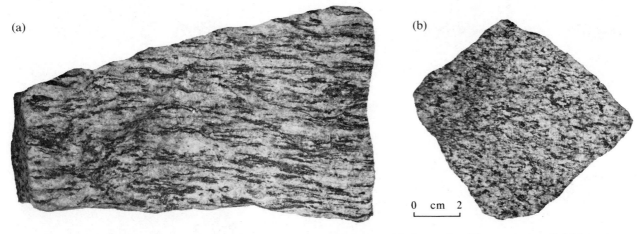

0   cm   2

Figure 11-4   Lineated and weakly foliated feldspar–quartz–biotite gneiss. Photos are parallel (a) and perpendicular (b) to lineation defined by bladelike aggregates of biotite.

extremes of perfectly oriented (which is never attained in nature) and random (isotropic). Some anisotropic fabrics are so weak and imperceptibly oriented that they can only be detected by careful measurements and statistical analysis.

There is a temptation on the part of the novice to denote any linear feature seen on the surface of a rock as a lineation. Traces, or intersections, of foliation on random, oblique broken surfaces (joints) or other nonpenetrative surfaces are not a true lineation. The lines on the sides of the blocks in Figure 11-2 are merely traces of foliation.

Influence of Composition on Imposed Fabric

The mineralogical composition of a rock body—reflecting *P, T,* and *X* of the system at equilibrium—has a strong influence on the expression of fabric because of the characteristic shapes of the mineral grains themselves.

Interlayered bodies of quartzite and schist metamorphosed under the same *P–T* conditions, for example, can display conspicuously different development of foliation; the schist may be strongly foliated because of a high degree of preferred orientation of abundant mica flakes, whereas the quartzite is only weakly foliated because of moderately discoidal quartz grains, or it may be sensibly nonfoliated if the quartz grains are equant. Rocks containing characteristically inequant mineral grains—micas, chlorite, amphibole—will be foliated if these grew under nonhydrostatic stress conditions. But rocks formed under the same stress conditions with an abundance of characteristically equant mineral grains—carbonates, quartz, feldspars, pyroxenes, garnet—may be only weakly foliated. Lower-grade metamorphic rocks will tend to be better foliated because of the generally greater amounts of platy phyllosilicates compared to higher-grade rocks.

Lineation also depends upon compositional factors for expression; marble is hence a poor rock for development of lineation, but rocks with amphiboles can readily express lineation, as can relatively fine-grained rocks with intersecting foliations or easily crenulated micaceous rocks.

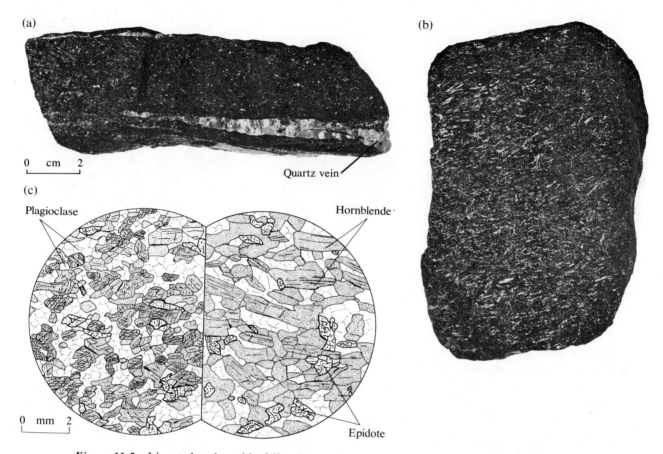

Figure 11-5   Lineated and weakly foliated amphibolite. Photographs are perpendicular (a) and parallel (b) to lineation and foliation. Microscopic scale drawings (c) are similarly oriented.

Summary of Metamorphic Fabrics

A wide variety of fabrics are found in metamorphic bodies. No attempt is made here to explain their origin completely. More than one type may occur in the same body—on the same scale in the same volume of rock, or on different scales in different volumes. If no specific name is available for a particular fabric, it may simply be described. *Blasto-* as a prefix refers to a relict fabric (blastoporphyritic, blastoophitic), whereas *-blastic* as a suffix refers to solid-state crystallization during metamorphism.

Specific metamorphic fabrics include the following:

Foliation   Any *penetrative* set of more or less parallel surfaces.

Lineation   Any *penetrative* set of more or less parallel lines.

The above two terms apply regardless of scale, whereas the following are textural and apply to grains in a metamorphic rock:

Phaneritic   Grains large enough to be identified by the naked eye.

Aphanitic   Grains too small to be identified by the naked eye.

Megacrystic   A useful nongenetic term to describe any rock, regardless of origin, or where origin is uncertain, containing relatively large grains—**megacrysts**—of any shape in a finer-grained matrix.

Blastoporphyritic   An inherited porphyritic fabric derived from a magmatic rock in which relict phenocrysts, even though replaced by aggregates of more stable metamorphic minerals, still show their characteristic habits (see Figure 1-9b).

Porphyroblastic or **metacrystic**   Larger grains that grew during metamorphism, as shown by shapes compatible with crystalloblastic processes or by their unique occurrence in metamorphic rocks (staurolite, kyanite, and so on), are embedded in a finer-grained matrix (see Figures 10-14 and 10-16).

Poikiloblastic   Grains, especially metacrysts, containing numerous small, randomly oriented inclusions of other minerals (see Figures 10-14a, 10-15, and 10-21).

Granoblastic   A mosaic of equidimensional anhedral grains; inequant grains, if present, are randomly oriented (see Figure 10-12).

Lepidoblastic   An abundance of platy mineral grains—micas, talc, chlorite, graphite—with strong planar preferred dimensional orientation; laminae of these platy grains, called folia, may be crenulated into small folds; the folia impart a **schistosity** to the rock (see Figures 10-20c,d) if it is phaneritic, or **slaty cleavage** if aphanitic (see Figure 10-20b).

Nematoblastic   An abundance of acicular or columnar grains displaying a linear preferred dimensional orientation; imparts a lineation to the rock (see Figure 11-5).

Mylonitic   A very fine-grained, usually aphanitic, streaked anisotropic fabric produced by intense deformation through cohesive solid-state ductile flow.

Flaser fabric   A type of mylonitic fabric in which ovoidal megacrysts that have survived deformation lie in the finer mylonitic matrix (see Figures 10-7c and 10-8c).

11.3 CLASSIFICATION OF METAMORPHIC ROCKS

Classification can be based upon (1) fabric, (2) field relations, in a broad sense, (3) chemical composition, and (4) inferred $P–T$ conditions of metamorphism. Metamorphic rocks may also be classified on the basis of surviving relict fabrics and mineralogical compositions, if obvious, simply by appending the prefix **meta-** to the name of the original rock. Examples include metasedimentary rocks, metavolcanic rocks, metaandesite (see Figure 1-9b), metagraywacke (see Figure 10-20a), metaconglomerate (see Figure 10-2), and metadiabase (see Figure 10-18a). The prefixes **para-** and **ortho-** are used specifically to denote sedimentary and igneous protoliths, respectively, where these can be ascertained. Examples include paragneiss and orthoamphibolite.

The classification most commonly used on the scale of the hand sample and outcrop is based chiefly upon how well a foliation is developed, with composition playing a subsidiary role. This reflects metamorphic conditions, especially the imposed fabric and the stress conditions under which it formed. Chemical composition is minimized because it is generally inherited from the protolith and hence may not be indicative of metamorphic conditions.

The basic scheme of classification of metamorphic rocks has three subdivisions: (1) strongly foliated rocks, which break readily along the foliation, usually because of abundant oriented mica or other phyllosilicate flakes; (2) weakly foliated

Table 11-1   The three basic fabric divisions for metamorphic rocks.

1 Strongly foliated rocks	2 Weakly foliated rocks	3 Nonfoliated to weakly foliated rocks
Slate	Gneiss	Granofels
Phyllite	Migmatite	Amphibolite
Schist	Mylonite	Serpentinite
		Greenstone
		Greisen
		Hornfels
		Quartzite
		Marble
		Argillite
		Skarn

rocks, in which the foliation is mechanically rather passive; and (3) nonfoliated rocks (see Table 11-1). Nomenclature can depend on scale because individual foliation surfaces may be more widely spaced than the size of a hand sample. For example, an outcrop of gneiss may be comprised of hand-sample-size layers of amphibolite and quartzo-feldspathic granofels.

Although the list of names in Table 11-1 may seem to be comprehensive and exhaustive, in practice this is not the case. Common misfits are relatively fine-grained rocks that lack a relict fabric and are composed of a weakly or nonfoliated aggregate of minerals not compatible with any of the above names. Such rocks may simply be called ''rock,'' with an appropriate compositional adjective. Examples would be actinolite rock, quartz–chlorite–epidote rock, and so on.

Because classification of metamorphic rocks depends less on specific boundary conditions and is certainly less quantitative than igneous rock nomenclature, it is common practice to prefix appropriate modifiers to the basic names listed

in Table 11-1. A few examples will suffice to illustrate how judicious use of terms can make the rock nomenclature quite informative:

Quartz–biotite–garnet schist   A strongly foliated rock whose major constituents are, in order of abundance, quartz, biotite, and then garnet.

Porphyroblastic (*metacrystic*) *feldspar schist*   Conspicuous, large porphyroblasts (metacrysts) of feldspar in schist.

Lineated hornblende–plagioclase schist   Strong alignment of acicular grains of hornblende imparts a lineation, as well as foliation, to the rock.

Crenulated phyllite   A phyllite whose foliation is intricately and pervasively folded on a small scale.

Phyllitic metatuff   A rock whose original nature as a tuff is conspicuous due to the presence of relict volcaniclastic texture but in which metamorphism has induced crystallization of strongly aligned, aphanitic micas.

Quartzitic (*or micaceous, or garnetiferous, and so on*) *schist*   The most conspicuous mineral is quartz (or mica, or garnet, and so on) in a strongly foliated rock.

Pelitic schist   Rock with strong foliation whose mineralogical composition denotes derivation from an aluminous sedimentary rock, such as shale.

Chloritic schist   Strongly foliated rock composed wholly or largely of chlorite.

Calc-silicate gneiss   A phaneritic rock whose weak foliation is expressed by segregated layers of calc-silicate minerals such as diopside, wollastonite, idocrase, and so on.

Foliated metaconglomerate   A rock whose relict texture is obviously that of a conglomerate but in which metamorphism has induced a foliation expressed by aligned mica flakes and/or flattened pebbles.

Schistose amphibolite A hornblende–plagioclase rock that displays foliation too weak for it to be placed, justifiably, in the class of a schist.

Other names can be built up in a similar manner.

11.4 DESCRIPTION OF METAMORPHIC ROCK TYPES

The purpose of this section is to present the salient fabrics and mineralogical compositions of metamorphic rock types so they may be more adequately recognized and described in hand samples and outcrops. The outline follows the basic subdivisions outlined in Table 11-1.

Strongly Foliated Rocks

Rocks in this category readily break along a foliation that is usually expressed by lepidoblastic phyllosilicate mineral grains, locally augmented by thin layers of contrasting mineralogical composition.

Slates are aphanitic, tougher than shale, and have a dull luster on well-developed slaty cleavages. Rarely, in metasedimentary rocks thin relict silty or sandy beds may lie parallel or slightly oblique to the cleavage and these can show relict grading and other primary sedimentary features. Relict clastic fabrics are evident in thin section. Even though clay minerals in shale protoliths have recrystallized to chlorites and micas, the angular outlines of the original quartz grains may be still evident. Distorted fossils are very rarely found. Weakly schistose metagraywackes are commonly interlayered with slates, as are various metavolcanic rocks such as greenstones and metatuffs that have relict magmatic fabrics. Slates are purple or red (with abundant hematite), green (chlorite) and, more commonly, gray to black (graphite). Some have metacrysts of euhedral pyrite. Low-temperature recrystallization of slate in contact aureoles creates metacrysts of poikiloblastic cordierite, andalusite, or mica, forming **spotted slates** (see Figure 10-14a); locally the cordierite or another metacrystic phase preferentially weathers out, leaving pits. Most slates are pelitic and derived from shale or mudstone protoliths, but some are metamorphosed tuffs and other magmatic rocks.

Phyllites are also aphanitic, but because of a slightly coarser grain size than slates, they have a lustrous sheen on foliated surfaces. They are transitional between slates and schists and share common associations and properties.

Schists are phaneritic, commonly lineated, and have a weak to well-developed segregation layering of felsic and mafic material that usually augments the already strong schistosity expressed by lepidoblastic phyllosilicate mineral grains (see Figures 10-20c,d, 10-14b, and 11-6). Locally, the oriented phyllosilicate grains may be oblique to the layering. Lineation is commonly expressed by elongate segregations of contrasting minerals or by folds and wrinkles in the foliation. Relative to slates and phyllites, the coarser-grained schists have better-developed imposed metamorphic fabrics that have modified and usually erased relict magmatic and sedimentary features. The common segregation layering in schists may have originated by metamorphic differentiation rather than being inherited bedding. All chemical classes are found in schists, but pelitic and quartzo-feldspathic compositions with prominant micas and chlorites are most common. Diagnostic minerals that indicate low- to intermediate-grade P–T conditions include metacrysts of kyanite, garnet (usually Fe–Mg–Mn–Al solid solutions), staurolite, and alkali feldspar; matrix minerals commonly include (besides quartz, biotite, muscovite, and chlorite), plagioclase, epidote group minerals, amphiboles, sillimanite, Fe–Ti oxides, carbonates, sphene, serpentine, and talc. Bodies of schist and some phyllite have conspicuous veins, pods, and other segregated masses composed chiefly of quartz. A

(a)   (b)

Figure 11-6   Mica–quartz–feldspar schist. (a) Large exposure showing strong planar foliation. (b) Close-up view of the foliation surface, on which is a well-developed lineation (parallel to hammer handle) expressed by streaks of alternating darker micaceous and lighter quartzo-feldspathic material. Note conjugate set of joints perpendicular to foliation and intersecting in an obtuse angle that is bisected by the lineation.

wide variety of other metamorphic rock types are associated with schists.

Weakly Foliated Rocks

Rocks in this category have only a weak parallelism of inequant mineral grains, or a mechanically passive compositional layering. Foliation in this class of rocks can be readily discerned, but the rock does not readily break along it.

Mylonite is a hard, fine-grained rock with mylonitic or flaser fabric that generally occurs as sheetlike bodies within coarser rocks. Quartz and feldspar are usually the chief constituents, hence when very fine grained, mylonite commonly appears as chertlike masses. However, the streaked, foliated, and/or lineated appearance, the presence of less deformed ovoidal relict grains (flaser fabric), and the intensely strained grains obvious in thin section all serve to indicate mylonite (see Figures 10-7 and 10-8). Though mylonites were once

believed to have formed by brittle cataclasis, involving crushing, pulverization, and milling of the original rock (the Greek word *mylon* means "to mill"), it is now apparent that most if not all mylonites form by cohesive ductile flow under nonhydrostatic stress.

Gneisses are composed chiefly of quartz and feldspar and are perhaps the most widespread rock type in regionally metamorphosed Precambrian shields. They are medium- to coarse-grained phaneritic, and granoblastic to lepidoblastic. Compositional layering expressed by varying modal proportions of quartz, feldspar, micas, and hornblende is usually evident (see Figures 11-7a, 11-8, and 11-9), but in some granitic gneisses layering is quite subtle or even absent and a weak foliation is expressed by preferred orientation of inequant mineral grains or grain aggregates (see Figure 11-7b). Gneisses with ovoidal megacrysts of feldspar are sometimes called **augen gneiss** (*Augen* is German for "eyes").

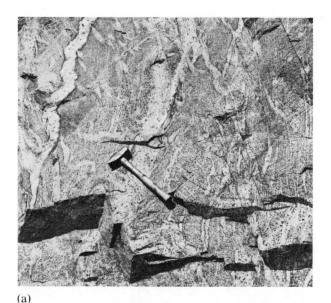

(a)

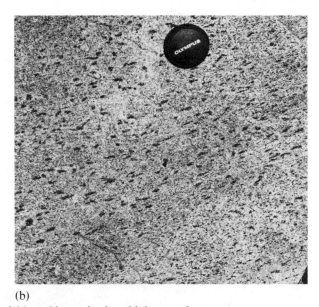

(b)

Figure 11-7   Rather massive migmatitic quartzo-feldspathic gneiss in which a weak, mechanically insignificant foliation is expressed by more felsic segregations that contain less black hornblende and biotite and by preferred orientation of these mafic minerals or aggregates of them. (a) Some segregations have very subtle definition and grade imperceptibly into more mafic rock; others, such as in the upper left corner, are almost wholly felsic and have thin borders of wholly mafic material. (b) Close-up view of more homogeneous area of exposure showing trace of weak foliation that is expressed by preferred orientation of mafic mineral grains and especially by elongate clots of them.

Both magmatic and sedimentary protoliths are possible for gneisses, and where this can be decided, the terms **orthogneiss** and **paragneiss,** respectively, can be used. Massive granites intruded pre- or syntectonically in catazonal environments are the parents of many bodies of quartzo-feldspathic orthogneiss. Deformation of porphyritic granite may produce augen orthogneiss, or the augen may develop by solid-state growth, perhaps in response to alkali metasomatism (see pages 515–517). Massive, structureless bodies of granite may develop gneissic segregation layers during syntectonic recrystallization. The possibilities are numerous.

Migmatite (which means "mixed" in Greek) is a composite, heterogeneous rock consisting of mafic metamorphic material mingled with leucocratic granitic rock in the form of planar to folded and contorted layers, criss-crossing veins, and irregular pods. The granitic component is sharply bounded to gradational with the more mafic host schist, gneiss, or amphibolite. Often the two components are so mingled as to suggest very ductile or even partially melted conditions (see Figure 11-10). Numerous illustrations by Mehnert (1968) show the great variety of fabrics in migmatites. Because the granitic and metamorphic members occur on scales of several centimeters or even meters, migmatites are recognizable in outcrops rather than hand samples. The distinction between quartzo-feldspathic gneisses and migmatites can be vague; some petrologists would call the rocks shown in Figures 11-7a, 11-8, and 11-9 migmatite.

Migmatites occur locally around some large mesozonal granitic plutons, but they are far more voluminous in deep regional metamorphic terranes

Figure 11-8   Migmatitic quartzo-feldspathic gneiss that is more strongly foliated than the body shown in Figure 11-7. Note the folded and transposed felsic layers that formed during cohesive ductile flow. Above and to left of hammer are wedges of finer-grained amphibolite that may represent an intrusive dike that was broken in a more brittle manner during flow.

where mineral assemblages are in the upper amphibolite facies. Their origin has long been the subject of heated controversy, and a bewildering terminology, partly descriptive and partly genetic, has grown during this debate (Mehnert, 1968). In some migmatites, granitic material has apparently been injected as magma from outside sources (see Figure 11-10b). However, in most cases, the mafic and leucocratic components are so intimately intermingled that some sort of *in situ* segregation

process seems required. This would be especially relevant in the case of regionally extensive migmatites not conspicuously associated with major granitic intrusions and where the leucocratic and mafic components seem to be segregations of an original material intermediate in composition between the two (see Figures 11-7a and 11-10a). But whether the leucocratic and mafic components segregate by subsolidus metasomatic processes or by partial melting is still debated. Perhaps both

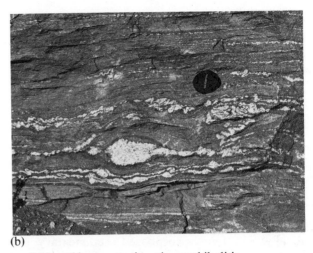

(a)

(b)

Figure 11-9 Migmatitic gneiss composed of quartzo-feldspathic segregations in amphibolitic schist. (a) Layers of felsic and mafic rock are fairly even and continuous. (b) More discontinuous segregations that are conspicuously coarser-grained than the moderately well-foliated, dark-colored hornblende–biotite feldspar schist.

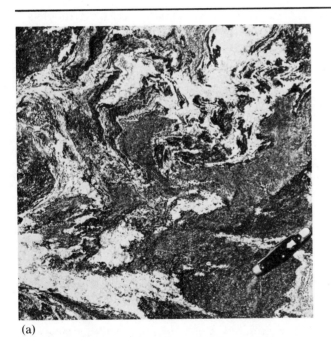

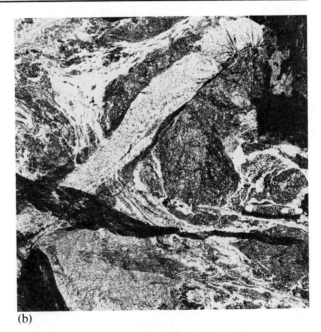

(a)

(b)

Figure 11-10 Migmatite. Pocketknife is 8 cm long. (a) Irregular, contorted veins and pods of coarse leucocratic granitic material lie in a finer-grained host containing abundant biotite, hornblende, and some plagioclase. Note that the major leucocratic area at the top center encloses material more mafic than the general host rock, as if the two resulted from segregation, or differentiation, of the original intermediate rock. (b) Similar to (a) except that a late granitic dike cuts across the migmatitic veins.

processes occur in different places at different times.

Nonfoliated to Weakly Foliated Rocks

Rocks with nonfoliated (isotropic) to very weakly foliated fabric are dominated by equant mineral grains; inequant grains, if present, are insignificant or randomly oriented so that the fabric is massive or granoblastic.

Quartzite is composed principally of quartz; when broken, the rock fractures through, not around, the grains. Rocks of this character occur locally in unmetamorphosed sedimentary sequences and may be called **orthoquartzite** to distinguish them from **metaquartzite** in metamorphic terranes. Massive quartzites, possibly with relict bedding structures, grade into laminated or slabby quartz schists in which microscopically thin folia of mica alternate with thicker lenses and layers of quartz. Marbles, schists, phyllites, and slates representing metamorphosed limestones and shales are commonly associated with quartzites and quartz–mica schists. Many quartzites are metacherts.

Marble, dolomite marble (dolomarble), and calc-silicate schists and gneisses are derived from limestones and dolomites with varying amounts of sandy and shaly material. Quartz, diopside, tremolite, talc, phlogopite, wollastonite, calcic plagioclase, idocrase, forsterite, and grossular–andradite appear together with calcite and dolomite. Gray coloration, often uneven and streaked, in otherwise pure marbles is produced by minute specks of graphite. Relict beds may be preserved.

Skarn (or tactite) is a rock composed of calcium, magnesium, and iron silicates that has been derived from nearly pure limestones or dolomite into which large amounts of Si, Al, Fe, and Mg were metasomatically introduced because of proximity to an igneous intrusion. Relict bedding may be apparent in outcrop. Skarns are commonly host rocks for economic deposits of magnetite and Cu sulfides.

Greisen is a granoblastic aggregate of quartz and muscovite (or lepidolite) with accessory amounts of topaz, tourmaline, fluorite, rutile, cassiterite, and wolframite formed by metasomatic hydrothermal alteration of granitic rock. Note the significant amount of fluorine in this assemblage.

Hornfels is a rock formed by thermal metamorphism and solid-state crystal growth within a contact aureole surrounding an igneous intrusion; aphanitic to fine phaneritic grain size and granoblastic fabric are typical but relict bedding may be present; hornfelses are commonly but not necessarily pelitic. Note that the field relations must be known in order to justify this term.

Argillite is a very low-grade, dark-colored, aphanitic, often conchoidally fracturing, firmly indurated rock that has undergone only slight recrystallization from the parent mudstone; relict bedding may be present.

High-grade quartzo-feldspathic rocks. **Granofels** is a term sometimes used for phaneritic, massive, granoblastic metamorphic rocks that, because of the dominance of equant grains, such as quartz and feldspar, lack a foliation. Mafic minerals are also equant grains of pyroxene and perhaps garnet. Such rocks are typical of the high-grade anhydrous granulite facies and are sometimes designated by the name **granulites.**

Generally, in Phanerozoic orogenic regional terranes, metamorphism culminates in sillimanitic or migmatitic rocks of the amphibolite facies. In contrast, Precambrian shields may locally expose older Proterozoic and Archean rocks transitional into, or lying wholly within, the granulite facies. Characteristic of this facies is the assemblage of augite (diopside) + hypersthene + plagioclase. As either monoclinic Ca-rich or orthorhombic Ca-poor pyroxene can occur at lower grades in certain types of rocks, a single pyroxene is not diagnostic. Other minerals stable in the granulite facies include garnet, kyanite, sillimanite, quartz, alkali feldspar, calcite, and forsterite. Hydrous phases, usually amphibole or biotite, are scarce.

Granulite facies assemblages apparently form under water-poor, catazonal conditions overlapping, at least in part, the *P–T* range of the amphibolite facies. Justification of this view lies in the occurrence of layers of amphibolite facies assemblages intermingled with granulite facies and in the occurrence of unusual mafic dikes with amphibolite facies assemblages along the margins and granulite in the core.

Granulite assemblages commonly occur in rocks with compositional layering and granoblastic textures; some rocks are locally mylonitic. In older texts on metamorphic rocks, ribbonlike grains (see Figure 10-7d) were said to define granulitic fabric, but this term has now fallen into disuse. Weak foliation expressed by compositional layering and mylonitic fabric may be subhorizontal over wide areas, and local variations in attitude are associated with large-scale recumbent folds. Lenses of granitic rock occur on many scales.

Feldspars in the granulite facies are generally dark colors—red, green, brown, and even black. Plagioclases have small amounts of exsolved potassium feldspar (antiperthite), and alkali feldspars are intergrowths of subequal amounts of K and Na end members (mesoperthites). These types of perthites indicate high temperatures of crystallization. Some quartz grains in hand sample show an unusual blue opalescence due to the presence of minute needles of rutile apparently formed by exsolution of TiO_2 originally dissolved in the quartz at high temperatures.

Familiar calc-alkaline granitic rocks have minerals (alkali feldspar + plagioclase + biotite + hornblende) stable in at least part of the amphibolite facies range. Rocks of similar bulk chemical composition with granulite facies assemblages (anhydrous mafic silicates, especially) and crystalloblastic fabric belong to the **charnockite suite.** The magmatic versus metamorphic origin of charnockites has long been debated.

Metamorphosed mafic rocks of low to intermediate grade. Basaltic, diabasic, and gabbroic and some andesitic protoliths composed chiefly of calcic plagioclase and pyroxene form greenstones and amphibolites with isotropic fabrics upon meta-

morphism. If nonhydrostatic stress is significant during recrystallization, mineralogically equivalent slates, phyllites, schists, and gneisses may form. **Greenstone** is an aphanitic, weakly metamorphosed mafic igneous rock composed of a haphazard isotropic array of fine-grained epidote group minerals, prehnite, pumpellyite, chlorite, actinolite, albite, sphene, and locally a carbonate mineral. This greenschist facies assemblage of minerals partially to completely replaces primary magmatic minerals (see Figures 10-18a,b); relict igneous fabrics are often preserved and monomineralogic veins and patchy segregations are common. With increase in metamorphic grade, albite and epidote react to form oligoclase or andesine and sodic aluminous hornblendes take the place of actinolite and chlorite so that the rock is composed chiefly of hornblende and plagioclase with possible minor amounts of biotite, sphene, quartz, and possibly pyroxene and almandine-rich garnet (see Figure 10-18c). Such **amphibolites** are phaneritic and show little if any relict magmatic fabrics.

Very possibly the greatest volume of greenstones and amphibolites lie buried beneath the oceans, where they have formed by sub-sea-floor metamorphism (alteration). Effects of recrystallization and metasomatism have been imprinted to a variable degree. Metamorphism is patchy, even on the scale of a thin section, and veins of secondary metamorphic mineral assemblages are common. Disequilibrium mineral assemblages and fabrics are widespread. Because of this, the change from pristine crustal magmatic rock is often referred to as submarine alteration rather than metamorphism. Original vesicles in submarine basaltic lavas have been filled with carbonates and zeolites, producing amygdules. Relict magmatic fabrics are commonly preserved in greenstones, such as pillow rinds, interpillow clastic fabrics, and variolitic, intergranular, ophitic, and porphyritic textures. Rocks with discernable primary fabrics can be classed as metabasalt, metagabbro, and so on where appropriate. Although most metamorphically imposed fabrics are isotropic, rare mylonitic samples have been dredged from the sea floor, and these presumably represent bodies from shear zones, especially related to transform faulting.

Cores of submarine pillow lavas are locally recrystallized to a greenschist facies assemblage of albite + epidote + chlorite + actinolite + sphene, whereas the original glassy rind is now a metasomatic chloritic aggregate. Ca, Na, and K seem to have been especially mobile during metasomatism.

Veins up to several centimeters in width are widespread and consist of combinations of quartz, chlorite, epidote, and calcite. Veins have more limited, even monomineralic, assemblages than host rocks, again reflecting the action of metasomatic processes and transport of dissolved ions in hydrothermal fluids. Locally, vein assemblages are out of equilibrium with the mineralogical composition of the host, indicating continuing percolation of hydrothermal solutions of different temperatures and composition after reconstitution of the rock.

It is of historical interest, in bridging pre– and post–plate tectonic geologic thinking, to consider the "spilite problem." A **spilite** is a greenstone of basaltic textural aspect composed of low-metamorphic-grade minerals including albite, chlorite, actinolite, sphene, calcite, various hydrated Ca–Al silicates (epidote, prehnite, laumontite), and relict pyroxene. Since their first recognition over 150 years ago as pillow lavas within ophiolitic sequences, spilites have held a tempestuous position in petrologic discussion. Despite the obvious relict basaltic textures and low-metamorphic-grade mineral phases, proponents of a unique spilite magma have only recently become silent. There is a close correspondence between the compositions of basalts and spilites, although the latter, as a group, are characteristically more variable in SiO_2 and slightly lower in FeO and CaO, and higher in Na_2O and H_2O. Transition metals such as Cu, Co, Ni, and so on are lower in spilites. We now view spilites as forming from sea-floor basalts near oceanic rifts by metasomatic interchange with heated sea water. Subsequent metamorphic imprints may develop during obduc-

tion of ophiolite onto continental margins, so the history of spilites can be complex.

Serpentinite.   The ultramafic members of ophiolite comprising peridotites and minor dunites and pyroxenites are variably serpentinized, forming **serpentinite.**

There are basically two fabrics in serpentinite. In the **massive** type, formed under essentially hydrostatic stress conditions, relict textures of the dunite or peridotite protolith are still evident in a meshlike aggregate of secondary serpentine minerals (see Figure 5-27). Veins of cross-fiber asbestos serpentine are common, and freshly broken surfaces are olive to black in color whereas weathered ones are yellow-orange to red-brown. In **sheared** serpentinite, a multitude of polished and slickensided slip surfaces are visible in hand sample. Colors range from black to green to pale yellow, commonly within a single hand sample. Relict fabrics have been obliterated by the penetrative deformation under nonhydrostatic stresses. Bodies of sheared serpentinite, as expected, are mechanically very weak.

Three serpentine minerals—antigorite, lizardite, and chrysotile—in addition to minor amounts of brucite, talc, secondary magnetite, and carbonate minerals constitute serpentinite. Subtle chemical as well as structural differences exist between the three serpentine phases, so they are not truly polymorphs with separate P–T stability fields. The first two have platy habits and the third is fibrous asbestos.

Most serpentinized alpine (ophiolite) ultramafics consist of lizardite + chrysotile + brucite + magnetite and probably formed between atmospheric temperatures and 300°C. Antigorite-bearing serpentinites are much less abundant and probably formed by metamorphism under greenschist- or amphibolite-facies conditions at temperatures as high as 550°C (Coleman, 1977).

There has been a good deal of uncertainty and resulting controversy regarding the specific hydration reaction involved in serpentinization. Four possible reactions involving chiefly olivine are as follows:

$$3Mg_2SiO_4 + 4H_2O + \underset{introduced}{SiO_2} \rightarrow 2Mg_3Si_2O_5(OH)_4 \quad (11.1)$$

olivine serpentine
131 cm³ 215 cm³

$$2Mg_2SiO_4 + 3H_2O \rightarrow Mg_3Si_2O_5(OH)_4 + Mg(OH)_2 \quad (11.2)$$

87 cm³ 108 cm³ brucite
24 cm³

$$Mg_2SiO_4 + MgSiO_3 + 2H_2O \rightarrow Mg_3Si_2O_5(OH)_4 \quad (11.3)$$

44 cm³ enstatite 108 cm³
31 cm³

$$5Mg_2SiO_4 + 4H_2O \rightarrow 2Mg_3Si_2O_5(OH)_4 + 4MgO + SiO_2 \quad (11.4)$$

219 cm³ 215 cm³ removed in
solution

The first three reactions demand large expansions in volume of the crystalline assemblages, whereas the last one maintains almost a constant volume but considerable MgO and SiO_2 are removed from the system. Although subtle expansion-related textures have been documented, common mesh-textured serpentinites with relict primary grain outlines do not seem to show expansion. Yet, voluminous Mg- and Si-metasomatism in neighboring rocks is not always apparent either.

In some serpentinized ultramafic bodies, element ratios, particularly Mg/Si, are the same as in a texturally inferred protolith. In such cases, it appears that the reaction shown in Equation (11.2) may have prevailed to serpentinize dunites, and combinations of Equations (11.2) and (11.3) for harzburgites (olivine + enstatite peridotite). These reactions ignore the small amounts of Fe present in these ultramafic systems, and an Fe-oxide phase should be listed as a product in every case.

Small amounts of Ca are held in solid solution in orthopyroxenes, and major amounts in clinopyroxene in peridotites. Although clinopyroxenes locally survive serpentinization where orthopyroxene and olivine are completely replaced, any Ca released from the ultramafic body can promote Ca-metasomatism as it migrates via a fluid phase into the surrounding country rocks. Such Ca-metasomatism, possibly overprinted upon earlier thermal contact metamorphic and possible tectonic effects related to emplacement of the peridotite, produces **rodingites** from gabbros and diabases composed of grossular and prehnite, sometimes with wollastonite, diopside, and hydrogrossular.

Subsolidus hydration reactions between water and magmatic olivine and pyroxene forming serpentine potentially occurs (1) in the oceanic environment along deep penetrating fractures, (2) during, and (3) after emplacement of the ophiolite on land. Insight into whether oceanic or connate water in the metamorphic system is involved can be gained by analysis of the hydrogen and oxygen isotopic composition of the serpentine (Coleman, 1977, p. 101).

Rocks of the high P/T facies series. A section briefly describing these unusual and rare rocks is warranted, especially to indicate their mineralogical nature. Though not present in all compositional rock types of the blueschist facies, blue sodic amphiboles with large concentrations of the glaucophane, $Na_2Mg_3Al_2(Si_8O_{22})(OH)_2$, end member are prominent and provide the name for the facies. In contrast, typical amphiboles of greenschist and amphibolite facies rocks are calcic actinolite and hornblende. Pyroxenes in blueschist facies rocks are also sodic and comprise monoclinic solid solutions containing appreciable concentrations of the jadeite, $NaAlSi_2O_6$, end member substituting for diopside, $CaMgSi_2O_6$, and other components in common pyroxenes of lower-P, higher-T environments. Calcium in blueschist assemblages is taken up in aragonite, sphene, and the hydrous Ca–Al silicates lawsonite, epidote, and pumpellyite. Lawsonite is a surprisingly dense (~ 3.1 g/cm³) hydrous mineral whose composition, $CaAl_2(OH)_2(Si_2O_7)H_2O$, is equivalent to anorthite (density 2.76 g/cm³) plus water. Pumpellyite is an epidotelike mineral with a composition roughly equivalent to epidote plus actinolite plus extra water.

As blue amphiboles have a broad P–T stability field extending into that of the greenschist facies, this phase alone has no particular significance, but in association with lawsonite, aragonite, and/or jadeitic pyroxene it is diagnostic of the blueschist (glaucophane–lawsonite schist) facies. Absence of feldspars other than albite; the absence of biotite, andalusite, and sillimanite; and the presence of the high-pressure $CaCO_3$ polymorph, aragonite, instead of calcite are noteworthy.

Graywackes and metagraywackes are widespread in high-P/T metamorphic belts such as in the Franciscan rocks of the California Coast Ranges, where they comprise $\sim$ 80% of the sequence. Originally, graywackes were composed of clastic grains of quartz, feldspar, mica, pyroxene, aphanitic volcanic rocks, and locally chert, all in a chloritic-clay matrix (see Figure 10-20a). Both sheared and undeformed massive metagraywackes with obvious epiclastic fabric contain zeolite veins and patchy replacements of detrital feldspar grains in lowest-grade rocks of the zeolite facies. Prehnite, pumpellyite, epidote, and albite appear in the higher-grade metagraywackes of the prehnite–pumpellyite facies. In highest-grade metagraywackes, still commonly with obvious relict epiclastic fabric and bedding (see Figure 11-11), lawsonite, jadeitic pyroxene, and sometimes glaucophane appear. Many of the critical minerals in all grades are easily overlooked even in thin section because of their fine grain size. Some syntectonically metamorphosed graywackes, are thoroughly recrystallized schists composed of quartz + albite + muscovite + chlorite $\pm$ glaucophane $\pm$ lawsonite $\pm$ aragonite.

Metamorphosed marine cherts rich in Mn and Fe^{+3} are thinly laminated granoblastic aggregates of quartz + hematite + Fe–Mn garnet (spessartine) $\pm$ riebeckite $\pm$ piemontite (Mn-epidote).

Basaltic rocks comprise approximately 10% of the Franciscan Complex. Most are weakly metamorphosed greenstones—isotropic aggregates of greenschist facies minerals commonly with relict magmatic minerals and fabrics, including pillows and variolitic, amygdaloidal, and ophitic textures.

Glaucophane-bearing rocks actually comprise less than 1% of the Franciscan rocks in the California Coast Ranges. In addition to its presence as disseminated grains in metagraywackes, glaucophane-rich blue amphibole also has three other prominent modes of occurrence: (1) in foliated and lineated schists, cut by quartz veins, comprised chiefly of lawsonite + glaucophane + quartz; (2) in metamorphosed basalts, now schists, composed of glaucophane $\pm$ lawsonite $\pm$ pumpellyite $\pm$ epidote $\pm$ almandiferous garnet $\pm$ jadeitic pyroxene $\pm$ aragonite; (3) as coarse-grained, rounded blocks up to a meter or so in diameter with sheared margins locally of actinolite and talc, lying in a matrix of serpentinite.

Eclogites (Coleman and others, 1965) are essentially bimineralic rocks comprised of almandine–pyrope garnet and jadeite–diopside pyroxene but with a bulk chemical composition essentially the same as that of basalt. Their relatively high density (3.4 to 3.5 g/cm³) suggests formation under high-pressure and probably dry conditions. Many outcrops of eclogite in the Coast Ranges are blocks within mélange (see Figure 12-24 and 12-25). Patches, veins, and layers of secondary hydrous assemblages, including glaucophane, white mica, epidote, chlorite, and actinolite are widespread (see Figure 10-31) and form by introduction of water, possibly at lower T than prevailed during formation of the primary anhydrous assemblage.

Figure 11-11 Bedded jadeitic metagraywacke exposed in roadcut near Pacheco Pass, 130 km southeast of San Francisco. [U.S. Geological Survey photograph courtesy of E. H. Bailey.]

11.5 COMPOSITION DIAGRAMS

Composition diagrams are widely used by metamorphic petrologists to show the interrelation between the bulk chemical composition of a rock and its constituent mineral assemblage, or grade. It may be surprising that so many components comprising most rocks can be adequately represented in two-dimensional diagrams, but, as we shall see, reasonable approximations are possible if certain simplifying assumptions can be made.

We begin our presentation with some fundamental concepts and then discuss specific types of diagrams.

Concepts

V. M. Goldschmidt pointed out in 1911 that most common metamorphic mineral assemblages that recur in many terranes will correspond to at least divariant equilibria (see pages 266–269). This fol-
lows from the fact that a given assemblage (mineralogical composition) in a rock of some fixed composition X is stable over a certain range of P and T, which can therefore be freely varied, within limits, without perturbing the equilibrium between the coexisting minerals. (Under very special P and T conditions, mineralogical reactions will occur, adding to the number of phases in the system and so reducing the degrees of freedom in the system to univariant or invariant conditions; see Section 14-2.) If a system is at least divariant ($F \geq 2$), the Gibbs phase rule (see page 269) $F = C + 2 - \varphi$ becomes $2 \leq C + 2 - \varphi$ or $\varphi \leq C$.

A composition or **chemographic diagram** can be drawn for a particular range of P–T conditions representing a metamorphic facies. In such a diagram, the number of phases, φ, will not exceed the number of components represented, C. To illustrate the topologic relations, consider a system composed of only three components ($C = 3$); call them h, k, and l and let them be represented

(a)

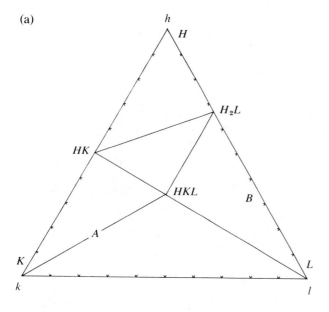

(b)

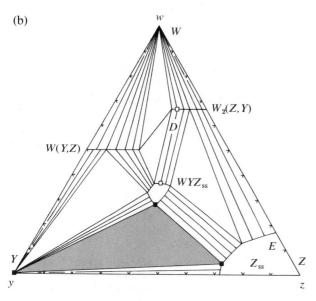

Figure 11-12 Compatible equilibrium phases in hypothetical metamorphic systems. (a) Only pure crystalline phases (K, HK, H, HKL, H_2L, L) are stable. (b) Two pure phases—W and Y—and four crystalline solid solutions—W, (Y,Z), $W_2(Z,Y)$, WYZ_{ss} and Z_{ss}—are stable. Only a few representative tie lines between coexistent phases are shown in the two-phase bands.

at the apices of a triangle (see Figure 11-12a). Only pure crystalline phases (no solid solutions) are stable in this hypothetical system. Tie lines connect phases that coexist stably together. With $C = 3$, the maximum number of phases that can compose any rock over a particular range of P and T is three. Rock A consists only of phases K and HKL, in equal proportions. B consists of the three-phase assemblage L, H_2L, and minor HKL. Figure 11-12b shows another hypothetical system composed of the components w, y, and z. Four of the crystalline phases are solid solutions. Z_{ss} and WYZ_{ss} have variable concentrations of all three components and are represented by an area in the triangle. $W(Y,Z)$ and $W_2(Z,Y)$ have variable amounts of only y and z and so are represented by lines. The extent of the line or area indicates the limits of solid solubility. W and Y are pure phases with no solid solution and so are represented by points. Because of the solid solution in the phases, the extent of stability fields of three-phase assemblages is reduced compared to those in Figure 11-12a, whereas two-phase assemblages, now represented by bands rather than lines, are more extensive.

Three-phase assemblages shown in Figure 11-12b are divariant ($F = 3 + 2 - 3 = 2$); the composition of each crystalline solid solution is uniquely fixed if P and T are uniquely specified. For example, the exact composition of each phase located at the apices of the three-phase triangle (shaded) $Y + WYZ_{ss} + Z_{ss}$ is indicated by the solid squares. Any rock within the same range of P and T whose bulk chemical composition lies within that three-phase triangle consists of exactly the same three phases represented by the solid squares; modal proportions of these phases vary, however, with changes in bulk composition of the rock.

Two phase assemblages, denoted by bundles of tie lines in Figure 11-12b, have an additional degree of freedom, that is, they are trivariant, over a particular range of P and T ($F = 3 + 2 - 2 = 3$). Specification of the mole fraction of just one component in one phase is sufficient to define completely the composition of both phases. For example, if a rock D consists of $WYZ_{ss} + W_2(Z, Y)$ and the mole fraction of component y in the latter is 0.10, then the mole fractions of w, y, and z in $(WYZ)_{ss}$ are 0.36, 0.31, and 0.33, respectively. In other words, there is a unique (for the particular range of P and T) tie line that connects this particular $W_2(Z, Y)$ phase (open square) to the $(WYZ)_{ss}$ in equilibrium with it, allowing its composition to be read from the diagram. The modal proportions of these two particular phases (open squares) in rock D are given by the lever rule (see Figure 8-15). A limited range of rocks can be composed of the same two phases, and variations in their chemical bulk composition are accomodated by variations in modal proportions of the two phases.

A one-phase assemblage over a particular range of P and T has four degrees of freedom. The composition of rock E, consisting entirely of Z_{ss}, can only be completely and uniquely defined by specifying the mole fractions of two of the three components. A variety of rocks can be composed exclusively of Z_{ss}, and their differences are reflected in the different compositions of Z_{ss}.

A variety of such triangular composition diagrams have been used to portray relationships in metamorphic rocks (Winkler, 1976). The most widely used are the AFM diagram for pelitic rocks and the ACF for mafic rocks (compare Figure 11-1).

AFM Diagram For Pelitic Rocks

Mineral assemblages in pelitic rocks have long served as valuable indicators of metamorphic conditions because of their sensitivity to changes in P and T. Metamorphosed shales and mudstones are chemically complex mixtures of many components (see Figure 11-1). Representing all of them on a two-dimensional or three-dimensional diagram is impossible. However, some reasonable assumptions can be made, following Thompson (1957), to simplify the representation. SiO_2 is generally present in sufficient concentrations so that quartz is present in all assemblages and is not totally consumed by any mineralogical reaction. Variations in the concentration of SiO_2 in such rocks are merely

reflected in variations in the modal proportion of quartz in them; this quartz-excess condition is implicit for the remainder of this section. H_2O is also generally present in excess during metamorphism. Fe_2O_3, MnO, CaO, Na_2O, and TiO_2 are generally in small enough concentration in pelitic rocks that they merely substitute for major components such as Al_2O_3, K_2O, FeO in various minerals; Na_2O may substitute for K_2O in biotite and muscovite, for example. If any one of their concentrations exceeds the solubility of the crystalline solid, then a separate mineral rich in that component will occur. Hence, the Mn-garnet spessartite may occur in rocks enriched in MnO, and rutile or ilmenite in rocks with abundant TiO_2; CaO and Na_2O in high concentrations cause the appearance of plagioclase. Four remaining major components, Al_2O_3, FeO, MgO, and K_2O, are responsible for much of the mineralogical variations observed in pelitic rocks, including compositions of index minerals such as chlorite, biotite, garnet, and so on in the classic Barrovian zones.

These four components can be represented at the corners of a tetrahedron, inside of which lies a point representing the bulk chemical composition of a pelitic rock (see Figure 11-13). But a three-dimensional tetrahedron is difficult to portray on a two-dimensional piece of paper. Thompson (1957) made the additional simplification of projecting all compositional points, lines, areas, and volumes occurring within the AFMK tetrahedron onto the AFM face, with muscovite as the point from which the projection is made. An ideal composition for muscovite, $KAl_2AlSi_3O_{10}(OH)_2$, is assumed, although real muscovites can deviate from this significantly. In projecting from muscovite, the AFM plane must be extended to catch all of the pertinent compositional features inside the tetrahedron. Choice of muscovite as a projection point is justified by the fact that it is a widespread mineral in nearly all low- to intermediate-grade pelitic rocks. In a sense, it has a status like quartz in being present in excess with all the assemblages represented in the diagram. For higher-grade rocks containing K-feldspar, that mineral may be used as a projection point. Choice of AFM as the plane of projec-

(a)

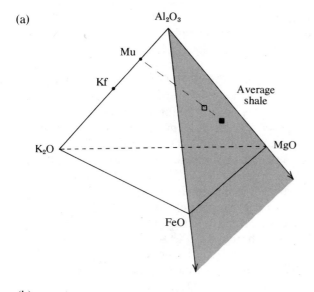

(b)

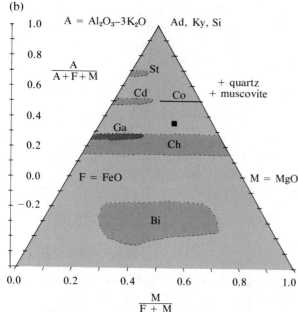

Figure 11-13 Thompson AFM projection. Abbreviations as in Figure 11-1. (a) Complete AFMK tetrahedron showing average shale from Table 11-2 (square), AFM projection face extended (shaded), and alternate ideal K-feldspar and muscovite projection points. (b) AFM projection from muscovite showing average shale and range of composition of aluminous silicates occurring in pelitic rocks.

Table 11-2 Chemical composition of average shale and calculation of A, F, and M plotted in Figure 11-13b.

	Weight percent	Molecular proportion	
SiO_2	58.1	0.967	A = mol. prop. Al_2O_3 −
Al_2O_3	15.4	0.151	3(mol. prop. K_2O)
Fe_2O	4.0	0.025	= 0.151 − 3(0.034)
FeO	2.4	0.033	= 0.049
MgO	2.4	0.060	M = mol. prop. MgO
CaO	3.1	0.055	= 0.060
Na_2O	1.3	0.021	F = mol. prop. FeO
K_2O	3.2	0.034	= 0.033
TiO_2	0.6	0.008	
P_2O_5	0.2	0.001	A/(A+F+M) = 0.35
MnO	trace	—	M/(F+M) = 0.65
CO_2	2.6	0.059	
SO_3	0.6	0.007	
C	0.8	0.067	
H_2O	5.0	0.278	
Total	99.7		

Source: H. Blatt, G. Middleton, and R. Murray, 1972, *Origin of Sedimentary Rocks* (Englewood Cliffs, N.J.: Prentice-Hall).

tion is justified by the fact that most critical minerals in pelitic rocks lie on it, or are close to it.

The detailed steps for plotting a rock or mineral onto an AFM projection are as follows:

1. Divide the weight percentage of each constituent oxide in the rock or mineral by its molecular weight to obtain the molecular proportion.

2. The amount of component A in the projection is the molecular proportion of Al_2O_3 minus three times the molecular proportion of K_2O. This subtraction is the arithmetic technique of projecting from the ideal muscovite composition point onto the AFM face of the AFMK tetrahedron, because in ideal muscovite there are three times as many moles of Al_2O_3 as K_2O.

3. The amount of component F in the projection is the molecular proportion of FeO. If ilmenite ($FeO \cdot TiO_2$) is a member of the mineral assemblage, then the molecular

Table 11-3 Chemical composition of basalt BCR − 1 in Table 2-2 and calculation of A, C, and F plotted in Figure 11-14.

	Weight percent	Molecular proportion
SiO_2	54.50	0.907
Al_2O_3	13.41	0.132
Fe_2O_3	3.28	0.021
FeO	9.17	0.128
MgO	3.41	0.085
CaO	6.98	0.124
Na_2O	3.23	0.052
K_2O	1.68	0.018
TiO_2	2.22	0.028
P_2O_5	0.36	0.003
MnO	0.19	0.003
CO_2	0.03	0.001
Total	98.45	

A = mol. prop. Al_2O_3 − mol. prop. Na_2O − mol. prop. K_2O
  = 0.132 − 0.052 − 0.018
  = 0.062

C = mol. prop. CaO − 10/3 (mol. prop. P_2O_5) − mol. prop. CO_2
  = 0.124 − 0.010 − 0.001
  = 0.113

F = mol. prop. FeO + mol. prop. MgO − mol. prop. TiO_2 − mol. prop. Fe_2O_3
  = 0.128 + 0.085 − 0.028 − .021
  = 0.164

A + C + F = 0.062 + 0.113 + 0.164
  = 0.339

%A = (0.62/0.339)100 = 18.3
%C = (0.113/0.339)100 = 33.3
%F = (0.164/0.339)100 = 48.4

proportion of TiO_2 in the rock must be subtracted from the molecular proportion of FeO to obtain F. This subtraction is the arithmetic technique of making ilmenite a part of the assemblage; alternatively, we could say that ilmenite is present in excess, or that we are projecting from ilmenite in the AFMK tetrahedron as well as from muscovite.

4. The amount of component M in the projection is simply the molecular proportion of MgO.

5. The amounts of A, F, and M from steps 2, 3, and 4 are plotted on the extended AFM projection plane using a grid of their ratios (see Figure 11-13b).

ACF Diagram for Mafic Rocks

In this diagram, which serves for metamorphosed mafic rocks as well as shaly limestones and dolomites, FeO and MgO are lumped together as one component, F, so that another component, C, essentially representing CaO, can be displayed.

In contrast to the accurate representation of minerals and rocks that can be made on an AFM projection, ACF diagrams are more commonly used for portraying generalized mineral assemblages. Serious overlaps occur, especially at the F apex of the triangle, where anthophyllite, cummingtonite, hypersthene, olivine, and so on all plot.

The values of the components A, C, and F are determined as follows (see Table 11-3):

1. Obtain molecular proportions of oxides comprising the rock or mineral.

2. Component A equals the molecular proportions of Al_2O_3 minus Na_2O minus K_2O. This formulation assumes that plagioclase is present and any K_2O and Na_2O present is in the form of K-feldspar ($K_2O \cdot Al_2O_3 \cdot 6SiO_2$) and albite ($Na_2O \cdot Al_2O_3 \cdot 6SiO_2$)

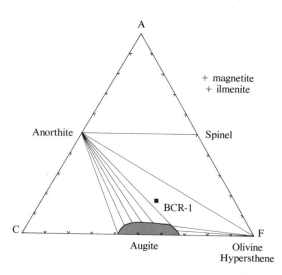

Figure 11-14 ACF diagram for basaltic and gabbroic mineral assemblages with the basalt BCR-1 from Table 2-2 plotted as a square. Note extensive solid solution in augite (shaded).

dissolved in the plagioclase; note the $1:1$ ratios of K_2O and Na_2O to Al_2O_3 in these alkali feldspars. If muscovite or biotite are present, this calculation for A is invalid (see Winkler, 1976).

3. Component C equals the molecular proportions of CaO minus $\frac{10}{3}(P_2O_5)$ minus CO_2. These subtractions allow for the presence of ideal apatite ($10CaO \cdot 3P_2O_5$) and calcite ($CaO \cdot CO_2$).

4. Component F equals the molecular proportions of FeO plus MgO. In some cases molecular proportions equal to TiO_2 and Fe_2O_3 may be subtracted to allow for the presence of ideal ilmenite ($FeO \cdot TiO_2$) and magnetite ($FeO \cdot Fe_2O_3$).

5. The sum of $A + C + F$ is found and the percentages of A, C, and F calculated for plotting, as in Figure 11-14.

Mafic Mineral Assemblages in the Metamorphic Facies

Significant variations in the mineral assemblages constituting mafic rocks occur across the *P–T* range of metamorphism. A convenient way to represent such variations is on a series of ACF diagrams keyed to six major metamorphic facies (see Figure 11-15; compare Figure 10-27).

SUMMARY

11.1. The enormous range of chemical composition of metamorphic rocks reflects their derivation from magmatic and sedimentary rocks as well as metasomatic modifications of these. The mineralogical composition of a metamorphic rock reflects its chemical composition and can be used to distinguish between sedimentary and igneous protoliths.

11.2. The geometric and chronologic aspects of fabric are essential to the description and classification of metamorphic rocks.

The geometry of the imposed metamorphic fabric reflects the nature of the stress field during metamorphism. Isotropic fabrics form under hydrostatic stress conditions. Imposed anisotropic fabrics form during ductile flow under nonhydrostatic stress conditions, and their geometric pattern—whether planar, linear, or planar–linear—reflects the pattern of flow. Most lineations lie on a foliation.

The mineralogical composition of a rock body influences the imposed fabric because of the characteristic shapes of the mineral grains themselves.

11.3 and 11.4. To emphasize metamorphic conditions, especially the nature of applied stress, rock nomenclature depends primarily on imposed fabric; three major categories are strongly foliated rocks, weakly foliated rocks, and essentially nonfoliated rocks.

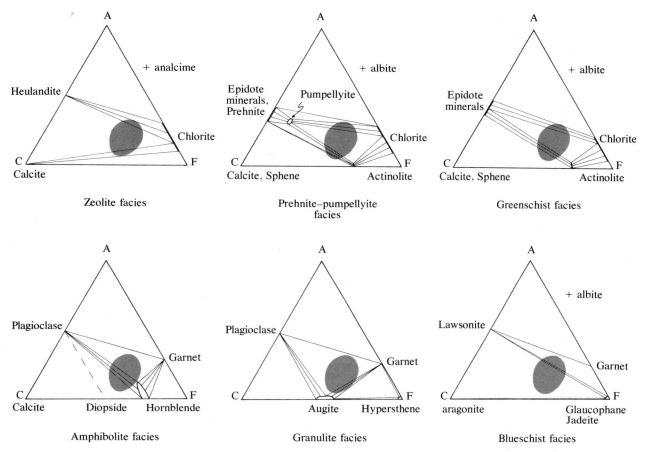

Figure 11-15   Simplified representation of mineral assemblages found in metamorphosed mafic igneous rocks (shaded area). A = Al_2O_3–K_2O–Na_2O; C = CaO; F = FeO + MgO–Fe_2O_3–TiO_2. Fe–Ti oxides and possible quartz are additional phases in all assemblages. Compare diagrams with the basaltic assemblage of Figure 11-14. [After W. S. Fyfe, N. J. Price, and A. B. Thompson, 1978, *Fluids in the Earth's Crust* (New York: Elsevier); and D. S. Coombs, 1961, Some recent work on the lower grades of metamorphism, *Australian Journal of Science* 24.]

11.5. By making certain simplifying assumptions, a single triangular diagram can show the correspondence between the bulk chemical composition of a rock and its constituent mineralogical composition at some particular *P* and *T,* or metamorphic grade. Over a restricted volume of a metamorphic system, the *P* and *T* can be constant, so that the mineralogical composition is dictated solely by the chemical composition. From the phase rule, the number of phases cannot exceed the number of components at some fixed values of *P* and *T*. In pelitic rocks, the four components Al_2O_3, FeO, MgO, and K_2O are chiefly responsible for most of the mineralogical variations found in typical Barrovian zones. By projecting from muscovite or K-feldspar, compositional relations within a four-component tetrahedron can be shown on an AFM triangular diagram. An ACF diagram is more useful for Ca-rich rocks, but FeO and MgO must be combined together.

STUDY EXERCISES

1. Plot the average shale listed in Table 11-2 on an AFM projection, assuming (a) that ilmenite (FeO · TiO$_2$), magnetite (FeO · Fe$_2$O$_3$), and muscovite are members of its mineral assemblage, and (b) that ilmenite and K-feldspar are found in its assemblage.

2. Plot a biotite and a Mg–Fe–Al garnet from Appendix C on an AFM diagram projecting first from ideal muscovite and then from K-feldspar.

3. Plot any one of the andesites or basalts in Chapters 3 and 5 on an ACF diagram. Assume that these rocks contain magnetite, ilmenite, and apatite but no calcite.

REFERENCES

Coleman, R. G. 1977. *Ophiolites.* New York: Springer-Verlag, 229 p.

Coleman, R. G., Lee, D. E., Beatty, L. B., and Brannock, W. W. 1965. Eclogites and eclogites. Their differences and similarities. *Geological Society of America Bulletin* 76:483–508.

Hobbs, B. E., Means, W. D., and Williams, P. F. 1976. *An Outline of Structural Geology:* New York: John Wiley and Sons, 571 p.

Mehnert, K. R. 1968. *Migmatites and the Origin of Granitic Rocks.* New York: Elsevier, 393 p.

Thompson, J. B. 1957. The graphical analysis of mineral assemblages in pelitic schists. *American Mineralogist* 42:842–858.

Winkler, H. G. F. 1976. *Petrogenesis of Metamorphic Rocks,* 4th Ed. New York: Springer Verlag, 254 p.

12

Field Relations of Metamorphic Bodies

Metamorphism is concentrated near lithospheric plate boundaries where dynamic processes are focused to change the states of geologic systems. Metamorphic rocks in continental margin orogenic belts have evolved over prolonged periods of time by episodic and more or less synchronous deformation and recrystallization of diverse rock sequences; the geometry and rates of plate convergence, plate character, and accompanying thermal processes in the asthenosphere and lithosphere all play a part in this metamorphic development. Prolonged subduction activity generates a pair of metamorphic belts along continental margins. Geothermal gradients are low in the outer belt adjacent to the oceanic trench, so that mineral assemblages of the high P/T metamorphic facies series are developed in ancient oceanic lithosphere (ophiolite) and arc-trench sediments that have been tectonically juxtaposed onto the continental margin in pervasively broken bodies of mélange. Farther inland, in the vicinity of the magmatic arc, where geothermal gradients are greater, intermediate to low P/T metamorphic facies series occur with granitic plutons. This inland metamorphic belt is generally regionally zoned from greenschist through amphibolite facies rocks, commonly with a culminating ultrametamorphic zone of migmatites.

Coupled thermal and mechanical (stress) changes in subduction-related regional metamorphic systems are expressed by syntectonic solid-state crystal growth and ductile flow, producing anisotropic fabrics. Where the action of nonhydrostatic stresses is locally more intense, as in deep-seated shear zones, localized ductile flow produces mylonitic rocks.

Deep burial, together with emplacement of hot magmas derived from the upper mantle or deeper regions of the crust, is responsible for much of the effect of elevated temperatures observed in metamorphic bodies. Contact metamorphic aureoles of thermally and metasomatically changed rocks form around epizonal and mesozonal intrusions. The thermal energy of the magma drives recrystallization of the aureole rocks as well as convective circulation of hydrothermal solutions that promote metasomatism.

Sub-sea-floor metamorphism (alteration) occurs near divergent oceanic plate margins as heated sea

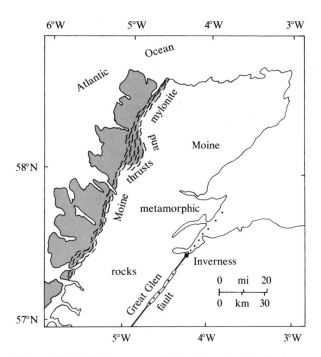

Figure 12-1   Moine zone of thrusts and mylonite in northwest Scotland. Shaded area is Cambro-Ordovician and older sedimentary and metamorphic rocks. [After M. W. Higgins, 1971, Cataclastic rocks, U.S. Geological Survey Professional Paper 687.]

water penetratively convects through the nascent basaltic crust. It can be studied in scraps of the oceanic lithosphere (ophiolite) emplaced tectonically onto continental margins.

The intimate association between different types of metamorphic systems and lithospheric plate boundaries provides a connecting link between geophysical and tectonic studies on the one hand and metamorphic petrology on the other.

12.1 MYLONITE ZONES

Sheets of **mylonite** that form by intense localized deformation are widespread in plutonic and metamorphic terranes. These zones of fine-grained rock, produced by reduction of grain size of an originally coarser-grained rock, are commonly braided and range in size from seams observable only in thin sections to belts several kilometers in width and hundreds of kilometers in length; displacements across mylonite zones range from barely perceptible to hundreds of kilometers. Mylonites have been frequently overlooked or misidentified (Higgins, 1971) because of their superficial resemblance to layers of chert, fine-grained quartzite, and some volcanic rocks. Careful study of fabric relations, especially in thin section, will reveal the intense undulatory extinction and other deformational features of typical mylonite (see Figures 10-7 and 10-8) formed by cohesive ductile flow. However, subsequent crystal growth processes may obliterate the deformational fabric.

Mylonites are important in showing displacement of deep crustal and mantle rock bodies. Especially in plutonic igneous terranes, stratigraphic markers that might denote zones of dislocation are lacking and mylonite zones may be their only conspicuous manifestation.

The classic mylonite zone, for which the term mylonite was coined by Lapworth in 1885, lies within the Moine thrust belt in northwest Scotland (see Figure 12-1). Mylonites are an intimate part of this belt, which extends for almost 200 km along the coast. Though generally only a few kilometers wide, locally the belt of thrust faults and mylonitic rocks is over 10 km wide. Detailed study of the zone by Christie (1963) and many others (see Hobbs and others, 1976, for additional references) discloses a complex fabric formed by multiple episodes of deformation. Fine-grained, intensely deformed mylonitic rocks occur throughout the belt (see Figure 13-18), but generally the most intensely deformed rocks comprise a gently east-dipping sheet less than 100 m thick with a major thrust fault separating rocks of different composition and age very near its base. Sheets of mylonite cut rocks ranging from greenschist to granulite facies; the higher-grade rocks are commonly retrogressively metamorphosed. More than a kilometer or so east of the Moine thrust-mylonite zone, the regionally metamorphosed Moinian rocks show no indication of mylonitization, but their foliation parallels that in the mylonites to the west.

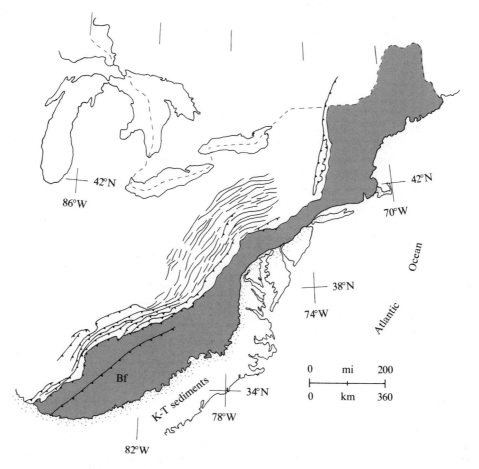

Figure 12-2   Simplified tectonic map of the Appalachian orogenic belt in the United States. Brevard fault-mylonite zone (Bf) lies within the regional metamorphic terrane (shaded) to the east of the belt of thrust-faulted and folded but essentially unmetamorphosed Paleozoic rocks. The metamorphic terrane is unconformably overlain on the east by postorogenic Cretaceous and Tertiary sediments of the Atlantic coastal plain. [After L. D. Harris and K. C. Bayer, 1974, Sequential development of the Appalachian orogen above a master decollement—A hypothesis, *Geology* 7.]

In current global plate reconstructions, the Caledonide orogenic belt of Great Britain and Scandanavia may be correlated at least in part with the Appalachian orogen of eastern North America. Convergence between the North America and European plates may have been responsible for the complex deformation in the Moine zone of thrusts and mylonites and in the Brevard zone, which extends in a northeasterly direction for at least 600 km in the southern Appalachian orogen (see Figure 12-2). This zone is comprised of mylonitic rocks as much as 6 km wide, although this width represents only the most intensely and pervasively deformed rocks in a belt of perceptible effects extending for several kilometers beyond. Amphibolite-facies rocks isoclinally folded during metamorphism east of the zone are cut by and become progressively more retrograded closer to the mylonitic zone; mostly low-grade rocks occur within the zone itself.

Multiple episodes of deformation of varying character are evident in the zone, as shown by the occurrence of brecciated mylonite, for example. The exact nature of the dislocation across the zone has been the subject of considerable debate. However, recent geological and geophysical investigations (Cook and others, 1979) suggest that the Brevard zone is the surface outcrop of but one of several thrust faults lying above a regional detachment surface (decollement) at depth (6 to 15 km) that have carried metamorphic rocks in the orogen at least 260 km to the west. Roper and Justus (1973) postulate that the Brevard zone has experienced a complex history during the Paleozoic, during which time an island arc and African continental segments may have been sutured onto the American continent.

The San Andreas transform in California and the Alpine in New Zealand are segments of the eastern and the western boundaries of the Pacific plate. Depths of exposure vary greatly along these two active faults, and for the Alpine zone there has possibly been more than 4 to 11 km of vertical uplift of the southeastern block in the past 5 My, although assessments vary. Consequently, the exposed rocks in these zones range from unconsolidated cataclastic fault breccia and gouge formed near the surface to cohesively deformed mylonitic rocks formed at elevated confining pressures and temperatures at greater depths.

The geology of the South Island of New Zealand clearly reflects 500 km of right lateral motion along the Alpine continental transform (see Figure 12-3). In the central region, metamorphic zones virtually parallel the transform. This pattern, together with an increase of K–Ar ages of schists from 4 to 120 My with increasing distance away from the fault, is interpreted by Scholz and others

(1979) to reflect generation of heat along the fault during its Mesozoic-Cenozoic movement. They believe that higher temperatures reset the radiochronometric systems, giving progressively younger ages closer to the fault, and producing the increase in metamorphic grade.

12.2 CONTACT METAMORPHIC AND METASOMATIC BODIES

Emplacement of bodies of hot magma into cooler country rocks thermally perturbs them. Explusion of aqueous solutions from the magma or convective mobilization of ground water containing dissolved materials in the country rocks due to the heat from the intrusion perturbs the wall rocks chemically as well as thermally. These perturbations may be sufficiently intense, depending upon the composition of the country rocks, to cause contact metamorphism and metasomatism in a contact aureole around the intrusion.

Contact effects are generally inconspicuous in relatively dry igneous rocks but can be pronounced in sedimentary rocks, especially shales and carbonates. The intensity and extent of contact metamorphism depends upon many factors. Some aureoles are only a few centimeters wide, whereas more extensive ones can be one to two kilometers or more wide. Aureoles are usually zoned with respect to fabric and mineralogical composition, reflecting thermal and chemical gradients around the intrusion in the wall rocks. Zonation in broad aureoles in pelitic rocks reflecting essentially isochemical thermal metamorphism tends to be fairly regular and more or less concentric to the intrusive contact, whereas metasomatic zones in calcareous rocks around granitic and dioritic intrusions are more irregular, perhaps reflecting the control exerted by the composition, structure, and especially the permeability of the country rock to migrating fluids. In all cases, it is important to remember that the exposed contact aureole is only one more or less horizontal section through the intrusion–wall rock system; significant variations can be expected in the vertical, or third, dimension.

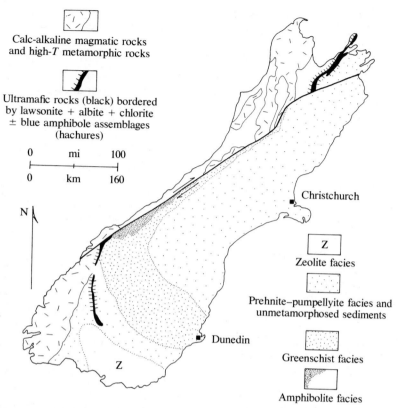

Calc-alkaline magmatic rocks
and high-*T* metamorphic rocks

Ultramafic rocks (black) bordered
by lawsonite + albite + chlorite
± blue amphibole assemblages
(hachures)

| 0 | mi | 100 |
| 0 | km | 160 |

N

Christchurch

Z
Zeolite facies

Prehnite–pumpellyite facies and
unmetamorphosed sediments

Greenschist facies

Dunedin

Amphibolite facies

Z

Figure 12-3 South Island of New Zealand, showing Alpine transform fault and generalized distribution of magmatic rocks and metamorphic zones. It is a beneficial learning experience in reading any map to color the different rock units lightly with colored pencils. [After C. A. Landis and D. G. Bishop, 1972, Plate tectonics and regional stratigraphic-metamorphic relations in the southern part of the New Zealand geosyncline, *Geological Society of America Bulletin* 83.]

Well-zoned aureoles are developed at mesozonal or epizonal crustal levels. In the epizone, country rocks were initially relatively cool and emplacement of the pluton occurs by uplift of the roof, by faulting, or by stoping so that the wall rocks do not suffer penetrative ductile deformation under nonhydrostatic stresses. Hornfelses with granoblastic fabrics are typical of aureoles in which rocks were simply thermally cooked or metasomatically soaked. In deeper mesozonal and especially catazonal environments, wall rocks are already hotter so that contact zonation is less pronounced and plutons are commonly emplaced by forceful shouldering aside of the ductile wall rocks. Metamorphosed wall rocks may then have anisotropic fabrics similar to those found in regional terranes, except that fabric patterns reflecting ductile flow will tend to correlate with the geometry of the enclosed intrusive contact.

Thermal Considerations

Temperatures in country rocks around cooling plutons depend upon the geometry, size, temperature, and heat content of the intrusion and the properties of the country rocks, especially their fluid content and permeability. If the water content of the country rocks is low or if their permeability, chiefly related to fractures and bedding planes, is insignificant, then heat flow is dominantly conductive. On the other hand, high water contents and permeabilities permit major convective transport not only of heat but also of the matter dissolved in the circulating fluids.

In models of purely conductive heat flow around a body of static magma (see pages 248–250; Turner, 1981, p. 18), the maximum T in the wall rocks at the contact is roughly half the initial magma temperature but depends upon their preintrusion T. This maximum T as well as the dimension, D, of the intrusion perpendicular to the contact govern the width and thermal history of the aureole. The maximum T ever attained in aureoles of initially similar rock at a distance X from the contacts of similar magma bodies is a function of X/D; taking a magma–country rock situation in which $X/D = 1$, then a pluton 1 km wide will produce a maximum T at a distance of 1 km from the contact identical to the maximum T a distance of 0.1 km from the contact of an intrusion 0.1 km wide. The duration of time that a particular T is maintained at some point in the aureole is proportional to D^2. Hence, doubling the dimension of a pluton increases the duration fourfold. Large intrusions thus produce broader thermal aureoles in wall rocks and maintain them for much longer times, assuring more complete approach to an equilibrium grain shape and mineralogical assemblage.

Convecting magmas can develop contact temperatures in wall rocks nearly that of the magma itself (Shaw, 1974). In a deeply buried, thick body of convecting basaltic magma, temperatures become very high in the roof as hotter magma from the base of the body moves upward and across the roof (Irvine, 1970). Substantial melting can occur if the solidus temperature of the roof rocks is less than the temperature of the basaltic magma. Temperatures in country rocks along the walls and especially the floor of the pluton are less than in the case of a static magma body.

The presence of fluids has an important bearing on the transfer of heat and matter between a pluton and its wall rocks. Conceivably, fluids could either augment or suppress heating of wall rocks. In the latter case, for example, intrusion of a dry, hot magma into wet country rocks could cause the water to flow into the pluton to even out the gradient in P_{H_2O} across the contact. Some of the thermal energy in the magma would be consumed in heating the connate water. Wall rocks would only be moderately heated and the pluton would cool quicker. On the other hand, relatively hydrous magma bodies, especially granitic ones in their late stages, may not take in water from wall rocks and indeed might lose water into surrounding permeable rocks, enhancing the heat transfer occurring by conduction. In this second instance, the pluton is a source of water. This phenomenon could explain relatively broad aureoles and metasomatic zones around many granitic intrusions.

Theoretical computer modeling by Norton and Knight (1977) has shown that large-scale fluid circulation is virtually inevitable in pluton–country rock systems with any realistic permeabilities and water contents. Water carrying dissolved matter is predicted to move several kilometers from the intrusive thermal source on time scales of a few hundred thousand years. Whereas conductive heat flow into roof rocks at the top of a pluton is greatest at the moment of emplacement and then decreases exponentially with time, the convective heat flow increases dramatically with time to values several times the maximum conductive transport before exponentially decaying. The calculated mass of water circulated in magma-driven circulation systems is stupendous; along the world's oceanic rifts, assuming a spreading rate of 5 cm/y, a mass equivalent to the entire mass of today's oceans would be circulated in 0.2 Gy, that is, since about the beginning of the Mesozoic. The validity of these theoretical models is greatly strengthened by the enrichment in light isotopes of oxygen and hydrogen documented in many pluton–country rock systems (see page 294).

Examples of Contact Aureoles

Many examples could be described (Turner, 1981), but we will cite only three.

(a)

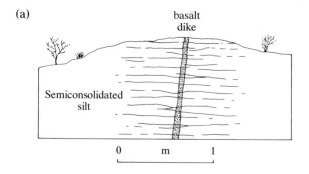

(b)

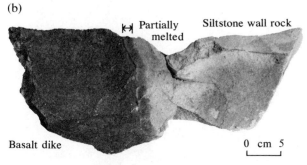

Figure 12-4 Contact effects adjacent to a thin basalt dike. (a) Schematic cross section showing prominent but irregular columnar joints cutting the dike and siltstone perpendicular to contact. (b) Segment of irregular columnar joint showing thin selvage of fused siltstone next to dike.

Contact effects next to basaltic dikes. Thin dikes generally produce very narrow contact effects; not uncommonly, there is only a barely conspicuous baking or discoloration of the wall rocks within a few centimeters of the contact. Dikes of basaltic magma, due to their relatively high *T* and thermal energy content, may nonetheless cause intense effects in a very narrow zone next to the contact if the wall rocks are particularly susceptible because of composition and high water content. Figure 12-4 shows contact effects next to a basaltic dike in northwestern Arizona. Temperatures were sufficient within a couple of centimeters to have partially fused the intruded material at the contact.

Onawa pelitic contact aureole. Thermal metamorphism of pelitic country rocks surrounding the Onawa granodiorite intrusion in Maine

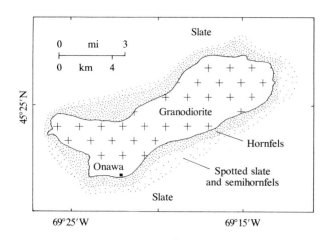

Figure 12-5 Generalized geologic map of the Onawa pluton and contact metamorphic aureole, Maine. Figure 12-6 shows representative slate (a), spotted hornfels (b), and hornfels (c) from this aureole. After Moore (1960).

(Moore, 1960) has created a zoned contact aureole (see Figure 12-5). The rocks into which the pluton was emplaced are slates formed during an earlier episode of regional metamorphism. Mineralogically, these country-rock slates consist of quartz + chlorite + phengite + graphite + Fe–Ti oxides (see Figure 12-6a). Phengite is a white mica with more SiO_2, FeO, Fe_2O_3, and MgO and less Al_2O_3 than "ideal" muscovite ($KAl_2(AlSi_3O_{10})$ $(OH)_2$). The first indication of thermal contact metamorphism, found as far as 2 km from the exposed margin of the pluton, is the more or less simultaneous development in the slates of biotite + cordierite + andalusite. Small ($\leq$ 1 mm), ill-defined spots manifest the poikiloblastic cordierite grains, now largely retrograded to extremely fine-grained phyllosilicates. Andalusite in these spotted slates appears as euhedral prisms up to 1 cm long and 1 to 2 mm across and is sometimes referred to as chiastolite (see Figure 10-14a). The exact reactions by which these three new phases were created are complex, but reaction between phengitic mica and chlorite would yield a more ideal muscovite plus biotite, cordierite, and andalusite. A minor amount of andalusite may also have appeared by breakdown of pyrophyllite, which appears very locally

(a)

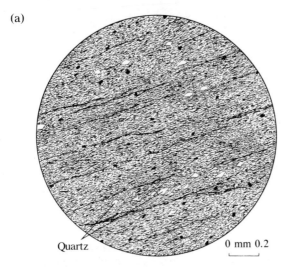

Quartz   0 mm 0.2

(b)

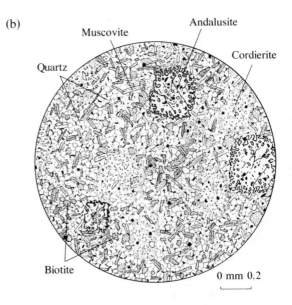

Muscovite   Andalusite

Quartz   Cordierite

Biotite   0 mm 0.2

(c)

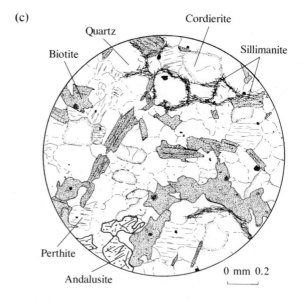

Quartz   Cordierite

Biotite   Sillimanite

Perthite

Andalusite   0 mm 0.2

Figure 12-6 Progressive thermal metamorphism of
pelitic rocks under hydrostatic conditions in the Onawa
contact aureole. See Figure 12-5 for map of zones. (a)
Slate country rock into which the Onawa pluton was
emplaced had experienced previous regional
metamorphism with development of a low-grade mineral
assemblage of Fe–Ti oxides + quartz + chlorite +
phengitic white mica + graphite; because of the
preferred orientation of platy grains, the rock has a slaty
cleavage. Angular quartz grains are relicts of the original
clastic fabric of the parent shale. (b) Spotted pelitic
hornfels. Solid-state crystal growth under hydrostatic
conditions has completely obliterated the clastic
sedimentary fabric and the cleavage of the original slate,
producing a coarser, isotropic granoblastic fabric. Most
of the rock is comprised of quartz + biotite +
muscovite + graphite (black opaque grains). Larger
poikiloblastic grains of andalusite and cordierite,
crowded with myriad tiny inclusions of the matrix
phases, stand out in relief (andalusite) or etch out
(cordierite) on weathered surfaces, forming spots. (c)
Pelitic hornfels. Coarser granoblastic rock from the
inner part of the contact aureole. Unstable andalusite
was decomposing when the metamorphic reactions
stopped. In the innermost part of the aureole, the
hornfelses contain myriad millimeter-scale lenticles of a
quartz–alkali feldspar aggregate that probably formed
by very limited partial melting.

in the slates. Because of the nearly simultaneous appearance of the new phases and some control on them of subtle variations in bulk chemical composition, it is not possible to map separate andalusite, cordierite, and biotite isograds and associated zones.

Closer to the intrusion, spotted slates give way to progressively more massive, spotted semihornfelses and hornfelses (see Figure 12-6b) where solid-state crystal growth under hydrostatic conditions coarsened the grain size and obliterated the slaty cleavage. Still closer to the pluton is a higher-grade assemblage of quartz + alkali feldspar + sillimanite + cordierite + biotite in coarser hornfelses (see Figure 12-6c). This zone is bounded by the alkali feldspar isograd governed by the breakdown reaction of muscovite with quartz

$$(K,Na)Al_2(AlSi_3O_{10})(OH)_2 + SiO_2 \rightarrow$$
$$\text{muscovite} \qquad\qquad \text{quartz}$$

$$(K,Na)AlSi_3O_8 + Al_2SiO_5 + H_2O$$
$$\text{alkali} \qquad\quad \text{sillimanite}$$
$$\text{feldspar}$$

The mineral assemblages found in the Onawa contact aureole are shown schematically in AFM projections in Figure 12-7.

Metasomatic skarn zones.   Extensive metasomatic zones occur around some intrusions (usually granitic to dioritic) due to significant gains and losses of matter as large volumes of hydrothermal solutions circulated through the country rock. Although the intrusion may not be the source of the fluids, it is at least the energy source to drive the flow of heat and matter. Metasomatic systems in calcareous wall rocks attempt to attain a stable thermodynamic equilibrium in the presence of gradients in T and X, forming mineralogical zones in skarns.

A clear example of metasomatic zones reflecting a gradient in X outward from an intrusion is found at Crestmore, California, where brucite marbles, formed during an earlier metamorphic event, were intruded by a small epizonal plug of quartz monzonite porphyry (see Figure 12-8;

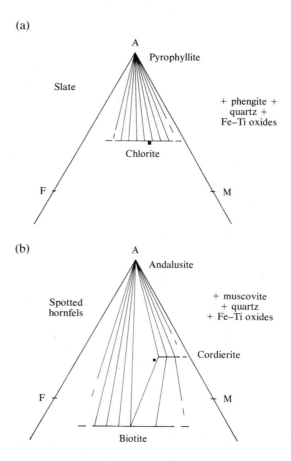

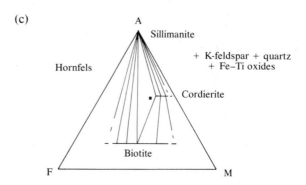

Figure 12-7   AFM projections for the three mineral assemblages in the Onawa aureole shown in Figure 12-6. The bulk chemical compositions of the rocks in the aureole are very uniform (Moore, 1960) and are represented by the filled square.

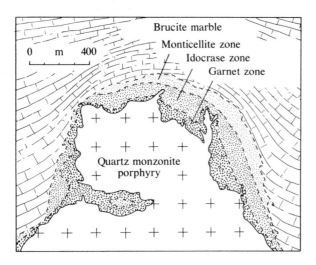

Figure 12-8 Schematic cross section through the quartz monzonite porphyry pluton and surrounding metasomatic zones at Crestmore, California. Forsterite zone lying between the original brucite marbles and the monticellite zone is omitted. [After C. W. Burnham, 1959, Contact metamorphism of magnesian limestones at Crestmore, California, *Geological Society of America Bulletin* 70.]

Table 12-1 Metasomatic zones and related mineral assemblages at Crestmore, California.

Metasomatic zone	Related mineral assemblages
Forsterite zone	Calcite + brucite + clinohumite + spinel
	Calcite + clinohumite + forsterite + spinel
	Calcite + forsterite + spinel + clintonite
Monticellite zone	Calcite + forsterite + monticellite + clintonite
	Calcite + monticellite + melilite + clintonite
	Calcite + monticellite + spurrite (or tilleyite) + melilite
	Monticellite + spurrite + merwinite + melilite
Idocrase zone	Idocrase + monticellite + spurrite + merwinite + melilite
	Idocrase + monticellite + diopside + wollastonite
Garnet zone	Grossular + diopside + wollastonite

Burnham, 1959). The wider parts of the meta-somatized aureole consist of three skarn mineral zones; from the original brucite marble (containing chiefly Ca, Mg, C, H, and O) inward to the porphyry, the mineral assemblages composing the zone are as shown in Table 12-1.

Considering only the major minerals containing CaO, MgO, and SiO_2, it is obvious from Figure 12-9 that this sequence of ten assemblages reflects increasing concentrations of Si derived from hydrothermal solutions emanating from the magma body. As the solutions migrated into the wall rock, Si was extracted by reaction with $CaCO_3$ and $Mg(OH)_2$. These metasomatic zones need not have required a large T gradient to develop. Significant amounts of Al were introduced into the monticellite, idocrase, and garnet zones, in addition to Si, thereby stabilizing melilite, idocrase, and grossular.

The transition from the brucite marble to the monticellite zone was possibly accomplished by a reaction such as

$$Mg(OH)_2 + SiO_2 + CaCO_3 \rightarrow$$
$$CaMgSiO_4 + CO_2 + H_2O$$

brucite + silica in hydrothermal solution
                    + calcite → monticellite

An important reaction in the development of idocrase from the monticellite zone mineral assemblage was likely

$$CaMgSiO_4 + 2(2Ca_2SiO_4CaCO_3) + 3Ca_3MgSi_2O_8$$
$$+ 4Ca_{10}Mg_2Al_6Si_7O_{35} + 15SiO_2 + 12H_2O \rightarrow$$
$$6Ca_{10}Mg_2Al_4Si_9O_{34}(OH)_4 + 2CO_2$$

monticellite + 2 spurrite + 3 merwinite
            + 4 melilite + 15 silica in
        hydrothermal solution → 6 idocrase

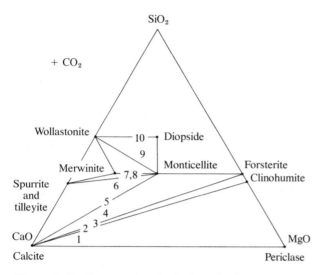

Figure 12-9 Compositional relations in the system CaO–MgO–SiO_2 with excess CO_2 at more or less fixed T and P. The path of progressive silication of the wall rocks at Crestmore is shown by the sequence of numbers representing the ten mineral assemblages that comprise the metasomatic zones. [After C. W. Burnham, 1959, Contact metamorphism of magnesian limestones at Crestmore, California, *Geological Society of America Bulletin* 70.]

A possible reaction for production of the garnet zone assemblage from idocrase was

$$Ca_{10}Mg_2Al_4Si_9O_{34}(OH)_4 + 3SiO_2 \rightarrow$$
$$2Ca_3Al_2Si_3O_{12} + 2CaMgSi_2O_6 + 2CaSiO_3 + 2H_2O$$

idocrase + 3 silica in hydrothermal solution →
    2 grossular + 2 diopside + 2 wollastonite

These reactions indicate progressive silication of the country rocks and, in the first two reactions, decarbonation. Other ions were involved in the metasomatism than just the ones indicated in these idealized reactions, including Fe^{+2}, Na, and K.

The exact nature of the chemical changes during metasomatism at Crestmore, and in most metasomatic bodies, is difficult to determine because the original rock is not available to serve as a standard for comparison. To estimate the changes,

it would be necessary to assume that unmodified rock some distance from the source of the metasomatizing solutions, if possible the same rock horizon, accurately represents the composition of the original rock.

12.3 BURIAL METAMORPHIC BODIES

Deeply buried rocks may suffer partial to complete mineralogical and textural reconstitution over a broad area independently of magmatic intrusions and tectonism. Absence of penetrative deformation and imposed anisotropic fabric indicates nonhydrostatic stresses were negligible during reconstitution. Burial metamorphism thus differs from syntectonic regional metamorphism, in which pervasive deformation is broadly concurrent with recrystallization, producing well-developed tectonite fabrics. Nonetheless, burial metamorphic bodies may have mineral assemblages similar to those that occur in lowest-grade regional terranes, and with increasing signs of penetrative deformation they must pass gradationally into typical regionally metamorphosed terranes.

Estimations of depth of burial during metamorphism based upon reconstruction of the overlying stratigraphic section indicate that a particular mineral assemblage may occur at significantly different depths in different buried bodies around the world. This suggests that temperature variations may be more important in determining the reconstituted mineral assemblage than pressure. In active geothermal areas (Salton Sea in southern California and Wairakei and Broadlands in northern New Zealand), metamorphic mineral assemblages are encountered in boreholes at relatively shallow depths (~ 2 km) and variations in mineral assemblages correlate with measured temperatures. There has been controversy over whether the reconstitution in these rocks should be called diagenesis or metamorphism, but the fact that it has occurred at P and T significantly different from that under which the rock was originally emplaced places it in the realm of metamorphism. Burial metamorphism is more evident in thick se-

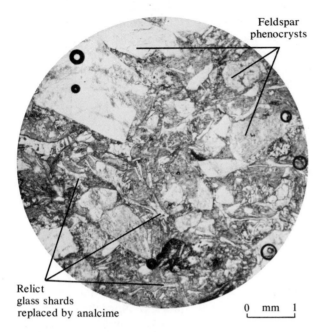

Feldspar phenocrysts

Relict glass shards replaced by analcime

0   mm   1

Figure 12-10 Metamorphosed vitric-crystal tuff from eastern Australia (see Wilkinson and Whetten, 1964). Shards of glass have been entirely replaced by analcime. Fine quartz, alkali feldspar, and analcime compose the very fine-grained matrix around the relict shards and phenocrysts. Black circles are air bubbles in thin section cement.

quences of volcaniclastic rocks, especially those containing unstable glass, but reconstitution can also occur in clay-rich sediments. The presence of abundant water is imperative for the production of the hydrous low-grade mineral assemblages characterizing these bodies.

Relict fabrics are usually preserved because solid-state grain growth occurs on a very small scale, chiefly within original grain outlines, forming a fine-grained replacement aggregate of new minerals (see Figure 12-10). Some of the zeolites and other low-*T* metamorphic minerals formed during burial metamorphism may appear in veins. In pelitic bodies imposed anisotropic fabrics may form mimetically after the bedding and fissility in the original clay-rich shales.

Extensive, thick basin sequences subjected to burial metamorphism may show zonation in mineral assemblages. The classic area for burial

metamorphism of this sort is in southernmost New Zealand, where zeolite facies assemblages have developed in a Triassic–Jurassic basin filled with about 9 km of volcanogenic sediments and tuffs (see Figure 12-3; Boles and Coombs, 1977). At an early stage in the alteration history, heulandite (or clinoptilolite) + montmorillonite + chlorite + quartz formed from unstable glass in tuffs; primary magmatic minerals are commonly unaltered, though they also were unstable. Some interstratified beds of tuff have altered to analcime + quartz. In the lower part of the section where higher temperatures prevailed, first quartz + analcime and then most heulandite-bearing assemblages disappear because of the dehydration reactions

$$NaAlSi_2O_6 \cdot H_2O + SiO_2 \rightarrow NaAlSi_3O_8 + H_2O$$
analcime   quartz   albite

and

$$CaAl_2Si_7O_{18} \cdot 6H_2O \rightarrow CaAl_2Si_4O_{12} \cdot 4H_2O$$
heulandite   laumontite
$$+ 3SiO_2 + 2H_2O$$
quartz

Adularia, a low-temperature K-feldspar, may form with albite.

Prehnite and pumpellyite form from the Ca-zeolites, or instead by a series of reactions depending on *P, T,* and concentrations of CO_2, Ca, and so on in the fluid phase. An example is the dehydration–decarbonation reaction

$$CaAl_2Si_4O_{12} \cdot 4H_2O + CaCO_3 \rightarrow$$
laumontite   calcite
$$Ca_2Al_2Si_3O_{10}(OH)_2 + SiO_2 + 3H_2O + CO_2$$
prehnite   quartz

Eventually, and especially under the influence of increasing temperature, all zeolite assemblages become unstable and are destroyed, marking the transition to the prehnite–pumpellyite facies.

In many metamorphic terranes around the world, lowest-grade mineral assemblages pass gradationally into the greenschist facies without the appearance of zeolites or prehnite + pumpellyite, or at least not to the extent observed in the volcanogenic basin-fill in southern New Zealand. Thus, metamorphosed miogeoclinal shales, sandstones, and carbonates in the northern Appalachians carry assemblages of kaolinite + calcite + dolomite. The reaction

$$2CaAl_2Si_4O_{12} \cdot 4H_2O + 2CO_2 =$$
$$\text{laumontite}$$
$$Al_4Si_4O_{10}(OH)_8 + 2CaCO_3 + 4SiO_2 + 4H_2O$$
$$\text{kaolinite} \qquad \text{calcite} \qquad \text{quartz}$$

indicates that in CO_2-rich systems the zeolite laumontite is unstable and the kaolinite + calcite + quartz assemblage is more stable; in CO_2-poor environments, zeolites are more stable. Similarly, the occurrence of prehnite, epidote, and pumpellyite is suppressed in CO_2-rich environments. The development of mineral assemblages specific to the lower part of the greenschist and all of the zeolite and prehnite–pumpellyite facies appears to depend upon concentrations of CO_2 and H_2O as well as temperature.

12.4 REGIONAL LOW TO INTERMEDIATE *P/T* METAMORPHIC BODIES

By far the most widespread products of metamorphism evident in both Phanerozoic orogenic belts and Precambrian shields are tectonites whose mineral assemblages testify to recrystallization under low to intermediate *P/T* conditions, spanning the entire range of metamorphic temperatures. Geothermal gradients were intermediate to high. Fabrics are typically, though not everywhere, anisotropic and reflect syntectonic recrystallization under nonhydrostatic stresses; generally there is evidence of multiple thermal and deformation episodes. Pervasive foliation, lineation, and folds are intimately associated.

Such regional metamorphic bodies evolve within orogenic belts during convergence of lithospheric plates. Due to more or less coeval magmatism, the metamorphic rocks are intimately associated with granitic and dioritic plutons. At higher crustal levels, particularly in the epizone, these plutons postdate tectonism and regional metamorphism; in the deeper catazone, pulses of metamorphism and magmatism may be essentially concurrent.

Regional Metamorphism in Orogenic Belts

Appalachian orogen. Metamorphic rocks are exposed over a large part of the now rather deeply eroded Appalachian orogenic belt from the southeastern United States northward through New England and the Canadian maritime provinces to Newfoundland. No mineral assemblages of the high *P/T* blueschist facies, including such phases as glaucophane, lawsonite, and aragonite, have been discovered. Assemblages of the zeolite and prehnite–pumpellyite facies are possibly more widespread than currently recognized. Reported mineral assemblages are chiefly restricted to the greenschist, amphibolite, and granulite facies.

Figure 12-11 shows the distribution of these metamorphic zones in the Appalachian orogen. Precambrian rocks now lying in a discontinuous belt in the western part of the orogen were first metamorphosed 1100 to 800 My ago and again during the Paleozoic. No assemblages of the greenschist facies are known in these Precambrian rocks except in some places where overprinted by the Paleozoic event(s). Granulite facies assemblages survive locally. These polymetamorphic rocks are unconformably overlain by sedimentary and volcanic rocks that bear only a Paleozoic imprint of metamorphism. Barrovian metamorphic zones and their delineating isograds related to Paleozoic metamorphism tend to be elongate parallel to the orogen, but a simple pattern of progressive metamorphism is by no means apparent. Though complete data are lacking, this complex pattern seems real and raises important questions. Is the distribution of zones complicated by multiple thermal events in space and time or by postmetamorphic tectonic disturbances? Recent evi-

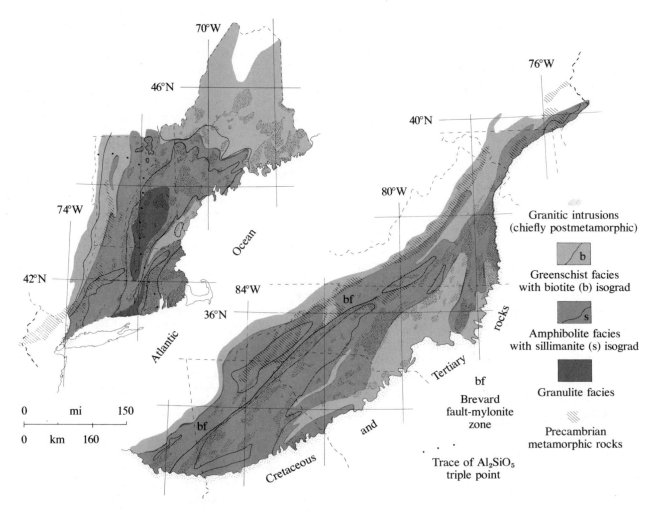

Figure 12-11   Simplified metamorphic map of the Appalachian orogen in the United States. Compare with Figure 12-2. The New England segment (top) joins onto the central and southern segment (bottom) along the Pennsylvania–New Jersey state line (dark dashed line). The western limit of metamorphic rocks is a thrust fault or is gradational into unmetamorphosed sedimentary rocks. Rocks are chiefly Paleozoic except where otherwise noted. [After B. A. Morgan, 1972, Metamorphic map of the Appalachians, United States Geological Survey Miscellaneous Geologic Investigations, Map I-724.]

dence (see Section 12.1) indicates that extensive thrust sheets that moved hundreds of kilometers occur throughout the orogen; the Brevard fault-mylonite zone is one such dislocation. Throughout the orogen, radiometric ages on metamorphic rocks fall into three episodes at 480–460, 380–360,

and 270–230 My, suggesting multiple thermal events. More research should clarify these questions.

Paleozoic metamorphism in northern New England was of the low-*P/T* type, as shown by the widespread occurrence of andalusite and cordierite in pelitic rocks. With increasing grade, andalusite gives way to sillimanite. The high geothermal gradient implied by these mineralogical features (see Figure 10-28) is supported by the relatively closely spaced isograds on the map (see Figure 12-11). Southward in the orogen, the metamorphism was of the intermediate P/T type; andalusite and cordierite are absent; kyanite, rather than andalusite, is the low-T aluminosilicate, and isograds are more widely spaced. The boundary between these contrasting P/T realms shown in Figure 12-11 represents the P and T of the aluminosilicate triple point during metamorphism.

Caledonide orogen. In the modern plate tectonic view, the Appalachian orogen continued northward beyond Newfoundland before the Mesozoic opening of the Atlantic Ocean. This northerly segment now lies in the northern British Isles, Scandinavia, and east Greenland, where it is known as the Caledonide orogen. The complex plate interactions recorded in Appalachian tectonic, magmatic, metamorphic, and sedimentologic features are duplicated to some degree in the Caledonide belt. Continental slices with European and African faunal assemblages are now attached to the east coast of North America, and North American fauna are found in the northern British Isles.

Figures 12-12 and 12-13 summarize the evolution of the Caledonide orogen in the northern British Isles. By the end of the Cambrian, a thick wedge of sediment as much as 1,000 My old and trending NE-SW had accumulated along the southeastern passive margin of a continental mass. Beginning in late Cambrian time and extending into the Ordovician (510 to 480 My ago), oceanic lithosphere was subducted beneath this wedge. Though only one locality of blueschist is known, widespread ophiolite sequences, regional metamorphism, deformation, and granitic magmatism firmly document the plate convergence. Polyphase deformation during this 30-My period was broadly concurrent with a thermal event that

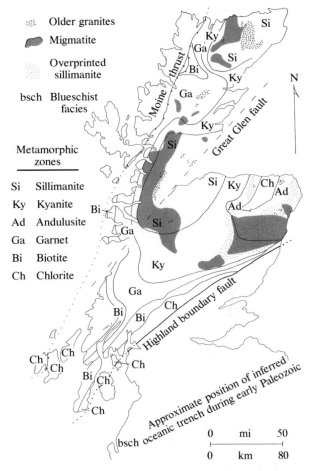

Figure 12-12 Simplified map of the Caledonide orogen in northern Scotland. The metamorphic zones shown here are a revision of earlier work (see Figure 10-25). Voluminous Newer granites are not shown. Postorogenic strike-slip movement along Great Glen fault has been eliminated to restore original position of segments of orogen. [After J. F. Dewey and R. J. Pankhurst, 1970, The evolution of the Scottish Caledonides in relation to their isotopic age pattern, *Royal Society of Edinburgh Transactions* 68.]

reached a culmination approximately between the third and fourth phases of deformation.

The distribution of metamorphic zones seen on the map in Figure 12-12 defines a southwestward-plunging thermal anticline with more deeply

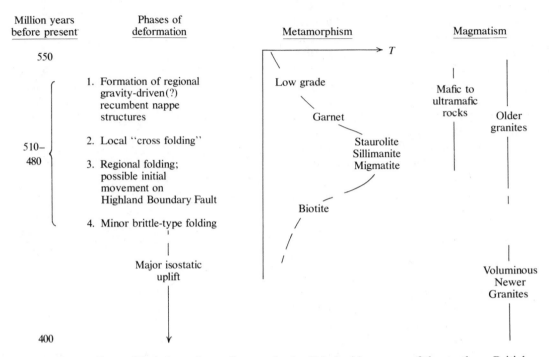

Figure 12-13 Generalized chronology of events in the Caledonide orogen of the northern British Isles. [After J. F. Dewey and R. J. Pankhurst, 1970, The evolution of the Scottish Caledonides in relation to their isotopic age pattern, *Royal Society of Edinburgh Transactions* 68.]

buried rocks undergoing higher grades of metamorphism lying generally toward the northeast. However, an area north of the east end of the Highland Boundary fault was apparently subjected to a low *P/T* type of metamorphism, as andalusite + cordierite are widespread; a chlorite zone is also present. These zones of andalusite and chlorite were apparently formed in response to a much higher geothermal gradient than prevailed to form the Barrovian zones to the north, west, and southwest.

Parts of the andalusite and kyanite zones north of the east end of the Highland Boundary fault were subjected at a later time to higher temperatures, so that these low-temperature aluminosilicates inverted to higher-temperature sillimanite. Local partial melting and formation of coarse

migmatitic gneisses accompanied this sillimanite overprint at higher temperatures.

Although a few so-called Older, usually foliated, granite plutons had been emplaced syntectonically and even pretectonically as much as 550 My ago, copious volumes of granitic magmas were emplaced after the 30 My period of recrystallization and deformation, forming chiefly massive Newer granites, commonly with well-developed contact metamorphic aureoles. During this ~100-My period of Newer granitic activity the orogen was isostatically uplifting and eroded, as shown by metamorphic and plutonic detritus in nearby late Ordovician and Silurian sedimentary rocks.

We can only speculate on challenging questions raised by the field relations in the Caledonide orogen. What was the cause of the major thermal event prompting metamorphism? Was it really a single event, or were there multiple pulses, the last of which everywhere erased earlier events, except in the area of the sillimanite–migmatite overprint? How is the cause of the thermal event(s) related to the forces behind the polyphase deformation? Why was the greatest volume of granite emplacement so much later than the peak of recrystallization and deformation? Were these Newer granitic magmas generated during the peak event(s) and merely took tens of millions of years to rise to the crustal level now exposed by erosion?

Nature of Regional Metamorphic Bodies

This section describes certain features of regionally metamorphosed rock bodies to supplement the description of rock types on the hand sample and outcrop scale given in Section 11.4.

Slate belts are elongate terranes that may be exposed over thousands of square kilometers (Hobbs and others, 1976, p. 403). Although slates formed by metamorphism of shale protoliths are prominent, other low-grade and fine-grained metasedimentary and metavolcanic rocks are widespread. Foliation is steeply inclined and subparallel to the belt. Mineral assemblages are generally of the prehnite–pumpellyite and greenschist facies.

Regionally extensive **migmatite complexes** are more or less confined to sillimanite or higher-grade zones of the upper amphibolite facies where muscovite is unstable and sillimanite coexists with K-feldspar (see Figure 12-12). Comparison of Figures 8-37b, 10-27a, and 9-11 shows that a water-deficient assemblage (see pages 289 and 312) of quartz + alkali feldspar + muscovite, such as might occur in many pelitic and quartzofeldspathic metamorphic rocks of the amphibolite facies, would begin to melt just below 700°C at 0.4 GPa, corresponding to a depth of ~ 15 km. Under water-excess conditions, the melting would begin

at somewhat lower temperatures, assuming in both cases that the muscovite is close to its ideal composition. The breakdown conditions for biotite and hornblende lie at much higher temperatures. Hence, local hydrous (but generally not water-saturated) granitic melts could form with a more refractory assemblage of biotite, hornblende, and calcic plagioclase. If the melts were not of sufficient volume to segregate, collect, and buoy out of the rock body to form a granitic pluton, they might eventually solidify into a coarse-textured granitic member of migmatite.

Despite these arguments for partial melting, subsolidus metamorphic differentiation and metasomatic processes continue to be invoked for specific migmatite bodies (see for example, Ashworth, 1979). Regardless of their origin, migmatites form by processes at $P–T$ conditions transitional between metamorphism and granitic magmatism. This transition is especially clear in the southern Black Forest of Germany, where Mehnert (1968) has documented a transition over several kilometers from ortho- and paragneisses through migmatites to more or less homogeneous granite (see Figure 12-14). He envisages (p. 273) an idealized sequence of processes as follows:

1. Solid-state crystal growth, chiefly of feldspars (metamorphism).

2. Formation of ternary (Kf–Ab–Q) minimum-temperature partial melts enriched in water, which segregate from the more mafic refractory residuum (migmatization).

3. Collection of predominantly molten material that includes the mafic components (magmatization).

4. Homogenization of the magma, intrusion to higher crustal levels, and possible differentiation.

Mehnert finds all transitions between distinctively magmatic and crystalloblastic textures in migmatites, indicating their mixed origin.

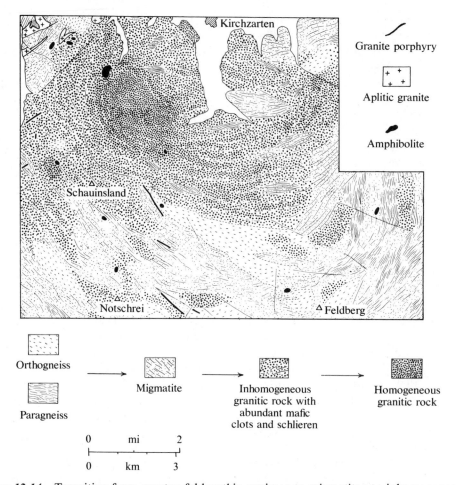

Figure 12-14 Transition from quartzo-feldspathic gneisses to migmatites to inhomogeneous gneissic granitic rocks to homogeneous granitic rock in the Black Forest. The sequence of rock types corresponds to the four processes listed in the text. [After K. R. Mehnert, 1968, *Migmatites and the Origin of Granitic Rocks* (New York: Elsevier).]

In the northern Appalachian orogen of New England, Thompson and Norton (1968, p. 325) note that muscovite–garnet-bearing S-type granites (see pages 140–141) "appear in most areas to be roughly contemporaneous with, or slightly younger than, the main regional metamorphism. Their distribution relative to the higher grade metamorphic zones shows a striking correlation . . . ." North and east of the trace of the Al_2SiO_5 triple point (see Figure 12-11), in the low P/T metamorphic terrane, the plutons have well-defined contact aureoles, which show that the magmas originated at greater depths and were then emplaced at cooler, shallower crystal levels after the culmination of regional metamorphism. In contrast, south and west of this triple-point trace, contact metamorphism around intrusions is minimal. At this deeper crustal level the plutons were syntectonic and contemporaneous with regional metamorphism.

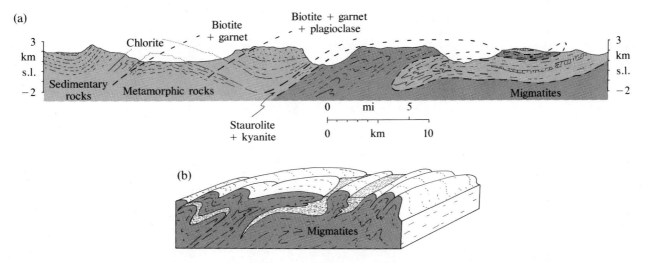

Figure 12-15 Migmatite diapirs in the Caledonide orogen in East Greenland. (a) Geologic section across the head of Kejser Franz Joseph Fjord. Note that horizontal and vertical scales are the same; s.l., sea level. (b) Interpretive model of migmatite diapirs. [After J. Haller, 1971, *Geology of the East Greenland Caledonides* (New York: Wiley Interscience).]

The question should be raised here whether the granitic rocks associated with migmatite in the Black Forest and more or less synmetamorphic granitic rocks in the northern Appalachians are the end products of ultrametamorphism or the cause of metamorphism in the surrounding rocks. But the greater areal extent of the metamorphism suggests that the granites and associated migmatites represent local regions where temperatures were high enough to produce partial melting.

Hot, partially melted migmatite bodies may be sufficiently less dense than surrounding, overlying rocks to buoy upward en masse into shallower crustal levels. Spectacular examples of such **migmatite domes** or diapirs can be seen in the rugged mountainous terrane of the East Greenland Caledonide orogen (see Figure 12-15; Haller, 1971, pp. 157–200). Nappelike sheets that spread out laterally enveloped surrounding lower-grade metamorphic rocks.

Rather similar domical bodies of gneiss mantled by essentially concordant layers of schists, amphibolites, and other metamorphic rocks were called **mantled gneiss domes** by P. Eskola in 1949 (see Hobbs and others, 1976). Belts of such domes together with associated nappes and recumbent folds in metamorphic rocks occur in many deeply eroded orogens, such as the northern Appalachians (see Figure 12-16; Thompson and others, 1968). More or less homogeneous gneisses in the cores of these domes are chiefly of granitic to quartz dioritic composition, and the mantling rocks include other gneisses, amphibolites, mica schists, quartzites, and calc-silicate rocks. Weak foliation in the core gneisses is concordant with the foliation in the mantling tectonites. The core gneisses are older than the mantling rocks and, due to their apparent buoyant mobility during regional metamorphism, mushroomed upward to higher crustal levels in the mantle complex.

12.5 SUB-SEA-FLOOR METAMORPHISM

It would be logical to discuss high *P/T* metamorphic bodies next, but because oceanic rocks figure so prominently in them, we must first briefly touch

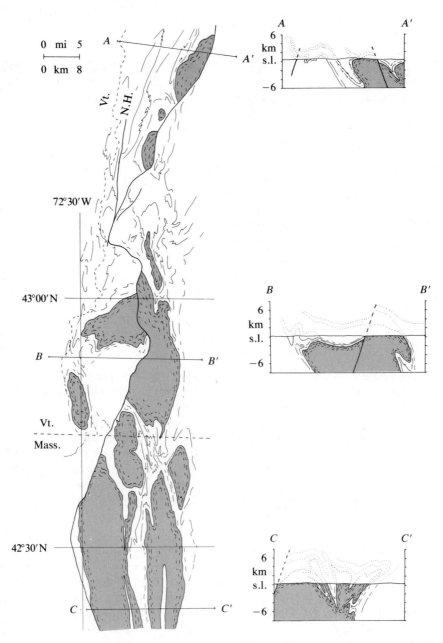

Figure 12-16   Generalized geologic map and cross sections showing gneiss domes (shaded) in the Appalachian orogen in Vermont (Vt.), New Hampshire (N.H.) and Massachusetts (Mass.). Trend lines show complexly folded metamorphic rocks mantling domes. Major north-south fault system shown by heavy lines is Triassic and postorogenic. [After J. B. Thompson, Jr., P. Robinson, T. N. Clifford, and N. J. Trask, Jr., 1968, Nappes and gneiss domes in west-central New England, in E. Zen, W. S. White, J. B. Hadley, and J. B. Thompson (eds.), *Studies in Appalachian Geology, Northern and Maritime* (New York: Wiley Interscience).]

upon metamorphic systems that prevail just beneath the sea floor. (Some petrologists prefer to call this particular process alteration).

During the intensive testing of the concept of sea-floor spreading in the late 1960s and early 1970s, it was found that significant amounts of the oceanic crust are metamorphosed. This realization came from the accelerated investigations of the sea floor itself and from studies of ophiolite complexes on land. Rocks dredged from the sea floor, especially in the vicinity of topographic scarps and transform faults, include greenstones, serpentinites, and rare amphibolites, some schistose. The heavy Sr-isotopic composition ($^{87}Sr/^{86}Sr > 0.704$) of these hydrated metamorphic rocks indicate that the water involved in reconstituting the original oceanic-rift magmatic rocks must have been sea water ($^{87}Sr/^{86}Sr$ for sea water is 0.709 compared to 0.703–0.704 for fresh rift rocks). Besides, the minute amounts of juvenile water in rift magmas could not accomplish the extensive hydration reactions necessary to produce greenstones and amphibolites. Drill cores recovered to date (1981) from the sea floor only show zeolitic alteration along cracks and as vug fillings. No more intensely recrystallized rocks have been penetrated. Analyses of radiometric isotopes in clay and calcite veins in drill cores indicate these secondary minerals formed only about 10 My later than their crustal host rock, which means the veins were emplaced within several hundred kilometers of the oceanic rift. Highly variable heat flows measured within several hundred kilometers of oceanic rifts suggest convective circulation of cooling sea water through newly emplaced, hot magmatic rocks (see pages 182–183). The heavy metal (V, Co, Ni, Cu) content of youngest sea-floor sediment increases one to two orders of magnitude in the vicinity of the crest of oceanic rifts, apparently because of the leaching of these metals from the hot crustal rocks by downward-percolating sea water and their precipitation in the sediment upon egress of the hot brines.

Ophiolite complexes are interpreted to be slices of ancient metamorphosed oceanic crust and uppermost mantle tectonically obducted onto the continental margin in subduction zones. Study of mineral assemblages, field relations, and fabric in these complexes on land adds further support to the concept of circulation of heated sea water through the oceanic crust near the crest of the oceanic rifts. Rapid increase in mineralogical grade stratigraphically downward into ophiolite sequences indicates geothermal gradients as much as several hundred degrees Celsius per kilometer, compatible with some of the high heat flows measured at oceanic rifts. These tightly telescoped mineralogical zones are underlaid by essentially unaltered oceanic crustal rocks. Secondary calcite is concentrated in the upper parts of ophiolite sequences, presumably because penetration of $CaCO_3$-saturated sea water into hot crustal rocks induces precipitation. Secondary hematite is also widespread in upper pillow lavas in ophiolites, and Fe_2O_3/FeO ratios in whole rocks and secondary epidotes decrease downward. The oxidative features are explained, once again, by penetration into the crust of highly oxygenated sea water, which subsequently becomes more reduced in transit as it reacts with oxygen-poor magmatic rocks (see Figure 12-17).

General Character of Rock Bodies

Before proceeding any further, we must strongly emphasize the incomplete state of knowledge regarding sea-floor metamorphism (alteration) and its products. The arguments just presented make a strong case for its existence; yet there are three reasons to urge caution in how the rocks are to be interpreted. First, only dredge samples of metamorphic rocks have so far been collected in the oceanic realm; we do not have any reasonable control on field relations. Second, on-land ophiolites may have suffered one or more metamorphic imprints subsequent to sea-floor activity and prior to and during emplacement along the continental margin. Third, only a decade of research effort has been devoted to the phenomenon, and new information comes to light almost monthly.

Mineral assemblages in altered oceanic crust are dominantly of the zeolite and greenschist

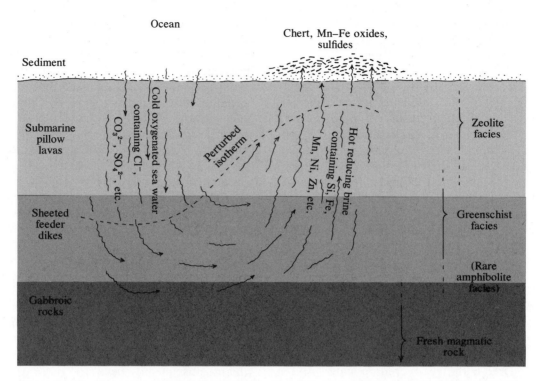

Figure 12-17   Conceptual model of convective circulation of sea water through hot magmatic crust. Influx of cold water depresses isotherms, whereas outflow of hot brines elevates isotherms. Temperature of alteration increases rapidly with depth into the pillow lavas and sheeted dikes, but because of only limited access of sea water below the dikes, virtually no reconstitution occurs in the underlying still hotter gabbroic and ultramafic rocks.

facies. Assemblages that are transitional or that rightfully belong to the amphibolite facies are less common and tend to occur in metamorphosed diabases and gabbros rather than basalts, as might be anticipated because of higher temperatures in the deeper oceanic crust. In a strict sense, it is inappropriate to refer to the mineral assemblages formed just beneath the sea floor as belonging to the zeolite, greenschist, and amphibolite facies because of the very low pressures involved (compare Figure 10-28).

Mineral assemblages and their spatial zonation in the Troodos and E. Liguria ophiolites (see Figure 12-18) furnish insight into the metamorphic conditions beneath the sea floor near oceanic rifts (Gass and Smewing, 1973; Spooner and Fyfe,

1973). It is assumed that the assemblages are due entirely to oceanic processes. Temperatures at the time of metamorphism may be inferred on the basis of experimental data and from measured temperatures and corresponding mineral assemblages observed in bore holes in geothermal fields in Iceland, New Zealand, and the Salton Sea of southern California. Pressure has a negligible effect on the stability of the assemblages in these systems. Inferred geothermal gradients in the relatively thin crustal segment represented in the E. Liguria ophiolite sequence were extremely high, in the range of 900 to 1300°C/km. These gradients are 5 to 8 times the highest gradients ($\sim$ 170°C/km) measured at active oceanic rifts and may, therefore, indicate a very intense and highly localized

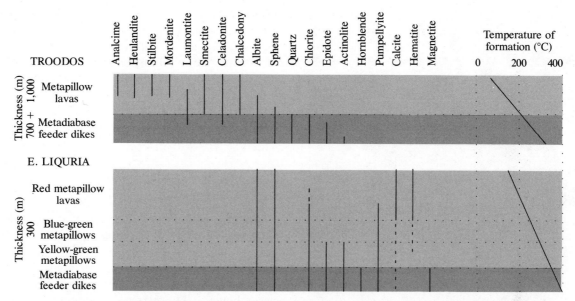

Figure 12-18 Mineral assemblages and inferred metamorphic temperatures in the pillow lava and sheeted feeder dike members of the Troodos and E. Liguria ophiolites (thicknesses not to scale). Note the greater thickness of the observed metamorphic effects in the Troodos ophiolite. Compare Figure 12-17. [After I. G. Gass and J. D. Smewing, 1973, Intrusion, extrusion and metamorphism at constructive margins: Evidence from the Troodos Massif, Cyprus, *Nature* 242; and E. T. C. Spooner and W. S. Fyfe, 1973, Sub-sea-floor metamorphism, heat and mass transfer, *Contributions to Mineralogy and Petrology* 42.]

thermal energy source (magma) along the rift. A more modest gradient of ~ 150°C/km is inferred for the Troodos metamorphic body.

Because of the low pressures and higher geothermal gradients of sub-sea-floor metamorphic (alteration) systems, a prehnite–pumpellyite facies between the zeolite and greenschist facies is absent (see Figure 10-27).

Hydrothermal Brine Systems

The exchange of chemical constituents in sub-sea-floor hydrothermal systems is significant (Spooner and Fyfe, 1973). Initial sea water is highly oxygenated with relatively large concentrations of O^{2-}, Cl^-, Br^-, HCO_3^-, CO_3^{2-}, and SO_4^{2-} anions. During penetrative convection into the relatively reduced magmatic rocks, chemical reactions such as the following oxidize the rock:

$$\underset{\text{in olivine solid}}{Fe_2SiO_4} + \tfrac{1}{2}O_2 \rightarrow \underset{\text{hematite}}{Fe_2O_3} + \underset{\text{silica}}{SiO_2}$$
solution

$$\underset{\substack{\text{in olivine} \\ \text{solid solution}}}{11Fe_2SiO_4} + 2SO_4^{2-} + 4H^+ \rightarrow$$

$$\underset{\text{magnetite}}{7Fe_3O_4} + \underset{\text{pyrite}}{FeS_2} + \underset{\text{silica}}{11SiO_2} + 2H_2O$$

Reaction between Ca (in silicates) and carbonate anions produces calcite:

$$\underset{\substack{\text{in pyroxene} \\ \text{solid solution}}}{CaSiO_3} + CO_3^{2-} + 2H^+ \rightarrow$$

$$\underset{\text{calcite}}{CaCO_3} + \underset{\text{silica}}{SiO_2} + H_2O$$

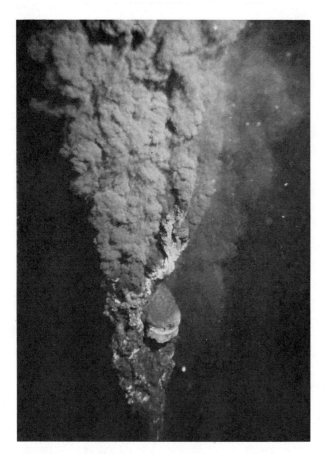

Figure 12-19   Plume of dark-gray brine discharging from a hot spring along the East Pacific Rise at a water depth of about 3 km. Temperatures in excess of 350°C have been measured at such orifices. [Photograph by Robert Ballard from the deep submersible Alvin, courtesy of the Woods Hole Oceanographic Institution.]

Note that large amounts of silica are liberated as a consequence of formation of metal oxides, carbonates, and sulfides. In some manner, this silica must eventually contribute to the formation of deep marine cherts so typically associated with ophiolite sequences.

Farther along its path of convective circulation through hot crustal rocks (see Figure 12-17), sea water becomes more reduced and large amounts of metals such as Mn, Fe, Co, Ni, Cu, Zn, Ag, and Au are leached from the mafic rocks and pass into solution as chloride complexes. Note the depletion of these elements in spilites (see p. 398). As these brines are eventually discharged from the basaltic crust into the overlying ocean (see Figure 12-19), the dissolved heavy metals are precipitated as colloidal, hydrated ferric and manganese oxides and as sulfides, which form coatings on lavas and sediment, creating potentially exploitable ore deposits (Bonatti and others, 1976).

12.6 HIGH-PRESSURE BLUESCHIST METAMORPHIC BELTS

Submarine pillow lavas and related gabbroic and ultramafic rocks of the ophiolite suite are intimately associated with arc-trench graywackes, deep-marine sediments, and, in many localities, metamorphic rocks containing high P/T mineral assemblages. These strikingly contrasted rock types are found tectonically juxtaposed and mingled together in elongate belts in young orogens, mostly Mesozoic and Cenozoic, around the margin of the Pacific Ocean (see Figure 12-20a) and in the Alpine-Mediterranean orogenic zone. This diverse petrotectonic association has puzzled geologists for decades.

Paired Metamorphic Belts and Plate Tectonics

In 1961, before plate tectonics had been conceived, A. Miyashiro (1973) recognized that in the Mesozoic-Cenozoic terranes of the circum-Pacific region there are **paired metamorphic belts,** as in Japan (see Figure 12-20b). The high P/T belt with its associated tectonically juxtaposed ophiolites and trench sediments generally lies closer to the ocean and adjacent to the oceanic trench. Farther inland is the second belt of intermediate to low P/T metamorphic rocks and associated calc-alkaline granitic rocks.

The formation of paired, regionally extensive, metamorphic belts with mineral assemblages characteristic of contrasting P/T gradients can be rationalized in the light of global plate tectonics. In regions of ocean–continental plate convergence,

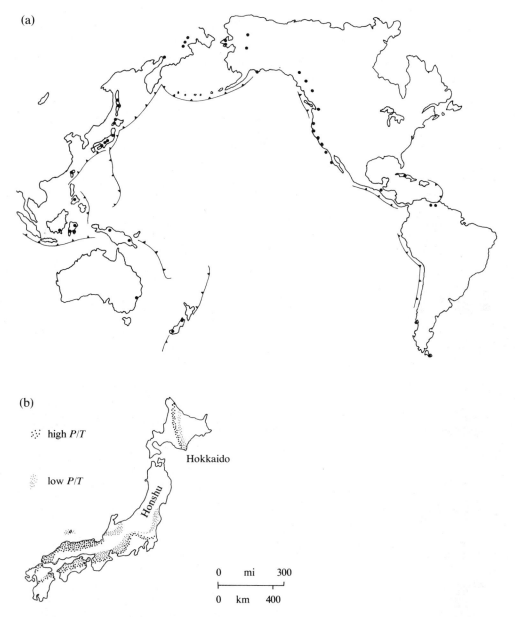

Figure 12-20 (a) High *P*/*T* metamorphic belts containing blueschists (solid circles) in the circum-Pacific region. Active subduction zones indicated by lines with teeth on overriding plate. (b) Paired metamorphic belts in Japan. On Honshu there are two belts, the one on the north being Permian-Jurassic and the younger on the south being Jurassic-Cretaceous. On Hokkaido, the polarity of the paired belts is reversed relative to the two belts on Honshu. [Both after A. Miyashiro, 1973, *Metamorphism and Metamorphic Belts,* (New York: John Wiley and Sons); part (a) also after W. P. Irwin and R. G. Coleman, 1972, Preliminary map showing global distribution of alpine-type ultramafic rocks and blueschists, U.S. Geological Survey Miscellaneous Field Studies, Map MF-340.]

Field Relations of Metamorphic Bodies

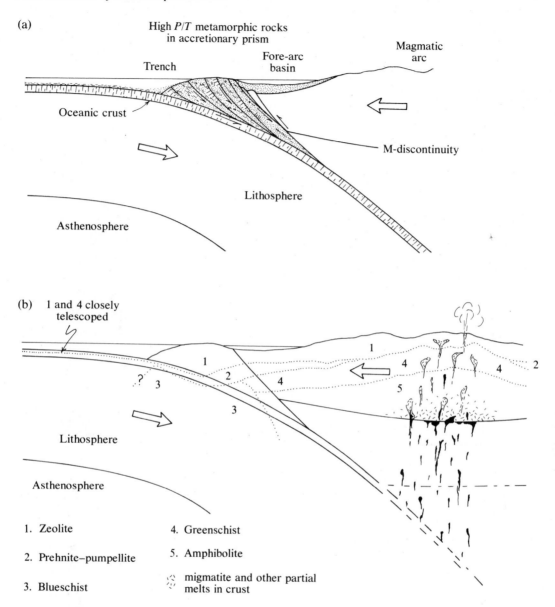

(a)

High *P/T* metamorphic rocks
in accretionary prism

Magmatic
arc

Fore-arc
basin

Trench

Oceanic crust

M-discontinuity

Lithosphere

Asthenosphere

(b) 1 and 4 closely
telescoped

1

1

4

4

2

2

3

2

4

5

?

3

3

Lithosphere

Asthenosphere

1. Zeolite
2. Prehnite–pumpellite
3. Blueschist

4. Greenschist
5. Amphibolite

migmatite and other partial
melts in crust

Figure 12-21 Schematic cross sections through a convergent oceanic–continental plate boundary. (a) Geologic model shows imbrication of slices of arc-trench sediments and scraped-off oceanic material as oceanic lithosphere is underthrust beneath the continental margin. This imbricate underplating forms an accretionary prism that builds the margin of the continent seaward and possibly causes the inclination of the subducting plate to decrease. (b) Approximate distribution of metamorphic facies assemblages and regions of partial melting in the upper mantle and lower crust. Compare with Figure 10-28. [Part (a) after W. G. Ernst (ed.), 1975, Subduction zone metamorphism, Benchmark Papers in Geology 19 (Stroundsburg, Pa.: Dowden, Hutchinson and Ross); part (b) after W. G. Ernst, 1976, *Petrologic Phase Equilibria* (San Francisco: W. H. Freeman and Company).]

there are two, but not necessarily distinct and separable, regions of potential metamorphism, which may be appreciated by examination of Figure 12-21. Near the oceanic trench at the immediate boundary of the two converging plates, arc-trench sediments and subducting lithosphere penetrate to increasing depths and higher pressures but experience only slightly increasing temperatures because of the slow rates of heat conduction into the descending cool slab. Note depressed isotherms in the slab beneath the trench in Figure 9-2. The descending lithosphere and overlying wedge of arc-trench sediment, therefore, follows a steep, high P/T trajectory. In California, geothermal gradients are interpreted to have been on the order of only 10 to 15°C/km (see Figure 10-28). This high P/T trajectory begins in the P/T field of the zeolite facies and runs into the realm of the blueschist facies, perhaps passing through the prehnite-pumpellyite facies field en route. Farther inland, such as in the Sierra Nevada region of California, there is another region of contemporaneous metamorphism more or less coincident within the calc-alkaline magmatic arc, where geothermal gradients are significantly higher and follow intermediate to low P/T trajectories.

The fact that Paleozoic blueschist belts are less common than Mesozoic and Cenozoic ones and that Precambrian ones are rare has puzzled many geologists. Slower rates of subduction in pre-Mesozoic time and possibly thinner plates that allowed higher temperatures to develop during subduction have been suggested. But another explanation (see pages 373–375) is that the more slowly a high-P/T metamorphic body is uplifted, and therefore the older it is when finally exposed at the surface, the greater is the opportunity for it to be heated and overprinted by higher-temperature mineral assemblages.

Field Relations

High-P/T blueschist belts record a complex interplay of more or less contemporaneous sedimentation, tectonism, metamorphism, and uplift during plate convergence. The geologic rapidity of these interrelated processes is indicated by the fact that depositional ages inferred from rarely preserved fossils are commonly not much greater than the radiometric age of metamorphic recrystallization. Before detailing a specific blueschist belt, we need to describe briefly a characteristic type of rock body in such belts, known as mélange, and to make further comments regarding ophiolite.

Mélange. Mélange (French for "mixture") is a body of rock mappable in the field that contains heterogeneous blocks within a pervasively deformed matrix. Mélange is widespread in the Upper Jurassic–Cretaceous Franciscan complex of the California Coast Ranges (see Figures 12-22 to 12-24). For an area north of San Francisco, Maxwell (1974, pp. 1200–1201) observed that

> typical Franciscan rocks...consist of broken and discontinuous masses of graywacke and silty mudstone, with zones of scattered exotic blocks in a matrix of graywacke or mudstone. The matrix is characterized by unsystematic, discontinuous folding and intense local shearing and rubbling. Exotic blocks include all members of the ophiolite suite—serpentinite, gabbro, diabase, extrusive lava and radiolarian chert—as well as large and small blocks of coarse blueschist....The association of broken and sheared matrix with dispersed exotic blocks is best described as a mélange . . . [and] gives rise to an "accidental" topography characterized by scattered blocks jutting out of a rounded or landslide surface.

Mélange in subduction zone–oceanic trench settings might form by faulting, shearing, diapirism, and squeezing of unconsolidated material and surficial landsliding on steep slopes. No mixing process is excluded in the formation of mélange (Silver and Beutner, 1980).

Ophiolite. Ancient oceanic lithosphere formed at divergent plate junctions is consumed into the mantle in subduction zones. But, locally in time and space, apparent tectonic accidents are believed by most geologists to have allowed less than 0.001% of this lithosphere to be obducted into the overriding plate, where it can eventually be ex-

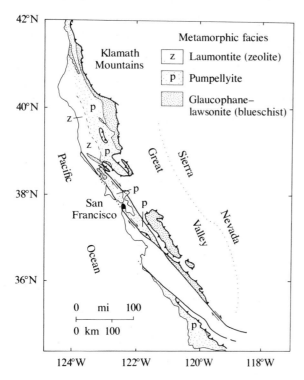

Figure 12-22 Simplified geologic map showing metamorphic zonation in the Franciscan Complex in the California Coast Ranges. Heavy lines with teeth are traces of the Coast Range Thrust; teeth are on upper plate rocks of the Great Valley Sequence (unpatterned). Postmetamorphic, mid- to late-Cenozoic faults of the San Andreas transform system are unpatterned heavy lines. For interpretive cross sections of the geologically controversial Coast Ranges, see Maxwell (1974) and Suppe (1979). [After W. G. Ernst (ed.), 1975, Subduction zone metamorphism, Benchmark Papers in Geology 19 (Stroundsburg, Pa.: Dowden, Hutchinson and Ross).]

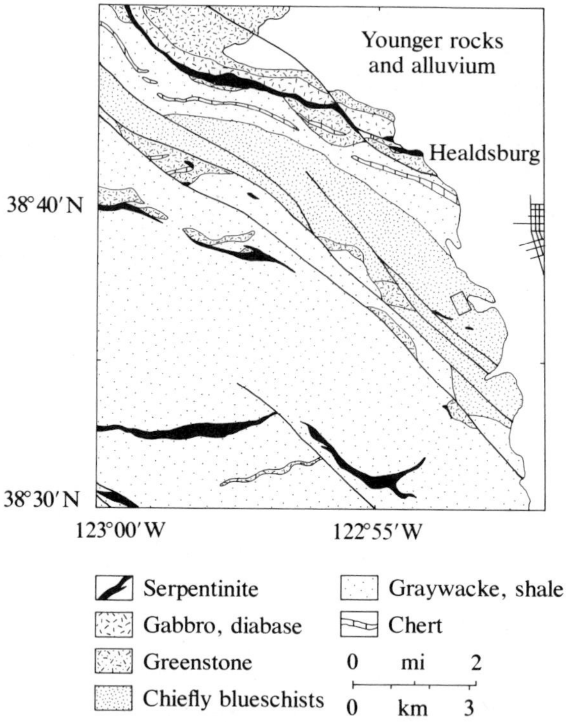

Figure 12-23 Geologic map of a part of the California Coast Ranges west of Healdsburg and 95 km north of San Francisco (see Figure 12-22) showing tectonically intermingled blueschist facies metamorphic rocks, dismembered ophiolite, and unmetamorphosed arc-trench sedimentary rocks of the Franciscan Complex. Designated rock types on the map are locally generalized. Heavy lines are faults. Rectangle is area shown in Figure 12-24. [After W. K. Gealey, 1950, Geology of the Healdsburg Quadrangle, California, *California Division of Mines Bulletin* 161.]

posed at the surface due to isostatic uplift and erosion as ophiolite (Coleman, 1977).

Where exposed on land, ophiolite commonly lies in terranes of mélange where it has been tectonically dismembered. Metamorphic rocks with high P/T mineral assemblages (such as lawsonite + albite + chlorite) locally tectonically underlie ophiolite. Sedimentary and low-grade metamor-

phic sequences may lie in depositional contact on ophiolite. In still other localities relatively intact sheets of ophiolite overlie zoned metamorphic rock bodies, and contact relations are clear enough to furnish insight into the mode of emplacement. In western Newfoundland, for example, Williams and Smyth (1973) and Jamieson (1980) report that syntectonic and locally mylonitic metamorphic

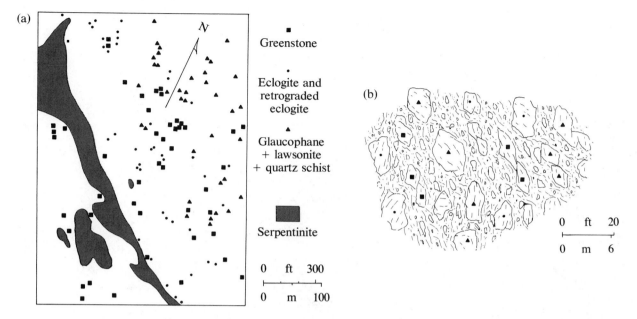

Figure 12-24 Franciscan mélange. (a) Distribution of rock types in the small rectangular area outlined in Figure 12-23. Each symbol represents an outcrop, usually several meters across. Other than for serpentinite, it is not possible to draw contacts between rock types because of their chaotic distribution. The terrane is interpreted to be a mélange that, if wholly exposed, might appear something like the sketch in (b), showing exotic heterogeneous blocks in a sheared matrix. Compare with Figure 12-25. [Part (a) after I. Y. Borg, 1956, Glaucophane schists and eclogites near Healdsburg, California, *Geological Society of America Bulletin* 67.]

rocks up to several hundred meters thick consist of pyroxene-bearing amphibolites, amphibolites, and greenschists with the intensity of metamorphic grade and of polyphase deformation decreasing structurally downward beneath the overlying early Paleozoic ophiolite sheets. Mineral geothermometers in the ophiolite peridotite and underlying syntectonic metamorphic aureole indicate that the former was emplaced at 900°C to 1050°C and that the aureole rocks recrystallized at gradationally lower temperatures away from the peridotites. Mineral geobarometers indicate the last equilibration in the peridotites was at 0.7 to 1.0 GPa, whereas pressures were substantially less in the underlying aureole rocks. To what extent shear heating during emplacement was involved is uncertain, but the evidence does suggest that a hot,

once deeply buried slab of ultramafic rock overrode continental margin rock sequences, syntectonically metamorphosing them.

Other oceanic slabs form caps on rock sequences with blueschist facies assemblages. But blueschists are not always in contact with ultramafics. Mélanges of serpentinite may carry inclusions of blueschist and eclogite, but as serpentinized ultramafics themselves have no phases that can serve as geobarometers, their metamorphic history is not easily reconstructed. Most on-land ophiolites have either escaped or at least now show no high-P/T metamorphism. Overprinting of higher-T, lower-P metamorphism is possible as progressive continental accretion brings earlier-obducted slices farther inland, closer to the hot magmatic arc.

Figure 12-25   Franciscan mélange exposed in wall of quarry near Greenbrae, 16 km north of San Francisco. Blocks of graywacke, greenstone, tuff, and chert lie in a matrix of sheared shale. Exposure partly covered by regolith and disturbed by local minor landslides. [U.S. Geological Survey photograph courtesy of E. H. Bailey.]

High-P/T metamorphic belt in California.   Blueschists, dismembered ophiolite, and associated arc-trench rocks comprise the Franciscan Complex of the Coast Ranges of California (see Figures 12-22 to 12-25). Ernst (1975, p. 3) summarizes:

> The relatively high pressure metamorphic progression in Franciscan rocks ranges from laumontite zone developed in the northwestern coastal belt of the California Coast Ranges through pumpellyite-bearing rocks to a lawsonitic glaucophane schist assemblage adjacent to the tectonic contact with virtually unmetamorphosed coeval rocks of the Great Valley Sequence

.... This profound structural break, the Coast Range Thrust..., juxtaposes the Franciscan, an inferred oceanic trench mélange, including dismembered ophiolites, on the west against Great Valley continental margin sediments laid down to the east on the stable American lithospheric plate....The inferred relative underflow was to the east. Few major internal structures have been recognized within the Franciscan terrane, but at least in northern California...more-or-less coherent tectonic slices dip to the east, as does the Coast Range Thrust itself.

The highest grade *in situ* blueschists, and eclogites and albite amphibolites that occur only as tectonic blocks, are disposed in general in the eastern part of the Franciscan terrane...; these coarse-grained schists possess latest Jurassic and Early Cretaceous radiometric ages, whereas farther to the west, mid-Cretaceous *in situ* depositional and recrystallization ages are typical....The feebly metamorphosed northwestern coastal belt of the Franciscan group contains strata at least as young as Paleocene in age....

During the same time interval (from late Jurassic through Cretaceous) that this high *P/T* assemblage was forming near the coast, granitic plutonism (following slightly earlier regional metamorphism) was occurring farther inland in the area of the present Sierra Nevada (see pages 142–143).

Unsolved Problems

We have alluded to the formation of blueschist facies assemblages under high *P/T* conditions, but this interpretation was not widely favored in the decades prior to the application of experimental and theoretical thermodynamic constraints to the problem and prior to plate tectonic concepts. One of the older explanations for the formation of glaucophane-bearing rocks invoked Na-metasomatism around serpentinite intrusions. Although this has been documented in local instances, detailed mapping elsewhere, as shown in Figure 12-24a, indicates blueschists are not necessarily related to serpentinite; moreover, many blueschists are chemically equivalent to common graywackes and basalts, so isochemical metamorphism is feasible. An alternate suggestion that blueschist assemblages form metastably in some manner has not received independent support.

The association of relatively dense jadeitic pyroxenes, lawsonite, and aragonite indicates that elevated pressures at low to moderate temperatures generate blueschist facies assemblages. Deep burial and tectonic overpressures have been postulated as causes of the high pressures. Ernst (1975) has advocated deep burial, 20 to 30 km, as the cause of blueschist metamorphism. In his subduction model (see Figure 12-21), oceanic crust and arc-trench deposits are underthrust as imbricate slices beneath the continental margin. Older, once more deeply buried rocks subjected to high *P/T* metamorphism emerge to the surface farther inland due to isostatic uplift and erosion of the thickened, buoyant accretionary wedge (see quotation by Ernst, above). Some high *P/T* terranes are now bounded by postmetamorphic strike-slip faults—the San Andreas transform in California, for example. These faults reflect a shift from compressive plate convergence to passive horizontal slip, which might facilitate buoyant rebound and erosion of the mélange wedge, exposing the deeper, high *P/T* metamorphic assemblages. A substantial decrease in the rate of plate convergence might have a similar effect along a boundary of continuing subduction. The dip of the descending plate may also play a role; shallow-dipping plates may encourage uplift of the overlying slab.

Some geologists have not been willing to accept the deep burial explanation for blueschist metamorphism and have appealed, instead, to **tectonic overpressures.** In bodies of tectonically stressed rocks, the mean pressure, $\bar{P}$, is greater than the confining pressure, P, which is due solely to the overburden. However, the amount of this overpressure ($\bar{P} - P$) is limited by the strength of the rock, which in turn is dictated partly by the pressure of fluids in pore spaces. In dry and hence stronger rock bodies, overpressures of a few hundred MPa could develop in massive graywackes, which could effectively decrease the required depth of burial for metamorphism by about half. But the hydrated blueschist mineral assemblages of the Franciscan rocks testify to the presence of water during metamorphism, which implies rocks were not so strong. This observation plus the presence of local, weak, shaly layers within the Franciscan suggest that, realistically, any overpressures are likely to have been less than 100 MPa, equivalent in depth to less than 4 km. Upside-down metamorphic zonation in Franciscan rocks beneath the exposed sole of the Central Valley Thrust in northwestern California is believed by some geologists to have been produced by abnormally high pore water pressures near the thrust, rather than by deep burial. Hence the debate between deep burial and tectonic overpressures to account for blueschists has not been wholly resolved.

Serpentinites

Serpentinites representing metamorphosed peridotites, pyroxenites, and dunites in ophiolite are as characteristic of high-*P*/*T* metamorphic belts as glaucophane–lawsonite rocks. Thus, in the compilation of Irwin and Coleman (1972) all blueschist occurrences are associated with serpentinite. However, many Paleozoic and Precambrian serpentinite belts such as in the Canadian and Brazilian shields, have no high *P*/*T* blueschists with them. Serpentinite belts are as valid an indicator of ancient subduction zones as blueschists.

A discontinuous belt of ophiolite fringes the western margin of the Sierra Nevada batholith in California (see Figure 4-43). Part of this ophiolite belt has been internally mixed to form a serpentinite mélange (Saleeby, 1979). Within the schistose serpentinite matrix are blocks up to 20 km long of peridotite (partly serpentinized), gabbro, basalt, chert, and breccias of serpentinite blocks cemented by carbonates. The serpentinite mélange is envisaged to have developed in a transform fracture zone near an oceanic rift. Accompanying the tectonic shearing of the crust, which enhanced access of sea water, serpentinizing ultramafic rocks were protruded onto the sea floor, forming detrital serpentinite. Larger clasts were cemented by carbonate to form **ophicalcite**, whereas finer detritus, together with very minor amounts of pelagic sediment, formed "sedimentary" serpentinite. More solid blocks of the disrupted ophiolite sequence were tectonically mixed into this soft, tractable serpentinite matrix. Later, this mélange body was sutured onto the margin of the American continent during late Paleozoic–early Mesozoic time, producing additional internal deformation.

SUMMARY

12.1. Mylonite is a sheetlike body ranging from microscopic dimensions to many kilometers that is more intensely deformed and finer-grained than adjacent rocks due to pervasive ductile flow. Zones of mylonitic rock in orogenic belts commonly show complex multiple episodes of deformation.

12.2. Country rocks surrounding magmatic intrusions are thermally and chemically perturbed, forming zoned contact aureoles of massive hornfelses or metasomatic rocks. The nature of the aureole depends upon many factors, including the size, *T*, and composition of the magmatic intrusion, whether it convects during cooling, and the *T*, composition, and fabric of the country rocks, especially their water content and permeability.

Major intrusions into pelitic rocks commonly form broad, zoned aureoles in which rocks spotted with cordierite or andalusite give way toward the intrusion to more thoroughly recrystallized hornfelses containing more anhydrous mineral assemblages.

Skarns of calc-silicate mineral assemblages are typically developed in carbonate rocks adjacent to granitic and dioritic intrusions as aqueous fluids containing dissolved Si, Al, Fe, and so on are expelled and react with calcite and dolomite.

12.3. Deeply buried rocks may suffer partial to complete mineralogical reconstitution over broad areas independent of magmatic intrusions and tectonism. During this burial metamorphism, zeolite facies assemblages commonly develop in water-rich glassy volcanogenic rocks, whereas clay–carbonate assemblages characterize rocks containing more CO_2.

12.4. Regionally extensive, foliated and lineated tectonites with mineral assemblages of the intermediate to low *P*/*T* facies series are widely exposed in deeply eroded orogenic belts and in Precambrian shields. Granitic to dioritic intrusions are ubiquitous in these terranes.

Metamorphic zonation in the Appalachian orogen of the United States is complicated in pattern and likely reflects multiple episodes of deformation and heating in response to complicated plate convergence over a 250-My interval of Paleozoic time. In the restricted segment of the Caledonide orogen in the northern British Isles, multiple episodes of deformation concurrent with essentially only one thermal pulse are compressed into ~ 30 My in Cambro-Ordovician time before a 100-My period of copious granitic intrusion and isostatic uplift.

Regional metamorphic terranes exposed in the roots of ancient orogens are zoned from greenschist (locally zeolite) through amphibolite and granulite facies. In the sillimanite zone, field and experimental observations suggest that localized partial melting in rocks of appropriate composition can form bodies of granite melt; this culminating ultrametamorphism yields migmatites and large synmetamorphic plutons capable of diapirically buoying upward through their ductile but more refractory and denser country rocks to shallower crustal levels. Buoyant diapirs of migmatite and granitic gneiss form migmatite and gneiss domes. High-T granulite facies terranes apparently originate from dry rocks that may previously have been dewatered by an episode of partial melting.

12.5. Convective penetration of sea water into pervasively fractured hot basaltic rocks of the oceanic crust near divergent rift junctures produces sub-sea-floor metamorphism (alteration). Original fabrics are commonly conspicuous but overprinted to a variable degree by hydrous replacement assemblages, which are generally of metasomatic character. Due to geothermal gradients of hundreds of degrees Celsius per kilometer, zeolite and greenschist mineral zones are tightly telescoped into one or two kilometers of pillow basalts and sheeted feeder dikes in the top part of the oceanic crust; relatively fresh magmatic rocks underlie these zones.

Highly oxygenated sea water carrying Cl^-,

HCO_3^-, CO_3^{2-}, and SO_4^{2-} forms ionic complexes with transition metals leached from the oceanic crust; as this brine is expelled from its convective circuit through the crust into the cold ocean, encrustations of potentially exploitable hydrated Mn- and Fe-oxides and bodies of Ni–Zn–Cu sulfides form on the sea floor.

12.6. Rocks of the high P/T metamorphic facies series, ophiolite, deep marine cherts, and arc-trench sediments are intimately associated and tectonically juxtaposed as mélange in Phanerozoic belts adjacent to ancient oceanic trenches; farther inland there is commonly an intermediate- to low-P/T metamorphic belt coincident with the calc-alkaline magmatic arc. These paired belts, found chiefly around the Pacific Ocean and in the Mediterranean region, are explained by a subduction model: Arc-trench sediments together with scraped-off oceanic crust and uppermost mantle are underthrust beneath the accretionary prism between the old continental margin and the oceanic trench into an environment where geothermal gradients are low, forming a high P/T facies series; subsequently, in some uncertain way, this mass of metamorphic rocks is uplifted and exposed at the surface. Farther inland, where rising magmas above the Benioff zone deflect geotherms upward, creating high geothermal gradients, intermediate to low P/T regional metamorphism develops. But not all orogens may follow this model.

Large parts of the high P/T belt are a chaotically intermingled mélange of high-P/T metamorphic rocks, including blueschists and eclogites, together with ophiolite rock types and arc-trench sediments. Ancient oceanic lithosphere (ophiolite) is almost always tectonically dismembered. Glaucophane-bearing rocks with coexisting high-P/T phases such as lawsonite, aragonite, jadeitic pyroxene, and pyropic garnet comprise but 1% of late Mesozoic Franciscan rocks in the Coast ranges. Clastic sedimentary rocks, serpentinite, and mafic igneous rocks, in that order, are far greater in volume.

These high-*P/T* mineral assemblages are probably formed as a result of deep burial in the vicinity of the descending oceanic lithosphere, rather than by tectonic overpressures, though the debate continues.

Alpine serpentinite belts formed by hydration of dunite and peridotite in ophiolite are valuable indicators of ancient convergent plate boundaries.

REFERENCES

Ashworth, J. R. 1979. Genesis of the Skagit Gneiss migmatites, Washington, and the distinction between possible mechanisms of migmatization: Discussion and reply. *Geological Society of America Bulletin* 90:887–888.

Boles, J. R. and Coombs, D. S. 1977. Zeolite facies alteration of sandstones in the Southland syncline, New Zealand. *American Journal of Science* 277:982–1012.

Bonatti, E., Zerbi, M., Kay, R., and Rydell, H. 1976. Metalliferous deposits from the Apennine ophiolites: Mesozoic equivalents of modern deposits from oceanic spreading centers. *Geological Society of America Bulletin* 87:83–94.

Burnham, C. W. 1959. Contact metamorphism of magnesian limestones at Crestmore, California. *Geological Society of America Bulletin* 70:879–920.

Christie, J. M. 1963. The Moine thrust zone in the Assynt region, northwest Scotland. University of California Publications in Geological Sciences 40:345–440.

Coleman, R. G. 1977. *Ophiolites.* New York: Springer-Verlag, 229 p.

Cook, F. A., Albaugh, D. S., Brown, L. D., Kaufman, S., Oliver, J. E., and Hatcher, R. D., Jr. 1979. Thin-skinned tectonics in the crystalline southern Appalachians: COCORP seismic-reflection profiling of the Blue Ridge and Piedmont. *Geology* 7:563–567.

Dewey, J. F., and Pankhurst, R. J. 1970. The evolution of the Scottish Caledonides in relation to their isotopic age pattern. *Royal Society of Edinburgh Transactions* 68:361–389.

Ernst, W. G. (ed.). 1975. Subduction zone metamorphism. Benchmark Papers in Geology/19. Stroundsburg, Pennsylvania: Dowden, Hutchinson and Ross, 445 p. (A collection of important papers by several geologists on high-*P/T* metamorphic belts).

Gass, I. G., and Smewing, J. D. 1973. Intrusion, extrusion and metamorphism at constructive margins: Evidence from the Troodos Massif, Cyprus. *Nature* 242:26–29.

Haller, J. 1971. *Geology of the East Greenland Caledonides.* New York: Wiley Interscience, 413 p.

Higgins, M. W. 1971. Cataclastic rocks. U.S. Geological Survey Professional Paper 687, 97 p.

Hobbs, B. E., Means, W. D., and Williams, P. F. 1976. *An Outline of Structural Geology:* New York: John Wiley and Sons, 571 p.

Irvine, T. N. 1970. Heat transfer during solidification of layered intrusions. I. Sheets and sills. *Canadian Journal of Earth Sciences* 7:1031–1061.

Irwin, W. P., and Coleman, R. G. 1972. Preliminary map showing global distribution of alpine-type ultramafic rocks and blueschists. U.S. Geological Survey Miscellaneous Field Studies, Map MF-340.

Jamieson, R. A. 1980. Formation of metamorphic aureoles beneath ophiolites—Evidence from the St. Anthony complex, Newfoundland. *Geology* 8:150–154.

Landis, C. A., and Bishop, D. G. 1972. Plate tectonics and regional stratigraphic-metamorphic relations in the southern part of the New Zealand geosyncline. *Geological Society of America Bulletin* 83:2267–2284.

Mason, R. 1978. *Petrology of the Metamorphic Rocks.* London: George Allen and Unwin, 254 p.

Maxwell, J. C. 1974. Anatomy of an orogen. *Geological Society of America Bulletin* 85:1195–1204.

Mehnert, K. R. 1968. *Migmatites and the Origin of Granitic Rocks.* New York: Elsevier, 393 p.

Miyashiro, A. 1973. *Metamorphism and Metamorphic Belts.* New York: John Wiley and Sons, 492 p.

Moore, J. M., Jr. 1960. Phase relations in the contact aureole of the Onawa pluton, Maine. Unpublished Ph.D. thesis. Cambridge: Massachusetts Institute of Technology.

Morgan, B. A. 1972. Metamorphic map of the Appalachians. United States Geological Survey Miscellaneous Geologic Investigations, Map I-724.

Norton, D., and Knight, J. 1977. Transport phenomena in hydrothermal systems: Cooling plutons. *American Journal of Science* 277:937–981.

Roper, P. J., and Justus, P. S. 1973. Polytectonic evolution of the Brevard Zone. *American Journal of Science* 273A:105–132.

Saleeby, J. 1979. Kaweah serpentinite mélange, southwest Sierra Nevada foothills, California. *Geological Society of America Bulletin* 90:29–46.

Scholz, C. H., Beavan, J., and Hanks, T. C. 1979. Frictional metamorphism, argon depletion, and tectonic stress on the Alpine fault, New Zealand. *Journal of Geophysical Research* 84:6770–6782.

Silver, E. A., and Beutner, E. C. 1980. Penrose Conference Report: Melanges. *Geology* 8:32–34.

Spooner, E. T. C., and Fyfe, W. S. 1973. Sub-sea-floor metamorphism, heat and mass transfer. *Contributions to Mineralogy and Petrology* 42:287–304.

Suppe, J. 1979. Structural interpretation of the southern part of the northern Coast Ranges and Sacramento Valley, California: Summary. *Geological Society of America Bulletin* 90:327–330.

Thompson, J. B., Jr., and Norton, S. A. 1968. Paleozoic regional metamorphism in New England and adjacent areas. In E. Zen, W. S. White, J. B. Hadley, and J. B. Thompson (eds.), *Studies of Appalachian Geology, Northern and Maritime*. New York: Wiley Interscience, pp. 319–327.

Thompson, J. B. Jr., Robinson, P., Clifford, T. N., and Trask, N. J., Jr. 1968. Nappes and gneiss domes in west-central New England. In E. Zen, W. S. White, J. B. Hadley, and J. B. Thompson (eds.), *Studies in Appalachian Geology, Northern and Maritime*. New York: Wiley Interscience, pp. 203–218.

Turner, F. J. 1981. *Metamorphic Petrology: Mineralogical, Field and Tectonic Aspects,* 2nd Ed. New York: McGraw-Hill, 524 p.

Wilkinson, J. F. G., and Whetten, J. T. 1964. Some analcime-bearing pyroclastic and sedimentary rocks from New South Wales. *Journal of Sedimentary Petrology* 34:543–553.

Williams, H., and Smyth, W. R. 1973. Metamorphic aureoles beneath ophiolite suites and Alpine peridotites: Tectonic implications with west Newfoundland examples. *American Journal of Science* 273:594–621.

13

Origin of Fabric in Metamorphic Systems

Imposed metamorphic fabrics evolve through thermally, chemically, and mechanically induced destructive and constructive processes. Cohesive ductile flow in response to nonhydrostatic stress destroys existing grains, and the mechanical deformational energy is partly converted to increased surface and internal energy in grains. Constructive, thermally activated processes form new, more stable grains by nucleation, growth, and grain boundary adjustment, reducing stored grain energy. The most widespread metamorphic fabrics, found in regional tectonites, appear to be essentially syntectonic and are formed by concurrent thermally activated and mechanical processes.

Given sufficient time, solid-state crystal growth produces the coarsest possible grain sizes. Grain shapes are dictated by the requirement to minimize grain-boundary energy in the aggregate; phases with low-energy anisotropy approach a configuration in which grain boundaries (seen in thin section) meet at approximately 120° triple junctions, whereas more anisotropic grains assume a more euhedral habit.

Metamorphic rock bodies subjected to nonhy-drostatic stress under elevated but subsolidus T and high P deform by cohesive ductile flow, especially if the time rate of strain is slow and water is present to weaken silicates hydrolytically. Grains and grain aggregates change shape (strain) in ductile flow by intracrystalline plastic slip and twinning and by pressure solution.

Plastic slip and twinning are experimentally well-documented phenomena and occur on favorably oriented crystallographic planes in a crystal. Slip is made possible by the presence of line imperfections, or dislocations, which allow parts of the crystal to translate past one another at much lower applied stress than for a theoretically perfect crystal. If thermally activated at high T, dislocations can climb away from obstacles in their slip planes, promoting recovery to a state where more slip can occur. More or less steady-state plastic flow by simultaneous slip and recovery and recrystallization can, therefore, occur at elevated temperatures under nonhydrostatic stress, resulting in large strains.

The pattern and symmetry of ductile flow under nonhydrostatic stress is reflected in the pattern of the anisotropic fabric, whether planar, linear, or planar–linear. The origin of segregation layering and preferred orientation of mineral grains in metamorphic tectonites is still controversial, but layer transposition, rigid grain rotation, pressure solution, and intracrystalline plastic slip and twinning have each been demonstrated to be pertinent in specific instances. Fluids liberated from a body during prograde metamorphism have an important bearing on the evolution of the anisotropic fabric that may be concurrently developing if the body is under nonhydrostatic stress.

13.1 SOLID-STATE CRYSTAL GROWTH

Crystal growth is a thermally (and sometimes chemically) activated response of a solid-state system to minimize its total energy by nucleation and growth of new mineral phases that are thermodynamically more stable and by adjustment of the boundaries and surface areas of existing mineral grains in the aggregate; both processes consume grains of relatively higher energy.

Solid-state crystal growth includes: (1) crystallization of new phases not already present by mineralogical reactions, nucleation, and subsequent crystal growth, as in the growth of garnet in a shale protolith experiencing prograde metamorphism; and (2) modification of boundaries of existing grains that may be strained or otherwise less stable than alternate ones of different shape or size; metamorphism of fine-grained chert to coarse quartzite and recrystallization of strained quartz grains are examples.

Nucleation

In metamorphic bodies, a nucleus of an entirely new mineral grain might form within an existing homogeneous solid grain. The fundamental concepts of homogeneous nucleation in silicate melts developed on pages 238–240 can also be applied to formation of nuclei in a solid medium. Local, transient fluctuations from the normal atomic configuration are assumed to yield a group of atoms with

the arrangement of a new phase. If the radius of this entity is sufficiently great that the free energy difference between the old and new phases is not counteracted by the greater surface energy of the nucleus, then the nucleus is viable and can grow.

The rate of homogeneous nucleation in solid-state systems is slow in comparison to that in silicate melts because of the lower T and consequently reduced thermal agitation of atoms in crystals. For this reason, many investigators (Fyfe and others, 1958; McLean, 1965; Spry, 1969) emphasize the role played by **heterogeneous nucleation** occurring along grain boundaries or along internal crystal discontinuities such as dislocations and deformation band boundaries (see Figure 10-10a) where there are arrays of atoms that are perturbed compared to the more normal parts of the grain. That grain boundaries are favored nucleation sites may be seen in the widespread occurrence of oriented subsolidus overgrowths of one mineral upon another (see Figure 10-17). Whether or not the outermost margin of a grain a few atoms in thickness is amorphous, like glass, or some sort of intermediate ordered lattice is a matter of controversy. But in either case, fluctuations in this grain boundary layer to form a viable nucleus would seem to be more likely than at sites within a homogeneous normal crystal.

The formation of porphyroblasts (metacrysts) of one mineral phase in a matrix of smaller crystals of other minerals appears to be a nucleation phenomenon. Common metacrysts are relatively dense aluminosilicates such as garnet, staurolite, kyanite, or andalusite. Their relatively high surface-related energies imply that only a very few abnormally large nuclei survive, and these few then grow to large size to consume available nutrient chemical constituents.

Crystal Growth

Two constraints on growth of crystals operate in metamorphic systems. First is the requirement of minimum free energy and stable thermodynamic equilibrium at the prevailing P–T–X conditions, met by growth of grains of new phases. Second is the requirement for minimum surface and internal strain energy, satisfied by adjustment of boundaries of grains.

Growth of grains of a new phase from a stable nucleus.   Growth processes in solid-state metamorphic systems are more complicated than in silicate melts because requisite chemical constituents for the growing crystal must be derived by decomposition of preexisting grains rather than from a surrounding homogeneous melt. Growth rates are also much slower. In magmas, unwanted atomic components diffuse away from the vicinity of the growing crystal face through the liquid melt, whereas in the case of a new grain of A ($= B + C$) growing in an aggregate of solid phases $B + C + D + E$ there must be some way of handling unwanted constituents ($D + E$). Although some metamorphic mineral grains are free of foreign inclusions, many are poikiloblastic, especially in low-grade assemblages (see Figure 10-15).

A critical factor governing the rate of crystal growth in solid metamorphic aggregates is the rate of diffusion. Diffusion of chemical species through silicate minerals is extremely slow, even at high temperatures. Measured rates of diffusion are, however, faster in an aggregate than in a single crystal, even when the aggregate is dry and has no measurable porosity, because of the presence of grain boundaries along which the diffusion rate might be intermediate between that through the crystal and that through its melt (Fyfe and others, 1978, p. 114). Textural evidence for this enhanced rate of diffusion is seen in some poikiloblastic metacrysts that seem to have grown by fingerlike extensions between engulfed matrix grains.

Prograde mineral reactions occurring during crystal growth can liberate large amounts of volatiles, especially water, from the metamorphic body. This water must escape from the aggregate somehow during these reactions (pockets of water are not found in rocks now exposed at the surface), possibly along grain boundaries. This film of migrating water would have a profound influence in enhancing diffusion rates.

Migration of existing grain boundaries.   Even though a rock may be composed of grains of thermodynamically stable phases with lowest possible Gibbs free energy, growth adjustments may take place to minimize the surface area of grains comprising the aggregate (Spry, 1969, chap. 3; Vernon, 1976, chap. 5). As discussed on page 238, surface energy is the energy associated with unbonded atoms on the surface of a solid in contact with a liquid or gas. In the case of solid in contact with solid, a **grain-boundary energy** must be minimized for stable equilibrium of the aggregate. The mutual boundary between two grains is a thin film of disarrayed atoms whose energy is greater than a similar volume of atoms within the interior of either grain in its stable state. Grain-boundary energies for a given mineral grain will, of course, depend upon its atomic structure and composition as well as upon the type of neighboring grain and its orientation.

Almost no reliable values for grain-boundary energies of silicate minerals exist, but they appear to be several orders of magnitude smaller than Gibbs free energies of the whole grain. Hence, grain-boundary adjustments must require the most favorable kinetic conditions of elevated T and long time periods, otherwise equilibrium grain shapes in the aggregate are not achieved, or only partly so. With respect to grain-boundary energy, the most stable state would be of a single crystal of each phase in the aggregate, but the driving force for grain-boundary adjustments becomes vanishingly small long before this minimum energy state is reached.

This small but significant driving force in a crystalline aggregate to minimize grain-boundary areas and maximize grain size also tends to form a hierarchy of grain shapes of least energy in aggregates of strongly anisotropic minerals.

Let us consider equilibrium grain shapes first. If a crystal made into a perfect sphere by grinding off its faces and edges is placed in a solution supersaturated with respect to the crystalline phase, crystal faces grow with rational, low Miller-index orientations. The surface energy of the crystal is minimized. These stable, low-energy crystal faces, according to the Law of Bravais (Spry, 1969, p. 142), tend to have the highest density of atoms and smallest interplanar spacing. But in a compact, pore-free aggregate, not every grain can enjoy the privilege of being bounded entirely by its lowest-energy crystal faces. A compromise must be realized.

In the ideal case of a monomineralic aggregate, such as pure marble or quartzite, minimum grain-boundary energy is approached for polyhedral grains with 14 faces, approximating a distorted tetrakaidodecahedron, or a truncated octahedron (see Figure 10-11f). The dihedral or interfacial angle between adjacent faces of such grains is about 120°. Clusters of bubbles in foaming liquids provide analogous examples. In thin sections of rocks, these polyhedral grains are intersected randomly and apparent grain boundary angles only approach 120° statistically.

However, all mineral grains are anisotropic to some extent with respect to grain boundary energy, even those with isometric symmetry. This anisotropy dictates the shape arrangement of grains in real metamorphic bodies. Minerals ranking low in the crystalloblastic series (see page 358), such as quartz, feldspars, and calcite, nonetheless tend to behave as if they were isotropic with respect to grain-boundary energy; in aggregates, there is a close approach to the 120° dihedral or interfacial angle in the granoblastic fabric. Minerals higher in the series tend not only to have greater overall grain-boundary energy, but also stronger energy anisotropy (Spry, 1969, p. 148). Hence, to minimize overall energy in polymineralic aggregates, grains of such minerals form with more or less rational crystal faces; any other boundary relationship—for example, euhedral grains ranking low in the series juxtaposed against anhedral high-series grains—would have greater overall aggregate energy.

With regard to grain size in an aggregate, the least energy is realized for the coarsest grains because, for a given volume, they have less aggregate surface area. Thus, very fine-grained rocks, such as chert or mylonite, will, if given the opportunity, recrystallize into coarser-grained aggregates. But the presence of a second mineral in an otherwise monomineralic aggregate can limit the size of the first grains because migrating grain boundaries become anchored to them (see Figure 10-23; Hobbs and others, 1976, p. 112).

13.2 STRESS AND DEFORMATION

Stress and Pressure

Development of imposed fabric represents a change in state of the rock body, so changes in P, T, and X are involved. Insofar as thermodynamic equilibrium in geologic systems is concerned, the intensive variable pressure, P, is generally taken to be equivalent to the confining pressure at depth due to the superincumbent load of rock. Confining pressure is considered to act with equal magnitude in all directions as if hydrostatic conditions prevailed, that is, as if the entire body were a static fluid. P increases with depth, Z, according to the geobaric gradient $\Delta P/\Delta Z = \rho g$ at the rate of about 27 to 33 MPa/km depending upon the mean density (2.7 to 3.3 g/cm^3) of the rock overburden (see box on page 25). Although rocks near the surface are strong and rigid under their own weight, as evidenced by steep natural cliffs, canyon walls, quarry walls, and tunnels, they become weaker at increasing depth as T increases, especially with respect to long intervals of geologic time. The well-established principle of isostasy presupposes that large segments of the upper mantle behave as fluids over long time periods.

Since the action of hydrostatic pressure is to exert an equal squeeze from all directions, rock is merely compressed to a smaller volume without any change in shape. However, the occurrence of folds and deformed pebbles, oolites, fossils, and mineral grains in rocks once deeply buried testifies to the existence of some other type of forces that operate in tectonic situations.

Any body acted upon by forces is in a **state of stress.** In rock bodies, forces arise because of gravity and the consequent weight of overlying rock and because of tectonism, chiefly related to lithospheric plate motion. The magnitude of a force on a

(a)

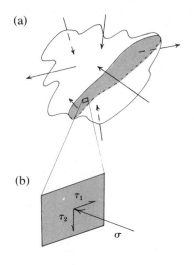

(b)

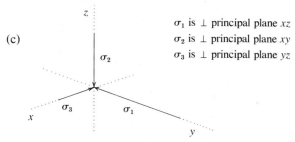

(c)

σ_1 is $\perp$ principal plane xz
σ_2 is $\perp$ principal plane xy
σ_3 is $\perp$ principal plane yz

Figure 13-1 State of stress in a body acted upon by forces. (a) Force vectors (arrows) on the body, with an arbitrary plane (shaded) through it. (b) Resolved components of normal, σ, and shear stresses, τ, on a segment of the plane. (c) The whole state of stress at any point in the body can be represented by three principal axes of stress, $\sigma_1 > \sigma_2 > \sigma_3$, normal to three orthogonal principal planes.

plane, divided by the area over which it operates, is the **stress** on that plane; its magnitude is expressed in units of pascals, like pressure. Although a rigorous mathematical analysis of stress is beyond our scope here (see Hobbs and others, 1976; Means, 1976), we can gain some insight into a state of stress by visualizing a body acted upon by forces as shown in Figure 13-1a. On any arbitrarily oriented *plane* in the body, there will be one normal stress, σ, perpendicular to the plane and two

shear stresses, τ, tangential or parallel to the plane (see Figure 13-1b); these three stresses are mutually perpendicular, or orthogonal, and constitute the components of the total stress on the plane. Theoretically, at any *point* in a body it is possible to find three suitably oriented orthogonal **principal planes** on which the stresses are only normal (the shear stresses on these planes are zero); these three **normal principal stresses** represent the total state of stress at that point in the body and are designated with subscripts according to their magnitude, $\sigma_1 > \sigma_2 > \sigma_3$ (see Figure 13-1c).

Another way to represent the total state of stress at a point in a body is to express it in terms of two components. One component, called the **mean stress**, $\overline{\sigma}$, consists of only normal stresses of equal magnitude given by

$$\overline{\sigma} = \frac{\sigma_1 + \sigma_2 + \sigma_3}{3}$$

Because of its equal magnitude in all directions, it is a hydrostatic pressure and so, to a first approximation for most geologic systems, we may consider it to be equivalent to the confining pressure, P. The second component of the total state of stress at a point is the **nonhydrostatic** or **deviatoric stress**, which embodies both normal and shear stresses. These two components are difficult to portray in a diagram, but their effects on the deformation of the body are more readily grasped.

Deformation

Individual particles within a stressed body are displaced to new positions until a state of equilibrium between the imposed forces and the body is attained. The body in its final deformed state has, in general, experienced rotation and distortion; the distortion, or **strain**, encompasses changes in both shape and size.

In metamorphic rocks we concentrate our attention on the strain caused by internal flow and measure this by some sort of internal reference frame in the body. Examples of such reference frames include outlines of grains or of grain aggregates and more specifically phenocrysts, oolites, pebbles, amygdules, lapilli, folded layers, and so on.

Strain, ϵ, is measured in terms of the change in some dimensional or angular aspect of the body relative to the original state. Therefore, volumetric strain $\epsilon_v = \Delta V/V_0$ and linear strain, or elongation, $\epsilon_l = \Delta l/l_0$. Another measure of linear strain is given by $\lambda = (l/l_0)^2 = (1 + \epsilon_l)^2$, where l is the new dimension. Strain is a dimensionless fraction, or a percent change, such as $\frac{1}{30}$, or 3.3%.

Because the mean or hydrostatic stress component of the total state of stress is of equal magnitude in all directions, it causes only a volumetric strain without change in shape; there is no rotational component of deformation either. The nonhydrostatic stress component, on the other hand, embodies shear stresses, causes changes in the shape of rock bodies, and is responsible for the solid-state flow and resulting imposed anisotropic fabric of tectonites. The box at right summarizes these relations.

Nonhydrostatic stress influences solid-state crystal growth by producing deformed grains with greater energy, which are more likely to yield nuclei that will grow into new, more stable grains. At one time it was believed that nonhydrostatic stress actually stabilized certain "stress" minerals, such as kyanite, which could not form in its absence. However, no evidence has been found in support of this concept and it has been abandoned for at least a couple of decades; the mineralogical composition of a rock body is dictated only by P, T, and X.

Response to Stress

Stress and strain are mathematical concepts that apply to any material. In petrology, we are concerned with the mechanisms of response of rocks and individual mineral grains (as well as silicate melts) to applied stress. Exactly how does the body behave as it experiences stress and deforms? What are the mechanisms, on the scale of atoms or of single crystals, that permit strain? The response

---

Stress and Strain

| total stress | = | mean or hydrostatic component (equal magnitude in all directions; no shear) | + | deviatoric or nonhydrostatic component (embodies shear stresses whose magnitude varies with direction) |

| total strain | = | change in volume due to hydrostatic component of stress | + | change in shape due to nonhydrostatic component of stress |

---

of rocks is complex but can be viewed initially in terms of ideal, mathematical end members. We will first briefly describe these ideals and then consider the deformational response of real rocks to stress.

Idealized types of behavior. Seismic waves passing through the Earth are an example of **elastic behavior.** The strain is very small (less than 1% to 2%) and reversible; that is, it is instantaneous upon application of stress and is totally recoverable, so the material returns instantaneously to its initial undeformed state when the stress is eliminated. There is a linear or direct proportionality (**Hooke's Law**) between stress and elastic strain, $\sigma = K\epsilon$, where K is a constant for the material (see Figure 13-2). The **strain rate**—the strain produced in unit time—for elastic behavior is large. Strain rate is $\Delta\epsilon/\Delta t$, or in differential form $d\epsilon/dt$ ($= \dot{\epsilon}$); the unit of strain rate is s^{-1}, or reciprocal seconds.

The isothermal elastic compressibility, $\beta = -\Delta V/V\Delta P$, of minerals is on the order of $10^{-5}/$MPa (Birch, 1966). Hence a mineral subjected to a confining pressure increment $\Delta P = 10^3$MPa corresponding to a depth of ~ 35 km in the crust would elastically dilate by only $-\Delta V/V = \beta\Delta P = (10^{-5}/\text{MPa})(10^3\text{MPa}) = 10^{-2}$ or a volumetric elastic strain of 1%!

IDEAL BEHAVIOR

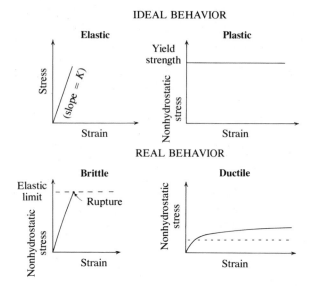

Figure 13-2   Types of response to applied stress. The dashed line in the diagram for ductile behavior represents an approximate plastic yield strength.

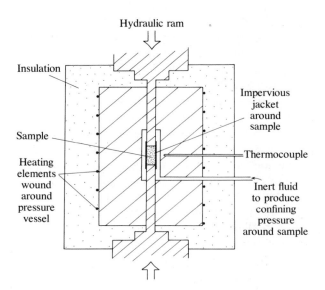

Figure 13-3   Investigations of the deformational behavior of materials under nonhydrostatic stress are made in laboratories using a hydraulic squeezer of the sort illustrated here in a highly schematic cross section. Elevated temperatures on the deforming sample are produced by the heating wires around the hardened steel pressure vessel and are monitored by the thermocouple positioned close to the sample. Steel is weak at high temperatures, so a small heater can be built immediately around the sample inside the pressure vessel, which may then be cooled by circulating water to maintain its strength; such heaters are deformed with the sample and cannot be reused. Confining pressures are created by pumping some inert fluid, such as argon, around the sample. If this fluid pressure exceeds the axially directed normal stress exerted on the ends of the sample cylinder by the hydraulically driven pistons, the sample will be axially extended (see Figure 13-23a). If the opposite, the sample is shortened, or compressed, as in Figure 13-5.

In the ideal **plastic behavior,** no strain occurs in response to applied nonhydrostatic stress until some critical value is reached, called the **plastic yield strength,** at which the material strains permanently (see Figure 13-2). **Steady-state flow,** sometimes called **creep,** is a type of plastic strain that proceeds isothermally at constant strain rate under a constant nonhydrostatic stress. This type of behavior may be compared to the Newtonian viscous flow described on page 233, in which the strain rate is the ratio of the applied shear stress to the coefficient of viscosity.

Response of real rock bodies.   Rocks near the surface of the Earth, where P and T are low, subjected to very high strain rates respond in a **brittle** manner to nonhydrostatic stress. Every geologist who has collected a hand sample from a rock outcrop by striking it with a hammer has witnessed brittle behavior. A more revealing exhibition is to squeeze (load) a small cylinder of rock in a hydraulic testing machine (see Figure 13-3). As the axially

directed load (the maximum principal stress, σ_1) is increased, the cylinder shortens proportionately by only a few percent in an elastic manner. Eventually, as the load reaches the **elastic limit,** the cylinder breaks—sometimes violently—along fractures oriented parallel to or at an acute angle to σ_1. Brittle deformation includes fracturing and faulting

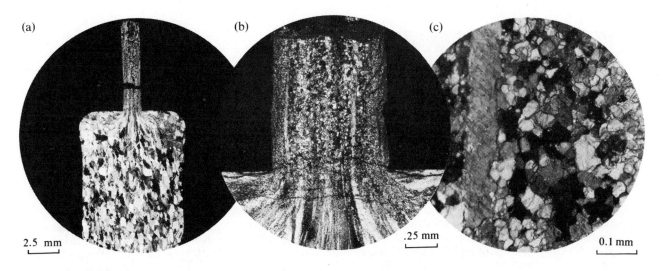

(a)   (b)   (c)

2.5 mm   .25 mm   0.1 mm

Figure 13-4   Solid-state ductile flow in marble produced in the laboratory. A small cylinder of marble at 500°C subjected to a confining pressure of 500 MPa was extruded from a hole in the loading piston. The rate of strain in the extruded region was 3×10^{-7}/s, or the approximately 100% strain was accomplished in about 35 days. (a) Photomicrograph shows a thin section of the deformed cylinder under cross-polarized light. Note deformation twin lamellae in the calcite grains, especially prominent in and near the extruded region. Extruded neck broke after termination of experiment. (b) Enlarged view of extruded region showing extreme elongation of twin lamellae and extensive recrystallization of grains after they were extruded, or perhaps during extrusion. (c) Enlarged view of granoblastic grain aggregate produced by recrystallization that replaced intensely strained grains. [Photographs and information courtesy of H. C. Heard from his unpublished Ph.D. dissertation, University of California at Los Angeles, 1962; see also H. C. Heard, 1963, Effect of large changes in strain rate in the experimental deformation of Yule marble, *Journal of Geology* 71.]

on multigrain scales and as well as **cataclasis,** or cataclastic flow, on a granular scale involving frictional sliding of broken fragments past one another, not unlike flow of loose, or cohesionless, sand grains. Brittle response produces **fault** or **tectonic breccia** and more finely pulverized rock flour, called **gouge.**

Two factors make brittle response strongly pressure-dependent, limiting it to environments of low P in the upper crust: (1) greater confining pressures increase the frictional resistance of fragments moving past one another, and (2) the internal cracking and rupturing of initially pore-free rock cause an increase in volume, called dilatancy.

Rocks once deeply buried in the crust or in the upper mantle, as testified by their mineralogical composition, display clear evidence of significant pervasive deformation in the form of folds and distorted but not broken relict pebbles, fossils, and mineral grains. These cohesive types of response to nonhydrostatic stress are expressions of **ductile behavior** (see Figure 13-4). There is a permanent change in shape without gross fracturing. It is the same sort of flow that is embodied in the deformation of glaciers (see box on page 359), salt diapirs, and pieces of metal.

In summary, nonhydrostatic stress produces reversible elastic strain up to the elastic limit, or elastic strength. Applied nonhydrostatic stress greater than this limit produces permanent or irreversible deformation such as ruptures, faults, or

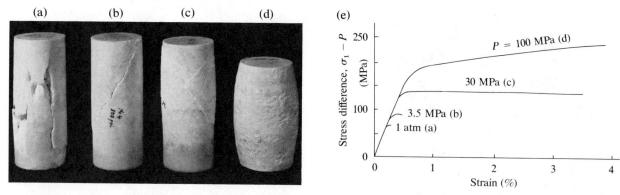

Figure 13-5 Cylinders of marble deformed in the laboratory showing a transition between brittle and ductile behavior as a function of confining pressure, P (= $\sigma_2 = \sigma_3$). (a) Extension fractures parallel to direction of applied maximum compression σ_1 typify brittle behavior under atmospheric P–T conditions. (b) Inclined fault formed at P = 3.5 MPa. (c) Conjugate set of faults combined with a little permanent ductile barreling at P = 30 MPa. (d) Ductile flow without any conspicuous faults at 100 MPa confining pressure. (e) Stress–strain curves for the marble cylinders, all of which were deformed at 25°C. Note the increasing strength with increasing P. Increasing T decreases the strength. Compare Figure 13-2. [From M. S. Paterson, *Experimental Rock Deformation—The Brittle Field.* Copyright 1978 by Springer-Verlag New York, Inc.]

permanent ductile flow. Ductile flow is the hallmark of metamorphic rock bodies deformed by nonhydrostatic stress at elevated P and T and slow strain rates, whereas faulting, cataclasis, and other types of brittle behavior typify sedimentary bodies deformed at low P and T and rapid rates of strain.

Laboratory experiments and field observations demonstrate that there is no sharp boundary between the conditions for and the fabrics produced by brittle and ductile behavior. Rather, there is a continuum between these two extremes, or a **brittle-to-ductile transition,** depending upon P, T, composition, and strain rate. The brittle to ductile transition occurs under different conditions for different materials; rock salt, for example, experiences ductile flow in shallow salt domes (diapirs) at low P and T, whereas silicate rocks require much higher P and T, such as prevail in the deep crust or upper mantle, to flow. For some particular P–T conditions, some rocks and minerals will be more ductile than others; relatively brittle material may rupture under nonhydrostatic stress while surrounding more ductile material experiences cohesive flow (see Figure 11-8). Other aspects of

the brittle to ductile transition are illustrated in Figure 13-5.

13.3 DUCTILE FLOW

Two ductile flow mechanisms that can act in unison, especially at high temperatures, have been identified in laboratory experiments and in real tectonites: intracrystalline plasticity and diffusive flow or pressure solution. Much research regarding these mechanisms is still current, so the following discussion presents the state of the art in 1981.

Crystal Plasticity

In unstrained crystals, the periodic array of atoms is maintained in an equilibrium state by interatomic forces of attraction and repulsion. During elastic strain, the atomic bonds are distorted, shortening or lengthening the interatomic distances (see Figure 13-6b). If the applied nonhydrostatic stress exceeds the plastic yield strength, then groups of atomic bonds can be systematically broken or rotated along particular crystallographic directions into a new, permanent equilibrium position where they remain upon relaxation of the stress. This permanent, irreversible **intracrystalline plastic deformation** (see Figure 13-6c,d) is especially favored where confining pressures are too great to permit brittle rupture or temperatures are high enough to facilitate rearrangement of atoms. Because the crystalline state is the lowest energy state of a solid in even the most intensely deformed condition, grains will retain their ordered atomic structure, albeit on very small scales, rather than form an amorphous (glass or melt) structure.

Modern engineering technology is strongly dependent upon plastic deformation in crystalline metals such as iron, aluminum, and copper and in various alloys, which can be formed into desired shapes by bending, extruding, and rolling. Most of our understanding of crystal plasticity came initially from research on metals (metallurgy), for only in the past couple of decades have experimentalists turned their attention to rock-forming minerals.

Plastic deformation of a single crystal occurs by two fundamental mechanisms—translation gliding and twin gliding—in favorably oriented crystallographic directions.

In **translation gliding,** or simply **slip** (see Figure 13-6c), discrete layers of the crystal slide past one another by integral multiples of fundamental crystallographic units. Afterward, the internal lattice is still continuous and the parts on either side of the slip panels are still similarly oriented; only the external outline of the crystal has changed. Obviously, the plane on which slip occurs and the direction of slip in that plane must be controlled by the nature of the atomic structure in the crystal. **Slip planes** in crystals tend to lie parallel to crystallographic directions of densest atomic packing and widest interatomic spacing, where interplane bonding is weakest. These same directions tend to parallel crystal faces with simple Miller indices (compare the Law of Bravais). But the nature of the atoms in the crystallographic planes is also important, making for many exceptions.

In halite (see Figure 13-7), a prominent slip

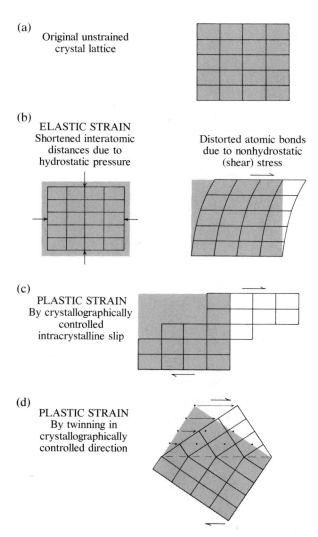

(a) Original unstrained crystal lattice

(b) ELASTIC STRAIN Shortened interatomic distances due to hydrostatic pressure

Distorted atomic bonds due to nonhydrostatic (shear) stress

(c) PLASTIC STRAIN By crystallographically controlled intracrystalline slip

(d) PLASTIC STRAIN By twinning in crystallographically controlled direction

Figure 13-6 Contrast between elastic and plastic strains shown by schematic sections through a hypothetical crystal lattice. (c) Intracrystalline slip can produce pronounced changes in the external shape of the crystal without distorting the internal lattice. (d) In the lattice strained by twinning, the original lattice points are indicated by dots and the arrows show they have moved progressively farther the more distant they are from the twin plane (dashed light line).

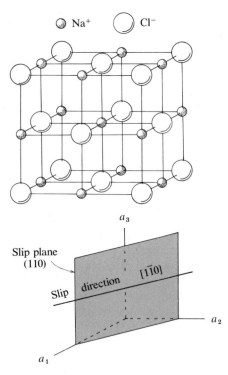

Figure 13-7   Atomic structure of halite and one of its six symmetrically related slip planes and directions.

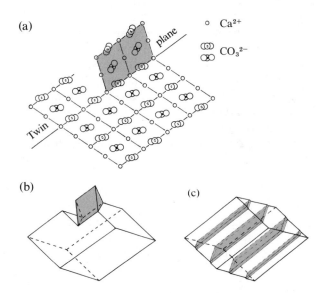

Figure 13-8   Twin gliding in calcite on $\{1\bar{1}02\}$. (a) Atomic structure across the twin plane. (b) A small wedge-shaped twin (shaded) can be produced by compressing the edge of a calcite cleavage rhombohedron with a knife perpendicular to its c-axis. (c) Lamallar twins (shaded) in a calcite rhomb. Compare with Figure 13-4a,b.

plane is in the orientation $\{110\}$, of which, because of symmetry, there are six planes. The **slip direction** in the one plane (110) is $[1\bar{1}0]$. Although $\{111\}$ is the orientation of densest atomic packing, the alternating layers of Na^+ and Cl^- ions have a strong attraction for one another, so it is not the easiest slip plane. Other **slip systems** (a plane plus the direction on it) have been documented in halite as well as in many of the important rock-forming carbonate and silicate minerals. Many crystals of rock-forming minerals possess several slip systems; certain of these will be favored under a particular range of *P, T,* and strain rate. See Nicolas and Poirier (1976, chap. 5) for an extensive review of the still-growing body of data on slip systems in minerals.

   In **twin gliding** (see Figures 13-6d and 13-8), sometimes called **deformation** or **mechanical twinning**, a portion of the crystal structure is sheared in such a way as to produce a mirror image of the original crystal. (Such deformation twinning should be contrasted with growth twins, such as those seen ubiquitously in magmatic plagioclases grown free-floating in a silicate melt.) The twin plane, which is the mirror across which the parts are symmetrically related, may be parallel to a plane of plastic slip in some crystals. Deformation twins may be thin lamellae and distributed throughout the crystal, or they can take over most, and possibly all, of the crystal, depending upon circumstances. Wedge-shaped or curved polysynthetic twins of mechanical origin are common in carbonates and plagioclase (see Figures 10-7a and 10-19b).

   Translation and twin gliding both enable a crystal under nonhydrostatic stress to change its shape by permitting part of the crystal to move with respect to another part. For translation or twin gliding to occur on any particular plane in a crystal, a **critical resolved shear stress** on that plane must be exceeded. This critical shear stress depends upon the whole state of stress in the rock body containing the crystal (see Figure 13-1b) as well as the orientation of the crystal and its potential slip or twin planes in relation to the stresses. In micas, for example, the only known slip plane is (001), and if this direction lies at a high angle to the maximum compressive principal stress (σ_1), the resolved shear stress on (001) is too small to allow slip.

Plasticity of crystalline aggregates. The concept of critical resolved shear stress helps to explain the well-known fact that metals, such as lead, gold, copper, and aluminum, are far more easily strained plastically than rocks. There are at least two reasons for this difference. First, in metal crystals, atomic bonds are weaker than in silicates and carbonates, hence the critical resolved shear stress necessary to produce slip is less. Second, because metal crystals (most of which are isometric) generally have higher symmetry than many common rock-forming minerals (typically triclinic, monoclinic, or orthorhombic), more slip planes and directions exist in metal crystals, so that in a nonhydrostatically stressed polygranular aggregate of randomly oriented crystals at least one slip plane in virtually every grain can be activated.

Rocks display varying plasticity. Hexagonal carbonates have enough slip and twin systems to allow for considerable plasticity in aggregates (marble) over a wide range of *T, P,* and strain rate. Hexagonal quartz is less easily strained plastically because greater nonhydrostatic stresses than for carbonates are required to rupture the stronger Si–O bonds. Triclinic and monoclinic feldspars in quartzo-feldspathic rocks are still less plastic because of the paucity of slip systems. So what actually happens in polyphase rocks that are dominated by strong, low-symmetry silicate minerals that have obviously experienced considerable permanent strain?

Nicolas and Poirier (1976, p. 42) conceived a "thought experiment" to cast light on this question. Grains in an aggregate are separated from one another and allowed to deform individually by slip in response to applied nonhydrostatic stress. After deformation, we attempt to reassemble the grains into a compact, pore-free aggregate, but find it impossible because there is a large proportion of voids and overlaps between deformed grains. In order to accomplish the reassembly, it would be necessary to distort virtually every grain by special stress frames and then fit them together. The stress state is now quite heterogeneous on the scale of grains. Strain in some grains may be only elastic, held there by the confinement of the neighboring grains. This elastic strain, which is only a few percent, will bend or otherwise distort grains, perhaps changing their volume (see Figure 13-6b). Most grains, however, will be bent (twisted) and kinked, both of which rely on a form of plastic slip.

In **kinking** (see Figure 13-9), differential slip (or twin gliding in some instances) is restricted to a **kink band** in which the slip planes are oriented several tens of degrees from the same planes in the essentially undeformed parts of the crystal on both sides. Kinking is most commonly found in crystals that have only one easy slip plane, such as micas, enstatite, and kyanite. **Bending** of a crystal lattice, which must occur at the margins of a kink band, or which can develop in a more distributed fashion throughout an entire crystal, offers an intriguing dilemma. Bending depends upon differential movement between layers of the lattice, as in distorting a pack of cards; but the bending shown in Figure 13-10 is not in accord with plastic slip (see Figure 13-6c), where the translation is only by integral multiples of lattice spacings, leaving "good" crystal continuously across the plane of movement. To maintain continuity throughout most of the bent crystal lattice it is necessary to postulate the existence of geometrically necessary stacking imperfections, or defects. Such defects are extremely important in plastic strain.

Crystal defects. Imperfections or irregularities in crystals are basically of two types: point defects and line defects.

The periodicity of a lattice can be broken at a single point by a **point defect.** The different types of point defects shown schematically in Figure 13-11

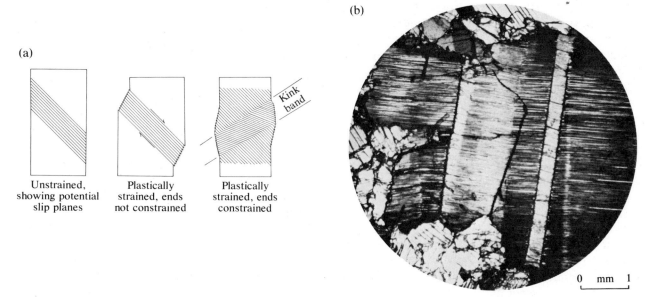

(a)

Unstrained,
showing potential
slip planes

Plastically
strained, ends
not constrained

Plastically
strained, ends
constrained

Kink
band

(b)

0   mm   1

Figure 13-9   Kink bands. (a) Unstrained and plastically strained cylinders of a single crystal with a set of parallel slip planes. The kink band developed in the third cylinder represents what happens in a crystal that is strained while confined within an aggregate. Extension of a confined crystal can also produce a kink band. [After D. V. Higgs and J. Handin, 1959, Experimental deformation of dolomite single crystals, *Geological Society of America Bulletin* 70.] (b) Photomicrograph shows kink bands in plastically strained enstatite crystals in a peridotite.

play roles in diffusion, the color of crystals, and their electrical and mechanical properties. Although defects are departures from the perfect lattice, there is an equilibrium concentration of them at a particular temperature.

Line defects, or **dislocations,** as they are more commonly called, are chiefly responsible for the plasticity of crystals. A dislocation is not simply a linear collection of point defects, but is instead a special category of irregularity that may be visualized as a line in the slip plane separating slipped from unslipped parts of the crystal. The nature of dislocations and how they relate to slip are shown in Figures 13-12 and 13-13.

Hobbs and others (1976, p. 77) describe the phenomenon of dislocation-enhanced slip as follows:

Microscopically, slip may be compared to the gliding of cards past each other as a deck of cards is sheared. On the atomic scale, however, slip does not take place simultaneously over an entire slip plane in a crystal, but nucleates in a small region of high stress concentration. This region of slip then spreads out in an expanding loop across the slip plane until it ultimately intersects a grain boundary where it produces a small step. The line that separates the slipped region from the unslipped region at any instant is a dislocation.

Propagation of an **edge dislocation** (see Figure 13-12) accompanying slip through a crystal has been likened to the movement of a caterpillar, which humps up the rear of its body by moving its tail forward, and then moves the hump progressively forward to the head. Another analogy is provided by the old parlor trick of how to remove the carpet from a room without first taking out all the furniture. The easy way is to ruck up the carpet along one edge and move the ruck across the room by lifting up corners of furniture one at a

(a)

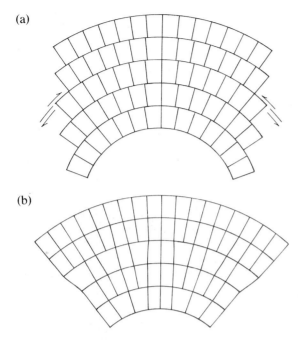

(b)

Figure 13-10　Bent crystal lattice. (a) The warped ''card deck'' model does not correspond to bending of a crystal by plastic slip because of the discontinuity of the lattice across the surfaces of differential movement. (b) Another model mostly preserves continuity but embodies geometrically necessary imperfections, or defects, which here may be viewed as half lattice planes extending part way through the lattice, perpendicular to the plane of the drawing.

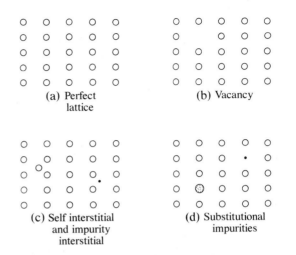

Figure 13-11　Types of point defects.

time as necessary. This procedure would have to be repeated several times to remove the entire carpet from the room, but it does not require much energy at any instant.

A **screw dislocation** may be likened to tearing a sheet of paper, starting at one edge and propogating the tear to the opposite edge.

In a crystal, if all the atoms along the entire slip plane were to be moved simultaneously to a new equilibrium position, a great deal of stress would be required to break all their bonds with neighboring atoms simultaneously. But slip by propagation of a dislocation requires far less shear stress, about one thousand times smaller than the theoretical, because only a line of atoms needs to be moved at any instant. This process is efficient because of the large number of dislocations possible in a crystal.

Dislocations are thermodynamically unstable because they represent a local region in a crystal where the atoms are not in proper position and atomic bonds are elastically distorted. But all crystals have them from the time of their growth; they are built into the crystal from the beginning, like the extra partial rows of kernels in some cobs of corn. In undeformed crystals of quartz, the ''density'' of dislocations, expressed as the total length of dislocation lines per volume, is about 10^3 cm/cm^3 (see Figure 13-14).

Not only does shear stress cause existing dislocations to move, it also produces new ones at grain boundaries, cracks, and other sources within the crystal (Nicolas and Poirier, 1976, p. 94); all such dislocations can propagate through the crystal, allowing it to slip plastically. An intensely deformed quartz crystal, for example, might contain 10^{12} cm/cm^3 dislocations! But as the dislocation density increases, greater shear stress is required to move dislocations and therefore to slip parts of the crystal plastically. This phenomenon of continued plastic strain requiring ever higher applied stress is called **strain** or **work hardening** and can be seen in the stress–strain curve for ductile behavior compared to the mathematical idealized plastic response of Figure 13-2. Work hardening and im-

(a)   Unstrained

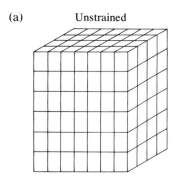

(b)   Edge dislocation

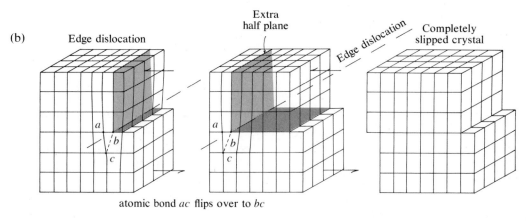

atomic bond *ac* flips over to *bc*

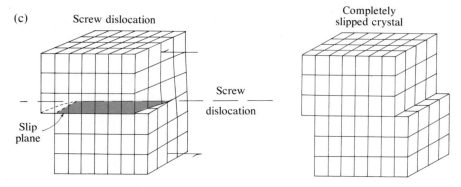

Figure 13-12   A dislocation may be either an edge (b) or a screw (c) and is a line in the slip plane (heavy shade) between slipped and unslipped parts of the crystal.

mobilization of dislocations is caused by (1) interaction of elastic forces around each dislocation in a slip plane, (2) extra dislocation segments lying between slip planes, (3) interaction of dislocations on intersecting oblique slip planes, and (4) obstacles such as point defects or second-phase inclu-

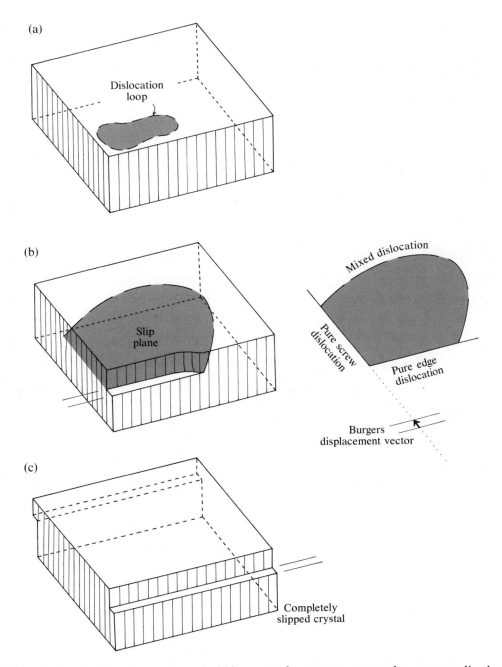

Figure 13-13 A dislocation nucleated within a crystal propagates outward as an expanding loop, ultimately intersecting the edge of the crystal in screw and edge dislocations. A dislocation oblique to the **Burgers displacement vector** is a mix of the pure screw and edge dislocations, which are parallel and perpendicular, respectively, to that vector.

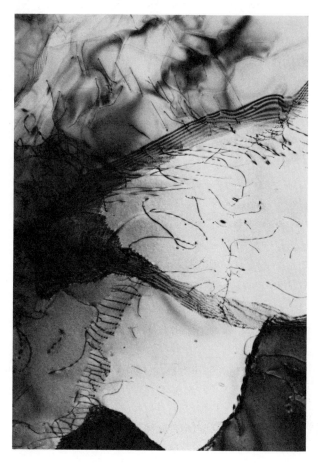

Figure 13-14 Transmission electron micrograph of moderately strained quartz showing dislocations. The central subgrain at its widest part is about 4 μm across. Compare Figure 13-13. [Photograph courtesy of M. S. Weathers.]

sions on which the abundant migrating dislocations get hung up.

Recovery. The energy of a crystal increases with increasing dislocation density. The mechanical work of deformation in a nonhydrostatic stress field is partly stored in crystal dislocations. Given the appropriate conditions, the crystal will, therefore, attempt to reduce its dislocation density. At low temperatures or rapid rates of strain, elimination of dislocations is difficult; under the opposite conditions, especially high temperatures, say, greater than 0.3 to 0.5 T_m, where T_m is the melting T of the crystal in degrees Kelvin, rates of atomic diffusion become appreciable so that dislocations can be mobilized and even eliminated, a process called **recovery.**

Recovery can be achieved by **dislocation climb,** where an edge dislocation moves perpendicular to the slip plane by the addition or removal of atoms, one at a time, by diffusion along the edge of the extra half-plane (see Figure 13-15). Two dislocations of opposite sign climbing toward each other are annihilated; this is the chief mechanism of reducing dislocation density. Dislocation climb, acting together with slip during **dynamic** or **syntectonic recovery,** enable dislocations to migrate up and around various types of obstacles.

In bent (or twisted) and kinked grains, syntectonic recovery allows dislocations to arrange themselves into **dislocation walls** lying at low angles to the more or less unstrained lattice on either side, which are regions of the crystal called **subgrains** (see Figure 13-16). Many of the geometrically unnecessary dislocations are annihilated and the geometrically necessary ones are stacked into walls, leaving virtually strain-free subgrains between. Subgrain configurations are more stable than a whole strained grain because the arrayed dislocations and lattice between them have lesser energy. Edge dislocation walls will be virtually perpendicular to slip planes, and if several are present, subgrains will be polygonal; hence subgrain development is often referred to as **polygonization** (not to be confused with the growth of randomly oriented polygonal grains in granoblastic aggregates; see Figure 10-11). In a sense, subgrain formation is an incipient recrystallization (see below) and often precedes it, but in contrast, no new nuclei are formed. Textural evidence for polygonization in thin section consists of small polygonal areas within a grain in each of which the optical extinction is fairly uniform but differs by a few

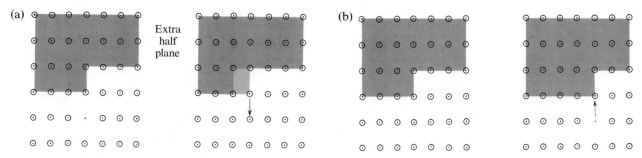

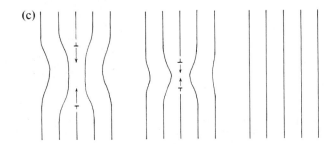

· Atom above extra half plane of atoms

○ Atom below

Figure 13-15 Recovery by dislocation climb and annihilation. (a) Dislocation climb is accomplished by losses or gains of atoms one at a time by diffusion along a jogged half-plane edge. An atom from the edge of the extra half plane of atoms (see Figure 13-12) has diffused into a nearby vacancy, causing the edge dislocation to climb upward. (b) The reverse occurs, extending the area of the half plane and causing the dislocation to climb downward. (c) Two edge dislocations, now viewed from their ends, climbing toward one another are annihilated and the two extra half planes become one whole. T symbolizes the end of an edge dislocation. [After A. Nicolas and J. P. Poirier, 1976, *Crystalline Plasticity and Solid State Flow in Metamorphic Rocks* (New York: John Wiley and Sons).]

degrees from immediate neighbors (see Figure 13-16b).

Static or posttectonic recovery may occur after release of applied stress, but its significance is probably not as great as that of syntectonic recovery, which permits a more or less steady-state plastic flow by counterbalancing the work hardening resulting from interaction of slip dislocations.

Recrystallization. High-energy strained grains may recrystallize by nucleation and subsequent growth of completely new, lowest-energy, strain-free grains. These grains differ significantly in orientation from the original strained parent and from each other. Recrystallization rather than recovery conceivably occurs at relatively high temperatures, perhaps $> 0.5 \, T_m$ in degrees Kelvin. An intensely strained grain, or parts of a grain, such as along a kink band where atomic bonds are most distorted, are favored sites for nucleation of new strain-free grains.

In practice, the timing of recrystallization is often difficult to ascertain from the rock fabric. Many fabrics interpreted in the past as posttec-

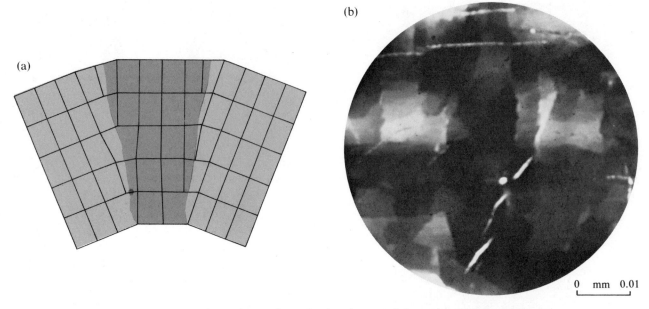

Fig. 13-16 Low-angle dislocation walls and subgrains. (a) Schematic of lattice with edge dislocations of the same sign (half planes all extend in the same direction) concentrated into low-angle walls between subgrains (shaded). Compare with Figure 13-10b. (b) Photomicrograph of quartz subgrains viewed in cross-polarized light under a petrographic microscope. For a view of subgrains under an electron microscope, see Figure 13-14, in which dislocations are concentrated at low-angle boundaries.

tonic may actually have originated at high temperatures syntectonically. Recovery and recrystallization, as well as plastic slip, may all function almost simultaneously in different parts of the body or even parts of one grain. In this syntectonic process, new strain-free grains may lie side by side with polygonal subgrains and older strained grains, all coexisting in a state of dynamic equilibrium during steady-state flow (White, 1977).

Hydrolytic weakening. For many years, quartz was one of the most recalcitrant of the common rock-forming minerals to deform plastically in the laboratory. At 1.5 GPa confining pressure, for example, the stress difference $\sigma_1 - \sigma_2 \, (= \sigma_3)$ required to induce plastic flow is about 4 GPa at 400°C. A still higher T of 900°C, which is about the upper limit for metamorphic conditions, only re-

duces the necessary stress difference to about 2 GPa. Yet there is widespread evidence, such as deformation bands and lamellae, in metamorphic tectonites, demonstrating that quartz has flowed plastically under what surely must have been smaller stress differences (see Figure 10-7).

In the mid-1960s, D. T. Griggs and J. D. Blacic resolved this dilemma by their discovery of the phenomenon of hydrolytic weakening (see references in Nicolas and Poirier, 1976, p. 206). Minerals such as quartz, feldspar, olivine are nominally anhydrous. Nonetheless, infrared spectroscopy reveals that significant concentrations of water as $(OH)^-$ ions reside in such phases, not just as inclusions of fluid H_2O. Griggs and Blacic found that quartz crystals grown synthetically in the presence of water and containing on the order of 0.13 wt.% $H_2O \, (\sim 10^{-2}$ H/Si atoms) deform plastically above

380°C at stress differences ten to twenty times lower than dry quartz crystals. Figure 13-17 is a schematic explanation of hydrolytic weakening in silicate minerals, using quartz as the prime example. For dislocations to move in a dry crystal, the strong Si–O bonds have to be broken. However, if some water is dissolved in the crystal as (OH)⁻, these hydroxyls might diffuse to dislocations, which can then be easily propagated through the crystal, allowing plastic slip, because only weaker hydrogen bonds need be broken in the slip process.

Applications to metamorphic tectonites. There is still much to be learned regarding plastic flow of rocks—the mechanisms at various temperatures, strain rates, and water concentrations in different minerals and the textural evidence for these mechanisms. It is not always easy to correlate dislocation features seen with transmission electron microscopy with the larger-scale textural features seen with an optical microscope.

Small amounts of strain are manifest under cross-polarized light in **undulatory optical extinction** of grains, especially obvious in quartz, so that as the microscope stage is rotated, a region of extinction sweeps fanlike across the grain. Strained quartz is no longer uniaxial and has an optic axial angle of up to 20°. Undulatory extinction is associated with bent cleavages and twin lamellae (see Figure 10-7a). Crystals can bend by formation of edge dislocations of one sign along slip planes (see Figure 13-16a).

As recovery begins to take place in bent crystals, dislocations group into low-angle walls separating subgrains (see Figure 13-16). Sometimes, these subgrains are elongate and are concentrated into ribbonlike **deformation bands,** whose orientation may differ conspicuously from one to another (see Figure 10-7d). Deformation bands differ from areas of undulose extinction in a grain only in the severity of the inhomogeneity of the strain. The exact nature and origin of deformation bands is still controversial, but in quartz intracrystalline slip on the basal plane (0001) and on

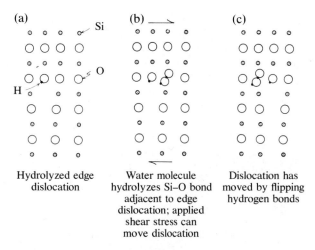

Figure 13-17 Enhanced mobility of dislocations in hydrous quartz is due to the ease of breaking weak hydrogen bonds, rather than strong Si–O bonds in dry crystals. [Schematic drawing after D. T. Griggs, 1967, *Geophysical Journal, Royal Astronomical Society* 14.]

prism planes parallel to the *c*-axis produces bands subparallel and subnormal to the *c*-axis, respectively (Vernon, 1976). Locally, bending of a crystal produces kink bands (see Figure 13-9), which may be considered a type of deformation band.

Commonly associated with deformation bands in strained quartz and generally oriented at a high angle to them are **deformation lamellae** (see Figure 10-7e). These are narrow (a few micrometers), planar features that have slight variations in refractive index. They probably originate in different ways, including slip and twin gliding.

At relatively low temperatures or high strain rates, grains show undulatory extinction, kink and deformation bands, deformation lamellae, and under the electron microscope, high dislocation densities. At higher temperatures or slower strain rates, the effects of thermally activated recovery (polygonization creating subgrains) and recrystallization are more apparent. Deformation bands and entire strained grains will have sutured (serrated) boundaries rather than smooth boundaries if incipient recrystallization has occurred. Or, small, new,

(a)

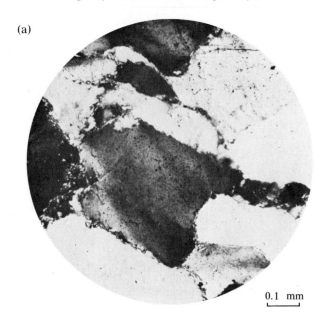

0.1 mm

(b)

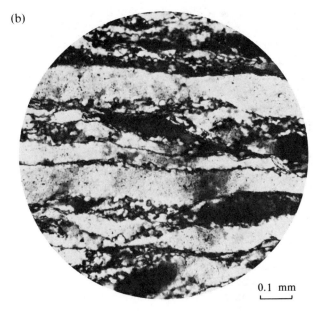

0.1 mm

(c)

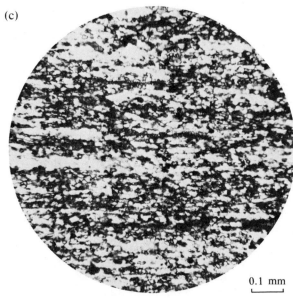

0.1 mm

Figure 13-18 Progressive mylonitization of the basal quartzite unit at the Stack of Glencoul, Moine thrust zone, Scotland, viewed under cross-polarized light. (a) Sample 110 m from the thrust contact between quartzite and overlying Moine schist has <5% new recrystallized grains along margins of relict grains. (b) Sample 5 to 10 m from the thrust contact has elongated ribbons of relict quartz grains, showing undulatory extinction, surrounded by small grains that formed by dynamic, or syntectonic, recrystallization during flow. (c) At the thrust contact, the quartzite is almost entirely composed of recrystallized grains. Note the progressive increase in proportion of recrystallized to relict quartz grains approaching the thrust even though the recrystallized grains remain essentially uniform in size. [Photographs courtesy of M. S. Weathers, from M. S. Weathers, J. M. Bird, R. F. Cooper, and D. L. Kohlstedt, 1979, Differential stress determined from deformation-induced microstructures of the Moine thrust zone, *Journal of Geophysical Research* 84.]

strain-free grains will replace strained areas showing undulatory extinction (Figures 10-10 and 13-4c).

Very large strains found in many metamorphic bodies with intensely folded or stretched markers were likely created by more or less steady-state plastic flow, so that the body behaved much like a very viscous fluid; syntectonic recovery and recrystallization continuously counterbalanced work hardening.

Experimental studies have shown that dislocation density and the size of syntectonic subgrains and recrystallized grains are a function only of the magnitude of shear stress. If, for example, the magnitude of nonhydrostatic stress driving steady-state flow increases, recovery processes will adjust to maintain a steady state by gathering the faster-forming dislocations into more closely spaced subgrain walls; the dislocations have a smaller distance to migrate and the recovery rate is increased. Weathers and others (1979) found that dislocation density and recrystallized grain size are essentially uniform in mylonites exposed across the strike of the Moine thrust in Scotland (see Figure 12-1), regardless of relict predeformation grain size and position within the thrust zone (see Figure 13-18). From this, they conclude that the nonhydrostatic stress was essentially uniform throughout the sampled mylonite zone and was on the order of 60 to 130 MPa, based upon laboratory studies of grain size in relation to plastic flow laws. Relict quartz grains became less abundant and more flattened toward the center of the fault zone, implying an increase in total strain. Because the shear stress was approximately constant, this increase in strain must have corresponded to an increased rate of strain toward the fault, possibly as a result of an increase in temperature due to strain heating. There may be a sort of feedback phenomenon (see page 236) where strain heating raises the T of the rock, making it more plastic and weaker so that further deformation is localized in the heated volume, and so on.

Although mylonites were originally defined a century ago as forming by milling, crushing, and grinding, it now appears that generally, if not always, such brittle processes are not relevant. Rather, Bell, and Etheridge (1973) have demonstrated that mylonites possess textural features typical of materials deformed in the laboratory by plastic flow, including translation and twin gliding, bending, kinking, and dynamic recovery and recrystallization.

Tullis and Yund (1977) find that if granite is deformed at low temperatures where recovery is sluggish, dislocations get tangled; due to this work hardening applied stress can rise high enough to initiate brittle cracks in mineral grains. At low confining pressures, these cracks line up to form through-going faults in the crystalline aggregate, whereas at higher P the cracks are confined to individual grains. At higher temperatures favoring easier diffusion, recovery is rapid and considerably more permanent cohesive strain is possible.

Experiments on flow of rocks in the laboratory cannot exactly duplicate conditions in real rock bodies because of the vast difference in time scales. Deformation in the laboratory is accomplished over time periods of weeks or months at strain rates of more than 10^{-7}/s. Judging from rates of lithospheric plate movement, strain in natural rock bodies is more likely to be on the order of 10^{-12} to 10^{-15}/s. In order to evaluate the effects of these much slower strain rates, the steady-state flow of experimentally deformed rocks is fitted to a rate equation (Heard, 1976) of the form

$$\dot{\epsilon} = A \ \sigma^n e^{-(E_a/RT)}$$

where A is the rate constant, E_a is the activation energy for steady-state flow, R is the gas constant, T is absolute temperature, σ is the nonhydrostatic stress, and n is a number between 2 and 8 depending upon the temperature and material. With this equation, it is possible to extrapolate experimental behavior to geologically reasonable rates of strain (see Figure 13-19). Monomineralic metamorphic rocks such as quartzite, marble, and dolomarble experience steady-state plastic flow under diminished nonhydrostatic stress as T increases, as expected. Marble flows under lower stress than dolomarble and quartzite, for a given T, although wet quartzite at 600°C and 800°C is weaker than marble due to hydrolytic weakening. Flow of rock salt (halite) in shallow crustal salt domes is a well-known phenomenon, and Figure 13-19 shows that halite at only 100°C has a strength in steady-

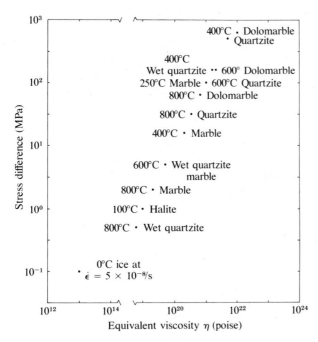

Figure 13-19 Relation between nonhydrostatic stress (difference), σ, and an equivalent viscosity, η, for steady-state plastic flow, assuming a geologically reasonable strain rate $\epsilon = 3 \cdot 10^{-14}$/s and a flow law $\sigma = 3\dot{\epsilon}\eta$. Data from Heard (1976, figs. 5 and 6).

state flow intermediate between wet quartzite and marble at 800°C. Also, for the sake of comparison, ice at 0°C flows at a still lower nonhydrostatic stress but for faster strain rates; rapid ductile flow of glacial ice is a well-known phenomenon (see box on page 359).

The absolute values of the nonhydrostatic stress (stress difference) required to cause flow are also of interest. Stresses in the range of 60 to 130 MPa for flow of quartz mylonites in the Moine thrust zone were found by Weathers and others (1979). This strength agrees well with the data plotted in Figure 13-19 for dry quartzite at 600°C or wet quartzite just above 400°C; these are reasonable metamorphic temperatures. The stress difference of only 1 MPa to produce flow in halite at 100°C is equivalent to that occurring at the base of a free-

standing column of halite only 50 m high—the halite flowing under its own weight.

These data emphasize that rocks at elevated temperatures and pressures strained very slowly at geologically reasonable rates support only very small stress differences; steady-state plastic flow occurs at very small stresses, relaxing applied forces (see box on page 467). Hence, rock bodies in the deep crust and mantle respond to geologically slow strain rates almost as if they were fluid.

Diffusive Flow or Pressure Solution

The preceding major section has dealt with the mechanisms and evidences for plastic flow, which is one of the two mechanisms of solid-state ductile deformation occurring in metamorphic bodies. Plastic flow involves intracrystalline glide, recovery, and recrystallization; these are all essentially constant volume and constant composition processes. In aggregates, some deformation may also occur by grain boundary sliding, where the whole grain moves past its neighboring grains. This process may be envisaged as slip along the film of misfit atoms constituting the grain boundary; evidence for it is not easily seen because of the lack of markers. Strain-induced recovery and recrystallization are thermally activated diffusive processes, but recrystallization cannot produce strain in a body because no change in overall shape occurs.

The second mechanism of ductile deformation is a thermally activated, stress-induced, diffusive recrystallization that can produce large strains by movement of material in a hydrous fluid phase. This process, variously called diffusive flow, solution transfer, **pressure solution,** diffusive creep, Riecke's principle, or Nabarro-Herring or Coble creep, can act in concert with intracrystalline plastic flow or quite independently in response to nonhydrostatic stresses less than the plastic limit. Although discussed sporadically in the geologic literature for over a century and investigated extensively by metallurgists, most of the petrologic information concerning this type of flow is observational or theoretical (McClay, 1977; Robin,

Rock Strength and Tectonic Overpressures

Some petrologists have advocated **tectonic overpressures** as an explanation for the puzzling formation of high-*P* blueschist facies assemblages (see page 439). This is not the same concept as that of a **stress mineral** stabilized by nonhydrostatic stress, which has been discredited by considerable experimental studies, but is a similar effort to find another factor that might indirectly influence mineral stability.

Tectonic forces can augment gravitational forces at depth so that the magnitude of the confining pressure can be assumed to be equal to the **mean pressure**, $\bar{P} = (\sigma_1 + \sigma_2 + \sigma_3)/3$, rather than the lesser, overburden-related confining pressure, $P = \rho gz$. The difference between these magnitudes ($\bar{P} - P$) is the tectonic overpressure and is dependent upon the strength of the rock. Under most metamorphic conditions, considering appropriately low natural strain rates and high

temperatures, rocks are quite weak and flow in a ductile manner under nonhydrostatic stresses (unrelaxed stress differences) of 100 MPa to perhaps 10 MPa. Hence, tectonic overpressures can be no more than these strengths, so that the state of stress is approximately hydrostatic because the absolute value of *P* is so much greater, say, 1 GPa in blueschist bodies. An unrelaxed stress difference of 100 MPa is equivalent to an uncertainty in the depth of overburden of ~ 4 km. However, in blueschist bodies, temperatures are relatively low and one could argue that strain rates might be greater, both of which make rock stronger.

Still another factor is the water pressure, P_{H_2O}, which pushes grains apart, reducing the effective confining pressure on them; this makes the rock more brittle but reduces its strength. Field and petrographic evidence indicates high P_{H_2O} prevailed in blueschists.

1979; Green, 1980), and its exact formulation and explanation lies in the realm of sophisticated thermodynamics. Unfortunately, confirmatory experiments are scanty.

Theoretically, a region of a mineral grain boundary under greater normal stress has greater free energy than other parts where the normal stress is less, which then ought to grow at the expense of the more unstable region, if diffusive transfer of matter is possible. Of three possible diffusion paths (see Figure 13-20), migration of matter through an intergranular fluid film is favored because rates of diffusion are faster even at relatively low temperatures and because fluids are normally present in rocks during prograde metamorphism. The driving force for diffusion is related to gradients in chemical potential caused by the nonhydrostatic stress (Nicolas and Poirier, 1976, p. 156–158; Green, 1980).

Apparent effects of pressure solution were recognized as early as 1863 by the father of micro-

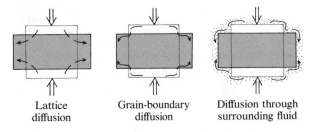

Lattice diffusion   Grain-boundary diffusion   Diffusion through surrounding fluid

Figure 13-20 Three paths of diffusion allowing particles to move from an unstable compressed area of a grain to an area where growth is possible. The path through a hydrous fluid solution surrounding the grain may be the most effective in rock bodies and is the basis for the term *pressure solution*.

scopic petrography, H. C. Sorby, and are widespread in sedimentary and low-grade metamorphic rocks (Durney, 1972; Elliot, 1973). Pebbles and cobbles in some unmetamorphosed conglomerates have conspicuous indentations where adjacent

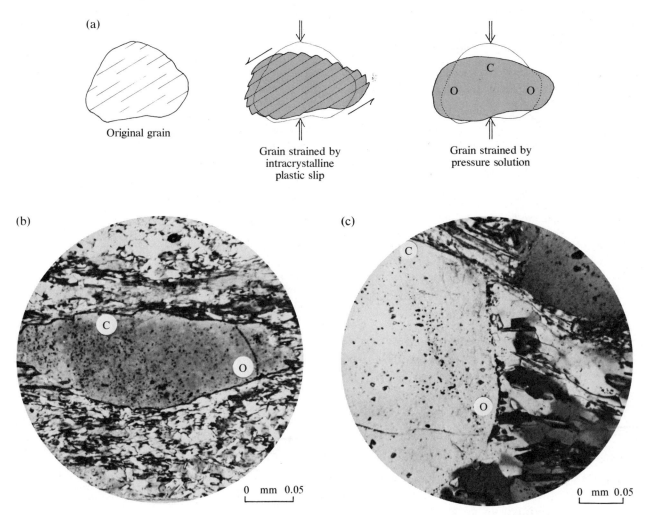

Figure 13-21   Quartz grains showing the effects of pressure solution that has caused them to grow in the plane of the imposed metamorphic foliation at the expense of dissolution perpendicular to it. (a) Schematic comparison of intracrystalline plastic slip and pressure solution, where C is corroded surface resulting from removal of material and O is a segment of the original grain surface marked by dustlike impurities onto which new grain material has been deposited. (b,c) Originally more equant clastic quartz grains have been flattened by growth of optically continuous quartz beyond the original grain boundary O and by substantial corrosion of surfaces labeled C. [From D. S. Elliott, 1973, Diffusion flow laws in metamorphic rocks. *Geological Society of America Bulletin* 84.]

clasts pressed and apparently caused removal of material by pressure solution. The well-known stylolites in carbonate rocks are a pressure-solution effect. Fossils and mineral grains are truncated in compressed areas by solution, commonly leaving an insoluble foreign residue on the dissolved surface or juxtaposing inert grains against the dissolved surface (see Figure 13-21). Optically continuous overgrowths or fibrous crystals extend outward from original grains in pressure shadows along the plane of flattening—the direction in which less work is needed to push against the matrix.

13.4 ROLE OF FLUIDS IN DEFORMATION

The significance of water in hydrolytic weakening of nominally anhydrous silicates has been discussed in relation to plastic flow. In pressure solution, intergranular fluids must play a very significant role in promoting diffusion of matter to change shapes of individual mineral grains. Integration of small, grain-scale effects over the entire volume of the rock body can produce significant strain.

Surprisingly, there is almost no experimental information on how rocks deform under nonhydrostatic stress while prograde fluid-release reactions are occurring (Fyfe, 1976). In one study, Raleigh and Paterson (1965) found that as serpentinite partially dehydrates with rising T to form talc + forsterite, it ceases to flow in a ductile manner, becomes embrittled, and fractures. High water pressures in pore spaces reduce the effective confining pressure, making the rock more brittle. In general, nothing is known on a grain scale of the effects of dehydration during deformation. Do neighboring hydrous grains detach from one another as water is liberated, causing local loss of cohesion and volumetric expansion? Could migration of this liberated water (or CO_2) through and eventually out of the rock body promote pressure solution? If so, at what rate, and to what extent in terms of overall strain? To what extent might the major strain of a body occur under high fluid pressure brittle conditions—cataclasis—with subsequent overprinting by ductile grain adjustment and solid-state crystal growth (Vernon, 1976, pp. 53–56)?

13.5 ORIGIN OF ANISOTROPIC FABRIC IN METAMORPHIC TECTONITES

Field and laboratory investigations have shown that imposed anisotropic fabrics in tectonites re-
sult from syntectonic ductile flow in nonhydrostatic stress fields. The large variation in the details of syntectonic flow fabrics found in metamorphic bodies implies a similarly varied range of factors affecting their origin. Relevant factors must include external ones such as the character of the applied nonhydrostatic stress, the magnitude of the overall strain, the rate of strain, T, and P and internal ones such as the inherited fabric of the body (whether anisotropic or isotropic), its grain size, and mineral and chemical composition, especially the availability of water or other fluids. Moreover, imposed fabrics can be inhomogeneous on different scales, probably reflecting inhomogeneities in one or more of these factors.

There are two general problems concerning the origin of the imposed anisotropic fabric in a body. First is the tectonic significance of the geometry of the fabric. For example, what pattern of ductile flow produced the lineation and foliation in a certain body of schist? And what stress field and resulting strain produced the flow? The second general problem concerns the origin, at the scale of individual grains, of the anisotropic fabric, such as the nematoblastic aggregate of hornblendes, or the lepidoblastic mica aggregate. The next two sections consider these two problems.

Patterns of Deformation and Flow: Fabric Geometry

Interpretation of the tectonic significance of an anisotropic fabric is more the concern of the structural geologist, yet petrologists dealing with metamorphic bodies can hardly be completely oblivious to this facet of the body if they are seeking a full understanding of its origin. Because of numerous factors controlling the geometric pattern of fabric—whether linear, planar or planar–linear—only general interpretive statements can be made here. The interconnected links in the chain of factors affecting the final fabric of a metamorphic tectonite are shown in Figure 13-22. If the initial rock body is perfectly homogeneous (every point within it is indistinguishable) and perfectly isotropic, then

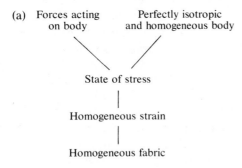

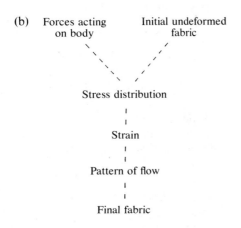

Figure 13-22 Chain of cause and effect relations in the development of imposed fabrics. (a) Geologically unrealistic case. (b) Situation in most real metamorphic rock bodies. [After F. J. Turner and L. E. Weiss, 1963, *Structural Analysis of Metamorphic Tectonites* (New York: McGraw-Hill).]

the state of stress within the body is perfectly correlated with any small applied forces; moreover, the resulting strain can be homogeneous (Hobbs and others, 1976, p. 22) and the final fabric correlates perfectly with the strain and stress. But real rock bodies are almost never isotropic nor homogeneous, although some may approach these qualities on certain scales (Turner and Weiss, 1963, Chap. 2); planar and linear discontinuities pervading real bodies (bedding, schistosity, and so on)

generally cause heterogeneous stress fields when forces are applied (see Figure 3-22b). Strains resulting from this uneven state of stress are generally not homogeneous nor are they directly predictable because of these mechanically significant discontinuities. The strain distribution in the metamorphic body is a consequence of the pattern of ductile flow, but as Turner and Weiss (1963, p. 371) point out, there can be an infinite variety of flow patterns, or movement pictures, for a given strain.

It appears that, in general, the pattern of ductile flow is reflected in the geometry of final imposed fabric, but the ways in which the flow pattern is influenced by the overall strain, by the distribution of stresses, and finally by the applied forces on the body are progressively more difficult to specify unambiguously.

Although strain and pattern of flow may not correlate closely, we may make tentative correlations between fabric and strain if the discontinuities in the body such as foliation surfaces, compositional layers or lenses, and so on are closely spaced and not mechanically very significant. By making a link between strain and flow pattern, we can overcome much of the frustration experienced by the novice in trying to find something tangible to say about the significance of the geometry of the imposed fabric. Keeping such reservations in mind, some simple types of homogeneous strain and the geometric types of anisotropic fabric that might result via ductile flow are shown in Figure 13-23. Note that planar–linear fabrics can develop by any of three types of strain, including plane strain, simple shear, and pure shear. This ambiguity in the tectonic significance of the most common type of anisotropic fabric emphasizes the inherent dangers in interpretation; facile conclusions can be very misleading.

Where strained oolites, fossils, or pebbles are present together with mineralogical foliation and lineation, they commonly, but not always, show similar geometry, reflecting similar patterns of deformation (Flinn, 1965). Strain and patterns of flow are, of course, not necessarily homogeneous

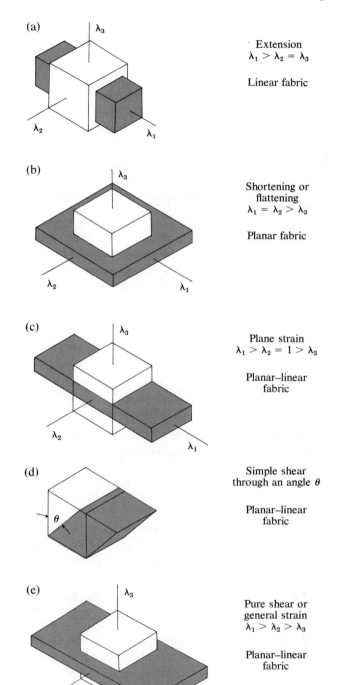

(a)

Extension
$\lambda_1 > \lambda_2 = \lambda_3$

Linear fabric

Figure 13-23 Simple types of homogeneous strain and the imposed anisotropic fabric that can result if the assumptions described in the text are satisfied. The cube represents the unstrained state and the shaded block the strained. $\lambda = (l/l_0)^2$ where l_0 is the initial length of the cube edge and l is the length after strain. [After B. E. Hobbs, W. D. Means, and P. F. Williams, 1976, *An Outline of Structural Geology* (New York: John Wiley and Sons).]

(b)

Shortening or
flattening
$\lambda_1 = \lambda_2 > \lambda_3$

Planar fabric

throughout the body, though this is common at least on some scale of observation. The statistically uniform orientation of foliation and lineation over large areas of some regional metamorphic terranes is remarkable (see Figure 10-6b).

More sophisticated correlations can be drawn between fabric, flow pattern, and strain with regard to their symmetries (Turner and Weiss, 1963, pp. 385–391). Fabric symmetries summarized in Table 13-1 can be compared to Figure 13-23.

(c)

Plane strain
$\lambda_1 > \lambda_2 = 1 > \lambda_3$

Planar–linear
fabric

Preferred Orientation of Mineral Grains

The origin of the preferred orientation of mineral grains in tectonites has intrigued geologists for well over a century. Like other interpretations of fabric, the problem is both petrological and structural in scope (Hobbs and others, 1976, p. 118–140, 231–266). Foliation commonly bears an axial plane relation with associated folds and parallels the direction of overall flattening in folded bodies. Lineation commonly parallels hinge lines of associated folds. In the following discussion, we will emphasize the grain-scale processes possibly involved in development of imposed anisotropic fabrics.

(d)

Simple shear
through an angle θ

Planar–linear
fabric

(e)

Pure shear or
general strain
$\lambda_1 > \lambda_2 > \lambda_3$

Planar–linear
fabric

In planar, linear, and planar–linear metamorphic fabrics, mineral grains show preferred orientation of their shape, or of their internal lattice, or both. Preferred dimensional shape orientation of platy, tabular, prismatic, or otherwise inequant grains is manifest in foliated and lineated fabrics. Phyllosilicates and single-chain silicates are most commonly involved, but locally discoidal or pencillike grains of carbonates, feldspars, and quartz

Table 13-1   Symmetries of imposed metamorphic fabrics.

Symmetry	Imposed fabric	Example of origin
Spherical	Isotropic or granoblastic	Static recrystallization under hydrostatic conditions
Axial	Planar (foliation)	Mimitic recrystallization of clay particles in shale during burial metamorphism, producing phyllite
	Linear (lineation)	Extensional strain and flow within tightly constricted fold hinge or in country rocks alongside forcefully emplaced pluton
Orthorhombic	planar–linear (foliation plus lineation)	Plane or general strain occurring in many metamorphic systems where there is compound extension and flattening
Monoclinic	planar–linear (foliation plus lineation)	Simple shear as in a thrust-fault zone, producing mylonite
Triclinic	Multiple planar–linear (more than one obliquely intersecting foliation and/or lineation)	Oblique overprinting during polymetamorphism

Source: After J. Verhoogen, F. J. Turner, L. E. Weiss, C. Wahrhaftig, and W. S. Fyfe, 1970, *The Earth* (New York: Holt, Rinehart and Winston).

may also define the anisotropic fabric. Although not obvious to the unaided eye, grains in many intermediate- to high-grade metamorphic tectonites also have their lattices or crystallographic axes preferentially oriented in a particular direction or pattern (see Figure 13-24). Crystallographic axes, such as c-axes in quartz, may show a statistically uniform direction of axial orientation, or they may be oriented statistically within a plane, and so on. Obviously, phyllosilicates, sillimanite, and other characteristically euhedral minerals will show a lattice preferred orientation if they are shape oriented. In contrast, minerals low in the crystalloblastic series such as quartz, carbonates, and feldspars can possess lattice preferred orientation without shape orientation, and vice versa.

The precise mechanism by which the preferred grain orientation in a specific tectonite originated is not easily determined. In some instances a particular mechanism may have left distinctive evidence, but in others we are forced to say only that certain ones were possible, including:

1. Intracrystalline plastic slip and twinning

2. Preferred nucleation and growth of favorably oriented grains

3. Rotation of rigid inequant grains

4. Pressure solution

Ambiguities arise from possible overprinting of one mechanism after another, from simultaneous operation of more than one, and from a lack of complete understanding of how the actual process operates in real rocks (Means, 1977).

Intracrystalline plastic slip and twinning.   Plastic slip and twinning are verifiable mechanisms of producing preferred orientation of grain shapes and lattices in rocks.

In simple shear by slip in a confined grain (see Figures 13-23 and 13-25), several hundred percent extension is possible. Not only does the grain change shape—note especially the external rotation of its faces—it also experiences a significant internal rotation of crystallographic planes and axes. Because of the nature of plastic slip, preserving essentially undistorted crystal across the slip planes, the crystallographic axes and planes serve

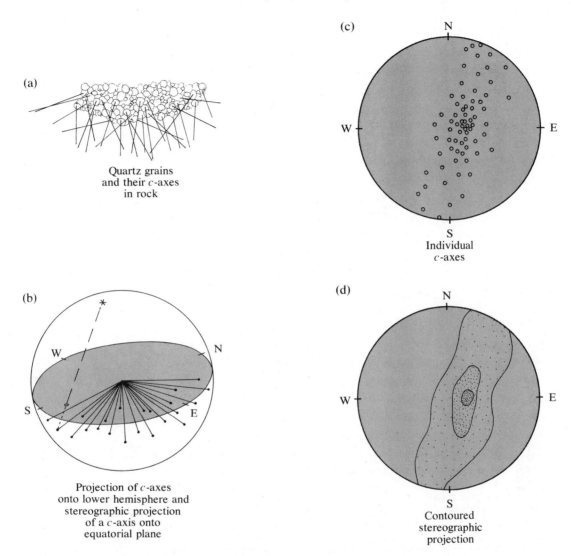

(a)

Quartz grains
and their *c*-axes
in rock

(b)

Projection of *c*-axes
onto lower hemisphere and
stereographic projection
of a *c*-axis onto
equatorial plane

(c)

Individual
c-axes

(d)

Contoured
stereographic
projection

Figure 13-24 Lattice preferred orientation in an aggregate of mineral grains, for example, *c*-axes
in quartz. The orientation of this crystallographic axis is determined in each quartz grain in a thin
section taken from an oriented hand sample by a universal stage attached to a petrographic
microscope (Phillips, 1971, chaps. 9–11). Alternatively, X-ray diffraction techniques can be used.
(a) Schematic of grain aggregate showing orientations of *c*-axes. (b) Orientation of the *c*-axis in
each grain is drawn from the center of a sphere whose equatorial plane (shaded) is oriented
horizontally. From the point (solid circle) on the lower hemisphere where the *c*-axis intersects it,
a projection line (dashed) is drawn to a point (*) on top of the sphere. The point where this
projection line intersects the equatorial plane (open circle) is called the **stereographic projection** of
the *c*-axis. (c) Stereographic projection of *c*-axes in a hypothetical quartz aggregate forms a
bandlike pattern. (d) Contoured diagram with each line bounding an area of some specified
minimum point density (compare Hobbs and others, 1976, pp. 120–121, appendix A; see also
Turner and Weiss, 1963, especially chap. 3).

(a)

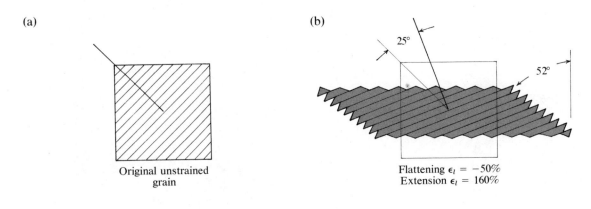

Original unstrained
grain

(b)

Flattening $\epsilon_l = -50\%$
Extension $\epsilon_l = 160\%$

(c)

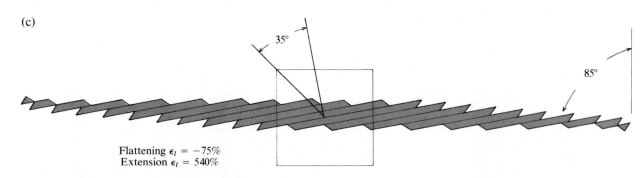

Flattening $\epsilon_l = -75\%$
Extension $\epsilon_l = 540\%$

Figure 13-25   Development of flattened grain with reoriented lattice due to intracrystalline plastic slip. Model grain is strained by simple shear as if confined between the wide jaws of a vise. (a) Original unstrained grain showing potential slip planes and perpendicular lattice direction. This grain model could represent quartz with basal (0001) slip planes and the crystallographic c-axis. (b) Flattening by $\epsilon_1 = (l - l_0)/l_0 = -50\%$ with a corresponding extension of 160% has resulted in an internal rotation of the crystallographic axis and slip planes by 25° and an external rotation of the edge of the grain by 52°. (c) More extreme strain and rotation. The constraints on this strain model are that the top and bottom faces of the grain remain parallel and constant in overall orientation and that the dimensions of the grain parallel to slip planes and the thickness of the slipped segments of crystal remain constant, that is, the strain is purely plastic. [After B. E. Hobbs, W. D. Means, and P. F. Williams, 1976, *An Outline of Structural Geology* (New York: John Wiley and Sons).]

as valuable markers for the overall deformation. In this idealized model, which represents but one mode of strain of an isolated yet confined grain, the normal to the slip plane rotates toward the direction of shortening with increasing strain. Of

course, in actual compact grain aggregates, plastic strain and orienting mechanisms differ somewhat.

Informative pictorial models of plastic orienting mechanisms in aggregates have been presented by Etchecopar (1977). A grain aggregate was modeled by a computer as a two-dimensional array of initially tightly fitting hexagons, each of which has one possible slip system randomly oriented over the whole array. The array was progressively strained in the pure shear and simple shear modes (see Figure 13-23) by allowing the hexagons to slip (as in Figure 13-25), rotate, translate, or even rupture, all the while minimizing gaps or overlaps between the grains. Certain features of these computer models shown in Figure 13-26 are of special interest: (1) intergrain voids and overlaps are considerable, even though the modelling procedure attempts to minimize them. As discussed on page 455, in a real rock undergoing deformation these voids and overlaps must be compensated by bending, kinking, and other strain effects and possibly by brittle rupturing as well as ductile processes. (2) Grains whose slip directions are more nearly parallel to the plane of flattening have low resolved shear stress on their slip planes and do not strain appreciably. (3) Increasing bulk strain of the aggregate results in a stronger preferred orientation of lattice slip directions and of grain shape (see Figure 13-26f).

Plastic flow in a real tectonite body has been documented by Compton (1980). The Raft River Mountains of northwest Utah is one of several metamorphic core complexes lying to the west of the belt of Cretaceous overthrust faults that run the length of the Cordillera from Alaska into Mexico. These complexes, which appear to have experienced a complex history of metamorphism, tectonism, and magmatism through Mesozoic and Cenozoic time, are commonly characterized by tectonites with subhorizontal foliation and lineation. In the Raft River Mountains complex, these lineations are oriented principally east-west, perpendicular to the strike of the nearby thrust belt, in an elongate east-west structural dome. Such anisotropic fabric might be explained by simple shear (see Figure 13-23), with the lineation developing in the direction of shear (like slickensides) associated with easterly transport of the thrust sheets. However, in the Raft River Mountains complex, the lineations consist not only of steaked

mineral aggregates but also of hinge lines of folds on a wide range of scales whose direction of overturning, locally forming recumbent isoclinal folds, is so consistent as to indicate tectonic transport at right angles to the lineations and hence parallel to the strike of the thrust belt.

Compton (1980) studied in detail the anisotropic fabric of lineated metasedimentary quartz–feldspar schists in the core complex. Quartz and minor feldspar grains are relict sand grains of fairly uniform size that have been deformed into crude triaxial ellipsoids with shortest axes perpendicular to foliation and longest axes parallel to lineation. Local relict conglomeratic beds in the quartzitic metasedimentary sequence have remarkable triaxial cobbles of similar shape and orientation (see Figure 10-2) as relict sand grains. Several features indicate the anisotropic fabric developed principally by plastic processes. Thus most quartz occurs as ribbonlike lenticular deformation bands with uneven optical extinction under cross-polarized light, or as elongate subgrains. Deformation lamellae, though not abundant, have subbasal orientations. A few relict quartz sand grains that are undeformed have c-axes oriented perpendicular to the foliation, an orientation unfavorable to either basal or prism slip under a compression acting normal to the foliation. Deformed quartz grains have volumes similar to those of less-deformed feldspar grains, supporting a constant volume process of plastic deformation acting on initially uniform size sand grains. Some relict feldspar sand grains have been broken and pulled apart by extension in the direction of the lineation, whereas enclosing, more ductile quartz is plastically deformed into deformation bands. Small, randomly oriented and equant quartz grains constituting 10% to 80% of the schists apparently formed by syntectonic recrystallization. Although some quartz samples have monoclinic or triclinic c-axis orientation patterns, orthorhombic symmetry patterns are more common (see Figure 13-27).

Experimental and computer-simulated deformation of quartzite by plane strain has yielded features similar to those observed in the Raft River

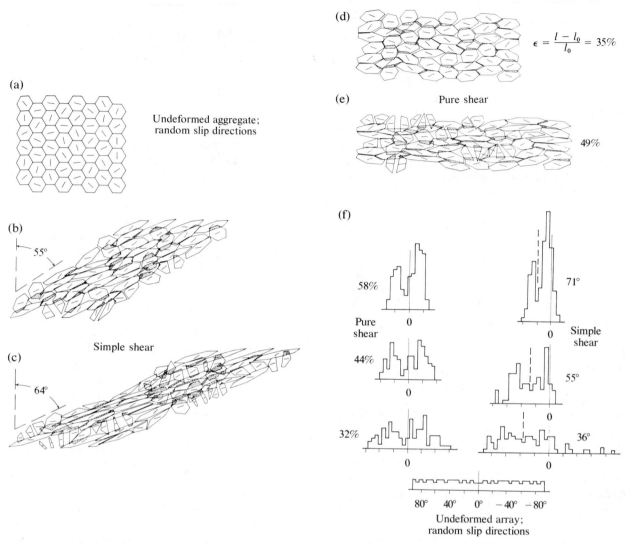

Figure 13-26 Computer models of deformation in grain aggregates. (a) Undeformed array of grains, each with one slip direction as shown. (b,c) Progressive simple shear. Stippled areas are grain overlaps; note gaps between grains and ruptured grains. (d,e) Progressive pure shear. (f) Histograms showing development of increasing preferred lattice (slip direction) orientation during progressive strain; dashed lines in histograms for simple shear are orientations of long diagonal of strained aggregate. [After A. Etchecopar, 1977, A plane kinematic model of progressive deformation in a polycrystalline aggregate, *Tectonophysics* 39.]

Mountains core complex. Patterns of *c*-axes are orthorhombic, and triaxial ellipsoidal grains that are similarly oriented with respect to the *c*-axis pattern have similar relative lengths of axes; the quartz grains deformed experimentally by dislocation glide and climb.

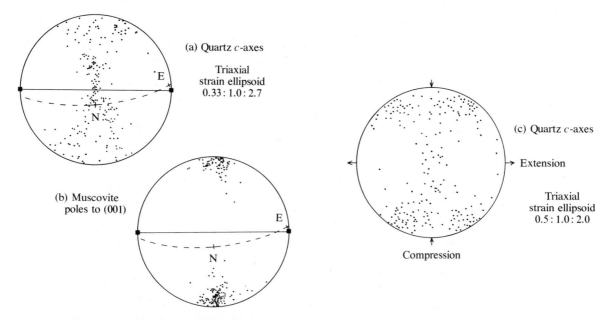

Figure 13-27 Stereographic (equal area) lower hemisphere plots of lattice orientation of grains in tectonites subjected to plane strain. Compare Figures 13-23 and 13-24 and Table 13-1. (a) Quartz *c*-axes in a sample of quartz (89%)–feldspar (8%)–muscovite (3%) schist from the Raft River Mountains metamorphic core complex show an orthorhombic symmetry. Solid line is the trace of the foliation, square is the lineation, and dashed arc is the trace of the horizontal plane with geographic north (N) and east (E). Numbers indicate average relative lengths of axes of triaxial ellipsoidal quartz grains, assuming an original sphere of radius 1. (b) Normals to (001) in muscovite from same sample as in (a). (c) Quartz *c*-axes in quartzite deformed experimentally in plane strain at 1.5 GPa, 800°C and $\dot{\epsilon} = 10^{-6}$/s. Arrows indicate direction of compression and extension. Note orthorhombic symmetry and resemblance to (a). [All after R. R. Compton, 1980, Fabrics and strains in quartzites of a metamorphic core complex, Raft River Mountains, Utah, Geological Society of America Memoir 153; part (c) also after J. Tullis, 1977, Preferred orientation of quartz produced by slip during plane strain, *Tectonophysics* 39.]

Compton (1980) concludes that the prominent planar–linear fabric of the Raft River Mountains complex was produced by east-west extension in essentially plane strain, modified by moderate axial flattening and locally by shear. Formation of contemporaneous mineral lineations and folds is speculated to have occurred in a heated and elongate east-west, up-domed body—the metamorphic core. Radially directed outward flow under gravity buckled and overturned relict beds into folds and, at the same time, caused stretching (extension) tangentially around the rising domical core.

Preferred nucleation and growth of favorably oriented grains. Of the four possible mechanisms of developing preferred grain orientation, this is the most difficult one to evaluate. Critical experiments and diagnostic petrographic criteria to assess its relevance are almost nonexistent. Some preliminary experiments show, as we might intuitively anticipate, that anisotropic fabrics produced by solid-state growth in a nonhydrostatic stress field have correlatable patterns of ductile flow and strain, but the specific mechanisms involved are uncertain.

(a)

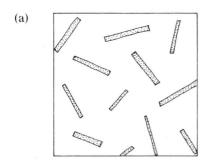

(b)

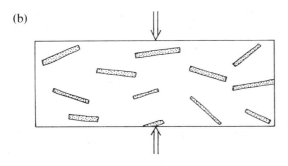

Figure 13-28 Rigid grains embedded in a softer or more ductile matrix can be rotated toward the plane of flattening during overall strain of the body.

Rotation of rigid inequant grains. Rigid platelike or rodlike grains embedded in a soft or ductile matrix conceivably might rotate toward the plane of flattening, for planar preferred orientation, or toward the axis of extension, for linear orientations. As a result, typically euhedral inequant grains, such as phyllosilicates, amphiboles, and sillimanite, will possess a lattice preferred orientation as well as a shape orientation. Very large strains are necessary to produce strong preferred orientations solely by rotation if the inequant grains are initially quite random (see Figure 13-28).

Rigid grain rotation during dewatering of soft muds under nonhydrostatic stress has been invoked by some geologists to explain the origin of cleavage in slates. They envisage that prior to con-

solidation, a thick sequence of water-rich muds, silts, and sands was flattened by tectonic forces, at the same time rotating phyllosilicate platelets into a preferred planar orientation parallel to the plane of flattening and to axial planes of folded beds. Expulsion of water during this event is postulated to have occurred along the cleavage, as indicated by sheets or dikes of sandstone alleged to be parallel to the cleavage, which locally penetrate into the slate from adjacent sandy beds.

The role of tectonic dewatering to produce slaty cleavage has been challenged by other geologists. Beutner (1980) finds that sandstone dikes are not actually parallel to the cleavage and suggests that they formed by soft sediment deformation prior to a regional metamorphic event during which the cleavage developed. Statistical analysis of phyllosilicate grain shapes and orientations in some Austrailian slates by Bell (1978) reveals patterns difficult to reconcile with purely soft-state grain rotation. A significant imprint from pressure solution and preferred growth of favorably oriented grains is apparent.

Thus, soft-state grain rotation producing slaty cleavage has not been satisfactorily demonstrated as a viable mechanism. It cannot be universal in any case because this fabric is also found in metamorphosed igneous rocks and as a secondary feature in polymetamorphic rocks.

Pressure solution. Differential pressure solution of grains such as quartz produces flattened shapes that impart a foliation to the body. No lattice preferred orientation is created unless certain unfavorably oriented grains are somehow entirely dissolved.

Differential pressure solution may also enable disseminated inert platy grains such as micas to rotate toward the plane of flattening. This could produce the preferred orientations of phyllosilicate flakes seen in some slates, whose very fine grain size and consequent short diffusion paths would appear to make pressure solution quite possible.

Origin of Segregation Layering

Some segregation layers, defined by variations in the proportions of constituent minerals, are locally demonstrably relict beds; lepidoblastic micas in them have formed mimetically after clays oriented in the beds. However, the complex deformational history of most metamorphic bodies precludes preservation of bedding to any recognizable degree. Other planar segregations, which may actually be closer to lenses than layers, result from mechanically dominated transposition processes (see page 352; Myers, 1978); preferred orientation of inequant grains in the segregations may also result from the transposition. Shear of initially coarse-grained rocks can also produce small-scale, lenticular segregations in resulting mylonites, simply because the originally large grains have been drawn out into extremely flattened shapes.

In a sense, it is inappropriate to consider the origin of preferred orientation of mineral grains, as was done in the last section, independently of the origin of segregation layering; the two are commonly intimately associated in deformed metamorphic bodies. Even in some low-grade phyllites, a lepidoblastic fabric is accompanied by a fine-scale segregation layering (best seen under the microscope), which together produce the foliation.

Studies of interbedded metagraywackes, metasiltstones, and slates in the Appalachian orogen by Groshong (1976) and in southeastern Australia by Williams (1972) have enlightened our understanding of how slaty cleavage and foliation in the metasediments originates. These foliations are oriented parallel to the axial surfaces of associated folds in bedding and are defined by alternating quartz-rich domains with weak or nonexistent preferred grain orientations and phyllosilicate-rich folia with stronger preferred orientation (see Figure 13-29). The lepidoblastic folia are interpreted to have formed from originally more quartz-rich material through selective pressure solution of the quartz by fluids migrating along surfaces of move-

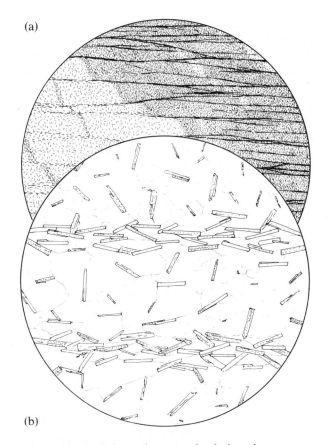

(a)

(b)

Figure 13-29   Schematic textural relations in a metagraywacke–slate sequence showing cleavage and segregation layering developed by selective pressure solution under nonhydrostatic stress. (a) Microscopic view under low magnification shows a relict bedding contact vertical between slate and coarser grained metagraywacke that is offset along dark phyllosilicate-rich domains; horizontal cleavage defined by these anastomosing micaceous films is better developed in the finer-grained, quartz-poor slate than in the metagraywacke. (b) Under higher magnification, the metagraywacke displays lepidoblastic segregation layers with flattened, selectively dissolved quartz grains lying between the more inert phyllosilicates. Quartz-rich layers show no evidence of pressure solution.

ment in the rock; in this metamorphic differentiation process, the volume decrease occurring in the lepidoblastic domains permitted the relatively inert phyllosilicate flakes to rotate into a preferred orientation. Textural evidence for these two conclusions is convincing. Quartz-rich domains have relatively equidimensional grains; many are obviously detrital in origin. Lepidoblastic domains contain fewer quartz grains, none obviously detrital and all conspicuously flattened; some quartz grains appear truncated by adjacent phyllosilicate flakes and local pressure-shadow overgrowths; compare, for example, Figure 13-21. Mica films displace markers, such as obliquely oriented relict bedding contacts. These textural relations are difficult to explain by selective introduction of quartz into, or removal of mica from, a homogeneous rock like the phyllosilicate domains to form quartz-rich domains.

Williams (1972) emphasizes that the quartz selectively removed did not travel only to immediately adjacent domains but rather migrated to different parts of the deformed body. Ubiquitous quartz veins attest to this migration. He states (p. 41) that

> the environment envisaged for these rocks, during the development of the layering is one in which water, initially connate in origin and then derived from metamorphic reactions in underlying rocks, was continually migrating through the rock into shallower parts of the crust.... It is reasonable to assume that by the time differentiation was taking place, the porosity and permeability of the rock were low. Now if this rock was deforming and the deformation was occurring close to the brittle/ductile transition, it is conceivable that movement zones would develop that would become "channel-ways" for the fluids migrating through the rock. Quartz would then be dissolved from the walls of these "channelways." In fact, at Bermagui, the deformation did give rise to what have been described above as movement zones; that some of these are faulted indicates that deformation was occurring close to the brittle/ductile transition. Thus, as deformation took place in these zones, many of the grains would be unable to adjust their shape to accommodate the intergranular movements, and voids would develop in the rock, thereby increasing its permeability. This

hypothesis provides a reasonable explanation of why the quartz is selectively removed from the "movement zones," that is, from the mica-rich domains.

Williams' hypothesis is a good illustration of the sort of link between prograde dehydration reactions and deformation envisaged by Fyfe (1976) and discussed in Section 13.4. Once selective dissolution has been initiated to produce phyllosilicate-rich folia, subsequent prograde reactions will liberate more water from them, facilitating additional movement of rock and corrosion of remaining soluble grains. This metamorphic differentiation could thus continue over a range of metamorphic conditions. Certainly, additional field and laboratory studies of solution-related anisotropic fabrics are needed, especially if the role of possibly concurrent fluid-release reactions and formation of veins can be evaluated. And to compliment such studies of natural rocks, we need a substantial program of experimental investigations under controlled conditions (Means, 1977).

The origin of the common segregation layering in higher-grade rocks—amphibolites, schists, gneisses, and so on—is still another problem. To what extent might it, too, be answered by some sort of selective solution process acting concurrently with deformation and flow of fluids liberated during prograde mineral reactions? Robin (1979) proposes that in domains of a rock body with higher modal proportions of phyllosilicates, the stress difference ($\sigma_1 - \sigma_3$) will be smaller than in neighboring more competent quartz or feldspar dominant regions because of the inherent differences in the strength of these minerals. The maximum principal stress σ_1 will be more or less uniform over some considerable volume of the body whereas σ_3 will vary, depending upon differences in the phyllosilicate/quartz ratio. In the more competent—large ($\sigma_1 - \sigma_3$)—domains, quartz grain boundaries oriented perpendicular to σ_3 will have a lower energy than any quartz grain boundary in the phyllosilicate-enriched domains where ($\sigma_1 - \sigma_3$) is small. Consequently, silica ought to migrate from the phyllosilicate-enriched domains to the low-energy quartz grain boundaries in the quartz-enriched domains. Robin envisages that even subtle compositional variations may become amplified in nonhydrostatic stress fields by means of solution transfer. A similar model for generation of compositional layering in alpine peridotites has been advocated by Dick and Sinton (1979).

SUMMARY

13.1. Constructional, thermally activated solid-state growth processes create new, more stable phases by nucleation and growth in the course of mineralogical reactions; solid-state growth also adjusts shapes and sizes of existing grains. Heterogeneous nucleation along grain boundaries and at intracrystalline structural irregularities is probably favored over homogeneous nucleation within normal grains.

Grain growth after nucleation is driven by the requirement to minimize internal and grain-boundary energies in the aggregate. Growth of new grains is greatly facilitated by the presence of intergranular fluids, which permit rapid diffusion of nutrient components through the grain aggregate. Given sufficient time, adjustment of grain boundaries tends to form grains of maximum feasible size whose boundaries meet at approximately 120° triple junctions or to form more euhedral outlines if the grain is highly anisotropic in surface energy.

13.2. Any body acted upon by forces is in a state of stress. The magnitude of stress is force divided by the area acted upon. Any general state of stress at a point consists of a hydrostatic component of equal magnitude in all directions, the confining P, and a nonhydrostatic component embodying shear stresses.

Rock bodies subjected to stress are deformed, an important component of which is strain—the change in shape (distortion) or size of a body; strain can be determined from internal markers. In the Earth, the hydrostatic component of stress, P, causes insignificant strain but dictates mineralogical equilibrium; nonhydrostatic stress causes solid-state deformational flow, which is manifest in anisotropic fabric.

Materials respond to applied stress in different ways. Elastic behavior involves small reversible strains over short time periods; that is, strain rate is large. Nonhydrostatic stress above an elastic limit causes permanent rupture (faulting); stress greater than a plastic limit leads to plastic flow. The response of real rock bodies in nature to nonhydrostatic stress is infinitely varied but can be viewed in terms of brittle and ductile end members. Most metamorphic bodies under elevated T and P strained at slow rates experience cohesive ductile flow.

13.3. Two mechanisms of ductile flow are intracrystalline plasticity and diffusive flow, or pressure solution.

Under certain conditions, especially where confining P is sufficiently great to inhibit brittle rupture, parts of a crystal yield to nonhydrostatic stress by slipping or twinning along special lattice orientations to a new equilibrium position. This intracrystalline plastic deformation changes the external shape of the crystal but preserves the order of the lattice across the slip or twin plane. Plastic slip commonly occurs on planes parallel to lattice directions of widest interplanar spacing and weakest bonding and is activated when the resolved shear stress on the plane exceeds a critical yield value.

Compared to rock-forming minerals, metals slip at lower resolved shear stress because of weaker atomic bonding and on more numerous planes because of higher crystal symmetry. Lower symmetry, covalent–ionic bonded silicate grains in rocks commonly bend and kink by local plastic slip in heterogeneous stress fields.

Movement of dislocations along slip planes permits intracrystalline slip under nonhydrostatic stress. Continued application of stress causes dislocations to tangle and pile up, causing strain hardening.

Thermally activated recovery at elevated T and slow strain rates is accomplished by movement of dislocations out of slip planes, counteracting strain hardening; such movement culminates in poly-gonization—the formation of more stable, less strained subgrains bounded by low-angle disloca-tion walls. Syntectonic recovery and recrystalliza-tion during intracrystalline slip and twinning permit large strains during steady-state flow in tectonites.

Dissolved water in silicates allows them to de-form plastically at lower stress; this hydrolytic weakening is a consequence of weak Si–OH–HO–Si linkages, which are more easily broken than Si–O–Si bridges.

Under relatively low T and/or high strain rate, silicate grains in plastically deformed rocks show undulatory optical extinction, kink and deforma-tion bands, and, under the electron microscope, high dislocation densities. At higher T and/or slower strain rates, thermally activated syntec-tonic recovery and recrystallization effects are more evident and flow can produce large strains in rocks at only small stress differences.

In diffusive flow, or pressure solution, grains become flattened due to differential dissolution, transfer, and precipitation of material in pressure shadows along the foliation surface or in nearby veins.

Deep crustal and mantle rock bodies subject to slow strain over long periods of geologic time are very weak and do not support large nonhydrostatic stress differences. They deform by ductile flow almost as if they were fluids with no shear strength.

13.4. Large volumes of water liberated from metamorphic rock bodies subjected to nonhydro-static stress while undergoing prograde reactions must influence mechanisms of deformation, such as by hydrolytic weakening and pressure solution.

13.5. The origin and significance of anisotropic fabrics is not always clear because of the multitude of relevant factors. The geometry of the imposed fabric, whether planar, linear, or planar–linear, is closely linked with the pattern of ductile flow pro-ducing it and is less directly, or ambiguously, re-lated to the type of strain.

Mineral grains in tectonites have preferred orientation of their inequant dimensions, or of their internal lattice, or both. Preferred orientation due to intracrystalline plastic slip and twinning has been verified in many instances and potentially of-fers considerable information regarding the history of deformation. The significance of preferred nucleation and growth of favorably oriented grains under nonhydrostatic stresses in real rock bodies remains uncertain. Pressure solution and rotation of rigid inequant grains appear to have operated in unison to produce the familiar slaty cleavage in many bodies. Differential pressure solution may explain the origin of some segregation layering.

STUDY QUESTIONS

1. List and discuss factors that promote brittle behavior of rocks under nonhydrostatic stress. Characterize brittle behavior.

2. List and discuss factors that promote ductile behavior of rocks under nonhydrostatic stress. Characterize ductility.

3. Contrast movement of a dislocation along a slip plane with dislocation climb.

4. Floating logs are linearly oriented parallel to a flowing stream of water. Cite other familiar examples of the way in which the geometry of a flow process correlates with the geometry of its effects.

5. Why are most metals more plastic than rocks?

6. Study the atomic structure of micas, enstatite, and kyanite in a reference book. Predict the easiest slip plane in these crystals.

7. Describe and sketch features in grains of cal-cite, mica, quartz, and feldspar that have been plastically strained. Contrast features in single grains and grain aggregates.

8. Describe and sketch the mechanisms that can cause change in grain shape. To what extent do these mechanisms also change the orienta-tion of the crystal lattice?

9. Discuss in a qualitative way the nonhydro-static stresses required for plastic slip in quartz

as a function of temperature, strain rate, and availability of water.

10. What factors inhibit the development in aggregates of grain boundaries with minimum energy?

11. Justify the commonly invoked assumption that states of stress at depth in the Earth are, to a first approximation, hydrostatic.

12. Why should pressure solution be a more appropriate term for metamorphic rocks than diffusive flow, which is used to describe a similar phenomenon in metals?

13. Would one anticipate any correlation between metamorphic grade and the mode of deformation producing preferred grain orientation in tectonites? Justify your answer. If an idealized correlation exists, describe it and specify any assumptions made.

14. Discuss the role of water in the formation of anisotropic fabric in metamorphic tectonites.

REFERENCES

Bell, T. H. 1978. The development of slaty cleavage across the Nackara arc of the Adelaid geosyncline. *Tectonophysics* 51:171–201.

Bell, T. H., and Etheridge, M. A. 1973. Microstructures of mylonites and their descriptive terminology. *Lithos* 6:337–348.

Beutner, E. C. 1980. Slaty cleavage unrelated to tectonic dewatering: The Siamo and Michigamme slates revisited. *Geological Society of America Bulletin* 91:171–178.

Birch, F. 1966. Compressibility; elastic constants. Geological Society of America Memoir 97:97–173.

Compton, R. R. 1980. Fabrics and strains in quartzites of a metamorphic core complex, Raft River Mountains, Utah. Geological Society of America Memoir 153:385–398.

Dick, H. J. B., and Sinton, J. M. 1979. Compositional layering in alpine peridotites: Evidence for pressure solution creep in the mantle. *Journal of Geology* 87:403–416.

Durney, D. W. 1972. Solution-transfer, an important geological deformation mechanism. *Nature* 235:315–317.

Elliot, D. 1973. Diffusion flow laws in metamorphic rocks. *Geological Society of America Bulletin* 84:2645–2664.

Etchecopar, A. 1977. A plane kinematic model of progressive deformation in a polycrystalline aggregate. *Tectonophysics* 39:121–139.

Flinn, D. 1965. On the symmetry principle and the deformation ellipsoid. *Geological Magazine* 102:36–45.

Fyfe, W. S. 1976. Chemical aspects of rock deformation. *Philosophical Transactions of the Royal Society of London,* Series A, 283:221–228.

Fyfe, W. S., Price, N. J., and Thompson, H. B. 1978. *Fluids in the Earth's Crust.* Amsterdam: Elsevier, 383 p.

Fyfe, W. S., Turner, F. J., and Verhoogen, J. 1958. Metamorphic reactions and metamorphic facies. Geological Society of America Memoir 73, 259 p.

Green, H. W., II. 1980. On the thermodynamics of nonhydrostatically stressed solids. *Philosophical Magazine,* A, 47A:637–647.

Groshong, R. H., Jr. 1976. Strain and pressure solution in the Martinsburg slate, Delaware Water Gap, New Jersey. *American Journal of Science,* 276:1131–1146.

Heard, H. D. 1976. Comparison of the flow properties of rocks at crustal conditions. *Philosophical Transactions of the Royal Society of London,* Series A, 283:173–186.

Hobbs, B. E., Means, W. D., and Williams, P. F. 1976. *An Outline of Structural Geology.* New York: John Wiley and Sons, 571 p.

McClay, K. R. 1977. Pressure solution and cobble creep in rocks and minerals: A review. *Journal of the Geological Society* (London) 134:57–70.

McLean, D. 1965. The science of metamorphism in metals. In W. S. Pitcher and G. W. Flinn (eds.), *Controls of Metamorphism.* Edinburgh: Oliver and Boyd, pp. 103–118.

Means, W. D. 1976. *Stress and Strain.* New York: Springer-Verlag, 339 p.

Means, W. D. 1977. Experimental contributions to the study of foliations in rocks: A review of research since 1960. *Tectonophysics* 39:329–354.

Myers, J. W. 1978. Formation of banded gneisses by deformation of igneous rocks. *Precambrian Research* 6:43–64.

Nicolas, A., and Poirier, J. P. 1976. *Crystalline Plasticity and Solid State Flow in Metamorphic Rocks.* New York: John Wiley and Sons, 444 p.

Oertle, G. 1970. Deformation of a slaty, lapillar tuff in the Lake District, England. *Geological Society of America Bulletin* 81:1173–1188.

Paterson, M. S. 1978. *Experimental Rock Deformation— the Brittle Field.* New York: Springer-Verlag, 254 p.

Phillips, W. R. 1971. *Mineral Optics: Principles and Techniques.* San Francisco: W. H. Freeman and Company, 249 p.

Raleigh, C. B., and Paterson, M. S. 1965. Experimental deformation of serpentinite and its tectonic implications. *Journal of Geophysical Research* 70:3965–3985.

Robin, P-Y. F. 1979. Theory of metamorphic segregation and related processes. *Geochimica et Cosmochimica Acta* 43:1587–1600.

Spry, A., 1969. *Metamorphic Textures.* New York: Pergamon, 350 p.

Tullis, J., and Yund, R. A. 1977. Experimental deformation of dry Westerly granite. *Journal of Geophysical Research* 82:5705–5718.

Turner, F. J., and Weiss, L. E. 1963. *Structural Analysis of Metamorphic Tectonites.* New York: McGraw-Hill, 545 p.

Verhoogen, J., Turner, F. J., Weiss, L. E., Wahrhaftig, C., and Fyfe, W. S. 1970. *The Earth.* New York: Holt, Rinehart and Winston, 748 p.

Vernon, R. H. 1976. *Metamorphic Processes.* London: Thomas Murby, 247 p.

Weathers, M. S., Bird, J. M., Cooper, R. F., and Kohlstedt, D. L. 1979. Differential stress determined from deformation-induced microstructures of the Moine thrust zone. *Journal of Geophysical Research* 84:7495–7509.

White, S. 1977. Geological significance of recovery and recrystallization processes in quartz. *Tectonophysics* 39:143–170.

Williams, P. F. 1972. Development of metamorphic layering and cleavage in low grade metamorphic rocks at Bermagui, Australia. *American Journal of Science* 272:1–47.

14
Mineralogical and Chemical Aspects of Metamorphic Systems

Thermodynamic concepts and experimentally determined mineral stabilities guide investigations of mineralogical reactions in metamorphic systems. They allow the petrologist to elucidate the $P-T-X$ conditions of metamorphism, assuming a state of equilibrium was attained. Reactions occur wherever there is a significant change in P, T, or X in the system. Increasing P on a body causes it to contract so that phases of lesser molar volume (greater density) are more stable, whereas increasing T expands the body so that phases of greater molar volume and greater entropy are more stable. Hydrous and carbonate-bearing solids that are stable at low temperatures break down with increasing T, liberating H_2O and CO_2. These endothermic prograde reactions can be represented in a $P-T$ diagram by breakdown curves that usually have positive slope. Stability fields of minerals are not influenced by the concentration of components in the system that cannot enter into the mineral or into any phase generated by its breakdown. In contrast, thermal stability field boundaries of crystalline solid solutions can shift hundreds of degrees depending on the concentrations of substitutive components in the system. The stability field of a phase is reduced in a system where it can react with another phase.

In metamorphic systems, P_{fluid} is commonly approximately equal to the confining P, but may be less. The maximum $P-T$ stability of a volatile-bearing phase is attained where the fluid has the greatest concentration of that particular volatile species.

Fluids influence mineral kinetics as well as equilibria. Without the catalytic influence of a fluid, rates of mineral reactions are generally very sluggish, and considerable overstepping of the equilibrium conditions is required for the reaction to proceed. Persistence of metastable phases plagues interpretations of natural and experimental systems. Fortunately, however, this same phenomenon tends to assure that culminating prograde mineral assemblages are preserved for study without being erased by retrograde reactions in the fluid-free body.

The intent of this chapter is to examine the phenomena of chemical reactions between minerals and virtually ubiquitous intergranular fluids caused by changes in P, T, or X. We first examine ideal types of mineralogical reactions, searching for simple fundamental thermodynamic principles that allow us to predict their direction of change. Actual reactions in real rocks are then considered in the light of these principles. Finally, having analyzed possible equilibrium states, we briefly survey kinetic factors controlling the rates at which such equilibrium states might actually be attained in changing systems.

14.1 INTENSIVE VARIABLES AND ENERGY

Change in P, T, or X of sufficient magnitude in a metamorphic system will cause a drift toward a new state of stable equilibrium. Any new mineralogical state is reached via one or more mineralogical reactions that consume older, less stable phases in favor of new, more stable ones.

As we will first be discussing essentially isochemical reactions triggered by changes in only P and T, we will begin with a brief review of these two intensive variables. P and normally T increase with depth in the Earth, so that their gradients are positive (see Figure 10-22); however, locally in subduction zones the T gradient is believed to be negative (see Figure 9-2). The gradient in confining pressure ($\Delta P/\Delta Z$) is approximately 27 MPa/km within the crust and 33 MPa/km in the upper mantle (see box on page 25). Such pressure gradients assume hydrostatic conditions, which are approximately valid in deep crustal and upper mantle environments where the rocks have small strengths—that is, they can only support small stress differences before flowing (see pages 465–466 and box on page 467). Geothermal gradients ($\Delta T/\Delta Z$) range from as low as about 5°C/km in oceanic trenches where blueschists form to hundreds of degrees per kilometer in oceanic rift environments.

It is sometimes wondered whether the heat produced mechanically by rock deformation is significant in raising the T in metamorphic systems. The amount of heat, q, in joules, produced in a cubic meter of rock per second as the result of deformation is $q = \tau \Delta\epsilon/\Delta t$ where τ (in pascals) is the shear stress difference causing the deformation and $\Delta\epsilon/\Delta t$ is the strain rate. From page 465, a representative geologic strain rate is $3 \cdot 10^{-14}$/s, and stress differences for steady state flow are between 0.1 and 1,000 MPa, depending upon the rock and the prevailing T. These values show q is $3 \cdot 10^{-5}$ to $3 \cdot 10^{-9}$ J/m$^3 \cdot$ s. For comparison, the amount of heat generated in an average granitic rock due to its radioactivity is about $3 \cdot 10^{-6}$ J/m$^3 \cdot$ s (see Table 9-1).

These values indicate that mechanically produced thermal energy could range from an order of magnitude greater than to as little as one-thousandth that due to radioactivity. However, several unstated factors and assumptions need to be considered. Rates of heat production in rocks due to radioactivity vary over three orders of magnitude, with most rocks being colder than granites. Rates of strain are very difficult to determine in real rock bodies (Fyfe and others, 1978, p. 242) and the representative geologic rate cited above is merely based on rates of isostatic uplift and of displacement on active faults such as the San Andreas in California. Strain rates could well vary in different parts of a body; along faults or major shear zones, rates could be high, causing large q's. Stress differences and strain rates range so widely that each system must be evaluated individually regarding possible heat generation by mechanical deformation.

Most dehydration and decarbonation reactions are endothermic; they consume heat and moderate otherwise rising temperatures. Exothermic hydration and carbonation reactions converting volatile-free rocks, such as dry igneous rocks like basalt, into volatile-bearing greenschist or zeolite facies assemblages produce heat. Providing CO_2 and H_2O are available, these exothermic reactions are self-accelerating and might possibly cause the system to move rapidly through lowest metamorphic grades. However, two important prograde dehydration reactions that mark the transition from the zeolite to the blueschist facies are exothermic: laumontite → lawsonite + quartz + H_2O and analcime → jadeite + H_2O. Zen and Thompson (1974) show that temperatures can increase as much as 50°C to 100°C if all the heat in a body is conserved during exothermic reactions.

14.2 OVERVIEW OF REACTIONS

The products of mineralogical reactions as well as any remaining relics of the initial reactant minerals in a rock body are important in elucidating the nature of changes in the system—its metamorphic history. Textural vestiges of preexistent phases that reacted together provide direct indication of the reaction.

In many rock bodies, the avilability of a catalytic intergranular fluid and rapid nucleation and diffusion rates have permitted reactions to go to completion with accompanying adjustment to relatively stable grain-boundary configurations. These processes obliterate prior mineralogical states and textures and thus limit any history that can be directly read in the rock body.

Great strides have been made over the past three decades in our understanding of the metamorphic conditions responsible for the formation of specific mineral assemblages. This understanding has come from extensive laboratory studies fostered by development of experimental apparatus and techniques that simulate P–T–X conditions in metamorphic systems (Bailey, 1976) and by application of thermodynamic concepts to mineral equilibria. Because of lower temperatures in metamorphic systems as contrasted with magmatic and because of small free energy differences between many products and reactants, laboratory experiments are commonly plagued by metastability problems. Some systems, especially those involving solid–solid reactions, have proven to be notoriously intractable, and investigations by different workers are seriously discrepant.

Experimentally or theoretically determined **stability fields** of individual minerals, or of assemblages of minerals, are represented on a two-dimensional phase (stability) diagram. Most of these plots are in P–T space, but others show P–X or T–X. A **boundary line** between two stability fields shows P–T–X conditions for a state of equilibrium between phases in the two fields. Change of P, T, or X may shift a system across an equilibrium boundary, causing a mineralogical reaction to occur; hence, boundary lines represent P–T–X conditions for reaction. Experimental determination of an equilibrium field boundary by monitoring only the P–T–X conditions for a one-way reaction $A + B \rightarrow C + D$ can be significantly in error because sluggish rates of nucleation and growth (see Section 14.7) will permit the reaction to proceed only at temperatures far removed from the equilibrium conditions, if at all. Due to this metastable persistence of reactant phases, boundaries so determined are **kinetic reaction curves** rather than boundaries between states of stable equilibrium. Most modern experimental determinations of stability boundaries seek P–T conditions for (1) reversal of a reaction $A + B = C + D$ going both up and down P or T, or (2) growth of crystals of a particular phase, or phase assemblage, at some P and T in the presence of seed crystals of all possible phases in the system; unstable seed crystals are consumed.

Thermodynamic data can be used to calculate mineral stabilities (Thompson, 1955; Ernst, 1976; Wood and Fraser, 1967; Helgeson and others, 1978; Powell, 1978; Robie and others, 1978; Turner, 1981). Some simple calculations will be shown later. These depend upon molar entropies, S, and volumes, V, which generally have to be measured experimentally. According to the concepts outlined on pages 256–257, states of greater relative randomness or entropy are more stable at higher T, whereas states of relatively smaller molar volume, V, are more stable at higher P.

Mineralogical reactions in subsolidus metamorphic systems are basically of three types:

Solid–solid reactions

Solid–(fluid + solid) reactions

Oxidation-reduction reactions

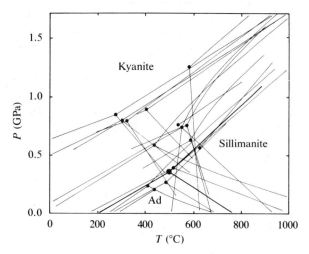

Figure 14-1   Boundary curves between the stability fields of the Al_2SiO_5 polymorphs as determined by many investigators; Ad, andalusite. Filled circles are triple points. [Light lines modified after E. Althaus, 1969, *Neues Jahrbuch für Mineralogie* 111; heavy lines from M. H. Holdaway, 1971, Stability of andalusite and the aluminum silicate phase diagram, *American Journal of Science* 271.]

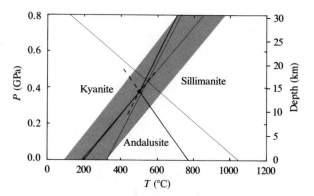

Figure 14-2 Al_2SiO_5 stability diagram. Heavy lines were experimentally determined (Holdaway, 1971), whereas light lines are those calculated in text. Shaded band encloses the calculated uncertainties for the kyanite → andalusite reaction line from Table 14-2. Dashed lines represent metastable extensions of equilibrium lines past the triple point (filled circle).

We will first discuss these reactions in closed systems where the fluid phase, if present, is pure H_2O or CO_2. Such idealized systems are not always applicable to natural rock bodies but furnish a useful starting point for discussion. Later, we will consider open systems, the important role played by dissolved constituents in fluids, and how movement of matter between the system and its surroundings affects mineralogical reactions and equilibria.

14.3 SOLID–SOLID REACTIONS

Breakdown and growth of new, more stable solids without any fluid involved directly in the reaction are considered here. (A fluid may, however, be involved in a catalytic sense, as will be discussed in Section 14.7.) Included are polymorphic transitions, unmixing of a solid solution, coupled reac-

tions between pure phases, and continuous reactions between crystalline solid solutions.

Polymorphic Transitions

Important polymorphic transitions in metamorphic systems are those involving the aluminosilicates (Al_2SiO_5) andalusite–sillimanite–kyanite and the calcium carbonates calcite–aragonite. Because of the common occurrence of the aluminosilicates in pelitic metamorphic rocks, their transitions and the way in which the boundaries of their stability fields can be calculated will be discussed to illustrate principles.

Numerous experimental determinations of Al_2SiO_5 polymorph stability fields have been published, but with very significant discrepancies (see Figure 14-1). Because of the small free energy differences between these polymorphs, factors such as grain size, strain (during grinding of the experimental mixtures), trace amounts of extraneous ions (such as Fe^{3+}), and order–disorder of Si–Al in tetrahedral sites in sillimanite impart sufficient energy to the experimental materials to shift the

Table 14-1 Thermodynamic data per mole for the aluminosilicate polymorphs at 500°C and 1 atm with estimated maximum uncertainties.

	V(cm³)	S(J/K)	H(MJ)
Andalusite	52.29 ± 0.05	245.1 ± 0.4	−2.51515 ± 0.00021
Sillimanite	50.23 ± 0.05	246.9 ± 0.4	−2.51278 ± 0.00021
Kyanite	44.69 ± 0.05	236.0 ± 0.4	−2.51931 ± 0.00021

Source: M. H. Holdaway, 1971, Stability of andalusite and the aluminum silicate phase diagram, *American Journal of Science* 271.

reaction conditions and equilibrium boundaries. Greenwood (1976, p. 217) and others have argued that the determination by Holdaway (1971) is probably the most accurate (see Figures 14-1 and 14-2).

Thermodynamic data at 500°C and 1 atm given by Holdaway (1971) and shown here in Table 14-1 can be used to approximate the stability diagram over a range of P and T (compare Wood and Fraser, 1976, p. 44). For a more accurate determination of the diagram, it would be necessary to take into account the variations in molar entropy and volume as a function of P and T. Qualitatively, we can say that kyanite, having the smallest molar volume, ought to be stable at highest pressures, whereas sillimanite ought to be stable at highest temperatures because its molar entropy is greatest. The Clapeyron equation (see page 258) can be used to approximate the slopes of boundary lines separating the stability fields of the three polymorphs. For the kyanite–andalusite boundary line, we have

$$\frac{dP}{dT} = \frac{\Delta S}{\Delta V} = \frac{(245.1 - 236.0)\ \text{J/K}}{(52.29 - 44.69)\ \text{cm}^3} = \frac{9.1\ \text{J/K}}{7.60\ \text{cm}^3}$$
$$= 1.20\ \text{MPa/K}$$

Taking the uncertainities (± values) of the data in Table 14-1 into account, this value becomes 1.20 ± 0.12 MPa/K. Slopes and uncertainties for the other two boundary lines can be similarly calculated;

Table 14-2 Slopes and T_e values at 1 atm for reaction lines delineating stability field boundaries of the aluminosilicate polymorphs calculated from the data in Table 14-1.[a]

	Slope (MPa/K)	T_e (°C)
Kyanite → andalusite	1.20 ± 0.12	184 $\begin{matrix} +95 \\ -79 \end{matrix}$
Kyanite → sillimanite	1.98 ± 0.17	326 $\begin{matrix} +89 \\ -77 \end{matrix}$
Andalusite → sillimanite	−0.87 ± 0.36	1044 $\begin{matrix} +1473 \\ -567 \end{matrix}$

[a] Note the large uncertainties, especially in the T_e values.

values are given in Table 14-2. Note that the andalusite–sillimanite boundary line has a negative slope.

To complete the construction of the stability diagram, we must determine the positions of the boundary lines whose slopes have just been calculated. From Equation (1.9), at constant P and T, we have $\Delta G = \Delta H - T\Delta S$, where the differentials dG and dS have been replaced by changes in free energy (ΔG) and entropy (ΔS) for the reaction and the heat differential is replaced by the enthalpy of the reaction, ΔH. At the equilibrium $T(= T_e)$, $\Delta G = 0$ and so $\Delta H = T_e \Delta S$. Solving for T_e for the reaction kyanite $\rightarrow$ andalusite at 1 atm using the data in Table 14-1,

$$T_e = \frac{\Delta H}{\Delta S} = \frac{(-2.51515 + 2.51931) \text{ MJ}}{(245.1 - 236.0) \text{ J/K}}$$
$$= \frac{4160}{9.1} \text{ K or } 184°\text{C}$$

This value could be as great as 279°C or as small as 105°C, taking the uncertainties in the data into account. Because the enthalpies are so similar for the polymorphs and uncertainties so large, the possible errors in the calculated T_e values are enormous (see Table 14-2). As H, V, and S are dependent upon P and T, our use of only single fixed values (at 500°C and 1 atm) introduces additional error.

The three equilibrium boundary lines in Figure 14-2 intersect at an invariant **triple point** on the P–T diagram for the Al_2SiO_5 system. Only at this unique P and T are all three polymorphs stable together. The low-pressure metastable extension of the kyanite–sillimanite boundary line (dashed) lies within the P–T field where andalusite is more stable than either of these two phases. Likewise, the dashed line extensions of kyanite–andalusite and sillimanite–andalusite show metastable equilibria. Several decades ago F. A. H. Schreinemakers devised an elegant and rigorous approach to the theoretical construction of phase diagrams; his method is recommended for the interested reader (see Zen, 1966).

If we are to believe Holdaway's stability diagram, any sequence of pelitic rocks showing a prograde metamorphic transition directly from kyanite into sillimanite must have been buried more than 14 km beneath the surface at the peak of metamorphism. Such sequences would comprise the intermediate P/T greenschist–amphibolite facies series. Conversely, other rock sequences containing andalusite must have formed at depths less than 14 km—the low P/T facies series of

Miyashiro (see pages 373–374). However, natural occurrences of two polymorphs in the same thin section are fairly common and there are even reports of all three coexisting together. Such rocks might possibly have been subject to a P and T coincident with the equilibrium boundary or the triple point, respectively. However, it is more likely that one or more of the polymorphs persisted metastably. It has been suggested that very small concentrations of Fe^{+3} dissolved in the polymorphs could make the equilibria between them divariant, rather than univariant (see below), so two could coexist stably; but this effect is probably not significant in this particular case (Holdaway, 1971).

The P–T stability boundaries of pure crystalline nonvolatile phases, whose compositions are not varied by solid solution, are not modified in any respect by compositional variations in the overall system. The occurrence in a rock of a pure phase with a known and limited P–T stability field, such as the Al_2SiO_5 and $CaCO_3$ polymorphs (Newton and Fyfe, 1976) can in principle, therefore, provide important insight into the conditions of metamorphism. This will be better appreciated as we consider additional mineralogical equilibria and learn how they can be severely influenced by the X of the system. Unfortunately, however, this potential usefulness of polymorphs is marred by unfavorable kinetics and metastability problems (see Section 14.7).

Unmixing and Coupled Solid–Solid Reactions

Some high-grade mafic metamorphic rocks of the granulite facies show evidence of changing mineral stabilities in the form of coronas and exsolution textures. Because of their dry composition and the consequent restricted sphere of effective diffusion, and possibly because of limited time, recrystallization may not completely obliterate primary coarse grains.

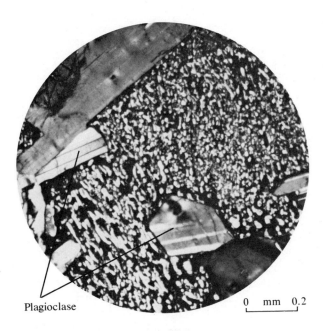

Plagioclase

0 mm 0.2

Figure 14-3 Exsolved blebs of augite (light areas) in an
orthopyroxene host (black) in gabbro viewed under
cross-polarized light. The primary orthopyroxene
precipitated from the magma contained appreciable
calcic pyroxene in solid solution, but as it cooled below
the solvus augite exsolved, forming the intergrowth. In
some instances, the exsolved phase forms regular planar
lamellae in the host grain.

Exsolution in pyroxenes. High-T magmatic pyroxenes in mafic and ultramafic rocks upon adjustment to lower subsolidus temperatures unmix
or **exsolve,** forming an intergrowth of two
pyroxenes within a primary grain (see Figure
14-3). The texture is similar to intergrowths of
K- and Na-rich feldspars in perthite and the
process of formation is the same in principle
(see page 280). With decreasing T, the extent
of mutual solid solubility of Ca-rich pyroxene
and Ca-poor pyroxene diminishes as the limbs
of the solvus expand outward toward pure
$Ca(\overline{Mg},Fe)Si_2O_6$ and $(Mg,Fe)_2Si_2O_6$ end members.
(The actual phase relations are complicated due to

(1) extensive solid solution of Al, Ti, Mn, and so
on, (2) incongruent melting relations, and (3)
polymorphism in the Ca-poor pyroxenes.)

High-T Ca-bearing hypersthene exsolves to a
hypersthene with less Ca plus a separate Ca-rich
pyroxene at lower T. Where favorable kinetic factors prevail, the two stable low-T pyroxenes may
form as distinct separate grains in a texturally stable crystalloblastic array forming two-pyroxene
gneisses and granofelses.

Coupled solid–solid reactions. Arrested incomplete solid–solid reactions in mafic rocks are well
displayed in various types of **coronas** where more
stable phases surround the grain of a phase unstable with respect to its neighboring minerals. In
Figure 14-4, an olivine grain originally in contact
with plagioclase has been replaced by hypersthene
and the excess Mg and Fe from this replacement
has diffused into the plagioclase to form spinel. Ca
from decomposed plagioclase must have diffused
in the opposite direction toward the olivine, where
diopside formed. Plagioclase became more sodic
during growth of the corona by uncertain and
probably complex reactions.

To a first approximation, corona development
in this metamorphosed gabbro can be modeled
by simple reactions between pure end-member
phases, ignoring the albite and Fe components.
For the replacement of olivine by orthopyroxene,

$$2Mg_2SiO_4 = 2MgSiO_3 + 2MgO \quad (14.1)$$
forsterite   enstatite   periclase

and for the formation of the rim of diopside +
spinel surrounding the orthopyroxene,

$$CaAl_2Si_2O_8 + 2MgO =$$
anorthite   periclase
$$CaMgSi_2O_6 + MgAl_2O_4 \quad (14.2)$$
diopside   spinel

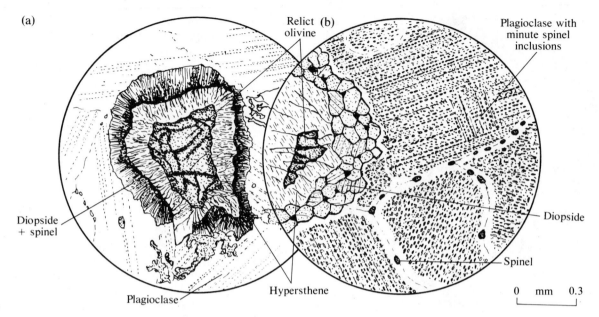

Figure 14-4 Coronas in metagabbros from Norway and New York consisting of hypersthene + diopside + spinel developed around an olivine grain originally in contact with calcic plagioclase. (a) Inner corona of fibrous hypersthene replacing original olivine is surrounded by fine intergrowth of diopside + spinel replacing plagioclase. (b) Texturally more stable corona, due to more advanced diffusion, of hypersthene and granoblastic diopside grains. Although a few small spinels occur with diopside, most of them cloud plagioclase grains and have grown preferentially along twin planes and grain boundaries. [From S. R. Nockolds, R. W. O'B. Knox, and G. A. Chinner, 1978, *Petrology for Students* (London: Cambridge University Press).]

Combining these two reactions, we obtain

$$2Mg_2SiO_4 + CaAl_2Si_2O_8 =$$
forsterite   anorthite

$$\qquad 2MgSiO_3 + CaMgSi_2O_6 + MgAl_2O_4 \quad (14.3)$$
$$\qquad \text{enstatite} \qquad \text{diopside} \qquad \text{spinel}$$

Molar entropies and volumes for these phases at 25°C and 1 atm are listed in Robie and others (1978) as follows:

$$S_{(2\,Fo+\,An)} = 389.7 \text{ J/K}$$
$$S_{(2\,En+Di+Sp)} = 359.4 \text{ J/K}$$
$$V_{(2\,Fo+\,An)} = 188.4 \text{ cm}^3$$
$$V_{(2\,En+Di+Sp)} = 168.7 \text{ cm}^3$$

If the relative values of these quantities do not change at high P and T, they would indicate that

the assemblage $2Fo + An$ ought to be stable at higher temperatures because of its greater entropy and at lower pressures because of its greater volume. The reaction boundary line with the assemblage $2En + Di + Sp$ has a positive slope on a $P–T$ diagram because $\Delta S/\Delta V$ is positive.

Anorthite has a relatively large molar volume, so with increasing P it reacts with MgO from breakdown of olivine into the denser assemblage diopside + spinel according to Equation (14.2). At still higher P, garnet is a denser repository for Ca and Al than spinel; garnet can form by the reaction

$$CaAl_2Si_2O_8 + 2(Mg,Fe)SiO_3 =$$
anorthite   hypersthene

$$\qquad Ca(Mg,Fe)_2Al_2Si_3O_{12} + SiO_2 \quad (14.4)$$
$$\qquad \text{garnet} \qquad\qquad \text{quartz}$$

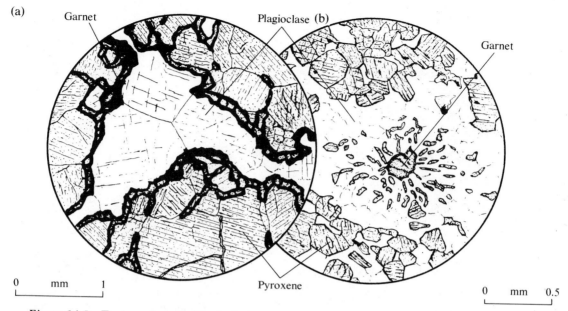

Figure 14-5 Textures in granulite-facies rocks from Australia and Germany modeled by the reaction anorthite + hypersthene = garnet + quartz. (a) Well-developed crystalloblastic fabric has been modified by growth of rims of garnet along grain boundaries between pyroxene and plagioclase. (b) The same reaction in reverse has produced a somewhat vermicular intergrowth of pyroxene + plagioclase from garnet, only a small remnant of which remains. [From S. R. Nockolds, R. W. O'B. Knox, and G. A. Chinner, 1978, *Petrology for Students* (London: Cambridge University Press).]

Figure 14-5 shows textures resulting from the forward and backward progress of this reaction.

Albite is stable to higher pressures than anorthite, but ultimately this phase also becomes unstable and decomposes to form the aluminous pyroxene, jadeite, $NaAlSi_2O_6$, according to

$$\underset{\text{albite}}{NaAlSi_3O_8} = \underset{\text{jadeite}}{NaAlSi_2O_6} + \underset{\text{quartz}}{SiO_2} \quad (14.5)$$

The reactions shown in Equations (14-3), (14-4), and (14-5) serve as models bounding the P–T regions where basaltic, granulitic, and eclogitic mineral assemblages are stable, shown in Figure 14-6. This figure is highly schematic and no values are placed on the P and T coordinate axes because of the chemical complexity of natural systems.

In summary, relatively large molar volume plagioclases stable in low-P basaltic–gabbroic rocks react at increasing P with pyroxenes and olivines to form lower molar volume aluminous pyroxenes, spinels, garnets, and quartz. Olivine disappears (by reaction with quartz) through this granulite facies transition into the high-pressure eclogite facies characterized chiefly by the bimineralic assemblage of aluminous pyroxene and (Mg,Fe) garnet.

Two fundamental concepts pertaining to mineralogical reactions in general are implicit in the foregoing discussions but should now be identified and emphasized. These have to do with coupled reactions and continuous reactions.

It is well known that olivine is stable to quite high P and T in the mantle, yet in the presence of plagioclase in mafic rocks these two phases react and are consumed in the deep crust. Albite in the

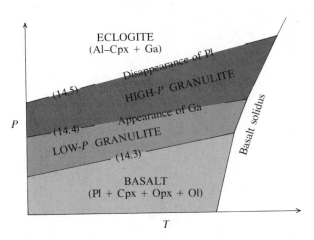

Figure 14-6 Schematic stability fields of basaltic, granulitic, and eclogitic mineral assemblages separated by reactions shown in Equations (14.3), (14.4), and (14.5). Boundary line slopes are schematic and assumed to be linear. Eclogite and granulite facies assemblages are stable in either the crust or upper mantle, depending upon temperature. Cpx, clinopyroxene including diopside; Opx, orthopyroxene including hypersthene; Pl, plagioclase; Ol, olivine; Ga, garnet.

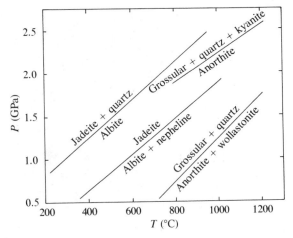

Figure 14-7 High-pressure breakdown reaction lines for albite and anorthite. [Experimental data for albite from R. C. Newton and W. S. Fyfe, 1976, High pressure metamorphism, in D. K. Bailey and R. Macdonald (eds.), *The Evolution of the Crystalline Rocks* (New York: Academic Press); data for anorthite from J. F. Hays, 1967, Lime-alumina-silica, *Annual Report of the Director, Carnegie Institution of Washington, Geophysical Laboratory, Yearbook* 65.]

absence of nepheline breaks down at significantly higher P than does albite in the presence of nepheline (see Figure 14-7). With respect to increasing P, the stability fields of olivine, anorthite, and albite are all reduced in the presence of other phases that can react with it.

In such **coupled reactions,** the stability field of a solid phase is reduced in the presence of another phase containing components that can react with it. The stability field of an isolated solid phase encompasses all other subfields representing stability limits of assemblages in which the phase is a member.

Continuous Reactions Between Crystalline Solid Solutions

The foregoing discussion pertains to pure phases whose chemical compositions are fixed and do not vary in spite of any compositional variations in the overall system. However, most rock-forming minerals are crystalline solid solutions whose compositions are a function of the composition, as well as P and T, of the system. In most rocks, only quartz is a pure phase (although it, too, may contain trace amounts of Ti, Al, and Fe^{3+} substituting for Si). The following examples will show that crystalline solid solutions have stability limits depending upon their composition; multivariant **continuous reactions** involving them shift over a range of P and T as X varies.

Aluminum in pyroxene solid solutions. In coupled solid–solid reactions at high pressure involving pyroxenes and an aluminous phase (garnet, spinel, plagioclase), Al dissolves in pyroxenes as the jadeite ($NaAlSi_2O_6$) and Ca-Tschermak ($CaAl(Al-Si)O_6$) end members. In jadeite, all of the Al is in the space-conserving octahedral coordination; this end member forms an extensive solid solution with diopside ($CaMgSi_2O_6$) in the pyroxene called omphacite, which is one of the two principal constituents of eclogite. In the Ca-Tschermak end member, which does not exist as a separate mineral, one Al is tetrahedrally coordinated and the other is octahedrally coordinated.

Thus, in the reaction between plagioclase + olivine previously modeled by Equation (14.3), the products might involve solid solution of the Ca-Tschermak molecule in pyroxenes. Considering only the solubility in clinopyroxene, the relevant reaction is

Mg_2SiO_4 + $CaAl_2Si_2O_8$ =
forsterite   anorthite

      $Mg_2Si_2O_6$ + $CaAl(AlSi)O_6$ (14.6)
      enstatite   in clinopyroxene
              solid solution

In the simplified system $CaO–MgO–Al_2O_3–SiO_2$, Herzberg (1978) has determined how P and T govern the mole fraction of $CaAl(AlSi)O_6$ in clinopyroxene and the partition of the $Mg_2Si_2O_6$ end member between clinopyroxene and orthopyroxene for various mafic and ultramafic mineral assemblages. Using this grid (see Figure 14-8), the equilibrium P and T of their subsolidus crystallization can be determined from the chemical composition of the two coexisting pyroxenes. The pyroxene pair hence serve as a **geobarometer** and a **geothermometer**. As FeO is an important component in many pyroxenes, corrections would have to be made for its presence, and possibly for minor amounts of Ti, Mn, Na, and so on as well.

Mg–Fe solid solution in cordierite and garnet. The most common solid solution in rock-forming minerals involves substitution between Fe^{2+} and Mg. An example of this is found in some high-grade pelitic rocks that show evidence for the reaction

$3Al_3(Mg,Fe)_2AlSi_5O_{18}$ = $2(Mg,Fe)_3Al_2Si_3O_{12}$
    cordierite   garnet
          + $4Al_2SiO_5$ + $5SiO_2$ (14.7)
          sillimanite   quartz

This divariant ($F = 2 + 4 - 4 = 2$) equilibrium extends over a range of P and T so that on a $P–T$ diagram (see Figure 14-9a) it is represented by a band, instead of a univariant line. Reaction is continuous through the band, rather than discontinuous at a unique P and T. In naturally coexistent cordierite and garnet, the former has a higher Mg/Mg + Fe ratio. This contrasting ratio can be shown in a $P–X$ diagram such as Figure 14-9b, which is topologically identical to the $T–X$ phase diagram for melting reactions in plagioclase and olivine solid solutions. These solid solutions melt incongruently, and crystallization (or melting) involves continuous reaction relations between melt and crystals as T or P changes (see pages 276 and 285). By analogy, we may refer to the **incongruent subsolidus breakdown** of cordierite with increasing P or T. In the divariant band region, there are continuous reactions gradually consuming cordierite and forming garnet, sillimanite, and quartz. Each reaction consumes some cordierite, leaving a little less that is more Mg-rich than originally, plus the additional garnet, sillimanite, and quartz.

Summary of Solid–Solid Reactions

Investigations of solid–solid reactions are guided by four fundamental concepts:

1. High temperatures stabilize high entropy phases whereas high pressures favor dense, low molar volume phases. Phases of high molar volume commonly—but not always—have higher entropy than chemically equivalent phases so that the reaction line between them on a $P–T$ diagram has a positive slope.

2. The $P–T$ transition curve between two pure polymorphs not accepting chemical components from their surroundings into solid solution is unaffected by the chemical composition of the system. Hence, pure polymorphs of $CaCO_3$ or Al_2SiO_5, if truly equilibrated, can serve as important indicators of P and T at the time of last equilibration.

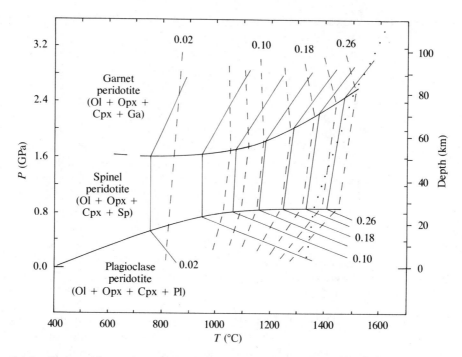

Figure 14-8 Compositions of crystalline solid solutions in the peridotitic part of the system $CaO-MgO-Al_2O_3-SiO_2$. Mole fraction of the Ca-Tschermak component ($X_{CaAl(AlSi)O_6}$) in clinopyroxenes is indicated by light solid lines; light dashed lines show values of the fraction of Mg in M_2 sites in clinopyroxene divided by the fraction in orthopyroxene. Note that with increasing P in the plagioclase peridotite field, clinopyroxenes become more aluminous (due to progressive plagioclase breakdown); through the spinel peridotite field, they remain essentially constant in Al content and then decrease in Al as garnet progressively forms in the garnet peridotite field. Dotted line is approximate location of peridotite solidus. Depths based on an average density of 3.2 g/cm³. Compare Figures 6-11 and 9-3b. [After C. T. Herzberg, 1978, Pyroxene geothermometry and geobarometry: Experimental and thermodynamic evaluation of some subsolidus phase relations involving pyroxenes in the system $CaO-MgO-Al_2O_3-SiO_2$, *Geochimica et Cosmochimica Acta* 42.]

3. In coupled reactions, the stability field of a phase is reduced in a system where it can react with another phase. A phase assemblage can only occur within the stability limits of the individual constituent phases. This concept applies to pure phases as well as solid solutions.

4. Crystalline solid solutions break down incongruently through continuous reactions over a range of P and T.

As we proceed in the next section to solid–(fluid + solid) reactions, it will be found that these same concepts still apply.

14.4 SOLID–(FLUID + SOLID) REACTIONS

Prograde reactions liberating fluids (H_2O and CO_2) and retrograde reactions consuming them are the most common reactions in metamorphic systems. Prograde metamorphism, essentially due to rising

(a) (b)

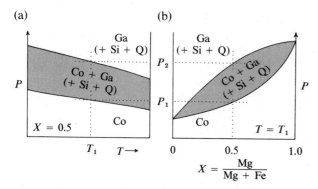

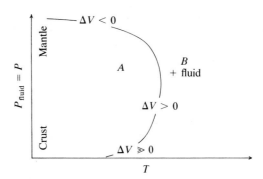

Figure 14-9 Schematic phase diagrams related to the divariant reaction 3 cordierite = 2 garnet + 4 sillimanite + 5 quartz. $X = Mg/(Mg + Fe)$. Co, cordierite; Ga, garnet; Si, sillimanite; Q, quartz. (a) Schematic P–T diagram showing divariant reaction band (shaded) in which cordierite with an equal proportion of Fe and Mg atoms (that is, $X = 0.5$) progressively breaks down to garnet with increasing P or T. (b) Schematic P–X diagram drawn at $T = T_1$, again showing divariant reaction band. Note that pure Mg- or Fe-cordierite reacts at a unique P, rather than continuously over a range of P as do intermediate solid solutions. [After B. J. Hensen, 1971, Theoretical phase relations involving cordierite and garnet in the system MgO-FeO-Al$_2$O$_3$-SiO$_2$, *Contributions to Mineralogy and Petrology* 33.]

Figure 14-10 Idealized univariant reaction curve for Equation (14.8).

T, involves a complex sequence of mostly continuous reactions progressively liberating H$_2$O and CO$_2$ from the assemblage of crystalline solid solutions.

Two prograde sequences of progressively less hydrous silicates, not strictly chemically equivalent, and involving various amounts of other reactants, are

clay minerals $\rightarrow$ chlorites $\rightarrow$ micas $\rightarrow$
amphiboles $\rightarrow$ anhydrous silicates

and

hydrated Ca–Al silicates $\rightarrow$ plagioclases

Prograde reactions liberate CO$_2$ according to

carbonates + quartz $\rightarrow$ Ca–Mg silicates

In typical solid–(fluid + solid) reactions of the form

$$E_t + A = B + \text{fluid} \qquad (14.8)$$

where E_t is thermal energy and A and B are solid phases or phase assemblages, there are large changes in entropy, volume, and hence free energy due to the liberation of the fluid phase.

Stability Relations

At low pressures and high T, CO$_2$ and H$_2$O are not very dense so ΔV for the reaction shown in Equation (14.8) is relatively large and positive and the slope of the reaction line, $dP/dT = \Delta S/\Delta V$, is small. At higher pressures nearer to continental geotherms, volatiles are more compressed so ΔV's are smaller and reaction line slopes are steeper. Ultimately, high pressures may sufficiently compress the fluid so that ΔV for a dehydration or decarbonation reaction is negative; as ΔS is still positive, the reaction curve bends back on itself with negative slope (see Figure 14-10). For the reaction

SiO$_2$ + NaAlSi$_2$O$_6$ · H$_2$O =
quartz   analcime
NaAlSi$_3$O$_8$ + H$_2$O (14.9)
albite

(a)

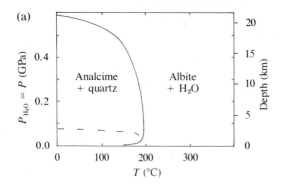

(b)

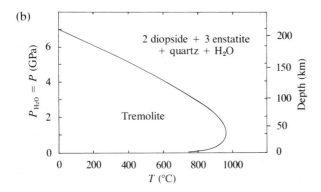

Figure 14-11 Univariant dehydration reaction curves for two extreme cases assuming $P = P_{H_2O}$. (a) Stability of the hydrous phase analcime is limited to crustal pressures. The dashed curve is the reaction curve for conditions where $P_{H_2O} = P/3$. (b) Stability of the hydrous phase tremolite extends into the mantle. [Part (a) after R. C. Newton and W. S. Fyfe, 1976, High pressure metamorphism, in D. K. Bailey and R. Macdonald (eds.), *The Evolution of the Crystalline Rocks* (New York: Academic Press); part (b) after W. S. Fyfe, N. J. Price, and A. B. Thompson, 1978, *Fluids in the Earth's Crust* (New York: Elsevier).]

this bendback occurs under crustal confining pressures, limiting the stability field of quartz + analcime to $P < 0.6\,\text{GPa}$ (see Figure 14-11a). Similar bendbacks under essentially crustal conditions occur in other high-P, low-T reactions including laumontite $\rightarrow$ lawsonite + quartz + H_2O and prehnite $\rightarrow$ grossular + zoisite + quartz + H_2O.

For most other fluid-release reactions, the bend-back of the reaction curve to negative slopes occurs under mantle conditions (see Figure 14-11b).

As with solid–solid coupled reactions, the presence of an additional, potentially reactive phase can shrink the stability field of a particular phase or phase assemblage. Pure calcite breaks down according to the reaction

$$CaCO_3 = CaO + CO_2 \qquad (14.10)$$
$$\text{calcite} \quad \text{lime}$$

As this reaction occurs at temperatures of over 1200°C, calcite alone, as in pure marbles, is stable throughout the realm of metamorphism and lime does not occur naturally. Because calcite is commonly coexistent with quartz, a coupled reaction can occur:

$$CaCO_3 + SiO_2 = CaSiO_3 + CO_2 \qquad (14.11)$$
$$\text{calcite} \quad \text{quartz} \quad \text{wollastonite}$$

The free energy change at, say, 600°C and 1 atm for this reaction is -54 kJ/mol in contrast to $+33$ kJ/mol for breakdown of pure calcite (Robie and others, 1978). Hence wollastonite + CO_2 is stable at 600°C whereas CaO + CO_2 is not. The coupled reaction involving quartz has thus substantially lowered the temperature of the breakdown of calcite; in the presence of another phase, such as rutile, in addition to quartz, calcite breaks down at still lower T (see Figure 14-12).

Another example of the lowering of a breakdown curve by reaction with another phase is afforded by comparison of the reaction curves for muscovite alone and for muscovite + quartz (see Figure 14-13). Most rocks containing muscovite also have quartz. Because of limited mutual solubility between muscovite and its Na-analog, paragonite, $NaAl_2AlSi_3O_{10}(OH)_2$, breakdown of these white micas is continuous over only a small range of P and T (see Figure 14-13). Incongruent breakdown of paragonite containing a little dissolved muscovite, for example, would initially yield as products a more K-enriched mica + Na-

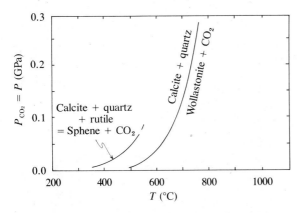

Figure 14-12 Univariant breakdown reaction curves for calcite-bearing assemblages where $P = P_{CO_2}$. [After H. J. Greenwood, 1976, Metamorphism at moderate temperatures and pressures, in D. K. Bailey and R. Macdonald (eds.), *The Evolution of the Crystalline Rocks* (New York: Academic Press).]

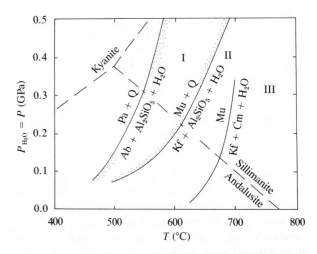

Figure 14-13 Breakdown conditions for muscovite and paragonite. The stippled divariant bands indicate conditions for incongruent decomposition of K-bearing paragonite (on the left) and Na-bearing muscovite. Dashed lines show stability limits of Al_2SiO_5 polymorphs from Figure 14-1. Pa, paragonite; Q, quartz; Ab, albite; Mu, muscovite; Kf, K-feldspar; Cm, corundum. Roman numerals refer to Figure 14-21. [After H. J. Greenwood, 1976, Metamorphism at moderate temperatures and pressures, in D. K. Bailey and R. Macdonald (eds.), *The Evolution of the Crystalline Rocks* (New York: Academic Press).]

rich feldspar + Al_2SiO_5 + H_2O. With increasing *T*, the residual mica completely decomposes to an alkali feldspar with the same Na/K ratio as the original mica. Fyfe and others (1978, p. 132) discuss situations in which incomplete solid solution exists in the hydrous and anhydrous phases (compare Figure 14-9).

Nature and Composition of Metamorphic Fluids

Although metamorphic systems are described as subsolidus or solid state, virtually all contain a fluid phase. What is the chemical composition of this phase and what is its state—liquid or gas?

The widespread presence of hydrous minerals in metamorphic rocks suggests that water is generally the dominant fluid in rock bodies. Carbonate-bearing rocks, however, may have significant concentrations of CO_2, and small amounts of sulfide minerals in some rocks indicate the presence of S. Metamorphic minerals commonly contain minute fluid inclusions, $<10^{-2}$ mm diameter, that provide important insight into the composition of the fluid present when the mineral grain formed (Fyfe and

others, 1978, p. 35; Greenwood, 1976, p. 198). In addition to a fluid, such inclusions also may contain a bubble of gas and one or more crystalline phases apparently formed as the fluid cooled from high temperatures. Heating the host mineral homogenizes the contents of the inclusion at some *T*, which can sometimes be taken as the *T* of formation of the host grain enclosing the inclusion, after appropriate correction for the pressure believed to have prevailed when the inclusion was trapped. Precipitated solids in inclusions are usually halite, but sylvite, calcite, and various fluorides have also been found. The fluid in these inclusions generally consists of H_2O with variable amounts of dissolved CO_2, Na^+, K^+, Ca^{2+}, Cl^-, SO_4^{2-}, and CO_3^{2-}.

The fluid phase in metamorphic systems has

Table 14-3   Critical points of H_2O and CO_2.

	P(MPa)	T(°C)
H_2O	21.8	371
CO_2	7.3	31

Source: From P. W. Atkins, 1978, *Physical Chemistry* (San Francisco: W. H. Freeman and Company). Copyright © 1978.

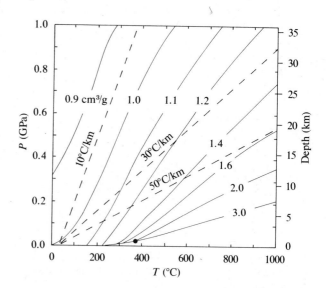

Figure 14-14   The specific volume of water in cm³/g in relation to geothermal gradients in metamorphic systems. Filled circle is the critical point. Compare with Figure 10-28a. [After W. S. Fyfe, N. J. Price, and A. B. Thompson, 1978, *Fluids in the Earth's Crust* (New York: Elsevier).]

been referred to as a vapor, gas, liquid, fluid, and supercritical fluid. Above the critical point, there is no sharp distinction, in the form of a meniscus, between gas (vapor) and liquid. Because of the modest P and T of critical points of the dominant fluid components in most metamorphic systems (see Table 14-3), the density, or specific volume, of the fluid in most real systems varies continuously as a function of T and P. Fyfe and others (1978, p. 21, 138) have pointed out that for most crustal regional metamorphic systems, increasing P with depth just about compensates for increasing T so that the density of water remains close to its familiar value at atmospheric conditions of 1 g/cm³ (see Figure 14-14). Carbon dioxide in such systems similarly has densities near unity. In view of these considerations, it is justified to refer to any free volatile phase in metamorphic systems as simply a **fluid.**

Fluid Release and Fluid Pressure

Simultaneously increasing T and P during burial of marine rock sequences along continental margins and in subduction zones liberates enormous volumes of aqueous fluid by mechanical compaction and by prograde chemical reactions. During compaction of sediments, most of the water initially present is forced out. Progressively less remains in increasingly more isolated pore spaces and as possible grain boundary films. A few percent bound structurally in clay minerals and zeolites is driven out of the rock with rising T during a complex series of prograde dehydration reactions ensuing over virtually the whole range of metamorphic conditions. As every phase in the sequence

clay minerals → chlorites → micas →
    amphiboles (+feldspars) → anhydrous silicates

is a solid solution, there must be more or less continuous breakdown of these hydrates with increasing depth of burial and T, liberating water all the time (see Figure 14-15).

In rock bodies with interconnecting pore spaces and cracks extending all the way to the surface, such as geothermal systems venting water to the surface in springs, the liberated water will flow out of the body; in any particular pore space $P_{H_2O} \sim P/3$ because rock is roughly three times denser than water. In effect, the interconnected pore spaces constitute a free-standing column of water extending all the way to the surface within the rock body. But as metamorphic grade increases, pore volume diminishes as a consequence of grain boundary adjustments during pressure solution and recrystallization; porosity becomes nil. Prograde dehydration reactions liberating water may locally cause P_{H_2O} to exceed P, but this condition may not persist for very long if H_2O eventually migrates out of the body along grain boundaries; or, if P_{H_2O} is sufficiently large relative to P, grains can be forced apart so that the rock ruptures. In either case, the excess water pressure is relieved and P_{H_2O} approaches the value of P ($P_{H_2O} = P$). All of these arguments also apply to P_{CO_2} in carbonate rocks, and to mixtures of H_2O and CO_2 in rocks containing both silicate and carbonate minerals.

An argument for the approximate equality $P_{fluid} \sim P$ lies in the worldwide consistency of mineral assemblages expressed in the metamorphic facies concept; such consistency and recurrence of assemblages would not be anticipated if in every system P_{fluid} was different (but less than P). This consistency implies a generally close correlation between T, P, and P_{fluid}, all three being mostly depth-dependent.

However, in some metamorphic systems, $P_{fluid} \ll P$. For example, a very large volume of dry igneous rock (such as gabbro) subjected to metamorphism might remain essentially fluid-free if pore spaces and fractures are unavailable to allow penetration of water from the distant surroundings.

Influence of Fluid Composition on Mineral Stability

Equilibria in most metamorphic systems are influenced by multicomponent fluids containing H_2O + CO_2 as well as other constituents (summary and references in Greenwood, 1976; Ernst, 1976, p. 229; Fyfe and others, 1978, chap. 7). Stability limits of carbonate phases are reduced where the fluid contains H_2O ($P_{CO_2} < P$), and hydrous silicates have smaller stability fields in the presence of CO_2 ($P_{H_2O} < P$). These relations are illustrated in Figure 14-16. In general, where the partial pressure of a volatile species is less than P, the stability field of phases bearing that volatile shrinks. The purpose of this section is to consider metamorphic reactions in systems in which $P_{H_2O} < P$ and $P_{CO_2} < P$, whereas we have heretofore assumed the particular fluid pressure to be equal to P (see Figures 14-10 to 14-13).

Such inequalities are possible in two situations. In the first, H_2O or CO_2 is the only volatile species present and its pressure is less than the confining pressure; this situation could occur where there is only a limited supply of the one volatile, or if the fluid somehow escapes from the vicinity of the reaction faster than the rate of reaction, or if the fluid is diluted by some inert constituent (such as argon) that does not react with any of the solid phases in the system. In the second situation, the fluid is a mixture of H_2O, CO_2, and possibly other volatile species in rocks containing both hydrous and carbonate minerals; symbolically, $P_{CO_2} + P_{H_2O} \sim P_{fluid}$. Conveniently, it turns out that both of these two situations ($P_{fluid} < P$ and $P_{CO_2} + P_{H_2O} \sim P_{fluid}$) influence mineral stability in a similar manner.

The stability field of analcime + quartz is greatly reduced in systems where $P_{H_2O} < P$ [see Figure 14-11a and Equation (14.9)]. This fact is of great importance in the development of the assemblage analcime + quartz in the zeolite facies. To illustrate how this comes about, consider a point at some T and $P = P_{H_2O} = P_{fluid}$ on the reaction curve (solid line in Figure 14-11a) where ΔG for the reaction is zero. Now, if the pressure on only the solids is increased, holding T and P_{H_2O} constant, then the free energy change for the reaction will become $\Delta G = \Delta V_{solids} (P - P_{H_2O})$, from Equation (8.3). For this particular reaction $\Delta V_{solids} \sim -13$ cm³/mol at 1 atm. Hence $\Delta G < 0$ and albite + H_2O is stable. The reaction boundary where $\Delta G = 0$ must lie at some lower increment of T ($-\Delta T$), an increment that would increase ΔG by an amount $\Delta S \Delta T$, from Equation 8.2, where $\Delta S = (S_{Ab} + S_{H_2O}) - (S_{Am} + S_Q)$. Because, for this reaction, ΔS is relatively small positive and ΔV_{solids} large, there is a large ΔT required to reach the reaction curve. Hence, the stability field of analcime + quartz shrinks drastically for $P_{H_2O} < P$ (see Figure 14-11a). For other reactions in which

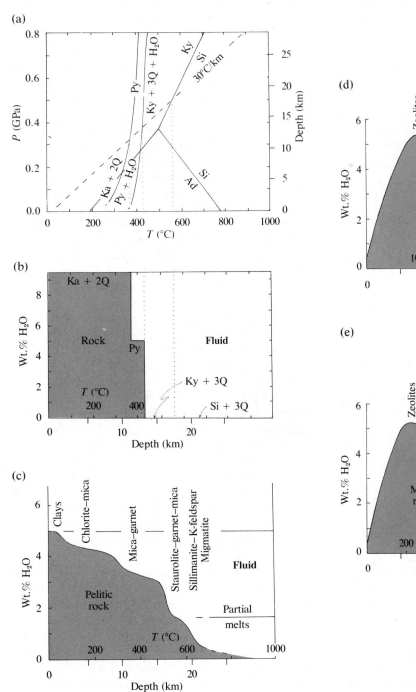

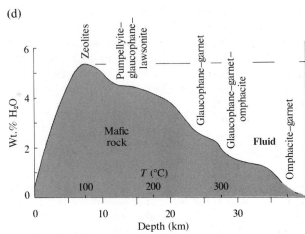

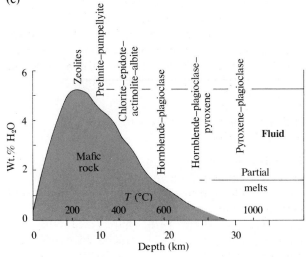

Figure 14-15 (*opposite*) Fluid-release curves for volatile-bearing assemblages subjected to prograde reactions. (a) Stability fields and breakdown curves for some phases in the system Al_2O_3–SiO_2–H_2O. Ka, kaolinite; Q, quartz; Py, pyrophyllite; Ky, kyanite; Si, sillimanite; Ad, andalusite. (b) Fluid-release curve from the assemblage kaolinite + 2 quartz along a 30°C/km geotherm. This assemblage contains ~9.5 wt.% water bound structurally in kaolinite. When the temperature along the geotherm reaches 370°C, corresponding to a depth of about 12 km, the reaction Ka + 2Q → Py + $2H_2O$ occurs, releasing ~4.5 wt.% free water into the system; 5 wt.% remains structurally bound in pyrophyllite. When the temperature of the rock system, now composed of pyrophyllite as the only solid phase, reaches 425°C, this phase decomposes into kyanite + 3 quartz + H_2O. All of the water in the rock has now been released and it consists of a wholly anhydrous crystalline assemblage, kyanite + quartz. At a little over 550°C, or about 18 km (note vertical dotted line), the kyanite undergoes a polymorphic transition to sillimanite. Because kaolinite and pyrophyllite are here considered to be pure phases with no solid solution, release of water occurs sharply at their univariant dehydration curves. (c) Generalized fluid-release curve for a shale along a 30°C/km geotherm. Release is continuous with increasing P and T because of the complex crystalline solid solutions comprising pelitic rocks that progressively break down. Changes in slope of the release curve are related to beginning and completion of dehydration of major hydrous silicates. Depending upon the exact composition of the system, especially the water concentration, partial melting may begin somewhere near 700°C to form migmatite. (d) Generalized fluid-release curve for mafic rock along a 10°C/km geotherm, assuming rock is initially dry and then hydrates to a zeolite facies assemblage with ~5 wt.% water. (On the sea floor palagonitization could produce a more hydrous rock at lower T.) (e) Same as (d) but along a 30°C/km geotherm. Note temperatures for partial melting and migmatite formation. [After W. S. Fyfe, N. J. Price, and A. B. Thompson, 1978, *Fluids in the Earth's Crust* (New York: Elsevier).]

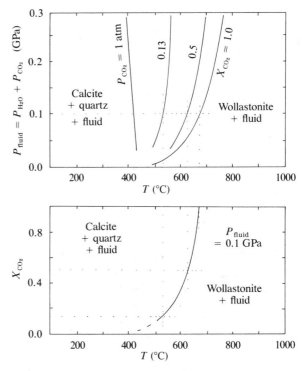

Figure 14-16 Univariant decarbonation curves as a function of composition of the fluid where $P = P_{fluid} = P_{HO} + P_{CO_2}$. Compare Figure 14-12. Fluid composition is denoted in terms of the mole fraction of CO_2, that is, X_{CO_2}. Dotted lines show correspondence between specific values of P_{fluid}–X_{CO_2}–T in the two diagrams. [From experimental work of H. J. Greenwood.]

ΔV_{solids} is smaller, the depression of the reaction curve is less.

From an alternate standpoint (see box on page 504), reducing the value of the activity or fugacity of the fluid less than unity sympathetically reduces the free energy or chemical potential of the assemblage of solids + fluid (relative to the volatile-bearing solids), making solids + fluid more stable at lower P and T.

The CO_2/H_2O ratio in pore fluids varies in space and time during the course of metamorphism, depending upon the mobility or amount of the fluid, the nature of reactions, and especially the assemblage of buffering solid phases. This concept can be illustrated by the phase relations in the system MgO–CO_2–H_2O (see Figure 14-17). Consider a pure magnesite rock in equilibrium with fluid in which $X_{CO_2} = 0.03$. With increasing T, magnesite begins to break down at 557°C with liberation of CO_2. Two different end-member processes can now be envisaged. If there is only a small amount

Activity, Fugacity, and Chemical Potential

As explained in the box on page 259, for systems in which any phase is not pure and can vary in composition, the Gibbs free energy function for the phase must incorporate a chemical potential energy term. At constant P and T, the chemical potential of some component i in an **ideal** liquid or solid **solution** depends only upon the mole fraction of i in the phase X_i. For an ideal or perfect gas (where for one mole the value of $P_{gas} V/RT$ is unity), the chemical potential of a component depends upon its equilibrium partial vapor pressure, which is an alternate expression for its mole fraction in the system. Although such ideal solutions and perfect gases are rare in nature, in many geologic systems phases may be so treated without serious error.

A more rigorous approach for **real solutions** and real gases, in which significant interatomic forces occur, is to define an activity and a fugacity (Wood and Fraser, 1976; Ernst, 1976). The **fugacity**, f_i, of some gas component i divided by the partial pressure of i, P_i, is the fugacity coefficient, γ_i, or $\gamma_i = f_i/P_i$. For ideal gases $\gamma_i = 1$ and $f_i = P_i$; for greater departures from the ideal behavior γ_i decreases to less than one. From Equation (8.3), at constant T, $dG = VdP$ and so for an ideal gas, $dG = (RT/P)dP = RTd(lnP)$ and for a real gas, $dG = RTd(lnf)$. For a gaseous component in a real solution, $d\mu_i = RTd(lnf_i)$.

In completely analogous fashion, we can define an **activity**, a_i, for components in real fluid and crystalline solid solutions. Because most metamorphic systems lie above the critical points of common fluid components (H_2O and CO_2), there is no discontinuity in physical properties of fluids and gases, so for constant T conditions, $dG = RTd(lna)$ and $d\mu_i = RTd(lna_i)$ if the solution is real or $d\mu_i = RTd(lnX_i)$ if it is ideal.

---

of this essentially aqueous ($X_{H_2O} = 0.97$) fluid entrapped in isolated pore spaces and along grain boundaries, the system is essentially closed and the composition of the immobile fluid will be initially determined by the reaction $MgCO_3 + H_2O = Mg(OH)_2 + CO_2$. The CO_2 concentration increases as H_2O is consumed, forming brucite, $Mg(OH)_2$, from magnesite. So long as these two solids coexist and T continues to rise, the composition of the fluid will move up the univariant curve to the invariant point where brucite, magnesite, and periclase, MgO, coexist at fixed P_{fluid}. Addition of heat at this point causes no rise in T; it is consumed in the two reactions

$$E_t + Mg(OH)_2 = MgO + H_2O$$

and

$$E_t + MgCO_3 = MgO + CO_2$$

As soon as all of the brucite is consumed, and some magnesite remains in addition to the newly formed periclase, the system moves up T along the curve as more heat is added. In so doing, progressively more periclase forms at the expense of magnesite and the fluid becomes more enriched in CO_2.

In contrast to this closed system, we can imagine another situation in which the mass of the fluid is much greater than the magnesite and its composition is fixed by conditions in a fluid reservoir external to the open magnesite system. As T rises, magnesite will again begin to decompose at 557°C if the fluid composition is fixed at $X_{CO_2} = 0.03$, but the CO_2 produced by reaction is carried away in the large mass of flowing, mobile fluid, which supplies the requisite amount of H_2O to make the brucite. Eventually, all of the magnesite will disappear at 557°C and the system moves into the divariant stability field of brucite. If the T continues to rise in this open system, the brucite is consumed at 655°C by the reaction $Mg(OH)_2 = MgO + H_2O$ and the liberated H_2O is removed by the fluid.

Thus, in a **closed system,** components are immobile and do not exchange with the surroundings, and the composition of the fluid is at all times determined by the solid phases or the reactions between them. In an **open,** or **fluid-dominated system,** one or more components are mobile and freely move between the system and its surroundings; fluid composition is externally controlled.

These concepts of component mobility are also exemplified by the role that oxygen and hydrogen play in oxidation-reduction equilibria and the stability of Fe-bearing solid phases. All metamorphic fluids contain free molecules of O_2 and H_2, but their concentrations are exceedingly small. In pure water at 1 atm, the equilibrium $P_{O_2} \sim 10^{-23}$ Pa; P_{H_2} is twice this value (Fyfe and others, 1978, p. 170). Because of these small concentrations in the fluid phase, the P_{O_2} (or more accurately, the fugacity of oxygen; see box on page 259) of the system is dictated by its bulk chemical composition as reflected in the mineralogical composition. Only if very large volumes of fluid move through an open system is the P_{O_2} governed by the fluid composition.

The way in which a given mineral assemblage controls the P_{O_2} can be appreciated from the model system Fe–Si–O. Consider the reaction (upper curve in Figure 8-34)

$$4Fe_3O_4 + O_2 = 6Fe_2O_3 \quad (14.12)$$
$$\text{magnetite} \qquad\qquad \text{hematite}$$

Provided magnetite and hematite are coexistent at some particular value of P and T, P_{O_2} is fixed. If oxygen is added (at the particular P and T), some magnetite is consumed to form more hematite and the P_{O_2} is maintained at the fixed value. The equilibrium assemblage hematite + magnetite (HM) thus constitutes an oxygen **buffer** that maintains a fixed P_{O_2} at some P and T no matter how much oxygen enters or leaves the system. However, should a very large amount of oxygen be introduced into the system, so that all of the magnetite is consumed in the reaction shown in Equation (14.12), leaving only hematite, the P_{O_2} is

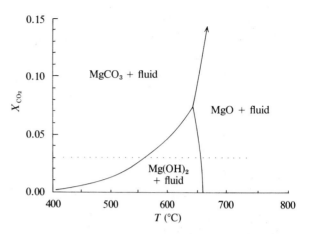

Figure 14-17 Phase relations of brucite, periclase, and magnesite at $P = P_{\text{fluid}} = P_{H_2O} + P_{CO_2} = 0.2$ GPa as a function of fluid composition and T. [After H. J. Greenwood, 1975, Buffering of pore fluids by metamorphic reactions, *American Journal of Science* 275; and H. J. Greenwood, 1976, Metamorphism at moderate temperatures and pressures, in D. K. Bailey and R. Macdonald (eds.), *The Evolution of the Crystalline Rocks* (New York: Academic Press).]

no longer buffered; the system is divariant. If T varies in the presence of both hematite and magnetite, the P_{O_2} can only vary along the upper curve in Figure 8-34. Another assemblage buffering P_{O_2}, but at lower values, is coexistent quartz + fayalite + magnetite (QFM) related by the reaction (see Figure 8-34)

$$3Fe_2SiO_4 + O_2 = 3SiO_2 + 2Fe_3O_4 \quad (14.13)$$
$$\text{fayalite} \qquad\quad \text{quartz} \quad \text{magnetite}$$

Highly reduced rocks may contain no Fe^{3+} and contain graphite and possibly sulfides. Many rocks, however, contain a buffering assemblage of magnetite plus an Fe-bearing silicate modeled by

$$3Fe^{2+}(\text{silicate}) + \tfrac{1}{2}O_2 = Fe_3O_4 +$$
$$\text{Fe-poor or Fe-free silicate} \pm \text{fluid} \quad (14.14)$$

A specific example is afforded by the reaction involving the ferrous-mica, annite:

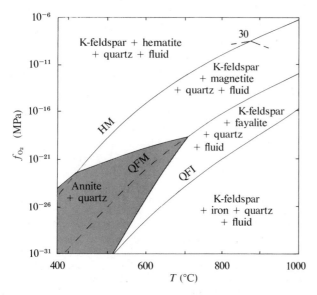

Figure 14-18 Stability field (shaded) of the Fe^{2+}-biotite, annite, $KFe_3AlSiO_3O_{10}(OH)_2$, at 207 MPa as a function of T and f_{O_2}. Compare Figure 8-34. The curve labeled QFI represents the buffer reaction $SiO_2 + 2Fe + O_2 = Fe_2SiO_4$ and the curves HM and QFM are the buffer reactions in Equations (14.12) and (14.13), respectively. The short dashed line segment labeled 30 represents a portion of the stability boundary in the quartz-deficient system of a biotite whose composition (mol %) is annite$_{30}$ phlogopite$_{70}$ under f_{O_2}–T conditions near the buffer assemblage hematite + magnetite. [After W. G. Ernst, 1976, *Petrologic Phase Equilibria* (San Francisco: W. H. Freeman and Company), from experimental data by D. R. Wones and H. P. Eugster.]

$$KFe_3^{2+}AlSi_3O_{10}(OH)_2 + \tfrac{1}{2}O_2 =$$
annite

$$\underset{\text{K-feldspar}}{KAlSi_3O_8} + \underset{\text{magnetite}}{Fe_3O_4} + H_2O \quad (14.15)$$

Figure 14-18 is an isobaric stability diagram for annite as a function of P_{O_2} (f_{O_2}) and T. Its greatest stability occurs in conjunction with the buffer assemblage magnetite + quartz + fayalite (QFM); with more oxidized or more reduced assemblages lacking fayalite, there is less or no Fe^{2+} available to stabilize annite, and its stability field is thereby reduced.

The stability of any Fe-bearing silicate is influenced by the oxygen pressure (fugacity) in the system, which is in turn usually buffered by the overall mineralogical composition unless very large amounts of fluid move through an open system. Fe-rich hydrous silicates have upper thermal stability boundaries at much lower temperatures than the corresponding Mg end members; compare curves 4 and 11 in Figure 14-19.

Ion Exchange Reactions in Open or Metasomatic Systems

We now consider additional concepts relating to reactions in open systems, foundations of which were laid about the turn of the century by Waldemar Lindgren, but more recently in the past three decades by J. B. Thompson and D. S. Korzhinskii.

An ideally closed system is probably rare in natural rock bodies, especially in situations where fluids are released during prograde reactions. As we find no textural evidence for pore spaces that might have stored fluids, we must assume the fluids have left the body, inevitably carrying dissolved constituents. Where the distance between the source and sink of the transported material is large, the process is conventionally called metasomatism. But, in principle, there is little difference between this situation and smaller scales of transport over a distance of a mineral grain. On the scale of grains, only polymorphic transitions might be truly isochemical (but see page 515); other reactions cause a change in the chemical composition of some volume. A kyanite grain, for example, uses only Si, Al, and O, yet the pelitic matrix in which it originally grew also contained Mg, Fe, K, and so on, and these must have been transported away from the growing grain.

For a system to be in complete thermodynamic equilibrium, all intensive variables—T, P, and chemical potentials of every component—must be uniform throughout. No incompatible phases, such as quartz and calcite at 800°C and 0.1 GPa

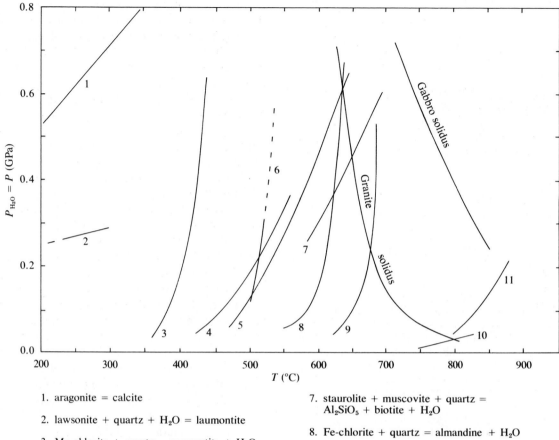

1. aragonite = calcite

2. lawsonite + quartz + H_2O = laumontite

3. Mn-chlorite + quartz = spessartite + H_2O

4. Fe-tremolite = fayalite + quartz + hedenbergite + H_2O

5. Mg-chlorite + muscovite + quartz = Mg-cordierite + phlogopite + H_2O

6. serpentine = forsterite + talc + H_2O

7. staurolite + muscovite + quartz = Al_2SiO_5 + biotite + H_2O

8. Fe-chlorite + quartz = almandine + H_2O

9. talc + forsterite = anthophyllite + H_2O

10. phlogopite + quartz = forsterite + K-feldspar + H_2O

11. tremolite = enstatite + diopside + quartz + H_2O

Figure 14-19   Univariant equilibria for some end-member systems in which $P_{H_2O} = P$ and for the granite and gabbro solidi. For additional reaction curves see Figures 14-2, 14-7, 14-11, 14-12, and 14-16. [After H. J. Greenwood, 1976, Metamorphism at moderate temperatures and pressures, in D. K. Bailey and R. Macdonald (eds.), *The Evolution of the Crystalline Rocks* (New York: Academic Press); and W. S. Fyfe, N. J. Price, and A. B. Thompson, 1978, *Fluids in the Earth's Crust* (New York: Elsevier).]

(see Figure 14-12), can exist together. However, as Thompson (1959, p. 430) points out, "though a large volume of rock commonly contains mutually incompatible phases and is thus not in internal equilibrium, it is generally possible to regard any part of such a thermodynamic system as substan-

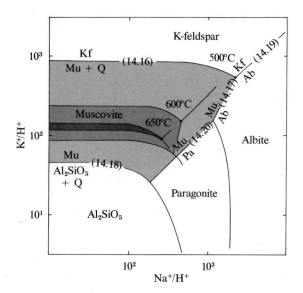

Figure 14-20   Phase relations in the system
Na–K–Al–Si–O–H as a function of T and the ratios of
alkali ions to hydrogen ions at $P = 0.5$ GPa and $a_{H_2O} =$
0.5. Curves represent ion-exchange equilibria listed in
text; see Equations (14.16) through (14.20). Note
shrinking of muscovite stability field with increasing T
(light shaded at 500°C, medium shaded at 600°C, and
dark shaded at 650°C) and absence of paragonite field at
600 and 650°C (compare Figure 14-13). [After R. P.
Wintsch, 1975, Solid-fluid equilibria in the system
$KAlSi_3O_8$-$NaAlSi_3O_8$-Al_2SiO_5-SiO_2-H_2O-HCl, *Journal
of Petrology* 16.]

tially in internal equilibrium if that part is made
sufficiently small.'' This statement is an expression
of the concept of **local equilibrium** (compare the
''mosaic'' equilibrium of Korzhinskii, 1970).

In fluid-dominated, open rock systems, because
the composition of the fluid is externally controlled, it may not everywhere and at all times be
in equilibrium with all solid phases in the system.
Adjustment toward a state of at least local equilibrium may involve **ion-exchange reactions.** Certain
ions in the fluid exchange with other ions in solids.
One type of exchange reaction is

$$AX \quad + \quad B^+ \quad = \quad BX \quad + \quad A^+$$
solid   in fluid   solid   in fluid

A familiar example of an ion exchange reaction is
the water-softening action of certain types of Na-
bearing zeolites in which dissolved Ca^{2+}, which
causes hardness in water, is taken into the zeolite
in exchange for two Na^+, which pass from the
zeolite into the water.

Most rock-forming silicate minerals and their
chemical components are only very slightly soluble in pure aqueous or carbonic fluids. This fact
limits the dissolved ion concentration gradients
that potentially might develop in a volume of static
fluid and thereby limits diffusion down any compositional gradient in the fluid. In a flowing fluid,
very large volumes are required to transport significant masses of dissolved silicates. However,
salts of Cl^-, SO_4^{2-}, and CO_3^{2-}, such as KCl,
$CaSO_4$ and Na_2CO_3, have significantly greater solubilities in aqueous fluids than silicates and consequently greater amounts of these materials can be
transported by diffusion in static fluids, or by fluid
flow. As carbonates, halides, and sulfates are the
very materials generally found in fluid inclusions in
metamorphic minerals (see page 499), they are
likely to have been involved in the solid–fluid
reactions occurring during metamorphism.

Alkali-ion exchange reactions can involve micas
and feldspars, as these are the chief repositories of
Na and K in metamorphic rocks. Many different
reactions can be written. For muscovite breakdown, possible reactions include:

$$KAl_2AlSi_3O_{10}(OH)_2 \quad + \quad 6SiO_2 \quad + \quad 2K^+ \quad =$$
muscovite   quartz   in fluid
141 cm³   136 cm³

$$3KAlSi_3O_8 \quad + \quad 2H^+ \quad (14.16)$$
K-feldspar   in fluid
327 cm³

$$KAl_2AlSi_3O_{10}(OH)_2 \quad + \quad 6SiO_2 \quad + \quad 3Na^+ \quad =$$
muscovite   quartz   in fluid
141 cm³   136 cm³

$$3NaAlSi_3O_8 \quad + \quad K^+ + 2H^+ \quad (14.17)$$
albite   in fluid
300 cm³

$$2KAl_2AlSi_3O_{10}(OH) + 2H^+ = 3Al_2SiO_5 +$$

muscovite   in fluid

282 cm^3 134–155 cm^3

$$3SiO_2 + 3H_2O + 2K^+ \quad (14.18)$$

quartz   in fluid

68 cm^3 54 cm^3

We may also write exchange reactions for Na–K micas and Na–K feldspars:

$$KAlSi_3O_8 + Na^+ =$$

K-feldspar   in fluid

109 cm^3

$$NaAlSi_3O_8 + K^+ \quad (14.19)$$

albite   in fluid

100 cm^3

$$KAl_2AlSi_3 10(OH)_2 + Na^+ =$$

muscovite   in fluid

141 cm^3

$$NaAl_2AlSi_3O_{10}(OH)_2 + K^+ \quad (14.20)$$

paragonite   in fluid

134 cm^3

Breakdown of albite can proceed according to

$$2NaAlSi_3O_8 + 2H^+ = Al_2SiO_5 +$$

albite   in fluid

201 cm^3 44–52 cm^3

$$5SiO_2 + H_2O + 2Na^+ \quad (14.21)$$

quartz   in fluid

114 cm^3 18 cm^3

Stability fields can be represented in terms of ratios of ion concentration (or activities; see box on page 504) at specified P and T, as in Figure 14-20.

14.5 INTERPRETIVE ANALYSES OF MINERAL ASSEMBLAGES

One of the important objectives of metamorphic petrology is to determine the P–T conditions of metamorphism of a body based upon its miner-alogical composition, assumed to represent a state of thermodynamic equilibrium (see box on page 510).

Petrogenetic Grid

Certainly, we are now in a much better position to approximate the P–T–X conditions of metamorphism than in 1940 when N. L. Bowen first proposed the concept of a **petrogenetic grid.** His idea was to construct a P–T diagram (grid) with univariant reaction curves bounding all conceivable divariant mineral assemblages for a given bulk chemical composition. Each mineral assemblage would lie in a unique P–T pigeonhole and inform the investigator immediately as to the conditions of metamorphism. But even after four decades of extensive experimental and theoretical effort, the goal of a wholly comprehensive petrogenetic grid has still not been achieved. Added to the long-standing difficulty in ascertaining stable equilibrium mineral assemblages in short-term laboratory experiments, there is the virtually infinite compositional complexity of natural assemblages. It would take many more decades of experimental work to evaluate stabilities of all natural combinations of chemical components.

However, univariant reaction lines delineating stability limits of many end-member minerals and mineral assemblages have been determined; some of the more important ones are shown in Figure 14-19. As more thermodynamic data become available, it will be easier to calculate mineral stabilities theoretically, and this will probably be the main thrust in future studies of metamorphic mineral assemblages (Helgeson and others, 1978).

Available univariant curves for simple end-member systems can be used in conjunction with principles governing mineral stabilities and reactions (discussed in previous sections) to provide important constraints on the P–T conditions of metamorphism in chemically complex systems.

Evidence for Stable Equilibrium in Metamorphic Mineral Assemblages

Unfortunately, there are no infallible criteria to verify that a particular mineral assemblage in a rock represents attainment of a state of stable equilibrium during crystallization of the assemblage (Zen, 1963; Zen and Thompson, 1974). The following criteria are for the most part only necessary and not sufficient:

1. Absence of known incompatible mineral pairs such as quartz + magnesian olivine, or hematite + graphite.

2. Limited number of different minerals in the rock.

3. All phases are in mutual contact with one another. A particular phase occurring only inside a corona or only as inclusions in a poikiloblast should be suspect.

4. Lack of zoning or compositional inhomogeneity in all grains. If a particular phase occurs as zoned grains, their edges should be of similar composition.

5. No irregularities in compositions of coexistant phases manifest in crossed tie lines in graphic representations of mineral assemblages, in irregular partition of trace elements between phases, or isotopic disequilibrium between phases.

The domain of a stable mineral assemblage is commonly taken to be a thin section, but in low-grade, fine-grained rocks a more restricted scale may be required because of the more limited sphere of diffusion. It should be realized that sophisticated laboratory investigations using mass spectrometer and electron microprobe analyzer may be used in an all-out effort to evaluate stable equilibrium.

Isograds

Graphic representations of multivariant stability fields and their bounding univariant equilibrium boundaries serve as models of mappable metamorphic zones and isograds, respectively, on the ground surface in real rock bodies. An isograd separating two zones is the line of intersection of the ground surface with a "fossil" plane of univariant equilibrium between multivariant assemblage volumes within the rock body.

In real metamorphic bodies, there are few univariant equilibria causing the appearance of a single index mineral because of extensive compositional variations in both minerals and fluids. Most real reactions are at least divariant. For example, to define an isograd in pelitic rocks on the first appearance of Mg–Fe garnet formed by breakdown of cordierite according to Equation (14.7), we would have to select only those rocks of a unique Mg/Fe ratio; otherwise, the isograd would be smeared over the ground surface because of the range of P–T conditions over which the continuous divariant reaction occurs in rocks of different Mg/Fe ratios. Theoretically, polymorphic transitions between pure phases unaffected by compositional variations in the system, or reactions between pure phases (such as quartz + calcite = wollastonite + CO_2) in a pure fluid, if such reactions actually exist, would be preferable to define a clean isograd reflecting "fossil" univariant P–T equilibria.

But there are even inherent pitfalls in the interpretation of any isograd based upon the first appearance of a single phase from a coupled reaction, such as that producing wollastonite from calcite + quartz. To illustrate, suppose a metamorphic terrane of pelitic rocks comprises a range of bulk compositions that can be represented in terms of the components Al_2O_3, SiO_2, and $KAlSi_3O_8$ (plus H_2O) in Figure 14-21. In a lower-grade zone (I) of the terrane, mineral assemblages in rocks C

and D contain muscovite + quartz ($\pm$ K-feldspar $\pm$ Al$_2$SiO$_5$), rock A contains muscovite + corundum (+ Al$_2$SiO$_5$), and rock B contains only muscovite. With increasing grade, K-feldspar forms by breakdown of muscovite; but there are two K-feldspar breakdowns and isograds at different P and T, one for quartz-bearing rocks and one for aluminous quartz-free rocks (see Figure 14-13). These two isograds can be distinguished by designating the first appearance of a pair of minerals: K-feldspar + Al$_2$SiO$_5$ in rocks containing muscovite + quartz at lower grade and K-feldspar + corundum in rocks initially containing no quartz. These two isograds occurring in the spectrum of pelitic compositions delineate three zones in Figure 14-21. Thus, isograds based upon the first appearance (or disappearance) of one index mineral can depend on the modal proportions of reactants, especially quartz.

In short, valid interpretation of any isograd based on a coupled reaction (which includes all nonpolymorphic transitions) hinges upon correct assessment of all the participating phases, their modal proportions, and the bulk chemical composition of the rock. The first appearance of a mineral pair is usually a more meaningful isograd than that of a single phase.

Isograds separating the classic Barrovian zones in pelitic rocks have been investigated by many petrologists for several decades, but the actual reactions governing them are still not thoroughly understood. This is particularly true in the lower grades (Mather, 1970) and is not surprising in view of the compositional complexity of pelitic rocks.

Analysis of isograds in higher-grade pelitic rocks around Whetstone Lake, Ontario by Carmichael (1970) furnishes a good example of the use of compositional projections. The lowest-grade rocks in this area contain chlorite + garnet, which reacted discontinuously with rising T to produce an assemblage of coexistent staurolite + biotite:

chlorite + muscovite + garnet =
    staurolite + biotite + quartz + water (14.22)

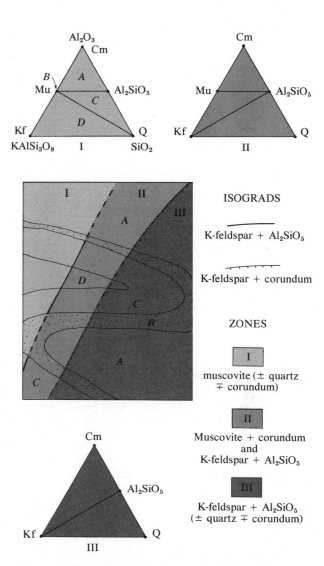

Figure 14-21 Hypothetical map of metamorphic isograds and zones graphically analyzed in terms of reactions with muscovite. A, B, C, and D represent different rock compositions. Compare with Figure 14-13. Note that K-feldspar is stable at lowest grade in rocks of suitable composition. [After R. H. Vernon, 1976, *Metamorphic Processes* (London: George Allen and Unwin).]

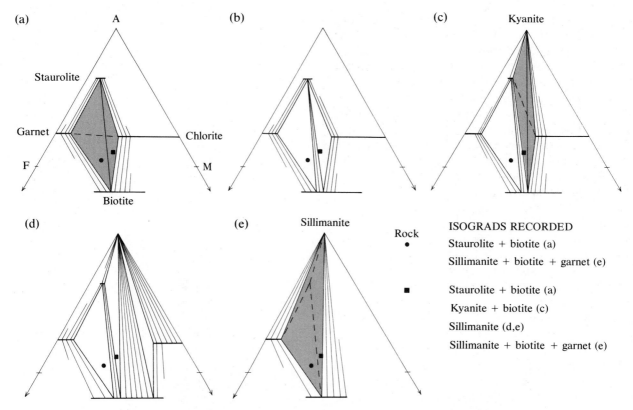

Figure 14-22 Sequence of continuous and discontinuous mineralogical reactions recorded in pelitic rocks around Whetstone Lake, Ontario. All assemblages include quartz + muscovite (± plagioclase ± Fe–Ti oxides) and are plotted on standard AFM diagrams (see pages 403–405). Any whole rock plotting in the shaded area of a diagram will record a discontinuous mineralogical reaction, that is, an isograd. [After D. M. Carmichael, 1970, Intersecting isograds in the Whetstone Lake area, Ontario, *Journal of Petrology* 11.]

Any rock plotting within the shaded area of Figure 14-22a will record this univariant reaction, manifest on the projection as a topologic discontinuity—an abrupt switch in mineral compatibilities; in Figure 14-22a the dashed line represents compatible chlorite + garnet and the full line compatible staurolite + biotite. The isograd mapped on the ground is not merely the first appearance of biotite, or of staurolite, as either one alone may be present in more mafic and more aluminous rocks, respectively, at lower grade. Rather, the isograd is the first appearance of the stable pair staurolite + biotite.

Though incompatible above this staurolite + biotite isograd, chlorite and garnet individually form stable three-phase assemblages with staurolite + biotite in the AFM diagram. Chlorite coexistent with staurolite + biotite (see Figure 14-22b) is more magnesian than the chlorite coexistent with garnet in the lower-grade zone because the stability of chlorites shrinks away from the less refractory Fe-rich chlorites. This contraction in chlorite stability, an example of a continuous reaction variation, is caused by the consumption of more Fe-rich chlorite according to

chlorite + muscovite = staurolite +
$$\text{biotite + quartz + water} \quad (14.23)$$

At higher grade, the staurolite + chlorite compatibility is broken by a discontinuous dehydration reaction:

staurolite + chlorite + muscovite + quartz =
$$\text{kyanite + biotite + water} \quad (14.24)$$

This reaction, expressed by the crossed tie lines in Figure 14-22c, is another topologic discontinuity, seen on the ground as the kyanite + biotite isograd. In the graphic representation of the zone above this isograd, the three-phase triangle kyanite + biotite + chlorite expands continuously with changing P and T at the expense of chlorite (see Figure 14-22d), according to the continuous reaction

chlorite + muscovite + quartz =
$$\text{kyanite + biotite + water} \quad (14.25)$$

further reducing the range of stable chlorite solid solutions.

At somewhat greater T, the polymorphic transition kyanite $\rightarrow$ sillimanite consumes kyanite; this first appearance of sillimanite constitutes a simple single mineral isograd.

The last isogradic reaction, forming coexistent sillimanite + biotite + garnet, involves breakdown of staurolite according to

staurolite + muscovite + quartz =
$$\text{sillimanite + biotite + garnet} + \text{water} \quad (14.26)$$

In Figure 14.22e, tie lines linking staurolite to garnet, biotite, and sillimanite all vanish with the discontinuous breakdown of staurolite and one new three-phase triangle replaces these three lower-grade compatibilities.

In summary, changes in P and T during metamorphism are manifest on compositional projections by (1) continuous variations in the orientations of tie lines associated with contracting and expanding stability fields; and (2) discontinuities in the basic pattern (topology) of connecting tie lines that define two- and three-phase regions because of the disappearance of a stable phase, emergence of a new one, or change in phase compatibilities. In the real rock body, continuous reactions occur within mappable metamorphic zones, separated by discontinuous reactions at isograds.

One rock type (pelitic in this case) can go through several prograde reactions with new minerals appearing while lower-grade ones survive. Not every mineral in the rock participates in every reaction. For example, in the Whetstone Lake terrane, biotite first appears at the biotite + staurolite isograd but is not a reactant phase and remains stable in the next two higher dehydration reactions.

Isograds delineated in pelitic rocks around Whetstone Lake are intersected by another isograd related to a combined decarbonation–dehydration reaction in carbonate-bearing rocks:

biotite + calcite + quartz = amphibole +
$$\text{K-feldspar} + CO_2 + H_2O \quad (14.27)$$

The T of isogradic reactions (except the kyanite–sillimanite transition) depends upon the relative concentrations of CO_2 and H_2O in the fluid phase (see Figure 14-23). Carmichael (1970), therefore, believes the intersecting pattern of isograds around Whetstone Lake was caused by a regional geothermal gradient, with higher temperatures to the northwest, combined with a gradient in X_{H_2O}, possibly related to H_2O expelled from a granitic intrusion in the northern part of the terrane.

What Are the Actual Reactions in Rocks?

It is not always possible to elucidate the actual reactions that occurred in a metamorphic body in response to changes in P, T, or X, but as shown in the previous section, reasonable models can be developed in many situations (see also examples in

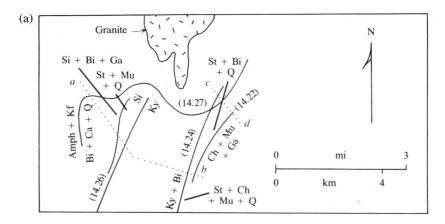

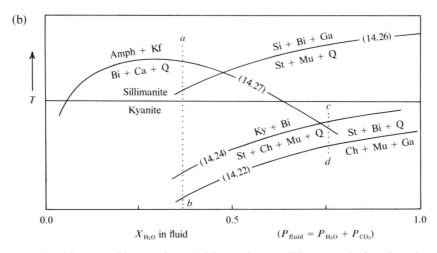

Figure 14-23 Metamorphic reactions and isograds near Whetstone Lake, Ontario. (a) Map of isograds, keyed to discussion in text. (b) Mole fraction–*T* plot for reactions. Dotted lines show how regional thermal gradients at different fluid compositions could intersect the sequence of univariant reaction curves to give the sequence of isograds on the ground shown in (a). [After D. M. Carmichael, 1970, Intersecting isograds in the Whetstone Lake area, Ontario, *Journal of Petrology* 11.]

Vernon, 1976, chap. 4). The investigator must carefully integrate textural and chemical properties of the mineral grains across the isograd and ask: What phases appeared and disappeared? Has the reaction gone to completion, producing a stable equilibrium assemblage (see box on page 510), or has it been arrested in transit, as in the corona textures of Figure 14-4, because of depletion of a reactant phase or for kinetic reasons? Textures can be deceptively simple and interpretations ambiguous. Even pseudomorphic replacements can be deceptive. For example, replacements of garnets and staurolites by chlorite are common, yet the actual hydration reaction is difficult to write; it is not simply chlorite = garnet (or staurolite) + $n\mathrm{H_2O}$. Commonly, textural evidence supporting a particular mineralogical reaction is lacking, even though a comparison of mineral assemblages in rocks on either side of an isograd indicates it must have occurred. For example, at the sillimanite isograd in typical Barrovian zonations, kyanite apparently disappears by the simple polymorphic transition kyanite → sillimanite. However, textural evidence indicating direct replacement of kyanite grains by sillimanite is rarely found. So what is the actual mechanism of transition?

Petrographic scrutiny reveals that small acicular sillimanites are commonly intimately intergrown with micas, as if these phyllosilicates provided a favorable site for nucleation of sillimanite. Carmichael (1969) has integrated this textural observation with a postulated cyclic sequence of local metasomatic ion-exchange reactions to explain the net reaction, kyanite = sillimanite. In the first local reaction

$$3\mathrm{Al_2SiO_5} + 3\mathrm{SiO_2} + 3\mathrm{H_2O} + 2\mathrm{K^+} =$$
kyanite   quartz   in fluid
$$2\mathrm{KAl_2AlSi_3O_{10}(OH)_2} + 2\mathrm{H^+} \qquad (14.28)$$
muscovite   in fluid

and in the second

$$2\mathrm{KAl_2AlSi_3O_{10}(OH)_2} + 2\mathrm{H^+} =$$
muscovite   in fluid
$$3\mathrm{Al_2SiO_5} + 3\mathrm{SiO_2} + 3\mathrm{H_2O} + 2\mathrm{K^+} \qquad (14.29)$$
sillimanite   quartz   in fluid

Obviously, by adding these two local reactions together, the net balanced polymorphic reaction is achieved. These two local reaction systems are envisaged to function simultaneously, though in different locations; hence, exchange of $\mathrm{H^+}$ and $\mathrm{K^+}$ ions via an aqueous fluid produces muscovite grains from kyanite in one local system and sillimanite grains from muscovite in the other. Relatively insoluble, more slowly diffusing $\mathrm{Al^{3+}}$ and $\mathrm{Si^{4+}}$ ions are immobile and these ions (plus oxygens), which make up the final sillimanite grains, are not the same ones in the original kyanite. However, overall, there is no transfer between the net system and its surroundings.

In such cyclic chemical reactions, the intermediate, transient phase may not leave evidence of its existence. However, in the situation under consideration here, muscovite is stable under P–T conditions above and below the kyanite → sillimanite transition; additional muscovite grains produced in reaction (14.28) are compensated by destruction of others in (14.29); any available muscovite grain may be utilized for nucleation of sillimanite. Other cyclic reactions in pelitic rocks have been examined by Carmichael (1969).

14.6 METASOMATISM

We have already considered metasomatic ion-exchange reactions in grain-scale situations and now consider in this section two larger-scale types of metasomatic phenomena, in part depending upon ion exchange: metamorphic differentiation and metasomatic zoning.

Metamorphic Differentiation

Metamorphic differentiation creates conspicuously heterogeneous rock, with layers or pods of contrasting mineralogical composition, from an initially more homogeneous body (Turner and Verhoogen, 1960, pp. 581–586; Vidale, 1974). Models of differentiation examined here involve metasomatism of alkali feldspars via ion-exchange reactions in chloride-bearing aqueous fluids (compare the deformation-related models of metamorphic differentiation on pages 479–481).

In dealing with equilibria between a silicate melt and plagioclase solid solution, we have seen (Figure 8-20) that Na is more highly concentrated in the melt phase and Ca in the crystals at any P and T. Similar unequal partitioning of elements occurs between coexisting crystalline solid solutions and fluid solutions in subsolidus metamorphic systems.

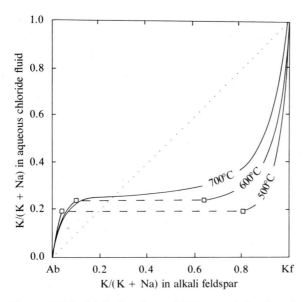

Figure 14-24 Mole fractions of K in coexisting 2 molar aqueous chloride fluid and alkali feldspar at 200 MPa. Coexisting feldspars (squares) delineate a segment of the alkali feldspar solvus. Dotted line indicates equal partition of K and Na between fluid and crystalline phases. [After P. M. Orville, 1963, Alkali ion exchange between vapor and feldspar phase, *American Journal of Science* 261.]

Orville (1963) has demonstrated experimentally that the concentration of K^+ relative to Na^+ in a chloride-bearing aqueous fluid coexisting in a state of univariant equilibrium with two alkali feldspars increases at higher temperatures (see Figure 14-24). This means that in an open rock body *with a significant gradient in T* and a freely communicating fluid, exchange equilibria will produce a more K-rich fluid in the higher-*T* region through replacement of K-feldspar by Na feldspar; in the lower-*T* region, the reverse occurs and increasingly K-rich feldspars coexist with increasingly Na-rich fluids at progressively lower temperatures. In currently active geothermal areas, such as Yellowstone National Park, near-surface, lower-*T* waters have very low concentrations of K but high Na because of the exchange equilibria in

which primary Na-feldspar in volcanic rocks is being replaced by K-feldspar. The large metacrysts of K-rich feldspar in some metamorphic country rocks surrounding deep-seated granitic intrusions are believed by Orville to have formed by alkali exchange metasomatism in the thermally zoned intrusion–wall rock system. The presence of Ca in these systems and the effect of different absolute concentrations of Cl^- in the fluids and of different P all modify the equilibria but do not change their general effects.

Ion-exchange equilibria may also have a bearing on the origin of layers of contrasting mineralogical composition in quartzo-feldspathic and pelitic rocks of the amphibolite facies. Part of the expression of some of these alternating layers lies in the modal abundance of plagioclase in one and K-rich feldspar in the other, though both layers contain both feldspars and their compositions are similar. Orville's experimental data and model show how this aspect of differentiated layers might originate under isothermal conditions (see Figure 14-25). Consider a body of calcareous shale or mudstone in which alternating thin beds have slightly greater and lesser concentrations of calcite. In response to an initial pulse of essentially isochemical metamorphism, the more calcic layers, *B* in Figure 14-25, form approximately equal proportions of plagioclase and alkali feldspar (open squares). The alternating layers, *A*, do the same, except the plagioclase in them is less calcic and the alkali feldspar in equilibrium with it is more sodic than in *B* layers. Subsequently, fluids from each layer may mix together and homogenize to an intermediate composition. Dissolved Na^+ and K^+ in the communicating fluid participate in ion-exchange reactions so that the bulk composition of layer *B* moves to *B'* and *A* moves to *A'*. Compositions of equilibrium plagioclase and alkali feldspar (solid squares) in both layers are now the same, except their modal proportions differ.

The situation in most natural rock systems is obviously much more complex than in these simplified models, but ion-exchange reactions in-

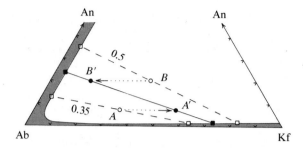

Figure 14-25 Equilibria between fluid and ternary feldspar solutions at 200 MPa. Compare Figure 8-30. Shaded area is one-feldspar field. Numbers are mole fractions of K in fluid, that is, K/(K + Na). Points *A* and *B* represent bulk feldspar composition of local systems containing plagioclase and alkali feldspar (open squares) at end of tie lines. If fluids from local systems *A* and *B* communicate and mix together, ion-exchange reactions involving Na⁺ and K⁺ in the open system metasomatize *A* and *B*, shifting their bulk compositions to *A'* and *B'*, respectively; note that the An content remains unchanged during this metasomatism. At equilibrium in the open system, *A'* consists mostly of K-rich alkali-feldspar and *B'* consists chiefly of plagioclase. The feldspars constituting each system are identical, however. [After P. M. Orville, 1963, Alkali ion exchange between vapor and feldspar phase, *American Journal of Science* 261.]

volving a fluid phase offer a promising approach to the origin of more complex differentiated metamorphic bodies (Vidale, 1974).

The Thompson Model of Metasomatic Zoning

A theoretical model of metasomatic zoning has been developed by Thompson (1959, see also Korzhinskii, 1970) and can be illustrated by an example from the binary system MgO–SiO₂. At atmospheric pressure four minerals are stable depending upon the composition of the local system (see Figure 14-26). Quartz and periclase constitute an unstable pair and, depending upon their proportions, can react to form stable one- or two-phase assemblages, as follows:

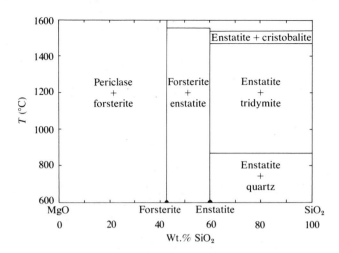

Figure 14-26 A portion of the system MgO–SiO₂. Stability fields involving silicate melt not shown or labeled. [After G. W. Morey, 1964, Phase-equilibrium relations of the common rock-forming oxides except water, U.S. Geological Survey Professional Paper 440-L.]

$$\underset{\text{periclase}}{MgO} + \underset{\text{quartz}}{2SiO_2} = \underset{\text{enstatite}}{MgSiO_3} + \underset{\text{quartz}}{SiO_2}$$

$$MgO + SiO_2 = MgSiO_3$$

$$3MgO + 2SiO_2 = MgSiO_3 + \underset{\text{forsterite}}{Mg_2SiO_4}$$

$$2MgO + SiO_2 = Mg_2SiO_4$$

$$3MgO + SiO_2 = Mg_2SiO_4 + MgO$$

Periclase + enstatite and forsterite + quartz also are unstable pairs and react to form stable forsterite and enstatite, respectively. These stable mineral compatibilities are implicit in the *T–X* diagram shown in Figure 14-26.

Thompson proposed a hypothetical situation (at fixed *P* and *T*) in which a region of pure periclase is separated from a region of pure quartz by three two-phase zones in which the bulk composition varies continuously and linearly (see Figure 14-27a). Within each of these two-phase zones, such

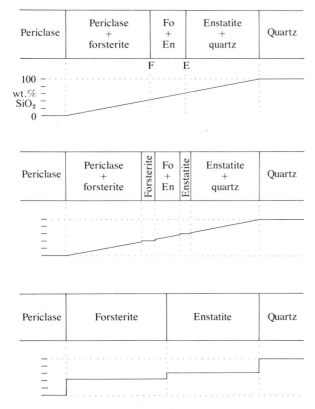

Figure 14-27 Hypothetical model of metasomatic zoning. Compare with Figure 14-26. [After J. B. Thompson, 1959, Local equilibrium in metasomatic processes, in P. H. Abelson (ed.), *Researches in Geochemistry,* Vol. 1 (New York: John Wiley and Sons).]

as periclase + forsterite, the modal proportions of each phase varies reciprocally and continuously from 0% to 100% across the zone to permit the variation in bulk composition. Within each zone, as well as in the two end regions, local equilibrium prevails. Each of the two-phase assemblages is stable and no reaction should occur within them. However, along boundary F, grains of periclase in the left zone may be in communication (say, via a migrating fluid) with enstatite grains in the right and, as this pair of phases is less stable than forsterite, an intermediate one-phase zone of that mineral may form, resulting in a local reduction of the

number of phases. Similarly, a zone of pure enstatite may form in place of the interface E. The extent to which these monomineralic forsterite and enstatite zones replace the neighboring two-phase zones would depend upon the available time for the diffusion of Si and Mg. Ultimately, the metasomatic diffusion process might produce the configuration of Figure 14-27c, in which only monomineralic forsterite and enstatite zones occur between the periclase and quartz regions; four areas of local one-phase equilibrium exist.

Two conclusions may be drawn from this hypothetical metasomatic model. First, transport of matter by diffusion (or by fluid flow) can produce discontinuities in bulk composition where none existed before. Second, metasomatic processes tend to reduce the number of phases in an assemblage.

Monomineralic veins, such as epidote in granitic rocks and quartz in pelitic schists, are examples of metasomatism in open, fluid-dominated systems (see box on page 519).

Special formulations of the Gibbs phase rule have been used by Thompson (1959) and Korzhinskii (1970) to assess phases and components in open metasomatic systems, although this approach is considered by Weill and Fyfe (1967; see especially their 1964 paper referred to therein) to be an incorrect application of the phase rule.

Joesten (1974) has described a sequence of zoned calc-silicate assemblages developed by metasomatic reaction between chert nodules and limestone host rock in a contact metamorphic aureole, and Brady (1977) discusses more complex metasomatic reactions around ultramafic bodies and between pelitic and calcareous layers in intermediate-grade regional metamorphic bodies.

14.7 KINETICS OF METAMORPHIC MINERAL REACTIONS

Throughout this chapter, it has been repeatedly assumed that mineral assemblages in metamorphic rocks represent states of stable equilibrium and, therefore, convey information on the intensive

Veins

Widespread veins in metamorphic bodies testify to the migration of fluids carrying dissolved ions.

Many veins are virtually monomineralic quartz but many other phases also occur, such as feldspar, epidote, and carbonates. Mineral assemblages in some veins are similar to assemblages in the host wall rock, suggesting local interchange of material via a fluid into an opening fissure; in this situation grains may extend more or less continuously from the wall rock into the interior of the vein. In other veins, the phase assemblage was not likely derived from the immediate wall rock and transport of dissolved constituents over considerable distances is implied; this would apply, for example, to veins of pure quartz in host marble and to very large massive bodies of quartz in almost any host.

The general simplicity of vein assemblages implies that all but a few components were mobile with chemical potentials externally controlled.

As quartz is a dominant vein phase, it is warranted to consider the solubility of quartz in water. The figure in this box shows that the solubility increases with increasing T and P. Deep-seated aqueous fluids saturated with respect to silica would precipitate quartz upon rising to shallower levels. The amount is conceivably great. Take, for example, a prograde metamorphic event liberating just 1 wt.% water due to dehydration reactions (see Figure 14-15). From 1 km³ of rocks, 2.7×10^{13} g of water is released, which could

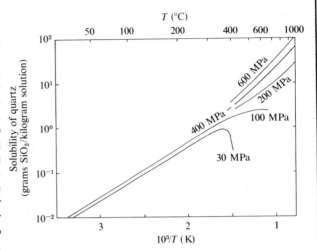

Solubility of quartz in water as a function of P and T. [After H. D. Holland, 1967, Gangue minerals in hydrothermal deposits, in H. L. Barnes (ed.), *Geochemistry of Hydrothermal Ore Deposits* (New York: Holt, Rinehart, and Winston).]

carry, if saturated at 400 MPa and 500°C, 2×10^{11} g of dissolved silica. Upon rising to a shallower depth where $P = 100$ MPa and $T = 300°C$, 3×10^{10} g of quartz would be deposited, equivalent to a vein 10 cm thick, 10 m wide, and 10 km long.

variables P, T, and X during metamorphism. Yet it must be realized that metastable assemblages can occur in both natural and laboratory systems, as demonstrated by the embarrassingly large number of discrepant phase boundaries determined for the Al_2SiO_5 system (see Figure 14-1) and the coexistence of two or more Al_2SiO_5 polymorphs in the same rock. Note also that most metamorphic assemblages exposed at the Earth's surface are metastable and have not equilibrated with atmospheric P and T; the good fortune of this metastability, which allows petrologists to study the

record of mineralogical adjustments in the deep crust, should not obscure the question as to the reason for the metastability in the first place. Some rocks show textural evidence for an arrested reaction, as both reactant and product phases are present; while the restricted availability of a reactant phase (such as water in a retrograde reaction) can be argued in some instances, other arrested reactions may depend upon unfavorable kinetics. These considerations justify a brief review of available information on reaction rates gained from experimental and theoretical studies.

Concepts of Rate Theory

The Eyring theory of absolute reaction rates (Fyfe and others, 1958) depends upon the existence of a transition state, or **activated complex,** which is a transient intermediate grouping of atoms as reactants transform into products. The complex could be represented by a chance encounter of reactant atoms in some intermediate configuration where two or more phases are reacting, or fluctuations in atomic configuration in a homogeneous phase due to its thermal energy in the case of a polymorphic transition. Obviously, the formation of an activated complex during the course of a reaction is enhanced at higher temperatures because of the increased agitation of atoms. Also, for a given reaction, there will be a particular frequency of chance encounters of reactant atoms that form the activated complex.

The free energy of the activated complex is greater than that of either reactants or products (see Figure 14.28). The **activation energy** is the amount of excess energy possessed by the activated complex relative to the reactants and may be viewed as an energy barrier impeding the formation of more stable products from reactants. The existence of this activation energy barrier is known from the persistance and survival of metastable states—for example, liquid water a few degrees below 0°C. If the activation energy is large, only a small proportion of encounters of reactant materials have sufficient energy to reach the activated state and the rate of the reaction is small. A smaller activation energy enables more encounters to be successful and the rate is greater. A **catalyst** is a substance that lowers the activation energy, allowing the reaction to proceed at a faster rate without disturbing the thermodynamic equilibrium.

The opportunity of forming the activated complex in a reaction is a measure of the reaction rate, K. Formally, the **reaction rate** is given by the Arrhenius equation

$$K = Ae^{-E_a/RT} \qquad (14.30)$$

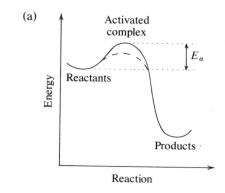

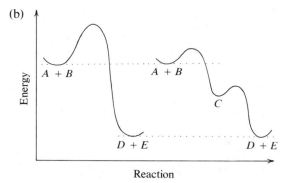

Figure 14-28 Energy levels controlling kinetics of mineral reactions. (a) Direct path via an activated complex. The activation energy barrier is lowered by a catalyst (dashed line). (b) Formation of an intermediate metastable assemblage, C, may offer a smaller activation energy barrier than in the direct reaction $A + B \rightarrow D + E$. [After R. H. Vernon, 1976, *Metamorphic Processes* (London: George Allen and Unwin).]

where A is the frequency factor, e is the base of natural logarithms, E_a is the activation energy, R is the gas constant, and T is the absolute temperature in degrees Kelvin. The expression $e^{-E_a/RT}$ is the fraction of encounters having energy E_a; such encounters are exponentially more numerous at higher T.

An experimental activation energy can be determined by measuring K as a function of T, taking the logarithm of the Arrhenius equation

$$\ln K = -E_a/TR + \ln A \qquad (14.31)$$

Table 14-4 Activation energies, E_a, in kJ/mol.

E_a (kJ/mol)	Type of reaction
40	zeolite dehydration
50–85	movement of alkali ions through open crystalline structures
85–100	alkali-ion exchange diffusion; migration of divalent cations
210–300	Si-Al exchange diffusion; dehydration reactions
420–840	breakup of Si-O bonds

Source: After E. D. Lacy, 1965, Factors in the study of metamorphic reaction rates, in W. S. Pitcher and G. W. Flynn (eds.), *Controls of Metamorphism* (Edinburgh: Oliver and Boyd).

Table 14-5 Comparison of reaction rates at different temperatures as a function of E_a, holding A constant.

E_a (kJ/mol)	$\dfrac{\text{Rate at 300°C}}{\text{Rate at 200°C}}$	$\dfrac{\text{Rate at 600°C}}{\text{Rate at 500°C}}$
63	16.2	3.1
272	1.7×10^5	1.3×10^2
419	1.2×10^8	1.7×10^3
628	1.3×10^{12}	7.2×10^4
837	1.3×10^{16}	3.0×10^6

Source: After E. D. Lacy, 1965, Factors in the study of metamorphic reaction rates, in W. S. Pitcher and G. W. Flynn (eds.), *Controls of Metamorphism* (Edinburgh: Oliver and Boyd).

and plotting $\ln K$ vs. $(-1/T)$. Equation (14.31) is a straight line whose slope times R is the desired activation energy and whose intercept on the $\ln K$ axis is $\ln A$. Such measurements are tedious and time-consuming; very few have been made for metamorphic reactions. Table 14-4 shows the approximate range of activation energies for different sorts of reactions.

It is a well-known fact that reaction rates are faster at higher T (see Table 14-5). For breakdown of a silicate crystal and formation of a new one with $E_a = 628$ kJ/mol, the rate is about 10^{12} times faster at 300°C than 200°C, or what would take 1 second at 300°C would require $\sim 10^5$ years at 200°C!

Rate theory shows that the frequency factor A increases exponentially with the **entropy of activation,** which is generally the same sign (positive or negative) as ΔS for the reaction. Large positive entropies of activation, such as occur in dehydration and decarbonation reactions, make A large, whereas small entropies (or negative ones) make A small. Hence, polymorphic transitions and hydration and carbonation reactions are slower than reactions releasing fluid.

Steps in Reactions

Most reactions proceed via a sequence of consecutive steps, each of which is governed by its own

rate. The rate of the overall reaction is the rate of the slowest step. Formation of new mineral grains from a reaction requires nucleation followed by crystal growth. Each of these processes requires consecutive steps, in a generalized way as follows:

NUCLEATION

1. Assembly of proper atoms to the nucleation site by diffusion.

2. Arrangement of atoms into a new configuration (the nucleus) via an unstable intermediate activated complex.

GROWTH

1. Disruption of atomic bonds in unstable reactant grains; this step must also procede step 1 under nucleation.

2. Diffusion of necessary atoms to viable nuclei of the new mineral grain.

Nucleation. There is little factual information regarding nucleation in metamorphic systems. As discussed on pages 238–239, an energy barrier to formation of a stable nucleus is related to its greater surface energy per volume as compared to a large crystal. Only nuclei of some critical size can survive and grow. A stable phase (or phases) whose free energy is only slightly lower than the existing unstable phase (or phases) may nucleate only with considerable overstepping in T (or P). This is especially critical in the case of solid–solid reactions, such as polymorphic transitions, where ΔS is very small (see Table 14-1); nuclei formation is sluggish and requires substantial overstepping in P or T beyond the lower equilibrium stability limits.

These comments apply essentially to homogeneous nucleation within a uniform media. But in rocks, there are boundaries between grains, perhaps with a fluid film between, as well as discontinuities within grains, such as dislocations, where heterogeneous nucleation may be favored. Indeed, textural evidence in metamorphic rocks commonly indicates nucleation has preferred grain boundaries or discontinuities within deformed grains, and grains of sillimanite may not directly replace kyanite.

Diffusion.   At equilibrium, the chemical potential of each component in a system must be the same in all phases. If not, the system is unstable and components migrate in such a way as to equalize potentials. In ideal solutions, components move down concentration gradients, whereas in real liquid or solid solutions, matter diffuses down activity gradients. This diffusion of matter is analogous to flow of heat from hotter to cooler bodies.

Rates of diffusion, governed by the diffusion coefficient, D (see pages 236–237), are very slow through crystalline solids. Data summarized in Fyfe and others (1978, pp. 115, 116) show D values on the order of 10^{-15} cm^2/s at 400°C for cations moving through silicates. Calculated average depths of penetration are on the order of 1 cm in 10 My. At 800°C, D values are on the order of 10^{-13} to 10^{-11} cm^2/s, so penetrations are ten to a hundred times greater. Experiments show diffusion is faster through crystals with substantial concentrations of point defects (see pages 455–457). Diffusion rates along disordered grain boundaries are greater than through grains, especially if a film of fluid is present. Fine-grained rocks would profit most from grain-boundary diffusion. And, of course, diffusion rates are considerably faster through fluids where penetrations are on the order of kilometers at 400°C over 10 My. In fluid-dominated systems, where the fluid itself is moving through the rock, dissolved material may be transported even faster than by diffusion in a stagnant fluid.

Application to Actual Metamorphic Reactions

Solid–solid reactions have observed activation energies of 400 kJ/mol or more, indicating that breakdown of Si–O bonds at the surface of grains is the limiting rate. However, reactions involving strained grains, or very small grains, have significantly lower E_a. Diffusion rates in solids are very sluggish. Nucleation rates are also very slow because of the small ΔS of these reactions, which requires substantial overstepping of the equilibrium boundary before any nuclei can persist and grow into crystals. Interpretation of metamorphic conditions on the basis of solid–solid reactions could, therefore, be misleading. Experiments show that completion of solid–solid reactions may take more time under metamorphic conditions than the probable duration of the metamorphic event itself (Fyfe and others, 1958 and 1978).

In contrast, activation energies for fluid-release reactions are on the order of 150–300 kJ/mol (Fyfe and others, 1958, p. 102), or smaller, which is significantly less than for solid–solid reactions. ΔS values are much greater, necessitating only small amounts of overstepping (no more than a few tens of degrees) beyond the reaction conditions for effective nucleation to occur.

Fluids are solvents (especially in the presence of dissolved Cl^- and other anions) that promote breakdown of reactant grains by carrying away surface material so as to expose more reactant; dissolved material in the fluid can diffuse rapidly to sites of growing product grains. Nucleation rates are likely faster in the presence of a fluid. Even though the reaction between MgO and SiO_2 to produce forsterite is a solid–solid reaction, it has been shown experimentally that the reaction rate is at least 10^8 times faster in the presence of water at 450°C than in the dry state at 1000°C (Fyfe and others, 1958, p. 85). At 300°C in a dry system, it would take $\sim 10^4$ years to convert just 10% aragonite to calcite, but in the presence of water, this same conversion takes only about one hour (Fyfe and others, 1978, p. 118).

Some mineral reactions appear to proceed faster by a path involving an intermediate transient phase, rather than going directly from reactant to product phases (see Figure 14-28b). Greenwood (1963) showed that at 100 MPa and $T \sim 820$°C talc decomposes initially to anthophyllite,

$$7Mg_3Si_4O_{10}(OH)_2 =$$
$$\text{talc}$$
$$3Mg_7Si_8O_{22}(OH)_2 + 4SiO_2 + 4H_2O$$
$$\text{anthophyllite} \qquad \text{quartz}$$

but that eventually in long-term experiments anthophyllite breaks down according to

$$3Mg_7Si_8O_{22}(OH)_2 = 21MgSiO_3 + 3SiO_2 + 3H_2O$$
$$\text{anthophyllite} \qquad \text{enstatite} \qquad \text{quartz}$$

Adding these two reactions together, we find:

$$Mg_3Si_4O_{10}(OH)_2 = 3MgSiO_3 + SiO_2 + H_2O$$
$$\text{talc} \qquad \text{enstatite} \qquad \text{quartz}$$

Apparently, the reaction path from unstable talc to stable enstatite + quartz runs faster via intermediate metastable anthophyllite than by direct conversion. In burial metamorphism of tuffs, Zen and Thompson (1974) wonder if the tetrahedral coordination of Al and Si in highly unstable glass might favor nucleation of zeolites rather than other minerals, such as calcite + kaolinite + quartz, which may under some conditions be more stable, notably where CO_2 concentrations are appreciable. It may be recalled from page 515 that Carmichael (1969) postulated that the transition from kyanite to sillimanite may proceed indirectly via an ion-exchange reaction with muscovite. He concludes (p. 255):

> Given a choice between a number of possible reaction paths, a natural metamorphic reaction will proceed to completion by means of that path for which the activation energy is lowest, provided that the over-stepping of the equilibrium temperature is too small to activate the other paths as well. This condition is likely to be met in regional metamorphic rocks, because of their necessarily slow rate of heating. It is safe to conclude that the activation energy for the reaction kyanite = sillimanite, by the reaction path (via intermediate transient muscovite)...is significantly lower than that for nucleation and growth of sillimanite directly from kyanite....all of the chemical species that participate in the various subreactions, without appearing in the balanced reaction that expresses the net change taking place in the system, are *catalysts*, in that they accelerate the reaction without changing its equilibrium conditions.

If dehydration and decarbonation reactions proceed on the time scale of only days to years under metamorphic conditions, then fluid release must keep pace with deformation and tectonic processes. This would have a profound, but as yet little investigated, effect on how the rock body deforms. Depending on the rate of flow of the water out of the body, two situations are possible (Fyfe and others, 1978, pp. 126–127):

> If flow rates are greater than reaction rates, then incomplete leaching will result in disequilibrium mass transport over varying distances, depending upon the dimensions of the fractures and reactive capacity of the fluid. This situation characterizes near-surface fluid-dominated regimes, such as some hydrothermal alteration systems and mineral deposition in open veins. If flow rates are less than reaction rates, the rock can easily buffer the composition of the fluid and fluid compositions will be rock-dominated.

We now return, finally, to the problem of why metastable, highest-grade mineral assemblages are preserved in metamorphic rocks now exposed at the surface. Do kinetic factors explain this irreversibility? Prograde fluid-release reactions with their large ΔS may well be inherently faster than the opposite retrograde reactions. It is significant, however, that retrograde reactions generally require H_2O or CO_2 to combine with the volatile-free, high-grade minerals and that occurrences of retrograde metamorphism are commonly seen only locally along grain margins. Inaccessibility of fluids to combine with high-grade volatile-free phases and to serve as catalysts for reactions may explain the rarity of retrograde metamorphism. The same explanation basically applies also to the preservation of high-T assemblages in magmatic bodies.

SUMMARY

14.1. New mineral assemblages are produced along generally positive P–T gradients in the Earth and wherever there are significant changes in chemical composition. Mechanical deformation may be a source of thermal energy in metamorphic bodies.

14.2. Higher temperatures stabilize higher entropy states and higher pressures stabilize more dense phases or phase assemblages. In many crustal reactions, the change in entropy and volume are of the same algebraic sign and accordingly, from the Clapeyron equation $dP/dT = \Delta S/\Delta V$, related reaction curves on a P–T diagram have positive slope.

14.3. With regard to metamorphic reactions:

1. The P–T stability field of a pure, crystalline, nonvolatile phase, whose composition is not varied by solid solution, is independent of any variations in the composition of the system.

2. The P–T stability field of a phase is reduced in the presence of another phase that can react in coupled manner with it.

3. The P–T stability field of crystalline solid solutions depends upon their composition; multivariant continuous reactions involving them slide over a range of P and T as the composition of the system varies.

4. A crystalline solid solution decomposes incongruently and continuously over a range of P and T to another composition in the solid solution series plus other breakdown phases.

14.4. Most prograde reactions in real metamorphic bodies are of the type solid → (solid + fluid) and have large positive ΔS because of the liberated fluid. The change in volume, ΔV, in such reactions is also positive under most crustal conditions, but because of the compressibility of the fluid, the reaction curve becomes steeper at higher P on a P–T diagram and at very high P and low T may bend back with negative slope.

Most metamorphic fluids are aqueous, but significant concentrations of CO_2 occur locally, as in some carbonate bodies, and other gaseous species such as CH_4, S, and so on may also be present. Generally, $P_{\text{fluid}} \sim P$. The density (in g/cm³) of fluids along most P–T gradients in the crust is close to unity.

In systems where, for any reason, $P_{\text{fluid}} < P$, the stability field of a phase or phase assemblage containing this fluid is reduced relative to its limits at $P_{\text{fluid}} = P$. In a closed, or rock-dominated, system, where components are immobile and do not communicate with the surroundings, the composition of the fluid, especially with respect to oxygen, is buffered by solid phases. In an open, or fluid-dominated, system, the fluid composition is generally externally controlled.

Ion activity in metamorphic fluids influences mineral stability via ion-exchange reactions.

14.5 An important objective of metamorphic petrology is to elucidate the P–T conditions of metamorphism of a body. This objective is ideally realized with petrogenetic grids in P–T space.

Multivariant stability fields and their bounding univariant equilibria curves on P–T diagrams model mappable metamorphic zones and isograds, respectively, in real metamorphic bodies. Continuous reactions occur within metamorphic zones and discontinuous ones at isograds. Valid interpretation of any isograd based on a coupled reaction relies on proper evaluation of all the participating phases, their modal proportions, and the bulk chemical composition of the rock.

Metamorphic reactions may not proceed directly from reactant to product phases but may use an indirect path and an intermediate phase.

14.6. In large-scale, fluid-dominated metasomatic systems, ion-exchange reactions partitioning a particular element unequally between coexistent phases at equilibrium may promote some types of metamorphic differentiation. Transport of mobile components in open systems can produce discontinuities in bulk composition where none previously existed and tend to reduce the number of the stable phases in the constituent assemblage.

14.7. There are valid concerns about whether metamorphic mineral assemblages represent states of stable equilibrium. For mineralogical reactions to progress, the equilibrium P–T conditions must be overstepped. Reactions are impeded by activation energy barriers. Most reactions proceed via a sequence of steps, and the rate of the slowest one governs the overall rate of reaction.

Reaction rates are faster and the amount of overstepping of equilibrium conditions is less for fluid-release reactions than for solid–solid reactions. Fluids are effective catalysts in metamorphic reactions. Some reactions may proceed faster via an indirect path involving a transient phase. Retrograde metamorphism is rare and metastable persistence of highest-grade volatile-poor mineral assemblages is common chiefly because of inaccessibility to fluids.

STUDY QUESTIONS

1. It is not uncommon to find coexisting andalusite and sillimanite within one thin section of a pelitic metamorphic rock. Discuss possible explanations for this coexistence.

2. Some Ca–Mg–Fe–Al garnets in granulite facies rocks show replacement by a fine-grained intergrowth, or symplectite, of aluminous hypersthene + Mg–Fe spinel + cordierite. Using S and V values in Robie and

others (1978), postulate how the symplectite might have originated.

3. Contrast the functions of a petrogenetic grid and a triangular composition projection.

4. Explain how it might be possible, from a kinetic standpoint, for a mineral to form outside of its stability field.

5. Propose two geologically realistic histories for the rocks shown in Figure 14-4. In other words, what changes in *geologic setting* might have created the coronas?

6. Discuss possible geologic histories of the rocks illustrated in Figure 14-5.

7. Figure 6-11 shows the equilibrium boundary for the graphite–diamond polymorphs. Discuss the validity of this boundary line in elucidating P and T in mantle systems of varying chemical composition.

8. Using Figure 14-8 as a guide, sketch compositional relations on a sequence of ACM diagrams at 0.4 GPa intervals and 1200°C for the peridotite whose chemical composition is as follows:

	Wt.%
SiO_2	44.7
Al_2O_3	4.7
Fe_2O_3	3.0
FeO	4.7
MgO	36.6
CaO	4.2
Na_2O	0.5
K_2O	0.1
TiO_2	0.1
MnO	0.1
Cr_2O_3	0.5
NiO	0.3
Total	99.5

Let $A = Al_2O_3$, $C = CaSiO_3$, and $M = MgSiO_3$ in the diagram. Devise a means of handling the abundant olivine in these assemblages.

9. a. Represent the continuous reaction cordierite = garnet + sillimanite + quartz in a *T–X* diagram.
   b. Represent the same reaction in a sequence of triangular AFM diagrams, making sure you show the changing stability limits with respect to F and M of the cordierite and garnet during the reaction.
   c. Sketch textural relations that would show the reaction had occurred in the backward direction.
   d. Under what geologic conditions would such a reaction occur in a forward direction? Backward direction?

10. Draw the approximate breakdown curve for pure calcite up to 0.5 GPa, given that the curve for Equation (14.10) lies at $P_{CO_2} = 3$ MPa at 1090°C (Ernst, 1976). Would the breakdown curve for dolomite → calcite + periclase + CO_2 lie at higher or lower *T*? Relative to the dolomite breakdown curve, would the curve for dolomite + 2 quartz = diopside + $2CO_2$ lie at higher or lower *T*?

11. Using the breakdown curves sketched in question 10, together with those in Figure 14-12, draw fluid-release (CO_2) curves similar to those in Figure 14-15 for (a) dolomite + 2 quartz and (b) 3 dolomite + quartz as a function of *T* at some constant *P*.

12. Write a possible reaction to consume annite in a sulfur-rich environment, producing pyrite and silicate phases. Discuss the possible geologic significance of such a reaction.

13. Discuss evidence for the existence of a separate fluid phase in metamorphic bodies at the time of their metamorphism.

14. From data in Figure 14-16, draw a perspective view of the univariant surface for the reaction calcite + quartz + fluid = wollastonite + fluid in a *P–T–X* block diagram.

15. Analyze the K-feldspar isograds in Figure 14-21 by writing a series of reactions with varying molecular ratios of muscovite/quartz.

16. Consider the reaction

$$NaCa_2(Fe,Mg)_4Al_3Si_6O_{22}(OH)_2 + (Fe,Mg)_3Al_2Si_3O_{12}$$
$$\text{hornblende} \qquad \text{almandiferous garnet}$$
$$+ \; 5SiO_2 \; = \; 7(Fe,Mg)SiO_3 \; +$$
$$\text{quartz} \qquad \text{orthopyroxene}$$
$$NaAlSi_3O_8 + 2CaAl_2Si_2O_8 + H_2O$$
$$\text{in plagioclase}$$

Comment on the validity of using this reaction to separate amphibolite- from granulite-facies rocks. Would the presence of highly variable modal amounts of quartz in the rock body influence your answer?

17. Account for the slopes of the reaction curves in Figure 14-23b.

18. Discuss the utility of $CaCO_3$ and Al_2SiO_5 polymorphs as indicators of *P–T* conditions. Consider both equilibrium and kinetic implications.

REFERENCES

Bailey, D. K. 1976. Experimental methods and the uses of phase diagrams. In D. K. Bailey and R. Macdonald (eds.), *The Evolution of the Crystalline Rocks*. New York: Academic Press, p. 3–97.

Brady, J. B. 1977. Metasomatic zones in metamorphic rocks. *Geochimica et Cosmochimica Acta* 41:113–125.

Carmichael, D. M. 1969. On the mechanism of prograde metamorphic reactions in quartz-bearing pelitic rocks. *Contributions to Mineralogy and Petrology* 20:244–267.

Carmichael, D. M. 1970. Intersecting isograds in the Whetstone Lake area, Ontario. *Journal of Petrology* 11:147–181.

Ernst, W. G. 1976. *Petrologic Phase Equilibria*. San Francisco: W. H. Freeman and Company, 333 p.

Eugster, H. 1959. Reduction and oxidation in metamorphism. In P. H. Abelson (ed.), *Researches in Geochemistry*. New York: John Wiley and Sons, pp. 397–426.

Fyfe, W. S., Turner, F. J., and Verhoogen, J. 1958. Metamorphic reactions and metamorphic facies. Geological Society of America Memoir 73, 259 p.

Fyfe, W. S., Price, N. J., and Thompson, A. B. 1978. *Fluids in the Earth's Crust*. New York: Elsevier, 383 p.

Greenwood, H. J. 1963. The synthesis and stability of anorthophyllite. *Journal of Petrology* 4:317–351.

Greenwood, H. J. 1976. Metamorphism at moderate temperatures and pressures. In D. K. Bailey and R. Macdonald (eds.), *The Evolution of the Crystalline Rocks*. New York: Academic Press, pp. 187–259.

Helgeson, H. C., Delany, J. M., Nesbitt, H. W., and Bird, D. K. 1978. Summary and critique of the thermodynamic properties of rock-forming minerals. *American Journal of Science* 278A:1–229.

Hensen, B. J. 1971. Theoretical phase relations involving cordierite and garnet in the system $MgO-FeO-Al_2O_3-SiO_2$. *Contributions to Mineralogy and Petrology* 33:191–214.

Herzberg, C. T. 1978. Pyroxene geothermometry and geobarometry: Experimental and thermodynamic evaluation of some subsolidus phase relations involving pyroxenes in the system $CaO-MgO-Al_2O_3-SiO_2$. *Geochimica et Cosmochimica Acta* 42:945–957.

Holdaway, M. H. 1971. Stability of andalusite and the aluminum silicate phase diagram. *American Journal of Science* 271:97–131.

Holland, H. D. 1967. Gange minerals in hydrothermal deposits. In H. L. Barnes (ed.), *Geochemistry of Hydrothermal Ore Deposits*. New York: Holt, Rinehart and Winston, pp. 382–436.

Joesten, R. 1974. Local equilibrium and metasomatic growth of zoned calc-silicate nodules from a contact aureole, Christmas Mountains, Big Bend region, Texas. *American Journal of Science* 274:876–901.

Korzhinskii, D. S. 1970. *Theory of Metasomatic Zoning* (translated by J. Agrell). Oxford: Clarendon Press, 162 p.

Lacy, E. D. 1965. Factors in the study of metamorphic reaction rates. In W. S. Pitcher and G. W. Flynn (eds.), *Controls of Metamorphism*. Edinburgh: Oliver and Boyd, pp. 140–154.

Mather, J. D. 1970. The biotite isograd and the lower greenschist facies in the Dalradian rocks of Scotland. *Journal of Petrology* 11:253–275.

Miyashiro, A. 1973. *Metamorphism and Metamorphic Belts*. New York: John Wiley and Sons, 479 p.

Newton, R. C., and Fyfe, W. S. 1976. High pressure metamorphism. In D. K. Bailey and R. Macdonald (eds.), *The Evolution of the Crystalline Rocks*. New York: Academic Press, pp. 181–186.

Nockolds, S. R., Knox, R. W. O'B., and Chinner, G. A. 1978. *Petrology for Students*. New York: Cambridge University Press, 435 p.

Orville, P. M. 1963. Alkali ion exchange between vapor and feldspar phase. *American Journal of Science* 261:201–237.

Powell, R. 1978. *Equilibrium Thermodynamics in Petrology: An Introduction*. New York: Harper and Row, 284 p.

Robie, Hemingway, B. C., and Fisher, J. R. 1978. Thermodynamic properties of minerals and related substances at $298.15°K$ and 1 bar (10^5 Pascals) pressure and at higher temperatures. *U.S. Geological Survey Bulletin* 1452, 456 p.

Thompson, J. B., Jr. 1955. The thermodynamic basis for the mineral facies concept. *American Journal of Science* 253:65–103.

Thompson, J. B. 1959. Local equilibrium in metasomatic processes. In P. H. Abelson (ed.), *Researches in Geochemistry*, Vol. 1. New York: John Wiley and Sons, p. 427–457.

Touret, J. 1977. The significance of fluid inclusions in metamorphic rocks. In D. G. Fraser (ed.), *Thermodynamics in Geology*. Boston: D. Reidel, pp. 203–227.

Turner, F. J. 1981. *Metamorphic Petrology: Mineralogical, Field and Tectonic Aspects*, 2nd Ed. New York: McGraw-Hill, 524 p.

Turner, F. J., and Verhoogen, J. 1960. *Igneous and Metamorphic Petrology*, 2nd Ed. New York: McGraw-Hill, 694 p.

Vernon, R. H. 1976. *Metamorphic Processes*. London: Thomas Murby, 247 p.

Vidale, R. 1974. Metamorphic differentiation layering in pelitic rocks of Dutchess County, New York: In A. W. Hofmann, B. J. Giletti, H. S. Yoder, Jr., and R. A. Yund (eds.), *Geochemical Transport and Kinetics*. Carnegie Institution of Washington Publication 634, pp. 273–286.

Weill, D. F., and Fyfe, W. S. 1967. On equilibrium thermodynamics of open systems and the phase rule (A reply to D. S. Korzhinskii). *Geochimica et Cosmochimica Acta* 31:1167–1176.

Wintsch, R. P. 1975. Solid-fluid equilibria in the system $KAlSi_3O_8$-$NaAlSi_3O_8$-Al_2SiO_5-SiO_2-H_2O-HCl. *Journal of Petrology* 16:57–79.

Wood, B. J., and Fraser, D. G. 1976. *Elementary Thermodynamics for Geologists.* Oxford: Oxford University Press, 303 p.

Zen, E-an. 1966. Construction of pressure-temperature diagrams for multicomponent systems after the method of Schreinemakers—a geometric approach. *U.S. Geological Survey Bulletin* 1225, 56 p.

Zen, E-an. 1963. Components, phases and criteria of chemical equilibrium in rocks. *American Journal of Science* 261:929–942.

Zen, E-an, and Thompson, A. B. 1974. Low-grade regional metamorphism: Mineral and equilibrium relations. *Annual Reviews of Earth and Planetary Sciences* 2:179–212.

Part IV

THE EARLY HISTORY OF THE EARTH AND OTHER PLANETARY BODIES

During the past two decades, the science of geology has experienced a remarkable revolution that overshadows its more or less steady development during the previous two centuries. This explosive growth in knowledge and ideas has been the result of a greatly accelerated program of research into the sea floor, which resulted in the development of the theory of plate tectonics, and the exploration of the Moon and planets by manned and unmanned space probes. These two new frontiers have stimulated geologists to ask many new questions about the early history of our planet. How far back in the geologic past has plate tectonics operated? What was the nature of geologic processes in the infancy of our planet?

As the sea floor on Earth is renewed via plate spreading and consumption on a time scale of only 0.2 Gy, we must turn to the continental masses, or to other planetary bodies, to find clues about the Earth's early development. Recent improvements in laboratory techniques for analysis of trace elements and isotopes have revealed fascinating new insights into the chronology and petrology of Precambrian continental rocks. Some of these old rock bodies are similar to younger rocks, but others are very different.

The aim of this last part of the text is to consider these similarities and differences and, with the additional insights provided by information on extraterrestrial rocks, to construct a model for the origin and early evolution of the Earth and other planetary bodies.

15

Precambrian Rock Bodies and Systems

In this chapter we deal with Precambrian rock bodies on Earth. Their nature is seriously obscured due to fragmentation and overprinting by more recent events, and their evolution may not be explicable by analogy with modern global geologic processes.

The oldest rocks on Earth are as old as 3.8 Gy and occur in widely scattered Archean terranes that are chiefly tonalitic gneisses and metamorphosed sequences of submarine basaltic lavas and minor calc-alkaline volcanic rocks. While some geologists interpret this twofold assemblage in terms of ancient vestiges of sialic continents and mafic ocean basins, few would venture to explain their formation by plate tectonic processes as we know them today. Unusually high geothermal gradients and probably unique tectonic regimes during the Archean and early Proterozoic are indicated by the occurrence of high-T peridotite lava flows and the general absence of highly alkaline magmatic rocks.

During the Proterozoic, oceanic mafic–ultramafic associations are virtually absent, yet swarms of basaltic dikes and several huge stratiform differentiated basaltic intrusions occur in continental settings. What tectonic and magmatic processes are implied by this record? How and why did massifs of the anorthosite–charnockite suite develop in the late Proterozoic, and at no time before or since? Elongate Proterozoic "mobile" belts of reworked older rocks are unlike the lenticular and reticulate pattern of Archean terranes and also unlike longer, sinuous Phanerozoic orogenic belts formed by lithospheric plate subduction. Is this still another facet of unique processes, or merely an artifact of depth of erosion or of our inability to read the rock record correctly?

Sialic continents have surely grown through the eons of recorded time, but at what rate? Was most of the growth during the Archean, with only very minor juvenile accretion since then? Continental growth by extraction and flotation of low-T partial melts enriched in incompatible elements has left the upper mantle source regions depleted in these constituents and chemically inhomogeneous. How do oceanic rift, oceanic island, and continental plateau magma systems tap different mantle source regions? And how have these regions evolved in relation to deeper and apparently less undifferentiated mantle?

15.1 ARCHEAN TERRANES ON EARTH

Every Precambrian shield on Earth contains scattered terranes of metamorphic and magmatic Archean rocks > 2.5 Gy old surrounded or transected by more or less linear belts of Proterozoic rocks 2.5 to 0.57 Gy old (see Figure 15-1). The oldest terranes are a montage built by multiple overprints of deformation, recrystallization, partial melting, and magmatic intrusion. Opening and reequilibration of isotopic systems hampers radiometric age determinations, which are the only means of establishing an absolute chronology in these nonfossiliferous rocks.

Archean terranes consist of two types of rock bodies: **supracrustal sequences,** or **greenstone belts,** comprised of metamorphosed basaltic and locally ultramafic lava flows, partly of submarine character, plus minor intermediate and silicic lavas and sedimentary deposits that are surrounded by a **gneiss complex** of quartzo-feldspathic, chiefly tonalitic orthogneisses and migmatites intruded by granitic plutons.

Field and Chronologic Relations

Three of the better-documented terranes are described below; they show the progressively greater development (or preserved extent?) of supracrustal sequences in younger terranes.

West Greenland. The oldest rocks authenticated on Earth occur in West Greenland between Godthab and Isua, where McGregor (1973; also in Barker, 1979) and others have mapped an extensive area of gneisses tectonically interleaved with an assortment of supracrustal rocks that were

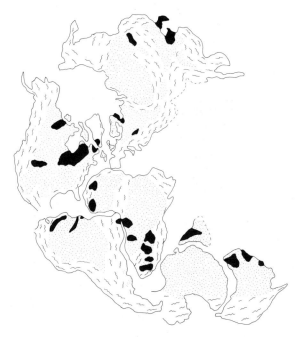

Figure 15-1 Permian predrift map of the continents showing generalized distribution of Archean (black), Proterozoic (stippled), and Phanerozoic crust (wavy lines). See also A. M. Goodwin in Windley (1976). [After K. C. Condie, 1976, *Plate Tectonics and Crustal Evolution* (New York: Pergamon).

emplaced chiefly on the Earth's surface (see Figure 15-2). There are two chronologic successions of supracrustal rocks and younger orthogneisses. The oldest Isua supracrustal sequence is found as scattered inclusions and rafts hundreds of meters long in the younger Amitsoq gneisses but is best preserved in an arcuate belt 32 km long and a few kilometers wide at Isua (see Figure 15-2b). This sequence, dated by Pb–Pb, Rb–Sr, and Sm–Nd methods at 3.71 to 3.77 Gy, consists of amphibolites of presumed volcanic origin with tholeiitic composition, pelitic schists, banded quartz–magnetite ironstones, thinly laminated quartzites (metacherts?), talc schists with relict masses of dunite, calc-silicate rocks, and quartzo-feldspathic schists of volcanic and conglomeratic derivation (Allart, in Windley, 1976). Inclusions within the

(a)

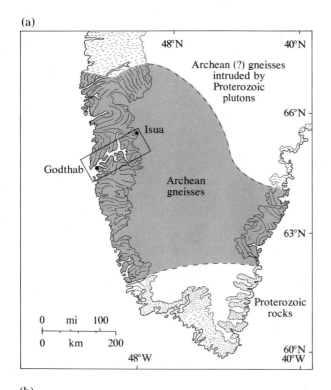

Figure 15-2   (a) Precambrian rocks in southern
Greenland. Proterozoic rocks include reworked
Archean(?) gneisses (short dashed lines) and
supracrustal and granitic rocks (dotted pattern). This
terrane is a part of the North Atlantic craton, which
includes the Labrador coast in eastern Canada
(Collerson and Bridgewater, in Barker, 1979) and the
northeastern tip of Scotland. (b) Enlarged map of the
Godthab–Isua area outlined in (a). [Part (a) after J. S.
Myers, The early Precambrian gneiss complex of
Greenland, in B. F. Windley (ed.), 1976, *The Early
History of the Earth* (New York: John Wiley and Sons);
part (b) after V. R. McGregor, Archean gray gneisses
and the origin of the continental crust, in F. Barker (ed.),
1979, *Trondhjemites, Dacites and Related Rocks* (New
York: Elsevier).]

(b)

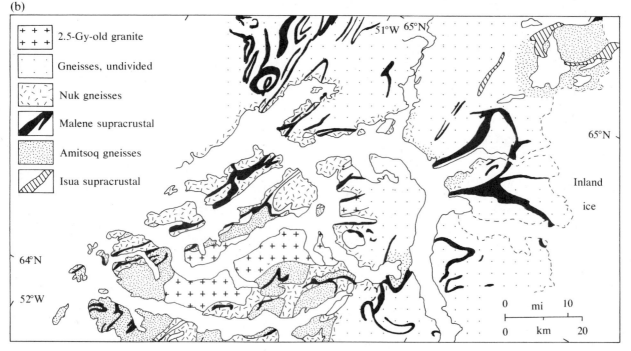

Amitsoq gneisses include amphibolites and pyroxenites with tholeiitic and komatiitic (see pages 539–540) affinities. Much of the Amitsoq gneiss complex appears to have been originally emplaced as syntectonic magmatic intrusions at least 3.6 Gy ago, although this record has been nearly obliterated by subsequent overprints. The Amitsoq gneisses were intruded by a swarm of Ameralik basalt dikes sometime before the second-generation supracrustal sequence, the Malene, was deposited about 3.0 Gy ago. Though generally intensely deformed, pillow structures are locally preserved in the Malene sequence, showing that the sheets of amphibolites up to 1 km thick were in part originally submarine basaltic lavas. Metasedimentary rocks in the sequence include pelitic schists (with garnet, sillimanite, and cordierite), marbles, quartzites, and banded iron formations. The Amitsoq gneisses, Ameralik dikes, Malene supracrustal rocks, and some local anorthositic leucogabbroic sheets were variably deformed (see Figure 15-3) and tectonically interleaved as imbricate thrust slices more or less during emplacement of tonalitic to granodioritic magmas that became the second-generation gneiss complex. In addition to numerous inclusions of older rock, these Nuk gneisses have compositional layering and clots, veins, and diffusely bounded masses of pegmatite, all drawn out into elongated lenses and streaks and intricately folded in ductile manner. Though compositionally similar to some calc-alkaline batholiths of Phanerozoic continental margin settings (see Chapter 4) and of similar volume, the form of the Nuk bodies is utterly different. Instead of pluglike diapiric masses up to several kilometers in horizontal diameter, they are mostly subconcordant sheets that greatly inflated the smaller volume of older rock into which they were emplaced.

The youngest major event in west Greenland, about 2.8 Gy ago, was a severe episode of major deformation and granulite to amphibolite facies metamorphism. The oldest rocks in many Archean terranes around the world are also of this same age.

Southeastern Africa. The Rhodesian and Kaapvaal Cratons in the southeastern part of the African continent harbor extensive belts of supracrustal rocks lying in a "sea" of quartzo-feldspathic rocks (see Figure 15-4).

The Ancient Gneiss Complex in the Kaapvaal Craton has suffered polyphase deformation and metamorphism, obscuring original fabrics and complicating chronologic relations (Hunter, in Barker, 1979). The oldest rocks are bimodal gneisses with complexly alternating layers of felsic rock (actually trondhjemite–tonalite with 70% SiO_2) and amphibolite (55% SiO_2) that appear to have formed by intense tectonic intermingling of basaltic and tonalitic volcanic or plutonic bodies rather than by metamorphic differentiation of an originally more homogeneous protolith. An apparently younger sequence of homogeneous, nonlayered hornblende tonalite gneisses has an age of 3.4 Gy. Both of these gneisses show evidence of at least three episodes of fold deformation. Younger migmatites, cordierite–garnet-bearing gneisses, metamorphosed ironstones, and lenses of quartz monzonite (3.2 Gy old) comprise the remainder of the Ancient Gneiss Complex. Granodioritic plutons are intrusive into this complex.

In contrast to the relatively small volume of supracrustal rocks in West Greenland, such sequences in younger Archean terranes are larger and are generally referred to as **greenstone belts** because the elongate exposures are dominated by mafic to ultramafic igneous rocks metamorphosed to more or less greenschist facies mineral assemblages. The lower 11 km of the 16-km-thick sequence in the Barberton greenstone belt in the Kaapvaal craton (Anhaeusser and others, 1969) includes alternating metaperidotite and metabasalt lava flows that are locally pillowed; minor intrusive sills of basalt and peridotite, some of which are layered and differentiated, and thin intercalations of felsic tuffs, cherts, and shales also occur (see Figure 15-5). All these are metamorphosed. Overlying this basal ultramafic–mafic group of rocks are several kilometers of mafic and

(a)

Figure 15-3   Progressive metamorphism and ductile flowage of Ameralik basalt dikes and their host Amitsoq gneisses in West Greenland. (a) Essentially undeformed dike cuts augen granodioritic gneiss. (b) Deformed dike and host gneiss; note the flattened feldspar augen, which define a folded foliation. (c) More intensely deformed body in which the slightly less ductile basalt dike rock has been broken into crude boudins encased in a swirled gneiss matrix; note intensely flattened feldspar augen and pegmatitic segregations in gneiss. [Geological Survey of Greenland photographs courtesy of V. R. McGregor, from V. R. McGregor, 1973, The early Precambrian gneisses of the Godthab district, West Greenland, *Philosophical Transactions of the Royal Society of London,* Series A, 273.]

felsic metavolcanic and metasedimentary rocks; at least five cyclic units are recognized, beginning with metabasaltic (and minor metaperidotitic) flows, then metamorphosed andesites, dacites, rhyodacites, and rhyolites, and capped by metamorphosed cherts and minor shales and carbonate rocks. In each younger cycle, the sediment/volcanic proportion increases and the lava flows become more felsic and more typically tholeiitic and calc-alkaline. Relict brecciated flow margins, pillows, varioles, vesicles, and spherulites are apparent. Increasing proportions of metavolcaniclastic rocks occur toward the top, and some of these are apparently subaqueous in origin. The uppermost part of the Barberton sequence is composed of a wide variety of metasedimentary rocks that reflects

(b)

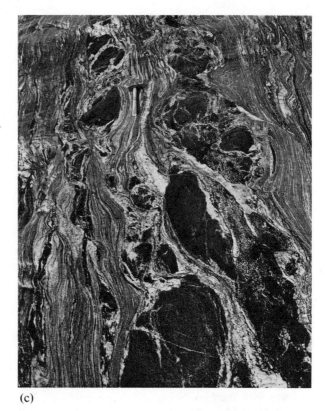

(c)

deep marine to shallow shelf depositional environments.

Many geologists envision greenstone belts, such as the Barberton, to be simple V-shaped synclines or synclinoria lying within the surrounding gneiss complex. Others interpret them as nappe or thrust sheets that have subsequently been infolded into the gneisses so that stratigraphic successions are locally inverted. These simplistic interpretations are challenged by Burke, Dewey, and Kidd (in Windley, 1976). They refer to a little known investigation by John G. Ramsay, who found that most of the Barberton belt is as strongly and complexly deformed as the surrounding gneiss complex or as any body in a Phanerozoic orogenic belt. Ramsay recognized three regionally significant imprints of

deformation involving horizontal shortening (up to 90%) as their principal strain component. Major mylonitic shear zones formed during a second deformation that produced a metamorphic foliation. The former designation of greenstone belts as ''schist belts'' better connotes their structural character. Burke and others (p. 125) emphasize ''that addition (or removal) of strata several kilometers thick, or juxtaposition of sequences originally deposited a hundred or more kilometers apart may take place along mylonite zones as little as a few centimeters thick and without causing any anomalous deformation outside them. Also, once one dislocation zone is recognized in any succession, it is almost certain that many are present.'' In view of the ductile shortening involved and the

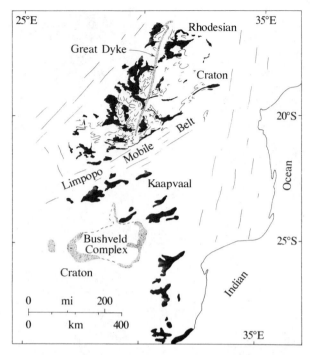

Figure 15-4   Southeast part of Africa showing Archean greenstone belts (black) in the Rhodesian and Kaapvaal Cratons with their bordering and transecting Proterozoic mobile belts (long lines). Trend lines in part of the Rhodesian Craton show form of foliation in "gregarious" gneiss domes between greenstone belts. Stippled areas are ultramafic rocks in the Great Dike and Bushveld Complex. Compare Figure 15-14. [After D. R. Hunter and P. J. Hamilton, 1978, The Bushveld Complex, in D. H. Tarling (ed.), *Evolution of the Earth's Crust* (New York: Academic Press); and A. E. Phaup, 1973, Geological Society of South Africa Special Publication 3.]

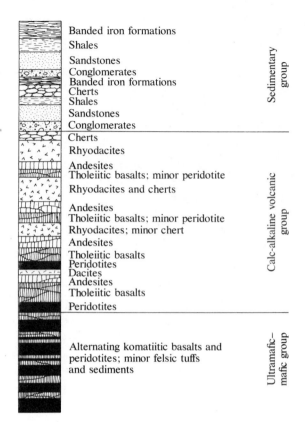

Figure 15-5   Hypothetical stratigraphic column of the volcanic–sedimentary sequence in Archean greenstone belts. This column is most representative of the Barberton belt. [After Anhaeusser, 1971, Cyclic volcanicity and sedimentation in the evolutionary development of Archean greenstone belts of shield areas, Geological Society of Australia Special Publication 3.]

possibility of multiple shear zones that repeat the stratigraphic section by imbricate stacking, we may wonder about certain allegedly unique properties of greenstone belts, namely, their extraordinary thickness and the cyclic nature of their depositional sequence.

Disagreements persist regarding the relative age of the Ancient Gneiss Complex and the Barberton greenstone sequence. Nowhere are the two in un-

faulted contact without intervening younger intrusions. Radiometric ages are roughly the same, about 3.4 Gy.

In the Rhodesian craton, greenstone–gneiss development appears to have been repeated three times during the Archean (Wilson and others, 1978). Highly deformed tonalitic gneisses intruded by ~ 3.5-Gy-old granite are associated with small remnants of a recumbently folded and inverted se-

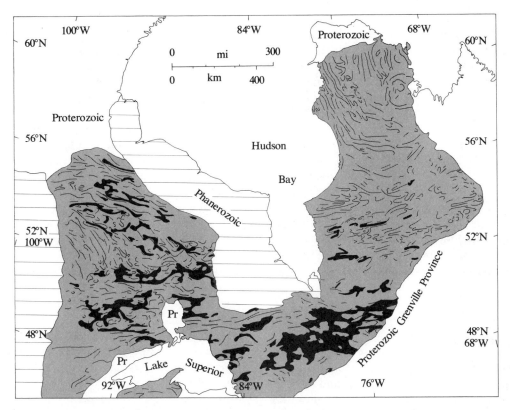

Figure 15-6 Archean gneiss complex (trend lines show foliation) and greenstone belts (black) in the Superior Province (shaded) of the south central Canadian shield. The Abitibi belt lies in the southeastern part of the province. Compare the outcrop area of these greenstone belts with the older African (see Figure 15-4) and Greenland (see Figure 15-2) supracrustal sequences. Pr, Proterozoic. [After Tectonic Map of Canada in *Geological Survey of Canada Economic Geology Report* 1.]

quence of metamorphosed basaltic and peridotitic lavas; this greenstone sequence is cut by a 3.4-Gy-old intrusion. An intermediate age greenstone–tonalitic gneiss terrane ~ 2.9 Gy old is poorly documented, whereas unconformably overlying this and the older terrane are remnants of a widespread greenstone sequence of volcanics and sediments; associated gneisses and later granitic intrusions are 2.7 to 2.6 Gy old.

Superior Province of Canada. Archean greenstone belts in the Canadian Shield (see Figure 15-6) are

mostly 2.7 to 2.5 Gy old. Large ones such as the Abitibi belt (Goodwin and Smith, 1980) are characterized by 5 to 8 km of tholeiitic and minor peridotitic flows overlaid by 4 to 7 km of intermediate and felsic calc-alkaline metavolcanic rocks. The commonly only very weakly metamorphosed flows have pillows and variolitic fabric that testify to subaqueous emplacement. The intermediate to felsic rocks are partly volcaniclastic, indicating shallow-water or even subaerial environments. These vast piles of now-deformed metavolcanic rocks are interpreted to represent huge volcanic edifices similar to modern Hawaiian shield volcanoes. As they ultimately emerged above water, erosional debris was shed into nearby deep troughs. Capping these volcanogenic metasediments are dominantly quartzo-feldspathic clastics, locally 10 km thick, that also appear to have been deposited in rapidly subsiding basins with proximal granitic sources.

The gneiss complexes surrounding the Superior greenstone belts consist of old tonalitic gneisses (+3.0 Gy) and younger metasedimentary–migmatite masses and tonalite–granite intrusions 2.8 to 2.7 Gy old that may be coeval with some parts of the greenstone sequences (Baragar and McGlynn, 1978). Field relations that might define the relative age of greenstone and gneiss are usually ambiguous but in some places favor an older gneissic basement.

Petrology of the Archean Association

Gneiss complexes.   Though comprised chiefly of feldspar, quartz, mica, and hornblende and averaging as a whole to sodic granodiorite or tonalite, the rocks of Archean gneiss complexes vary in fabric, mode of emplacement, and ultimate origin. Undeformed intrusions with well-preserved magmatic fabric, augen gneiss with megacrysts of K-feldspar, gneisses with alternating mafic–felsic layers, homogeneous but foliated orthogneisses, and migmatites are all represented.

In some terranes, widespread rocks with high Na and plagioclase as the sole or chief feldspar clearly predate K-feldspar-bearing granitic bodies. These earlier rocks have the composition of granodiorite, tonalite, and trondhjemite but are usually foliated, finer-grained, gray, layered to homogeneous orthogneisses. Later, more K-rich rocks are massive porphyritic quartz monzonites and granites and compositionally equivalent augen gneisses. The orthogneiss appellation, implying magmatic ancestry, is supported by certain chemical features (Tarney, in Windley, 1976) and

by relict intrusive relations observed in many terranes.

Amitsoq and Nuk gneisses have similar compositions (McGregor, in Barker, 1979), making derivation of the one from the other by melting unlikely because the melting range of tonalitic rocks is considerable. Before completion of melting, mobile, low-density, minimum-temperature granite melts would have tended to segregate and form separate intrusions. (Interestingly, the 2.5-Gy-old granite in the Godthab area, sketched in Figure 15-2b, has a composition appropriate to this petrogenic model and the initial $^{87}Sr/^{86}Sr$ ratio of 0.708 is significantly greater than the 0.701–0.703 ratios for the gneisses.) Generation of Nuk gneisses by partial melting of older gneisses therefore seems unlikely, and it is equally difficult to appeal to a melting process as having created the Amitsoq gneisses, for no older gneisses are exposed anywhere. The overwhelming volume of exposed quartzo-feldspathic gneisses relative to possible genetically complementary ultramafic–mafic rock types in West Greenland makes a process of magmatic differentiation difficult to believe, unless it is argued that these complementary differentiates are unseen below the present erosion surface.

McGregor (in Barker, 1979) and O'Nions and Pankhurst (1978), among others, conclude that Archean tonalitic orthogneisses were originally derived as partial melts from sources containing amphibole and garnet and with overall tholeiitic basalt composition. This conclusion is based upon patterns of trace element concentrations that show, among other features, depleted heavy REE in the gneisses. Trace element concentrations in the more K-rich granitic bodies rule out derivation as magmas from the same source as tonalitic rocks, from the tonalitic rocks themselves, or from peridotite, but favor graywacke or intermediate composition calc-alkaline source rocks.

Metamorphism in gneiss complexes ranges from amphibolite to granulite facies; charnockites are common. Inferred *P–T* conditions of metamorphism in Archean terranes lie in the range of 0.7 to

1.2 GPa and 700°C to 1200°C (Tarney and Windley, 1977). This implies depths of 25 to 42 km appropriate to geothermal gradients of 18°C/km to 35°C/km in the intermediate *P/T* facies series.

Many petrologists have found that granulite facies rocks are depleted in K, Rb, Th, and U relative to lower-grade amphibolite facies rocks, conceivably because of their preferential incorporation into hydrous silicate partial melts or hydrous fluids that migrated out of the bodies (Tarney and Windley, 1977). This depletion has a significant bearing on the rate of radiometric heat generation in the lower crust as well as the $^{87}Sr/^{86}Sr$ ratio. With Rb/Sr ratios mostly below 0.02, Sr-isotope ratios will lie in the same range as mantle peridotites; hence it is not possible to distinguish on this basis between partial melts derived from mantle or lower crustal, granulite-facies source rocks.

Komatiite. Almost simultaneously, petrologists working on the Barberton (Viljoen and Viljoen, 1969), Western Australia (Nesbitt, 1971), and Superior (Pyke and others, 1973) greenstone belts documented the first unequivocal occurrences of peridotitic magmas. The unusual fabric and composition of these unique Archean rocks warrant a special name—**komatiite,** taken from the Komati River, which passes through the Barberton greenstone belt—and receive special consideration here.

Relict vesicles, pillows, and breccias indicate emplacement as extrusive lava flows, partly in subaqueous environments; other komatiite bodies are sills. Relict glass, spinifex texture, and the thinness of some flows (see Figure 15-7) indicate the ultramafic magmas had very low viscosity and undercooled rapidly upon extrusion. **Spinifex fabric** (named after a spiky Australian grass) is an array of criss-crossing sheafs of subparallel blade- or platelike feathery olivine (Fo_{90-95}) or aluminous clinopyroxene crystals (see Figure 15-7b). Between such crystals, and comprising about half of the spinifex zone of the flow, is a finer-grained aggregate of devitrified glass, skeletal to feathery

clinopyroxenes, and skeletal chromites. Olivines are largest at the base of the spinifex zone and are length oriented subperpendicular to flow margins, that is, they grew normal to the cooling surface in the direction of maximum supersaturation. Similar textures occur in metallurgical slags and in silicate melts in the laboratory that have been undercooled drastically below the range of normal nucleation and crystal growth (see pages 239–245). Many komatiite magmas were more or less wholly liquid at the time of extrusion and were at or above their liquidus temperature.

Spinifex-textured komatiites range from peridotitic to basaltic in composition. Peridotitic komatiites share chemical features of peridotites in general (see Table 15-1) and must therefore be distinguished from them texturally (Arndt and others, 1979). Basaltic komatiites can be distinguished from tholeiitic basalts, with which they are commonly intimately associated, on the basis of lower Fe/(Fe + Mg) ratios, K and Ti contents, and higher Mg, Ni, and Cr (see Table 15-1). The variation diagram in Figure 15-8 suggests that suites of associated komatiitic to tholeiitic rocks may have evolved essentially by olivine fractionation; clinopyroxene and perhaps plagioclase fractionation are reflected in the basaltic range.

Phenocryst-free, noncumulate, spinifex-textured peridotitic komatiites with almost 30 wt.% MgO must have solidified from very high-*T* melts formed by extensive partial melting of peridotite source rocks. Melt temperatures in excess of 1600°C and degrees of partial melting as high as 80% have been considered possible (Green, 1975). Lesser degrees of melting at lower temperatures of similar source rock conceivably yield more tholeiitic primary melts. Thus the tholeiitic basalts (and andesites) commonly associated with komatiites could have a dual origin—crystal fractionation and partial melting.

Disagreements persist regarding the exact definition of komatiite, but it is apparent that spinifex-textured *peridotitic* komatiites that formed from very high-*T* , highly magnesian primary (?) melts

(a)

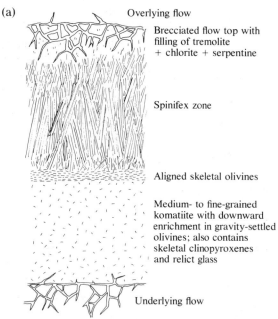

Overlying flow

Brecciated flow top with filling of tremolite + chlorite + serpentine

Spinifex zone

Aligned skeletal olivines

Medium- to fine-grained komatiite with downward enrichment in gravity-settled olivines; also contains skeletal clinopyroxenes and relict glass

Underlying flow

Figure 15-7   Spinifex texture in komatiitic peridotite flows in the Archean greenstone belt of the Canadian Superior Province. (a) Idealized cross section through a flow. Other flows are entirely similar to the lower part of the one shown here and still others are pillowed (Arndt and others, 1979). Flows are 0.5 to 15 m thick and average 3 m. (b) Coarse spinifex texture enhanced by weathering of exposed flow. (c) In thin section somewhat skeletal blades of olivine are arrayed among feathery augite and devitrified glass. [Part (a) after D. R. Pyke, A. J. Naldrett, and O. R. Ekstrand, 1973, Archean ultramafic flows in Munro township, Ontario, *Geological Society of America Bulletin* 84; photographs in parts (b) and (c) courtesy of N. T. Arndt.]

(b)

(c)

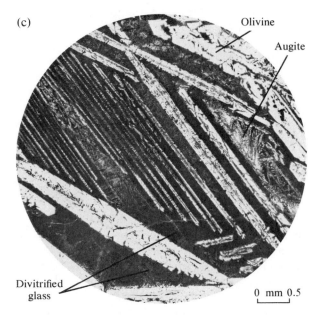

Olivine

Augite

Divitrified glass

0   mm   0.5

Table 15-1 Representative chemical and normative compositions of komatiitic and tholeiitic rocks from the Archean Abitibi greenstone belt in the Superior Province of northeast Ontario, Canada (in wt.%).

	1	2	3	4
SiO_2	44.4	42.26	53.0	51.7
TiO_2	0.35	0.63	0.83	1.43
Al_2O_3	7.40	4.23	14.4	12.8
Cr_2O_3	0.47	—	0.03	0.01
Fe_2O_3	13.3	3.61	2.67	14.3
FeO		6.58	8.49	
MnO	0.26	0.41	0.18	0.25
MgO	27.5	31.24	7.76	5.24
NiO	0.02	—	0.0	0.0
CaO	6.33	5.05	11.2	12.2
Na_2O	0.06	0.49	1.83	1.25
K_2O	0.08	0.34	0.02	0.79
Q	0.0	0.0	3.19	3.44
Or	0.43	2.02	0.12	4.83
Ab	0.55	4.15	16.61	11.52
An	18.9	8.32	31.28	28.29
Di	9.0	11.22	19.78	27.46
Hy	19.3	15.79	27.86	22.41
Ol	51.4	46.39	0.0	0.0
Mt	—	5.23	1.16	2.05
Il	0.47	1.19		

1. Spinifex-textured peridotitic komatiite recalculated water free.
2. Average peridotite (see Appendix D).
3. Basaltic komatiite lava recalculated water free.
4. Tholeiitic basalt recalculated water free.
Source: N. T. Arndt, A. J. Naldrett, and D. R. Pyke, 1977, Komatiitic and iron-rich tholeiitic lavas of Munro Township, northeast Ontario, *Journal of Petrology* 18.

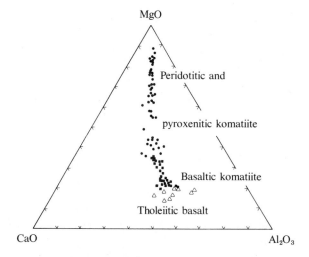

Figure 15-8 Compositional spectrum of peridotitic to basaltic rocks from Munro Township in the Archean Abitibi greenstone belt, Ontario. [After N. T. Arndt, A. J. Naldrett, and D. R. Pyke, 1977, Komatiitic and iron-rich tholeiitic lavas of Munro Township, northeast Ontario, *Journal of Petrology* 18.]

are restricted to the Archean. Several bodies of basaltic komatiite of Proterozoic age have been reported, and there is one occurrence of Phanerozoic (Ordovician) age in ophiolite in Newfoundland (see Arndt and others, 1979).

Significance and origin of volcanic rocks in greenstone belts. Many petrologists (for example, O'Nions and Pankhurst, 1978) allege that, with the exception of komatiitic rocks, Archean metavolcanic rocks are broadly similar in chemical composition to younger magmatic rocks. There are, however, some differences that may have real significance or that may merely reflect biases in sampling and analyses or the effects of secondary recrystallization. Similarities between Archean and younger mafic magmas have been interpreted to mean that Archean upper mantle source regions were similar to younger upper mantle and that similar tectonic settings prevailed. But the fact that some Archean rocks have compositions similar to those in modern island arcs, or oceanic rifts, or whatever tectonic setting is not in itself any indication of the Archean tectonic setting; only field relations can verify that. Compositional similarity only indicates similarity in the combination of variables relating to magma source, generation, and diversification. It must also be realized that because of migration of K, Ca, Si, Fe, and so on in open metamorphic systems, a chemical analysis of a metaigneous rock may tell nothing of its original magmatic composition. All it may indicate is the present mineralogical composition. Elements not tolerated in metamorphic minerals will have been removed. However, in a more positive sense, it has been found that, at least in low-grade metamorphic systems, certain minor and trace elements are relatively immobile (Condie in Windley, 1976), including Ti, Y, Zr, Cr, V, Sr, Sc, Hf, Co, Nb, Ni, and rare earth elements.

In Canadian Archean metavolcanic rocks, Goodwin and Smith (1980) note a chemical discontinuity between basalts with depleted LIL elements and unfractionated REE abundance patterns and andesites with greater LIL-element concentrations and fractionated REE patterns showing HREE depletions (see page 55). More felsic rocks show similar but more extreme features as andesites. They postulate that this discontinuity marks a shift from wholly mantle derivation for basaltic rocks to magma-generating processes involving sialic crustal material.

Speculations on the Evolution of the Archean Crust

Because of the multitude of variables and too few facts, geologists have proposed many different hypotheses regarding the early development of the Earth's crust; all are highly speculative and should be considered as guides to future data gathering and thinking. Basically, these multiple working hypotheses fall between two extremes: those that adopt a uniformitarian approach based upon modern plate tectonic processes, casting Archean evolution in a framework of sea-floor spreading and lithospheric subduction, and those that adopt a more unique approach to Archean processes. Allied questions are concerned with the composition of the first crust—whether mafic or sialic—and with the rate of continental growth during the Archean.

Thermal considerations.   There is general consensus that the Earth was much hotter in its infancy than at present. Although accreting material that formed the proto-Earth may not have been very hot, subsequent segregation of the metallic core (see Section 15-5) and probable catastrophic meteorite bombardment together with a much greater rate of radiogenic heat production in the Earth (see Figure 15-9) likely produced sufficient thermal energy to cause melting of its outer-periphery. Near-surface geothermal gradients would have been much greater than at present. Convective flow within this hotter Earth probably occurred at two levels: small-scale currents in a shallow layer just below a thin crust, and deeper larger cells that might have encompassed the whole mantle (McKenzie and Weiss, 1975).

We may speculate that the thin, scumlike crust over this double convective system was warped into undulating furrows and rises and may have suffered episodic breakage and piecemeal engulfment. Or, if this scum were a low-density differentiate of an underlying molten layer, it may have become thickened into floating rafts over descending limbs of convection cells. Intense meteorite bombardment terminating ~ 3.9 Gy ago on the Moon was likely also experienced by other nearby planetary bodies, including Earth (see Section 16.2). Impacting presumably created significant breakup of the thin crust and certainly would have generated a good deal of thermal energy.

Whatever happened in the early history of the Earth, the thermal regime must have been such as to preclude the formation, or at least the preservation, of any crust prior to 3.8 Gy ago. O'Nions and Pankhurst (1978, p. 212) note that "the earliest crust which was formed may have been completely destroyed, even within a short space of time, and the oldest rocks preserved may represent a fairly advanced stage of crustal evolution."

Is the oldest preserved crust mafic or sialic?   Advocates of a mafic crust point out that no early sialic crust has been positively identified; gneiss complexes seem to be composed of magmatic proto-liths intruded into older dominantly mafic supra-crustal rocks. For the West Greenland Archean, McGregor (in Barker, 1979) postulates that early oceanic crust composed chiefly of basaltic vol-canic rocks and minor sediments (Isua and Malene supracrustal sequences) was swept together, form-ing piles of imbricate thrust slices over descending mantle currents. As these piles thickened, the deeper portions were converted to amphibolite and after further depression began to melt, forming to-nalitic magmas. Ascending magmas penetrated and consolidated along thrust planes in the grow-ing pile, forming syntectonic sills. Subsequently, some of these gneisses were deformed by ductile flow and depressed deeply enough to be partially melted, forming granodioritic to granitic melts.

In view of the probably greater geothermal gra-dients in the crust of the early Earth, it came as a surprise that Archean metamorphism was not of the low-P/T type and that the inferred thickness of the Archean crust was so great. However, the crustal imbrication in sialic rafts overlying de-scending mantle currents envisaged by McGregor could explain the more modest thermal regime re-corded in Archean metamorphic bodies.

Fyfe (1976), Hargraves (1976), and Baragar and McGlynn (1978) argue instead that a world-encircling ocean and sialic crust formed early in Earth history; remnants of this crust are now seen in the widespread gneiss complexes that make up the major part of Archean terranes. Sialic partial melts collected and froze above more refractory, mafic and denser source regions in a protomantle. The maximum thickness of this sialic layer is con-strained by its minimum melting temperature, which may be taken as roughly 750°C for water-rich systems. Where temperatures exceeded this value, melting ensued and liquids buoyed upward, thus adjusting the base of the layer at the 750°C isotherm. Because of the theoretically greater geothermal gradients 3 to 4 Gy ago, this 750°C isotherm must have been shallower than at pres-ent. Proportionately, if the radiogenic heat pro-

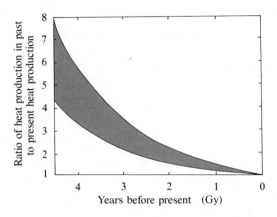

Figure 15-9 Ratio of radiogenic heat production in the past to present heat production. Shaded band encloses the ratios according to different models, all of which, however, reflect the exponential decay of radioactive elements with time. [After W. R. Dickinson and W. C. Luth, 1971, A model for plate tectonic evolution of mantle layers, *Science* 174.]

duction had been at least 3 to 4 times greater, then the sialic crust would have been roughly that much thinner. If the volume of this sial was similar to that of the present time, this early thinner crust could have covered most, if not all, of the globe. Due to its buoyancy and low melting T, this uni-versal sialic crust would not easily be subducted back into the interior and there could be no plate tectonics, at least not as we know it today.

The early universal ocean overlying the sialic crust is thought to have formed by condensation of steam escaping from the hot interior via vol-canism.

For the Archean terranes in Canada where gneissic complexes are recognizably older than the greenstone belts they enclose, Goodwin and Smith (1980) hypothesize the existence of "numerous small, thin pre-continental sialic nuclei" onto which mantle-derived basaltic extrusions accumu-lated into large volcanic edifices. Extensive melting accompanying rapid upwelling and decompression in small mantle convection cells may locally have produced komatiitic magmas. As the accumulations of dense basaltic to komatiitic lavas thickened, they conceivably sagged downward, causing the basal layers to become hotter, and eventually partially melting them, thus forming the more felsic lavas that cap the greenstone sequences.

Uniformitarian plate tectonic models. Some geologists have interpreted the formation of greenstone belts by more uniformitarian, even plate tectonic models.

Burke, Dewey, and Kidd (in Windley, 1976) envisage numerous small sialic island arcs and microcontinents that collided, squeezing basaltic oceanic crust between them to form greenstone belts. The microcontinents are believed to have been up to a hundred times smaller than at present, but no thinner and not as rigid. During this Archean **permobile** (the term preferred by Burke and others, rather than plate tectonic) activity, plates were believed to be moving two to three times more rapidly than at present away from ocean ridges whose total length probably exceeded that of today because of higher geothermal gradients.

Advocates of a form of plate tectonics during the Archean favor the greenstone sequences forming in a marginal basin–island arc setting rather than a large oceanic area. This follows from the postulated abundance of small sialic plates with nearby young ocean floor catching sialic sediments in local basins. Tarney, Dalziel, and deWit (in Windley, 1976; see also Windley, 1977, pp. 50–64) suggest the Cretaceous Rocas Verdes marginal basin off the southern tip of South America furnishes a close analog for the Archean. (Note the irony in this alleged analog: *rocas verdes* is Spanish for "green stone.") Now exposed there in a synclinal succession are pillow basalts, breccias, and tuffs overlaid by shales and graywackes with a high proportion of volcanogenic detritus and iron-rich cherts. Andesites are sparse and sheeted dike complexes are common. The succession locally overlies older granitic basement and is intruded by younger tonalite–granodiorite plutons.

DeWit and others (1980) have recently documented the first occurrence, in the Barberton greenstone belt, of a sheeted dike complex flanked by pillow flows and capped by a thick deposit of sediments similar to those characterizing Phanerozoic ocean ridge sequences. This discovery provides support for the view that at least one Archean greenstone belt may be an ancient counterpart of Phanerozoic ophiolites.

Despite the many significant uncertainties regarding the nature and development of the earliest crust, there can be little doubt that the Archean permobile tectonic activity that formed small lenticular to reticulate greenstone belts surrounded by gneiss complexes (see Figures 15-2b, 15-4, and 15-6) was unlike Phanerozoic plate tectonic activity that formed orogenic belts tens of thousands of kilometers long. Phanerozoic orogenic belts have spacings related to breadths of their associated lithospheric plates of thousands to tens of thousands of kilometers, in contrast to spacings between adjacent Archean greenstone belts of ~ 100 km. This contrast in scale can be viewed as a shift toward larger convection cells within the cooler Earth in more recent geologic time (McKenzie and Weiss, 1975).

15.2 PROTEROZOIC TERRANES

Toward the end of the Archean, 2.5 Gy ago, major changes in global geologic processes seem to have occurred that may have been more significant than any differences between Precambrian and Phanerozoic activity. The demise of Archean processes and the singular resulting rock bodies—apparently occurring over a period of a few hundred million years on different continents—reflects gradual stabilization of large continental slabs, or **cratons,** presumably due to a decreasing global geothermal gradient. This cooling would have the compound effect of allowing for a thicker, more rigid lithosphere—that is, depressing the solidus-temperature isotherm to greater depth—and possibly allowing for larger-scale convective cells in the underlying mantle. Energy transfer, vertical and horizontal motions, and consequent geologic processes tended to be focused at more widely spaced wavelength nodes. There was a closer approach to contemporary plate tectonics.

The Proterozoic magmatic–metamorphic rock record is dominated by three types of bodies: (1) swarms of basaltic dikes and huge stratiform basaltic intrusions; (2) anorthosite suite massifs; and (3) linear tectonic–metamorphic mobile belts overprinted onto earlier Archean terranes.

Basaltic Intrusions

The continental crust bears a complex and very long record of recurring extension and emplacement of mantle-derived mafic magmas, beginning at least as early as the intrusion of the Amerilik basalt dike swarm into Amitsoq gneisses in west Greenland more than 3.0 Gy ago. Proterozoic dike swarms and major stratiform intrusions are especially common and suggest widespread horizontal extensional stresses that failed to rift apart the continental terranes completely.

Dike swarms, some containing hundreds of dikes, are exposed over large areas, particularly in Greenland, Canada, and Africa, and were likely feeders for extensive basaltic plateaus that are now eroded away. Tabulation of major swarms of all ages from around the world (Windley, 1977) reveals that most were emplaced 2.5 to 1.2 Gy ago, that is, during the first two-thirds of the Proterozoic. In the Canadian shield (see Figure 15-10), basaltic volcanism migrated toward the periphery of the craton with time, apparently in response to a broadening region of continental stabilization (Baragar, 1977).

Dikelike bodies of layered basaltic differentiates of earliest Proterozoic age crop out in Western Australia and Zimbabwe. The largest and best known of these, the Great Dyke of Zimbabwe, 480 km long and averaging 6 km wide (see Figure 15-4), is apparently not a true dike but an elongate, graben-controlled outcropping of four funnel-shaped

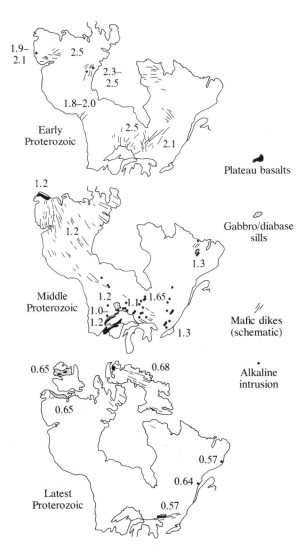

Figure 15-10 Continental basaltic flood lavas and dike swarms and alkaline intrusions in the Canadian shield. Ages in Gy. [After W. R. A. Baragar, 1977, Volcanism of the stable crust, Geological Association of Canada Special Paper 16.]

masses, which may once have been interconnected (Worst, 1958). Present marginal contacts with Archean rocks are mostly steep faults that cut the gently inward-dipping layers of the intrusions. The major part of the body is a layered sequence of ultramafic cumulates, including layers of chromite capped by gabbros and anorthosites that show well-developed cryptic mineral variations. The unusually magnesian composition of the earliest cumulus olivines (Fo_{94}) and enstatites (En_{93}) suggests an unusually mafic and hot parental magma.

Many other stratiform differentiated basaltic intrusions occur in Africa as well as in North America. The Duluth anorthositic gabbro complex (~ 1.2 Gy old) in the Great Lakes area is comparable in outcrop area to the Great Dyke. It and the smaller Muskox intrusion (see Figure 5-32) of similar age in the Northwest Territories of Canada are associated in space and time with some of the oldest known continental flood basalt provinces that are still preserved—Keweenawan and Coppermine River, respectively (see Figure 15-10). The Stillwater complex (actually late Archean, 2.7 Gy old) in Montana is small in comparison. Best known for their economic importance are the Sudbury body (1.7 Gy old) in Ontario, which is of intermediate size and a major source of nickel, and the colossal Bushveld complex (2.1 Gy old) in South Africa (see Figure 15-4). This complex is one to two orders of magnitude larger in area than the other stratiform intrusions just mentioned and is reputedly the world's largest repository of magmatic ore deposits of Cu, Ni, V, Cr, and Pt (Willemse, 1969; Visser and von Gruenewald, 1970).

The structure of the Bushveld complex is poorly understood, but to a first approximation it has a lobate form defined by five adjoining basin-shaped masses of magmatic rock. The rhythmically layered basal ultramafic cumulates show fluctuations in cryptic mineral variations interpreted to have been produced by recurring injection of fresh draughts of basaltic magma into the residual magma body. This mixing process, chiefly responsible for multiple layers of economically important oxides and sulfides, produced over 2 km of ultramafic rocks. The overlying 5 km of differentiates, also rhythmically layered, display a regular cryptic variation toward fayalite–hedenburgite granitic rocks at the top. Overlying these differentiates is a roof of felsites with hornfelsic fabric

and micropegmatite, whose origin is highly controversial (Hunter and Hamilton, 1978); magmatic differentiation, partial melting, and metamorphism of older intruded sedimentary or volcanic deposits have been suggested. Still another problem concerns the origin of a generally coarse-grained granite that is intrusive into the roof rocks and has too great a volume to be a magmatic differentiate.

How can we account for the unusual development of basaltic dike swarms and especially the large stratiform intrusions in the Proterozoic? The longer duration of the Proterozoic, almost four times the Phanerozoic, cannot be the only factor, as *large* stratiform bodies are not found in younger rocks. Do these intrusions reflect special magmatic and tectonic conditions during a long transition between the highly mobile and hot crustal conditions of the Archean, with its greenstone–gneiss terranes, and the cooler, plate tectonic Phanerozoic Earth? Or could even more unique circumstances have prevailed? Dietz (1964) has contended, for example, that the Sudbury mafic body formed as a result of meteorite impact, and W. Hamilton (in Visser and von Gruenewald, 1970) has advocated the same mechanism for the Bushveld.

Massifs of the Anorthosite Suite

Nature of bodies.   Anorthosite, a rock comprised chiefly of plagioclase, occurs as thin sheets in Archean terranes and more commonly as layers in stratiform intrusions ranging from Precambrian to Tertiary in age. By far the largest volume, however, appears in mid-Proterozoic **massifs**— laterally extensive bodies cropping out over hundreds or thousands of square kilometers and surrounded by high-grade, usually granulite-facies metamorphic rocks. Their form at depth is commonly obscure, but some appear from geophysical data to be sheets a few kilometers thick that are domical, basinlike, or have the form of a widely flared funnel.

The massifs actually comprise a suite of plutonic igneous rock types emplaced in part as separate intrusions. Although anorthosite is generally the principal rock type, other associated types include gabbro, troctolite, norite, and leucocratic relatives of these rock types, together with diorite, granodiorite, monozonite, quartz monzonite, and granite. Rocks with greater than about one-third mafic minerals in the mode are generally not found in massifs, and negative gravity anomalies over many preclude the chance of substantial buried volumes of very mafic rock. Unlike Phanerozoic arc plutonic batholiths of dioritic to granitic rocks characteristically containing hornblende and biotite, mafic phases in anorthosite suite massifs are anhydrous minerals including pyroxenes, olivine, Fe–Ti oxides, and rarely garnet. Because of their anhydrous mineralogical composition, monzonitic rocks are referred to as mangerite and felsic ones in general as charnockite (see de Waard, in Isachsen, 1968). The anhydrous mafic assemblage denotes crystallization under low P_{H_2O}, as indicated by the model reaction

hornblende + biotite + quartz =
    hypersthene + K-feldspar + plagioclase + H_2O

K-feldspar is usually found as a phase within perthitic feldspar. Figure 15-11 shows the modal compositions of the anorthosite suite in three representative massifs in eastern North America, the area of most voluminous development of such bodies (see Figure 15-12). The plagioclase in anorthosite is andesine or labradorite, averaging close to An_{50}, in contrast to the more calcic plagioclase, approximately bytownite, of Archean and stratiform-intrusion anorthosites. Grains are large, measurable in centimeters, rarely in meters. Where primary magmatic fabrics are preserved, subophitic and poikilitic relations exist between the subhedral, locally oscillatory zoned plagioclases and somewhat sporadically distributed pyroxenes. Tabular grains commonly have a preferred orientation.

Although the more mafic rock types commonly grade into one another, the felsic members of the suite usually show a late intrusive relationship. Compositional layering on various scales together

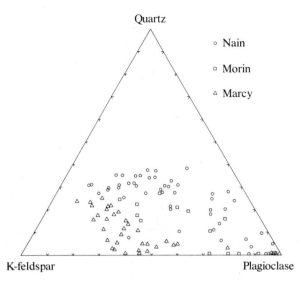

Figure 15-11 Modal proportions of quartz, K-feldspar, and plagioclase in samples from three of the massifs of the anorthosite suite shown in Figure 15-12. Most samples also contain one or two pyroxenes, olivine, Fe–Ti oxides, and rarely hornblende and biotite. [After D. de Waard, The anorthosite problem: the problem of the anorthosite–charnockiite suite of rocks, in Y. W. Isachsen (ed.), 1968, *Origin of Anorthosite and Related Rocks,* New York State Museum and Science Service Memoir 18.]

with cryptic mineral variations have been noted in the more mafic parts of undeformed massifs in central Labrador studied by Emslie (1975). He also notes that they have high-T contact metamorphic aureoles consistent with confining pressures less than 0.5 GPa. As this is lower than those generally considered for granulite facies metamorphism, it suggests the high-grade terranes were partially uplifted and eroded prior to emplacement of the anorthosite suite massifs.

Post emplacement metamorphism of the massifs is widespread in the Grenville Province of eastern North America (see Figure 15-12). Much of the confusion over the years regarding the nature of massifs and their origin has no doubt stemmed from this overprint. Pervasive deformation has obscured contact relations and has resulted in gneissic and mylonitic fabrics. Strained plagio-

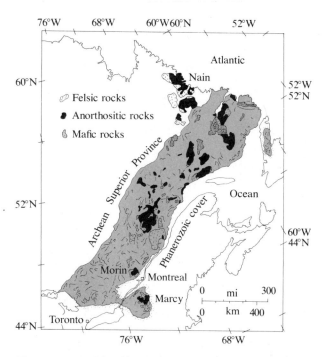

Figure 15-12 Massifs of the anorthosite suite in central Labrador and the Grenville Province (shaded) in the eastern Canadian Precambrian Shield. Compare with Figure 15-6. [After R. F. Emslie, 1978, Anorthosite massifs, rapakivi granites, and late Proterozoic rifting of North America, *Precambrian Research* 7; and the Tectonic Map of Canada in *Geological Survey of Canada Economic Geology Report* 1.]

clase megacrysts lie in a finer-grained plagioclase matrix of plastically strained or recrystallized granoblastic grains. Accompanying this metamorphism is a color change in the plagioclase from dark gray to purple (caused by abundant inclusions of Fe–Ti oxides) to pale gray or green. Pyroxenes in some metamorphosed anorthosites have been replaced by amphibole, indicating hydrous amphibolite facies conditions.

Origin. A variety of petrogenic models have been advocated for massifs of the anorthosite suite. At one extreme is the suggestion that they originated by metasomatic homogenization of interbedded

evaporites and shales during regional metamorphism. Although metamorphic fabrics are widespread in the Grenville Province, there is also clear manifestation of primary magmatic fabrics and intrusive relationships (Emslie, 1975). Pyroxene geothermometry in magmatic textured rocks yields crystallization temperatures of 1200°C to 1000°C—surely a magmatic range. High positive europium anomalies in the trace element pattern (see page 55) of anorthosites indicate that plagioclase crystals equilibrated with some sort of silicate melt. These and other arguments (Romey and others, 1979) support a magmatic origin.

That anorthosites are accumulative rocks is compatible with the virtual absence in the rock record of anorthositic volcanic rocks and with the relatively uncommon occurrence of anorthosite dikes. It also agrees with petrographic observations, such as expressed by deWaard (in Isachsen, p. 81, 1968): "Anorthosite is essentially a cumulate of plagioclase megacrysts with interstices filled with rock of noritic composition. With an increase in interstitial norite or a decrease of plagioclase crystals, the anorthosite grades into norite." Trace element studies by Duchesne and Demaiffe (1978) on Norwegian massifs indicate the possibility of fractional crystallization linking anorthosite with noritic rocks on the one hand and felsic rocks on the other. However, Simmons and Hanson (1978) looked at trace element patterns in eastern North American massifs and concluded that felsic rocks also have positive europium anomalies and cannot, therefore, be residual liquids of a more plagioclase-rich magma.

Yoder (in Isachsen, 1968), among others, has noted that elevated confining pressure enlarges the composition field in basaltic magmas where plagioclase is the first crystalline phase to precipitate upon cooling; this would allow it to accumulate and produce the observed ophitic to poikilitic fabrics. However, plagioclase could also be favored as the first crystallizing phase in highly aluminous (> 17 wt.%) basaltic magmas. Emslie (1978) has found such compositions in the chilled margins and allegedly parental magmas in unmetamorphosed massifs in central Labrador.

Opinion has been divided about whether the felsic charnockitic rocks so persistently associated with anorthosite in massifs are cogenetic. The close chronologic and spatial associations argue for some sort of genetic relation. However, it appears that the felsic and anorthositic/gabbroic members of the suite may not be comagmatic, or formed from the same parent magma, because of differing $^{87}Sr/^{86}Sr$ ratios. Where ratios have been determined for both members, the felsic rocks have higher initial ratios, suggesting old sialic crustal material might have been involved in their ancestry (Emslie, 1978, Table 1).

In a recent comprehensive petrogenetic model of anorthosite suite massifs, Emslie (1978) begins with mantle-derived olivine tholeiite magmas generated in an essentially orogenic regime beneath thick stable continental crust. During ascent, such magmas become buoyantly blocked near the base of the sialic crust and undergo fractionation of olivine and orthopyroxene, producing highly aluminous residual magmas. These magmas rise higher into the crust where they crystallize as masses of anorthosite with lesser amounts of gabbroic rock. Rare megacrysts of aluminous orthopyroxene in anorthosite are believed to have been derived from the fractionating parent magma near the base of the crust. The generally younger felsic rocks of the anorthosite suite are postulated to represent partial melts of the lower crust formed around the fractionating parent magma bodies. These crustal rocks were presumably of the granulite facies and had likely experienced prior partial melting or metasomatism (see page 539) that had depleted them in Fe, Rb, U, and most importantly H_2O. Felsic rocks in anorthosite suite massifs are accordingly low in these constituents.

An intriguing aspect of anorthosite massifs, first emphasized by Herz (1969), is their apparent limited occurrence in space and time. Despite considerable difficulties in radiometric dating, their ages appear to be limited to 1.7 to 1.1 Gy, with a tighter clustering near 1.4 Gy in North America. Geographically, massifs extend across North America

° Anorthosite suite massifs >1.1 Gy old in the Grenville Province
· Anorogenic anorthosite suite massifs ~1.45 Gy old

▲ Anorogenic rapakivi granites and anorthosites ~1.7 Gy old

Figure 15-13   North American–European occurrences of anorthosite and associated rocks on a predrift continent reconstruction. [After R. F. Emslie, 1978, Anorthosite massifs, rapakivi granites, and late Proterozoic rifting of North America, *Precambrian Research* 7.]

and Eurasia in a poorly defined and diffuse zone (see Figure 15-13), and in a minor belt across the Southern Hemisphere continents. Herz suggested that anorthosites were the product of a unique cataclysmic "anorthosite event" occurring only during the Earth's early history. This possibility was brought into sharper focus shortly later when it was found that a significant part of the crust of the Moon is comprised of anorthosite (see page 571). However, the age of the lunar crust turned out to be much older, 4.6 to 4.4 Gy.

The question of why anorthosite massifs are so concentrated in space and time in North America must be considered in the light of the tectonic history and evolution of the Grenville Province (see pages 550–554).

Other Proterozoic Magmatic Bodies

Another prominent rock type of the Proterozoic that locally forms the felsic member of anorthosite suite massifs is **rapakivi granite.** In this textural type of granite large anhedral grains of potassic feldspar are surrounded by rims of sodic pla-gioclase (Turner and Verhoogen, 1960, p. 368). Through the North America–Eurasia region, bodies of rapakivi granite range in age from 1.8 to 1.1 Gy but are most voluminous in Finland and Sweden as 1.7-Gy-old plutons (see Figure 15-13) whose individual outcrop area may exceed 15,000 km². Some rapakivi granites clearly were emplaced at shallow crustal levels, as indicated by their porphyritic texture, miarolitic cavities, and sharp and locally brecciated contact with host rocks. During shallow emplacement, *P–T–X* condi-tions within the magma must have shifted in such a way as to allow early precipitation of potassic feldspar followed by partial resorption and precipi-tation of oligoclase rims. Locally, rapakivi granites are associated with extrusive rocks, some of which (in Sweden) have relict vitroclastic and eutaxitic textures and are the oldest known ash-flow tuffs.

Although a few alkaline magmatic bodies of older age have been found, they begin to appear in quantity in North America–Eurasia about 1.7 Gy ago (Windley, 1977).

Proterozoic Mobile Belts

Belts of rock with consistent tectonic, metamor-phic, magmatic, and chronologic features are typi-cal of Proterozoic terranes, particularly in the Af-rican continent, where they are referred to as **mobile belts.** These differ from Phanerozoic orogenic belts or cordillera such as the Appala-chians (see pages 421–423) in that no calc-alkaline volcanism with asymmetric compositional pat-terns (see Figure 9-13) is present and sedimentary sequences (locally several kilometers thick) rest upon sialic continental basement and nowhere upon mafic oceanic crust; structures in older ter-ranes on either side are commonly found to pass virtually undeflected, albeit overprinted, through the younger mobile belt. The rocks of mobile belts consisting of **reworked** (reactivated or re-mobilized) older sialic material, together with any younger supracrustal deposits present, have been deformed in a ductile manner, recrystallized, and even partially melted to form magmatic bodies.

Southern Africa.   Kroner (1977) has summarized specific evidence for the concept that Proterozoic mobile belts of southern Africa are **ensialic—** evolved by polyphase reworking of older sialic continental material. For example, the Limpopo mobile belt separates the Archean (3.6- to 3.5-Gy-old) Rhodesian and Kaapvaal Cratons (see Figure 15-4), and was developed from this older terrane by deformation and granulite-facies metamorphism. The 1.0 Gy old Kibaran belt cuts the > 1.8 Gy old Ubendian belt, which in turn cuts the still older Zambia Craton (see Figure 15-14).

Ensialic reworking involves no new crustal ma-terial and is thus incompatible with concepts of crustal accretion and orogeny accompanying oceanic plate subduction. Mobile belts may have formed by welding or suturing of small sialic blocks butting together (compare the "cratoniza-tion" of Clifford, 1970), or alternatively they might have developed by segmentation of a once larger sialic block, with the parts jostling alongside each other to produce the reworked material.

Grenville Province.   Belts of volcanic and sedimentary Proterozoic rocks in Canada are mainly peripheral to Archean terranes and usually lie on sialic basement, that is, they are ensialic (Moore, 1977). Basal units in rock successions are submarine basalts, giving way upward to interme-diate and silicic volcanics of calc-alkaline affinity and then to clastics derived in part from the ear-lier volcanic terranes. Subsequent regional metamorphism and deformation of these se-quences has produced folds, which tend to be overturned toward the Archean terrane, and metamorphism, whose grade increases away from it. The stratigraphic, structural, and compositional features of some early Proterozoic volcanic se-quences are more similar to Archean than to Phanerozoic bodies.

Among the Proterozoic terranes of the Cana-

(a)

(b)

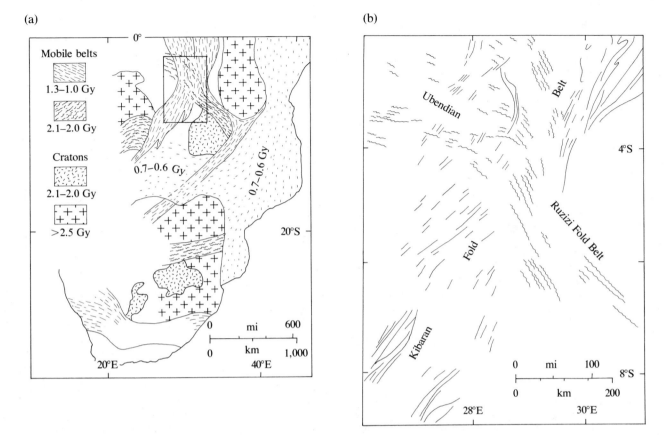

Figure 15-14 Mobile belts (showing trends of folds) that cut craton segments in southeastern Africa. Location of larger-scale map in (b) is shown by rectangular area in (a). [After A. Kroner, 1977, Precambrian mobile belts of southern and eastern Africa—ancient sutures or sites of ensialic mobility? *Tectonophysics* 40; and R. M. Shackleton, 1973, Correlation of structures across Precambrian orogenic belts in Africa, in D. H. Tarling and S. K. Runcorn (eds.), *Implications of Continental Drift to the Earth Sciences,* Vol. 2 (New York: Academic Press).]

dian Precambrian Shield, none has been more controversial than the Grenville Province, which lies to the southeast of the Archean Superior Province and extends under Phanerozoic cover into the eastern and southern United States (see Figure 15-15). The controversial facets of the Grenville include the voluminous enigmatic massifs of the annorthosite suite; the nature and origin of the boundary with the adjacent Superior Province, or Grenville Front; the conflicting ages of rocks (K–Ar ages are generally ≳ 1.0 Gy but other meth-

ods give older ages); the origin of the extensive high-grade metamorphism culminating in granulite facies assemblages that indicate a crustal thickness during metamorphism of 60 to 65 km (the sum of the present 35 km depth of the crust based on seismic data plus 25 to 30 km for the depth of burial necessary to form granulite facies assemblages; see Baer and others, 1974); and finally, but by no means least, the origin of the Province itself.

The oldest rocks recognized by most geologists

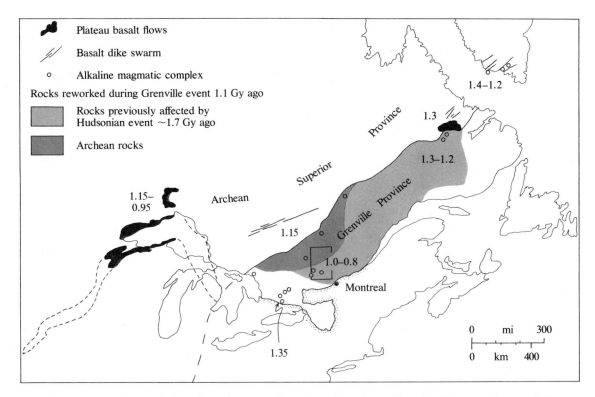

Figure 15-15   Precambrian of southeastern Canada and southern Greenland (restored to predrift position) showing magmatic assemblages compatible with pre-Grenvillian rifting. Short dashed lines indicate buried extension of Keweenawan basaltic rocks beneath Phanezoic cover as discerned by gravity and magnetic data; long dashed line indicates buried extension of Grenville front. Numbers are ages in Gy. Compare with Figure 15-12. Rectangle is area shown in Figure 15-16. [After W. R. A. Baragar, 1977, Volcanism of the stable crust, Geological Association of Canada Special Paper 16; and J. F. Dewey and K. C. A. Burke, 1973, Tibetan, Variscan, and Precambrian basement reactivation: products of continental collision, *Journal of Geology* 81.]

(for example, Edwards, 1972) on the basis of field relations and radiometric dating are a widespread complex of migmatites, gray leucocratic quartzofeldspathic gneisses, and mafic gneisses. These give Rb–Sr and Pb–U zircon ages clustering near 2.6 and 1.8 Gy, corresponding to intense metamorphic imprints during the Kenoran and Hudsonian "Orogenies," respectively. Measured ages in different areas range widely about these values. Overlying this complexly deformed basement is a sequence of metasedimentary and metavolcanic rocks, the Grenville Supergroup, originally several kilometers thick, that was deposited perhaps 1.4 to 1.2 Gy ago. At the southwestern end of the Province in Ontario, a basal sequence 4 km thick of low-K, pillowed tholeiitic basalts is overlaid by 3 km of calc-alkaline, chiefly andesitic rocks; locally, this apparent arc volcanic association lies on an ultramafic–mafic complex considered to represent oceanic crust. Overlying the volcanics are metamorphosed carbonate and clastic sequences. Massifs of the anorthosite suite appear to have been emplaced partly along the unconformity at the base of the Grenville Supergroup.

The igneous rock record suggests (Baragar, 1977; Emslie, 1978) that an episode of continental rifting essentially parallel to the Grenville Front heralded the beginning of its evolution (see Figure 15-15). Plateau basalts and dike swarms with northeast strikes are exposed north of the Province, where they apparently escaped the later severe metamorphic and deformational imprint. Highly alkaline complexes, some now appearing as feldspathoidal gneisses, lie along a similar northeasterly trend through the Province. Radiometric ages on these mafic and alkaline rocks are widely scattered from 1.4 to 0.8 Gy, very possibly because of resetting by multiple deformational and thermal events. Possibly, arc volcanism and sedimentation that followed this rifting produced the Grenville Supergroup in a seaway southeast of the Archean Superior block. Radiometric ages of about 1.1 Gy mark an intense and widespread episode of deformation and metamorphism ranging from greenschist to granulite facies that affected all of the aforementioned rocks. Late Archean and early Proterozoic rocks imprinted by the Kenoran and Hudsonian events were later reworked during this Grenvillian "Orogeny." K–Ar ages ranging down to 0.9 Gy reflect closing isotopic systems as the rocks were uplifted and cooled.

Many geologists have envisaged, in a plate tectonic context, the Grenville Front to be an ancient continental margin and suture line. However, Baer and others (1974) and Brocoum and Dalziel (1974) reject this interpretation because Superior radiometric ages occur within the Grenville Province within a few tens of kilometers of the front and because fold structures, apparently Hudsonian, are continuous across the Front. Grenvillian folding begins to be apparent to the southeast of the Front (see Figure 15-16), whereas at the Front itself there is commonly a pervasive, strong mylonitic foliation dipping steeply beneath the Grenville Province with a downdip lineation. Terranes

Hinge lines of Grenvillian folds

Hinge lines of Hudsonian folds

Figure 15-16 Structural map of the area in the Grenville Province northwest of Montreal outlined in Figure 15-15, showing metamorphic foliation that was folded during the Hudsonian event 1.7 Gy ago along a more or less north-south trend and then overprinted by northeast-trending folds during Grenvillian deformation 1.1 Gy ago. [After H. R. Wynne-Edwards, 1969, Tectonic overprinting in the Grenville province, southwestern Quebec, Geological Association of Canada Special Paper 5.]

on either side of the Front have experienced different degrees of metamorphism at different times.

The evolution of the Grenville Province can be viewed, in the broadest sense, in terms of an opening and closing ocean—a familiar theme from Phanerozoic plate tectonic activity. The closure that produced the Grenvillian deformation and metamorphism ("Orogeny") roughly 1.1 Gy ago (following the earlier rift episode ~ 1.3 Gy ago) has been likened to the collision of the Indian and Asian continents during the Cenozoic (Dewey and Burke, 1973). The resulting Tibetan Plateau with its thick crust (60 to 80 km) formed as the Indian continental plate and was thrust under the Asian continental plate. Such a collision and crustal thickening during the Proterozoic could have produced the Grenvillian deformation and the locally high-grade metamorphism.

Anorthosite–rapakivi granite magmatism in North America that peaked around 1.45 Gy ago is believed by Emslie (1978) to be an expression of the same sort of igneous bimodality common in younger Phanerozoic continental rifts (see pages 166–167 and Section 6.6), but in a deep plutonic setting. The existence of a thick, stable, cratonic crust is viewed as a critical factor in the evolution of anorthosite suite massifs. Yet, it is puzzling why similar major belts of massifs are not found elsewhere and of different ages.

15.3 PRECAMBRIAN CONTINENTAL GROWTH AND EVOLUTION

Previous sections have painted pictures of Archean and Proterozoic continental terranes that contrast to varying degrees with rock associations and tectonic features of Phanerozoic age. What has been the course of evolution of the continents through time? If plate tectonic processes as we now know them were not strictly operative in the Archean, when did they begin? At what rate have the continents grown—steadily throughout the 3.8-Gy recorded history, or rapidly in the Archean with only reworking since then?

Precambrian Continental Evolution

The small lenticular to reticulate greenstone belts surrounded by more expansive gneiss complexes of Archean terranes gave way to different tectonic styles and rock associations during Proterozoic time. During the Proterozoic, highly alkaline magmatic complexes, including carbonatites, became relatively abundant for the first time; such rocks in Phanerozoic settings require relatively thick and tectonically stable continents with low geothermal gradients (see Section 6.5). Numerous swarms of Proterozoic basaltic dikes emplaced in continental terranes indicate extensional stresses at work, possibly over upwelling mantle; such failed rifts indicate a relatively thick and resistive continental crust. Proterozoic mobile belts generally appear to have been ensialic—that is, overprinted onto older crust; reworking involved recrystallization, ductile deformation, and resetting of certain radiometric systems but allowed vestiges of the older crust to be preserved.

Figure 15-17 indicates that mafic–ultramafic rock associations have formed throughout recorded geologic history but at rates and in volumes that are difficult to estimate. It is quite apparent, nonetheless, that the crustal environments in which they were emplaced varied significantly. Although continental associations are especially evident in the Proterozoic, oceanic mafic–ultramafic bodies of this age are virtually absent. Jackson and Thayer (1972) have stressed the absence of alpine peridotite bodies older than 1.2 Gy old, although ophiolite sequences may be present in the Grenville Supergroup (1.4 to 1.2 Gy) and in the Barberton greenstone belt (~ 3.3 Gy), as previously indicated. Is the absence of mafic–ultramafic oceanic crust in the early Proterozoic real, or only an artifact of subsequent reworking and subduction, or the fault of our limited abilities to recognize it? This absence is disconcerting, especially if one espouses models of continental growth that claim substantially less continental area during this period than at present (Glikson, 1979).

Piper (1976) has amassed evidence, mainly

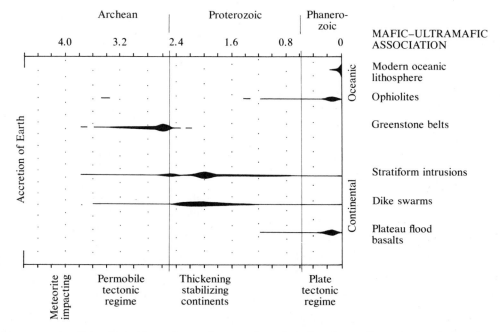

Figure 15-17 Schematic chronology of expression of mafic–ultramafic rock associations in relation to crustal tectonic regimes. The width of each line as a function of time is intended to show the relative development of that one association, not relative volumes between associations.

paleomagnetic, consistent with a global model in which the bulk of the Precambrian shields were aggregated together as a single supercontinent during much of Proterozoic time.

The Proterozoic appears to have been a time during which global processes were in transition from the small-scale, highly mobile, high-temperature activity of the Archean to the much larger-scale, Wilson-cycle plate tectonism of the Phanerozoic. The transition was diachronous, that is, took place at different times in different places. There was overall cooling of the interior of the Earth and control of tectono-magmatic processes by progressively larger, slower-moving convection cells associated with lower geothermal gradients.

Continental Growth

Contrasts in models of continental growth through recorded geologic history are still another example of the uncertainties involved in interpreting the early evolution of the Earth. Depending upon the initial constraints—such as isostasy, sediment record, isotopic composition, thermal regime, and so on—the proportion of continental area 2.5 Gy ago relative to the present claimed by different models ranges from 10% to at least 80% (Veizer and Jansen, 1979; Hargraves, 1976; Wise, 1974).

Some geochemists have noted that initial $^{87}Sr/^{86}Sr$ ratios from many Archean and some Proterozoic gneisses are within the range of 0.700 to 0.703 and fall close to and a little above the presumed ratio in the upper mantle at the corresponding age (see Figure 2-9). The inference is drawn that the protoliths of these gneisses were derived from the mantle and constitute juvenile additions to the continental crust and are not melted or highly mobilized older crust rich in Rb. However, trace element data (see pages 538–539) seem to favor a hydrous mafic tholeiitic source for protoliths of Archean gneisses rather than peridotite. Also, the relative depletion of Rb in deep crustal granulite facies rocks (see page 539) imparts initial Sr isotope ratios to partial melts that are no different from mantle-derived melts. Hence, magmas derived from either upper mantle or lower crustal sialic sources could have similar initial $^{87}Sr/^{86}Sr$ ratios.

Sm–Nd isotopes (see page 557) may resolve some of the ambiguities in the Rb–Sr data concerning continental accretion or reworking. For example, McCulloch and Wasserburg (1978) conclude from such data that most of the Canadian Shield, except the Grenville Province, formed from juvenile material 2.7 to 2.5 Gy ago.

Previous discussion indicated that widespread Proterozoic mobile belts represent reworked older rock; Rb–Sr isotopic data seem to support this (Glikson, 1979).

15.4 EVOLUTION OF THE MANTLE

In considering the origin and evolution of the continents, we cannot ignore the role played by the underlying mantle. Significant flow of energy and matter occurs across the M-discontinuity between these two layers of the Earth. On the one hand, the mantle is a source of thermal energy and matter in the form of partial melts accreting into and onto the crust. On the other, it is a sink, because of subduction of oceanic plates into the mantle; perhaps 10 to 100 times as much plate material descends into the mantle as is added to the continents. How much of the mantle is actually involved in this interchange? Has all of the mantle, comprising 68% of the Earth's total mass, been involved in creating the present day crust, which comprises only 0.4% of the Earth? To what extent is crustal, and especially continental, material recycled back into the mantle? Do mantle-derived mafic magmas that build the crust come only from pristine "fertile" regions of the mantle not previously melted?

Resolution of these major questions would appear to be an impossible task because of the relative inaccessibility of the mantle. However, mafic magmas that are essentially chemically unmodified since departure from their mantle sources as well as mantle-derived xenoliths in these magmas serve as poor man's "probes" into at least the upper reaches of that important Earth domain. Exciting answers are beginning to appear from geochemical studies of trace elements and isotopic tracers in these mantle-derived mafic magmas and xenoliths.

Evidence for a Chemically Heterogeneous Mantle

The growth of sialic continents beginning at least 3.8 Gy ago by extraction of low-T incongruent partial melts has generally depleted the upper mantle in incompatible large-ion-lithophile (LIL) elements (see pages 55–56 and Figure 15-18).

Sr isotopes. Selective extraction from mantle peridotite of LIL incompatible elements such as Rb during partial melting will be recorded in isotopic tracers such as ^{87}Sr measured within mantle samples. Because ^{87}Rb decays radioactively to ^{87}Sr, removal of any Rb in a partial melt from a mantle source region will cause the $^{87}Sr/^{86}Sr$ ratio to be less at some future time than in a region of the mantle not melted (see Figure 2-9). Any liquids formed during a second episode of partial melting of the Rb-depleted source rock and in isotopic equilibrium with it will have a low $^{87}Sr/^{86}Sr$ signature.

Oceanic mantle sources will be considered first because there is no complicating possibility of contamination with high $^{87}Sr/^{86}Sr$ continental material as the mantle-derived mafic magmas rise toward the surface. It has been known for some time (Gast, 1968) that the voluminous sea-floor or mid-ocean ridge basalts (MORB) have relatively low $^{87}Sr/^{86}Sr$ ratios, mostly between 0.702 and 0.704. Additionally, and surprisingly, they do not even contain sufficient ^{87}Rb to have generated their measured ^{87}Sr contents, assuming a chondritic Earth 4.6 Gy old (see Chapter 16). The upper mantle source of MORB must therefore have experienced a prior episode of partial melting. The nature and time of this episode is obscure.

Compared to MORB, volcanic rocks of oceanic islands have generally higher, but quite variable $^{87}Sr/^{86}Sr$ ratios, ranging from 0.7028 to 0.7066 (see Figure 15-19), together with complimentary greater but variable LIL concentrations. The source of these island magmas cannot be the same as MORB; hence, beneath oceanic areas of the globe there must be two source regions of basaltic magmas, the one previously depleted in LIL and the other having somewhat greater but more heterogeneous concentrations.

Duncan and Compston (1976) found that the mantle sources for the lavas that built the mid- to late-Cenozoic islands of French Polynesia in the southwest Pacific were heterogeneous in Rb/Sr, so that there is a strong positive correlation between measured $^{87}Sr/^{86}Sr$ and $^{87}Rb/^{86}Sr$. A particular source region with a higher Rb/Sr yielded a higher $^{87}Sr/^{86}Sr$. The slope of the linear relationship between these ratios, an isochron (see Figure 2-8), gives the time since the last homogenization of Sr isotopes by some sort of major melting or massive diffusion event. The age of this event, from the measured slope, is ~1 Gy, and the initial $^{87}Sr/^{86}Sr$ ratio at the time of this event was between 0.7005 and 0.7024.

Other workers using similar but not as well-constrained isochrons have found isolated mantle source regions with ages as great as 3 Gy (Pankhurst, 1977). Some of these sources underlie continental areas where their validity is open to question because possible contamination of old sialic material can lead to false isochrons. Nonetheless, heterogeneities in the oceanic mantle that last for a billion years or more appear to pose a problem within the context of plate tectonics because the accompanying convection would seem to exert a stirring action on the mantle.

Nd isotopes. As discussed on pages 58–59, the ratio $^{143}Nd/^{144}Nd$ serves as a petrogenetic tracer in much the same way as the $^{87}Sr/^{86}Sr$ ratio because

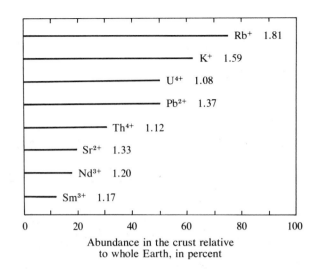

Figure 15-18 Abundances of some large-ion-lithophile (LIL) elements in the crust relative to the whole Earth. Ionic radii in Å (see Table 2-8). Because the crust comprises only 0.4% of the Earth, all the elements shown are greatly enriched in the crust. Ions most enriched in the crust are most incompatible in peridotitic crystalline phases (see Table 2-7). [After data of S. R. Taylor in R. K. O'Nions, P. J. Hamilton, and N. M. Evensen, 1980, The chemical evolution of the Earth's mantle, *Scientific American* 242.]

^{143}Nd is a decay product of ^{147}Sm and the Sm/Nd ratio is governed to some extent by petrologic processes. Because of their relative ionic radii (see Figure 15-18), Nd is a slightly more incompatible element than Sm in peridotitic systems, preferring the liquid state more than Sm, so that partial melts have lesser Sm/Nd, but greater Rb/Sr, than the coexisting olivines and pyroxenes. Accordingly, values of $^{143}Nd/^{144}Nd$ and $^{87}Sr/^{86}Sr$ ratios in the whole family of mantle-derived magmas vary inversely and lie along the negatively sloping mantle correlation line in Figure 15-19. As chondritic meteorites serve as a good first approximation to the bulk chemical composition of the Earth, their present-day isotopic ratios are used as standards of reference to decide whether recent mantle materials are relatively depleted or enriched in incompatible elements.

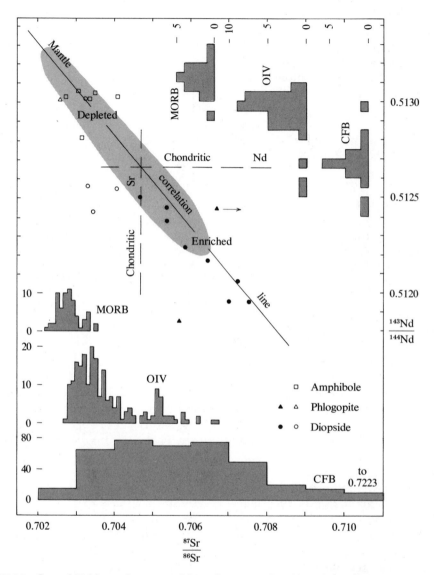

Figure 15-19 Sr and Nd isotopic composition of young volcanic rocks and minerals in xenoliths in basaltic rocks (open symbols) and in kimberlite (solid symbols). Shaded area along mantle correlation line is where MORB and oceanic island volcanic rocks (OIV) plot. Histogram of Sr isotope ratios for continental flood basalts (CFB) includes numerous older, less precise values. Present isotopic ratios for a chondritic bulk Earth divide the diagram into quadrants; the upper left one represents incompatible-element-depleted compositions, and the lower right, enriched. [After A. W. Hofmann and S. R. Hart, 1978, An assessment of local and regional isotopic equilibrium in the mantle, *Earth and Planetary Science Letters* 38; M. Menzies and V. R. Murthy, 1980, Enriched mantle: Nd and Sr isotopes in diopsides from kimberlite nodules, *Nature* 283; and D. J. DePaulo and G. J. Wasserburg, 1979, Neodymium isotopes in flood basalts from the Siberian platform and inferences about their mantle sources, *Proceedings of the National Academy of Sciences, U.S.A.,* 76. Histogram of Sr isotope ratios for continental flood basalts after F. Faure, 1977, *Principles of Isotope Geology* (New York: John Wiley and Sons).]

The distinction between mantle sources for oceanic islands and MORB based on Sr-isotopic ratios is strengthened by contrasts in their Nd-isotopic ratios (see Figure 15-19). Heterogeneous mantle sources for oceanic island volcanics are chiefly depleted in Rb and Nd but not to the degree of the MORB source, which stands apart as a more uniform, highly depleted entity. Depleted oceanic mantle source regions as a whole are complementary to the greatly enriched continental crust.

The composition of the subcontinental mantle differs for the most part from oceanic mantle source regions (see Figure 15-19). Sr ratios of continental flood basalts are highly variable and skewed off the mantle correlation line toward high $^{87}Sr/^{86}Sr$ ratios, apparently because of contamination with old ^{87}Rb-rich sialic material, according to DePaolo and Wasserburg (1979). Although these workers contend that the source of continental flood basalts has been essentially chondritic with regard to Nd isotope ratios, the Nd data in Figure 15-19 are dispersed significantly on either side of the chondritic reference line.

Mineral grains in ultramafic xenoliths in continental basalts and kimberlites are highly variable in Nd–Sr isotopic composition (see Figure 15-19), a fact that supports an underlying heterogeneous mantle.

In contrast to the heterogeneities in basaltic source regions, Basu and Tatsumoto (1979) found that the groundmasses of seven kimberlite samples from the United States, India, and South Africa with ages of 90, 377, 940, and 1,300 My have initial $^{143}Nd/^{144}Nd$ ratios identical to those of chondrite, calculated for the same times of formation. This remarkable correspondence of Nd isotopic compositions indicates the kimberlite magmas, which had deeper mantle sources than basalts, tapped more pristine material not previously melted or enriched by melts (but compare pages 216–218). It is highly unlikely that a combination of uniquely compensating depletion and enrichment events could have occurred on three continents at four different times.

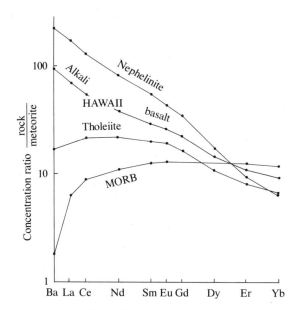

Figure 15-20 Chondrite normalized Ba and REE patterns for three Hawaiian lavas and MORB. [After G. N. Hanson, 1977, Geochemical evolution of the suboceanic mantle, *Journal of the Geological Society* (London) 134.]

Trace elements. Hanson (1977) has reviewed trace element data on oceanic basaltic magma sources and confirms the significant differences observed by Gast (1968) a decade earlier on the basis of less information. MORB have K/Rb, K/Ba, Rb/Sr, and Zr/Nb ratios significantly less than oceanic island rocks. MORB have Ba and light REE depletion contrasted with enrichment in oceanic island rocks (see Figure 15-20). On the basis of available distribution coefficients (see page 55) and quantitative model calculations, Hanson concludes that the difference between MORB and oceanic island magmas is unlikely to have originated by partial melting, crystal liquid fractionation, or zone melting. However, MORB magmas could be derived from mantle peridotite from which very small amounts of incompatible-element-enriched melts had previously been extracted.

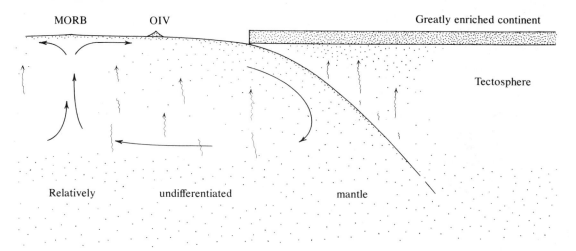

Figure 15-21 Highly schematic model of variable enrichment and depletion of crust and mantle regions in incompatible elements due to migration of metasomatic fluids (sinuous arrows). Density of stippling indicates concentration of incompatible elements. The lithosphere– asthenosphere boundary is not shown in this petrologic model because penetration of metasomatizing fluids across it blurs any compositional distinction. OIV, oceanic island volcanic rocks.

Speculations. To account for the observed iso- topic and trace element differences between the four mantle source regions of basalt and kimber- lite, we could postulate a simple two-layer mantle model. A fairly uniform but highly depleted upper layer immediately below the oceanic crust could be the source of MORB; the underlying mantle, extending up to the M-discontinuity beneath the thicker continents, could be the source of flood basalts and kimberlites. Oceanic island basalts would be derived from a mixture of the deep and shallow oceanic mantle regions. But how would this mixing be achieved? And how do we reconcile the existence of such layers with plate tectonic processes and the accompanying mantle convec- tion?

A better model (compare Hanson, 1977; Tat- sumoto, 1978; and others cited in Chapter 6) em- bodies the phenomenon of mantle metasomatism (see pages 216–218), in which incompatible- element-enriched hydrous melts or fluids from deeper sources variably penetrate the upper levels

of the mantle, forming amphibole- and phlogopite- rich veins. In this model (see Figure 15-21), a con- vecting layer of mantle that is variably enriched and depleted in incompatible elements due to the action of ascending metasomatic liquids overlies a deeper, less-differentiated mantle. Mantle moving in the convective circuit toward the mid-ocean ridge becomes depleted as ascending metasomatic liquids sweep through it. Upon ascent and decom- pression beneath the ridge, this depleted mantle partially melts to form MORB. Away from the ridge, mantle becomes more enriched due to pene- tration of ascending metasomatic liquids, but source regions of oceanic island magmas are still generally depleted relative to the chondritic com- position.

The subduction of MORB and other oceanic material in the descending lithospheric plate is an important link in the cycle of irreversible chemical differentiation of continents from mantle. Fyfe and others (1978; reference at the end of Chapter 14) note that the water incorporated into the subducting oceanic sediments and metamorphosed MORB is of such magnitude that the entire mass of the oceans would have been subducted into the mantle in ~ 1 Gy. The reason this has not happened is the progressive dewatering of the oceanic crust during descent and heating of the plate. Escaping water generates metasomatic liquids that are believed to sweep through the overlying wedge of mantle peridotite, depleting it of incompatible elements (Best, 1975; Mysen, 1979). Hence, continental material is extracted from the oceanic crust and from the mantle overlying the descending plate in the process of subduction zone magmatism.

The fate of the depleted lithosphere formed as a result of production of magmas at oceanic rifts has been an enigma. Some geologists have postulated that it must sink deep into the mantle at subduction zones, whereas Jordan (1978) and Oxburgh and Parmentier (1978) have noted that because of its lower relative density, it must tend to float toward the top of the mantle. The lower density of depleted mantle is a result of the preferential incorporation into liquids of low-T, Fe-end-member components of olivine and pyroxene solid solutions and possibly also the complete extraction of dense garnet during partial melting, leaving a lighter Mg-enriched crystalline residuum of dunitic to harzburgitic composition. The less dense depleted lithosphere is envisaged to rise in multiple diapirs as it is heated during descent in the subduction zone. These buoyant diapirs are believed to collect at the top of the mantle to form a gravitationally stable, "infertile," subcontinental **tectosphere** that may be more than 400 km thick (Jordan, 1978). The tectosphere concept answers the geophysical dilemma, based on heat flow investigations, of how a necessarily cooler subcontinental mantle can be in hydrostatic equilibrium with a hotter suboceanic mantle. If these regions were chemically the same, they could not be in gravitational equilibrium because of the more expanded and hence less dense suboceanic mantle. Because of its higher solidus T and therefore greater viscosity, shear deformation associated with plate mo-

tion will be deflected below the tectosphere, assuring its firm attachment to the base of the continent. It is hypothesized that tectosphere development has been the principal factor in stabilizing the continents over the eons of geologic time and, in particular, was the reason for the transition from Archean permobile activity to the more rigid continent-mobile belt activity of the Proterozoic.

In the metasomatic model presented above, the depleted lithosphere formed at an oceanic ridge may become infiltrated by LIL-enriched metasomatic fluids as it moves away from the ridge. Upon reaching a subduction zone, it may no longer be much depleted and so could not contribute to a tectosphere. However, a subcontinental tectosphere could also grow by metasomatic depletion of the wedge of mantle above a subducting plate during arc magmatism; in long-lived subduction zones, the volume of mantle that is partially melted and depleted could be enormous, particularly if some sort of convective flow continuously feeds more mantle into the wedge-shaped region above the descending plate.

The source of continental flood basalts is not likely to lie within the underlying depleted tectosphere but may possibly involve pristine chondritic mantle rising into it from deeper levels during extensional rifting (DePaolo and Wasserburg, 1979). Alternatively, the dispersion of Nd and Sr isotopic ratios in flood basalts seen in Figure 15-19 may originate by metasomatism of the tectosphere to the extent that the average source is approximately chondritic.

Summary of Critical Points in Crust–Mantle Differentiation

Continental and mantle evolution are intimately linked. The volume of the mantle from which continental material has been extracted is likely only a fraction (less than half, perhaps one-fourth) of the whole. The deeper mantle, which must remain separate from the overlying convecting part associated with the plate tectonic system, may be approximately chondritic with respect to some

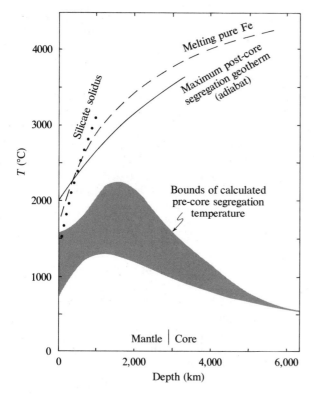

Figure 15-22 Melting temperatures and postulated geotherms in the Earth before and after segregation of the core. Melting temperature of metallic core material is somewhat lower than the melting curve of pure iron because of possible solid solution of FeO and FeS. The maximum post-core geotherm assumes that heating of the interior due to core segregation is sufficient to induce convection in the subsolidus mantle, which establishes an adiabatic T gradient. The calculated pre-core T distribution (shaded) ranges widely because of uncertainties in heating effects of accreting planetesimals of different size and infall rate (see Section 16.4); the calculations also incorporate heating effects due to self-compression and phase transformations in the interior. [After A. E. Ringwood, 1979, *Origin of the Earth and Moon* (New York: Springer-Verlag).]

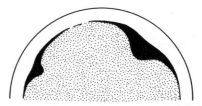

Figure 15-23 Model of core segregation according to W. M. Elsasser. Dense droplets of molten metal in the outermost periphery of the Earth where temperatures exceeded the solidus of both metal and silicate (see Figure 15-22) descend to the top of the region where silicate material is at subsolidus temperatures. Coalescence of droplets forms a layer of dense molten metal that sags into solid silicate and eventually falls to the center of the Earth as a negative diapir, scavenging additional metal particles enroute. Diapirs a few hundred kilometers in diameter would fall to the center through solid silicate whose coefficient of viscosity $\eta = 10^{20}$ Pa · s on a time scale of ~0.1 My, if descent by a simple Stokes' Law (see page 318) is assumed. Actually, the time for descent may be less because the effect of viscous heating due to loss of gravitational potential energy is felt immediately around the drop, locally reducing η. [After W. M. Elsasser, 1963, Early history of the Earth, in J. Geiss and E. Goldberg (eds.), *Earth Science and Meteoritics* (Amsterdam: North Holland Publishing).]

constituents, such as Sm and Nd. Archean basalts have incompatible element concentrations (O'Nions and Pankhurst, 1978) and Nd-isotopic compositions that indicate somewhat less differentiated mantle sources compared to Phanerozoic MORB sources. This suggests that continental crust has been growing since the Archean.

Over the eons of geologic time, continental volume has been increasing while the thickness of continents has also increased due to the decreasing geothermal gradient and general cooling of the Earth. A depleted-mantle tectosphere has been accreting onto the base of the continents. However, as the Earth cools still more and the region in the mantle of stability of amphiboles and micas increases, more volatiles, LIL, and other incompatible elements of continental character may be fixed into the mantle as they ride down in subducting lithosphere. The mantle will "undifferentiate" to some degree. Eventually, the asthenosphere will cool and harden, arresting plate motion.

The importance of water in crust-mantle evolution cannot be emphasized enough. Without water to help make the asthenosphere as ductile as it is, plate motion, and thus all the geologic processes accompanying it, might be impossible. Yet the way in which water moves through the mantle and where and how it resides there can only be conjectured. Does water in the subducted lithosphere somehow become entrained into the convective mantle circuit toward the oceanic rift, facilitating metasomatism en route (see Figure 15-21)? Does water-induced metasomatism of mantle promote plate motion by depleting a region that then buoys upward at oceanic rifts?

15.5 CORE SEGREGATION

The currently most acceptable model for the origin of the Earth (see Section 16-4) involves more or less homogeneous accretion from a solar nebula of chondritic meteorite composition. But the concentration of Fe in mantle-derived alpine peridotites and in xenoliths in kimberlite is much lower than chondritic concentrations (see Table 16-4). Segregation of an iron core from the mantle is thus invoked to account for this discrepancy. Calculations indicate that the thermal energy created due to viscous resistance of the settling metal as it lost gravitational potential energy would have heated the whole Earth over 2000°C (see Table 16-3). On geochemical grounds, Ringwood (1979, p. 54) argues that core segregation must have occurred rapidly within no more than a few hundred million years after accretion. Considering these time and energy factors for core segregation, together with other thermal energy sources, there must have been significant melting in at least part of the Earth early in its history.

Calculated temperature estimates for the Earth's interior immediately after accretion but before core segregation are shown in Figure 15-22. Most of the interior at this time may have been too cold to cause melting of Mg-rich silicate. However, because the melting temperature for metallic iron is slightly lower than the solidus for silicate, iron particles in the chondritic material from which the Earth accreted may have melted, forming drops of molten metal that coalesced and settled through the solid but plastically deformable protomantle (see Figure 15-23). The resulting generation of thermal energy raised the temperature in the Earth so that a more positively sloping geothermal gradient resulted and the solid mantle could begin to convect. The rising geotherm could well have intersected the silicate solidus so that partial melting of the upper few hundred kilometers occurred.

SUMMARY

15.1. The oldest terranes in Precambrian shields are a montage of complex rock bodies multiply overprinted by magmatic, deformation, and recrystallization events. Archean terranes consist of quartzo-feldspathic gneiss complexes and supracrustal or greenstone belt sequences.

The relative age of the gneiss complexes and supracrustal rocks differs in any particular area and is commonly ambiguous. The oldest authenticated rocks on Earth, 3.8 Gy old, are in a small supracrustal sequence in West Greenland. Younger sequences up to roughly 2.5 Gy in age tend to be larger in volume and comprise thick metamorphosed accumulations of chiefly submarine mafic and ultramafic lava flows of tholeiitic and komatiitic affinities succeeded by calc-alkaline andesitic and more felsic volcanic rocks and some sediments.

Quartzo-feldspathic rocks in gneiss complexes are predominantly tonalitic and trondhjemitic orthogneisses with high concentrations of Na, although a variety of metasedimentary rocks with granulite to amphibolite facies assemblages are also present. The magmatic protoliths of the tonalitic gneisses appear to have been derived as partial melts of tholeiitic mafic source rocks containing amphibole and garnet. Basaltic and komatiitic lavas in greenstone belts appear to have been derived from peridotitic sources, the latter representing unusually high degrees of melting under apparently high geothermal gradients.

Consideration of all possible sources of thermal energy in the early history of the Earth almost surely indicates much higher near-surface temperatures and geothermal gradients, which may have precluded formation of much, if any, crust prior to 3.8 Gy ago. Whether the earliest crust was mafic or sialic is controversial. Although some geologists advocate the operation of plate tectonics during the Archean, most would characterize global processes as uniquely permobile.

15.2. The Proterozoic rock record can be interpreted in terms of a transition from earlier, Archean permobile activity, engendered by vigorous convection in high geothermal gradients beneath thin lithosphere, to a plate tectonic regime of thicker lithosphere and more stable continents overlying longer-wavelength convective systems in a cooler Earth. Principal Proterozoic rock bodies are basaltic dike swarms and large stratiform intrusions, anorthosite-suite massifs, and reworked mobile belts.

Widespread basaltic intrusions into continental rocks are about 2.5 to 1.2 Gy old. Thick intrusive sheets or massifs of anorthosite and pyroxene-bearing felsic rocks are surrounded by high-grade metamorphic rocks; the restriction of such bimodal anorogenic massifs almost entirely to the middle of Proterozoic time raises the possibility that some unique set of circumstances caused their formation. Ash-flow tuffs and highly alkaline rocks including carbonatites appear in substantial volumes in the Mid-Proterozoic.

Ensialic mobile belts composed of reworked (deformed and recrystallized) older continental rocks typify Proterozoic tectonic style. The controversial Grenville Province, a mobile belt with numerous slightly older anorthosite suite massifs in eastern Canada, is believed by some geologists to have formed by a continent–continent collision similar to that which produced today's Himalaya and Tibetan Plateau.

15.3. Details regarding the pattern in space and time of continental growth and evolution remain elusive, particularly for the Precambrian, although a general trend of global activity is discernable. After the short-wavelength premobile activity that produced Archean greenstone-belt gneiss complexes, Proterozoic continents stabilized and thickened so that highly alkaline magmas could form and many swarms of basaltic dikes could be emplaced as they experienced extensional stresses during aborted rifting. Jostling of continental segments in a single supercontinent may have created many of the known mobile belts. In contrast to the widespread continental record during the Proterozoic, the oceanic one is virtually missing, for reasons not understood.

Controversy continues regarding the rate at which continents have grown, although radiometric data seem to favor formation of juvenile magmatic material during the Archean, culminating about 2.7 to 2.5 Gy ago, followed by reworking of this sial during the Proterozoic.

15.4 The mantle, comprising 68% of the Earth's mass, is both a sink and a source of matter and energy associated with the continents, which possess less than 1% of Earth's mass. The growth of sialic continents beginning at least 3.8 Gy ago by extraction of low-T partial melts has variable depleted the upper mantle in incompatible LIL elements. Migration of aqueous metasomatic fluids and melts carrying LIL incompatible elements in a convecting mantle can account for much of the variable depleted upper mantle. The volume of the mantle from which continental material has been extracted is likely only a fraction of the whole. The thickening of continental sial due to lessening geothermal gradients in the cooling Earth may be paralleled by thickening of a subcontinental tectosphere of cold depleted mantle peridotite.

Water must play a paramount role in mantle metasomatism, segregation of LIL elements from the mantle into the crust, and motion of lithospheric plates because of an underlying ductile asthenosphere. But the details of how water moves and where and how it resides in the mantle are largely unknown. As the Earth continues to cool, micas and amphiboles will become more widespread as stable phases in the mantle, fixing some continental and hydrospheric material back into the mantle.

15.5. Although there are uncertainties regarding how the metallic core segregated from a postulated initially homogeneous Earth, there is no question that this event would have generated a substantial amount of thermal energy as the falling iron material lost gravitational potential energy.

REFERENCES

Anhaeusser, C. R., Mason, R., Viljoen, M. J., and Viljoen, R. P. 1969. A reappraisal of some aspects of Precambrian shield geology. *Geological Society of America Bulletin* 80:2175–2200.

Arndt, N. T., Francis, D., and Hynes, A. J. 1979. The field characteristics and petrology of Archean and Proterozoic komatiites. *Canadian Mineralogist* 17:147–163.

Arndt, N. T., Naldrett, A. J., and Pyke, D. R. 1977. Komatiitic and iron-rich tholeiitic lavas of Munro Township, northeast Ontario. *Journal of Petrology* 18:319–369.

Baer, A. J., Emslie, R. F., Irving, E., and Tanner, J. C. 1974. Grenville geology and plate tectonics. *Geoscience Canada* 1:54–61.

Baragar, W. R. A. 1977. Volcanism of the stable crust. Geological Association of Canada Special Paper 16:377–405.

Baragar, W. R. A., and McGlynn, J. C. 1978. On the basement of Canadian greenstone belts: Discussion. *Geoscience Canada* 5:13–15.

Barker, F. (ed.). 1979. *Trondhjemites, Dacites and Related rocks*. New York: Elsevier, 659 p.

Basu, A. R., and Tatsumoto, M. 1979. Samarium-neodymium systematics in kimberlites and in the minerals of garnet lherzolite inclusions. *Science* 205:398–401.

Best, M. G. 1975. Migration of hydrous fluids in the upper mantle and potassium variation in calc-alkaline rocks. *Geology* 3:429–432.

Brocoum, S. J., and Dalziel, I. W. D. 1974. The Sudbury basin, the Southern province, the Grenville front, and the Penokean orogeny. *Geological Society of America Bulletin* 85:1571–1580.

Clifford, T. N. 1970. The structural framework of Africa. In T. N. Clifford and I. G. Gass (eds.), *African Magmatism and Tectonics*. Edinburgh: Oliver and Boyd, pp. 1–26.

DePaolo, D. J., and Wasserburg, G. J. 1979. Neodymium isotopes in flood basalts from the Siberian platform and inferences about their mantle sources. *Proceedings of the National Academy of Sciences, U.S.A.,* 76:3056–3060.

Dewey, J. F., and Burke, K. C. A. 1973. Tibetan, Variscan, and Precambrian basement reactivation: Products of continental collision. *Journal of Geology* 81:683–692.

deWit, M., Hart, R., Stern, C, and Barton, C. M. 1980. Metallogenesis related to seawater interaction with 3.5 B.Y. oceanic crust. *American Geophysical Union Transactions* 61:386.

Dietz, R. S. 1965. Sudbury structure as an astrobleme. *Journal of Geology* 72:412–434.

Duchesne, J-C., and Demaiffe, D. 1978. Trace elements and anorthosite genesis. *Earth and Planetary Science Letters* 38:249–272.

Duncan, R. A., and Compston, W. 1976. Sr-isotope evidence for an old mantle source region for French Polynesian volcanism. *Geology* 4:728–732.

Emslie, R. F. 1975. Nature and origin of anorthosite suites. *Geoscience Canada* 2:99–104.

Emslie, R. F. 1978. Anorthosite massifs, rapakivi granites, and late Proterozoic rifting of North America. *Precambrian Research* 7:61–98.

Fyfe, W. S. 1976. Heat flow and magmatic activity in the Proterozoic. *Philosophical Transactions of the Royal Society of London,* Series A, 280:655–660.

Gast, P. W. 1960. Limitations on the composition of the upper mantle. *Journal of Geophysical Research* 65:1287–1297.

Gast, P. W. 1968. Trace element fractionation and the origin of tholeiitic and alkaline magma types. *Geochimica et Cosmochimica Acta* 32:1057–1086.

Glikson, A. Y. 1979. The missing Precambrian crust. *Geology* 7:449–454.

Goodwin, A. M., and Smith, I. E. M. 1980. Chemical discontinuities in Archean metavolcanic terrains and the development of Archean crust. *Precambrian Research* 10:301–311.

Green, D. H. 1975. Genesis of Archean peridotitic magmas and constraints on Archean geothermal gradients and tectonics. *Geology* 3:15–18.

Hanson, G. N. 1977. Geochemical evolution of the suboceanic mantle. *Journal of the Geological Society* (London), 134:235–253.

Hargraves, R. B. 1976. Precambrian geologic history. *Science* 193:363–371.

Herz, N. 1969. Anorthosite belts, continental drift and the anorthosite event. *Science* 164:944–947.

Hunter, D. R., and Hamilton, P. J. 1978. The Bushveld Complex. In D. H. Tarling (ed.), *Evolution of the Earth's Crust.* New York: Academic Press, pp. 107–173.

Isachsen, Y. W. (ed.). 1968. *Origin of anorthosite and related rocks.* New York State Museum and Science Service Memoir 18, 466 p.

Jackson, E. D., and Thayer, T. P. 1972. Some criteria for distinguishing between stratiform, concentric and alpine peridotite-gabbro complexes. *24th International Geologic Congress, Section 2,* p. 289–296.

Jordan, T. H. 1978. Composition and development of the continental tectosphere. *Nature* 274:544–548.

Kroner, A. 1977. Precambrian mobile belts of southern and eastern Africa—ancient sutures or sites of ensialic mobility? A case for crustal evolution towards plate tectonics. *Tectonophysics* 40:101–135.

McCulloch, M. T., and Wasserburg, G. J. 1978. Sm-Nd and Rb-Sr chronology of continental crust formation. *Science* 200:1003–1009.

McGregor, V. R. 1973. The early Precambrian gneisses of the Godthab district, West Greenland. *Philosophical Transactions of the Royal Society of London,* Series A, 273:343–358.

McKenzie, D., and Weiss, N. 1975. Speculations on the thermal and tectonic history of the earth. *Geophysical Journal of the Royal Astronomical Society* 42:131–174.

Moore, J. M. 1977. Orogenic volcanism in the Proterozoic of Canada. Geological Association of Canada Special Paper 16:127–148.

Muehlberger, W. R., Denison, R. E., and Lidiak, E. G. 1967. Basement rocks in continental interior of United States. *American Association of Petroleum Geologists Bulletin* 51:2351–2380.

Mysen, B. O. 1979. Trace element partitioning between garnet peridotite minerals and water-rich vapor: Experimental data from 5 to 30 kbar: *American Mineralogist* 64:274–287.

Nesbitt, R. W. 1971. Skeletal crystal forms in the ultramafic rocks of the Yilgarn block, western Australia: Evidence for an Archean ultramafic liquid. Geological Society of Australia Special Publication 3:331–347.

O'Nions, R. K., Evensen, N. M., Hamilton, P. J., and Carter, S. R. 1978. Melting of the mantle past and present: Isotope and trace element evidence. *Philosophical Transactions of the Royal Society of London,* Series A, 288:547–559.

O'Nions, R. K., and Pankhurst, R. J. 1978. Early Archean rocks and geochemical evolution of the Earth's crust. *Earth and Planetary Science Letters* 38:211–236.

Oxburgh, E. R., and Parmentier, E. M. 1978. Thermal processes in the formation of continental lithosphere. *Philosophical Transactions of the Royal Society of London,* Series A, 288:415–429.

Pankhurst, R. J. 1977. Strontium isotope evidence for mantle events in the continental lithosphere. *Journal of the Geological Society* (London), 134:255–268.

Piper, J. D. A. 1976. Paleomagnetic evidence for a Proterozoic super-continent. *Philosophical Transactions of the Royal Society of London,* Series A, 280:469–490.

Pyke, D. R., Naldrett, A. J., and Eckstrand, O. R. 1973. Archean ultramafic flows in Munro township, Ontario. *Geological Society of America Bulletin* 84:955–978.

Ringwood, A. E. 1979. *Origin of the Earth and Moon.* New York: Springer-Verlag, 295 p.

Romey, W. D., Hudson, M. R., and Elberty, W. R. 1979. Comment on evaporites as precursors of massif anorthosite. *Geology* 7:3–5.

Simmons, E. D., and Hanson, G. N. 1978. Geochemistry and origin of massif-type anorthosites. *Contributions to Mineralogy and Petrology* 66:119–135.

Tarney, J., and Windley, B. F. 1977. Chemistry, thermal gradients and evolution of the lower continental crust. *Journal of the Geological Society* (London), 134:153–172.

Tatsumoto, M. 1978. Isotopic composition of lead in oceanic basalt and its implication to mantle evolution. *Earth and Planetary Science Letters* 38:63–87.

Turner, F. J., and Verhoogen, J. 1960. *Igneous and Metamorphic Petrology,* 2nd Ed. New York: McGraw-Hill, 694 p.

Veizer, J., and Jansen, S. L. 1979. Basement and sedimentary recycling and continental evolution. *Journal of Geology* 87:341–370.

Viljoen, M. J., and Viljoen, R. P. 1969. The geology and geochemistry of the lower ultramafic unit of the Onverwacht Group and a proposed new class of igneous rock. Geological Society of South Africa Special Publication 2:55–85.

Visser, D. J. L., and von Gruenwald, G. 1970. Symposium on the Bushveld igneous complex and other layered intrusions. Geological Society of South Africa Special Publication 1, 763 p.

Willemse, J. 1969. The geology of the Bushveld igneous complex, the largest repository of magmatic ore deposits in the world. Economic Geology Monograph 4:1–22.

Wilson, J. F., Bickle, M. J., Hawkesworth, C. J., Martin, A., Nisbet, E. G., and Orpen, J. L. 1978. Granite-greenstone terrains of the Rhodesian Archean craton. *Nature* 271:23–27.

Windley, B. F. (ed.). 1976. *The Early History of the Earth.* New York: John Wiley and Sons, 619 p.

Windley, B. F. 1977. *The Evolving Continents.* New York: John Wiley and Sons, 385 p.

Wise, D. U. 1974. Continental margins, freeboard and the volumes of continents and oceans through time. In C. A. Burk and C. L. Drake (eds.), *The Geology of Continental Margins.* New York: Springer-Verlag, pp. 45–58.

Worst, B. G. 1958. The differentiation and structure of the Great Dyke of Rhodesia. *Transactions of the Geological Society of South Africa* 61:283–354.

Wynne-Edwards, H. R. 1972. The Grenville Province. Geological Association of Canada Special Paper 11:263–334.

16

Extraterrestrial Petrology and Evolution of the Planets

Radiometric dating shows that the most ancient rocks on Earth are only 3.8 Gy old, yet the age of the solar system is 800 My older. What transpired on Earth during the missing part of the rock record, which is longer than all of Phanerozoic time? Why is there no record from 4.6 to 3.8 Gy ago? How did the Earth originate and become so strongly differentiated? This chapter briefly reviews the exciting discoveries of the past decade or so regarding the nature and origin of other bodies of rock in the solar system and the bearing they have on these questions, which are otherwise unanswerable.

Other planetary bodies are also chemically differentiated. The Moon experienced widespread melting of at least its outer part 4.6 to 4.4 Gy ago, which produced a differentiated anorthositic crust and olivine–pyroxene mantle. Intense meteorite cratering culminated 3.9 Gy ago. This same episode of meteorite bombardment is assumed to have been the cause of widespread craters on Mercury, Mars, and possibly Venus. Could this bombardment have also occurred on Earth and

have been so violent as to obliterate the earlier rock record? If so, what factors prevented post-bombardment plate tectonics on these other planetary bodies, thus preserving craters, while the Earth's surface has been and continues to be vigorously recycled by the effects of global plate motions and erosive processes?

Some meteorites are fragments of small planetary bodies that were differentiated into silicate and metallic parts. Other less common meteorites called carbonaceous chondrites are nearly undifferentiated and have essentially the same bulk chemical composition for nonvolatile elements as does the Sun; they appear to be a disequilibrium aggregation of solar nebula material from which the Sun and planets accreted 4.6 Gy ago.

16.1 THE MOON

Because of the unmanned and manned landings by the American and Russian space programs of the 1960s and 1970s, there is now an enormous volume of data concerning the Moon. In addition to thou-

Figure 16-1 Telescopic view of the Moon showing contrast between lighter-colored, intensely cratered highlands and the less cratered, topographically lower darker maria. Note sprays of ejecta from large craters in maria. [Photograph courtesy of Lick Observatory.]

Figure 16-2 Outcrop of basalt flow about 50 m long near the top of the 350 m deep Hadley Rille. About 5 m of regolith overlies the flow, and beneath its exposure the wall of rille is covered by talus. [National Aeronautics and Space Administration photograph AS15-89-12116.]

sands of journal articles and several books (for example, Taylor, 1975), there are the *Proceedings of the Lunar Science Conference,* published annually, and a symposium volume (Brown and others, 1977). The books by King (1976) and Wood (1979) and review articles by Smith (1979) and Marvin, Wasson, and others (1979) consider all extraterrestrial rock bodies. The purpose of this section is to review briefly those aspects of the structure of the Moon, the petrology of its rocks, and its history that bear upon the largely unrecorded earliest history of the Earth.

Structure of the Moon

The lunar surface, seen from a distance (see Figure 16-1), is readily divisible into an intensely cra-

tered, light-colored **highlands** and smoother, less cratered, darker-colored lowlands, or **maria,** that comprise about 17% of the lunar surface. Covering the entire surface of the moon is a **regolith** several meters thick of churned rock fragments that has formed as a result of billions of years of meteorite impacting; outcrops of coherent rock as seen on Earth are very rare (see Figure 16-2).

Seismic investigations using moonquakes, impacting meteorites, and spent spacecraft parts for energy sources reveal a layered Moon (see Figure 16-3). The proportionately thicker lid above a postulated partially melted layer indicates the Moon has cooled more than Earth, which is not surprising in view of the lack of major volcanism since possibly 2.5 Gy ago on the lunar surface. The possible existence of a metallic iron core is controver-

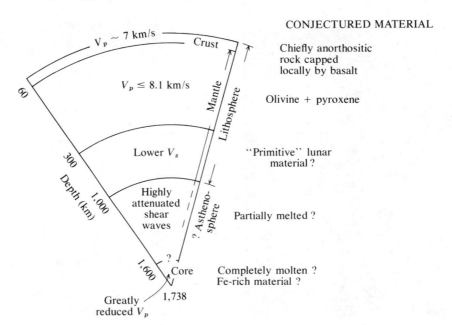

Figure 16-3   Seismic layering of the Moon's interior. V_p and V_s are velocities of compressional and shear waves, respectively. The olivine + pyroxene assemblage of the lunar mantle is not labeled peridotite so as to discourage any notions that it is necessarily the same rock as that comprising the upper mantle of the Earth. [After D. R. Lammlein, 1977, Lunar seismicity, structure, and tectonics, *Philosophical Transactions of the Royal Society of London*, Series A, 285.]

sial, and if one exists, its volume is only $\sim 1\%$ of the whole Moon.

Petrology and Chronology

Lunar petrologic processes are dominated by anhydrous magmatism and shock metamorphism. The volatile-free and highly reduced character of lunar rocks, together with the total absence of hydrothermal alteration, fluid-involved metamorphism, and chemical weathering so widespread in terrestrial systems, are striking contrasts.

Shock metamorphism—the regolith. Impacting meteorites, ranging in size from projectiles perhaps 100 km in diameter down to microscopic dust, have acted as both destructive and constructive rock-forming processes. Their kinetic energy has deformed bedrock and ejected particulate material from impact craters. In a constructional sense, new material has been added to the Moon and some of the energy from larger meteorites is consumed in melting the impacted material; slight fusion merely welds the impacted fragments into a cohesive breccia, whereas more intensive or even complete melting creates small droplets that splash away from the crater, forming glass spheres. The regolith resulting from these impact processes is quite heterogeneous, although core samples reveal the existence of successive layers that may represent material ejected during discrete major cratering events.

Much of the effort expended by geologists who have studied samples returned by the Apollo and Luna missions has been devoted to unraveling the

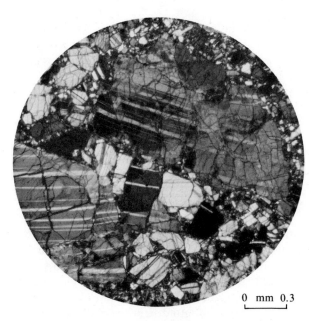

0 mm 0.3

Figure 16-4 Lunar anorthosite breccia. [National Aeronautics and Space Administration photograph 60215,13.]

Table 16-1 Chemical compositions of lunar rocks, with terrestrial comparisons (in wt.%).

	Anorthosite		Basalt	
	1	2	3	4
SiO_2	45.0	53.40	37–49	50.72
TiO_2	0.56	0.77	0.3–13	1.96
Al_2O_3	24.6	23.96	7–14	14.98
Fe_2O_3	—	0.91	—	3.51
FeO	6.6	3.02	18–23	8.22
MgO	8.6	1.88	6–17	7.38
CaO	14.2	9.85	8–12	10.35
Na_2O	0.45	4.17	0.1–0.5	2.44
K_2O	0.075	0.80	0.02–0.3	0.45
Total	99.2	98.76	—	100.00
Normative				
An	95	57		

1. Average composition of the lunar highlands.
2. "Primary" anorthosite magma of Adirondack Mountains, New York.
3. Range in composition of mare basalts.
4. Average tholeiitic basalt.
Source: Column 2 after A. F. Buddington, 1939, Geological Society of America Memoir 7; columns 1 and 3 after S. R. Taylor, 1975, *Lunar Science: A Post-Apollo View* (New York: Pergamon); column 4 after R. W. LeMaitre, 1976, The chemical variability of some common igneous rocks, *Journal of Petrology* 17.

nature and history of the highlands and maria from regolith samples and trying to distinguish pristine igneous rocks from impact-generated magmatic material.

Highlands. The pervasively shocked samples recovered from the lunar highlands are chiefly anorthositic rocks (see Figure 16-4 and Table 16-1) in which the plagioclase, by terrestrial standards, is unusually calcic—usually calcic bytownite and anorthite. Lesser amounts of plagioclase-rich norites, olivine gabbros, and very rare dunite with Mg-rich olivines and orthopyroxenes have been found. Among the minor plagioclase-rich rocks are those called KREEP because of their high concentrations of K, rare earth elements, and P. The origin of these minor rock types is controversial, but there is considerable agreement that the major anorthositic rocks represent an accumulation of buoyant plagioclase crystals that floated to the top of a lunar magmatic "ocean." The thick mantle beneath the anorthositic crust (see Figure 16-3),

inferred to be an olivine–pyroxene rock, is envisaged to represent a complementary settled crystal accumulation. Radiometric ages of 4.6 to 4.4 Gy for highland samples indicate that this magmatic segregation occurred during the initial accretion of planetary bodies in the solar system, or very shortly afterward. Several factors, including energy from accretion of particles in the primordial solar nebula, energy of impacting planetesimals (that is, small primitive planets), and radioactive decay of short-lived isotopes (such as [26]Al to [26]Mg) might have conspired to produce the high temperatures required to cause melting of the outer few hundred kilometers of the Moon.

Ringwood (1979) points out that complete melting of the whole Moon or even its outer perimeter was less likely from an energy standpoint than par-

Figure 16-5 Basalt flows about 30 m thick on Mare Imbrium. [National Aeronautics and Space Administration photograph AS-15-1556.]

tial melting. Furthermore, this partial melt must have been roughly of basaltic composition in order to have the segregating plagioclase as a near-liquidus phase; an ultramafic melt would have insufficient Ca and Al to allow much precipitation of plagioclase. See Ringwood (1979, pp. 168–177) for an extended discussion of attempts to evaluate the overall bulk composition of the lunar magma "ocean," which he finds is similar to oceanic rift basalts.

An epoch of massive meteorite bombardment culminating ~ 3.9 Gy ago (resetting Ar radiometric systems) excavated major basins and may have generated zones of weakness in the cooling and thickening crust and mantle through which mare basalt magmas were later extruded.

Mare basalts. Following the massive meteorite bombardment, excavated basins and lowlands were flooded with innumerable thin flows of very fluid basalt lava (see Figure 16-5). The extreme length of mare flows, some apparently in excess of 1,000 km, reflects their low apparent viscosities and high rates of extrusion. Although locally accumulating to thicknesses of perhaps several kilometers, the overall volume of flows must be only a few percent, at most, of the whole lunar crust. Isotopic analyses of returned samples yield ages in the range of 4.0 to 3.2 Gy, but crater densities on some maria suggest that volcanism may have persisted to 2.5 Gy ago.

Lunar basalts display the same types of fabrics as terrestrial basalts (see Figure 16-6). Some returned samples are somewhat vesicular; the composition of the responsible gas is not known but was not likely to be water. Compositionally, the lavas extruded over the hundreds of millions of years of mare activity are quite variable; nomenclature problems abound. But, overall, they differ in significant compositional respects from basalts on Earth. The major minerals are strongly zoned Ti-rich clinopyroxene and very calcic plagioclase, averaging between 80% and 90% An; contrast terrestrial basalts, which generally contain labradorite plagioclase. Olivine is not always present; late-crystallizing tridymite or cristobalite are common. All mare basalts contain Fe–Ti oxides, usually ilmenite, and in some samples these oxides are a major phase (ranging to 25 modal percent). The basalts are highly reduced, lack Fe^{3+}, and may even contain accessory amounts of metallic iron. Relatively volatile chemical species—H_2O, CO_2, S, Cl, Pb, As, Bi, Hg, Na, K—have exceedingly small concentrations with respect to terrestrial basalts. While it may be thought possible for Na and K, for example, to volatilize from the hot lava flows extruded into the vacuum of the lunar surface, this is not the reason for their low concentrations because: (1) all mare basalts are poor in these elements relative to terrestrial basalts, rather than just some and to a variable degree; and (2) at the very small oxygen fugacities (see pages 504–506) of the lunar basalt magmas, Na would volatilize as the metal, oxidizing Fe^{2+} to Fe^{3+} in the lava, but

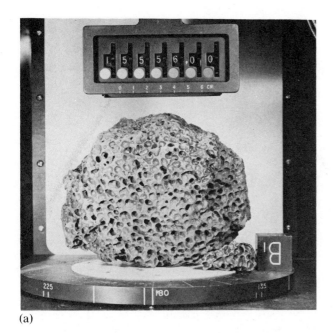

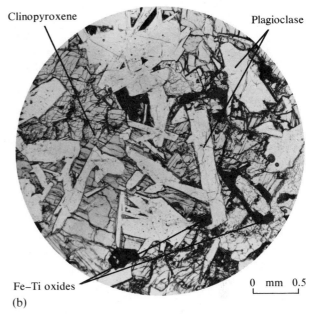

Clinopyroxene

Plagioclase

Fe–Ti oxides

0 mm 0.5

(a)

(b)

Figure 16-6 Lunar basalts. (a) Vesicular basalt from Hadley Rille (see Figure 16-2). (b) Ophitic basalt composed of calcic plagioclase, titaniferous clinopyroxene, and Fe–Ti oxides. [National Aeronautics and Space Administration photographs S-71-45477 and S-69-47907.]

ferric iron is not found. Mare basalts tend to have relatively greater abundances of refractory elements such as Ti, Sc, Zr, and Y; this is manifest in the presence of the unusual minerals armalcolite $(FeMg)Ti_2O_5$, tranquillityite $Fe_8(Zr,Y)_2Ti_3Si_3O_{24}$, and zirconolite $CaZrTi_2O_7$.

Still another significant aspect of mare basalts is their universal depletion in europium relative to other REE (see Figure 16-7). Several interpretations have been proposed (Taylor, 1975, p. 156), but the most attractive follows as a consequence of the postulated segregation of a plagioclase-rich crust, now exposed as the lunar highlands. As explained on page 55, the Eu^{2+} ion strongly partitions into plagioclase crystals in equilibrium with a silicate melt, leaving it depleted. Lunar highlands samples are enriched in europium and hence postulated complementary olivine–pyroxene differentiates in the underlying lunar mantle segregated 4.6 to 4.4 Gy ago are presumably depleted.

Partial melting of this olivine–pyroxene mantle assemblage would, therefore, yield melts also depleted in europium—the mare magmas (Brown, 1977).

The source of the thermal energy required for mare magma formation remains a puzzle.

Implications of Lunar History for the Evolution of the Earth

The evolution of the Moon almost entirely before ~ 3 Gy ago complements in an extraordinary way the Earth's recorded history mostly since that time. It is possible that the brief, earlier development of the Moon is related to its smaller size, which would have allowed it to cool faster, and to its more refractory and essentially volatile-free overall composition, as manifest in the Ca–Al-rich highlands and Ti–Zr-enriched mare basalts. This compositional contrast implies that solidus tem-

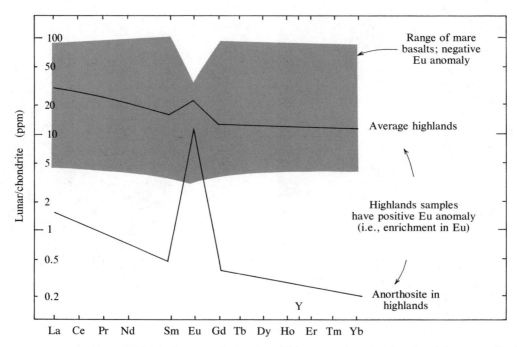

Figure 16-7   Ratios of REE in lunar rocks to chondritic meteorite, that is, chondrite-normalized REE pattern. All mare basalts fall in shaded band and are depleted in Eu relative to other REE. Normalizing to chondritic values eliminates the irregular sawtooth abundance pattern that otherwise results as a reflection of the fact that elements with even atomic numbers are more abundant than those with odd atomic numbers. [After S. R. Taylor, 1975, *Lunar Science: A Post-Apollo View* (New York: Pergamon).]

peratures were reached sooner in the cooling Moon, effectively shutting off magmatic activity. Heat sources in these two bodies may also have differed.

Are there any parallels between lunar and terrestrial evolution, albeit on different time scales, cooling rates, and compositional parameters? Can we view the segregation of a plagioclase-rich scum on the lunar magmatic "ocean" 4.6 to 4.4 Gy ago as analogous to the generation of tonalitic crustal rocks in the Archean roughly 3.8 to 2.5 Gy ago on Earth? Would the greater H_2O, K, and Na concentration in the Earth cause the difference in the composition of these earliest crusts on the two bodies? Could mare basaltic activity and mid-ocean rift basaltic volcanism be similar responses

to fundamentally similar circumstances in their interiors? Or should mare basalts be compared to greenstone belt volcanics? The catastrophic bombardment of the lunar surface terminating ~ 3.9 Gy ago occurred just before the oldest authenticated (3.8 Gy) rocks formed on Earth. Are these two dates merely coincidental? Or could the Earth also have suffered bombardment, completely obliterating any vestiges of a pre-3.8-Gy-old crust?

16.2   COMPARATIVE PLANETOLOGY

Study of the other rocky planetary bodies in the inner part of the solar system is providing fruitful insight into their origin, individually and collectively. In addition, comparisons of processes and products on different planets may answer some of the questions posed above regarding the Earth's evolution.

Early Meteorite Bombardment

The discovery that Mars and Mercury, as well as the Moon, possess impact craters with diameters of as much as 200 km constitutes a strong argument that the Earth, whose orbit lies between them, also suffered a similar massive bombardment. We can only surmise that the termination of this intense bombardment $\sim$ 3.9 Gy ago was the same on all four bodies. Vigorous geologic activity on Earth since then would have obliterated the postulated impact products. (Only slightly more than 100 impact craters on Earth have been recognized, ranging up to 140 km in diameter; the oldest is only $\sim$ 2 Gy old.)

Scaling calculations for the effects of the early meteorite bombardment on Earth using the Moon for calibration have been carried out by Frey (1980) and Grieve (1980). At least one-third to one-half of the Earth's surface would have been cratered. Several tens of basins in excess of 1,000 km in diameter and 3 to 4 km deep would have formed by projectiles perhaps hundreds of kilometers in diameter excavating any crust that might have been present to depths of 15 km or so. Not to be ignored, however, is the thermal energy imparted to the Earth by the massive impact event. Although this energy, calculated to have been at least 10^{28} to 10^{29} J, is two to three orders of magnitude less than that calculated for core segregation (see Section 15.5), it was focused in only the outer few kilometers of the Earth's surface. If all of this energy were accumulated in an instant, the surface temperature would have risen many thousands of degrees. However, the bombardment probably occurred over some extended time period, but of unknown duration, so that heat generated by each impact could at least partially dissipate before the next impact.

Figure 16-8 Mariner 10 view of the planet Mercury. [National Aeronautics and Space Administration photograph.]

Differentiated Inner Planets

All of the inner rocky planets—Mercury, Venus, Earth and Moon, Mars—in the solar system appear to be differentiated; that is, different parts have different chemical compositions (King, 1976). Some cosmologists believe this differentiation came about by an initial inhomogeneous accretion from the solar nebula, creating metallic cores covered by silicate mantles. However, to anticipate a subsequent discussion on pages 586–588, there are more difficulties with this model than with one of homogeneous accretion, followed by subsequent chemical differentiation into core, mantle, and possibly crust.

Evidence that Mercury is a differentiated body is the weakest for the inner planets. Although most closely resembling the Moon because of extensive cratering (see Figure 16-8), its greater uncompressed density relative to the Moon (see Table 16-2) suggests that Mercury must be composed mostly of Fe–Ni in contrast to the silicate-dominant Moon.

Venus, with a similar radius and density as Earth, may have a similar gross internal structure and composition. But little is known of its surface geology because of a thick atmosphere of chiefly H_2SO_4 and CO_2 that is opaque to the visual part of the electromagnetic spectrum. Recent radar mea-

Table 16-2 Relevant properties of the planets and the Moon.

	Mass (10^{27} g)	Radius (10^3 km)	Overall real density (g/cm^3)	Estimated density at 1 GP (g/cm^3)	Distance to Sun (10^8 km)
Mercury	0.33	2.44	5.44	5.3	0.58
Venus	4.87	6.05	5.27	3.9	1.08
Earth	5.98	6.38	5.52	4.0	1.50
Moon	0.07	1.74	3.34	3.4	1.50
Mars	0.65	3.39	3.96	3.9	2.28
Jupiter	1900	70	1.3		7.78
Saturn	570	60	0.7		14.3
Uranus	87	26	1.2		28.7
Neptune	100	25	1.7		45.0
Pluto	0.7?	3?	3?		59

Source: After J. V. Smith, 1979, Mineralogy of the planets: A voyage in space and time, *Mineralogical Magazine* 43.

surements from Earth and especially from the Pioneer-Venus spacecraft have provided a topographic map of most of the surface with elevations accurate to about 200 m. Certain conclusions may be drawn from such a map (Masursky and others, 1980). In contrast to the bimodal topography of Earth, in which elevations are dominated by continental cratons and ocean basins, Venus has a unimodal relief. Rolling upland plains constitute nearly two-thirds of the surface and have a total relief of only 2 km. Numerous circular features may be lava-filled impact basins, and their size frequency distribution is reminiscent of the ancient cratered terranes on the Moon and Mars. K, Th, and U concentrations obtained by the gamma-ray spectrometer on the USSR Venera 8 lander suggest a granitic composition for the upland plains.

About 8% of the Venusian surface lies at higher elevations, as much as 9 km more than the most elevated upland. These highlands appear to be in isostatic equilibrium with the surrounding upland plains, judging from measurements of gravity by spacecraft, presumably because of crustal thickening or lateral compositional variations in the periphery of Venus. One highland region resem-

bles a chain of shield volcanoes along a discontinuous fault zone; Venera landers 9 and 10 gamma-ray spectrometry indicate a basaltic composition. About 27% of the planet is classified as lowlands and may be covered by basaltic lavas. No integrated global pattern of troughs and ridges that might be associated with plate subduction and accretion have been identified although some youthful-appearing, local ridge and trough complexes do exist.

Mars apparently has a metallic core, judging from its density. Viking lander photography shows a surface of extensive dune fields littered with abundant boulders whose pitted surfaces suggest vesicular lava flows (see Figure 16-9). X-ray fluorescence spectrometry of Martian wind-blown soil at the lander site reveals a chemical composition resembling basalt, perhaps with an admixture of chondritic meteorite (Clark and Baird, 1979). On a larger scale, the Martian surface consists of intensely cratered terranes dominating the southern hemisphere and relatively uncratered plains that are prominent in the northern hemisphere. This dicotomy resembles the lunar highlands and maria, but until rock samples can be returned and dated, we can only assume that the chronology of development of Mars parallels that of the Moon.

There is an enormous bulge in the Tharsis plains region in the Martian crust ~ 7 km high and 4,000 km across, on top of which are four huge shield volcanoes (see Figures 16-10 and 16-11). Radiating outward as far as 3,000 km are numerous fractures. Impact crater density calibrated against crater chronology on the Moon suggests to Carr (1974) that the fracturing associated with the bulge terminated ~ 1.2 Gy ago, whereas the shield volcanoes formed later, perhaps 0.8 to 0.1 Gy ago. Carr further hypothesizes that the bulge, radial fractures, and capping volcanoes formed over an enormous upwelling column of mantle that must have been unique in the evolution of Mars. This sort of activity is analogous to intraplate processes on Earth that are commonly cited as precursors of complete plate rifting and eventual spreading and subduction. However, these aspects of plate tectonics are absent on Mars; perhaps Mars tried but didn't quite make it for some unknown reasons.

The much greater size of Martian shield volcanoes compared to terrestrial ones, such as those of Hawaii (see Figure 5-11), is believed by Carr to be due to the fixation of the Martian crust over the magma-generating source. Whereas movement of the Pacific plate over a stationary magma source for tens of millions of years produced the long Emperor-Hawaiian chain of volcanoes (see pages 201–202), the crust and source were fixed relative to one another on Mars for tens and possibly hundreds of millions of years, so that the extruded magma just kept piling up over the source. Alternatively, the greater volume and, therefore, height of Martian shields may simply reflect deeper sources beneath thicker lithosphere, as argued from hydrostatic principles on pages 334–336.

Evolution of the Differentiated Planets

The evolution of the inner rocky planets into differentiated bodies after their accretion from the solar nebula some 4.6 Gy ago has apparently followed interesting parallels, punctuated by significant differences (Kaula, 1975; Lowman, 1976; Wood, 1979, pp. 177–179).

The principal factor governing evolution of the inner planets after their accretion appears to their size and how it influences the thermal energy budget. This budget in turn is a function of relative gains and losses. First, consider the credit side of the thermal ledger. Planets gained thermal energy during accretion as infalling planetesimals (ranging down to small bodies of perhaps meteorite size) imparted their kinetic energy. Accretion into larger planets yielded more energy (see Table 16-3). After accretion, or possibly in the latter stages of it, segregation of a metallic core (see Section 15.5) heated the body still more; this assumes, of course, that sufficient metallic Fe was present. Intense bombardment by meteorites before 3.9 Gy ago, radioactive decay, and possible tidal interactions—the latter two persisting to the present—are the remaining heat sources.

In terms of heat losses, the smaller planets suffer more because of their larger ratio of surface area to volume. Internal heat is dissipated into space by radiative and conductive processes, both a function of the surface area of the body. Another important factor in heat loss is the location of the source. Thermal energy produced by accreting planetesimals and impacting meteorites is rapidly lost from the thin surface layer it affects most; energy from deep internal sources such as core segregation is more effectively retained, except as convection brings it near the surface.

Figure 16-12 is a schematic portrayal of the similarities and differences in the evolution of the planets after accretion as a function of their temperature. The Moon and Mercury died geologically at a young age; Mars is either dead or a slumbering adolescent, while Earth is still vigorously active and in full flower even though 4.6 Gy old. The small Moon experienced melting to form a global magma "ocean" that was the first step in differentiation into a lighter crust and denser mantle; after massive meteorite bombardment, there followed a period of mare basaltic volcanism.

Figure 16-9   Martian surface viewed from Viking Lander I showing apparently vesicular volcanic rocks. [NASA.]

Moon and Mercury, the smaller bodies, then died geologically. Mars subsequently experienced, on a grand scale in one locale, crustal uplift, fracturing, and volcanism that may be viewed as an aborted attempt at plate tectonics. It has been geologically dormant (or dead?) since perhaps a few hundred million years ago.

Although little is known for certain about Venus, it does not appear to have experienced plate tectonics and has large areas of cratered terrane. If size and thermal energy content are the determining factors in planetary evolution, then why hasn't Venus behaved more like Earth in having a prolonged and still active program of global geologic activity, especially plate tectonics? In addition to size, the composition of the planet must have some influence on its evolution, and in this regard an obviously unique feature of the Earth is its water content. Is plate tectonics somehow permitted only in a water-rich planet? Is water from the oceans somehow fed back into the mantle in subduction zones to maintain the asthenosphere

and plate motion during a particular global thermal regime?

16.3 METEORITES

Rock samples from the Earth and Moon and the surface terranes of the other inner planets all reflect chemical differentiation, either during or after accretion of the body. Certain types of undifferentiated meteorites hold significant clues regarding the earliest history of planets and their accretion in the primordial solar nebula. Other meteorites show evidence of chemical differentiation, collisional deformation, and metamorphism that are insightful for our ideas concerning subsequent planetary evolution.

Meteorites are pieces of extraterrestrial stony or metallic material captured by the gravity field of the Earth (Wood, 1979; Wasson, 1974). They are believed to have resided in the asteroid belt between Mars and Jupiter but were bumped out of

that heliocentric orbit by mutual collisions combined with gravitational perturbations from Jupiter and other planets.

Types of Meteorites

Carbonaceous chondrites. The cosmological significance of **carbonaceous chondrites** far exceeds their relatively minor abundance (see Figure 16-13). They are composed of as much as 50% chondrules and inclusions in a gray to black aphanitic matrix (see Figure 16-14). **Chondrules** are spheroidal objects a millimeter or so in diameter composed of olivine and pyroxene and sometimes Fe–Ni metal, troilite (FeS), and glass in a variety of textures (see Figure 16-15). They seem to have formed by rapid cooling of droplets of silicate melt in a reducing environment, the origin of which is unknown. Generally light-colored **inclusions** are aggregates of Ca–Al–Mg–Ti-rich and Si-poor minerals such as olivine, diopside, spinel ($MgAl_2O_4$), Al–Ti clinopyroxene, melilite (($Ca,Na,K)_2(Mg$,

$Fe,Al,Si)_3O_7$), grossular ($Ca_3Al_2Si_3O_{12}$), and perovskite ($CaTiO_3$). These minerals are all relatively refractory and are believed to have formed by condensation of a cooling hot vapor. The matrix surrounding the inclusions and chondrules is a messy, semiopaque mixture of ill-defined phyllosilicate minerals, magnetite, Fe–Ni sulfide, carbonates and sulfates of Ca and Mg, and as much as 5% complex tarlike organic compounds. Clearly, carbonaceous chondrites are mechanical mixtures of materials formed at high to low temperatures that are grossly out of mineralogical equilibrium. Note especially the coexistence of three valence states of iron.

There is a striking similarity between the bulk chemical composition of carbonaceous chondrites richest in C, H_2O, and volatile trace elements and the composition of the Sun, determined spectroscopically. (This particular class of carbonaceous chondrites, called type I, is essentially lacking in chondrules, for some unknown reason.) Scrutiny of the values in Table 16-4 shows that, within

Figure 16-10 Map of a part of Mars showing old cratered terrane, smoother younger plains, and fractures radiating from the Tharsis bulge, which is capped by four enormous shield volcanoes, including Olympus Mons to the northwest (see Figure 16-11). Topographic contour lines showing extent of bulge are 1 km apart in elevation. [After M. H. Carr, 1974, Tectonism and volcanism of the Tharsis region of Mars, *Journal of Geophysical Research* **79**; and *U.S. Geological Survey Atlas of Mars* M 25M 2R, 1975, I-940.]

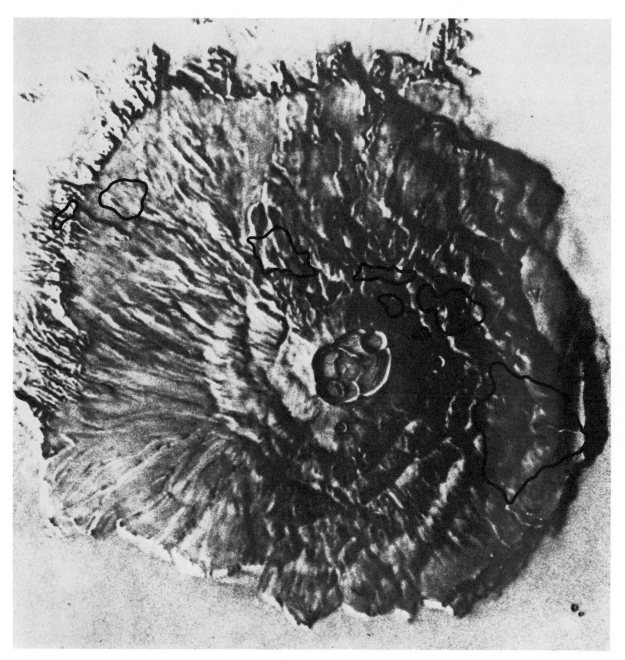

Figure 16-11   Mariner 9 photograph of the Martian shield volcano, Olympus Mons (previously called Nix Olympica). Note the summit caldera complex. The enormous size (24 km high, 600 km in diameter) is apparent when contrasted with the outlines of the Hawaiian Islands, the largest of which consists of five coalescing shield volcanoes (see Figure 5-11). [After R. Greeley, 1974, *Geologic Guide to the Island of Hawaii—A Field Guide for Comparative Planetary Geology,* NASA Cr 152416.]

Table 16-3 Calculated temperature rise (ΔT) of the inner planets due to initial accretion of planetismals and to core segregation.[a]

	Earth	Venus	Mars	Mercury	Moon
Radius (10^3 km)	6.38	6.05	3.39	2.44	1.74
Accretion (ΔT°C)	29,000	25,000	5,900	4,100	1,300
Core segregation (ΔT°C)	2,300	?	300	700	~ 20

[a] Calculations assume all thermal energy accruing from transformation of kinetic and gravitational potential energies is retained in the planet. *Source:* After J. A. Wood, 1979, *The Solar System* (Englewood Cliffs, N.J.: Prentice-Hall); and S. C. Solomon, 1980, Differentiation of crusts and cores of the terrestrial planets: Lessons for the early Earth, *Precambrian Research* 10.

analytical error, the concentrations of most elements (see also Wasson, 1974, p. 78) are identical, except for the lightest and most volatile elements H, He, C, N, and O, which have anomalously low concentrations. These primitive meteorites with the presumed generalized chemical composition of the primordial solar nebula have escaped the subsequent chemical differentiation via melting and fractionation processes suffered by many other meteorites as well as by the inner rocky planetary bodies of the solar system. Their primitive nature is also indicated by radiometric ages of ~ 4.6 Gy.

As a group, carbonaceous chondrites show variable depletion in the more volatile constituents and commensurate enrichment in refractory constituents. They grade compositionally into ordinary chondrites.

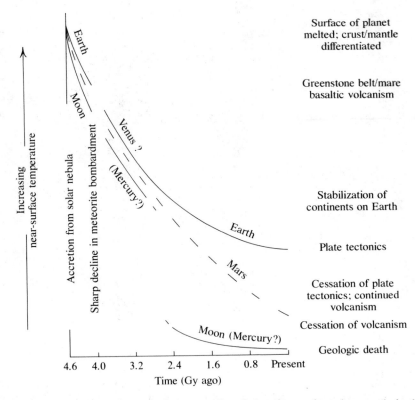

Figure 16-12 Highly schematic evolutionary paths of the planets based upon their thermal energy content. [After J. A. Wood, 1979, *The Solar System* (Englewood Cliffs: Prentice-Hall).]

Ordinary chondrites.   The most abundant meteorites that are observed falling to Earth (see Figure 16-13) are harder and lighter in color than are the carbonaceous type. These **ordinary chondrites** lack the water and organic compounds typical of the carbonaceous group and are significantly depleted in volatile elements such as S, Bi, Pb, Hg, and Cl. More refractory elements—Mg, Si, Fe, Ti, and so on—are enriched (see Table 16-4) and are lodged in olivine, pyroxenes, Ni–Fe alloys, troilite, and minor plagioclase and chromite. Textures are decidedly crystalloblastic and boundaries between chondrites and matrix are commonly blurred (see Figure 16-16). Clearly, ordinary chondrites have experienced an imprint of thermal metamorphism, probably at temperatures exceeding 500°C, so that equilibrium grain shapes and mineral compositions

were produced. It is not certain, however, whether this metamorphism drove the volatile constituents out of a material that was initially similar to carbonaceous chondrite, or whether the protolith was already volatile-poor owing to special circumstances of condensation from the solar nebula. The date of metamorphism seems to vary almost any time after 4.6 Gy ago, judging from the wide range in K–Ar ages observed (Wood, 1979, p. 122).

Achondrites.   These are stony meteorites more similar in composition and texture to differentiated terrestrial magmatic rocks than the chondrites. Indeed, many **achondrites** are essentially basalts with ophitic and vesicular fabrics; others are gabbros and allied phaneritic ultramafic rocks. The magmas from which they crystallized could have been

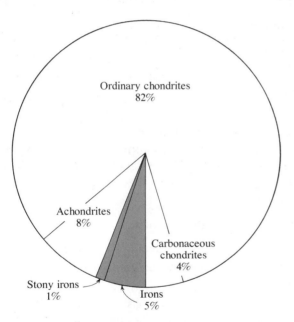

Figure 16-13   Proportions of the five major categories of meteorites observed to fall to Earth. Proportion of iron meteorites is shaded; that of stony meteorites is unshaded. A far greater proportion of iron meteorites is found because of their striking contrast to terrestrial rocks; many stony meteorites resemble rocks, and others, such as the carbonaceous chondrites, weather rapidly after falling to Earth. [After J. A. Wood, 1979, *The Solar System* (Englewood Cliffs: Prentice-Hall).]

Figure 16-14   Fragment of the Allende (Mexico) carbonaceous chondrite. The dark upper surface is a fusion crust formed by frictional heating during flight through the Earth's atmosphere. Note abundant light-colored Ca–Al-rich inclusions and spherical chondrules. [Smithsonian Astrophysical Observatory photograph courtesy of John A. Wood.]

Chondrule

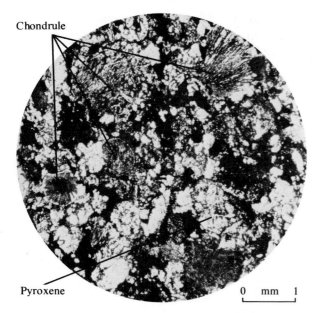

Pyroxene

0   mm   1

Figure 16-15   Chondrules in the chondritic meteorite Gilgoin. Textures within chondrules vary from radiating to barred. [A sample of this meteorite was kindly loaned to the author by Carleton Moore.]

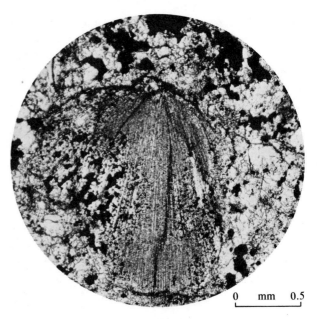

0   mm   0.5

Figure 16-16   Recrystallization in the Lumpkin ordinary chondrite. Needlelike radiating grains in the chondrules have begun to recrystallize into more equant grains, obliterating the definition of the chondrule. [Smithsonian Astrophysical Observatory photograph courtesy of John A. Wood.]

formed by impact processes on the differentiated crust of a large planetary body or by melting in its interior. Many achondrites are breccias, suggesting impacting in their history, as if they were some sort of regolith (see Figure 16-17).

Ages of achondrites range downward from 4.6 Gy by more than a billion years, indicating the parent body (or bodies) were differentiating well after initial accretion.

Iron meteorites.   This class of meteorites essentially complements in a chemical sense the achondrite differentiates of parent chondrite bodies. Iron meteorites originated as high-temperature Fe–Ni alloys, but due to subsequent slow cooling they have unmixed into coarse blades of Ni- and Fe-rich metal; this is the familiar Widmanstätten intergrowth of taenite and kamacite (Wood, 1979, p. 105). The kinetics of unmixing and the phase com-

positions are well known from laboratory studies and indicate that the exsolution occurred over a T range of 600°C to 400°C at cooling rates of only a few degrees per million years.

Parent Bodies of Meteorites

Wood (1979, p. 107) succinctly summarizes current evidence regarding the parent bodies of meteorites as follows:

> The slow rate of cooling registered by metal alloys in most of the meteorites, and the fact that some of them (achondrites) are the products of igneous fractionation (which requires a gravitational field to separate liquids from crystals), indicates that the meteorites once resided in one or more planets. These properties could not have developed if meteorites had always been small objects orbiting at large in the solar system. The interiors

Table 16-4 Chemical composition of the solar photosphere, least differentiated carbonaceous chondrite (type I), and ordinary chondrite.

	Solar spectrum atomic[a]	Carbonaceous meteorite		Ordinary chondrite Wt. %	Pyrolite[c] Wt. %
		Atomic[b]	Wt. %		
H	12	8.36	2	—	—
C	8.62 ± 0.12	7.43 ± 0.20	3.1	0.2	—
N	7.94 15	6.28 40	0.26	0.005	—
O	8.84 07	8.47	46.0	35	43.7
Na	6.28 05	6.36 04	0.51	0.58	0.3
Mg	7.60 15	7.61 01	9.56	14.2	23.0
Al	6.52 12	6.51 03	0.85	1.01	1.75
Si	7.65 08	7.59 01	10.5	17.0	21.1
P	5.50 15	5.57 20	0.11	0.12	—
S	7.20 15	7.28 07	5.90	2.0	—
Cl	5.5 04	5.17 30	0.05	0.01	—
K	5.16 10	5.15 06	0.054	0.08	0.02
Ca	6.35 10	6.44 10	1.06	1.19	2.2
Ti	5.05 12	4.97 03	0.043	0.062	0.12
Cr	5.71 14	5.68 06	0.24	0.34	0.3
Mn	5.42 16	5.55 07	0.19	0.22	0.12
Fe	7.50 08	7.53 01	18.4	27.6	6.2
Ni	6.28 09	6.25 13	1.0	1.8	0.2

[a] Logarithm of atomic abundance, normalized so that atomic proportion of $H = 10^{12}$.
[b] Logarithm of atomic abundance, normalized so that atomic proportion of $Mg + Si = 10^{15.2}$, the proportion of these two elements in the Sun.
[c] Pyrolite is a hypothetical or model upper mantle peridotite conceived by Ringwood (1979, Table 2.1).
Source: After J. V. Smith, 1979, Mineralogy of the planets: A voyage in space and time, *Mineralogical Magazine* 43.

of these bodies must have become hot, just as Earth's interior is; this had the effect of metamorphosing some primitive planetary material (producing ordinary chondrites) and causing melting elsewhere. Components of the melted material separated gravitationally, yielding zones of pure metal (the source of iron meteorites) and pure silicates (achondrites).

The rate at which a rock cools depends upon how deeply it is buried: the less overburden, the faster it dissipates heat to space. The cooling rates found for...ordinary chondrites correspond to depths of 100 to 200 km in a small planet. The parent meteorite planets were probably not much larger than a few hundred kilometers in radius, that is, of asteroidal dimensions. Origin in smallish planets is further indicated by the absence of high-pressure minerals....

Had meteorites been derived from larger parent bodies, tightly packed crystalline phases such as pyrope or jadeite would occur, but they have never been found.

Altogether, the textural and chemical properties of meteorites indicate derivation from many small parent bodies that had accreted from the primordial solar nebula and subsequently, at least in

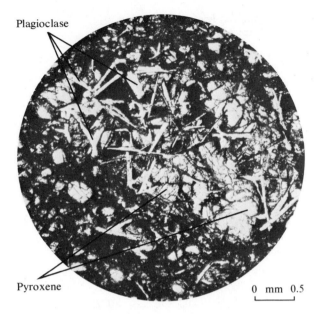

Plagioclase

Pyroxene

0 mm 0.5

Figure 16-17   Brecciated basaltic achondrite
Pasamonte. A clast with subophitic texture composed of
Ca-poor pyroxene and plagioclase is surrounded (top
and bottom third of view) by finer fragmental material of
the same composition. [A sample of this meteorite was
kindly loaned to the author by Carleton Moore.]

many bodies, experienced large-scale differentiation like other planetary terrestrial bodies. Why
such small bodies melted is a puzzle, but decay of
short-lived radioactive isotopes, such as $^{26}Al \rightarrow$
^{26}Mg with a half-life of 0.7 My, may have contributed to the melting.

16.4 ACCRETION OF THE EARTH AND OTHER PLANETARY BODIES

Modern theories of the birth of the planets and Sun
in the solar system begin with a disc-shaped solar
nebula composed of dispersed hot gas and dust.
The origin of the nebula and the reasons why it
collapsed and condensed into discrete bodies are
problems left to astronomers and cosmologists
(Wood, 1979, chaps. 6, 7). However, it is interesting to note that anomalously high concentrations of
^{16}O, ^{26}Mg, and other exotic isotopes in carbonaceous meteorites are believed to have resulted from

the explosion of a nearby supernova; the resulting
shock wave may have triggered the condensation
and collapse of the solar nebula (Marvin, Wasson,
and others, 1979).

The actual processes of condensation and accretion are highly speculative but do have a bearing
on the nature of the planets and so will be considered very briefly here. At some point in the history
of the nebula, solid particles—perhaps small
clumps of dust, chondrules, and Ca–Al-rich
inclusions—were drawn gravitationally toward its
midplane (Ringwood, 1979; Wood, 1979). With increasing concentration of particles in the midplane
region, still larger clumps would grow by mutual
gravitational attraction, coagulating into planetary
nuclei, or **planetesimals**, perhaps at most a few
kilometers to tens of kilometers in diameter.
Larger planetesimals grew at the expense of
smaller ones as they, together with any remaining
stray small dust particles, were swept up into the
greater gravitational field of the larger bodies.
Eventually, still larger protoplanets (hundreds of
kilometers in diameter) grew by accretion of
smaller planetesimals and stray dust. All of this
activity may have occupied roughly 100 My.

It is readily apparent from Table 16-2 that the
planets have grossly different densities and therefore chemical compositions. Mercury, for example, must have a proportionately large metallic
iron core, whereas the other inner rocky planets
have smaller cores. The outer planets, such as Jupiter, have the equivalent of several Earth masses
of silicate material but are proportionately more
gaseous bodies. Is it possible that some sort of
chemical fractionation occurred in the condensing
solar nebula to produce these differences?

Condensation of the Solar Nebula

The basic sequence of condensation in cooling gas
of solar composition has been calculated from
thermodynamic data by Grossman and Larimer
(1974; see also Ringwood, 1979, p. 98) and is
shown in simplified form in Figure 16-18. As in
melting or crystallization processes of silicate
rocks, the condensation process is incongruent and

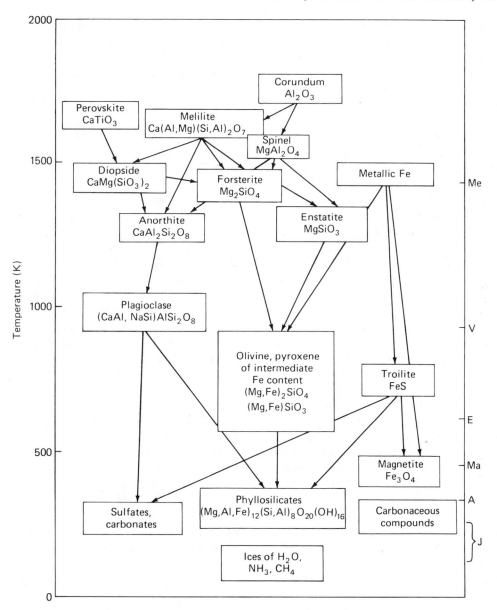

Figure 16-18 Simplified sequence of condensation of minerals from a nebula of solar composition. Arrows show falling temperature reaction between residual gas and mineral(s) in the upper box to produce mineral(s) in lower box. Note the temperatures are in K. Bracket at lower right margin indicates range of temperatures at which the inner planets and satellites of Jupiter (J) are believed to have accreted according to the equilibrium condensation model. Carbonaceous chondrites represent a disequilibrium collection of high- through low-temperature condensates; some of the Ca–Al-rich inclusions in them are concentrically layered with a higher temperature assemblage enclosed by lower. Me, Mercury; V, Venus; E, Earth; Ma, Mars; A, asteroids. [From John A. Wood, *The Solar System,* copyright © 1979, p. 162. Reprinted by permission of Prentice-Hall, Inc., Englewood Cliffs, New Jersey.]

sequential—crystalline phases of different composition appear at different temperatures and react (perhaps incompletely) with any residual gas to form new phases as T falls. At least three different models of condensation can be conceived, as described below (Wood, 1979, pp. 172–177).

Equilibrium condensation. In this model, a more or less specific temperature was assumed to have prevailed for most of the period of slow accretion of particles and planetesimals into each particular planet; this T essentially determines the overall bulk composition of the planet (see Figure 16-18). Where Mercury is presumed to have accreted, temperatures were ~ 1400 K and thus it should have formed solid condensates of perovskite, melilite, spinel, metallic iron, and some diopside and forsterite; highest-T corundum had reacted to form melilite and spinel before 1400 K was attained in the cooling nebula. Accretion of Earth at about 600 K involved troilite, Mg–Fe olivines, pyroxenes, and plagioclase (and K-feldspar).

This model would provide, after subsequent internal differentiation, for a metallic core in Mercury and a Ca–Ti–Al–Mg silicate mantle. Earth's core would contain troilite. The major difficulty with this model is that the Earth would have no volatile constituents—H_2O, CO_2, and so on—nor any ferric iron; although this is certainly a serious deficiency, various secondary mechanisms have been proposed to overcome it.

Fractional condensation. A fractional condensation model can be proposed that produces from the outset a highly differentiated and layered planet analogous to a zoned plagioclase crystal formed in a rapidly cooled magma (see the discussion on pages 277–278 of the contrasting modes of crystallization of magma). In this model, Earth accreted with a Ca–Al–Ti silicate inner core and an outer metallic Fe core overlaid by olivines and pyroxenes and finally light, low-T phyllosilicates and other volatile-rich phases. There is no evidence today of a Ca–Al–Ti silicate core, but due to the rapid accretion assumed in this model, temper-atures may have been sufficiently high to melt the interior of such an Earth and allow this material to buoy upward and let the denser, metallic Fe sink into the core.

An important feature of this model is the generalized overall bulk chemical composition of the planet, equivalent to the solar or carbonaceous chondrite composition. However, major differences between planets, such as between Mercury and Earth, are not accounted for.

Combined condensation. This model combines favorable aspects of the previous two. Accreting planetesimals are assumed to have been more or less like carbonaceous chondrites, representing disequilibrium mechanical mixtures of high- to low-T condensates of the nebula. Density differences among the planets reflect the dominant T of condensate—fairly high in the case of Mercury, with its large iron core. Besides this contribution from an equilibrium type of condensation, it is held that a fractional type of condensation added a late veneer onto the planets of low-T volatile-rich material.

It is impossible to determine a very detailed model of accretion of the Earth with available information (compare, for example, Ringwood, 1979, chap. 8; and Smith, 1979, p. 77). Major uncertainties include: (1) the rate and size of infalling planetesimals into the proto-Earth and the consequent rate at which energy of impact converted into heat was dissipated; and (2) the composition of early- and late-arriving planetesimals, whether reduced or oxidized, chondritic or enriched in volatiles, refractories or metal, and so on.

Early Degassing of the Earth

Since the classic investigation of Rubey (1951; see also Fanale, 1971), there has been widespread belief that the present atmosphere and oceans of the Earth are not relics of an early atmosphere but instead have accumulated over eons of time by degassing of the interior via magmatism. The evidence for this is the extremely low concentrations of the inert gases Ne, ^{36}Ar (not a radiogenic daughter product), Kr, and Xe relative to H_2O and CO_2 in the Earth's atmosphere, hydrosphere, and lithosphere. In terms of solar abundances, these inert gases are depleted by factors of hundreds to millions of times. Apparently, any chemically uncombined gases were lost during accretion, or were not in the accumulating planetesimals. Only gaseous components locked in crystalline phases, chiefly CO_2 and H_2O in carbonates and phyllosilicates, stable to at least moderate temperatures, seem to have been retained in the Earth. These combined gases have been coming out of the interior through melting processes since early in the Earth's history.

Origin of the Moon

The origin of the Earth's single satellite, so large in relation to its parent planet, has intrigued mankind from the beginning of history. Three basic explanations have been offered: (1) capture—the Moon initially formed in another orbit in the early solar system and was somehow captured into its present orbit around Earth; (2) binary planet—Moon and Earth accreted from a common cluster of nebular material, which segregated into a gravitationally coupled doublet; (3) fission—Moon was torn out of the Earth at some early stage of its evolution by unknown but necessarily mighty forces.

Ringwood (1979; see also Brown, 1977) has recently marshalled evidence in support of the fission model, specifically advocating derivation of the Moon from the Earth's mantle. The Moon has an overall density (see Table 16-2) significantly less than those of the other inner rocky planets, which have substantial metallic cores; but its density is close to an uncompressed Earth mantle density. It was noted on page 572 that the hypothetical lunar magmatic "ocean" from which highland anorthositic rocks were differentiated might have had a composition not unlike oceanic rift basalts, suggesting similar sources. The fission event must have predated segregation of the lunar highlands 4.6 to 4.4 Gy ago yet must have postdated segrega-

tion of the Earth's core. Tidal interactions attending the fission process would have been a major source of thermal energy to reduce the overall volatile content of the Moon, contributing to the widespread melting of the lunar magma "ocean" and also contributing to the high early temperatures of the Earth.

SUMMARY

16.1. The two basic surface features of the Moon—the older intensely cratered highlands and the younger, darker colored, and less cratered maria—are everywhere covered by a regolith of impacted rock fragments. The lunar lithosphere is very thick; the existence of a thin underlying asthenosphere and a small metallic core is uncertain.

Lunar petrologic processes are dominated by anhydrous magmatism and shock metamorphism due to meteorite impact. The lunar highlands consist chiefly of highly calcic anorthosite that apparently formed by accumulation of floating plagioclases in a magma "ocean" soon after initial accretion of the Moon from the solar nebula 4.6 Gy ago. An epoch of massive meteorite bombardment culminated ~ 3.9 Gy ago. Highly fluid flows of lunar basalt flooded maria basins from ~ 4.0 Gy to perhaps 2.5 Gy ago. These basalts are generally more enriched in refractory elements (Ti, Zr, and so on) but depleted in volatile elements (H, O, Na, S, Bi, and so on) relative to terrestrial basalts. They apparently originated as partial melts of the olivine–pyroxene cumulates of the magma "ocean."

16.2. Comparisons of features and processes on other planetary bodies furnishes insight into the evolution of the Earth.

Meteorite bombardment culminating 3.9 Gy ago would have been energetic enough on Earth to have obliterated any prior crust.

The composition and especially the size of planetary bodies has strongly influenced their evolution. Smaller bodies such as Moon and Mercury lost much of their thermal energy during infancy and have been geologically dead for a couple of billion years. Larger Mars experienced a localized but intense episode of magmatism roughly one billion years ago. The evolution of Venus, which is about the same size as Earth, is uncertain but does not appear to have involved global plate tectonism. Only Earth of the inner rocky planets of our solar system has experienced plate tectonism and is still very much alive geologically. It is argued that the abundance of water in Earth may have permitted this activity.

16.3. Meteorites captured by Earth's gravitational field are chiefly stony but include some of metallic composition.

Rare carbonaceous chondrites consist of mechanical disequilibrium mixtures of high-T chondrules, Ca–Al–Mg–Ti-rich inclusions, and low-T phases including organic compounds. Carbonaceous chondrites richest in volatiles are strikingly similar in composition to the sun and are believed to represent relatively undifferentiated samples of the primordial solar nebula from which all bodies in the solar system evolved.

Ordinary chondrites and achondrites are more differentiated meteorites that formed by metamorphism and by partial melting of more primitive bodies, followed by crystallization of these melts.

Textural and chemical properties of meteorites indicate derivation from many small parent bodies that accreted from the primordial solar nebula and subsequently, at least in many bodies, experienced differentiation.

16.4 Modern theories concerning the origin of planets involve gravitational collapse and condensation of a hot dispersed cloud of gas and dust—the solar nebula—followed by accretion of small planetesimals into larger bodies. Sequential and differential condensation occurred to produce some fundamental differences in the bulk composition of the planets.

The anomalously low concentrations of inert gases on Earth indicate a major loss of uncombined gases during accretion or shortly thereafter followed by a slow accumulation of the present hydrosphere through degassing of the interior via volcanism.

There is some evidence that the Moon split off from the Earth early in their lifetime, rather than being gravitationally captured or forming as part of a gravitationally coupled binary body.

REFERENCES

Brown, G. M. 1977. Two major igneous events in the evolution of the moon. *Philosophical Transactions of the Royal Society of London*, Series A, 286:439–451.

Brown, G. M., Eglinton, G., Runcorn, S. K., and Urey, H. C. (eds.). 1977. The Moon—A new appraisal from space missions and laboratory analyses. *Philosophical Transactions of the Royal Society of London*, Series A, 285:1–606.

Carr, M. H. 1974. Tectonism and volcanism of the Tharsis region of Mars. *Journal of Geophysical Research* 79:3943–3949.

Clark, B. C., and Baird, A. K. 1979. Is the Martian lithosphere sulfur rich? *Journal of Geophysical Research* 84:8395–8403.

Fanale, F. P. 1971. A case for catastrophic early degassing of the Earth. *Chemical Geology* 8:79–105.

Frey, H. 1980. Crustal evolution of the Earth: The role of major impacts. *Precambrian Research* 10:195–216.

Grieve, R. A. F. 1980. Impact bombardment and its role in protocontinental growth on the early Earth. *Precambrian Research* 10:217–247.

Grossman, L., and Larimer, J. W. 1974. Early chemical history of the solar system. *Reviews of Geophysics and Space Physics* 12:71–101.

King, E. A. 1976. *Space Geology*. New York: John Wiley and Sons, 349 p.

Lowman, P. D., Jr. 1976. Crustal evolution in silicate planets: Implications for the origin of continents. *Journal of Geology* 84:1–26.

Marvin, U. B., et al. 1979. Extraterrestrial samples: Progress and prospects. *Geotimes* 24(10):22–28.

Masursky, H., Eliason, E., Ford, P. G., McGill, G. E., Pettengill, G. H., Schaber, G. G., and Schubert, G.

1980. Pioneer-Venus radar results: Geology from images and altimetry. *Journal of Geophysical Research* 85:8232–8260.

Ringwood, A. E. 1979. *Origin of the Earth and Moon.* New York: Springer-Verlag, 295 p.

Rubey, W. R. 1951. Geologic history of sea water. *Geological Society of America Bulletin* 62:1111–1148.

Smith, J. V. 1979. Mineralogy of the planets: A voyage in space and time. *Mineralogical Magazine* 43:1–89.

Solomon, S. C. 1980. Differentiation of crusts and cores of the terrestrial planets: Lessons for the early Earth. *Precambrian Research* 10:177–194.

Taylor, S. R. 1975. *Lunar Science: A Post-Apollo View.* New York: Pergamon, 372 p.

Wasson, J. T. 1974. *Meteorites.* New York: Springer-Verlag, 360 p.

Wood, J. A. 1979. *The Solar System.* Englewood Cliffs, N.J.: Prentice-Hall, 196 p.

APPENDIXES

Appendix A
Units

The following standard international or SI units (Le Systeme International d'Unites) were approved in 1974 and are discussed in National Bureau of Standards Special Publication 330.

BASE UNITS

Quantity	Name of unit	Symbol
Length	meter	m
Mass	gram	g
Time	second	s
Temperature	kelvin	K
Amount	mole	mol

DERIVED UNITS

Quantity	Name of unit	Symbol	In base units
Force	newton	N	$m \cdot kg \cdot s^{-2}$
Stress, pressure	pascal	Pa	$m^{-1} \cdot kg \cdot s^{-2}$
Energy	joule	J	$m^2 \cdot kg \cdot s^{-2}$
Power	watt	W	$m^2 \cdot kg \cdot s^{-3}$
Viscosity	pascal second	Pa $\cdot$ s	$m^{-1} \cdot kg \cdot s^{-1}$
Heat capacity, entropy	joule per kelvin	J/K	$m^2 \cdot kg \cdot s^{-2} \cdot K^{-1}$

SI PREFIXES

Factor	Prefix	Symbol
10^9	giga	G
10^6	mega	M
10^3	kilo	k
10^{-2}	centi	c
10^{-3}	milli	m
10^{-6}	micro	μ

Other units have been conventionally used in the past and are encountered in the geologic literature. Some of these and their equivalent SI units include the following:

Quantity	Name of conventional unit	Symbol	Value in SI units
Length	angstrom	Å	10^{-10} m
Stress, pressure	bar	bar	0.1 MPa
Stress, pressure	kilobar	kbar	0.1 GPa
Stress, pressure	standard atmosphere	atm	101,325 Pa
Viscosity	poise	P	0.1 Pa · s
Energy	calorie	cal	4.186 J
Time	million years	m.y.	My
Temperature	degree Celsius	°C	K
Heat flow	heat flow unit	HFU (μ cal · cm^{-2} · s^{-1})	0.0418 W/m^2

Appendix B
Petrographic Techniques

B.1 PROCEDURE FOR STAINING K-RICH FELDSPAR

1. Cut slab of rock and polish with 400 grit, or finer if rock is very fine grained.

2. Immerse polished surface in 40% HF for about 20 seconds to etch silicate minerals. CAUTION: HYDROFLUORIC ACID IS EXTREMELY DANGEROUS. Never use it outside a fume hood. Use rubber gloves and tongs when handling rocks.

3. Wash HF off slab with tap water.

4. Immerse slab in a concentrated solution of sodium cobaltinitrite for 1 to 2 minutes. Potassium from decomposed K-rich alkali feldspar will form a bright canary-yellow precipitate of potassium cobaltinitrite.

5. Optimum results may require etching and/or staining for periods of time other than specified above.

6. A useful byproduct of this procedure is that plagioclases after etching are chalky white, but not stained, whereas quartz is only very slightly etched and appears gray (see Plate V).

7. Thin sections of rocks may be stained, but it is necessary to etch them in HF vapor, not by total immersion in liquid.

8. Calcium feldspars may be stained red by dipping in $BaCl_2$ solution after etching with HF as in step 2, then rinsing and treating with potassium rhodizonate solution.

B.2 MODAL ANALYSIS OF ROCKS BY POINT COUNTING

The modal composition, or mode, of a rock—the percentage of actual minerals present—may be estimated roughly by eye (see Figure B-1) or determined more accurately by point-counting techniques.

Point counting is a method of ascertaining the areal proportions of constituent minerals in a rock by superposing a grid upon a planar section through the rock. If the fabric is isotropic, a rep-

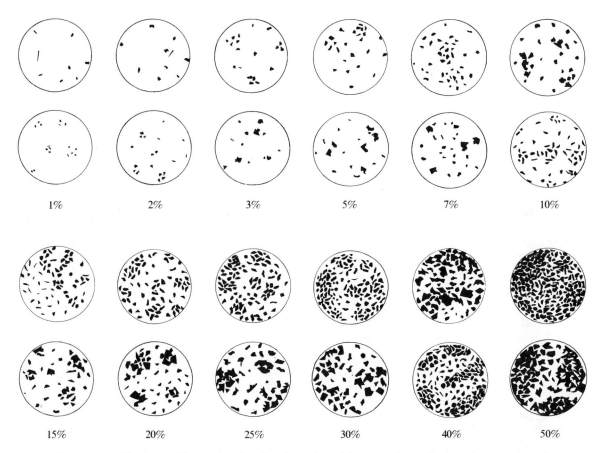

1%   2%   3%   5%   7%   10%

15%   20%   25%   30%   40%   50%

Figure B-1   Charts to aid the visual estimation of modal proportions of minerals in rocks [After
R. D. Terry and G. V. Chilingar, American Geological Institute Data Sheet 6.]

resentative planar section can yield a valid indication of the volumetric proportions of minerals in
the rock. Two or more orthogonally oriented sections are required for anisotropic fabrics.

If, in a grid of N points, n_1 points fall upon mineral 1, then the areal (volumetric) percentage of
that mineral in the rock is $n_1/N \times 100$; if n_{11} points
fall on mineral 11, then its percentage in the rock is
$n_{11}/N \times 100$; and so on. Obviously, $\Sigma n = N$.

In any determination of the sort described here,

consideration must be given to the accuracy—the
degree to which the determined value differs from
the true value. This difference, called error, can be
determined empirically or by statistical theory; it
must be minimized in order for the measurements
to be meaningful. There are three sources of error
in point counting, which are not easily separated:

1. *Operator Error*   Misidentification of minerals, biasing the tally in the case of grid

points that fall on boundaries between grains, and so on.

2. *Counting error*   Arising from the estimation of areal percentages in a planar section using too few grid points.

3. *Sampling error* Evaluating volumetric percentages of constituent minerals from planar sections of the rock that are not representative of the whole.

Operator error can only be minimized by careful work and the experience of the petrologist. The other two errors depend upon the area sampled, the grain size and shape, the grid spacing, and the percentage of the mineral sought. To illustrate, sampling an area of only 8 or 10 grains in a coarse granitic rock will obviously give a large sampling error. One measure of the error is the variance, σ^2, where σ is the standard deviation. At a 67% confidence level, the error is less than $\pm \sigma$, at 95% $\pm 2\sigma$, and at 99% $\pm 3\sigma$. The theoretical variance, according to Solomon (1963), resulting from errors 1 and 2 above in determining the modal percentage of a mineral is approximately equal to or less than

$$\frac{44 \, ps^3}{RA} \left[1 \, + \, 5.8 \left(\frac{R}{s} \right)^3 \right]$$

where p is the modal percentage of the mineral, s is the grid spacing (the mean spacing if rectangular), R is the mean grain diameter, and A is the area over which the grid extends.

To set up a modal analysis, one visually estimates p, measures R [see Solomon (1963) for methods and comments on irregularly shaped grains], and selects appropriate A and s such that the value of the theoretical variance is small, generally no more than 2. For expediency, the number of grid points to be counted (a function of s and A)

should be made small (unless the petrologist has lots of time). The following table shows some data for different grain dimeters, where $p = 25\%$ and the variance is 2.

s (mm)	A (mm²)	Number of points	R (mm)
0.2	75	1,800	0.1
0.5	720	1,250	0.1
0.5	950	3,650	0.5
2.0	10,000	1,250	0.5
2.0	25,000	6,400	2.5
4.0	33,800	2,100	2.5
25	10^6	1,600	12.5

Various types of attachments for the stage of a petrographic microscope are available to facilitate point counting of thin sections. For coarse rocks, it is convenient to use a grid on a transparent overlay placed on a polished slab.

REFERENCES

Bailey, E. H., and Stevens, R. E. 1960. Selective staining of K-feldspar and plagioclase on rock slabs and thin sections. *American Mineralogist* 45:1020–1025.

Heinrich, E. W. 1965. *Microscopic Identification of Minerals.* New York: McGraw-Hill, 414 p. (See p. 12–14 for staining techniques).

Jackson, E. D., and Ross, D. C. 1956. A technique for modal analyses of medium- and coarse-grained (3–10 mm) rocks. *American Mineralogist* 41:648–651.

Solomon, M. 1963. Counting and sampling errors in modal analysis by point counter. *Journal of Petrology* 4:367–382.

Solomon, M., and Green, R. 1965. A chart for designing modal analysis by point counting. *Geologische Rundschau* 55:844–848.

Appendix C
Review of Some
Rock-Forming Minerals

The purpose of this appendix is to provide a brief overview of some major rock-forming minerals found in igneous and metamorphic rocks to aid in the study of hand samples and in preliminary observations using the petrographic microscope and thin sections. For additional information, refer to the five-volume reference by W. A. Deer, R. A. Howie, and J. Zussman, 1962–1963, *Rock-Forming Minerals* (London: Longman); or to W. R. Phillips and D. T. Griffen, 1981, *Optical Mineralogy: The Nonopaque Minerals* (San Francisco: W. H. Freeman and Company). Most chemical analyses cited here are from Deer, Howie, and Zussman (1962–1963), with their permission.

FELDSPARS

Color Colorless, white, pink, gray; also black, irridescent, and various shades of brown and green; colorless in thin section.

Hardness 6.

Cleavage {001} and {010} approximately at right angles is perfect.

Habit Tabular to anhedral.

Alteration Fine-grained phyllosilicates and hydrated Ca–Al silicates.

Optical relief Nil.

Interference colors Low first order.

Feldspars are the most abundant of the rock-forming minerals, occurring in most igneous rocks and many metamorphic rocks. With few exceptions, feldspars consist of solid solutions (see Figure C-1) between three end members (see Table C-1):

K-feldspar (Kf) $KAlSi_3O_8$

Albite (Ab) $NaAlSi_3O_8$

Anorthite (An) $CaAl_2Si_2O_8$

Essentially two solid-solution series can be recognized in hand specimen and by simple optical examination of thin sections: triclinic plagioclase (Ab–An) and triclinic–monoclinic alkali feldspar (Ab–Kf). Most feldspars in igneous rocks are zoned, or chemically inhomogeneous, and possess

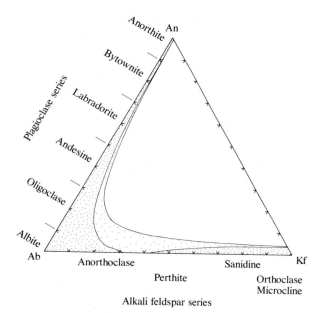

Figure C-1 Extent of solid solution and nomenclature of the feldspars at about 1000°C (light stippling) and 600°C (dark stippling).

Table C-1 Chemical compositions of select feldspars (in wt.%)

	1	2	3	4	5	6
SiO_2	64.50	65.86	66.84	63.49	53.44	43.54
TiO_2	—	—	—	0.00	0.02	tr
Al_2O_3	20.25	20.66	19.62	22.76	29.58	35.66
Fe_2O_3	0.47	0.29	0.57	0.14	0.13	0.58
FeO	—	0.10	—	0.04	0.14	0.00
MgO	—	0.12	—	0.00	0.06	0.06
BaO	—	0.12	—	—	—	—
CaO	0.48	1.50	0.58	3.51	11.83	19.53
Na_2O	4.72	8.17	11.53	9.46	4.51	0.26
K_2O	9.60	3.88	0.10	0.22	0.26	
Total	100.02	100.70	99.24	99.62	99.97	99.63

1. Orthoclase microperthite in pegmatite.
2. Anorthoclase in trachyte.
3. Albite in chlorite schist.
4. Oligoclase in pegmatite.
5. Labradorite in gabbro.
6. Anorthite in gabbro.

differing compositions in usually concentric zones that are evident in hand sample as contrasting colors (see Figure 4-1) and in thin section as bands of contrasting optical extinction (see Figure 7-15).

Plagioclase can be recognized by any one of the following features:

1. Polysynthetic lamellar twinning, usually albite twinning, is visible as perfectly even striae on {001} cleavages (see Figure C-2). Light can be reflected first from one then from the alternate set of {001} cleavage striae by rotation of the sample by only a few degrees. These twinning striae appear only on the {001} cleavage, which can usually be recognized by its elongate rectangular (or lath-shaped) outline. The {010} cleavage, usually recognized by its rather more square outline, gives no indication of twinning. Therefore, in a rock composed wholly of plagioclase but in random orientation, statistically only one-half of the cleavages appearing on a broken surface display twinning. Polysynthetic twinning in thin section is easily recognized under cross-polarized light as alternating light and dark bands (see Figures 4-3b and 4-12b). Polysynthetic twinning in plagioclases formed in metamorphic rocks by solid-state recrystallization is poorly developed or absent. This characteristic, together with their commonly anhedral shape, causes plagioclases in metamorphic rocks to resemble quartz.

2. Association can be helpful in distinguishing plagioclases but is less reliable than twinning. If feldspar grains are found in a rock containing olivine, pyroxene, or abundant (> 30%) amphibole, they are

(a) Carlsbad

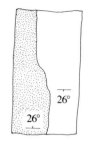

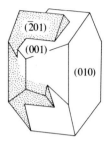

Ideal
twinned
crystal

Typical appearance
of {001} cleavage
in a rock

26°

26°

(b) Albite

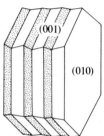

(001)

(010)

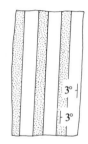

3°

3°

Figure C-2 Common twinning in feldspars that aid in recognition. Dip symbols show orientation of {001} cleavage. (a) Carlsbad twin in monclinic K-rich feldspar consists of two individuals, joined along an irregular composition plane, whose perfect {001} cleavages are tilted toward each other to form an angle of about 52°. (b) Albite twinning, the most common type of polysynthetic twinning in triclinic plagioclase, consists of multiple individuals joined along planar composition planes so that {001} cleavages appear to be thin lamellae of only slightly different orientation.

likely to be (at least in large part) of plagioclase.

3. Color and alteration can also be helpful but are again not as reliable as twinning. When plagioclase and alkali feldspar occur together in the same rock, as is commonly the case where micas, sparse amphibole, and/or quartz are associated, then the two can sometimes be distinguished as follows:

a. Two colors of feldspars, both fresh with vitreous luster: Plagioclase is generally white and alkali feldspar is pink, pale brown, or lavender.

b. Some feldspar grains are dull, chalky, pale green, or otherwise lacking vitreous luster, remainder are fresher and more vitreous: Plagioclase is generally altered and alkali feldspar is fresh.

K-rich alkali feldspars may be recognized as follows:

1. By the absence of polysynthetic twinning on the {001} cleavage. This is a necessary, not a sufficient, condition for identification of alkali feldspar. Untwinned plagioclase, which often occurs in metamorphic rocks, cannot be distinguished from alkali feldspar on this basis. K-rich alkali feldspars are commonly twinned on the Carlsbad Law (see Figure C-2).

2. By weak negative relief in thin section.

3. By perthic intergrowths (see Figure 4-12a), usually found in phaneritic plutonic rocks that cooled slowly after growth of an homogeneous alkali feldspar, allowing it to unmix into discrete, Na-rich and K-rich phases.

Sanidines and anorthoclases have a preserved high-*T*, poorly ordered Al–Si structure and occur in quickly cooled volcanic rocks. Orthoclase and microcline are more Al–Si-ordered phases and occur in phaneritic rocks; distinction between them relies on sophisticated optical or X-ray diffraction properties, except that the grid twinning of microcline is sometimes evident in thin section (see Figure C-3).

QUARTZ   SiO₂   trigonal

Color: Generally colorless, gray, or white; also shades of brown, red, and purple; colorless in thin section.

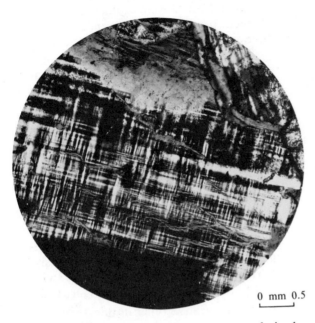

0 mm 0.5

Figure C-3 Photomicrograph under cross-polarized light of grid twinning in triclinic microcline.

Table C-2 Chemical compositions of select feldspathoids (in wt.%).

	1	2
SiO_2	43.61	54.62
TiO_2	0.00	0.00
Al_2O_3	33.05	22.93
Fe_2O_3	0.85	0.26
FeO	—	0.26
MgO	0.05	0.00
CaO	0.53	0.08
Na_2O	16.09	0.66
K_2O	4.92	21.02
H_2O^+	0.70	0.12
Total	99.81	99.95

1. Nepheline in nepheline syenite.
2. Leucite in leucitite.

Hardness: 7.

Cleavage: None; conchoidal fracture.

Habit: Anhedral in metamorphic or phaneritic igneous rocks; phenocrysts in volcanic rocks are bipyramidal but commonly embayed or corroded.

Alteration: None.

Optical relief: Nil

Interference colors: Low first order. Strained quartz grains show undulatory, or wavy, extinction.

FELDSPATHOIDS

These are relatively uncommon minerals occurring in silica-undersaturated igneous rocks and include nepheline and leucite (see Table C-2).

Nepheline   (Na,K)AlSiO₄   hexagonal

Color: Colorless, white, gray; colorless in thin section.

Hardness: 6.

Cleavage: Poor.

Habit: Typically anhedral, rarely rectangular or hexagonal outlines as phenocrysts in rare lavas.

Alteration: Fine-grained phyllosilicates and zeolites.

Optical relief: Nil.

Interference colors: Very low first order.

Nepheline is easily confused with quartz in hand sample and in thin section. In hand specimen, it lacks the vitreous luster of quartz and the conchoidal fracture. A weathered surface can be very helpful in distinguishing quartz from nepheline, since the latter decomposes more readily than most minerals and stands in negative relief below the adjacent grains. In some phaneritic plutonic rocks, the purple and yellow feldspathoids sodalite and cancrinite can occur with nepheline, aiding in its identification.

Leucite (K,Na)AlSi$_2$O$_6$ tetragonal

Color: White, gray, or pale yellow-brown; colorless in thin section.

Hardness: 6.

Cleavage: Very poor.

Habit: Rare phenocrysts may appear as trapezohedrons.

Alteration: As for nepheline, but more common.

Optical relief: Nil.

Interference colors: Very low first order, complex twinning usually obvious under cross-polarized light.

Leucite occurs only in volcanic rocks and very shallow dikes and sills of appropriate silica-undersaturated composition.

Fe–Ti OXIDES

Although there are numerous minerals in the system Fe–Ti–O, only the hematite, ilmenite, and magnetite end-members will be considered here (see Table C-3).

Hematite Fe$_2$O$_3$ trigonal

Color: Red brown, or gray with metallic luster; streak is always red brown.

Hardness: 5 to 6, earthy varieties crumble very easily.

Cleavage: None, but sometimes basal parting on metallic specularite flakes.

Habit: Metallic specularite is platy; more typically hematite occurs as very fine earthy aggregates.

Alteration: Limonite.

Optical properties: In thin section all but thinnest flakes are opaque.

Hematite is rare in igneous rocks except where they have experienced alteration or heating in the atmosphere, as in freshly extruded lavas. In such rocks hematite occurs as a finely disseminated coloring pigment throughout the rock. In metamorphic rocks, hematite may be identified as specularite flakes, sometimes with high concentrations of quartz, as in metamorphosed iron formations.

Table C-3 Chemical compositions of select Fe–Ti oxides (in wt.%).

	1	2
SiO$_2$	0.14	0.48
TiO$_2$	49.89	2.98
Al$_2$O$_3$	0.02	0.02
V$_2$O$_3$	0.18	—
Fe$_2$O$_3$	6.26	63.40
FeO	40.39	32.25
MnO	0.41	0.00
MgO	2.27	0.39
CaO	0.34	0.23
Total	99.90	99.75

1. Ilmenite in gabbro.
2. Magnetite in gneiss.

Ilmenite FeTiO$_3$ trigonal

Color: Gray to black with metallic luster; opaque in thin section.

Hardness: 5 to 6.

Cleavage: None, but parting on a basal plane is sometimes observed.

Habit: Platy grains are common in igneous and metamorphic rocks.

Magnetite Fe$_3$O$_4$ isometric

Color: Gray to black with metallic luster; opaque in thin section.

Hardness: 7 to 8.

Cleavage: None, although octahedral parting may be present.

Habit: Typically cubes or octahedrons.

Alteration: Limonite.

Magnetite and ilmenite are widespread in igneous rocks and are also found in metamorphic rocks, where they can often be distinguished on the basis of their contrasting habits. Iron-bearing chromite $(Cr,Fe)_3O_4$ occurs in subalkaline ultramafic rocks.

OLIVINE $(Mg,Fe)_2SiO_4$ orthorhombic

Color: Pale yellow green to green to black with increasing Fe.

Hardness: 7.

Cleavage: None, conchoidal fractures well expressed.

Habit: Typically anhedral grains with little crystal form evident in hand specimen; in thin section more euhedral grains with prisms and pinacoids are commonly evident.

Alteration: Fe-oxides, clay minerals, serpentine, talc, carbonates, brucite.

Optical relief: Moderately high.

Interference colors: High second order to third order.

There is complete solid solution between the forsterite (Fo) Mg_2SiO_4 and fayalite (Fa) Fe_2SiO_4 end members (see Table C-4); only limited substitution of Mn, Ca, Cr, Ni occurs in some olivines. Forsteritic olivines occur in ultramafic igneous rocks and in some metamorphosed rocks rich in magnesium. Slightly more iron-rich olivines (Fo_{80-50}) occur in gabbros, basalts, and rarely andesites. Olivines close to fayalite in composition occur in some felsic igneous rocks formed from dry magmas.

Olivines are commonly altered but the generally pseudomorphic replacement products differ in different rocks. As phenocrysts in mafic lavas subject to slight oxidation, olivines become irridescent;

Table C-4 Chemical compositions of select olivines (in wt.%).

	1	2	3
SiO_2	40.96	39.31	30.15
TiO_2	0.01	0.06	0.20
Al_2O_3	0.21	1.68	0.07
Fe_2O_3	0.00	—	0.43
NiO	0.25	—	—
FeO	7.86	19.84	65.02
MnO	0.13	0.17	1.01
MgO	50.45	34.74	1.05
CaO	0.15	0.87	2.18
Na_2O	0.01	—	—
K_2O	0.00	—	—
Total	100.31	99.67	100.11

1. Olivine (Fo_{92}) in dunite.
2. Olivine (Fo_{77}) in basalt.
3. Olivine (Fo_3) in late ferrogabbro differentiate in Skaergaard intrusion.

with more intense alteration, they are replaced by very fine-grained mixtures of red-brown iron oxides and phyllosilicates, sometimes called iddingsite. In plutonic rocks, olivines are hydrated to fine aggregates of talc, brucite, serpentine, or mixtures of these; in some instances, the pseudomorphic replacement is magnesite.

PYROXENES monoclinic and orthorhombic

Color: Black and various shades of green and brown; pyroxenes are generally a darker color than any associated olivine; almost colorless in thin section, but where associated with olivine, appear to be pale brown relative to more colorless olivine; sodic pyroxenes are pleochroic green in thin section.

Hardness: 5 to 6.

Cleavage: Prismatic {110} at right angles but generally poorly expressed in hand speci-

Table C-5 Chemical compositions of select pyroxenes (in wt.%).

	1	2	3	4	5
SiO_2	57.73	50.92	52.34	51.92	59.38
TiO_2	0.04	1.18	0.18	0.77	0.14
Al_2O_3	0.95	2.90	2.72	1.85	25.82
Fe_2O_3	0.42	0.47	1.31	31.44	0.45
FeO	3.57	11.11	1.77	0.75	tr
MnO	0.08	0.33	0.24	—	0.00
MgO	36.13	15.63	16.15	—	0.12
CaO	0.23	17.28	24.23	—	0.13
Na_2O	—	0.12	0.57	12.86	13.40
K_2O	—	0.12	0.08	0.19	0.02
Total	99.15	100.06	99.59	99.78	99.37

1. Enstatite in pyroxenite (includes 0.46 Cr_2O_3; 0.35 NiO).
2. Augite in basalt.
3. Diopside in calc-silicate gneiss.
4. Aegirine in riebeckite–albite granite.
5. Jadeite in albite–glaucophane schist.

men; more obvious in some grains in thin section; a {100} parting is seen in some magnesian pyroxenes.

Habit: Stubby prisms, also anhedral grains.

Alteration: Fe-oxides, clay minerals, chlorite, amphiboles, hydrated Ca–Al silicates.

Optical relief: Moderately high.

Interference colors: High first order to second order colors.

Pyroxenes are single chain silicates whose general formula (see also chemical compositions in Table C-5) can be represented by

$$(W)_{1-p}(X, Y)_{1+p}Z_2O_6$$

where
$$W = \text{Ca, Na}$$
$$X = \text{Mg, Fe}^{2+}, \text{Mn, Ni, Li}$$
$$Y = \text{Al, Fe}^{3+}, \text{Cr, Ti}$$
$$Z = \text{Si, Al}$$
$$p = 0 \text{ to } 1$$

Alkaline rocks can have a single monoclinic pyroxene that is a solid solution between diopside, $CaMgSi_2O_6$, and aegirine, $NaFe^{3+}Si_2O_6$, sometimes with significant concentrations of Ti and Al substituting for Si.

Subalkaline magmatic rocks commonly have two stably coexistent pyroxenes: (1) a calcic monoclinic pyroxene (clinopyroxene) that is essentially a solid solution between diopside and hedenbergite, $CaFe^{2+}Si_2O_6$, but contains substantial Al and Ti and other ions, as in augite; and (2) a Ca-poor monoclinic or orthorhombic pyroxene (orthopyroxene) that is a solid solution between the enstatite, $Mg_2Si_2O_6$, and ferrosilite, $Fe^{2+}Si_2O_6$, end members with only limited substitution of Ti, Al, and Ca. Increasingly more mafic rock types contain increasingly more magnesian pyroxenes. Calcic clinopyroxenes range from black to pale green in hand sample with increasing Mg, whereas orthopyroxenes range from black to gray-brown.

In addition to the above pyroxenes, metamorphic rocks formed at high P and low T can contain clinopyroxenes with significant amounts of the jadeite end member, $NaAlSi_2O_6$.

AMPHIBOLES monoclinic and orthorhombic

Color: White to darker shades of green and brown to black with increasing Fe; all but Fe-free tremolite are pleochroic in shades of green and brown in thin section.

Hardness: 5 to 6.

Cleavage: Prismatic {110} at about 60° is perfect; nearly always visible in hand specimen and serves as an important distinguishing feature from somewhat similar-appearing pyroxenes.

Habit: Columnar to acicular but in some rocks rather anhedral.

Alteration: Amphiboles generally are more resistant to alteration than olivines and pyroxenes, but sometimes are replaced by chlorite.

Table C-6   Chemical compositions of select amphiboles (in wt.%).

	1	2	3	4	5	6
SiO_2	59.46	56.06	46.88	57.73	52.59	40.96
TiO_2	—	0.12	0.84	—	0.91	3.92
Al_2O_3	0.49	2.14	8.54	12.04	1.64	15.35
Fe_2O_3	0.00	2.22	0.99	1.16	7.69	4.82
FeO	0.07	8.97	17.18	4.51	11.80	3.30
MnO	0.38	0.23	0.28	—	0.60	0.05
MgO	25.19	16.11	10.89	13.02	9.32	15.24
CaO	11.88	10.28	11.22	1.04	3.41	10.42
Na_2O	—	1.36	1.03	6.98	6.79	2.56
K_2O	—	0.00	0.17	0.68	2.06	1.74
H_2O^+	2.27	2.42	1.87	2.27	1.44	1.00
Total	99.73	99.91	99.98	100.33	98.25	99.36

1. Tremolite in skarn.
2. Actinolite in chlorite–quartz–albite–sphene schist.
3. Hornblende in amphibolite.
4. Glaucophane in schist (includes 1.20 F).
5. Arfvedsonite in syenite (includes 2.05 F).
6. Kaersutitic amphibole megacryst in basanite lava; includes 0.27 Cr_2O_3, 0.02 F, and 0.12 Cl.
Source: Column 6 after M. G. Best, 1974, Mantle-derived amphibole in xenoliths in alkalic basaltic magmas, *Journal of Geophysical Research* 79.

Table C-7   Chemical compositions of select micas (in wt.%).

	1	2	3	4
SiO_2	48.42	44.41	40.16	38.30
TiO_2	0.87	0.22	0.45	3.60
Al_2O_3	27.16	40.09	2.65	13.99
Fe_2O_3	6.57	1.72	5.52	3.98
FeO	0.81	0.28	4.39	20.24
MnO	—	0.02	0.03	0.09
MgO	tr	0.16	22.68	7.96
CaO	tr	0.67	0.24	0.90
Li_2O	tr	—	—	—
Na_2O	0.35	5.80	—	0.50
K_2O	11.23	2.22	9.62	8.31
F	tr	0.08	—	0.32
H_2O^+	4.31	4.45	3.00	1.63
Total	99.72	100.12	98.74	99.84

1. Muscovite in schist.
2. Paragonite in chlorite–sericite schist.
3. Phlogopite in kimberlite.
4. Biotite in granodiorite.

Optical relief: Moderate.

Interference colors: Second order.

The structure of amphiboles (linked tetrahedra of $(Si, Al)O_4$ forming double chains) permits a very wide range of substitutive ions and thus the chemical compositions (see Table C-6) and nomenclature of amphiboles are quite complex. The general formula is

$$X_{2-3}Y_5Z_8O_{22}(OH)_2$$

where X is Ca, Na, and K
   Y is Mg, Fe^{2+}, Fe^{3+}, Al, Ti, Mn, Cr, Li, and Zn
   Z is Si and Al

O, F, and Cl may substitute for OH in varying amounts. Tremolite, $Ca_2Mg_5Si_8O_{22}(OH)_2$, is characteristic of low- to intermediate-grade metamorphic rocks. Actinolite, the green, Fe-bearing but low-Al amphibole, occurs in a variety of low- and intermediate-grade mafic metamorphic rocks. Hornblende is the most widespread of all the amphiboles and occurs in medium- to high-grade metamorphic rocks and in many subalkaline igneous rocks. Glaucophane is an alkalic aluminous amphibole occurring in some high-P, low-T metamorphic rocks of the blueschist facies. Arfvedsonite and riebeckite are typical of alkaline rocks.

MICAS   monoclinic

Muscovite, biotite, and phlogopite are the principal rock-forming micas (see chemical compositions in Table C-7).

Hardness: 2 to 3.

Cleavage: Basal {001} perfect.

Habit: Almost always pseudohexagonal flakes or tablets.

Optical relief: Moderate.

Interference colors: Second to third order but characteristically speckled rather than uniform throughout the grain.

Muscovite $(K,Na)Al_2(AlSi_3)O_{10}(OH,F)_2$

Color: Pale silver gray to green, rarely pale rose.

Alteration: Clay minerals.

Muscovite is widespread in low-grade metamorphic rocks formed from shales and igneous feldspar-bearing protoliths, in which it may be confused with the less common paragonite $(Na,K)Al_2(AlSi_3)O_{10}(OH,F)_2$. Muscovite also occurs in some granites that are highly aluminous and poor in Ca, and especially in pegmatites, where the Li-mica lepidolite may be found instead. Sericite is a name for fine-grained white mica (muscovite or paragonite).

Phlogopite–Biotite
$K(Mg,Fe^{2+})_3(AlSi_3)O_{10}(OH,F,Cl)_2$

Color: Phlogopite is brown and pleochroic orange-brown in thin section. Biotite is dark brown to black and pleochroic brown or green in thin section; slightly decomposed biotites may appear bronze-colored in hand sample and so falsely resemble phlogopite.

Alteration: Chlorite, clay minerals, Fe-oxides.

Phlogopite, the rare Mg-rich mica, occurs in some alkaline ultramafic rocks, such as kimberlite, and in some metamorphosed shaly dolomites. Biotite, the more Fe-rich mica, is very widespread in fairly low- to moderate-grade metamorphic rocks and in many magmatic rocks.

SERPENTINE $Mg_3Si_2O_5(OH)_4$ monoclinic

Color: White, yellow, various shades of green; generally highly variable in hand specimen and may appear black because of finely disseminated magnetite; in thin section generally colorless.

Hardness: 2 to 3.

Cleavage: Basal {001} perfect, but generally the small grain size precludes recognition of this cleavage.

Habit: Platy or micaceous in the variety known as antigorite; fibrous in chrysotile.

Optical relief: Low.

Interference colors: First order.

Serpentine forms by hydration of olivine and pyroxene at submagmatic temperatures.

TALC $Mg_3Si_4O_{10}(OH)_2$ monoclinic

Color: White to pale green; colorless in thin section.

Hardness: 1.

Cleavage: Basal {001} perfect, but generally difficult to see because of very fine grain size.

Habit: Micaceous.

Optical relief: Low to moderate.

Interference colors: Second and third order.

Talc has only very limited substitution of Al for Si and Fe for Mg (see Table C-8). Talc is a low-temperature hydration product of olivines and pyroxenes and also occurs in low-grade metamorphosed siliceous dolomites.

CHLORITE
$(Mg,Al,Fe)_6(Si,Al)_4O_{10}(OH)_8$
monoclinic

Color: Various shades of green, colorless to pleochroic pale green in thin section.

Hardness: 2 to 3.

Table C-8 Chemical compositions of select phyllosilicates other than micas (in wt.%).

	1	2	3
SiO_2	44.70	62.47	25.62
TiO_2	0.00	0.00	0.88
Al_2O_3	0.50	0.47	21.19
Fe_2O_3	0.07	—	3.88
FeO	0.29	0.79	21.55
MnO	—	0.00	0.35
MgO	42.05	31.76	15.28
CaO	0.12	0.00	0.16
Na_2O	—	—	0.00
K_2O	—	—	0.00
H_2O^+	12.43	4.70	10.87
Total	100.16	100.19	99.78

1. Serpentine in serpentinite.
2. Talc.
3. Chlorite in epidote–albite schist.

Cleavage: Basal {001} perfect, cleavage flakes inelastic.

Habit: Micaceous.

Optical relief: Nil to low.

Interference colors: Low first order.

Chlorite is widespread in low-grade metamorphic rocks. In igneous rocks it forms by alteration of primary pyroxenes, amphiboles, and biotites. Chlorite forms in low-temperature environments, as do serpentine and talc, but requires considerable amounts of aluminum.

EPIDOTE GROUP
$Ca_2(Al,Fe^{3+}Mn^{3+})_3O(SiO_4)(Si_2O_7)(OH)$
monoclinic and orthorhombic

Color: Gray, yellow-green (pistachio green); in thin section colorless to pale yellow

Hardness: 6.

Cleavage: {001} perfect.

Habit: Prismatic, often in granular aggregate form.

Optical relief: Moderate.

Interference colors: Low first order, also second and third orders.

Minerals of the epidote group are found in low- to intermediate-grade metamorphic rocks and in altered igneous rocks that originally contained Ca–Al minerals such as plagioclase and hornblende. Fracture coatings of yellow-green epidote are common in mafic to felsic rock bodies.

CORDIERITE
$Al_3(Mg,Fe^{2+})_2AlSi_5O_{18}$ orthorhombic

Color: Gray to blue, colorless in thin section.

Hardness: 7.

Cleavage: Poor.

Habit: Anhedral grains.

Alteration: White micas or other phyllosilicates.

Optical relief: Low.

Interference colors: Low first order.

Cordierite is recognized only with considerable difficulty in hand specimen, but its former presence on weathered surfaces may be indicated by pits where the cordierite has decomposed more readily than the adjacent minerals. In thin section it resembles quartz and untwinned feldspars. Inclusions of various other minerals are common. Cordierite is typical of low- to moderate-grade pelitic metamorphic rocks and is associated with quartz, micas, and polymorphs of Al_2SiO_5.

STAUROLITE
$(Fe^{2+},Mg)_2(Al,Fe^{3+})_9O_6(SiO_4)_4(O,OH)_2$
monoclinic

Color: Brown; pleochroic yellow in thin section.

Hardness: 7.

Cleavage: Poor.

Habit: Columnar, prismatic.

Optical relief: Fairly high.

Interference colors: First order.

Staurolite has a restricted mode of occurrence, appearing in medium-grade metamorphic rocks enriched in Fe and Al.

Al_2SiO_5 POLYMORPHS

These include andalusite, sillimanite, and kyanite—all of which are characteristic of aluminous metamorphic rocks. Little atomic substitution occurs in their structures, although Fe^{3+} and Mn^{3+} may substitute in minor proportions for Al.

Andalusite   orthorhombic

Color: Colorless, pale pink, lavender, or gray; usually colorless in thin section.

Hardness: 6 to 7.

Cleavage: {110} good.

Habit: Prismatic; in slaty rocks andalusite prisms with graphitic inclusions forming a crosslike pattern are known as chiastolite.

Alteration: White micas; pyrophyllite.

Optical relief: Moderate.

Interference colors: First order and low second order.

Andalusite occurs mostly in low- to medium-grade, low-pressure, metamorphosed pelitic rocks with quartz, micas, chlorite, and cordierite.

Sillimanite   orthorhombic

Color: Colorless to pale yellow; colorless in thin section.

Hardness: 6 to 7.

Cleavage: {010} good; irregular transverse fractures.

Habit: Prismatic, acicular, and in very fine-grained fibrous almost hairlike aggregates that appear milky white in hand specimen.

Optical relief: Moderate.

Interference colors: Second order.

Sillimanite occurs in higher-temperature pelitic metamorphic rocks in both high- and low-pressure environments.

Kyanite   triclinic

Color: White to pale blue, gray, green, and pink; colorless in thin section.

Hardness: 6 to 7.

Cleavage: {100} perfect, {010} good.

Habit: Bladed, prismatic, generally much coarser grain size than sillimanite.

Optical relief: Moderately high.

Interference colors: Second order.

Kyanite occurs typically in medium- to high-grade pelitic metamorphic rocks.

GARNET GROUP

Color: Variable, depending on composition (see below); generally colorless in thin section.

Hardness: 6 to 7.

Cleavage: None; conchoidal fracture.

Habit: Typically dodecahedrons and trapezohedrons.

Alteration: Chlorite pseudomorphs sometimes form.

Optical relief: High.

Interference colors: Isotropic, rarely very low first-order colors.

The garnet group of minerals comprises an extensive solid-solution series whose general formula (see also Table C-9) can be expressed as

$$X_3 Y_2(SiO_4)_3$$

where $X = Fe^{2+}$, Ca, Mg, Mn
$Y = Al, Fe^{3+}, Ti, Cr$

Table C-9 Chemical composition of select aluminous metamorphic minerals (in wt.%).

	1	2	3	4	5	6
SiO_2	36.52	47.69	28.89	36.59	38.69	41.52
TiO_2	0.00	tr	0.81	1.68	0.55	—
Al_2O_3	20.97	32.52	52.61	22.42	18.17	23.01
Fe_2O_3	17.22	0.63	2.95	0.04	5.70	1.22
FeO	0.45	8.04	10.78	32.11	3.78	12.85
MnO	0.00	0.04	0.09	1.42	0.64	0.33
MgO	0.00	7.56	2.09	5.41	0.76	16.64
CaO	23.05	0.52	—	0.54	31.76	4.71
Na_2O	—	0.53	—	—	0.13	0.00
K_2O	—	0.42	—	—	0.06	0.00
H_2O^+	1.98	1.85	1.78	—	—	—
H_2O^-	—	0.55	—	—	—	—
Total	100.19	100.35	100.00	100.21	100.24	100.51

1. Epidote, vein in diabase.
2. Cordierite in pelitic hornfels.
3. Staurolite in muscovite–biotite–quartz–feldspar schist.
4. Almandine in biotite–sillimanite gneiss.
5. Grossular in anorthite–epidote gneiss.
6. Pyrope in eclogite.

The following end members are recognized:

End member	Color	Mode of occurrence of garnet enriched in end member
Almandine $Fe_3^{2+}Al_2$	Red-brown	Metamorphosed pelitic rocks and in some amphibolites; rarely in some granites and rhyolites
Andradite $Ca_3Fe_2^{3+}$	Brown	In calc-silicate metamorphic rocks, especially skarns
Grossular Ca_3Al_2	Apple-green	In calc-silicate metamorphic rocks, especially skarns
Pyrope Mg_3Al_2	Red to red-brown	In mafic to ultramafic rocks formed at high pressures
Spessartite Mn_3Al_2	Yellow to red	In low-grade metamorphosed rocks rich in Mn, rare in granites and rhyolites

Almandine forms an extensive solid solution with spessartite and with pyrope; grossular and andradite form extensive solid solutions.

VOLCANIC GLASS

Although not a mineral because of its amorphous nature, naturally occurring glass is widespread in a variety of volcanic materials from rhyolitic to

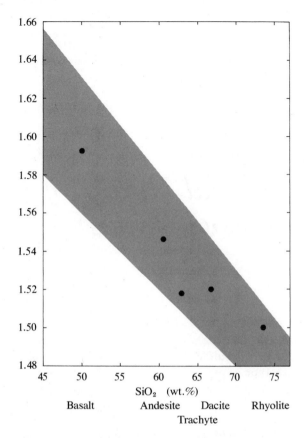

Figure C-4 Range of refractive indices of glasses as a function of their silica content is shown by shaded band. The extreme variation for a given SiO_2 is a result of wide variation in the concentrations of other constituents. Filled circles are average values for the indicated major rock types. [After B. N. Church and W. M. Johnson, 1980, Calculation of the refractive index of silicate glasses from chemical composition, *Geological Society of America Bulletin* 91.]

basaltic in chemical composition. In thin chips, rhyolitic glass is colorless except where pigmented by minute mafic or oxide minerals, but with increasing concentrations of Fe, glass becomes darker brown; basaltic glass (sideromelane) is brown even in thin section.

There is a rough correlation between the refractive index of volcanic glass and the concentration of the dominant constituent, silica. However, variable concentrations of other constituents—H_2O, Fe, Ti, Ca, and so on—between glasses at a particular SiO_2 content affect the index significantly (see Figure C-4), so that accurate determination of the composition must be made by chemical analysis.

Appendix D

Average Chemical Compositions and Norms of Some Major Rock Types

These data are drawn from the paper by R. W. LeMaitre (1976, The chemical variability of some common igneous rocks, *Journal of Petrology* 17). His paper should be consulted for unlisted rock types and for information regarding the chemical variability within each type. For some types, the variability is considerable.

	Nepheline syenite	Phonolite[a]	Syenite	Trachyte	Granite	Rhyolite
SiO_2	54.99	56.19	58.58	61.21	71.30	72.82
TiO_2	0.60	0.62	0.84	0.70	0.31	0.28
Al_2O_3	20.96	19.04	16.64	16.96	14.32	13.27
Fe_2O_3	2.25	2.79	3.04	2.99	1.21	1.48
FeO	2.05	2.03	3.13	2.29	1.64	1.11
MnO	0.15	0.17	0.13	0.15	0.05	0.06
MgO	0.77	1.07	1.87	0.93	0.71	0.39
CaO	2.31	2.72	3.53	2.34	1.84	1.14
Na_2O	8.23	7.79	5.24	5.47	3.68	3.55
K_2O	5.58	5.24	4.95	4.98	4.07	4.30
H_2O^+	1.30	1.57	0.99	1.15	0.64	1.10
H_2O^-	0.17	0.37	0.23	0.47	0.13	0.31
P_2O_5	0.13	0.18	0.29	0.21	0.12	0.07
CO_2	0.20	0.08	0.28	0.09	0.05	0.08
Total	99.69	99.86	99.74	99.94	100.07	99.96
Q	—	—	0.83	5.00	29.06	32.87
C	—	—	—	—	0.92	1.02

	Nepheline syenite	Phonolite[a]	Syenite	Trachyte	Granite	Rhyolite
Or	32.98	30.96	29.29	29.41	24.50	25.44
Ab	29.45	35.48	44.34	46.26	31.13	30.07
An	3.78	1.50	7.24	7.05	8.04	4.76
Ne	21.77	16.50	—	—	—	—
Di	4.53	6.89	5.35	2.14	—	—
Hy	—	—	4.16	2.06	3.37	1.34
Ol	0.28	—	—	—	—	—
Mt	3.27	4.05	4.41	4.33	1.75	2.14
Il	1.13	1.18	1.60	1.34	0.58	0.54
Ap	0.30	0.41	0.70	0.49	0.28	0.17
Cc	0.45	0.17	0.64	0.20	0.12	0.17

	Adamellite	Grandodiorite	Rhyodacite	Dacite	Tonalite	Diorite
SiO_2	68.65	66.09	65.55	65.01	61.52	57.48
TiO_2	0.54	0.54	0.60	0.58	0.73	0.95
Al_2O_3	14.55	15.73	15.04	15.91	16.48	16.67
Fe_2O_3	1.23	1.38	2.13	2.43	1.83	2.50
FeO	2.70	2.73	2.03	2.30	3.82	4.92
MnO	0.08	0.08	0.09	0.09	0.08	0.12
MgO	1.14	1.74	2.09	1.78	2.80	3.71
CaO	2.68	3.83	3.62	4.32	5.42	6.58
Na_2O	3.47	3.75	3.67	3.79	3.63	3.54
K_2O	4.00	2.73	3.00	2.17	2.07	1.76
H_2O^+	0.59	0.85	1.09	0.91	1.04	1.15
H_2O^-	0.14	0.19	0.42	0.28	0.20	0.21
P_2O_5	0.19	0.18	0.25	0.15	0.25	0.29
CO_2	0.09	0.08	0.21	0.06	0.14	0.10
Total	100.05	99.90	99.79	99.78	100.01	99.98
Q	25.17	22.36	22.67	22.73	16.62	10.28
C	0.28	0.26	0.25	—	—	—
Or	23.66	16.11	17.72	12.82	12.24	10.42
Ab	29.36	31.73	31.05	32.07	30.67	29.96
An	11.55	17.34	15.04	20.01	22.58	24.40
Ne	—	—	—	—	—	—
Di	—	—	—	0.11	1.49	4.67
Hy	5.66	7.40	6.19	5.73	9.68	12.56
Ol	—	—	—	—	—	—
Mt	1.79	2.00	3.08	3.53	2.66	3.63
Il	1.03	1.03	1.14	1.09	1.40	1.80
Ap	0.44	0.42	0.59	0.34	0.58	0.68
Cc	0.20	0.19	0.47	0.14	0.33	0.23

	Andesite	Monzonite	Latite[b]	Trachyandesite	Trachybasalt	Gabbro
SiO_2	57.94	62.00	61.25	58.75	49.21	50.14
TiO_2	0.87	0.78	0.81	1.08	2.40	1.12
Al_2O_3	17.02	15.65	16.01	16.70	16.63	15.48
Fe_2O_3	3.27	1.92	3.28	3.26	3.69	3.01
FeO	4.04	3.08	2.07	3.21	6.18	7.62
MnO	0.14	0.10	0.09	0.16	0.16	0.12
MgO	3.33	2.02	2.22	2.57	5.17	7.59
CaO	6.79	4.17	4.34	4.96	7.90	9.58
Na_2O	3.48	3.73	3.71	4.35	3.96	2.39
K_2O	1.62	4.06	3.87	3.21	2.55	0.93
H_2O^+	0.83	0.90	1.09	1.25	0.98	0.75
H_2O^-	0.34	0.19	0.57	0.58	0.49	0.11
P_2O_5	0.21	0.25	0.33	0.41	0.59	0.24
CO_2	0.05	0.08	0.19	0.08	0.10	0.07
Total	99.93	99.53	99.83	99.97	100.01	99.15
Q	12.37	14.02	14.26	7.80	—	0.71
C	—	—	—	—	—	—
Or	9.60	24.00	22.85	19.00	15.06	5.49
Ab	29.44	31.56	31.38	36.80	29.39	20.26
An	26.02	13.97	15.64	16.58	20.10	28.60
Ne	—	—	—	—	2.23	—
Di	4.84	3.78	2.05	3.95	11.85	13.70
Hy	9.49	6.01	4.57	6.06	—	22.13
Ol	—	—	—	—	8.28	—
Mt	4.74	2.78	3.91	4.73	5.36	4.36
Il	1.65	1.48	1.54	2.07	4.55	2.13
Ap	0.50	0.60	0.79	0.97	1.38	0.56
Cc	0.11	0.17	0.43	0.17	0.23	0.17

	Norite	Diabase	Basalt	Basanite	Nephelinite[c]	Hawaiite
SiO_2	50.44	50.14	49.20	44.30	40.60	47.48
TiO_2	1.00	1.49	1.84	2.51	2.66	3.23
Al_2O_3	16.28	15.02	15.74	14.70	14.33	15.74
Fe_2O_3	2.21	3.45	3.79	3.94	5.48	4.94
FeO	7.39	8.16	7.13	7.50	6.17	7.36
MnO	0.14	0.16	0.20	0.16	0.26	0.19
MgO	8.73	6.40	6.73	8.54	6.39	5.58
CaO	9.41	8.90	9.47	10.19	11.89	7.91
Na_2O	2.26	2.91	2.91	3.55	4.79	3.97
K_2O	0.70	0.99	1.10	1.96	3.46	1.53
H_2O^+	0.84	1.71	0.95	1.20	1.65	0.79
H_2O^-	0.13	0.40	0.43	0.42	0.54	0.55
P_2O_5	0.15	0.25	0.35	0.74	1.07	0.74
CO_2	0.18	0.16	0.11	0.18	0.60	0.04
Total	99.86	100.14	99.95	99.89	99.89	100.05

	Norite	Diabase	Basalt	Basanite	Nephelinite[c]	Hawaiite
Q	—	0.99	—	—	—	—
C	—	—	—	—	—	—
Or	4.15	5.85	6.52	11.61	3.16	9.03
Ab	19.14	24.65	24.66	12.42	—	33.61
An	31.49	24.98	26.62	18.38	7.39	20.62
Ne	—	—	—	9.55	21.95	—
Di	10.58	13.50	14.02	21.03	32.36	10.94
Hy	26.38	19.24	15.20	—	—	0.10
Ol	0.34	—	1.50	12.38	2.32	9.32
Mt	3.21	5.00	5.49	5.72	7.95	7.16
Il	1.90	2.83	3.49	4.77	5.05	6.13
Ap	0.36	0.58	0.82	1.74	2.51	1.75
Cc	0.41	0.36	0.26	0.40	1.37	0.08

	Anorthosite	Pyroxenite	Peridotite	Harzburgite	Dunite
SiO_2	50.28	46.27	42.26	39.93	38.29
TiO_2	0.64	1.47	0.63	0.26	0.09
Al_2O_3	25.86	7.16	4.23	2.35	1.82
Fe_2O_3	0.96	4.27	3.61	5.48	3.59
FeO	2.07	7.18	6.58	6.47	9.38
MnO	0.05	0.16	0.41	0.15	0.71
MgO	2.12	16.04	31.24	33.18	37.94
CaO	12.48	14.08	5.05	2.90	1.01
Na_2O	3.15	0.92	0.49	0.31	0.20
K_2O	0.65	0.64	0.34	0.14	0.08
H_2O^+	1.17	0.99	3.91	4.00	4.59
H_2O^-	0.14	0.14	0.31	0.24	0.25
P_2O_5	0.09	0.38	0.10	0.13	0.20
CO_2	0.14	0.13	0.30	0.09	0.43
Total	99.80	99.83	99.46	95.63	99.58
Q	—	—	—	—	—
C	—	—	—	—	0.80
Or	3.86	3.75	2.02	0.83	0.47
Ab	23.16	7.76	4.15	2.60	1.69
An	49.71	13.54	8.32	4.17	1.17
Ne	1.89	—	—	—	—
Di	8.61	42.09	11.22	6.93	—
Hy	—	6.26	15.79	21.13	14.48
Ol	2.01	15.12	46.39	46.22	67.38
Mt	1.40	6.20	5.23	7.94	5.20
Il	1.22	2.79	1.19	0.50	0.18
Ap	0.21	0.90	0.23	0.30	0.47
Cc	0.31	0.28	0.67	0.21	1.00

[a] Includes 0.73 *Wo* [b] Includes 0.58 *Hm* [c] Includes 13.57 *Lc*

Appendix E

Calculation of the CIPW Norm

In the calculation, constituents from the chemical analysis of the rock are allocated in a prescribed sequence to form normative minerals. The allocation is analogous to the sequential crystallization of minerals from a magma, except in the case of the norm (1) all of the normative minerals are anhydrous, (2) less solid solution is permitted (pyroxenes, for example, contain no TiO_2, Al_2O_3, or Fe_2O_3), (3) pyroxenes and olivines have the same Mg/Fe ratios, and (4) normative minerals are always formed in the same sequence, regardless of the composition of the rock. However, as in real magmas, incompatibilities such as nepheline–quartz and olivine–quartz are built into the calculation.

The normative minerals, their abbreviations, chemical formulas, and formula weights are given in Table E-1.

Most norms today are calculated by computer using the steps below (adapted from C. H. Kelsey, 1965, Calculation of the C.I.P.W. norm, *Mineralogical Magazine* 34:276–282). It is a useful exercise, however, to calculate a few by hand. Two are shown in Tables E-2 and E-3 that can be found by following the steps. Note that not every step applies to every rock, but that the ones that do must follow the prescribed sequence.

1. Calculate the amounts (molecular proportions) of the oxides and any elements listed in chemical analysis of the rock by dividing their weight percentages by their appropriate formula weights.

2. The amounts of MnO and NiO are added to that of FeO, and those of BaO and SrO to that of CaO.

3a. An amount of CaO equal to 3.33 times that of P_2O_5 (or 3.00 P_2O_5 and 0.33 F, if the latter is present) is allotted for apatite (*Ap*). To do this, the P_2O_5 is written in the *Ap* column opposite P_2O_5 and 3.33 times this amount is entered in the *Ap* column opposite CaO.

3b. An amount of Na_2O equal to twice that of Cl is allotted for halite (*Hl*).

3c. An amount of Na_2O equal to that of the SO_3 is allotted for thenardite (*Th*). It must be remem-

Table E-1   Abbreviations, chemical formulas, and formula weights for the normative minerals.

Name and abbreviation	Chemical formula	Formula weight
Quartz (Q)	SiO_2	60.1
Corundum (C)	Al_2O_3	102
Zircon (Z)	$ZrO_2 \cdot SiO_2$	183
Orthoclase (Or)	$K_2O \cdot Al_2O_3 \cdot 6SiO_2$	556
Albite (Ab)	$Na_2O \cdot Al_2O_3 \cdot 6SiO_2$	524
Anorthite (An)	$CaO \cdot Al_2O_3 \cdot 2SiO_2$	278
Leucite (Lc)	$K_2O \cdot Al_2O_3 \cdot 4SiO_2$	436
Nepheline (Ne)	$Na_2O \cdot Al_2O_3 \cdot 2SiO_2$	284
Kaliophilite (Kp)	$K_2O \cdot Al_2O_3 \cdot 2SiO_2$	316
Halite (Hl)	$NaCl$	58.4
Thenardite (Th)	$Na_2O \cdot SO_3$	142
Sodium carbonate (Nc)	$Na_2O \cdot CO_2$	106
Acmite (Ac)	$Na_2O \cdot Fe_2O_3 \cdot 4SiO_2$	462
Sodium metasilicate (Ns)	$Na_2O \cdot SiO_2$	122
Potassium metasilicate (Ks)	$K_2O \cdot SiO_2$	154
Diopside (Di)	$CaO \cdot (Mg,Fe)O \cdot 2SiO_2$	217–248[a]
Wollastonite (Wo)	$CaO \cdot SiO_2$	116
Hypersthene (Hy)	$(Mg,Fe)O \cdot SiO_2$	100–132[a]
Olivine (Ol)	$2(Mg,Fe)O \cdot SiO_2$	141–204[a]
Dicalcium silicate (Cs)	$2CaO \cdot SiO_2$	172
Magnetite (Mt)	$FeO \cdot Fe_2O_3$	232
Chromite (Cm)	$FeO \cdot Cr_2O_3$	224
Ilmenite (Il)	$FeO \cdot TiO_2$	152
Hematite (Hm)	Fe_2O_3	160
Sphene (Tn)	$CaO \cdot TiO_2 \cdot SiO_2$	196
Perovskite (Pf)	$CaO \cdot TiO_2$	136
Rutile (Ru)	TiO_2	79.9
Apatite (Ap)	$3.3CaO \cdot P_2O_5$	310
Fluorite (Fl)	CaF_2	78.1
Pyrite (Pr)	FeS_2	120
Calcite (Cc)	$CaO \cdot CO_2$	100

[a] These two numbers represent the weights of the pure Mg- and Fe-end members, respectively.

bered here that the SO_3 stated in the analysis usually represents the S of pyrite, so that step 3c is applicable only when the rock contains haüyn.

3d. An amount of FeO equivalent to half of S is allotted for pyrite (Pr).

3e. An amount of FeO equal to that of the Cr_2O_3 is allotted for chromite (Cm).

3f. An amount of FeO equal to that of the TiO_2 is allotted for ilmenite (*Il*). If there is an excess of TiO_2, an equal amount of CaO is to be allotted to this excess for provisional sphene (*Tn'*), but only after the allotment of CaO to Al_2O_3 for anorthite (step 4d). If there is still an excess of TiO_2, this latter excess is calculated as rutile (*Ru*).

3g. An amount of CaO equal to half that of the remaining F is allotted for fluorite (*Fl*).

3h. If the rock contains calcite, an amount of CaO equal to that of the CO_2 is allotted for calcite (*Cc*). (In most magmatic rocks, any modal calcite present is secondary.)

3i. Set aside the ZrO for Zircon (*Z*).

4a. An amount of Al_2O_3 equal to that of the available K_2O is allotted for provisional orthoclase (*Or'*).

4b. If there is an excess of K_2O over Al_2O_3 (an excess extremely rare), this excess is calculated as potassium metasilicate (*Ks*).

4c. Any excess of Al_2O_3 from 4a is combined with an equal amount of available Na_2O for provisional albite (*Ab'*). If there is insufficient Al_2O_3, see step 4g.

4d. If there is an excess of Al_2O_3 over the $K_2O + Na_2O$ used in 4a and 4c, this excess is allotted to an equal amount of CaO for anorthite (*An*).

4e. If there is an excess of Al_2O_3 over this CaO, this excess is calculated as corundum (*C*).

4f. If there is an excess of CaO over the Al_2O_3 of 4d, it is reserved for diopside (*Di*) and wollastonite (*Wo*) (see steps 7a and 7b).

4g. If in step 4c there is an excess of Na_2O over Al_2O_3, it is to be reserved for acmite and possibly for sodium metasilicate (see steps 5a and 5b). There is then no anorthite in the norm.

5a. To an amount of Fe_2O_3 equal to that of the excess of Na_2O over Al_2O_3 (see step 4g) is allotted an equal amount of Na_2O for acmite (*Ac*).

5b. If there is still an excess of Na_2O over Al_2O_3 (a rare excess), it is calculated as sodium metasilicate (*Ns*).

5c. If, as usually happens, there is an excess of Fe_2O_3 over remaining Na_2O, it is assigned to magnetite (*Mt*), an equal amount of available FeO being allotted to it.

5d. If there is still an excess of Fe_2O_3, it is calculated as hematite (*Hm*).

6. All of the MgO and FeO remaining from the previous allotments are placed in a *Remainder* column and their relative proportions are ascertained.

7a. To the amount of CaO remaining after Rule 4d is allotted an equal amount of MgO + FeO to form provisional diopside (*Di'*), *the relative proportions of MgO and FeO, as they occur in the Remainder, being preserved.*

7b. If there is an excess of CaO, it is reserved for provisional wollastonite (*Wo'*).

7c. If there is an excess of MgO + FeO over that needed for diopside (step 7a), they are allotted for provisional hypersthene (*Hy'*) *in the ratio as they occur in the Remainder.*

All the oxides except SiO_2 have now been assigned to actual or provisional normative minerals, and we have next to consider the distribution of the silica.

8. Allot the necessary amount of silica to ZrO_2 to form zircon (step 3i in the ratio 1 : 1), to CaO to

form provisional sphene (step 3f, 1 : 1), to excess Na_2O to form acmite (step 5a, 4 : 1), to excess K_2O and Na_2O to form potassium and sodium metasilicates (steps 4b, 5b, 1 : 1), to K_2O for provisional orthoclase (step 4a, 6 : 1), to Na_2O for provisional albite (step 4c, 6 : 1), to CaO for provisional anorthite (step 4d, 2 : 1), to CaO + (Mg + Fe)O for diopside (step 7a, 1 : 1), to excess CaO for wollastonite (step 7b, 1 : 1), and to (Mg + Fe)O for provisional hypersthene (Step 4c, 1 : 1).

9. Set Y equal to the amount of silica required for all the normative and provisional normative minerals so far formed in step 8. If $SiO_2 > Y$ set $Q = SiO_2 - Y$ and go to step 17. If $SiO_2 < Y$ the deficiency $D = Y - SiO_2$ and further calculations have to be made. These calculations must be continued until the deficiency of SiO_2 has been reduced to zero, at which point go to step 17.

10. If $D < Hy'/2$ set $Ol = D$ and $Hy = Hy' - 2D$. The silica deficiency is now zero. If $D > Hy'/2$ set $Ol = Hy'/2$ and $Hy = 0$; put $D_1 = D - Hy'/2$. Be sure to allot MgO and FeO in the proportions as they occur in the *Remainder*.

11. If $D_1 < Tn'$ set $Tn = Tn' - D_1$ and $Pf = D_1$. The silica deficiency is now zero. If $D_1 > Tn'$ set $Pf = Tn'$ and $Tn = 0$; put $D_2 = D_1/4 - Tn'$. Skip this step if no Tn' was formed, but set $D_2 = D_1$.

12. If $D_2 < 4Ab'$ set $Ne = D_2/4$ and $Ab = Ab' - D_2/4$. The silica deficiency is now zero. If $D_2 > 4Ab'$ set $Ne = Ab'$ and $Ab = 0$, put $D_3 = D_2 - 4Ab'$.

13. If $D_3 < 2Or'$ set $Lc = D_3/2$ and $Or = Or' - D_3/2$. The silica deficiency is now zero. If $D_3 > 2Or'$ set $Lc' = Or'$ and $Or = 0$; put $D_4 = D_3 - 2Or'$.

14. If $D_4 < Wo'/2$ set $Cs = D_4$ and $Wo = Wo' - 2D_4$. The silica deficiency is now zero. If $D_4 > Wo'/2$ set $Cs = Wo'/2$ and $Wo = 0$; put $D_5 = D_4 - Wo'/2$. Skip this step if no Wo' was formed, but set $D_5 = D_4$.

15. If $D_5 < Di'$ add an amount equal to $D_5/2$ to the amounts of Cs and Ol already in the norm; set $Di = Di' - D_5$. The silica deficiency is now zero. If $D_5 > Di'$ add an amount equal to $Di'/2$ to the amounts of Cs and Ol already in the norm; put $Di = 0$ and $D_6 = D_5 - Di'$.

16. Set $Kp = D_6/2$ and $Lc = Lc' - D_6/2$. The silica deficiency is now zero.

17. The fixed and provisional molecules of steps 8 and 9 are calculated into their weight percentages by multiplying oxide amounts by the formula weights of the normative minerals listed above. For example, the weight percentage of Or in the hornblende andesite calculation in Table E-2 is $0.031 \times 556 = 17.2$. The amount of orthoclase is the same as the number of Al_2O_3 or K_2O molecules. Note that the final weight percentages are given to only three significant figures, the same as the original weight percentages of the oxides in the rock and the formula weights of the normative minerals. The total of the normative mineral weight percentages should be the same as the total of the oxides in the rock, except for exclusion of H_2O and rounding-off errors.

The oxidation state of Fe in an analyzed rock profoundly affects the degree of silica saturation in the norm. If two rocks are identical except one has a high Fe_2O_3/FeO ratio due to secondary oxidation, it will have more normative magnetite and therefore less mafic silicates so that there will be relatively more SiO_2 in step 9. Oxidation can shift an Fe-rich rock such as a basalt from *Ne*-normative to *Hy*-normative or even *Q*-normative.

This aberrancy can be compensated for by taking the lowest Fe_2O_3/FeO ratio in the group of rocks analyzed or by simply adopting a standard value for the molecular ratio Fe_2O_3/FeO, such as 0.15 (see Cox and others, 1979, p. 413; reference at end of Chapter 2).

Table E-2 Normative calculation of hornblende andesite, east of Francis, Utah.

Oxide	Wt.%	1,2 Mol. prop.	3a Ap	3f Il	4a Or'	4c Ab'	4d An	5c Mt	5d Hm	6 Remainder	7a Di'	7d Hy'	9 Q
SiO_2	57.3	0.953			0.186	0.342	0.156				0.044	0.046	0.179
TiO_2	1.02	0.013		0.013									
Al_2O_3	16.9	0.166			0.031	0.057	0.078						
Fe_2O_3	4.58	0.029		0.013				0.024	0.005				
FeO	2.55	0.035						0.024					
MnO	0.11	0.002											
MgO	2.74	0.068								0.068	0.022	0.046	
CaO	6.11	0.109	0.010				0.078				0.022		
BaO	0.16	0.001											
Na_2O	3.52	0.057				0.057							
K_2O	2.91	0.031			0.031								
P_2O_5	0.42	0.003	0.003										
H_2O^+	1.05												
H_2O^-	0.65												
Total	100.02												

Q	10.8	Mt	5.57
Or	17.2	Il	1.98
Ab	29.9	Hm	0.80
An	21.7	Ap	0.93
Di	4.77	Total	98.25
Hy	4.60		

Note: This rock appears to have suffered secondary oxidation, because of the high Fe_2O_3/FeO ratio, and this has produced normative *Hm*, iron-free normative *Di* and *Hy* and considerable *Q*. Reduction of Fe_2O_3 so that Fe_2O_3/FeO (mol) = 0.15 gives Q = 7.33, Di = 5.08, Hy = 11.8, Mt = 2.79, and Hm = 0.

Table E-3 Normative calculation of mica peridotite, Western Uinta Mountains, Utah.

Oxide	Wt.%	1,2 Mole. Prop.	3a Ap	3f Il	4a Or'	4c Ab'	4d An	5c Mt	6 Re-mainder	7a Di'	7c Hy'	10 Ol	12 Ne	13 Lc'	10,15 Ol	15 Cs	16 Kp	16 Lc
SiO₂	37.7	0.627			0.342	0.066	0.008			0.398	0.357	0.179	0.022	0.228	0.278	0.100	0.008	0.212
TiO₂	0.64	0.008		0.008														
Al₂O₃	7.33	0.072			0.057	0.011	0.004						0.011	0.057			0.004	0.053
Fe₂O₃	3.99	0.025						0.025										
FeO	5.36	0.075		0.008				0.025	0.045	0.016	0.029	0.029			0.045			
MnO	0.19	0.003																
MgO	20.6	0.511							0.511	0.183	0.328	0.329			0.511			
CaO	12.2	0.217	0.017				0.004			0.199						0.199		
SrO	0.30	0.003																
BaO	0.04	0.000																
Na₂O	0.71	0.011				0.011							0.011					
K₂O	5.33	0.057			0.057									0.057				
P₂O₅	0.78	0.005	0.005															
H₂O⁺	3.44																	
H₂O⁻	1.24																	
Total	99.85																	

An	1.11		Cs	17.2
Lc	23.1		Mt	5.80
Ne	3.21		Il	1.22
Kp	1.26		Ap	1.55
Ol	40.6		Total	94.96

Step 9. $Y = 1.171$ so $SiO_2 < Y$; $D = 0.544$

Step 10. $D > Hy'/2$; set $Ol = 0.179$ and $Hy' = 0$; $D_1 = 0.544 - 0.179 = 0.365$

Step 11. Skip, but set $D_2 = D_1 = 0.365$

Step 12. $D_2 > 4 Ab'$; set $Ne = 0.011$ and $Ab = 0$; $D_3 = 0.365 - 0.044 = 0.321$

Step 13. $D_3 > 2 Or'$; set $Lc' = Or'$ and $Or = 0$; $D_4 = 0.321 - 0.114 = 0.207$

Step 14. Skip, but set $D_5 = D_4 = .207$

Step 15. $D_5 > Di'$; additional $Ol = Di'/2 = 0.099$; $Di = 0$; $D_6 = 0.207 - 0.199 = 0.008$

Step 16. $Kp = 0.004$ and $Lc = 0.053$

Index

Page numbers in **boldface** type refer to a definition. Page numbers followed by an asterisk indicate an illustration.